THE ELEMENTS

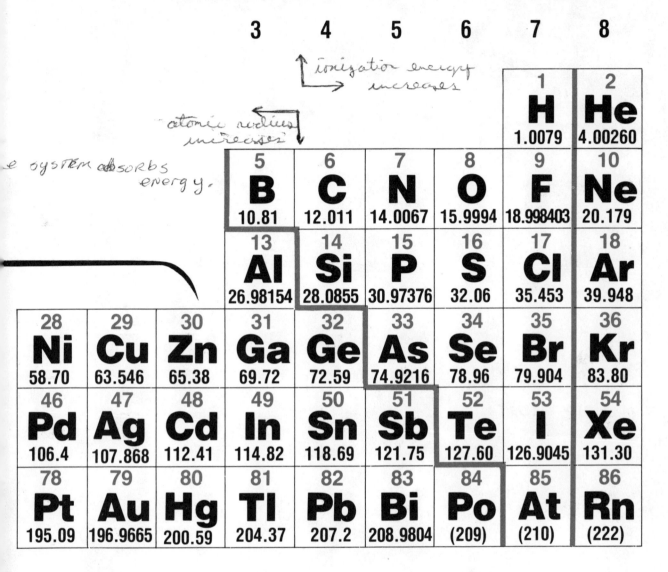

ionization energy increases

atomic radius increases

e system absorbs energy.

	3	4	5	6	7	8
					1 **H** 1.0079	2 **He** 4.00260
	5 **B** 10.81	6 **C** 12.011	7 **N** 14.0067	8 **O** 15.9994	9 **F** 18.998403	10 **Ne** 20.179
	13 **Al** 26.98154	14 **Si** 28.0855	15 **P** 30.97376	16 **S** 32.06	17 **Cl** 35.453	18 **Ar** 39.948
28 **Ni** 58.70	29 **Cu** 63.546	30 **Zn** 65.38	31 **Ga** 69.72	32 **Ge** 72.59	33 **As** 74.9216	34 **Se** 78.96	35 **Br** 79.904	36 **Kr** 83.80
46 **Pd** 106.4	47 **Ag** 107.868	48 **Cd** 112.41	49 **In** 114.82	50 **Sn** 118.69	51 **Sb** 121.75	52 **Te** 127.60	53 **I** 126.9045	54 **Xe** 131.30
78 **Pt** 195.09	79 **Au** 196.9665	80 **Hg** 200.59	81 **Tl** 204.37	82 **Pb** 207.2	83 **Bi** 208.9804	84 **Po** (209)	85 **At** (210)	86 **Rn** (222)

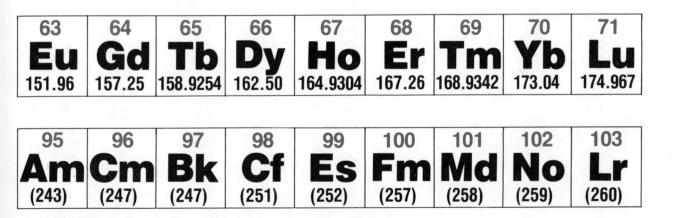

63 **Eu** 151.96	64 **Gd** 157.25	65 **Tb** 158.9254	66 **Dy** 162.50	67 **Ho** 164.9304	68 **Er** 167.26	69 **Tm** 168.9342	70 **Yb** 173.04	71 **Lu** 174.967
95 **Am** (243)	96 **Cm** (247)	97 **Bk** (247)	98 **Cf** (251)	99 **Es** (252)	100 **Fm** (257)	101 **Md** (258)	102 **No** (259)	103 **Lr** (260)

CONVERSION FACTORS:

1.609 km/mi   4.184 J/cal

1.057 qt/l.

3.281 ft/m.

TABLES:

Misc.

Atomic Radius and Ionization Energies, Pp. 186-7.
Geometry of Molecules with Expanded Octets, Fig 12.10, p. 294.

Covalent Molecule Geometry, Fig. 10.6, p. 232

Colloids

$$E_V = l_i + \gamma \frac{A}{V}$$

l_i = E per unit Volume
γ = E per unit Area

COARSE $\longrightarrow$ COLLOID $\longleftarrow$ ATOMIC

dispersion   condensation

coagulation   dissolution

MEDIUM	DISPERSED MATTER	COLLOID NAME
GAS	GAS	———
	LIQUID	AEROSOL
	SOLID	"
LIQUID	GAS	FOAM
	LIQUID	EMULSION
	SOLID	SOL
SOLID	GAS	SOLID FOAM
	LIQUID	" " (Rare)
	SOLID	SOLID SOL

SAUNDERS COLLEGE PUBLISHING COMPLETE PACKAGE FOR TEACHING GENERAL CHEMISTRY WITH CHEMICAL PRINCIPLES

Masterton, Slowinski & Stanitski: **CHEMICAL PRINCIPLES,** 5th edition. Instructors Manual to accompany

Masterton, Slowinski & Stanitski: **CHEMICAL PRINCIPLES,** 5th Edition. S.I. Version. Instructors Manual to accompany

Masterton & Slowinski: **CHEMICAL PRINCIPLES WITH QUALITATIVE ANALYSIS,** Instructors Manual to accompany

Slowinski, Masterton & Wolsey: **CHEMICAL PRINCIPLES IN THE LABORATORY,** 3rd edition. Instructors Manual to accompany

Slowinski, Wolsey & Masterton: **CHEMICAL PRINCIPLES IN THE LABORATORY WITH QUALITATIVE ANALYSIS,** revised reprint. Instructors Manual to accompany

Boyington: **STUDY GUIDE FOR CHEMICAL PRINCIPLES,** 5th edition and S.I. Version

Boyington: **TEST BANK FOR CHEMICAL PRINCIPLES,** 5th edition

Masterton, Slowinski & Stanitski: **OVERHEAD PROJECTUALS TO ACCOMPANY CHEMICAL PRINCIPLES,** 5th edition

Shakhashiri, Schreiner & Meyer: **GENERAL CHEMISTRY AUDIO-TAPE LESSONS,** 2nd edition. Transcripts & Instructors Manual to accompany

Shakhashiri, Schreiner & Meyer: **WORKBOOK FOR GENERAL CHEMISTRY AUDIO-TAPE LESSONS,** 2nd edition

Masterton & Slowinski: **ELEMENTARY MATHEMATICAL PREPARATION FOR GENERAL CHEMISTRY**

Masterton & Slowinski: **MATHEMATICAL PREPARATION FOR GENERAL CHEMISTRY**

Slowinski & Masterton: **QUALITATIVE ANALYSIS AND THE PROPERTIES OF IONS IN AQUEOUS SOLUTION**

Peters: **PROBLEM SOLVING FOR CHEMISTRY,** 2nd edition

Rochow: **MODERN DESCRIPTIVE CHEMISTRY**

 SAUNDERS GOLDEN SUNBURST SERIES

William L. Masterton

Professor of Chemistry
University of Connecticut
Storrs, Connecticut

Emil J. Slowinski

Professor of Chemistry
Macalester College
St. Paul, Minnesota

Conrad L. Stanitski

Professor of Chemistry
Randolph-Macon College
Ashland, Virginia

Chemical Principles **Fifth Edition**

SAUNDERS COLLEGE PUBLISHING
Philadelphia

 SAUNDERS GOLDEN SUNBURST SERIES

SAUNDERS COLLEGE PUBLISHING/CBS
Educational and Professional
Publishing, A Division of CBS Inc.
Address Orders to: 383 Madison Ave.,
New York, NY 10017
Address Editorial correspondence to:
West Washington Square
Philadelphia, PA 19105

This book was set in Times Roman by Hampton Graphics, Inc.
The editors were John J. Vondeling, M. Lee Walters, and Carol Field.
The art director, text and cover designer was Nancy E.J. Grossman.
The production manager was Thomas J. O'Connor.
The artwork was drawn by George W. Laurie, Pat Morrison, John Hackmaster,
  Linda Savalli, Mario Neves, Larry Ward, Sharon Iwanczuk,
  Dimitri Karetnikov, and Karen McGarry.
R.R. Donnelley & Sons Company was printer and binder.

Chemical Principles

ISBN 03-057804-3

1234   39   98765432

CBS COLLEGE PUBLISHING

To the author, a newly published textbook is a perfect specimen, infallible in all respects. However, as time passes, nagging doubts arise. Perhaps the development on p. 52 could be made clearer. Possibly some of the statements in Chapter 5 are misleading; conceivably, they could even be wrong. The passage of time makes certain sections of the text seem dated, perhaps even obsolete. Of such doubts are new editions born, encouraged by publishers who may have other motives.

In this revision of *Chemical Principles* we have retained the basic framework of previous editions. As in the past, the early chapters (1 – 7) emphasize the quantitative, experimental aspects of chemistry. These are followed by a treatment of electronic structure and chemical bonding (Chapters 8 – 12). The basic ideas of chemical equilibrium and kinetics are introduced near the midway point (Chapters 13 and 14). Later chapters in the text emphasize the chemistry of reactions in water solution (Chapters 16 – 22).

The most important change in this edition is the introduction of several descriptive chapters of a review nature. These include Chapters 5, 9, 12, 15, 20, and 22. The first of these (Chapter 5: Sources of the Elements) is perhaps typical. It presents the descriptive chemistry of the processes used to extract the more familiar elements from their natural sources. Throughout the chapter, principles covered in Chapters 1 to 4 are applied in text discussions, examples, and problems. In particular, Chapter 5 reviews the use of the mole (Chapter 2), chemical formulas (Chapter 3), and mass relations in reactions (Chapter 4).

The manner in which these new chapters are used will depend upon the wishes of the instructor and the needs of the class. They can serve to introduce a considerable amount of descriptive chemistry, beyond that covered in principles-oriented chapters. On the other hand, they may be used primarily for review purposes, giving students another chance to apply principles they may have had trouble grasping upon first exposure. Finally, one or more of these chapters (5, 9, 12, 15, 20, 22) may be omitted; they introduce no new principles that are required in later chapters.

Beyond this innovation, there are several organizational changes in this edition. The Periodic Table is now introduced in Chapter 1 to serve as a framework for organizing the descriptive chemistry of the elements. Concentrations and solution stoichiometry are presented in Chapter 4, much earlier than in previous editions. This allows for an early discussion of the principles of reactions in aqueous solution. For example, in thermochemistry (Chapter 6), we can now deal with ΔH for reactions involving ions in solution. Two topics have been moved back in the text: organic chemistry to Chapter 25 and chemical thermodynamics to Chapter 23. By postponing ΔG until ionic equilibria and electrochemistry have been covered, the various ways of expressing reaction spontaneity can be tied together rather neatly.

The number of solved examples and end-of-chapter problems is about 10% greater than in the last edition. Each example is now followed by an "exercise" which illustrates the same principle. The student can test his or her understanding of that principle by solving the exercise and checking against the answer given in the text. As in the past, the problems are arranged in matched pairs. Answers to one set, those in the column at the right of the page, are provided in Appendix 5. Comments or complaints about these answers should be addressed to WLM, who worked each problem three times as compared to only once for EJS and CLS.

In this edition, we have been influenced by the growing acceptance of the International System of Units. We now use the joule rather than the calorie as an energy unit. Atomic dimensions are expressed in nanometers rather than Angstroms. Atomic and molecular "weights" have become "masses." We have, however, retained two non-SI pressure units which are most commonly used in the United States: the atmosphere and the millimeter of mercury. In the alternate version of this text entitled *Chemical Principles Using the International System of Units,* SI units are used consistently throughout.

Many of the changes in this edition reflect the comments and suggestions of instructors and students who have used *Chemical Principles*. We are grateful to the following individuals who have provided us with written reviews:

Jane Bibler of the University of South Carolina, John Carpenter of St. Cloud State University, Ronald Clark of Florida State University, Jefferson Davis Jr. of The University of South Florida, Dan Decious of California State University, Sacramento, Dwaine Eubanks of Oklahoma State University, Peter Gold of Penn State University, Dorothy Goldish of California State University at Long Beach, Russell Grimes of the University of Virginia, Arnulf Hagen of the University of Oklahoma, Daniel Haworth of Marquette University, Edward Mellon of Florida State University, Edward Mercer of the University of South Carolina, Ted Musgrave of Colorado State University, Glen Nichols of Golden West College, L. C. Pedersen of the University of North Carolina, Gary Riley of Georgia Institute of Technology, and Wilmer Stratton of Earlham College.

We particularly appreciate the thoughtful, detailed review of the final manuscript by Dwaine Eubanks. CLS extends thanks to Kathryn and Sherman Hill and Dorothy and Ansley Nystrom for their warm hospitality during the initial stages of this work. It has been a pleasure, as always, to work with the effective and dedicated staff at Saunders. We owe a great deal to Tom O'Connor, a superb production manager, to John Hackmaster, a first-rate artist, and to John Vondeling, our editor. Without John's support and encouragement, we would never have finished the second edition of this text, let alone the fifth. We are happy to welcome back to the team Lee Walters, a delightful person of many talents which may even extend to the squash court.

WILLIAM L. MASTERTON
EMIL J. SLOWINSKI
CONRAD L. STANITSKI

CONTENTS

MATTER AND MEASUREMENTS

You are beginning a course in chemistry, one of the basic scientific disciplines of the modern world. At this point, you may well wonder what chemistry is all about. Perhaps we can obtain some idea of what chemistry is by reflecting for a moment on what chemists do. In the most general sense, chemists are people who study the properties and composition of matter, the "stuff" that makes up the universe. They are particularly interested in the interactions between matter and energy. These result in chemical reactions, in which the composition of matter changes.

The materials chemists study come from a variety of sources, some exotic, others more mundane. From the analysis of a moon rock or a sample of Martian soil, a chemist may draw conclusions about the origin of the universe or the possibilities of extraterrestrial life. Closer to home, chemists may look at the composition of a sample of water taken from the Hudson River or a sample of air from the Los Angeles Basin. Here, they are searching for substances which, in trace amounts, are responsible for the pollution of our environment.

A central goal of chemists throughout this century has been to isolate or synthesize new substances. "Miracle drugs" for the treatment of disease and synthetic fabrics such as nylon and Dacron were made by chemists. Chemists are actively involved in the search for new sources of energy and new approaches to conserving such "old" energy sources as petroleum and natural gas, upon which modern civilization depends.

In all their investigations, chemists work closely with people in other disciplines. A chemist doing research on solar cells or batteries for electric cars may be a member of a team that includes physicists and engineers. Drug design or the synthesis of new materials for heart valves requires active cooperation between chemists, biologists, and medical scientists. The search for the molecular basis of learning occurs at an interface between chemistry and psychology. Research on photosynthesis and chemical methods of birth control leads to results that are of interest to all those in the social sciences.

Discoveries in chemistry, as in all other areas, come from a variety of approaches. A sudden inspiration or an educated guess may persuade a chemist to follow a new path of research. With a certain amount of luck and a lot of hard work, he or she may succeed where others in the past have failed. Two requirements must be met if there is to be a high likelihood of success in any such effort. The people involved must be receptive to new ideas, willing to challenge the accepted wisdom of their time. Equally important, they must have available the means to test their ideas in the laboratory.

Regardless of where a new idea comes from, the method of testing it is one that has been used in all the sciences for at least two centuries. The "scientific method" starts with carefully designed experiments. These are carried out in a systematic way under controlled conditions. Usually, the system a chemist works with is a rather simple one.

In chemistry the experimentalist is king

1

It may consist of a single pure substance or perhaps a mixture containing two or three such substances. Data obtained with such systems, when properly interpreted, can lead to conclusions that apply to the more complex world outside the laboratory.

In this text, we will focus upon the basic principles of chemistry and their applications. Since chemistry is an experimental science, this means that we will be talking about a great many different kinds of experiments. All of these involve measurements of one type or another. It is important that you understand how these measurements are made and how they are expressed in scientific language. This topic is introduced in the first section of this chapter. Later in this chapter we will see how, on the basis of rather simple experiments, it is possible to separate and identify the components of matter.

1.1 MEASUREMENTS

Many experiments carried out in chemistry are quantitative. That is, they involve assigning numbers to such quantities as length, volume, mass, and temperature. Let us now consider the instruments used to measure those quantities and the units in which they are expressed.

Length

Most of us are familiar with a simple measuring device found in every general chemistry laboratory. This is the meter stick, which reproduces, as accurately as possible, the basic unit of length in the metric system, the **meter** (m). A meter stick is divided into one hundred equal parts, each one *centimeter* (cm) in length ($1 \text{ cm} = 10^{-2} \text{ m}$). A centimeter, in turn, is divided into ten equal parts, each one *millimeter* (mm) long ($1 \text{ mm} = 10^{-3} \text{ m}$). A much larger unit, familiar to runners, is the kilometer ($1 \text{ km} = 10^3 \text{ m}$).

In the metric system, units are related one to another by powers of 10

The prefixes "kilo," "centi," and "milli" are used in the metric system to designate units obtained by multiplying by 1000, 0.01, and 0.001, in that order. Another unit that we will use to express the dimensions of tiny particles such as atoms is the *nanometer* (nm): $1 \text{ nm} = 10^{-9} \text{ m}$.

Volume

Units of volume in the metric system are simply related to those of length. The *cubic centimeter* (cm³) represents the volume of a cube one centimeter on an edge. A larger unit is the *liter* (ℓ), which is exactly 1000 cm³:

$$1 \ \ell = 1000 \text{ cm}^3 \tag{1.1}$$

A milliliter, 1/1000 of a liter, has the same volume as a cubic centimeter.

The device most commonly used to measure volumes in general chemistry is the graduated cylinder. With this, we can measure out a known volume of a liquid, accurate to perhaps 0.1 cm³. When greater accuracy is required, we use a pipet or buret (Fig. 1.1). A pipet is calibrated to deliver a fixed volume of liquid (e.g., 25.00 ± 0.01 cm³) when filled to the mark and allowed to drain normally. Variable volumes can be delivered with about the same accuracy from a buret. With a buret, final and initial volumes must be read carefully to calculate the volume of liquid withdrawn. A volumetric flask is shown at the right of Figure 1.1. It is designed to contain a specified volume of liquid (e.g., 50, 100, . . . 1000 cm³) when filled to a level marked on the narrow neck.

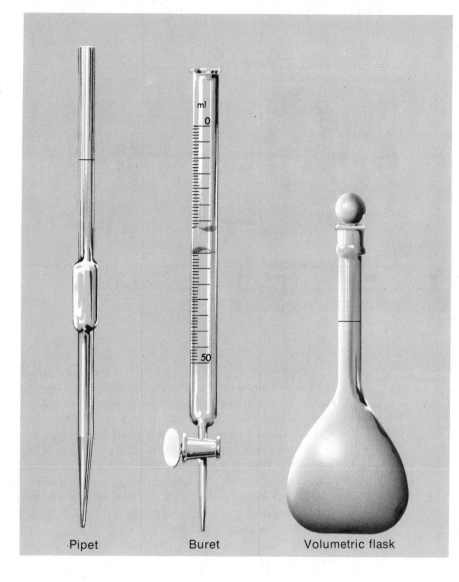

Figure 1.1 Instruments used with liquids to deliver a fixed volume (pipet), deliver a variable volume (buret), or contain a fixed volume (volumetric flask).

Pipet Buret Volumetric flask

Mass

The mass of a sample is a measure of the amount of matter it contains. In the metric system, mass may be expressed in *grams* (g), *kilograms* (kg), or milligrams (mg): $1 \text{ kg} = 10^3 \text{ g}$, $1 \text{ mg} = 10^{-3} \text{ g}$. Chemists measure mass on a balance (Fig. 1.2). To show what is involved in weighing ("massing") an object, consider the two-pan balance shown at the left of the figure. With nothing on either pan, the balance comes to rest with the two pans at the same height. To weigh an object, we place it on the left-hand pan. Pieces of metal of known mass are then added to the right-hand pan to restore balance, bringing the pans to the same height again. Under these conditions, the masses on the two pans are equal*:

$$\text{mass sample} = \text{mass metal}$$

---

*Strictly speaking, it is the weights that are equal when the pans are at the same height. However, weight = k(mass), where k is a proportionality constant that has a fixed value at a given location. So, k(mass sample) = k(mass metal), and we see that the masses must be equal.

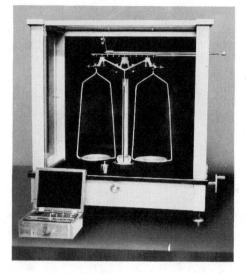
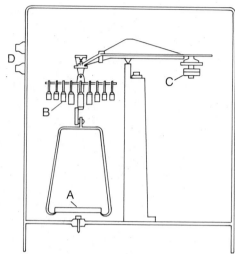

Figure 1.2 Two-pan balance *(left)* and schematic drawing of a single-pan balance *(right)*.

Nowadays most teaching and research laboratories use single-pan balances of the type shown at the right of Figure 1.2. Here a pan (A) and a set of movable masses (B) are suspended from one arm of the balance. Their mass is exactly balanced by that of a fixed mass (C) attached to the other arm. Adding a sample to the pan deflects it downwards. Balance is almost, but not quite, restored by turning the dials (D) to remove one or more movable masses. At this point, the beam is slightly tilted from a level position. An optical system is used to translate this deflection into a small mass correction. This appears as a number on a brightly lighted scale. The mass of the sample is found by adding this mass to that read on the dials. Fortunately. it takes less time to use this instrument than it does to describe it. You can carry out weighings to ± 0.001 g in a few seconds on a single-pan balance.

Temperature

The concept of temperature is familiar to all of us. This is because our bodies are so sensitive to temperature differences. When we pick up a piece of ice, we feel cold because its temperature is lower than that of our hand. After drinking a cup of coffee, we may refer to it as "hot," "lukewarm," or "atrocious." In the first two cases at least, we are describing the extent to which its temperature exceeds ours. From a slightly different viewpoint, temperature is the factor that determines the direction of heat flow. Anyone brave enough to swim in a Minnesota lake in January feels cold because heat is absorbed from his body. If he takes a hot shower afterward, which he certainly will, heat flows in the reverse direction. In general, whenever two objects at different temperatures touch each other, heat flows from the one at the higher to the one at the lower temperature.

To measure temperature we can use a mercury-in-glass thermometer. Here, we take advantage of the fact that mercury, like other substances, expands as temperature increases. When the temperature rises, the mercury in the thermometer expands up a thin tube. The total volume of the tube is only about 2 per cent of that of the bulb at the base. In this way, a rather small change in volume is made readily visible (Fig. 1.3).

Thermometers used in chemistry are marked in degrees *Celsius* (centigrade) after the Swedish astronomer, Anders Celsius (1701–1744). On this scale, the freezing point of

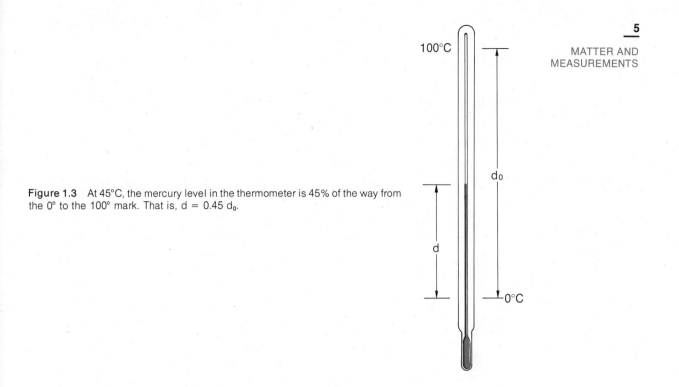

Figure 1.3 At 45°C, the mercury level in the thermometer is 45% of the way from the 0° to the 100° mark. That is, d = 0.45 d_0.

water is taken to be 0°C. The boiling point of water at one atmosphere pressure is 100°C. When we place a mercury-in-glass thermometer in a beaker containing crushed ice and water, the mercury comes to rest exactly at the 0° mark. In a beaker of boiling water, the mercury rises to the 100° mark. The distance between these two marks is divided into 100 equal parts. Each of these corresponds to a Celsius degree. Thus, a temperature of 45°C corresponds to a mercury level 45 per cent of the way from the 0° to the 100° mark.

A temperature scale in common use in the United States today is based on the work of Daniel Fahrenheit (1686–1736). He was a German instrument maker who was the first to use the mercury-in-glass thermometer. On this scale the normal freezing and boiling points of water are taken to be 32° and 212°. That is,

$$32°F = 0°C; 212°F = 100°C$$

As you can see from Figure 1.4, the Fahrenheit degree is smaller than the Celsius degree. The distance between the freezing and boiling points of water is 100 degrees on the Celsius scale and 180 degrees on the Fahrenheit scale. The relation between temperatures expressed on the two scales is

$$°F = 1.8(°C) + 32° \qquad (1.2)$$

Another temperature scale that we will find particularly useful in working with gases is the absolute or Kelvin scale. The relation between temperatures in K and °C is

$$K = °C + 273 \qquad (1.3)$$

The ° sign is omitted for Kelvin temperatures

This scale is named after Lord Kelvin, an English physicist. He showed, by a mathematical proof based upon both theory and experiment, that it is impossible to reach a temperature lower than 0 K.

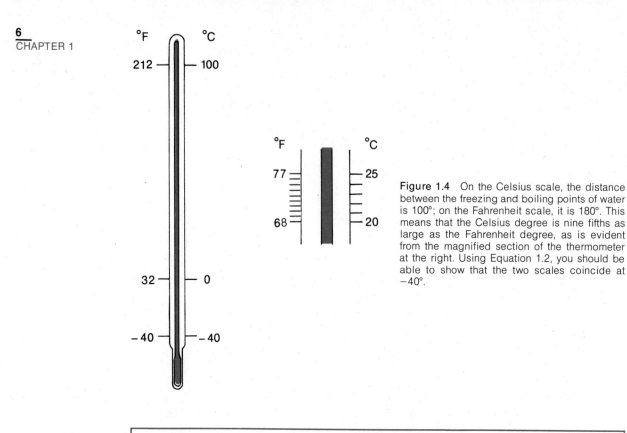

Figure 1.4 On the Celsius scale, the distance between the freezing and boiling points of water is 100°; on the Fahrenheit scale, it is 180°. This means that the Celsius degree is nine fifths as large as the Fahrenheit degree, as is evident from the magnified section of the thermometer at the right. Using Equation 1.2, you should be able to show that the two scales coincide at −40°.

Example 1.1   Express normal body temperature, 98.6°F, in °C and K.

Solution   Substituting in Equation 1.2, we obtain

$$98.6 = 1.8(°C) + 32$$

Solving:

$$1.8(°C) = 98.6 - 32 = 66.6$$

$$°C = 66.6/1.8 = 37.0$$

Applying Equation 1.3:

$$K = 37.0 + 273 = 310$$

Exercise   Convert 25°C to °F and to K. Answer: 77°F; 298 K.

Uncertainties in Measurements. Significant Figures

Every measurement we make carries with it a degree of uncertainty, or error. How large this error is depends upon the nature of the measuring device and the skill with which we use it. Suppose, for example, we try to measure out 8 cm³ of liquid using a 100-cm³ graduated cylinder. Here, the volume is likely to be in error by at least 1 cm³. With such a crude measuring device, we will be fortunate to obtain a volume closer to 8 than to 7 or 9 cm³. To obtain greater accuracy, we might use a narrow 10-cm³ cylinder, on which the divisions are much farther apart. The volume we measure now may be within 0.1 cm³ of the desired value of 8 cm³. That is, it is likely to fall in the range 7.9 to 8.1 cm³. Using a buret we can do even better. If we are very careful, we may reduce the uncertainty to 0.01 cm³.

The person who makes a measurement such as this should indicate the uncertainty associated with it. Such information is vital to anyone who wants to repeat an experiment or judge its accuracy. There are many ways to do this. We might report the three volume measurements referred to above as

8 ± 1 cm³ (large graduated cylinder)
8.0 ± 0.1 cm³ (small graduated cylinder)
8.00 ± 0.01 cm³ (buret)

In this text, we will drop the $\pm$ notation and simply write

8 cm³; 8.0 cm³; 8.00 cm³

When we do this, it is understood that there is an *uncertainty of one unit in the last digit* (1 cm³, 0.1 cm³, 0.01 cm³).

This method of citing the degree of confidence in a measurement is often described in terms of **significant figures.** We say that in 8.00 cm³ there are three significant figures. Each of the three digits in 8.00 has experimental meaning. Similarly, there are two significant figures in 8.0 cm³ and one significant figure in 8 cm³.

Frequently we need to know the number of significant figures in a measurement reported by someone else. Usually, this is easy to find. For example, in "12.3 g" and "6.04 g" there are clearly three significant figures. Sometimes it is not so obvious. Suppose you are told that a human hair weighs about 0.002 g. How many significant figures are there in this quantity? A moment's reflection should convince you that there is only one (the "2" at the end). The zeros serve only to fix the position of the decimal point. This might be a bit more obvious if the mass were expressed in exponential notation (Appendix 4). Here, we would write 0.002 g as

$$2 \times 10^{-3} \text{ g}$$

Now, clearly, there is only one significant figure. The uncertainty is $\pm 1 \times 10^{-3}$ g.

Sometimes the number of significant figures in a reported measurement is ambiguous. Suppose you are told that a coin weighs "50 g." You cannot be sure how many of these digits are meaningful. Perhaps the coin was weighed to the nearest gram (50 ± 1 g). If so, both the "5" and the "0" are known; there are two significant figures. Then again, the coin might have been weighed only to the nearest ten grams (50 ± 10 g). In this case, only the "5" is known accurately; there is only one significant figure. About all we can do in situations like this is to wish that the person who carried out the weighing had used exponential notation. The mass should have been reported as either

5.0×10^1 g (2 significant figures)
or 5×10^1 g (1 significant figure)

Example 1.2   An instructor asks a class in general chemistry to weigh a thimble. Three students report values of
   a. 20.03 g   b. 20.0 g   c. 0.02003 kg
How many significant figures should be assumed in each case?

Solution
   a. 4.
   b. 3. The zero after the decimal point is significant. It indicates that the thimble was weighed to the nearest 0.1 g.

c. 4. The zeros at the left are not significant. They are there only because the mass was expressed in kilograms rather than in grams. Note that "20.03 g" and "0.02003 kg" represent the same quantity.

Exercise   Give the number of significant figures in 2.6×10^2 cm³; 2.40×10^{-3} cm³. Answer: 2; 3.

Uncertainties in Calculated Quantities

Most of the quantities that we measure are not end results in themselves. Instead, they are used to calculate other quantities. We might, for example, measure the mass and volume of a sample in order to determine its density (Example 1.3). The precision of any such derived result is limited by those of the measurements upon which it is based. **When experimental quantities are multiplied or divided, the number of significant figures in the result is that in the least precise measurement.**

Example 1.3   A student checks the purity of a water sample by determining its density (mass per unit volume) at 25°C. (The value given for pure water in a handbook is 0.9970 g/cm³.) He measures out 25 cm³ of the sample from a cylinder and determines its mass to be 25.624 g. What should he report for the density?

Solution   The precision of the density determination is limited by that of the volume measurement, which has only two significant figures.

$$\text{density} = \frac{\text{mass}}{\text{volume}} = \frac{25.624 \text{ g}}{25 \text{ cm}^3} = 1.0 \text{ g/cm}^3$$

The five significant figures in the mass measurement are wasted. The student might as well have used a crude balance weighing to the nearest gram.

Exercise   Suppose the volume in Example 1.3 were known to be 25.0 cm³. What should the student report for the density? Answer: 1.02 g/cm³.

When measured quantities are added or subtracted, the uncertainty in the result is found in a different way. It is determined by the **absolute uncertainty** (rather than the number of significant figures) in the **least precise measurement**. Suppose, for example, we wish to calculate the total mass of a solution containing 10.21 g of instant coffee, a "pinch" of sugar, 0.2 g, and 256 g of water. The uncertainties in these masses are:

	Mass		Uncertainty
Instant coffee	10.21	±	0.01 g
Sugar	0.2	±	0.1 g
Water	256	±	1 g
Total mass	266	±	1 g

The sum of the masses cannot be more precise than that of the water (±1 g). The total mass should be reported as 266 g rather than 266.4 g or 266.41 g.

Frequently, in subtracting one quantity from another, we "lose" significant figures. To illustrate, suppose a student finds that a crucible weighs 56.245 g. She then adds a piece of magnesium ribbon to the crucible and determines the total mass to be 57.148 g. The mass of the magnesium is obtained by difference:

$$57.148 \text{ g} - 56.245 \text{ g} = 0.903 \text{ g}$$

Note that the uncertainty in the mass of the magnesium is the same as that in the two measured masses, 0.001 g. However, the quantity 0.903 g contains only three significant figures. This is true even though each measured mass contains five significant figures (57.148 g, 56.245 g).

In applying the rules given above, you should keep in mind one important point. Certain numbers involved in calculations are exact rather than approximate. If you were asked to express in feet a measured length in inches, 6.10 in, using the relation

If there are 25 students in a class, that number is exact

$$1 \text{ ft} = 12 \text{ in}$$

your answer should be given to three significant figures. This is the number of significant figures in the measured quantity, 6.10 in. The "12" in the equation is a defined quantity. There are exactly twelve inches in exactly one foot. A similar situation applies with the equation relating Fahrenheit and Celsius temperatures:

$$°F = 1.8(°C) + 32°$$

The numbers 1.8 and 32 are exact. Hence, they do not affect the accuracy of any calculation involving a temperature conversion.

1.2 CONVERSION OF UNITS

Frequently, we want to convert measurements expressed in one unit (e.g., grams) to another unit (milligrams or kilograms). Conversions within the metric system, where the units are related to each other by powers of ten, are readily carried out. The arithmetic is somewhat more complex with conversions within the English system where, for example:

$$1 \text{ ft} = 12 \text{ in}; \qquad 1 \text{ gallon} = 4 \text{ qt}; \qquad 1 \text{ lb} = 16 \text{ oz}$$

Because of its simple, decimal relations, the metric system is used by scientists. Indeed, it is in common use in almost every country in the world except the United States.

To change measured quantities from one set of units to another, we follow what is known as a conversion factor approach. To illustrate, suppose we want to convert a length of 25.6 in to feet. We know that 1 ft = 12 in. Dividing both sides of this equation by 12 in gives a quotient equal to unity.

$$\frac{1 \text{ ft}}{12 \text{ in}} = \frac{12 \text{ in}}{12 \text{ in}} = 1$$

We multiply 25.6 in by the quotient 1 ft/12 in, which is called a *conversion factor*. Since the conversion factor equals 1, this does not change the value of the length. However, it does accomplish the desired conversion of units.

$$25.6 \text{ in} \times \frac{1 \text{ ft}}{12 \text{ in}} = 2.13 \text{ ft}$$

The relation 1 ft = 12 in can be used equally well to convert a length in feet, say 4.00 ft, to inches. In this case, we divide both sides of the equation by 1 ft:

$$\frac{12 \text{ in}}{1 \text{ ft}} = 1$$

Multiplying 4.00 ft by the quotient 12 in/1 ft converts the length from feet to inches:

$$4.00 \text{ ft} \times \frac{12 \text{ in}}{1 \text{ ft}} = 48.0 \text{ in}$$

Notice that a single relation (1 ft = 12 in) gives us two conversion factors, 1 ft/12 in and 12 in/1 ft. Both of these are equal to unity. In making a conversion, we choose the factor that cancels out the unit we want to get rid of.

initial quantity × conversion factor(s) = desired quantity

Conversions between English and metric units are made in a similar way. The relations required can be obtained from Table 1.1. Suppose we find from a road map that the distance from St. Louis to Chicago is 259 miles. To convert this to kilometers, we note from Table 1.1 that 1 mile = 1.609 km. Since we want to go from miles to kilometers, we use the conversion factor

$$\frac{1.609 \text{ km}}{1 \text{ mi}}$$

Multiplying by the given distance, 259 miles, we have

$$259 \text{ mi} \times \frac{1.609 \text{ km}}{1 \text{ mi}} = 417 \text{ km}$$

TABLE 1.1   RELATIONS BETWEEN LENGTH, VOLUME, AND MASS UNITS

METRIC		ENGLISH		METRIC-ENGLISH	
Length					
1 km	$= 10^3$ m	1 ft	$= 12$ in	1 in	$= 2.54$ cm
1 cm	$= 10^{-2}$ m	1 yd	$= 3$ ft	1 m	$= 39.37$ in
1 mm	$= 10^{-3}$ m	1 mile	$= 5280$ ft	1 mile	$= 1.609$ km
1 nm	$= 10^{-9}$ m				
Volume					
1 m^3	$= 10^6$ cm$^3 = 10^3 \ell$	1 gallon	$= 4$ qt $= 8$ pt	1 ft^3	$= 28.32 \ell$
1 cm^3	$= 1$ ml $= 10^{-3} \ell$	1 qt (Can.)	$= 69.35$ in^3	1 ℓ	$= 0.8799$ qt (Can.)
		1 qt (U.S. liq.)	$= 57.75$ in^3	1 ℓ	$= 1.057$ qt (U.S. liq.)
Mass					
1 kg	$= 10^3$ g	1 lb	$= 16$ oz	1 lb	$= 453.6$ g
1 mg	$= 10^{-3}$ g	1 short ton	$= 2000$ lb	1 g	$= 0.03527$ oz
1 metric ton	$= 10^3$ kg			1 metric ton	$= 1.102$ short ton

Frequently we need to carry out more than one conversion to work a problem. This can be done by setting up successive conversion factors (Example 1.4).

Example 1.4   Analysis of an air sample taken from the Holland Tunnel in New York City shows that it contains 3.5×10^{-6} g/ℓ of carbon monoxide. Express the concentration of carbon monoxide in lb/ft³.

Solution   Two conversions are required, one from grams to pounds using the factor

$$1 \text{ lb} = 453.6 \text{ g}; \text{ (Table 1.1)}$$

the other from liters to cubic feet

$$1 \text{ ft}^3 = 28.32 \text{ } \ell; \text{ (Table 1.1)}$$

We can set up the arithmetic in a single expression

$$3.5 \times 10^{-6} \frac{g}{\ell} \times \frac{1 \text{ lb}}{453.6 \text{ g}} \times \frac{28.32 \text{ } \ell}{1 \text{ ft}^3} = 0.22 \times 10^{-6} \text{ lb/ft}^3$$

or, in standard exponential notation (Appendix 4)

$$= 2.2 \times 10^{-7} \text{ lb/ft}^3$$

Exercise   Convert a density of 3.50 g/cm³ to kg/m³. Answer: 3.50×10^3 kg/m³.

Experiments in general chemistry frequently involve measured quantities other than those given in Table 1.1 (length, volume, mass). You may, for example, measure the **time** required to carry out a reaction. This might be expressed in days (d), hours (h), minutes (min), or seconds (s).

$$1 \text{ d} = 24 \text{ h}; \qquad 1 \text{ h} = 60 \text{ min}; \qquad 1 \text{ min} = 60 \text{ s}$$

In experiments involving gases, it is important to measure the **pressure.** Many different units are used to express pressure. The more common ones are listed in Table 1.2, p. 12, along with the relations between them. In this text we will most commonly use the two units

—*millimeter of mercury* (mm Hg). This is the pressure exerted by a column of mercury one millimeter in height.

—*atmosphere* (atm). An atmosphere (760 mm Hg) is approximately the pressure of the atmosphere on a "normal" day at sea level.

Pascals (Pa) and kilopascals (1 kPa = 10^3 Pa) will be used less frequently.

In every reaction, there is an energy change whose magnitude can be measured. Chemists use a variety of energy units (Table 1.2). You may be most familiar with the calorie, which is the amount of heat required to raise the temperature of one gram of water by 1°C. Increasingly, nowadays, scientists are using a different unit, the **joule.** This is the preferred unit of energy in the International System of Units (p. 12). We will use the joule extensively throughout this text. The basic definition of this energy unit is given in Appendix 1. Notice from Table 1.2 that one calorie is about four joules; more exactly, 1 cal = 4.184 J. This means that an energy change expressed in joules will have a magnitude about four times that for the same energy change in calories. For example, it takes about four joules to raise the temperature of one gram of water 1°C.

The joule is actually a rather small energy unit. The energy released when a match

As you work with the various units in this section, you will feel more comfortable with them

TABLE 1.2   RELATIONS BETWEEN PRESSURE AND ENERGY UNITS
PRESSURE
1 atm = 760 mm Hg = 1.013×10^5 Pa = 14.70 lb/in^2
1 torr = 1 mm Hg
1 bar = 10^5 Pa
ENERGY
1 cal = 4.184 J = 4.129×10^{-2} $\ell \cdot$atm = 2.612×10^{19} electron volts
1 J = 10^7 ergs

burns is of the order of 2000 J. Most often, in chemical reactions, we will express energy changes in **kilojoules:**

$$1 \text{ kJ} = 10^3 \text{ J}$$

Conversions involving these units are carried out in the ordinary way, using the relations given in Table 1.2. (A more extensive list of relations between units of all types is given in Appendix 1, p. A.1.) Example 1.5 illustrates the conversion from calories to joules.

Example 1.5   When one gram of gasoline burns in an automobile engine, the amount of energy given off is about 1.03×10^4 cal. Express this in joules.

Solution   From Table 1.2 we see that 1 cal = 4.184 J. To convert calories to joules, we use the conversion factor 4.184 J/1 cal.

$$1.03 \times 10^4 \text{ cal} \times \frac{4.184 \text{ J}}{1 \text{ cal}} = 4.31 \times 10^4 \text{ J} \quad \text{(3 significant figures)}$$

Exercise   Convert 836.8 J to calories. Answer: 2.000×10^2 cal.

Using "proportions" is not as effective as using conversion factors

The conversion factor approach shown in Examples 1.4 and 1.5 will be used throughout this text. If this is your first contact with it, it may seem awkward or artificial. You will find, however, that it is the best way to solve a wide variety of problems in chemistry. It is particularly useful when multiple conversions are required (Example 1.4) or when the units may be unfamiliar to you (Example 1.5).

SI Units

As indicated by the large number of entries in Tables 1.1 and 1.2, many different units are often used to express a single quantity. Pressure may be expressed in atmospheres, millimeters of mercury, torrs, pascals. . . . This proliferation of units has long been of concern to scientists. In 1960 the General Conference of Weights and Measures recommended a self-consistent set of units based upon the metric system.

SI is pronounced ess eye

TABLE 1.3 SI UNITS	
QUANTITY	UNIT
Length	meter (m)
Volume	cubic meter (m³)
Mass	kilogram (kg)
Temperature	kelvin (K)
Time	second (s)
Pressure	pascal (Pa)
Energy	joule (J)

In the International System of Units (SI), a single base unit is used for each measured quantity. An extended discussion of this system is presented in Appendix 1. Table 1.3 lists the recommended SI units for each of the quantities referred to in this chapter.

For the most part, we will use SI units in this text. For example, we will express atomic dimensions in nanometers rather than angstroms (1 nm = 10^{-9} m = 10 Å). Energy changes will be expressed in joules rather than calories (1 cal = 4.184 J). However, in dealing with pressures, we will more commonly use the atmosphere or the millimeter of mercury as opposed to the pascal or kilopascal (1 kPa = 10^3 Pa). If desired, non-SI units such as the millimeter of mercury can readily be converted to SI units (Example 1.6).

Example 1.6   On a certain day, the pressure read on a barometer is 742 mm Hg. Express this in kilopascals, the pressure unit commonly used in Canada and other countries using SI.

Solution   From Table 1.2 we obtain the relation between millimeters of mercury and pascals: 760 mm Hg = 1.013×10^5 Pa. One kilopascal is 10^3 Pa. Setting up the conversion in two steps:

$$742 \text{ mm Hg} \times \frac{1.013 \times 10^5 \text{ Pa}}{760 \text{ mm Hg}} \times \frac{1 \text{ kPa}}{10^3 \text{ Pa}} = 98.9 \text{ kPa}$$

(Atmospheric pressure is usually close to 100 kPa.)

Exercise   Express 96.8 kPa in atmospheres. Answer: 0.956 atm.

1.3   KINDS OF SUBSTANCES

Chemists, by methods which we will consider shortly, have isolated many thousands of pure substances. These substances can be divided into two classes. Some of them, called **elements,** cannot be broken down by chemical means into two or more pure substances. All other pure substances are **compounds.** A compound, by definition, is a pure substance that can be broken down into two or more elements.

Elements and the Periodic Table

There are, at latest count, 106 known elements. Of these, 91 occur naturally. Many elements are familiar to all of us. The charcoal used in outdoor grills is nearly pure

In nature most elements occur in chemical combinations with other elements

carbon. Electrical wiring, jewelry, and water pipes are often made from copper, a metallic element. Another such element, aluminum, is used in many household utensils. The shiny liquid in the thermometers you use is still another metallic element, mercury.

In chemistry, an element is identified by its symbol. This consists of one or two letters, usually derived from the name of the element. Thus the symbol for carbon is C; that for aluminum is Al. Sometimes the symbol comes from the Latin name of the element or one of its compounds. The two elements copper and mercury, which were known in ancient times, have the symbols Cu *(cuprum)* and Hg *(hydrargyrum)*. Other examples include

Antimony	Sb	Lead	Pb	Sodium	Na
Gold	Au	Potassium	K	Tin	Sn
Iron	Fe	Silver	Ag		

You are probably familiar with a device used to organize the properties of elements. This is the Periodic Table, shown on the inside cover of this text. In later chapters we will consider how the Table is constructed and how it is used for a variety of purposes. At the moment, we need only be concerned with three general features of the Periodic Table.

1. The horizontal rows are referred to as **periods.** Thus the first period consists of the two elements hydrogen (H) and helium (He). The second period starts with lithium (Li) and ends with neon (Ne), and so on.

2. The vertical columns are known as **groups.** The groups at the far left and the right of the Table are numbered: Groups 1 and 2 are at the left, Groups 3 to 8 at the right. Elements that fall in these groups are often referred to as *main-group* elements. The elements in the center of periods 4 through 6 are called *transition* elements (e.g., Sc through Zn in the fourth period).

3. The diagonal line or stairway that runs from the upper left to the lower right of the Periodic Table separates metals from nonmetals. Elements below and to the left of this line are metals. All metals possess certain common properties. In particular, they are good conductors of electricity. All the metals, with one exception (Hg), are solids at 25°C. As you can see from the location of the diagonal line, most elements (about 75% of them) are metals.

Elements above and to the right of the diagonal line are nonmetals. With a few exceptions, these elements do not conduct an electrical current. Several nonmetals are gases at 25°C and 1 atm. The gaseous nonmetals include nitrogen (N), oxygen (O), fluorine (F), chlorine (Cl), and all the elements in the vertical column at the far right of the Table (He, Ne, Ar, Kr, Xe, Rn). One nonmetal, bromine (Br), is a liquid. The remaining elements of this type are solids at room temperature and atmospheric pressure. Included among these are the familiar nonmetals carbon (C), phosphorus (P), sulfur (S), and iodine (I). Although there are only a few nonmetals, they are very important in chemistry.

Air contains elementary nitrogen, oxygen, and argon

Along the diagonal line in the Periodic Table are several elements which are difficult to classify as metals or nonmetals. They have properties in between those of elements in the two classes. In particular, their electrical conductivities are intermediate between those of metals and nonmetals. The six elements

B	Si	Ge	As	Sb	Te
boron	silicon	germanium	arsenic	antimony	tellurium

are often called *metalloids*. Certain metalloids, notably Si and Ge, are used in semiconductor devices (Fig. 1.5).

As this discussion implies, metallic character varies in a systematic way with

Figure 1.5 A semiconductor chip made of extremely pure silicon. (The contact lenses show the relative size of the chip.) (Courtesy of Texas Instruments.)

position in the Periodic Table. Metallic character *decreases* as one moves *across* from left to right within a given period. For example, the third period starts with three elements which are distinctly metallic (Na, Mg, Al), moves on to a metalloid (Si), and is completed by four nonmetals (P, S, Cl, Ar). Metallic character *increases* as one moves *down* a given group. Consider, for example, Group 4. Carbon (C) is a nonmetal; silicon (Si) and germanium (Ge) are metalloids; tin (Sn) and lead (Pb) have the properties we associate with metals.

Compounds

As you might suppose, there are a great many more compounds than elements. The elements can combine with one another in many different ways. Thousands of different compounds have been prepared in the laboratory or extracted from natural sources. Each year hundreds of new compounds are reported.

Every compound contains two or more elements in fixed proportions by mass. Water, by far our most abundant compound, contains the two elements hydrogen and oxygen. In a 100-gram sample of pure water, there are 11.19 g of hydrogen and 88.81 g of oxygen. We would say that water contains 11.19% by mass of hydrogen and 88.81% by mass of oxygen. The chemical composition of sugar (sucrose) is somewhat more complex. This compound contains the three elements carbon (42.10%), hydrogen (6.48%), and oxygen (51.42%).

The properties of compounds are very different from those of the elements they contain. Ordinary table salt, sodium chloride, is a white, unreactive solid. As you can guess from its name, it contains the two elements sodium and chlorine. Sodium (Na) is a shiny, extremely reactive metal. Chlorine (Cl) is a poisonous, greenish-yellow gas. Clearly, when these two elements combine to form sodium chloride, a profound change takes place.

Many different methods can be used to resolve compounds into their elements. Sometimes, but not often, heat alone is sufficient. Mercury(II) oxide, a compound of mercury and oxygen, decomposes to its elements when heated to 600°C. Joseph Priestley, an English chemist, discovered oxygen 200 years ago by carrying out this reaction. He exposed a sample of mercury(II) oxide to an intense beam of sunlight focused by a powerful lens. A more common method of resolving compounds is called electrolysis. This involves passing an electric current through a compound, usually in the liquid state. In this way it is possible to separate water into its two elements, hydrogen and oxygen.

We can make sodium and chlorine from molten table salt by this method

1.4 IDENTIFICATION OF PURE SUBSTANCES

The chemist who isolates a pure substance has the further problem of identifying it and checking its purity. If the substance has never been prepared before, the task is a difficult one indeed. More frequently, the substance is one whose properties have been established and recorded in chemical reference books. In this case the problem is simpler but by no means trivial.

Chemical and Physical Properties

Substances are sometimes identified on the basis of their *chemical properties*. These are properties observed when the substance undergoes a chemical change, converting it to one or more other substances. You might, for example, show that a red solid is mercury(II) oxide by heating it in air and showing that it decomposes to mercury, a silvery liquid, and oxygen, a colorless gas.

More commonly, elements or compounds are identified by measuring their *physical properties*. These are properties which can be determined without changing the chemical identity of a substance. They include:

density =
$\dfrac{\text{mass of sample}}{\text{volume of sample}}$

—*density,* the mass-to-volume ratio of a substance. Densities of liquids or gases can be found by measuring, independently, the mass and volume of a sample. For solids, density is a bit more difficult to determine. A common approach is shown in Figure 1.6. The mass of the solid sample is first measured, using a balance. Its volume is found indirectly by determining the volume of liquid displaced by the solid.

—*melting point,* the temperature at which a pure substance changes from the solid to the liquid state. The melting point of a pure substance is identical with its *freezing point.*

—*boiling point,* the temperature at which a liquid boils, forming a gas.

—*solubility,* the extent to which a substance dissolves in a particular solvent. This may be expressed in various ways. A common method is to state the number of grams of the substance that dissolves in 100 g of solvent at a given temperature. Thus at 25°C, 36.5 g of sodium chloride dissolves in 100 g of water. In contrast, benzene is almost insoluble in water (0.07 g benzene per 100 g water).

If a substance has all of the physical properties of benzene, it *is* benzene

By measuring several physical properties and comparing them to known values, it is usually possible to identify a substance. Suppose, for example, that you find that a certain liquid boils at 81°C, is nearly insoluble in water, freezes at 5°C, and has a density of 0.882 g/cm³ at 25°C. Most likely, you are dealing with a sample of slightly impure benzene. Certainly the sample is neither water nor methyl alcohol.

Figure 1.6   One way to determine the density of a solid is to first find its mass, m. The solid is then added to a flask of known volume, V. The volume of water, V_w, required to fill the flask is determined. The density of the solid must then be $m/(V - V_w)$.

	Benzene	Water	Methyl alcohol
Boiling point	80°C	100°C	65°C
Melting point	5.5°C	0°C	− 98°C
Density (g/cm³)	0.879	1.00	0.787

Color. Absorption Spectrum

Some substances can be identified, at least tentatively, on the basis of their color. Gaseous nitrogen dioxide has a brown color. Bromine vapor is red and iodine vapor is violet. A water solution of copper sulfate is blue, a solution of potassium permanganate is purple, and so on.

The colors of gases and liquids are due to the absorption of visible light. Sunlight is a mixture of radiation of various wavelengths (λ). Light falling within a certain wavelength range is associated with a particular color. For example, light in the range 400–450 nm appears violet; for blue light, λ = 450–490 nm (see Fig. 1.7 and color plate 6, center of book). Bromine absorbs light in these regions and transmits light of other wavelengths. The subtraction of the violet and blue components of sunlight accounts for the red color of bromine (Table 1.4). In contrast, a water solution of potassium permanganate absorbs light in the green region, midway through the visible range. Light emitted from such a solution is rich in the blue (shorter wavelength) and red (longer wavelength) components. Hence, it appears purple, a mixture of blue and red (see color plate 1, center of book).

Substances which do not absorb visible light are colorless (or white, if they are solids). These substances often absorb radiation outside the visible region. Sometimes this occurs in the ultraviolet, at wavelengths below 400 nm. Benzene, a colorless liquid,

A black substance absorbs all colors

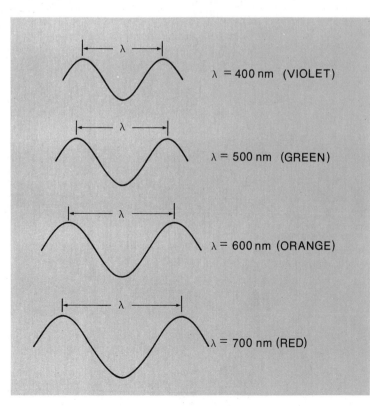

Figure 1.7 Relation between wavelength and color of visible light. The wavelength is the distance between successive crests of the wave. As wavelength increases, the color changes from violet (400 nm) to green (500 nm), orange (600 nm), and red (700 nm). Sunlight is a mixture of all wavelengths in the visible region as well as ultraviolet (<400 nm) and infrared (>700 nm) radiation.

λ = 400 nm (VIOLET)

λ = 500 nm (GREEN)

λ = 600 nm (ORANGE)

λ = 700 nm (RED)

TABLE 1.4 COLOR OF SUBSTANCES WHICH ABSORB LIGHT IN THE VISIBLE REGION

Wavelength Region	Color Absorbed	Color Transmitted
<400 nm	Ultraviolet	Colorless
400–450	Violet	Red, orange, yellow
450–490	Blue	Red, orange, yellow
490–550	Green	Purple
550–580	Yellow	Purple
580–650	Orange	Blue, green
650–700	Red	Blue, green
>700 nm	Infrared	Colorless

absorbs strongly around 255 nm. Ozone, a gaseous form of the element oxygen, is another substance that absorbs in the ultraviolet. Much of the harmful ultraviolet radiation of sunlight is taken up by ozone in the upper part of the earth's atmosphere.

Absorption in the infrared region, at wavelengths greater than 700 nm, is quite common. Water and carbon dioxide are among the substances that absorb in the infrared. Their presence in the atmosphere has an insulating effect. The earth, like other warm bodies, gives off heat in the form of infrared radiation. Much of this is absorbed by water vapor and carbon dioxide, preventing excessive loss of heat to outer space.

> To be colored, a sample must absorb visible light

Most organic compounds absorb at well-defined wavelengths in the infrared region. These substances can often be identified by exposing them to infrared radiation covering a wide range of wavelengths. The absorption can then be measured as a function of wavelength. In this way, an *absorption spectrum* of the type shown in Figure 1.8 is obtained. Such a spectrum serves as a "fingerprint" to identify the substance when it is compared with the spectrum of a known substance. Organic chemists use this technique routinely. One of its major advantages is that it can be applied to samples weighing a milligram or less.

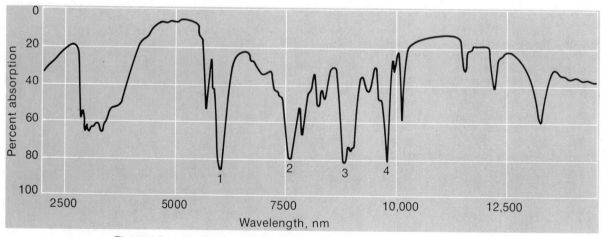

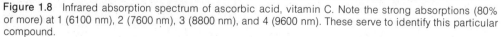

Figure 1.8 Infrared absorption spectrum of ascorbic acid, vitamin C. Note the strong absorptions (80% or more) at 1 (6100 nm), 2 (7600 nm), 3 (8800 nm), and 4 (9600 nm). These serve to identify this particular compound.

1.5 SEPARATION OF MIXTURES

Very few elements and compounds occur in nature in the pure state. Most often, they are found in mixtures with other substances. Air is a *homogeneous* (uniform) mixture (a **solution**) of several different gases. These include the elements nitrogen, oxygen, and argon and the compounds water and carbon dioxide. Seawater is a solution of various salts, including sodium chloride, in water. Most of the rocks and minerals in the earth's crust are *heterogeneous* (nonuniform) mixtures of several different substances.

Chemists, in carrying out reactions, ordinarily work with pure substances. To obtain a pure substance, it is often necessary to separate it from a mixture containing impurities. Such separations are based on differences in properties between the components of a mixture. Sometimes the process is a very simple one. Iron filings can be separated from powdered sulfur by using a magnet to extract the iron. Sand can be removed from sugar by shaking the mixture with water. The sugar goes into solution; the sand remains behind and can be filtered off.

There are several separation techniques which are more generally useful than those we have just mentioned. In this section, we will consider two processes commonly used in chemistry. One of these is distillation, used to separate the components of a liquid solution. The other is fractional crystallization, used to extract a pure solid from a mixture. Later, in Section 1.6, we will discuss a still more versatile method, chromatography, which can be applied to gaseous, liquid, or solid mixtures.

Chemists like pure substances so they can know what they are working with

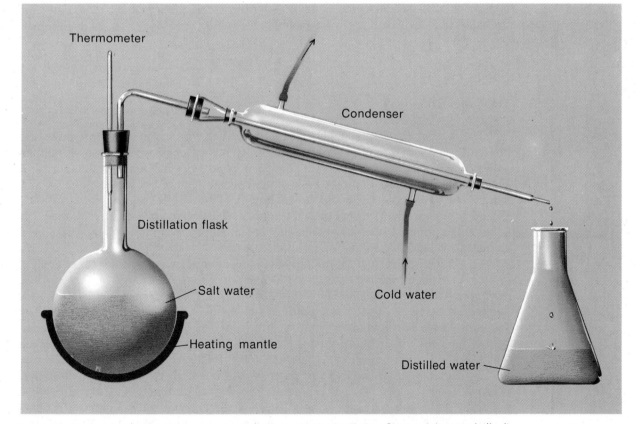

Figure 1.9 Pure water can be obtained from salt water by simple distillation. Since salt is not volatile, it remains in the distillation flask.

Distillation and Fractional Distillation

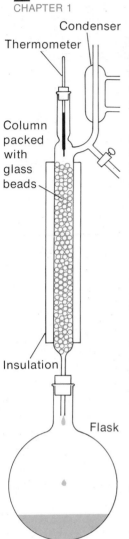

Condenser

Thermometer

Column packed with glass beads

Insulation

Flask

Figure 1.10 A simple fractionating column. Temperature decreases as one moves up the column. The less volatile component tends to condense back into the flask, while the more volatile component distills off and is recovered.

A mixture of two substances, only one of which is volatile, can be separated by distillation. A simple distillation apparatus which can be used to separate sodium chloride from water is shown in Figure 1.9. When the solution is heated the water boils off, leaving a residue of solid sodium chloride in the distilling flask. The water may be collected by passing the vapor down a cold tube and thus condensing it as a liquid. In many arid areas of the world, distillation is used to obtain fresh water from seawater.

If both components of a solution are volatile liquids, simple distillation does not give a complete separation. Consider, for example, a 50-50 mixture of ether (bp = 35°C) and benzene (bp = 80°C). If this solution is heated, it begins to boil somewhat above 35°C. The vapor formed is richer than the solution in the more volatile component, ether. However, the condensed vapor, called the *distillate,* still contains some benzene. The *residue* in the distilling flask becomes richer in benzene but still contains some ether. If we continue heating until half of the solution has distilled, we might find that the distillate contains 70% ether while the residue is 70% benzene.

To improve upon this separation, we might repeat the process, using the two fractions obtained in the first distillation. This would give a distillate still richer in ether (perhaps 85%) and a residue richer in benzene. A more effective way to carry out this process of **fractional distillation** is to use a fractionating column (Fig. 1.10). The glass beads in the column provide a surface on which much of the benzene condenses and falls back into the distilling flask. The more volatile ether tends to pass through the column without condensing. If the rate of heating and the temperature of the column are carefully controlled, a good separation can be achieved. Fractional distillation is used routinely in the petroleum industry. There the columns used are much larger and more complex than those shown in Figure 1.10. By this technique, crude oil is separated into fractions such as gasoline (bp 40–200°C), kerosene (bp 175–325°C), and diesel fuel (bp > 275°C).

Fractional Crystallization

The reagent grade solid chemicals that you work with in the laboratory have been purified, usually by fractional crystallization. In this process, we start with an impure solid, A, which we wish to purify. It contains a relatively small amount of impurity, B. The mixture of A and B is dissolved in a minimum amount of hot solvent, often boiling water. The solution is then cooled to room temperature or below. If all goes well, the solid that separates out on cooling will be pure A. This can be filtered off and dried. The solution that remains (the filtrate) should contain all of the impurity along with a small amount of A. This filtrate is ordinarily discarded.

To obtain a pure solid by fractional crystallization, two conditions must apply:

1. *The major component, A, must be much more soluble at high than at low temperatures.* Otherwise, much of it will be lost. That is, it will remain in solution upon cooling. Looking at Figure 1.11, you can see that it would be futile to try to purify sodium chloride by fractional crystallization from water. Its solubility is nearly independent of temperature. In contrast, this process works quite well with potassium nitrate, which is about eight times more soluble at 100°C than at 20°.

2. *The amount of impurity, B, must be rather small.* Otherwise, it will tend to crystallize out on cooling, contaminating the desired product, A. This is likely to happen with a mixture containing 10–20% B. Several recrystallizations might be required to obtain pure A. With a 50-50 mixture, the technique is likely to be completely ineffective.

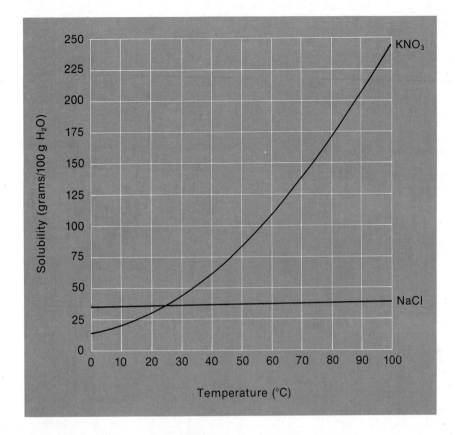

Figure 1.11 The solubility of potassium nitrate (KNO_3) in water increases rapidly with temperature, while that of sodium chloride (NaCl) remains nearly constant. This means that potassium nitrate can readily be purified by fractional crystallization while sodium chloride cannot.

1.6 CHROMATOGRAPHY

The separation techniques described in Section 1.5 have been in use for at least two centuries. In contrast, the process called chromatography is relatively new. It has become popular only within the past 25 years. The word chromatography is derived from the Greek word *chroma,* meaning color. In most of the early experiments the separated components were identified by their colors.

Chromatography has several advantages over more classical methods of separation. For one thing, it can be applied to very complex mixtures. As many as 20 amino acids can be isolated from a protein sample by this method. Moreover, it can be used for very small samples or for substances present at very low concentrations. Chromatographic techniques have been developed to detect air pollutants at concentrations of one part per million or less.

To illustrate the use of this method, let us consider a typical experiment in *paper chromatography,* which is often used in general chemistry. The sample is a liquid solution containing two or more dissolved solids. It is applied as a spot to a long strip of filter paper, near the lower edge. The spot is allowed to dry and the paper hung in a stoppered flask. Enough solvent is used to almost, but not quite, reach the spot. As the solvent rises by capillary action, the components of the sample travel along with it. A component which is very soluble tends to move farther up the paper. Another component that is less soluble, or more strongly absorbed by the paper, moves a small distance. In this way, a separation occurs. If the substances are colored, they are easily visible (see color plate 2, center of book). If not, some process must be used to detect them and make their positions clear.

Chromatography is among the most powerful techniques now available for resolving a mixture into its components

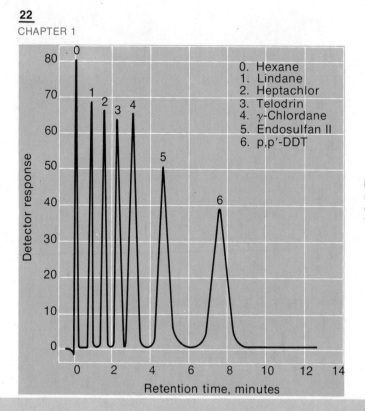

0. Hexane
1. Lindane
2. Heptachlor
3. Telodrin
4. γ-Chlordane
5. Endosulfan II
6. p,p'-DDT

Figure 1.12 A mixture of six different insecticides dissolved in hexane can be separated by vapor phase chromatography. The hexane comes out of the column first, while DDT is retained longest.

If desired, the components can be recovered by cutting the paper into sections and extracting them with a solvent.

The process just described is suitable for separating two or more solids. Many other chromatographic techniques are available. One of the most widely used is vapor phase chromatography (VPC). Here, the components are separated as vapors. The mixture is injected into a heated glass tube at one end. The tube is packed with a finely divided solid, often coated with a high-boiling viscous liquid. An unreactive "carrier" gas, often helium, is passed through the tube. The components of the sample gradually separate as they vaporize into the helium or adsorb onto the packing. Usually the more volatile components move faster and come out of the column first. As successive fractions leave the column, they activate a detector and recorder. The end result is a plot such as that shown in Figure 1.12.

SUMMARY

Chemistry is the study of matter and its composition, properties, structure, and reactions. Matter is composed of elements and compounds, which may occur as pure substances or as mixtures. A compound is a substance in which two or more elements are combined chemically. An element is unique; no two elements have the same symbol or exactly the same properties. Elements can be classified roughly as metals or nonmetals by their electrical conductivity. Metals are good conductors while most nonmetals are nonconductors. The Periodic Table is a convenient tabulation of the elements; we will refer to it frequently in future chapters. In the Table, metals are to the left and below the step-like dividing line beginning at boron (B). Nonmetals are located to the right and above this line.

Substances can be identified and classified by measuring their properties. Included among these are density, melting point, boiling point, and solubility. Still another property of a substance is the extent to which it absorbs radiant energy. Absorption in the visible region (400–700 nm) produces color. Infrared (>700 nm) and ultraviolet (<400 nm) radiation may also be absorbed.

Differences in properties such as boiling point or solubility permit components to be separated from a mixture. Distillation, fractional distillation, and fractional crystallization are among the most common separation techniques. Another separation method, chromatography, was discussed in a special section at the end of this chapter.

Many different kinds of measurements have been discussed in this chapter. All measured quantities have an uncertainty whose magnitude depends upon the instrument used and the skill of the person using it. Significant figures indicate the degree of uncertainty in a measurement (Examples 1.2 and 1.3). Measured quantities may be expressed in various units (Table 1.1). The International System of Units (SI) is a self-consistent set of units based on the metric system. To convert a quantity from one unit or set of units to another, we use conversion factors. Their use is illustrated in Examples 1.4, 1.5, and 1.6. We will use the conversion factor approach over and over again in future chapters, so you should become familiar with it now.

KEY WORDS AND CONCEPTS

These are listed in the order in which they appear in the chapter. They are defined in the Glossary in the back of the book.

meter	Kelvin scale	group	absorption spectrum
centimeter	significant figure	period	wavelength
millimeter	conversion factor	main group	ultraviolet
nanometer	millimeter of mercury	metal	infrared
cubic centimeter	atmosphere	nonmetal	mixture
liter	joule	metalloid	solution
mass	kilojoule	chemical property	homogeneous
gram	SI unit	physical property	heterogeneous
kilogram	element	density	distillation
milligram	compound	melting point	fractional distillation
Celsius scale	symbol	boiling point	fractional crystallization
Fahrenheit scale	Periodic Table	solubility	chromatography

QUESTIONS AND PROBLEMS

The questions and problems listed here are typical of those at the end of each chapter. Some involve discussions, others calculations. The topic emphasized in each question or problem is indicated in the Catalog below. Those in the "general" category may involve more than one concept. Problems indicated by an asterisk, listed at the end of each set, require extra skill and/or effort.

Most of the questions and problems are arranged in "matched pairs," opposite one another (e.g., 1.2 and 1.24; 1.3 and 1.25). The two problems within the pair illustrate the same concept and are of comparable difficulty. Answers are given in Appendix 5 for the member of the pair in the *right* column (e.g., 1.24, 1.25).

Catalog

Units and Measurements: 1.2–1.5, 1.24–1.27
Significant Figures: 1.6, 1.7, 1.28, 1.29
Conversion Factors: 1.8–1.13, 1.30–1.35
Elements and the Periodic Table: 1.14–1.16, 1.36–1.38

Properties (Density, Color): 1.17–1.20, 1.39–1.42
Separation of Mixtures: 1.21–1.23, 1.43–1.45
General: 1.1, 1.46–1.49

1.1 Review and know the meanings of the key words and concepts introduced in this chapter. Use the Glossary if necessary.

1.2 Classify each of the following as units of mass, volume, length, density, energy, or pressure:

    a. mg   b. mm Hg   c. m   d. kJ
    e. g/ℓ f. cm^3 g. km

1.3 Select the larger member of each pair:

    a. 2.54 cm or 2.54 m
    b. 20 g or 2×10^3 mg
    c. 10 mm or 10 km
    d. $10\ mm^3$ or $10\ cm^3$

1.4 The boiling point of liquid oxygen is $-183°C$. Express this in

    a. °F   b. K

1.5 The Ideal Gas Law states that PV = nRT. If P is in atm, V in ℓ, n in mol, and T in K, what are the units of R?

1.6 How many significant figures are there in each of the following?

    a. 26.2 cm   b. 10,000 m
    c. 0.00014 g   d. $14.0\ cm^3$
    e. 2.98×10^{10} cm/s   f. 2.04 g

1.7 Calculate each of the following to the correct number of significant figures:

    a. $x = \dfrac{12.2\ g}{50.3\ cm^3}$

    b. $x = \dfrac{39.314\ g}{14\ cm^3}$

    c. $x = \dfrac{6.54\ g + 1.12\ g}{3.91\ cm^3}$

    d. $x = 16.0\ g + 125\ g$

1.8 Using Table 1.1, find the number of cubic inches in one cubic meter.

1.9 When filled with helium, the Goodyear (or is it Goodrich?) blimp Columbia is 51 ft wide, 192 ft long, and 59 ft high. Assuming its volume is the product: width × length × height, express the volume in

    a. cubic feet   b. liters

1.10 One of the authors of this text has a height of 5 ft, 10.5 in, and weighs 132 lb. Express these measurements in meters and kilograms.

1.11 A world-class sprinter runs the 100.0-yd dash in 9.10 s. What is his speed in miles per hour?

1.24 Classify each of the following as units of mass, volume, length, density, energy, or pressure:

    a. cm   b. ℓ c. mg/cm^3 d. kPa
    e. J   f. kg   g. atm

1.25 Select the smaller member of each pair:

    a. 50 m or 0.50 km
    b. 500 mg or 0.0500 g
    c. $150\ cm^3$ or $150\ m^3$
    d. 1500 m or 1500 nm

1.26 The coldest nonlaboratory temperature recorded on earth was $-129.6°F$. Express this in

    a. °C   b. K

1.27 Consider the equation $\Delta E = hc/\lambda$. If ΔE is in J, c in m/s, and λ in m, what are the units of h?

1.28 How many significant figures are there in each of the following?

    a. 5.60 cm   b. 0.00560 m
    c. 6.02×10^{23} atoms   d. $5.06\ \ell$
    e. 2.1103 g   f. 0.050 km

1.29 How many significant figures are there in the values of x obtained from the following expressions?

    a. $x = 21.00\ cm^3 - 20.19\ cm^3$
    b. $x = (12.0\ cm)^3$
    c. $x = (6.19\ cm)\ (2.4\ cm)$
    d. $x = \dfrac{16.2\ g + 1.4\ g}{11.2\ cm^3}$

1.30 Using Table 1.1, determine the number of ounces in one metric ton.

1.31 A textbook is 27 cm long, 21 cm wide, and 4.4 cm thick. What is its volume in

    a. liters?   b. cubic inches?

1.32 Another author weighs 182 lb (never mind how tall he is). Express this mass in the old British unit "stones." One stone is 6.35 kg.

1.33 Bill Rodgers won the 1979 Boston Marathon (26.2 mi) in 2 h, 9 min, 27 s, finishing well ahead of our junior author. What was Rodgers' speed in kilometers per hour?

1.12 When the Pharmacopoeia of London was compiled in 1618, the troy system of measure was used to prepare medicine. Among the units used were: 20 grains = 1 scruple; 3 scruples = 1 drachm; 8 drachms = 1 ounce; 12 ounces = 1 pound. Hence

    a. 6.00 pounds = _____ drachms
    b. 12.0 drachms = _____ pounds; = _____ ounces

1.13 A gold atom has a diameter of 0.288 nm. (1 nm = 10^{-9} m; 1 mi = 1.609 km.) How many miles long would a single row of 6.022×10^{23} gold atoms be?

1.14 Referring to the table of elements inside the back cover of this text, name all the elements whose symbols are found in these names:

    a. Connecticut   b. your surname
    c. Macalester

1.15 Consider Group 4 of the Periodic Table. Which of the elements in that group are

    a. nonmetals?   b. metals?
    c. metalloids?

1.16 Classify each of the following elements as a metal, nonmetal, or metalloid: barium, arsenic, tungsten, chlorine, carbon, germanium.

1.17 Which of the following are physical properties of aluminum? chemical properties?

    a. density = 2.70 g/cm³
    b. reacts with oxygen to form a metal oxide
    c. melting point = 660°C
    d. good electrical conductor

1.18 The density of air at 20°C and 1 atm is 1.20 g/ℓ. How many kilograms of air are there in a balloon with a volume of one cubic meter under these conditions?

1.19 Blood plasma volume for adults is about 3.1 ℓ. Its density is 1.027 g/cm³. About how many pounds of blood plasma are there in your body?

1.20 What is the color of light having a wavelength of

    a. 800 nm?   b. 600 nm?
    c. 500 nm?   d. 300 nm?

1.21 Describe how you would separate each of the following from a mixture with water:

    a. sand   b. sugar
    c. gasoline   d. ethyl alcohol

1.34 During earlier times in England, land was measured in units such as fardells, nookes, yards, and hides: 2 fardells = 1 nooke; 4 nookes = 1 yard; 4 yards = 1 hide. Thus,

    a. 4.00 hides = _____ fardells
    b. 23 nookes = _____ hides; = _____ fardells.

1.35 A platinum atom has a diameter of 0.276 nm. How many platinum atoms would have to be aligned in a single row to form a line 4.6 mi long?

1.36 Referring to the table of elements inside the back cover of this text, name all the elements whose symbols are found in these names:

    a. Randolph-Macon   b. your college

1.37 Of the elements in Group 5 of the Periodic Table, which are

    a. nonmetals?   b. metalloids?
    c. metals?

1.38 Use the term metal, nonmetal, or metalloid to describe each of the following elements: sulfur, magnesium, bromine, antimony, lithium, osmium.

1.39 The following data refer to the element oxygen. Classify each as a physical or chemical property.

    a. forms a compound with iron
    b. boils at 90 K at 1 atm
    c. slightly soluble in water
    d. reacts with carbon to form carbon dioxide

1.40 The density of water is about 1.00 g/cm³. What is the mass of water, in kilograms, required to fill a flask with a volume of 650 mm³?

1.41 A water bed filled with water has the dimensions 8.0 ft × 6.5 ft × 0.75 ft. Taking the density of water to be 1.00 g/cm³, determine the number of kilograms of water required to fill the water bed.

1.42 Classify each of the following statements as true or false:

    a. A substance that absorbs at 450 nm is colored.
    b. UV radiation has a wavelength below 400 nm.
    c. IR radiation has a wavelength below 700 nm.

1.43 Describe how you would separate each of the following from a mixture with potassium chromate, a water-soluble solid:

    a. sand   b. water
    c. sugar   d. sodium chloride

1.22 A certain solid is to be purified by fractional crystallization. You have a choice of three solvents—A, B, or C. The solubility of the solid in each solvent is 10 g/100 g at 20°C. At 80°C the solubilities are 10 g/100 g of A, 30 g/100 g of B, and 15 g/100 g of C. Which solvent would you choose, and why?

1.23 Use Figure 1.11 to answer the following questions:

a. How much NaCl will dissolve in 50 g of water at 10°C? how much KNO_3?
b. At about what temperature does the solubility of KNO_3 equal that of NaCl at 20°C?
c. If you want to crystallize 53 g of KNO_3 from a hot solution containing 83 g KNO_3 in 100 g of water, to what temperature should you cool the solution?

1.44 You are given a solid mixture containing 60 g of A and 30 g of B, and you are asked to obtain pure A from the mixture. You carry out a fractional crystallization and obtain 40 g of A and 20 g of B. What would you do at this point (besides cursing your instructor)?

1.45 Use Figure 1.11 to answer the following questions:

a. How much KNO_3 will dissolve in 60 g of water at 40°C? how much NaCl?
b. At about what temperature does the solubility of KNO_3 equal that of NaCl at 10°C?
c. A hot solution contains 65 g of KNO_3 in 100 g of water. To what temperature must this solution be cooled to crystallize 35 g of KNO_3?

*1.46 At what temperature would a thermometer read the same whether it was marked in °F or in K?

*1.47 Oil spreads on water to form a film about 100 nm thick (2 significant figures). How many square kilometers of ocean will be covered by the slick formed when one barrel of oil is spilled? (1 barrel = 31.5 U.S. gal.)

*1.48 A flask containing 100.0 cm³ of ethyl alcohol (d = 0.789 g/cm³) is placed on one pan of a two-pan balance. A larger flask, weighing 11.0 g more than the first one when empty, is placed on the other pan. What volume of turpentine (d = 0.870 g/cm³) must be added to this flask to bring the two pans into balance?

*1.49 The solubilities in water at 0°C of the two amino acids L-alanine and L-tryptophan are 12.7 g/100 g and 0.82 g/100 g, in that order. A mixture of 25.0 g of L-alanine and 2.50 g of L-tryptophan is dissolved in 200 g of water at a higher temperature. When this solution is cooled to 0°C, how much of each amino acid crystallizes out of solution? What percentage of each amino acid is recovered by this fractional crystallization?

ATOMS, MOLECULES, AND IONS

In Chapter 1, we looked at some of the macroscopic properties of matter. These are properties which can be measured using samples large enough to see and weigh. Such properties include density, melting point, boiling point, and chemical reactivity. In this chapter we will look at chemical substances from a different point of view. We will examine the submicroscopic building blocks which make up elements and compounds. These include atoms, molecules, and ions.

For our purposes at this point, certain properties of atoms, molecules, and ions are of great importance. We need to know how these tiny particles compare to one another in size, charge, and mass. We will also examine the way in which they are put together. In later chapters we will return to this topic and discuss in more detail the principles of atomic structure (Chapter 8) and chemical bonding (Chapter 10).

2.1 ATOMIC THEORY

The notion that matter consists of discrete particles is an old one. About 400 B.C. this idea appeared in the writings of Democritus, a Greek philosopher. He had been introduced to it by his teacher, Leucippus. The idea was rejected by Plato and Aristotle, who had a great deal more influence than Democritus. Not until 1650 A.D. was the concept of atoms suggested again, this time by the Italian physicist Gassendi. Sir Isaac Newton (1642–1727) supported Gassendi's arguments with these words:

. . . it seems probable to me that God, in the Beginning, formed Matter in solid, massy, hard, impenetrable, movable Particles, of such Sizes and Figures, and with such other Properties, and in such Proportions to Space, as most conduced to the End for which he formed them. . . .

Prior to 1800 the concept of the particle nature of matter was based largely on intuition. Then, in 1808, an English schoolteacher, John Dalton, developed an explanation of several of the laws of chemistry. This became known as the **atomic theory.** Some of Dalton's ideas had to be discarded as chemists learned more about the structure of matter. However, the essentials of his theory have withstood the test of time. Three of the main postulates of modern atomic theory, all of which came from Dalton, are given below with examples to illustrate their meaning.

1. *An element is composed of tiny particles called atoms. All atoms of a given element show the same chemical properties.* The element oxygen is made up of oxygen atoms. These atoms are much too small to be seen or weighed (see, however, Fig. 2.1). All oxygen atoms behave chemically in the same way.

Figure 2.1 Electron microscopy image of a complex organic molecule. Four mercury atoms (dark spots) are clearly visible. (Courtesy of F. P. Ottensmeyer.)

2. *Atoms of different elements have different properties. In an ordinary chemical reaction, no atom of any element disappears or is changed into an atom of another element.* The chemical behavior of oxygen atoms is different from that of hydrogen atoms or any other kind of element. When hydrogen and oxygen react, all the hydrogen and oxygen atoms that react are present in the water formed. No atoms of any other element are formed.

3. *Compounds are formed when atoms of two or more elements combine. In a given compound, the relative numbers of atoms of each kind are definite and constant. In general, these relative numbers can be expressed as integers or simple fractions.* In the compound water, hydrogen atoms and oxygen atoms are combined with each other. For every oxygen atom present, there are always two hydrogen atoms.

The atomic theory explains two of the basic laws of chemistry:

1. The **Law of Conservation of Mass.** This law was first proposed by the French chemist Antoine Lavoisier in 1777. In modern form, it states that *there is no detectable change in mass in an ordinary chemical reaction.* If atoms are "conserved" in a reaction (Postulate 2), mass will also be conserved.

2. The **Law of Constant Composition.** This tells us that *a compound always contains the same elements in the same proportions* by mass. If the atom ratio of the elements in a compound is fixed (Postulate 3), their proportions by mass must also be fixed.

The validity of this law became generally recognized at about the same time that Dalton's theory appeared. Prior to 1808 many people agreed with the French chemist, Berthollet. He believed that the composition of a compound could vary over wide limits, depending on how it was prepared. Joseph Proust, a Frenchman working in Madrid, refuted Berthollet by showing that the "compounds" Berthollet had cited were actually mixtures.

It is now known that in certain compounds, particularly metal oxides and sulfides, the atom ratio may vary slightly from a whole number ratio. Careful analyses of samples of nickel oxide prepared by heating nickel with oxygen at high temperatures give an atom ratio of 0.97:1.00 rather than the expected 1:1 ratio. Compounds of this type are sometimes referred to as "Berthollides" or, more frequently, *nonstoichiometric* compounds. The deviations from constant composition arise because of defects in the crystal structure.

The third postulate of the atomic theory is in many ways the most important. Among other things, it led Dalton to formulate the **Law of Multiple Proportions.** This law applies

In a chemical reaction, mass of products = mass of reactants

to the situation in which two elements form more than one compound. It states that, in these compounds, *the masses of one element which combine with a fixed mass of the second element are in a ratio of small whole numbers* (for example, 2:1). To derive the Law of Multiple Proportions, Dalton reasoned along the following lines. Suppose elements A and B form two different compounds. In one of these (AB) one atom of A might be combined with one atom of B. The second compound (AB$_2$) might contain two atoms of B per atom of A. If this is true, the mass of B combined with a fixed mass (such as one gram) of A would be twice as great in the second compound. In other words, the masses of B per gram of A in the two compounds would be in a 2:1 ratio.

The mass of an atom is one of its properties. Like the mass of a billiard ball, it is fixed

---

Example 2.1   The elements carbon and oxygen form two different compounds. The first contains 42.9% by mass of carbon and 57.1% by mass of oxygen. In the second compound these percentages are 27.3 for carbon and 72.7 for oxygen.
  a. Show that these data are consistent with the Law of Multiple Proportions.
  b. Suggest an explanation in terms of atomic theory.

Solution
  a. According to the Law, the masses of oxygen in the two compounds that combine with a fixed mass of carbon should be in a whole-number ratio. To show that this is the case, let us base our calculations on one gram of carbon. In 100 g of the first compound there are 57.1 g O and 42.9 g C. Therefore the mass of O per gram of C is

$$\frac{57.1 \text{ g O}}{42.9 \text{ g C}} = 1.33 \text{ g O/g C}$$

In 100 g of the second compound there are 72.7 g O and 27.3 g C. The mass of oxygen per gram of carbon is

$$\frac{72.7 \text{ g O}}{27.3 \text{ g C}} = 2.66 \text{ g O/g C}$$

Clearly the masses of oxygen that combine with one gram of carbon are in a 2:1 ratio, since 2.66/1.33 = 2:1.
  b. A simple explanation happens to be the correct one. In the first compound (carbon monoxide) there is one atom of oxygen per atom of carbon. In the second compound (carbon dioxide) there are two oxygen atoms for every carbon atom.

Exercise   Calculate the mass of carbon which combines with one gram of oxygen in carbon dioxide; in carbon monoxide. Answer: 0.376 g; 0.751 g (a 1:2 ratio).

Dalton used data from other scientists to test the Law of Multiple Proportions. He also did some rather crude experiments himself which tended to confirm it. A few years later, the Swedish chemist Berzelius made some very careful experiments with several pairs of elements that formed more than one compound. He showed that the law was obeyed in all cases to within 0.1%.

The discovery of this law and its verification convinced most scientists that Dalton's atomic theory was valid. However, not everyone agreed. Some chemists maintained that it was a waste of time to speculate on the particle nature of matter. This minority point of view persisted throughout the nineteenth century. In 1900, a well-known German chemist, Ostwald, in a textbook of general chemistry, carefully avoided all mention of atoms.

Some scientists are slower than others to accept new ideas

2.2 COMPONENTS OF THE ATOM

Like any useful scientific theory, the atomic theory raised more questions than it answered. Scientists wondered whether atoms, tiny as they are, could be broken down into still smaller particles. Nearly one hundred years passed before the existence of subatomic particles was confirmed by experiment. Three physicists did pioneer work in this area. J. J. Thomson was an Englishman working at the Cavendish laboratory at Cambridge. Ernest Rutherford, a native of New Zealand, carried out his research at McGill University in Montreal and at Manchester and Cambridge in England. The third of these was an American, Robert A. Millikan, who did his research at the University of Chicago.

Electrons

The first evidence for the existence of subatomic particles came from studies of the conduction of electricity through gases at low pressures. A sketch of the apparatus used is shown in Figure 2.2. When the tube is partially evacuated and connected to a spark coil, an electric current flows through it. Associated with this flow are colored rays of light, spreading out from the negative electrode (cathode). The properties of these *cathode rays* were studied during the last three decades of the nineteenth century. It was found that they were bent by both electric and magnetic fields. From a careful study of this deflection, J. J. Thomson showed in 1897 that the rays consist of a stream of negatively charged particles. He called these particles **electrons.** Thomson went on to measure the mass-to-charge ratio of the electron, finding that

$$m/e = 5.69 \times 10^{-9} \text{ g/coulomb}$$

The fact that this ratio is the same, regardless of what gas is in the tube, implies that the electron is a basic particle, common to all atoms.

In 1909, Millikan determined the charge on the electron, using the apparatus sketched in Figure 2.3. He measured the effect of an electric field on the rate at which charged oil drops fall between two plates. From his data Millikan calculated the charges on the drops. He found that these charges were always an integral multiple of a smallest charge. Taking the smallest charge to be that of the electron, he arrived at a value of 1.60×10^{-19} coulomb. Combining this value with the mass-to-charge ratio, we obtain the mass of the electron:

The modern TV tube is a descendant of the cathode ray tube

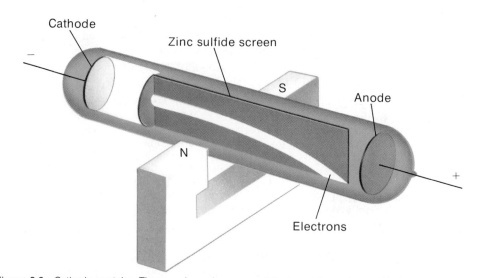

Figure 2.2 Cathode ray tube. The ray, shown here as a white beam, is made up of fast-moving electrons. In an electric or magnetic field, the beam is deflected in such a way as to indicate that it carries a negative charge.

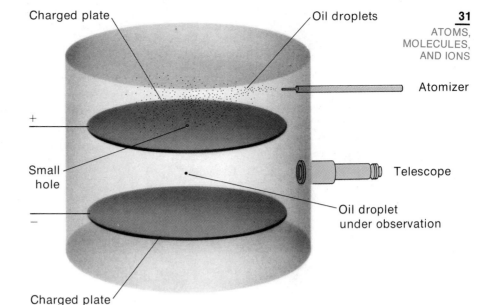

Charged plate   Oil droplets

Atomizer

+

Small hole

Telescope

Oil droplet under observation

Charged plate

Figure 2.3 Oil drop experiment. At a certain voltage, the electrical and gravitational forces on a charged drop between the two plates will be exactly balanced, and the drop will not move. Knowing this voltage and the mass of the drop, it is possible to calculate the charge on the drop. This charge must be a whole-number multiple of the charge on the electron.

$$m = (1.60 \times 10^{-19} \text{ coulomb}) \times (5.69 \times 10^{-9} \text{ g/coulomb})$$
$$= 9.11 \times 10^{-28} \text{ g}$$

The electron is the lightest weighable particle

This is roughly 1/2000 of the mass of the lightest atom, that of the element hydrogen.

Every atom contains a definite number of electrons. This number, which runs from one to over 100, is characteristic of a neutral atom of a particular element. All hydrogen atoms contain one electron. All atoms of the element uranium contain 92 electrons. We will have more to say in Chapter 8 about how these electrons are arranged relative to one another. At this time, we need only point out that electrons are found in the outer regions of atoms. They comprise what amounts to a cloud of negative charge about the nucleus of an atom.

The Atomic Nucleus. Protons and Neutrons

In 1911 Ernest Rutherford and his students carried out a series of experiments that shaped our ideas about the nature of the atom. They bombarded a piece of thin gold foil (Fig. 2.4) with α-particles (helium atoms stripped of their electrons). With a fluorescent screen, they observed the extent to which the α-particles were scattered. Most of them went through the foil unchanged in direction. A few, however, were reflected back from the foil at acute angles. The relative numbers of α-particles reflected at different angles were counted. By a mathematical analysis of the forces involved, Rutherford showed that the scattering was caused by a positively charged center within the gold atom. He estimated that almost all the mass of the atom was contained within this center. However, its diameter (about 10^{-5} nm) is only 1/10,000 of that of the atom. The experiment was repeated, with similar results, with foils of many different elements. In this way Rutherford showed that an atom contains a small, positively charged, massive center, called the **atomic nucleus.**

Since electrons and nuclei are exceedingly small, atoms consist mainly of empty space

Since the time of Rutherford, we have learned a great deal about the properties of atomic nuclei. For our purposes in chemistry, we can consider the nucleus of an atom to consist of two different types of particles:

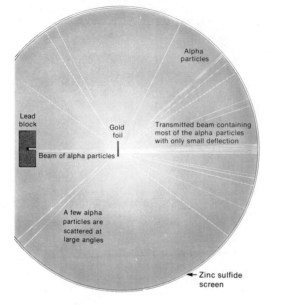

Figure 2.4 Rutherford scattering experiment. Most of the α-particles pass through the gold foil almost undeflected. However, a few collide with the nuclei of gold atoms, causing a large deflection. This experiment provided the first evidence for the existence of a small, positively charged nucleus in an atom.

1. The **proton,** which has a mass nearly equal to that of the lightest atom, the element hydrogen. Reflecting that fact, we assign the proton a *mass number* of 1. The proton carries a unit positive charge (+1), equal in magnitude to that of the electron (−1).

2. The **neutron,** an uncharged particle with a mass about equal to that of a proton. The neutron, like the proton, has a mass number of 1.

Atomic Number and the Periodic Table

All the atoms of a particular element have the same number of protons in the nucleus. This number is a basic property of an element, called its **atomic number.**

$$\text{atomic number} = \text{number of protons} \tag{2.1}$$

In a neutral atom the number of protons in the nucleus is exactly equal to the number of electrons outside the nucleus. Consider, for example, the elements hydrogen (at. no. = 1) and helium (at. no. = 2). All hydrogen atoms have one proton in the nucleus; all helium atoms have two. In a neutral hydrogen atom there is one electron outside the nucleus; in a helium atom there are two.

Atomic numbers of the elements are given in the Periodic Table (inside front cover). They are printed in color directly above the symbol of the element. Notice that atomic number increases steadily as we move across the Table. Indeed, the location of an element in the Table is determined by its atomic number. *The horizontal rows (periods) are just long enough to make elements with similar chemical properties fall in the same vertical column (group).* Each period ends with a very unreactive element called a noble gas (He, Ne, Ar, Kr, Xe, Rn). To make these elements in Group 8 fall directly beneath each other, the periods have the following lengths:

1st Period	2 elements	H (at. no. = 1) and He (at. no. = 2)
2nd Period	8 elements	Li (at. no. = 3) through Ne (at. no. = 10)
3rd Period	8 elements	Na (at. no. = 11) through Ar (at. no. = 18)
4th Period	18 elements	K (at. no. = 19) through Kr (at. no. = 36)
5th Period	18 elements	Rb (at. no. = 37) through Xe (at. no. = 54)
6th Period	32 elements	Cs (at. no. = 55) through Rn (at. no. = 86)

As with protons and neutrons, we can assign mass numbers to atoms. Recall that the mass number is 1 for both a proton and a neutron. Hence, we can find the mass number of an atom by adding up the number of protons and neutrons in the nucleus:

Electrons don't contribute appreciably to the masses of atoms

$$\text{mass number} = \text{number of protons} + \text{number of neutrons} \qquad (2.2)$$

For an atom with 17 protons and 20 neutrons in the nucleus,

$$\text{mass number} = 17 + 20 = 37$$

As we have noted, all atoms of a given element have the same atomic number (number of protons). They may, however, differ from one another in mass and hence in mass number. This can happen because, although the number of protons in the nucleus of a given kind of atom is fixed, the number of neutrons is not. It may vary and often does. Consider the element hydrogen (at. no. = 1). There are three different kinds of hydrogen atoms. They all have 1 proton in the nucleus. A "light" hydrogen atom (the most common type) has no neutrons in the nucleus (mass no. = 1). Another type of hydrogen atom (deuterium) has 1 neutron (mass no. = 2). Still a third type (tritium) has 2 neutrons (mass no. = 3).

Atoms which contain the same number of protons but a different number of neutrons are called isotopes. The three kinds of hydrogen atoms just described are isotopes of that element. They have masses which are very nearly in the ratio 1:2:3. Among the isotopes of the element uranium are the following:

Isotope	Atomic Number	Mass Number	Number of Protons	Number of Neutrons
Uranium-235	92	235	92	143
Uranium-238	92	238	92	146

The composition of a nucleus is shown by its nuclear symbol. Here, the atomic number appears as a subscript at the lower left of the symbol of the element. The mass number is written as a superscript at the upper left. The nuclear symbols of the isotopes of hydrogen are

Remember, the nature of an element is fixed by its atomic number

$$^1_1\text{H}, \ ^2_1\text{H}, \ ^3_1\text{H}$$

For the isotopes of uranium we write the nuclear symbols

$$^{235}_{92}\text{U}, \ ^{238}_{92}\text{U}$$

Example 2.2
    a. Write nuclear symbols for three isotopes of oxygen (atomic no. = 8) in which there are 8, 9, and 10 neutrons, respectively.
    b. One of the most harmful species in nuclear fallout is a radioactive isotope of strontium, $^{90}_{38}\text{Sr}$. How many protons are there in this nucleus? how many neutrons?

Solution
    a. The mass numbers must be: $8 + 8 = 16$; $8 + 9 = 17$; $8 + 10 = 18$. Thus we have

$$^{16}_8\text{O}, \ ^{17}_8\text{O}, \ ^{18}_8\text{O}$$

b. The number of protons is given by the atomic number, 38. To obtain the number of neutrons we subtract the number of protons from the mass number:

$$\text{number of neutrons} = 90 - 38 = 52$$

Exercise   Write the nuclear symbol for an atom that contains 32 protons and 38 neutrons (use the Periodic Table to find the symbol of the element). Answer: $^{70}_{32}\text{Ge}$.

2.3 MOLECULES AND IONS

Free atoms are usually very reactive chemically

Isolated atoms rarely occur in nature. They are found as such only in a few elements, namely the noble gases in Group 8 of the Periodic Table. Most elements and all compounds are built up of more complex structural units. Two of the most important of these are molecules and ions.

Molecules

An H—H molecule is much more stable than two separate H atoms

The basic structural unit in most volatile substances is the molecule. *A molecule is a group of two or more atoms held together by strong forces called chemical bonds.* The atoms which form the molecule may be the same or different. The gaseous element hydrogen consists of diatomic molecules, in which two hydrogen atoms are bonded to each other. The molecules of molecular compounds contain atoms of at least two elements. Hydrogen chloride is a simple example. In this gaseous compound, the basic structural unit is a molecule consisting of a hydrogen atom bonded to a chlorine atom.

The composition of a molecule can be shown in various ways. Sometimes we write what is known as a **structural formula.** Here the symbols of the atoms in the molecule are joined by straight lines which represent chemical bonds. The structural formulas of the two molecules discussed above are

H—H   H—Cl

hydrogen   hydrogen chloride

More simply, we can represent a molecule by its **molecular formula.** In the molecular formula, the number of atoms of each type is indicated by a subscript written after the symbol. The molecular formulas of the two substances above are

hydrogen: H_2 (2 H atoms per molecule)
hydrogen chloride: HCl   (1 H atom, 1 Cl atom per molecule)

(Note that no subscript is used when there is only one atom of a particular type present.)

Most molecular substances are composed of molecules more complex than those just cited. The water molecule, for example, consists of a central oxygen atom bonded to two hydrogen atoms. In the ammonia molecule, a central nitrogen atom is bonded to three hydrogen atoms. Methane, the major component of natural gas, has as a structural unit a molecule in which a carbon atom is bonded to four hydrogen atoms. The structural formulas of these molecules can be written as

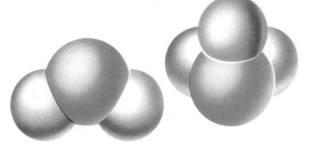

Figure 2.5 Space-filling models of three simple molecules. The atoms shown in gray are hydrogen atoms. They are held by strong chemical bonds to a central, nonmetal atom (O, N, or C).

Water (H_2O)   Ammonia (NH_3)   Methane (CH_4)

water   ammonia   methane

Their molecular formulas are

water:	H_2O	(2 H atoms, 1 O atom per molecule)
ammonia:	NH_3	(3 H atoms, 1 N atom per molecule)
methane:	CH_4	(4 H atoms, 1 C atom per molecule)

Two-dimensional structural formulas do not, in general, show the geometry of molecules. Space-filling models of the type shown in Figure 2.5 give a better picture of the relative positions of the atoms. We will have more to say about molecular geometry when we discuss chemical bonding in Chapter 10.

Ions

If enough energy is available, one or more electrons can be removed from a neutral atom. This leaves a positively charged particle somewhat smaller than the original atom. Electrons may be added to certain atoms to form negatively charged species larger than the original atom. Charged particles are called **ions.** An example of a positive ion (**cation**) is the Na^+ ion, formed from a sodium atom by the loss of a single electron. Another cation is Ca^{2+}, derived from a calcium atom by removing two electrons.

In a neutral atom, the number of electrons = number of protons = atomic number

$$\text{Na atom} \rightarrow Na^+ \text{ ion} + e^-$$
$$(11\,p^+, 11\,e^-) \quad (11\,p^+, 10\,e^-)$$

$$\text{Ca atom} \rightarrow Ca^{2+} \text{ ion} + 2\,e^-$$
$$(20\,p^+, 20\,e^-) \quad (20\,p^+, 18\,e^-)$$

Two common negative ions (**anions**) are the chloride ion, Cl^-, and the oxide ion, O^{2-}. These are formed when atoms of chlorine or oxygen acquire electrons:

$$\text{Cl atom} \quad + \quad e^- \quad \rightarrow \quad \text{Cl}^- \text{ ion}$$
$$(17\,p^+, 17\,e^-) \qquad\qquad\qquad (17\,p^+, 18\,e^-)$$

$$\text{O atom} \quad + \quad 2\,e^- \quad \rightarrow \quad O^{2-} \text{ ion}$$
$$(8\,p^+, 8\,e^-) \qquad\qquad\qquad (8\,p^+, 10\,e^-)$$

Notice that when an ion is formed the number of protons in the nucleus is unchanged. It is the number of electrons which increases or decreases. Negative ions contain more electrons than protons, while positive ions have an excess of protons. The charge of the ion is indicated by a superscript at the upper right. The O^{2-} ion has a negative charge equal to that of two electrons. The Na^+ ion has a positive charge equal to that of a proton.

Example 2.3   Give the number of protons and electrons in the Sc^{3+} ion.

Solution   Using the Periodic Table, we find that Sc (scandium) has an atomic number of 21. Hence, the Sc^{3+} ion contains 21 protons and

$$21 - 3 = 18 \text{ electrons}$$

Exercise   Give the symbol of an ion which has $10\,e^-$ and $7\,p^+$; an ion which has $10\,e^-$ and $12\,p^+$. Answers: N^{3-}; Mg^{2+}.

Many compounds are made up of ions. They are called ionic compounds. Since a bulk sample of matter must be electrically neutral, ionic compounds always contain both cations (+ charge) and anions (− charge). Ordinary table salt, sodium chloride, is made up of an equal number of Na^+ and Cl^- ions. The structure of a sodium chloride crystal is shown in Figure 2.6. Notice that there are no discrete molecules. Positive and negative ions are bonded together in a continuous network. Calcium oxide (quicklime), which is made up of Ca^{2+} and O^{2-} ions, has a similar structure.

In the ionic compounds sodium oxide and calcium chloride there are unequal numbers of cations and anions. To maintain electrical neutrality, there must be two Na^+ ions

At 25°C ionic compounds are always solids

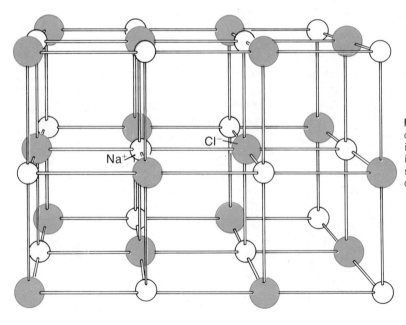

Figure 2.6 Ball-and-stick model of a crystal of NaCl (the Na^+ ions actually touch the Cl^- ions). Each Cl^- ion is surrounded by six Na^+ ions, and each Na^+ ion by six Cl^- ions. A single crystal of NaCl contains many billions of ions arranged in this pattern.

for every O^{2-} in sodium oxide. A similar situation applies with calcium chloride. Two Cl^- ions are required to balance one Ca^{2+} ion.

The composition of an ionic compound is indicated by writing its formula. Here, subscripts are used to indicate the relative number of ions of each type. For the four compounds just discussed:

Compound	Formula	Interpretation
Sodium chloride	NaCl	1 Na^+ ion for 1 Cl^- ion
Calcium oxide	CaO	1 Ca^{2+} ion for 1 O^{2-} ion
Sodium oxide	Na_2O	2 Na^+ ions for 1 O^{2-} ion
Calcium chloride	$CaCl_2$	1 Ca^{2+} ion for 2 Cl^- ions

2.4 RELATIVE MASSES OF ATOMS AND MOLECULES

A very important property of an atom or molecule is its mass. As you can imagine, measuring these masses is not an easy task. Indeed, to chemists of Dalton's time, it must have seemed impossible. How do you determine the mass of a particle so small that it can't be seen, let alone weighed?

Fortunately, it turns out that there is an approach to this problem which simplifies it considerably. For most purposes in chemistry, we don't need to know the *absolute* masses of atoms or molecules. Their *relative* masses will serve just as well. If we can find out how heavy one atom is compared to another, that will do for the time being.

Relative means "as compared to"

The problem of determining the relative masses of atoms and molecules is by no means a trivial one. In Section 2.7 we will describe some of the experiments that can be used for this purpose. Here we will consider only the results of these experiments.

Atomic Masses. The Carbon-12 Scale

Relative masses of atoms of different elements are expressed in terms of their atomic masses. **The atomic mass of an element is a number. It tells us how heavy, on the average, an atom of that element is compared to an atom of another element.** Consider, for example, the two elements hydrogen and oxygen. The atomic mass of hydrogen is slightly greater than 1; that of oxygen is slightly less than 16. This tells us that hydrogen atoms are about 1/16 as heavy as oxygen atoms. The atomic mass of sulfur is about 32. This means that sulfur atoms are about 32 times as heavy as hydrogen atoms and 32/16 or twice as heavy as oxygen atoms. In general, for two elements X and Y,

$$\frac{\text{mass of an atom of X}}{\text{mass of an atom of Y}} = \frac{\text{atomic mass of X}}{\text{atomic mass of Y}}$$

To set up a scale of atomic masses, it is necessary to establish a standard value for one species. Over the years, several different standards have been used. In 1961 it was agreed to assign an atomic mass of exactly 12 to the most common isotope of carbon, $^{12}_6C$. Atomic masses used in this text are based on the carbon-12 scale. When we say that the atomic mass of sodium (Na) is 22.98977, we mean that it weighs 22.98977/12.00000 = 1.915814 times as much as an atom of $^{12}_6C$.

Atomic masses are listed in the Periodic Table (inside front cover). They are given in black below the symbol of the element (H = 1.0079, He = 4.00260, and so on). Notice that, with a few exceptions, atomic mass increases in the same order as atomic number. Another table of atomic masses is given on the inside back cover of this text. Here the elements are arranged alphabetically.

Atomic masses range from about 1 to about 250

Atomic Masses and Isotopic Abundances

You may have noticed that the atomic mass of carbon is a bit larger than 12, 12.011 to be exact. Recalling that $^{12}_{6}C$ has an atomic mass of exactly 12, you might wonder where the "extra" 0.011 comes from. The explanation is a simple one. The element carbon, as it occurs in nature, is a mixture of two isotopes. About 99% of all carbon atoms are of the $^{12}_{6}C$ type. About 1% represent the $^{13}_{6}C$ isotope. The presence of these heavier atoms of mass number 13 explains why the atomic mass of carbon is slightly greater than 12.

The situation just described is quite common. Most elements, as they occur in nature, consist of two or more isotopes in fixed proportions. In such cases, the atomic mass of the element is a weighted average of those of its isotopes. To calculate the atomic mass we must know

—the masses of the isotopes on the carbon-12 scale.

—the isotopic *abundances* (percentages).

For an element X which consists of isotopes $X_1, X_2, \ldots$

$$\text{atomic mass X} = (\text{atomic mass } X_1) \times \frac{(\% \text{ of } X_1)}{100} + (\text{atomic mass } X_2) \times \frac{(\% \text{ of } X_2)}{100} + \ldots \quad (2.3)$$

The percentages of the several isotopes must, of course, add up to 100:

$$\% \text{ of } X_1 + \% \text{ of } X_2 + \ldots = 100$$

Example 2.4 The element boron consists of two isotopes, $^{10}_{5}B$ and $^{11}_{5}B$. Their masses, on the carbon-12 scale, are 10.01 and 11.01, in that order. The abundance of $^{10}_{5}B$ is 20.0%. What is

    a. the abundance of $^{11}_{5}B$?   b. the atomic mass of boron?

Solution

    a. The sum of the abundances must add to 100%. Hence

        abundance $^{11}_{5}B$ = 100.0 − 20.0 = 80.0%

    b. Substituting in Equation 2.3:

$$\text{atomic mass B} = 10.01 \times \frac{20.0}{100} + 11.01 \times \frac{80.0}{100} = 2.00 + 8.81 = 10.81$$

Note that this is the value listed for boron in the Periodic Table. The atomic mass of boron is closer to 11 than to 10. This reflects the fact that the heavier isotope is more abundant.

Exercise Taking the abundances of $^{12}_{6}C$ and $^{13}_{6}C$ to be 98.9% and 1.1% and their masses to be 12.00 and 13.00, calculate the atomic mass of carbon. Answer: 12.01.

In chemical reactions the various isotopes of an element behave identically

For most elements, isotopic abundances remain constant no matter where the element comes from or how it is prepared. Slight variations in abundance have been detected for a few elements. One of these is lead; the accuracy of the listed atomic mass (207.2 ± 0.1) is limited by this effect.

Molecular Masses

Relative masses of molecules can be expressed in much the same way as with atoms. The *molecular mass*, like the atomic mass, is a number. **It is found by adding the atomic**

masses of all the atoms in the molecule. Taking atomic masses from the Periodic Table (inside the front cover) we have

H_2: molecular mass = 2(at. mass H) = 2(1.008) = 2.016
H_2O: molecular mass = 2(at. mass H) + at. mass O = 2.016 + 16.00 = 18.02
NH_3: molecular mass = at. mass N + 3(at. mass H) = 14.01 + 3.024 = 17.03

Molecular masses are interpreted in the same way as atomic masses. When we say that the molecular mass of H_2O is about 18, we mean that a water molecule is

18/12 as heavy as a $^{12}_{6}C$ atom
18/16 as heavy as an oxygen atom (at. mass = 16)
18/17 as heavy as an NH_3 molecule (molecular mass = 17)

As you can see, molecular masses and atomic masses are both based on $^{12}_{6}C$

Example 2.5   What is the molecular mass of sugar (sucrose), which has the molecular formula $C_{12}H_{22}O_{11}$?

Solution   To find the molecular mass, we add the atomic masses of all the atoms in the molecule. Rounding off atomic masses to four significant figures, we have

molecular mass $C_{12}H_{22}O_{11}$ = 12(12.01) + 22(1.008) + 11(16.00) = 342.30

A sugar molecule weighs about 19 times as much as a water molecule.

Exercise   What is the molecular mass of glucose, $C_6H_{12}O_6$? Answer: 180.16.

2.5 AVOGADRO'S NUMBER

In color plate 3 (center of book) samples of three different elements are shown. One of these contains 12.01 g of carbon, another 32.06 g of sulfur, and the third 63.55 g of copper. The question is: what do these samples have in common? They don't have the same mass and clearly they don't look the same. However, they are comparable in one important way. Can you see what this is? (It may help you to realize that the atomic masses of C, S, and Cu are 12.01, 32.06, and 63.55, in that order.)

If you think about it for a moment, you should be able to answer this question. Look at it this way. A sulfur atom is 32.06/12.01 times as heavy as a carbon atom. If N carbon atoms weigh 12.01 g, then the same number, N, of sulfur atoms must weigh

$$\frac{32.06}{12.01} \times 12.01 \text{ g} = 32.06 \text{ g}$$

To understand this, it might help to first let N = 2, then 3, then 10, . . . up to any value N

In other words, 32.06 g of sulfur contains the same number of atoms as does 12.01 g of carbon. By the same argument, if there are N carbon atoms in 12.01 g of C, there must be the same number, N, of copper atoms in 63.55 g of Cu. In summary, there are the same number of atoms in each of the three samples referred to above.

The answer to this question suggests another, more difficult question. How many atoms are there in each of these samples (12.01 g of C, 32.06 g of S, 63.55 g of Cu)? As it happens, this problem is one that has been studied for at least a century. Several ingenious experiments have been designed to determine this number, known as **Avogadro's number** (see Problems 2.45 and 2.46). As you can imagine, it is huge. (Remember that atoms are tiny. There must be a lot of them in 12.01 g of C, 32.06 g of S, . . .) To four significant figures, Avogadro's number is

To get some idea of how large this number is, suppose the entire population of the world were assigned to counting the atoms in 12.01 g of carbon. If each person counted one atom per second and worked a 48-hour week, the task would take more than three billion years.

The importance of Avogadro's number in chemistry should be clear. *It represents the number of atoms in X grams of any element, where X is the atomic mass of the element.* Thus there are

6.022 × 10²³ C atoms in 12.01 g of carbon (at. mass C = 12.01)
6.022 × 10²³ S atoms in 32.06 g of sulfur (at. mass S = 32.06)
6.022 × 10²³ Cu atoms in 63.55 g of copper (at. mass Cu = 63.55)
6.022 × 10²³ O atoms in 16.00 g of oxygen (at. mass O = 16.00)

Avogadro's number is also useful in dealing with molecular substances. Consider, for example, a water molecule (molecular mass H_2O = 18.02). As we have seen, this molecule is 18.02/12.01 times as heavy as a carbon atom, 18.02/16.00 times as heavy as an oxygen atom, and so on. It follows that 18.02 g of water must also contain Avogadro's number of H_2O molecules. More generally, *in X grams of any molecular substance, where X is the molecular mass, there are Avogadro's number of molecules.* Thus, there are

6.022 × 10²³ H_2 molecules in 2.016 g of hydrogen (molecular mass H_2 = 2.016)
6.022 × 10²³ H_2O molecules in 18.02 g of water (molecular mass H_2O = 18.02)
6.022 × 10²³ NH_3 molecules in 17.03 g of ammonia (molecular mass NH_3 = 17.03)

and so on.

Example 2.6   Calculate the mass in grams of
a. a C atom.   b. 2.5 × 10⁹ H_2O molecules.

Solution
a. We know that 6.022 × 10²³ C atoms weigh 12.01 g. This relation gives us a factor to "convert" a carbon atom to grams:

$$\text{mass C atom} = 1 \text{ C atom} \times \frac{12.01 \text{ g}}{6.022 \times 10^{23} \text{ C atoms}} = 1.994 \times 10^{-23} \text{ g}$$

b. Here we know that 6.022 × 10²³ H_2O molecules weigh 18.02 g. So,

$$\text{mass} = 2.5 \times 10^9 \text{ } H_2O \text{ molecules} \times \frac{18.02 \text{ g}}{6.022 \times 10^{23} \text{ } H_2O \text{ molecules}}$$

$$= 7.5 \times 10^{-14} \text{ g}$$

Exercise   How many H_2O molecules are there in a snowflake weighing 1 mg? Answer: 3 × 10¹⁹.

2.6   THE MOLE

People in different professions often use special counting units. You and I eat eggs one at a time, but farmers sell them by the *dozen*. We spend dollar bills one at a time but

Congress distributes them by the *billion*. Chemists have their own counting unit, Avogadro's number. Since atoms and molecules are so small, it is convenient to talk about groups of 6.022×10^{23} of them. This counting unit is used so often in chemistry that it is given a special name, the **mole** (abbreviated as **mol**). **A mole, to a chemist, means Avogadro's number of items.** Thus

$$
\begin{aligned}
\text{One mole of C atoms} &= 6.022 \times 10^{23} \text{ C atoms} \\
\text{One mole of } H_2O \text{ molecules} &= 6.022 \times 10^{23} \text{ } H_2O \text{ molecules} \\
\text{One mole of electrons} &= 6.022 \times 10^{23} \text{ electrons} \\
\text{One mole of pennies} &= 6.022 \times 10^{23} \text{ pennies}
\end{aligned}
$$

(One mole of pennies is a lot of money. It's enough to pay all the expenses of the United States government for the next billion years or so.)

We can readily find the mass in grams of one mole of any chemical substance. To see how this is done, recall that there are

6.022×10^{23} atoms in 12.01 g of C, 32.06 g of S, . . .
6.022×10^{23} molecules in 2.016 g of H_2, 18.02 g of H_2O, . . .

But, as we have just seen, a mole always contains 6.022×10^{23} items. Hence

1 mol C weighs 12.01 g; 1 mol S weighs 32.06 g
1 mol H_2 weighs 2.016 g; 1 mol H_2O weighs 18.02 g

In these cases, and in all others, **one mole of a substance weighs Y grams, where Y is the formula mass, i.e., the sum of the atomic masses of the atoms in the formula.** Thus we have

Formula	Formula Mass	Mass of One Mole
C	12.01	12.01 g
H_2O	$2(1.008) + 16.00 = 18.02$	18.02 g
NaCl	$22.99 + 35.45 = 58.44$	58.44 g
$CaCl_2$	$40.08 + 2(35.45) = 110.98$	110.98 g

We really define the mole so that these relationships are valid

Mole-Gram Conversions

Very often in chemistry we have to convert from moles to grams, or vice versa. Thus we might want to know the number of moles of CO_2 in a piece of dry ice weighing one pound (454 g). In another case, we might need to obtain the mass in grams of 3.60 mol of sulfuric acid, H_2SO_4. Conversions of this type are readily made (Example 2.7).

Example 2.7   Determine
  a. the number of moles of CO_2 in 454 g.
  b. the mass in grams of 3.60 mol of H_2SO_4.

Solution
  a. The atomic masses of C and O are 12.01 and 16.00. The formula mass of CO_2 is $12.01 + 2(16.00) = 44.01$. Hence

$$1 \text{ mol } CO_2 \text{ weighs } 44.01 \text{ g}$$

This relation gives us a conversion factor to go from grams to moles, 1 mol/44.01 g:

This example shows how we go from our ordinary mass unit, grams, to the chemical unit, moles

$$\text{moles } CO_2 = 454 \text{ g} \times \frac{1 \text{ mol}}{44.01 \text{ g}} = 10.3 \text{ mol}$$

b. The formula mass of H_2SO_4 is

$$2(1.008) + 32.06 + 4(16.00) = 98.08$$

Hence 1 mol H_2SO_4 weighs 98.08 g.
In this case we want to go from moles to grams. So, the conversion factor to use is 98.08 g/1 mol.

$$\text{grams } H_2SO_4 = 3.60 \text{ mol} \times \frac{98.08 \text{ g}}{1 \text{ mol}} = 353 \text{ g } H_2SO_4$$

Exercise   What is the mass in grams of 2.50 mol of CO_2? How many moles are there in 49.0 g of H_2SO_4? Answer: 110 g; 0.500 mol.

Conversions of the type just carried out come up over and over again in chemistry. Indeed, they will be required in nearly every chapter of this text. Clearly, you must know what is meant by a mole. Remember, a mole always represents a certain number of items, 6.022×10^{23}. However, its mass differs with the substance involved. A mole of H_2O, 18.02 g, weighs considerably more than a mole of H_2, 2.016 g. A mole of sugar weighs still more, 342.30 g. A mole of pennies weighs much, much, much more (would you believe, about 10^{24} g?).

2.7 ATOMIC MASSES FROM EXPERIMENT

The problem of finding the relative masses of different atoms occupied the time of a great many famous chemists of the nineteenth century. We will not attempt to describe the tortuous path that led to the modern table of atomic masses. However, it may be helpful to describe a couple of the approaches that were used. Both of these led to approximate values for atomic masses, accurate to perhaps the nearest whole number.

The first direct approach to the determination of atomic masses was proposed in 1819 by two Frenchmen, Pierre Dulong and Alexis Petit. They suggested that the amount of heat required to raise the temperature of an atom of a solid element by, let us say, 1°C, should be independent of the type of atom. In their words, "the atoms of all simple bodies have the same capacity for heat." Since one mole of every element contains the same number of atoms, a fixed amount of heat should then be required to raise the temperature of one mole of a solid element by 1°C.

The Law of Dulong and Petit is expressed most simply in terms of the property called specific heat. This is the amount of heat required to raise the temperature of one gram of a substance by 1°C. The specific heat can be measured rather easily in the laboratory. You may do this later in this or other courses. By the Law of Dulong and Petit, the product of the molar mass of a solid element multiplied by its specific heat is about 25 J/°C.

$$\text{(molar mass)} \times \text{(specific heat)} \approx 25 \text{ J/°C} \qquad (2.4)$$

To show how this equation is used, consider the element iron. Its specific heat, as measured in the laboratory, is 0.448 J/g·°C. Substituting in Equation 2.4:

$$\text{molar mass Fe} = \frac{25 \text{ J/°C}}{0.448 \text{ J/g·°C}} = 56 \text{ g}$$

This leads to an atomic mass of 56 for iron, close to the true value, 55.847.   Not bad

The discovery of the Periodic Table by Mendeleef in 1869 suggested another way of estimating atomic masses. Recall that atomic mass increases in a more or less regular fashion as we move across the Table. This suggests that the atomic mass of an element should be close to the average of the two elements to its left and right. Consider, for example, the element scandium, which was unknown in 1869. Its neighbor on the left, calcium, has an atomic mass of about 40. To the right is titanium, with an atomic mass of 48. On this basis, the atomic mass of Sc should be about

$$\frac{40 + 48}{2} = 44$$

The true value is 44.9559, about one unit higher than this prediction.

The Mass Spectrometer

Today, classical methods of measuring atomic masses are largely of historical interest. Atomic masses can now be determined with great accuracy using a mass spectrometer (Fig. 2.7). Here, a gas such as helium, at very low pressures, is bombarded by high-energy electrons. A few helium atoms are converted to He^+ ions. These ions are focused into a narrow beam and accelerated by a voltage of 500 to 2000 V toward a magnetic field. The field deflects the ion beam out of its straight line path, toward the collector.

For ions of a given charge, let us say $+1$, the extent of deflection depends upon the mass of the ion. The heavier the ion, the less it will be deflected. At a given magnetic field strength and accelerating voltage (V_1), a $^{12}_{6}C^+$ ion might arrive at point A (Fig. 2.7). Under the same conditions, $^{4}_{2}He^+$ ions might be deflected to point B. By changing the accelerating voltage to V_2, it is possible to bring $^{4}_{2}He^+$ ions to point A instead. The accelerating voltages needed to bring the two ions to the same point are related by the equation

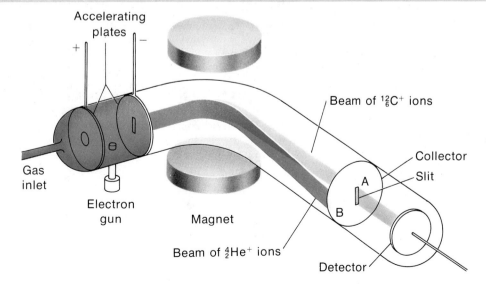

Figure 2.7 The mass spectrometer. A beam of gaseous ions is deflected in the magnetic field toward the collector plate. Light ions such as $^{4}_{2}He^+$ are deflected more than heavy ions such as $^{12}_{6}C^+$. By comparing the accelerating voltages required to bring the two ions to the same point, A, it is possible to determine the relative masses of the ions (Eq. 2.5).

Accelerating plates

Beam of $^{12}_{6}C^+$ ions

Collector

Slit

Gas inlet

Electron gun

Magnet

A

B

Beam of $^{4}_{2}He^+$ ions

Detector

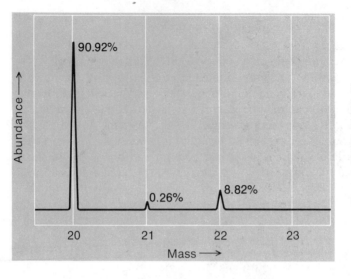

Figure 2.8 Mass spectrum of neon. Neon consists of three isotopes with atomic masses of 20.00 (90.92%), 21.00 (0.26%), and 22.00 (8.82%). Hence, the atomic mass of neon is 20.00(0.9092) + 21.00(0.0026) + 22.00(0.0882) = 20.18.

$$\frac{\text{mass } {}_{2}^{4}\text{He}^{+}}{\text{mass } {}_{6}^{12}\text{C}^{+}} = \frac{V_1}{V_2} \tag{2.5}$$

By this method atomic masses of isotopes can be found to within ±0.00001 units

By carefully measuring V_1 and V_2, we can calculate the relative masses of ${}_{2}^{4}$He and ${}_{6}^{12}$C. This way, we find that ${}_{2}^{4}$He has a mass 0.3336 times that of ${}_{6}^{12}$C. Hence, on the carbon-12 scale, the mass of ${}_{2}^{4}$He is

$$(0.3336)(12.00) = 4.003$$

But, naturally occurring helium consists almost entirely of this isotope. So, the atomic mass of the element helium is about 4.003.

Finding the atomic mass of an element with more than one isotope is more difficult. Consider neon, which has three isotopes, ${}_{10}^{20}$Ne, ${}_{10}^{21}$Ne, and ${}_{10}^{22}$Ne. A beam of neon ions is split into three parts when it passes through the magnetic field. To obtain the atomic mass of neon, we must know the abundance as well as the mass of each isotope. The abundances can be obtained by measuring the relative peaks produced on a recorder when the three beams reach the detector (Fig. 2.8). Knowing the mass of each isotope and its abundance, we can readily obtain the atomic mass of neon (recall Example 2.4).

SUMMARY

The three basic components of atoms are protons, neutrons, and electrons. Protons and neutrons are in the small, positively charged, nucleus of an atom. Electrons, which carry a negative charge, surround the nucleus. The proton carries a positive charge equal in magnitude to that of an electron. A neutron has zero charge. In a neutral atom, there are the same number of electrons as protons. The number of protons in the nucleus (atomic number) is characteristic of a particular element. In the Periodic Table, elements are arranged in order of increasing atomic number as we move from left to right in a horizontal row (period). Vertical columns (groups) contain elements with similar chemical properties.

The mass number of an atom is found by adding the number of neutrons and protons. Atoms of the same element (same number of protons) that differ in the number of neutrons are called isotopes (Example 2.2). The atomic mass is a number that tells us how heavy an atom is relative to a $^{12}_{6}C$ atom, which is assigned an atomic mass of exactly 12. The atomic mass of an element reflects the masses and relative abundances of the component isotopes (Example 2.4).

When an atom loses one or more electrons, a positive ion (cation) is formed. The gain of electrons by an atom leads to the formation of a negative ion (anion). Ionic compounds consist of positive and negative ions held together by strong attractive forces. In such a compound, the sum of positive charges equals the sum of negative charges. Certain elements and compounds contain discrete structural units called molecules. Molecules consist of two or more atoms joined by strong chemical bonds. The molecular mass of a substance is the sum of the atomic masses of the component atoms.

A mole represents Avogadro's number (6.022×10^{23}) of items. The "items" we refer to most frequently in chemistry are atoms or molecules. The mass in grams of a mole of a species can be found from its formula. It is numerically equal to the sum of the atomic masses of the element(s) present. Thus, we have

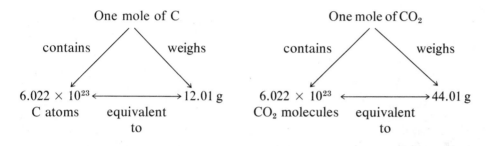

Relations such as these can be used to relate number of items to mass in grams (Example 2.6), number of moles to mass in grams (Example 2.7), or number of items to number of moles.

The basic concepts described above evolved from the atomic theory first proposed by John Dalton. Atomic theory leads directly to three of the basic laws of chemistry: (1) Law of Conservation of Mass; (2) Law of Constant Composition; and (3) Law of Multiple Proportions (Example 2.1).

KEY WORDS AND CONCEPTS

atom	proton	isotope	ionic compound
Law of Conservation of Mass	neutron	molecule	atomic mass
Law of Constant Composition	atomic number	molecular formula	molecular mass
Law of Multiple Proportions	period	ion	Avogadro's number
electron	group	cation	mole
alpha particle	mass number	anion	formula mass
nucleus			

QUESTIONS AND PROBLEMS

Catalog

Atomic Theory and Laws: 2.2, 2.3, 2.23, 2.24
Composition of Atoms and Ions: 2.4, 2.5, 2.25, 2.26
Atomic Masses, Isotopes: 2.6–2.8, 2.27–2.29
Molecular Masses: 2.9, 2.10, 2.30, 2.31

Avogadro's Number: 2.11–2.14, 2.32–2.35
Mole Concept and Conversions: 2.15–2.20, 2.36–2.41
General: 2.1, 2.21, 2.22, 2.42–2.47

2.1   Review and know the meanings of the key words and concepts in this chapter.

2.2   State, in your own words, the three basic postulates of the atomic theory of matter.

2.3   Phosphorus and chlorine form two compounds. In compound 1 the mass % of P and Cl are 22.54 and 77.46, respectively. In compound 2 these values are 14.88 and 85.12.

    a. Calculate the mass in grams of chlorine combined with one gram of phosphorus in each compound.
    b. Use the two values obtained in (a) to illustrate the Law of Multiple Proportions.

2.4   Consider the isotope of technetium used in brain scanning, $^{99}_{43}Tc$. How many

    a. protons are present in the nucleus?
    b. neutrons are present in the nucleus?
    c. electrons are in a Tc atom?
    d. electrons and protons are in a Tc^{3+} ion?

2.5   Complete the table below, using the Periodic Table where necessary to find the symbol of an element.

Symbol	Charge	Protons	Neutrons	Electrons
	0	9	10	___
$^{34}_{16}S^{2-}$		___	___	___
___	+3	23	28	___
___	+1	55	78	___

2.6   The atomic masses of Ni and S are 58.7 and 32.1, respectively.

    a. What is the ratio of the mass of a Ni atom to that of a S atom?
    b. If the atomic mass of sulfur were taken to be 50.0, what would be the atomic mass of nickel?

2.7   Bromine consists of two isotopes of atomic masses 78.92 and 80.92 with relative abundances, in the same order, of 50.54% and 49.46%. Calculate the atomic mass of bromine.

2.8   Copper, which has an atomic mass of 63.55, consists of two isotopes with masses of 62.96 and 64.96. What are the abundances of these isotopes?

2.9   Determine, to four significant figures, the ratio of the mass of an H_2O molecule to that of a

    a. $^{12}_{6}C$ atom   b. CO_2 molecule

2.10   The molecular formula of the insecticide DDT is $C_{14}H_9Cl_5$.

    a. What is the molecular mass of DDT?
    b. Is a DDT molecule heavier or lighter than a molecule of the insecticide lindane, $C_6H_6Cl_6$?

2.23   State, in your own words, how the Law of Multiple Proportions is explained in terms of atomic theory.

2.24   Using the data given in Problem 2.3, calculate the mass in grams of phosphorus combined with one gram of chlorine in each compound. Show that the two values obtained this way illustrate the Law of Multiple Proportions.

2.25   An isotope of strontium produced by nuclear fission is $^{90}_{38}Sr$. How many

    a. protons are in the nucleus?
    b. neutrons are in the nucleus?
    c. electrons are in a Sr^{2+} ion?
    d. protons are in a Sr^{2+} ion?

2.26   Complete the table below, using the Periodic Table where necessary to find the symbol of an element.

Symbol	Charge	Protons	Neutrons	Electrons
$^{64}_{30}Zn$	___	___	___	___
___	−3	15	16	___
$^{79}_{35}Br^-$	___	___	___	___
___	+3	26	30	___

2.27   At one time it was suggested that the atomic mass scale be set by taking the atomic mass of $^{19}_{9}F$ to be exactly 19. (On the C-12 scale, this isotope has a mass of 18.9984.) If this choice had been made, what would be the atomic mass of

    a. the element oxygen?   b. $^{12}_{6}C$?

2.28   Naturally occurring lithium consists of two isotopes with atomic masses of 6.015 and 7.016. The abundance of the heavier isotope is 92.58%. Calculate the atomic mass of lithium.

2.29   Magnesium, with an atomic mass of 24.31, consists of three isotopes of masses 23.99, 24.99, and 25.99. The lightest isotope has an abundance of 78.6%. Estimate the abundances of the other two isotopes.

2.30   A certain molecule contains one P atom and x Cl atoms. It is 17.35 times as heavy as a $^{12}_{6}C$ atom. What is the value of x, the number of Cl atoms in the molecule?

2.31   The controversial artificial sweetener saccharin has the molecular formula $C_7H_5O_3NS$.

    a. What is its molecular mass?
    b. Is a saccharin molecule heavier or lighter than a molecule of sucrose, $C_{12}H_{22}O_{11}$?

2.11 A penny has a mass of about 3.1 g. What is the mass, in metric tons, of Avogadro's number of pennies?

2.12 An electron has a charge of 1.60×10^{-19} coulombs and a mass of 9.11×10^{-28} g. What is the charge (in coulombs) and the mass (in grams) of Avogadro's number of electrons?

2.13 Silver has an atomic mass of 107.87. Calculate

    a. the mass of a silver atom in grams.
    b. the number of atoms in one gram of silver.

2.14 How many electrons are there in

    a. a silicon atom (at. no. Si = 14)?
    b. 2.0×10^6 Si atoms?
    c. 1.00 g of Si (at. mass = 28.09)?

2.15 Determine

    a. the number of moles in 2.25 g of Ca.
    b. the mass in grams of 2.61 mol of Ca.

2.16 Give the mass in grams of 1.20 mol of

    a. NaCl   b. CH_4 c. Si

2.17 Complete the following table for pyruvic acid, $C_3H_4O_3$:

Grams	Moles	Molecules	C Atoms
0.0880	___	___	___
___	0.00500	___	___
___	___	1.0×10^9	___
___	___	___	1.0×10^{20}

2.18 Arrange the following in order of increasing mass:

    a. an F atom
    b. 1×10^{-20} mol of F
    c. 1×10^{-20} g of F
    d. F_2 molecule

2.19 A fatal concentration of CO gas occurs if 2.48 g of CO is released into a closed garage having a volume of 1.50×10^3 ft^3. Under these conditions, how many molecules are there per cm^3?

2.20 In making semiconductor devices, silicon crystals can be prepared in which one atom out of 10^8 is an impurity.

    a. If such a crystal weighs 1 mg, how many atoms of the impurity does it contain?
    b. If the impurity is boron, what mass of boron is present?

2.32 A golf ball has a volume of about 20 cm^3. What is the volume, in cubic kilometers, of Avogadro's number of golf balls?

2.33 Genes are considered to be composed of a substance known as DNA. In a particular sample, the average mass of a DNA molecule is 1.0×10^{-15} g. Calculate the molecular mass of DNA.

2.34 Gold has an atomic mass of 196.97. Calculate

    a. the mass of 6.0×10^{15} Au atoms.
    b. the number of atoms in one ounce of gold.

2.35 Give the total number of protons and neutrons in

    a. an atom of lead (Pb) with a mass number of 208.
    b. a mole of $^{208}_{82}Pb$.
    c. 52.0 g of this isotope (at. mass = 208).

2.36 Hexane has the molecular formula C_6H_{14}.

    a. What is the mass in grams of one mole of hexane?
    b. How many moles of hexane are there in 1.00 kg?

2.37 Give the number of moles in 1.20 g of

    a. KNO_3 b. $C_6H_{12}O_6$ c. Fe

2.38 Complete the following table for lactic acid, $C_3H_6O_3$:

Grams	Moles	Molecules	O Atoms
0.00450	___	___	___
___	0.0125	___	___
___	___	3.0×10^{30}	___
___	___	___	1.2×10^{69}

2.39 Arrange the following in order of increasing number of atoms:

    a. 1×10^{-10} mol of C
    b. 1×10^{-10} g of C
    c. 1×10^{-10} mol of CO_2
    d. 5×10^{-11} mol of NH_3

2.40 DDT, a compound with a molecular mass of 354, has one of the lowest water solubilities of all known substances. Its solubility is estimated to be 1×10^{-6} g/ℓ. How many molecules of DDT are there in 1 cm^3 of a saturated solution?

2.41 When light strikes a photographic film covered with silver bromide, some of the Br^- ions are converted to bromine atoms which escape from the film. How many Br^- ions would have to be "lost" before one could detect a change in mass using the most sensitive balance, which weighs to 1.0×10^{-6} g?

2.21 Which of the following statements is always true? never true?

    a. A Cu^{2+} ion has more protons than a neutral copper atom.
    b. A mole of carbon has a greater mass than a mole of helium.
    c. A -2 ion is heavier than the atom from which it is derived.
    d. The mass of a mole of methane, CH_4, is the mass of a methane molecule.

2.22 Briefly describe an experiment to demonstrate

    a. the Law of Conservation of Mass.
    b. the Law of Constant Composition.
    c. that an electron has a negative charge.

2.42 Criticize each of the following statements:

    a. Atoms are lighter than molecules.
    b. Ionic compounds contain equal numbers of cations and anions.
    c. The mass of a carbon-12 atom is exactly 12 g.
    d. There are 6.02×10^{23} molecules in one mole of NaCl.

2.43 Write a brief description of an experiment to

    a. show that $SnCl_2$ and $SnCl_4$ follow the Law of Multiple Proportions.
    b. determine the mass of an electron.
    c. show that NaCl obeys the Law of Constant Composition.

*2.44 Taking the mass of an electron to be 9.11×10^{-28} g, determine the mass, to seven significant figures, of one mole of

    a. Na^+ ions   b. F^- ions   c. NaF

*2.45 One way to determine Avogadro's number is to measure the number of electrons required to plate out a known mass of a metal. It is found that 3037 coulombs are required to form one gram of copper from Cu^{2+} ions. Using the known atomic mass of copper and the charge of the electron in coulombs given in this chapter, calculate the number of atoms in one mole of Cu.

*2.46 By x-ray diffraction, it is possible to determine the geometric pattern in which atoms are arranged in a crystal and the distances between atoms. In a crystal of gold, four atoms effectively occupy the volume of a cube 0.407 nm on an edge. Taking the density of gold to be 19.4 g/cm³, calculate the number of atoms in one mole of Au.

*2.47 Cannizzaro, in 1858, suggested a method of deducing the atomic mass of an element such as hydrogen, based on the following reasoning:

    It is known from Avogadro's Law (Chapter 7) that in one liter of any gas at 0°C and 1 atm, there is a certain definite number of molecules x. In at least one of the gaseous compounds of hydrogen, there should be one atom of hydrogen per molecule. Consequently, by studying the masses of hydrogen in one liter of several of its gaseous compounds, it should be possible to deduce the mass of x hydrogen atoms. A similar study applied to a series of gaseous compounds of oxygen should yield the mass of x oxygen atoms.

    Applying Cannizzaro's reasoning to the data given below, obtain the atomic mass of hydrogen, taking that of oxygen to be 16.0.

Hydrogen Compounds	Mass One Liter (0°C, 1 atm)	%H	Oxygen Compounds	Mass One Liter (0°C, 1 atm)	%O
Hydrogen bromide	3.636 g	1.246	Nitrous oxide	1.975 g	36.35
Acetylene	1.170 g	7.743	Oxygen difluoride	2.422 g	29.63
Ammonia	0.765 g	17.76	Carbon dioxide	1.975 g	72.71
Hydrogen sulfide	1.532 g	5.915	Nitrogen dioxide	2.071 g	69.55

CHEMICAL FORMULAS

Chemical formulas, mentioned briefly in Chapter 2, are basic to the language of chemistry. They serve to identify substances and indicate their structures. When we write the formula H_2 to represent hydrogen, we indicate that this element consists of diatomic molecules. The formula NaCl identifies the compound we call sodium chloride.

In this chapter we will examine various types of chemical formulas. We will be interested in
—what they mean (Section 3.1).
—how they are determined (Sections 3.3–3.5).
—how they are used to find the percentages by mass of the elements in a compound (Section 3.2) or its name (Section 3.6).

3.1 TYPES OF FORMULAS

The **simplest** (*empirical*) formula of a compound gives the simplest whole-number ratio between the numbers of atoms of different elements in the compound. An example is the simplest formula of water, H_2O. This tells us that there are twice as many hydrogen atoms as oxygen atoms in water. In the compound potassium chlorate, there are three elements: potassium, chlorine, and oxygen. These are present in an atom ratio of 1 K:1 Cl:3 O. Hence, the simplest formula of potassium chlorate is $KClO_3$.

The **molecular** formula indicates the number of atoms of each type in a molecule. The molecular formula may be the same as the simplest formula. This is the case with water, H_2O; there are two atoms of hydrogen and one oxygen atom in a water molecule. In other cases, the molecular formula is a whole-number multiple of the simplest formula. Consider, for example, the compound of hydrogen and oxygen known as hydrogen peroxide. Here we write the molecular formula H_2O_2 to indicate that two hydrogen atoms are combined with two oxygen atoms in the hydrogen peroxide molecule. The simplest formula of this compound would be HO.

Molecular formulas can represent elements as well as compounds. Several nonmetals, at ordinary temperatures and pressures, consist of molecules. One of these, H_2, was mentioned in Chapter 2; a more complete list is given in Table 3.1. Elements in which there are no small, discrete molecules are represented simply by their symbols. Such elements include several nonmetals (C and Si in Group 4 of the Periodic Table; all the noble gases in 8) and all of the metals (Na, Mg, Fe, . . .).

Sometimes, to represent a compound, we go beyond the simplest or molecular formula. We write its formula in such a way as to suggest the structure of the compound. Consider, for example, the compounds called *hydrates*. These are solids, usually ionic in nature, which contain water molecules within the crystal lattice. An example is hy-

Where we can, we use molecular, rather than simplest, formulas

49

TABLE 3.1   FORMULAS OF MOLECULES FORMED BY ELEMENTS

GROUP 5	GROUP 6	GROUP 7
N_2 P_4*	O_2 S_8**	H_2 F_2 Cl_2 Br_2 I_2

*In white phosphorus, a reactive, highly toxic solid.
**In solid sulfur and in the liquid near the melting point. For simplicity, we will ordinarily represent the element by its symbol, S.

---

drated barium chloride. This compound contains two moles of water, H_2O, for every mole of barium chloride, $BaCl_2$. Its formula is written as

$$BaCl_2 \cdot 2\ H_2O$$

The water is loosely bound in hydrates (A dot is used to separate the formulas of the two compounds, $BaCl_2$ and H_2O.) We interpret the formula $CuSO_4 \cdot 5\ H_2O$ in a similar way. In this hydrate, there are five moles of water for every mole of copper sulfate, $CuSO_4$.

3.2   PER CENT COMPOSITION FROM FORMULA

As we have seen, the formula of a compound tells us the numbers and kinds of atoms present. It can also be used to determine the percentages by mass of the elements in the compound. These percentages are obtained by finding the mass of each element in one mole of the compound (Example 3.1).

Example 3.1   Sodium hydrogen carbonate, commonly called "bicarbonate of soda," is used in many commercial products to relieve an upset stomach. It has the simplest formula $NaHCO_3$. What are the mass % of Na, H, C, and O in sodium hydrogen carbonate? (Atomic mass Na = 22.99, H = 1.01, C = 12.01, O = 16.00.)

Solution   In one mole of $NaHCO_3$, there are

22.99 g (1 mol) of Na	1.01 g (1 mol) of H
12.01 g (1 mol) of C	48.00 g (3 mol) of O

The mass of one mole of $NaHCO_3$ is

$$22.99\ g + 1.01\ g + 12.01\ g + 48.00\ g = 84.01\ g$$

The mole concept really helps one deal with problems like this one

$$\text{mass \% Na} = \frac{22.99\ g}{84.01\ g} \times 100 = 27.36 \qquad \text{mass \% H} = \frac{1.01\ g}{84.01\ g} \times 100 = 1.20$$

$$\text{mass \% C} = \frac{12.01\ g}{84.01\ g} \times 100 = 14.30 \qquad \text{mass \% O} = \frac{48.00\ g}{84.01\ g} \times 100 = 57.14$$

The per cents add up to 100, as they should:

$$27.36 + 14.30 + 1.20 + 57.14 = 100.00$$

Exercise   What are the mass % of carbon and oxygen in CO_2? Answer: 27.29% C; 72.71% O.

The same reasoning can be applied to determine the percentage of water in a hydrate (Example 3.2).

Example 3.2   The hydrate known as Glauber's salt, $Na_2SO_4 \cdot 10\ H_2O$, is used in some types of solar heating units. Heat from the sun decomposes the hydrate to Na_2SO_4 and H_2O. Later, heat is evolved when these two compounds combine again to form the hydrate. Determine
   a. the percentage of water in $Na_2SO_4 \cdot 10\ H_2O$.
   b. the mass in grams of water obtained from 1.000 kg of the hydrate.

Solution
   a. In one mole of $Na_2SO_4 \cdot 10\ H_2O$, there are 2(22.99 g) = 45.98 g Na, 32.06 g of S, 4(16.00 g) = 64.00 g O, and 10(18.02 g) = 180.2 g of H_2O. The mass of one mole of the hydrate is

$$45.98\ g + 32.06\ g + 64.00\ g + 180.2\ g = 322.2\ g$$

$$\text{mass \% } H_2O = \frac{180.2\ g}{322.2\ g} \times 100 = 55.93$$

   b. Since the hydrate contains 55.93 mass % water:

$$\text{mass } H_2O = \frac{55.93}{100} \times 1000\ g = 559.3\ g\ H_2O$$

Exercise   What mass in grams of $Na_2SO_4 \cdot 10\ H_2O$ is required to produce 1.00 g of H_2O?
Answer: 1.79 g.

3.3 SIMPLEST FORMULA FROM PER CENT COMPOSITION

We saw in Section 3.2 that, given the formula of a compound, the percentages by mass of the elements present can be calculated. As you might suppose, this process can be reversed. If the mass % of the elements are known, the formula of a compound can be found. To do this, we start by determining the number of moles of each element in a convenient amount of compound, usually 100 g. Knowing the number of moles of each element present, we can obtain the mole ratio between these elements. This ratio must of necessity equal the atom ratio. From the atom ratio, the simplest formula follows directly.

Example 3.3   The element tin occurs in nature as the mineral cassiterite, a compound of tin and oxygen. Chemical analysis of this compound shows that the mass % of tin and oxygen are 78.8 and 21.2, in that order. From this information and the atomic masses (Sn = 118.7, O = 16.00), determine the formula of the compound.

Solution
   We need to know the relative numbers of atoms of tin and oxygen. To obtain these numbers, let us calculate the relative numbers of moles of Sn and O in a given mass of the compound. For convenience, we consider a 100-g sample, so that the per cents give us directly the masses of tin and oxygen in the sample.
   In 100 g of this compound there are 78.8 g of Sn and 21.2 g of oxygen. Converting to moles, we have

$$\text{moles Sn} = 78.8 \text{ g Sn} \times \frac{1 \text{ mol Sn}}{118.7 \text{ g Sn}} = 0.664 \text{ mol Sn}$$

$$\text{moles O} = 21.2 \text{ g O} \times \frac{1 \text{ mol O}}{16.00 \text{ g O}} = 1.33 \text{ mol O}$$

These quantities represent not only the relative numbers of moles of Sn and O, *but also the relative numbers of atoms of these elements*. This must be the case, since the number of atoms in one mole (6.022×10^{23}) is the same for all elements. Since the mole ratio in this compound is 0.664:1.33, the Sn:O atom ratio must also be 0.664:1.33.

To find the simplest formula, we need to know the simplest whole-number ratio between these two numbers. This is obtained by dividing each number by the smallest, 0.664:

$$\text{Sn:} \quad \frac{0.664}{0.664} = 1.00 \qquad \text{O:} \quad \frac{1.33}{0.664} = 2.00$$

We see that for every Sn atom, there are 2 oxygen atoms. Hence, the simplest formula of the compound must be SnO_2.

Exercise In another oxide of tin, the mass % of Sn and O are 88.1 and 11.9 in that order. What is the simplest formula of this compound? Answer: SnO.

This same approach can be used to determine the simplest formula of a compound containing three or more elements. The arithmetic is a bit longer but the principle is the same (Example 3.4).

Example 3.4 Vitamin C, which may or may not be effective in preventing colds, contains the three elements carbon, hydrogen, and oxygen. Analysis of a sample of vitamin C gives the following mass %:

$$C = 40.9; \ H = 4.58; \ O = 54.5$$

From these data, determine the simplest formula.

Solution Again, for convenience, we will work with a 100-g sample. In this sample, there are 40.9 g of C, 4.58 g of H, and 54.5 g of O. We first find the number of moles of each element:

$$\text{moles C} = 40.9 \text{ g C} \times \frac{1 \text{ mol C}}{12.01 \text{ g C}} = 3.41 \text{ mol C}$$

$$\text{moles H} = 4.58 \text{ g H} \times \frac{1 \text{ mol H}}{1.01 \text{ g H}} = 4.53 \text{ mol H}$$

$$\text{moles O} = 54.5 \text{ g O} \times \frac{1 \text{ mol O}}{16.00 \text{ g O}} = 3.41 \text{ mol O}$$

The numbers we have just calculated are in the same ratio as the numbers of atoms of C, H, and O in vitamin C. To find the simplest whole-number ratio, we divide by the smallest number, 3.41:

$$\text{C:} \quad \frac{3.41}{3.41} = 1.00 \qquad \text{H:} \quad \frac{4.53}{3.41} = 1.33 \qquad \text{O:} \quad \frac{3.41}{3.41} = 1.00$$

We see that for every carbon atom there is one oxygen atom. At the same time, there is $1.33 = 4/3$ H atom. The atom ratio could be expressed as 1 C:4/3 H:1 O. Multiplying by three to obtain the simplest whole-number ratio, we arrive at 3 C:4 H:3 O. The simplest formula of vitamin C must then be $C_3H_4O_3$.

Exercise Butane is a compound of carbon and hydrogen. The mass % of C is 82.7. What is the mass % of H? the simplest formula of butane? Answer: 17.3; C_2H_5.

Determination of Per Cent Composition by Experiment

Once the mass % of the elements in a compound are known, the calculation of the simplest formula is straightforward. Finding these percentages, however, is not a simple matter. They must be determined by experiment. Many different methods are possible. To illustrate the principle involved, consider the oxide of tin referred to in Example 3.3. To *analyze* this compound—that is, to determine the mass % of tin and oxygen—we might use the apparatus shown in Figure 3.1. Here we start with a weighed sample of the oxide, let us say 0.800 g. The sample is then heated in a stream of hydrogen gas. At high temperatures a reaction occurs. The oxygen in the compound is converted to water vapor, which escapes with the excess hydrogen. The solid residue that remains is pure tin; its mass is 0.630 g. The mass % of tin in the oxide must then be

$$\text{mass \% Sn} = \frac{0.630 \text{ g}}{0.800 \text{ g}} \times 100 = 78.8$$

The mass % of oxygen is found by subtracting from 100:

$$\text{mass \% O} = 100.0 - 78.8 = 21.2$$

Many simple compounds containing only two elements can be analyzed by a procedure similar to this. The general approach is to carry out a reaction in which one of the elements is produced in the pure state. Here, an oxide of tin is reacted with hydrogen to form tin metal. In another case, we might cause an oxide of iron to react with carbon, forming pure iron as one of the products.

The percentage composition of an organic compound such as vitamin C can be found using the apparatus shown in Figure 3.2. A weighed sample of the compound, usually

If it is feasible, it's safest to analyze for both elements

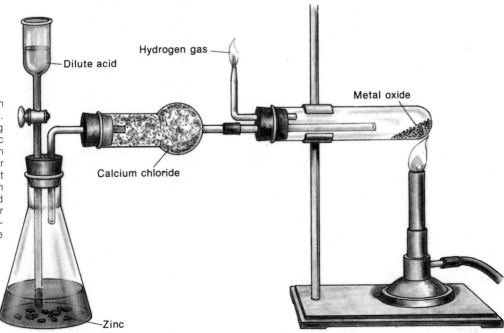

Figure 3.1 Reduction of oxides by hydrogen. Acid from the dropping funnel reacts with zinc to produce $H_2(g)$, which is dried by passing over $CaCl_2$. In the test tube at the right, the hydrogen reacts with the heated metal oxide. The water produced by the reaction passes off with the excess hydrogen.

Dilute acid

Hydrogen gas

Metal oxide

Calcium chloride

Zinc

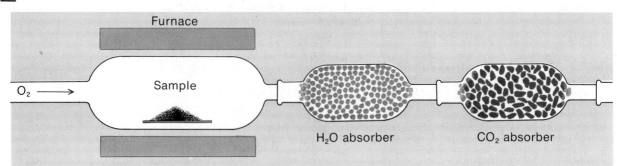

Figure 3.2 Combustion train used for carbon-hydrogen analysis. The absorbent for water is magnesium perchlorate, $Mg(ClO_4)_2$. Carbon dioxide is absorbed by finely divided sodium hydroxide supported on asbestos. Only a few milligrams of sample are needed for an analysis.

only a few milligrams, is burned in oxygen. The carbon in the sample is converted to carbon dioxide; the hydrogen present ends up as water. The amounts of CO_2 and H_2O produced are determined by measuring the increases in mass of the two absorption tubes. From these amounts, the masses of carbon and hydrogen in the sample and hence their percentages can be calculated. The percentage of oxygen, the only other element present in vitamin C, is found by subtraction from 100.

Note the small mass that is needed

Example 3.5   The mass % of C, H, and O in vitamin C (Example 3.4) are determined by burning a sample weighing 2.00 mg. The masses of CO_2 and H_2O formed are 3.00 mg and 0.816 mg, in that order. What are the mass % of C, H, and O?

Solution   Let us first calculate the mass of carbon in 3.00 mg of CO_2. Since one mole of CO_2, 44.0 g, contains one mole of carbon, 12.0 g:

$$12.0 \text{ g C} \simeq 44.0 \text{ g CO}_2*$$

$$\text{milligrams C} = 3.00 \text{ mg CO}_2 \times \frac{12.0 \text{ g C}}{44.0 \text{ g CO}_2} = 0.818 \text{ mg C}$$

Similarly, since there are 2.02 g of hydrogen in 18.0 g of H_2O:

$$\text{milligrams H} = 0.816 \text{ mg H}_2O \times \frac{2.02 \text{ g H}}{18.0 \text{ g H}_2O} = 0.0916 \text{ mg H}$$

Noting that the sample weighed 2.00 mg, we have

$$\%C = \frac{0.818}{2.00} \times 100 = 40.9\% \qquad \%H = \frac{0.0916}{2.00} \times 100 = 4.58\%$$

We obtain the percentage of oxygen by difference:

$$\%O = 100.0 - (40.9 + 4.6) = 54.5\%$$

Exercise   The combustion of 1.00 mg of butane gives 3.03 mg of CO_2 and 1.55 mg of H_2O. What are the mass % of C and H in butane? Answer: 82.6% C; 17.4% H.

*The symbol $\simeq$ means "chemically equivalent to." In conversions, it can be treated as an equals sign.

As pointed out in Section 3.1, the molecular formula is a whole-number multiple of the simplest formula. That multiple may be 1 as in H_2O, 2 as in H_2O_2, 3 as in C_3H_6, ... To find out what the multiple is, we need one more piece of data, the molecular mass. By comparing the molecular mass to the mass of the simplest formula, we can determine whether the multiple is 1, 2, 3, . . .

Example 3.6   The simplest formula of vitamin C was determined in Example 3.3 to be $C_3H_4O_3$. From another experiment, its molecular mass is found to be about 180. What is the molecular formula of vitamin C?

Solution   Let us first calculate the sum of the atomic masses for $C_3H_4O_3$. We have

$$3(12.0) + 4(1.0) + 3(16.0) = 88.0$$

Thus, the formula mass for $C_3H_4O_3$ is 88.0. The approximate molecular mass, 180, is twice this (180/88 = 2.0). Hence, the simplest formula must be multiplied by 2 to obtain the molecular formula:

$$\text{molecular formula vitamin C} = 2 \times C_3H_4O_3 = C_6H_8O_6$$

Note that an approximate molecular mass is sufficient here. An experiment that shows that the molecular mass of vitamin C is between 160 and 200 would be satisfactory. That would tell us that the simplest formula should be multiplied by 2 rather than 1, 3, or some other whole number.

Molecular masses can often be accurately found with a mass spectrometer

---

Exercise   The simplest formula of butane is C_2H_5. Its molecular mass is about 60. What is the molecular formula of butane? Answer: C_4H_{10}.

3.5 FORMULAS OF IONIC COMPOUNDS

In the past two sections, we have seen how the formulas of molecular substances can be determined by experiment. Formulas of ionic compounds can also be found in the laboratory. It is also possible to predict, "on paper," the formulas of many simple ionic compounds. This can be done using the principle of electrical neutrality, cited in Chapter 2. Consider, for example, the ionic compound calcium chloride. Knowing that the ions present are Ca^{2+} and Cl^- we can predict, correctly, that its formula must be $CaCl_2$. Two Cl^- ions, each with a -1 charge, are required to balance a single Ca^{2+} ion.

To use this principle, we must know the charges of the ions involved. Here the Periodic Table can be very helpful. We find that:

Elements which are close to a noble gas in the Periodic Table tend to form ions which have the same number of electrons as the noble gas atom.

Consider, for example, the noble gas neon. Since it has an atomic number of 10, a neon atom contains 10 protons and 10 electrons. The nonmetals N (at. no. = 7), O (at. no. = 8), and F (at. no. = 9) all form ions with 10 electrons. This requires that these ions have charges of -3, -2, and -1, in that order.

$$N\ (7\ p^+, 7\ e^-) + 3\ e^- \rightarrow N^{3-}\ (7\ p^+, 10\ e^-)$$
$$O\ (8\ p^+, 8\ e^-) + 2\ e^- \rightarrow O^{2-}\ (8\ p^+, 10\ e^-)$$
$$F\ (9\ p^+, 9\ e^-) + 1\ e^- \rightarrow F^-\ (9\ p^+, 10\ e^-)$$

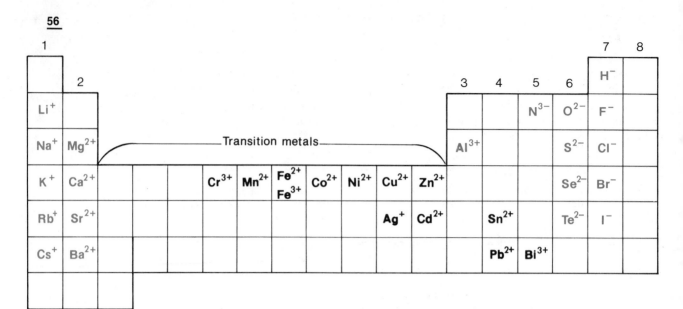

Figure 3.3 Ions with noble-gas structures are shown in color. Two other such ions, which are much less common, are Sc^{3+} and P^{3-}. Can you locate these in the Periodic Table? The transition metals and post-transition metals (Sn, Pb, Bi) form monatomic cations which may have charges of +1, +2, or +3. Several of these metals, notably iron, form two different cations (Fe^{2+}, Fe^{3+}).

The metals Na (at. no. = 11), Mg (at. no. = 12), and Al (at. no. = 13) behave in a similar way. They form ions which contain 10 electrons. This means that these ions have charges of +1, +2, and +3, in that order.

$$\text{Na (11 p}^+\text{, 11 e}^-\text{)} \rightarrow \text{Na}^+ \text{(11 p}^+\text{, 10 e}^-\text{)} + \text{1 e}^-$$
$$\text{Mg (12 p}^+\text{, 12 e}^-\text{)} \rightarrow \text{Mg}^{2+} \text{(12 p}^+\text{, 10 e}^-\text{)} + \text{2 e}^-$$
$$\text{Al (13 p}^+\text{, 13 e}^-\text{)} \rightarrow \text{Al}^{3+} \text{(13 p}^+\text{, 10 e}^-\text{)} + \text{3 e}^-$$

A list of common ions that have the same number of electrons as the nearest noble gas atom is given in Figure 3.3. Note that:

> *The ions of the Group 1 metals have +1 charges.*
> *The ions of the Group 2 metals have +2 charges.*
> *The ions of the Group 6 nonmetals have −2 charges.*
> *The ions of the Group 7 nonmetals have −1 charges.*

Several metals which are far removed from the noble gases in the Periodic Table form positive ions. The structures of these ions are not related in any direct way to those of noble gas atoms. Indeed, there is no simple way to predict their charges. Some of the more important of these ions are listed in Figure 3.3. They are derived from the **transition metals** (those in the groups near the center of the Periodic Table) and the heavier metals in Groups 4 and 5. The most common charge among these ions, you will note, is +2. Charges of +1 (Ag^+) and +3 (Cr^{3+}, Fe^{3+}, Bi^{3+}) are less common.

Using Figure 3.3, it is possible to predict the formulas of a large number of ionic compounds (Example 3.7). Notice that, *in writing the formula of an ionic compound, the positive ion is always given first.*

Example 3.7   Predict the formulas of the ionic compounds formed by
    a. lithium and oxygen (Li and O).   b. magnesium and chlorine (Mg and Cl).
    c. nickel and sulfur (Ni and S).   d. bismuth and fluorine (Bi and F).

Solution

    a. Li_2O; 2 Li^+ ions required to balance 1 O^{2-} ion

    b. $MgCl_2$; 1 Mg^{2+} ion required to balance 2 Cl^- ions

    c. NiS; 1 Ni^{2+} ion required to balance 1 S^{2-} ion

    d. BiF_3; 1 Bi^{3+} ion required to balance 3 F^- ions

These formulas follow
from charge balance:
Total + = Total −

Exercise   Give the formulas of zinc iodide, silver sulfide, and aluminum oxide. Answer: ZnI_2; Ag_2S; Al_2O_3.

All the ions considered so far have been *monatomic,* derived from a single atom. However, many familiar ionic compounds contain polyatomic ions. These are charged species containing more than one atom. Examples include

    —the hydroxide ion, OH^-, found in sodium hydroxide (lye), NaOH.

    —the carbonate ion, CO_3^{2-}, found in calcium carbonate (limestone), $CaCO_3$.

    —the ammonium ion, NH_4^+, found in ammonium chloride, NH_4Cl.

Table 3.2 lists some of the more important polyatomic ions, along with their charges and names.

TABLE 3.2   SOME COMMON POLYATOMIC IONS

+1	−1	−2	−3
NH_4^+ (ammonium)	OH^- (hydroxide)	CO_3^{2-} (carbonate)	PO_4^{3-} (phosphate)
	NO_3^- (nitrate)	SO_4^{2-} (sulfate)	
	ClO_3^- (chlorate)	CrO_4^{2-} (chromate)	
	ClO_4^- (perchlorate)	$Cr_2O_7^{2-}$ (dichromate)	
	MnO_4^- (permanganate)		
	HCO_3^- (hydrogen carbonate)		

You should learn these
formulas (including
charges!) and names

The formulas of compounds containing polyatomic ions can be predicted in much the same way as with monatomic ions. The principle of electrical neutrality is used to find the relative numbers of positive and negative ions. A minor complication arises when the number of polyatomic ions in the formula is two or greater. Here, the polyatomic ion is enclosed in parentheses to avoid confusion. Thus we have

    $Ca(OH)_2$ (1 Ca^{2+} ion, 2 OH^- ions)

    $(NH_4)_2S$ (2 NH_4^+ ions, 1 S^{2-} ion)

Ionic charges do not
appear in the formulas
of ionic compounds

Example 3.8   Predict the formulas, using Figure 3.3 and Table 3.2, of

    a. barium hydroxide   b. potassium sulfate   c. ammonium phosphate

Solution

    a. $Ba(OH)_2$; 1 Ba^{2+} ion required to balance 2 OH^- ions

    b. K_2SO_4; 2 K^+ ions required to balance 1 SO_4^{2-} ion

    c. $(NH_4)_3PO_4$; 3 NH_4^+ ions required to balance 1 PO_4^{3-} ion

Exercise   What is the formula of potassium hydroxide? barium phosphate? ammonium sulfate? Answer: KOH; $Ba_3(PO_4)_2$; $(NH_4)_2SO_4$.

3.6 NAMES OF COMPOUNDS

There are several million known compounds, all of which have names. Most of these are carbon compounds that fall in the area of organic chemistry (Chapter 25). At the moment, we need only be concerned with naming inorganic compounds, a somewhat simpler task. Moreover, we will restrict our attention here to two types of inorganic substances:

—simple ionic compounds of the type described in Section 3.5.

—binary compounds of the nonmetals (two different kinds of nonmetal atoms).

We have referred by name to many of these compounds in this and previous chapters. Quite possibly, you have deduced the main features of the system used. Their names follow in a straightforward way from the formulas written to represent them.

Ionic Compounds

The name of an ionic compound consists of two words. The first word gives the name of the positive ion, the second that of the negative ion. Thus we have

Compound	Cation	Anion	Name
$NaCl$	Na^+	Cl^-	Sodium chloride
K_2SO_4	K^+	SO_4^{2-}	Potassium sulfate
$Zn(NO_3)_2$	Zn^{2+}	NO_3^-	Zinc nitrate

To assign names to individual ions, note that

1. Monatomic positive ions take the name of the metal from which they are derived. Examples include

Na^+ *sodium* K^+ *potassium* Zn^{2+} *zinc*

There is one complication. Certain metals form more than one type of positive ion. An example is iron, where we have the Fe^{2+} and Fe^{3+} ions. To distinguish between these ions, the charge must be indicated in the name. This is done by giving the charge as a Roman numeral in parentheses after the name of the metal:

Fe^{2+} *iron*(II) Fe^{3+} *iron*(III)

(An older system used the suffixes -*ic* for the ion of higher charge and -*ous* for the ion of lower charge. These were added to the stem of the Latin name of the metal. The Fe^{3+} ion was referred to as ferric, the Fe^{2+} ion as ferrous.)

2. Monatomic negative ions are named by adding the suffix -*ide* to the stem of the name of the nonmetal from which they are derived. Thus we have

N^{3-} nitride	O^{2-} oxide	H^- hydride
	S^{2-} sulfide	F^- fluoride
	Se^{2-} selenide	Cl^- chloride
	Te^{2-} telluride	Br^- bromide
		I^- iodide

3. Polyatomic ions are given special names (Table 3.2, p. 57).

Example 3.9   Name the following ionic compounds:
   a. CaS   b. Al(NO$_3$)$_3$ c. FeCl$_2$

Solution
   a. calcium sulfide
   b. aluminum nitrate
   c. Recall that the chloride ion has a −1 charge (Cl$^-$). The positive ion must then be Fe^{2+}. The name is iron(II) chloride.

Exercise   Name Al$_2$O$_3$, Fe(NO$_3$)$_3$, and Ag$_2$Se. Answer: aluminum oxide; iron(III) nitrate; silver selenide.

Binary Compounds of the Nonmetals

When a pair of nonmetals forms only one compound, that compound is named very simply. The name of the element whose symbol appears first in the formula is written first. The second part of the name is formed by adding the suffix -*ide* to the stem of the name of the second nonmetal. Examples include

The element farther left in the Periodic Table usually comes first

HCl   hydrogen chloride
H$_2$S   hydrogen sulfide
NF$_3$ nitrogen fluoride

More often, a pair of nonmetals forms more than one compound. In this case, the name is a bit more complex. The Greek prefixes *di* = two, *tri* = three, *tetra* = four, *penta* = five, *hexa* = six, *hepta* = seven, and *octa* = eight are used to show the number of atoms of each element. The several oxides of nitrogen have the names

N$_2$O$_5$ dinitrogen pentoxide
N$_2$O$_4$ dinitrogen tetroxide
NO$_2$ nitrogen dioxide
N$_2$O$_3$ dinitrogen trioxide
NO   nitrogen oxide
N$_2$O   dinitrogen oxide

Example 3.10   Give the names of
   a. SO$_2$ b. SO$_3$ c. PCl$_3$ d. PCl$_5$

Solution
   a. sulfur dioxide   b. sulfur trioxide   c. phosphorus trichloride
   d. phosphorus pentachloride

Exercise   Give the names of SF$_4$ and SF$_6$. Answer: sulfur tetrafluoride; sulfur hexafluoride.

Many of the best-known binary compounds of the nonmetals have acquired common names. These are widely and, in some cases, exclusively used. Examples include

H_2O	water	PH_3	phosphine
H_2O_2	hydrogen peroxide	AsH_3	arsine
NH_3	ammonia	NO	nitric oxide
N_2H_4	hydrazine	N_2O	nitrous oxide

Water is never called
dihydrogen oxide

SUMMARY

The chemical composition of a compound is shown by its formula. The subscripts in the formula tell us the number of moles of each element present in one mole of the compound. From the formula, the mass % of each element can be calculated in a straightforward way (Example 3.1). Alternatively, if the mass % are known, the simplest (empirical) formula can be obtained (Examples 3.3, 3.4). Combustion analysis is often used to determine mass % of carbon and hydrogen in organic compounds (Example 3.5). The molecular mass is used to relate the simplest formula to the molecular formula (Example 3.6).

The simplest formula of an ionic compound can be predicted if the charges of the ions are known (Examples 3.7, 3.8). Elements in Groups 1, 2, 6, and 7 of the Periodic Table form ions with the same number of electrons as the neighboring noble gas atom. Hence, the charges of these ions are related in a simple way to their position in the Table ($+1$ for Group 1, $+2$ for Group 2, -2 for Group 6, -1 for Group 7). Charges of some important transition metal cations are given in Figure 3.3. The names, formulas, and charges of some common polyatomic ions are listed in Table 3.2.

In naming and writing formulas for ionic compounds, the cation is cited first. For metals which form two or more cations, the charge of the cation is indicated by a Roman numeral after the name of the metal (Example 3.9). Binary compounds of nonmetals are molecular and are named in a somewhat different way. If two nonmetals form more than one binary compound, Greek prefixes are used to indicate the number of atoms of each element in a molecule (Example 3.10).

KEY WORDS AND CONCEPTS

empirical formula   per cent composition
molecular formula   polyatomic ion
hydrate

QUESTIONS AND PROBLEMS

Catalog

Per Cent Composition from Formula: 3.2–3.4, 3.19–3.21
Per Cent Composition from Analysis: 3.5, 3.6, 3.22, 3.23
Simplest Formula from Experiment: 3.7–3.12, 3.24–3.29

Molecular Formula from Simplest Formula: 3.13, 3.14, 3.30, 3.31
Names and Formulas of Compounds: 3.15–3.18, 3.32–3.35
General: 3.1, 3.36–3.39

3.1   Review and know the meanings of the key words and concepts in this chapter.

3.2   Determine the mass % of the elements in

    a. $CaCO_3$ b. $C_6H_{12}O_6$ c. $Na_2CO_3 \cdot 10\ H_2O$

3.3   Consider the mineral $Ca_3(PO_4)_2$.

    a. What is the mass % of each element in this compound?
    b. How many grams of phosphorus can be obtained from one metric ton of this mineral?

3.4   Garlic salt contains NaCl in addition to garlic. Analysis of a 3.50-g sample of garlic salt yields 1.05 g of Cl.

    a. How many grams of NaCl are in the sample?
    b. What per cent, by mass, of garlic salt is NaCl?

(Assume there is no other Cl-containing compound in garlic salt.)

3.5   Hexane contains only the two elements carbon and hydrogen. Combustion of a 1.000-g sample of hexane gives 3.064 g of CO_2. What are the mass % of carbon and hydrogen in hexane?

3.6   A 2.000-g sample of a compound containing only C, H, and S is reacted with chlorine. The products are CCl_4 (9.903 g), HCl (7.042 g), and SCl_2 (3.315 g). What are the mass % of C, H, and S in the compound?

3.7   What are the simplest formulas of the following compounds (mass % given)?

    a. 59.0% Na; 41.0% O
    b. 15.1% K; 10.5% Al; 24.8% S; 49.6% O
    c. 20.2% Ni; 9.62% N; 4.16% H; 66.0% O

3.8   A hydrate of magnesium iodide has the formula $MgI_2 \cdot x\ H_2O$. To determine a value for x, a student heats a hydrate sample until all the water is removed. A 1.055-g sample of hydrate is heated until a constant mass of 0.695 g is reached. What is x?

3.9   Iron reacts with sulfur to form a sulfide. 2.236 g of iron reacted with 1.926 g of sulfur. What is the simplest formula of the sulfide?

3.10   A 1.997-g sample of a copper oxide, when heated in a stream of H_2 gas, yields 0.452 g of H_2O. What is the formula of the copper oxide?

3.11   Isopropyl alcohol, commonly called rubbing alcohol, contains only C, H, and O. When a 1.50-mg sample of the alcohol is burned completely in oxygen, 3.30 mg of CO_2 and 1.80 mg of H_2O are formed. What is the simplest formula for isopropyl alcohol?

3.12   The insecticide DDD contains only C, H, and Cl. When a 3.000-g sample is burned in oxygen, 5.78 g of CO_2 and 0.844 g of H_2O are formed. What is the simplest formula of DDD?

3.19   What is the mass % of carbon in

    a. propane, C_3H_8?   b. lactic acid, $C_3H_6O_3$?
    c. DDT, $C_{14}H_9Cl_5$?

3.20   How many grams of iron can be obtained from one kilogram of magnetite ore, Fe_3O_4?

3.21   A 500-mg sample of a commercial headache remedy contains 266 mg of aspirin, $C_9H_8O_4$.

    a. What is the mass % of aspirin in the product?
    b. How many grams of carbon are there in the aspirin contained in a tablet of this product weighing 0.611 g?

3.22   Naphthalene, like hexane, contains only carbon and hydrogen. Combustion of a 1.000-g sample of naphthalene gives 0.562 g of H_2O. What are the mass % of C and H in naphthalene?

3.23   A compound containing only C, H, and O is reacted with chlorine. The hydrogen is converted to HCl, the carbon to CCl_4. A 5.00-mg sample forms 10.7 mg of HCl and 30.2 mg of CCl_4. What are the mass % of C, H, and O in the compound?

3.24   Determine the simplest formulas of the following compounds (mass % given).

    a. 89.6% Pb; 10.4% O
    b. 37.0% C; 2.22% H; 42.3% O; 18.5% N
    c. 15.7% Cr; 14.5% S; 4.57% H; 65.2% O

3.25   The compound $CrSO_4 \cdot x\ H_2O$ is analyzed for per cent water by mass. A 1.370-g sample of the hydrate is heated to remove all the water; 0.740 g of $CrSO_4$ remains after heating. What is the mass % of water in the compound?

3.26   4.17 g of tin reacted directly with fluorine to form 6.83 g of the metal fluoride. What is the formula of the compound?

3.27   A 2.350-g sample of a copper oxide is heated in a stream of H_2 gas and 0.296 g of H_2O is formed. What is the formula of the copper oxide?

3.28   Glycerol contains the three elements C, H, and O. When a sample weighing 1.010 mg is burned in oxygen, 1.448 mg of CO_2 and 0.790 mg of H_2O are formed. What is the simplest formula of glycerol?

3.29   The insecticide lindane contains only C, H, and Cl. When a 2.463-g sample is burned in oxygen, 2.236 g of CO_2 and 0.458 g of H_2O are formed. What is the simplest formula of lindane?

3.13 In a certain hydrocarbon, there are twice as many H atoms as C atoms per molecule. The molecular mass of the compound is about 56. What is the molecular formula?

3.14 A certain compound has the simplest formula $C_4H_5ON_2$. Which of the following molecular masses are possible?

a. 48   b. 61   c. 156   d. 194   e. 220

3.15 Write the formula of each of the following ionic compounds. Use the Periodic Table and, if necessary, Figure 3.3 and Table 3.2.

a. zinc dichromate
b. tin(II) oxide
c. cobalt(III) hydroxide
d. barium nitrate

3.16 Complete the following table of ionic compounds. Use the Periodic Table and, if necessary, Figure 3.3 and Table 3.2.

Name	Formula
a. _____	$AgNO_3$
b. aluminum selenide	___
c. lithium oxide	___
d. ammonium iodide	
e. _____	$Fe(OH)_3$
f. _____	$KMnO_4$
g. _____	$Ca(ClO_3)_2$

3.17 Complete the following table of molecular compounds:

Name	Formula
a. nitrogen triiodide	___
b. _____	CCl_4
c. bromine trifluoride	___
d. bromine pentafluoride	___
e. _____	S_4N_4

3.18 How many moles of ions are in

a. 0.500 mol of barium chloride?
b. 0.125 mol of magnesium hydroxide?
c. 2.90 g of iron(II) carbonate?

3.30 The simplest formula of a certain hydrocarbon is found to be C_5H_4. Its molecular mass lies between 120 and 130. What is its molecular formula?

3.31 The molecular mass of a certain compound of sulfur and oxygen is 80. Which of the following are possible simplest formulas?

a. SO   b. SO_2 c. SO_3 d. S_2O

3.32 Name the following ionic compounds.

a. $(NH_4)_2Cr_2O_7$ b. CaI_2
c. $FeCO_3$ d. $KClO_4$

3.33 Complete the following table of ionic compounds:

Name	Formula
a. magnesium fluoride	___
b. _____	CdTe
c. aluminum sulfide	___
d. _____	$Fe(NO_3)_3$
e. ammonium sulfate	___
f. calcium phosphate	___
g. _____	$PbCrO_4$

3.34 Complete the following table of molecular compounds:

Name	Formula
a. _____	XeF_4
b. _____	XeF_2
c. iodine monochloride	___
d. iodine trichloride	___
e. _____	S_4N_2

3.35 How many moles of

a. anions are in one mole of cadmium phosphate?
b. cations are in 0.250 mol of ammonium phosphate?
c. ions are in 0.200 mol of zinc sulfide?

*3.36 A sample of an oxide of vanadium weighing 3.530 g was heated with hydrogen gas to form water and another oxide of vanadium weighing 2.909 g. The second oxide was treated further with hydrogen until only 1.979 g of vanadium metal remained.

a. What are the simplest formulas of the two oxides?
b. What is the total mass of water formed in the successive reactions?

*3.37 When 1.632 g of a metal oxide M_2O_5 is treated with HCl, 3.459 g of MCl_4 is formed. What is the atomic mass of M?

*3.38 A mixture of NaCl and NaBr weighing 1.234 g is heated with chlorine, which converts the mixture completely to NaCl. The total mass of NaCl is now 1.150 g. What are the mass % of Cl and Br in the original sample?

*3.39 A sample of cocaine, $C_{17}H_{21}O_4N$, is diluted with sugar, $C_{12}H_{22}O_{11}$. When a 1.00-mg sample of this mixture is burned, 1.75 mg of CO_2 is formed. What is the per cent of cocaine in the sample?

CHEMICAL REACTIONS
AND EQUATIONS

A major use of chemical formulas, discussed in Chapter 3, is to write chemical equations. An equation represents a chemical reaction. It identifies, by its formulas, the substances taking part in the reaction. These include the starting materials, called *reactants,* and the substances formed in the reaction, the *products.* In Section 4.1, we will consider how to write and balance chemical equations.

Chemical equations serve many purposes. In the most general sense, they describe what happens in a reaction. Here, we will be interested in one aspect of chemical equations: how to use them to relate amounts of reactants and products. These amounts and the relations between them can be expressed in moles (Section 4.2) or in grams (Section 4.3).

Chemical reactions can occur between pure substances, which may be either solids, liquids, or gases. We will consider several such reactions in this chapter. Other reactions take place in solution. Reactions in water solution are particularly common, both in the laboratory and in the world around us. This chapter concludes with a brief survey of such reactions. They will be considered in greater detail later (Chapters 16–22).

4.1 WRITING AND BALANCING EQUATIONS

Beginning students are sometimes led to believe that writing a chemical equation is a simple, mechanical process. Nothing could be further from the truth. One point that seems obvious is often overlooked. *You cannot write an equation unless you know what happens in the reaction it represents.* All the reactants and all the products must be identified.

To illustrate how an equation is generated, let us refer back to a reaction described in Chapter 3. Recall Figure 3.1, in which an oxide of tin, SnO_2, was heated with hydrogen, H_2. As pointed out in the discussion on p. 53, there are two products in this reaction. One is tin metal, Sn. The other is water vapor, H_2O. To write a balanced equation for this reaction, we follow a three-step path:

1. *Write an unbalanced equation.* Here the formulas of reactants appear on the left and those of products on the right. Reactants are separated from products by drawing an arrow between them. In this case we have

$$SnO_2 + H_2 \rightarrow Sn + H_2O$$

2. *Balance the equation.* That is, make it conform to the Law of Conservation of Mass. This requires that there be the same number of atoms of each element on each side of the equation. In the equation above, the oxygen atoms are not balanced. There are two

You might call this the Law of Conservation of Atoms

63

on the left and only one on the right. To correct this, we put a *coefficient* of 2 in front of H_2O:

$$SnO_2 + H_2 \rightarrow Sn + 2\ H_2O$$

Now the hydrogen atoms are out of balance. The coefficient we wrote multiplies the formula H_2O by 2, so we have $2 \times 2 = 4$ H atoms on the right, as opposed to only 2 on the left. To obtain 4 H atoms on the left, we write a coefficient of 2 for H_2.

$$SnO_2 + 2\ H_2 \rightarrow Sn + 2\ H_2O$$

The equation is now balanced. We have 1 atom of Sn, 2 atoms of O, and 4 atoms of H on both sides.

Note that equations are balanced by adjusting coefficients as we have shown. Subscripts within formulas must not be changed. In this case, you might have been tempted to write "H_2O_2" instead of "H_2O" on the right. That would have balanced the equation in a single step. However, it does not correspond to reality. The gaseous product in this reaction is water, H_2O, not hydrogen peroxide, H_2O_2.

H_2O and H_2O_2 are very different substances

3. *Indicate the physical states of reactants and products.* We will do this by using
(g) for substances in the gas state
(s) for solids
(l) for pure liquids
(aq) for species in water solution (Section 4.4)
For this reaction we have

$$SnO_2(s) + 2\ H_2(g) \rightarrow Sn(s) + 2\ H_2O(g) \tag{4.1}$$

This is the final balanced equation for the reaction.

Example 4.1   In a certain rocket motor the reactants are hydrazine, N_2H_4, and dinitrogen tetroxide, N_2O_4. When these two liquids are mixed, they react vigorously to form nitrogen gas and liquid water. The energy given off in the reaction is great enough to lift the rocket and its payload into space. Write a balanced equation for this reaction.

Solution   Start by writing the unbalanced equation. To do this, you must know the formulas of the products. By this time, you should know that the element nitrogen consists of diatomic molecules. Of course you know the formula of water. The unbalanced equation is

$$N_2H_4 + N_2O_4 \rightarrow N_2 + H_2O$$

To balance this equation, you might first write a coefficient of 4 for H_2O. This puts four oxygen atoms on both sides.

$$N_2H_4 + N_2O_4 \rightarrow N_2 + 4\ H_2O$$

Hydrogen must now be balanced by writing a coefficient of 2 for N_2H_4. There are then 8 H atoms on each side:

$$2\ N_2H_4 + N_2O_4 \rightarrow N_2 + 4\ H_2O$$

Now only the nitrogen atoms remain to be balanced. On the left we have 4 in the two N_2H_4 molecules and 2 more in N_2O_4 for a total of 6. To obtain 6 nitrogen atoms in the products, write a coefficient of 3 for N_2:

$$2 N_2H_4 + N_2O_4 \rightarrow 3 N_2 + 4 H_2O$$

To complete the job, indicate the physical states of reactants and products. These are given in the statement of the problem. The final balanced equation is

$$2 N_2H_4(l) + N_2O_4(l) \rightarrow 3 N_2(g) + 4 H_2O(l)$$

Exercise   Natural gas consists largely of methane, CH_4. When you light a Bunsen burner, the methane combines with oxygen of the air to form two products. These are liquid water and the gas carbon dioxide. Write a balanced equation for this reaction. Answer: $CH_4(g) + 2 O_2(g) \rightarrow CO_2(g) + 2 H_2O(l)$

By following the approach described in Example 4.1, you can readily balance most equations. In doing so, it is best to start with an element which appears in only *one* reactant and one product. Thus, in Example 4.1, hydrogen or oxygen would be a reasonable element to start with. Nitrogen, on the other hand, would be a poor choice because it appears in two different species (N_2H_4 and N_2O_4) on the left side of the equation.

4.2   MOLE RELATIONS IN BALANCED EQUATIONS

We can use a balanced equation to relate the amounts of substances that take part in a reaction. To do this we make use of the fact that **the coefficients of a balanced equation represent the relative numbers of moles of reactants and products.**

To show that this is the case, consider the reaction between CH_4 and O_2. As you just found, the balanced equation for this reaction is

$$CH_4(g) + 2 O_2(g) \rightarrow CO_2(g) + 2 H_2O(l) \tag{4.2}$$

Perhaps the simplest way to interpret the coefficients of this equation is to say that they represent numbers of molecules. That is,

1 molecule CH_4 + 2 molecules O_2 → 1 molecule CO_2 + 2 molecules H_2O

However, a balanced chemical equation behaves much like an algebraic equation. It remains valid if each of the coefficients is multiplied by the same number. That number might be 1/2, 100, . . . or, *in particular, Avogadro's number N.* That is,

N molecules CH_4 + 2 N molecules O_2 → N molecules CO_2 + 2 N molecules H_2O

But a mole represents Avogadro's number of items. In other words,

N molecules = 1 mol

So we can write

1 mol CH_4 + 2 mol O_2 → 1 mol CO_2 + 2 mol H_2O

Note how easy it is to get this equation from the balanced chemical equation

which is the relation we set out to demonstrate.

The argument we have just gone through can be applied to any reaction. For the reaction between SnO_2 and H_2

$$SnO_2(s) + 2 H_2(g) \rightarrow Sn(s) + 2 H_2O(g)$$

we can write

$$1 \text{ mol } SnO_2 + 2 \text{ mol } H_2 \rightarrow 1 \text{ mol } Sn + 2 \text{ mol } H_2O$$

Relations such as those just obtained can be expressed in a slightly different way. For this reaction we can say that

$$1 \text{ mol } SnO_2 \simeq 2 \text{ mol } H_2 \simeq 1 \text{ mol } Sn \simeq 2 \text{ mol } H_2O$$

where the symbol $\simeq$ means "is chemically equivalent to." Again, for the reaction considered in Example 4.1:

$$2 \text{ mol } N_2H_4 \simeq 1 \text{ mol } N_2O_4 \simeq 3 \text{ mol } N_2 \simeq 4 \text{ mol } H_2O$$

These relations can be interpreted to mean that, for example,

2 mol N_2H_4 reacts with 1 mol N_2O_4 1 mol N_2O_4 produces 3 mol N_2
2 mol N_2H_4 produces 3 mol N_2 1 mol N_2O_4 produces 4 mol H_2O

Example 4.2   For the reaction $2 \text{ N}_2\text{H}_4(l) + \text{N}_2\text{O}_4(l) \rightarrow 3 \text{ N}_2(g) + 4 \text{ H}_2\text{O}(l)$ determine
    a. the number of moles of N_2O_4 required to react with 3.62 mol of N_2H_4.
    b. the number of moles of N_2 produced when 1.24 mol of N_2H_4 reacts.

Solution
    a. We need a relation between moles of N_2H_4 and moles of N_2O_4. That relation is given by the coefficients of the balanced equation

$$2 \text{ mol } N_2H_4 \simeq 1 \text{ mol } N_2O_4$$

This gives us the conversion factor we need, 1 mol N_2O_4/2 mol N_2H_4. Thus,

$$\text{moles } N_2H_4 = 3.62 \text{ mol } N_2H_4 \times \frac{1 \text{ mol } N_2O_4}{2 \text{ mol } N_2H_4} = 1.81 \text{ mol } N_2O_4$$

    b. Here the appropriate relation is

$$2 \text{ mol } N_2H_4 \simeq 3 \text{ mol } N_2$$

Since we want to go from moles of N_2H_4 to moles of N_2, we use the conversion factor 3 mol N_2/2 mol N_2H_4.

$$\text{moles } N_2 = 1.24 \text{ mol } N_2H_4 \times \frac{3 \text{ mol } N_2}{2 \text{ mol } N_2H_4} = 1.86 \text{ mol } N_2$$

Exercise   For the reaction between SnO_2 and H_2 (Eq. 4.1) calculate the number of moles of H_2 required to produce 0.231 mol Sn. Answer: 0.462 mol H_2.

Using conversion factors like this, we can find amounts of substances that are equivalent to one another

4.3 MASS RELATIONS IN REACTIONS

We have seen how to obtain mole relations in chemical reactions using the coefficients of the balanced equation. The approach we have followed is easily extended to relations involving masses in grams of reactants and products. To do this we need only convert moles to grams, knowing the formula of the reactant or product involved. In this way it is possible to relate

—moles of one substance to grams of another (Example 4.3).
—grams of one substance to grams of another (Example 4.4).

Example 4.3   The ammonia used to make fertilizers for lawns and gardens is made by reacting nitrogen of the air with hydrogen. The balanced equation for the reaction is

$$N_2(g) + 3\ H_2(g) \rightarrow 2\ NH_3(g)$$

Determine
    a. the mass in grams of ammonia, NH_3, formed when 1.34 mol of N_2 reacts with H_2.
    b. the number of moles of ammonia formed when 24.0 g of N_2 is consumed in the reaction.

Solution
    a. We need a relation between grams of NH_3 and moles of N_2. To obtain it, we start with the mole relation

$$1\ \text{mol}\ N_2 \simeq 2\ \text{mol}\ NH_3$$

But, since the atomic masses of N and H are 14.0 and 1.0, one mole of NH_3 weighs 17.0 g. Hence, the relation we need is

$$1\ \text{mol}\ N_2 \simeq 2(17.0\ \text{g})\ NH_3$$

$$1\ \text{mol}\ N_2 \simeq 34.0\ \text{g}\ NH_3$$

$$\text{mass}\ NH_3 = 1.34\ \text{mol}\ N_2 \times \frac{34.0\ \text{g}\ NH_3}{1\ \text{mol}\ N_2} = 45.6\ \text{g}\ NH_3$$

    b. Again, we start with the relation 1 mol $N_2 \simeq$ 2 mol NH_3. This time, we want a relation between grams of N_2 and moles of NH_3. Since 1 mol $N_2 = 2(14.0\ \text{g}) = 28.0$ g, the relation is

$$28.0\ \text{g}\ N_2 \simeq 2\ \text{mol}\ NH_3$$

$$\text{moles}\ NH_3 = 24.0\ \text{g}\ N_2 \times \frac{2\ \text{mol}\ NH_3}{28.0\ \text{g}\ N_2} = 1.71\ \text{mol}\ NH_3$$

Exercise   For this reaction, calculate the mass in grams of H_2 required to form 1.0 mol of NH_3. Answer: 3.0 g H_2.

In conversions like this, the first term is the amount we are given (e.g., 1.34 mol N_2).

Example 4.4   For the synthesis of ammonia, $N_2(g) + 3\ H_2(g) \rightarrow 2\ NH_3(g)$, calculate
    a. the mass in grams of NH_3 that can be formed when 64.0 g of N_2 reacts.
    b. the mass in grams of N_2 required to form 1.00 kg of NH_3.

Solution
    a. As always, we start with the mole relation

$$1\ \text{mol}\ N_2 \simeq 2\ \text{mol}\ NH_3$$

Here, we need a relation between grams of N_2 and grams of NH_3. Since one mole of N_2 weighs 28.0 g and one mole of NH_3 weighs 17.0 g,

$$28.0\ \text{g}\ N_2 \simeq 2(17.0\ \text{g})\ NH_3$$

$$28.0\ \text{g}\ N_2 \simeq 34.0\ \text{g}\ NH_3$$

The approach in this example is a powerful one and can be applied to any chemical equation

$$\text{mass NH}_3 = 64.0 \text{ g N}_2 \times \frac{34.0 \text{ g NH}_3}{28.0 \text{ g N}_2} = 77.7 \text{ g NH}_3$$

b. We can use the relation obtained in (a). However, this time we want to go from grams of NH_3 to grams of N_2. So, the conversion factor is 28.0 g N_2/34.0 g NH_3.

$$\text{mass N}_2 = 1.00 \times 10^3 \text{ g NH}_3 \times \frac{28.0 \text{ g N}_2}{34.0 \text{ g NH}_3} = 824 \text{ g N}_2$$

Exercise For this reaction, calculate the mass in grams of H_2 required to form 1.0 g of NH_3. Answer: 0.18 g H_2.

Calculations of the type shown in Example 4.4 are very important in chemistry. Whenever a reaction is carried out in the laboratory, we need to know how much of each reactant is required and how much product we are likely to get. In practice, we may choose to use an excess of one reactant (Chapter 5). However, in designing an experiment, we always use the balanced equation as a guide to relate the amounts of reactants and products.

4.4   SOLUTION CONCENTRATION: MOLARITY

So far, all of the reactions we have considered have involved pure substances. In Section 4.5, we will discuss some of the types of reactions that occur in water solution. The principles developed in Sections 4.1 to 4.3 remain valid for these reactions. However, to apply these principles in a useful way, we need to specify the concentrations of species in solution. That is, we need to know how much solute is present in a given amount of solution.

Concentrations of species in solution can be expressed in many different ways. For our purposes here the most useful concentration unit is **molarity.** Molarity tells us how many moles of solute there are per liter of solution. Using the symbol M to stand for molarity, we have

$$\text{molarity (M)} = \frac{\text{moles solute}}{\text{liters solution}} \tag{4.3}$$

Thus a 1 M solution of NaCl would contain

	1 mol (58.44 g) of NaCl in one liter of solution
or	10 mol (584.4 g) of NaCl in ten liters of solution
or	0.1 mol (5.844 g) of NaCl in one tenth of a liter (100 cm³) of solution

and so on.

Using Equation 4.3, we can readily find the molarity of a solution knowing the mass of solute and the volume of solution (Example 4.5).

Example 4.5 What is the molarity of a solution prepared by dissolving 20.0 g of NaOH in enough water to give 482 cm³ of solution?

Solution To determine the number of moles of NaOH we note that

1 mol NaOH weighs 23.0 g + 16.0 g + 1.0 g = 40.0 g

Hence

$$\text{moles NaOH} = 20.0 \text{ g} \times \frac{1 \text{ mol}}{40.0 \text{ g}} = 0.500 \text{ mol}$$

Since $1 \ \ell = 1000 \text{ cm}^3$

$$\text{liters solution} = 482 \text{ cm}^3 \times \frac{1 \ \ell}{1000 \text{ cm}^3} = 0.482 \ \ell$$

Substituting in Equation 4.3:

$$\text{molarity} = \frac{0.500 \text{ mol}}{0.482 \ \ell} = 1.04 \text{ mol}/\ell$$

One liter of the solu-
tion contains 1.04 mol
NaOH

Exercise   What is the molarity of a solution containing 80.0 g of NaOH in 5.00×10^2 cm^3? Answer: 4.00 mol/ℓ.

Using the definition of molarity, Equation 4.3, it is possible to calculate two other quantities:

1. *The number of moles of solute in a given volume of solution:*

$$\text{moles solute} = (\text{molarity}) \times (\text{liters of solution}) \qquad (4.3a)$$

2. *The volume of solution containing a given amount of solute:*

$$\text{liters solution} = \frac{\text{moles solute}}{\text{molarity}} \qquad (4.3b)$$

Figure 4.1   The three flasks at the top of the figure have the same volume of solution, but their molarities differ. They contain 0.1 mol, 0.2 mol, and 0.3 mol of solute, in that order. The three flasks at the bottom of the figure also contain 0.1 mol, 0.2 mol, and 0.3 mol of solute, from left to right. Here, however, the volume increases rather than the molarity.

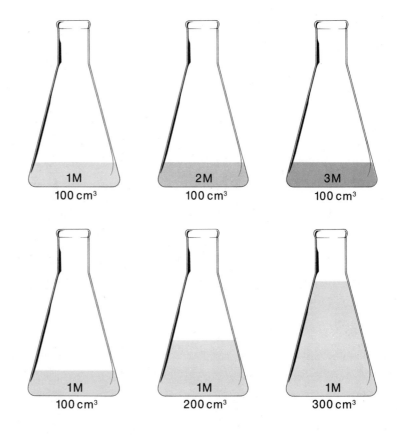

Example 4.6

a. How would you prepare 25 ℓ of 0.10 M BaCl$_2$ solution, starting with solid BaCl$_2$?

b. What volume of the solution in (a) would you take to get 0.020 mol of BaCl$_2$?

Solution

a. Using Equation 4.3a, we can calculate the number of moles of BaCl$_2$ required:

$$\text{moles BaCl}_2 = (\text{molarity BaCl}_2) \times (\text{liters solution})$$

$$= 0.10 \, \frac{\text{mol}}{\ell} \times 25 \, \ell = 2.5 \, \text{mol}$$

To find out how many grams of BaCl$_2$ we should weigh out, we note that one mole weighs 137 g + 2(35.5 g) = 208 g.

$$\text{mass BaCl}_2 = 2.5 \, \text{mol} \times \frac{208 \, \text{g}}{1 \, \text{mol}} = 520 \, \text{g}$$

We should then weigh out 520 g of BaCl$_2$ and stir with enough water to give a final volume of 25 ℓ (Fig. 4.2).

b. Using Equation 4.3b:

$$\text{liters solution} = \frac{\text{moles BaCl}_2}{\text{molarity BaCl}_2}$$

$$= \frac{0.020 \, \text{mol}}{0.10 \, \text{mol}/\ell} = 0.20 \, \ell \, (200 \, \text{cm}^3)$$

Exercise   How many moles of BaCl$_2$ are there in 4.0 ℓ of the solution in (a)? how many grams? Answer: 0.40 mol; 83 g.

 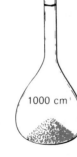

Figure 4.2   To prepare a solution of a desired molarity, the calculated amount of solute is weighed out and transferred to a volumetric flask. Enough water is added so that all the solid is dissolved by shaking. More water is then added to bring the level up to the mark on the neck. The flask is then shaken repeatedly until a homogeneous solution is formed.

1.
Take a
volumetric flask.

2.
Add carefully the
weighed amount
of solid.

3.
Add some water.
shake, and
dissolve solid.

4.
Fill flask to
1000-cm^3 mark
and shake until
homogeneous
solution is
obtained

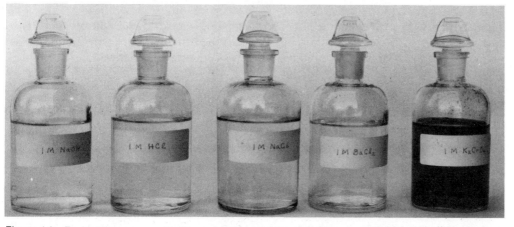

Figure 4.3 Each of these reagent bottles contains an ionic solute at a concentration of 1 M. The solutions of NaOH, HCl and NaCl each contain 2 mol/ℓ of ions. In $BaCl_2$ and K_2CrO_4, there are 3 mol/ℓ of ions.

Many of the water solutions that you work with in general chemistry contain ionic solutes (Fig. 4.3). The compounds NaOH and $BaCl_2$ used in Examples 4.5 and 4.6 are typical examples. The processes by which these solids dissolve in water can be represented by the equations

$$NaOH(s) \rightarrow Na^+(aq) + OH^-(aq) \tag{4.4}$$

$$BaCl_2(s) \rightarrow Ba^{2+}(aq) + 2\ Cl^-(aq) \tag{4.5}$$

Here the symbol (aq) is used to indicate that the ions are in water solution. As the equations indicate, NaOH and $BaCl_2$ are completely dissociated into ions in water. If we dissolve 1 mol of NaCl in water, we get 1 mol of Na^+ ions and 1 mol of OH^- ions. With 1 mol of $BaCl_2$, we have 1 mol of Ba^{2+} ions and 2 mol of Cl^- ions. Hence,

—a 1 M solution of NaOH is 1 M in Na^+ and 1 M in Cl^-.
—a 1 M solution of $BaCl_2$ is 1 M in Ba^{2+} and 2 M in Cl^-.

There are no NaOH or $BaCl_2$ molecules in the solutions

In the laboratory, we indicate the concentrations of solutions in terms of the reagent used to prepare them. Thus, a solution made by dissolving one mole (40.0 g) of NaOH per liter would be labeled "1.00 M NaOH." The concentrations of species in the solution would be better shown by writing on the label "1.00 M Na^+, 1.00 M OH^-" but that is not the way we do it.

Example 4.7 What is the concentration, in moles per liter, of each ion in
   a. 1.0 M $Al(NO_3)_3$?   b. 0.20 M K_2CrO_4?

Solution
   a. When 1 mol of $Al(NO_3)_3$ dissolves in water, 1 mol of Al^{3+} and 3 mol of NO_3^- are produced: $Al(NO_3)_3(s) \rightarrow Al^{3+}(aq) + 3\ NO_3^-(aq)$.   Hence

   conc. Al^{3+} = 1.0 M; conc. NO_3^- = 3.0 M

   b. One mole of K_2CrO_4 produces 2 mol of K^+ and 1 mol of CrO_4^{2-}. Hence

   conc. CrO_4^{2-} = 0.20 M; conc. K^+ = 2(0.20 M) = 0.40 M

Exercise Using Equation 4.3a, calculate the number of moles of K^+ in 2.0 ℓ of 0.20 M K_2CrO_4. Answer: 0.80 mol.

A few molecular solutes react with water to form ions. Among these are the compounds called acids. They ionize in water, forming H^+ ions. An example of a substance of this type is hydrogen chloride, HCl. When added to water, it completely ionizes, forming H^+ and Cl^- ions. The solution formed is referred to as hydrochloric acid. The process can be shown most simply as

On dissolving, all the HCl molecules break apart

$$HCl(g) \rightarrow H^+(aq) + Cl^-(aq) \quad\quad (4.6)$$
$$\text{hydrogen chloride} \quad\quad \text{hydrochloric acid}$$

Thus

—a 0.1 M solution of HCl is 0.1 M in H^+ and 0.1 M in Cl^-.
—a 1 M solution of HCl is 1 M in H^+ and 1 M in Cl^-.

4.5   REACTIONS IN WATER SOLUTION

Many different kinds of reactions take place in water solution. Most of them involve ions as reactants. Such reactions will be discussed in detail later in this text. However, it will be helpful at this point to consider briefly certain aspects of three rather simple types of reactions in water solution. In particular, we will be interested in using the equations for these reactions to find mole/mass relations.

A common type of reaction in water solution is that of **precipitation.** Here, an insoluble ionic compound comes out of solution (precipitates) when a cation and an anion come in contact with each other. An example is the reaction that occurs when water solutions of $AgNO_3$ and NaCl are mixed. The Ag^+ ions of the $AgNO_3$ solution and the Cl^- ions of the NaCl solution combine to give a white precipitate of silver chloride, AgCl (Fig. 4.4). The equation for the reaction is

The NO_3^- and Na^+ ions remain in the solution

$$Ag^+(aq) + Cl^-(aq) \rightarrow AgCl(s) \quad\quad (4.7)$$

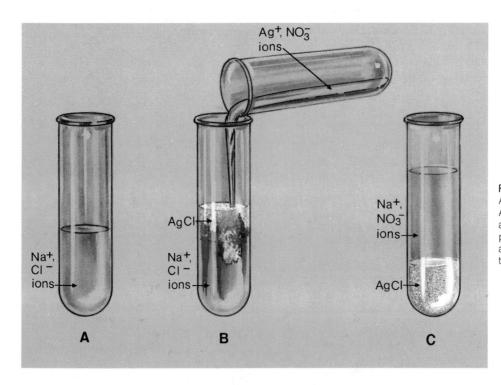

Figure 4.4 Precipitation of AgCl by mixing solutions of $AgNO_3$ and NaCl. The Na^+ and NO_3^- ions do not take part in the reaction, and so are not included in the equation.

Another important type of reaction in water solution is known as an **acid-base reaction.** As an example, consider what happens when solutions of HCl and NaOH are mixed. The H^+ ions of the hydrochloric acid solution react with the OH^- ions of the sodium hydroxide solution. Water molecules are formed. The equation is

$$H^+(aq) + OH^-(aq) \rightarrow H_2O \qquad (4.8)$$

In this reaction, HCl is acting as an **acid,** furnishing **H^+ ions.** The NaOH acts as a **base,** furnishing **OH^- ions.**

A third type of reaction in water solution is **oxidation-reduction.** Here, there is an exchange of electrons between two reactants. One species **loses electrons** and is said to be **oxidized.** The other species **gains electrons** and is **reduced.** An example of an oxidation-reduction reaction is that between hydrochloric acid and zinc metal (Fig. 4.5). The Zn atoms lose electrons; they are oxidized to Zn^{2+} ions:

$$Zn(s) \rightarrow Zn^{2+}(aq) + 2\ e^-$$

The H^+ ions of the HCl gain electrons. They are reduced to H atoms, which immediately combine to form H_2 molecules:

$$2\ H^+(aq) + 2\ e^- \rightarrow H_2(g)$$

Figure 4.5 An oxidation-reduction reaction occurs when hydrochloric acid is added to zinc metal. Zn atoms are oxidized to Zn^{2+} ions; H^+ ions are reduced to H_2 molecules.

In a reaction like this, the electrons produced in one step are used up in the other

The hydrogen escapes from the solution as a gas. The equation for the reaction is

$$Zn(s) + 2\,H^+(aq) \rightarrow Zn^{2+}(aq) + H_2(g) \tag{4.9}$$

We have now written three equations for reactions taking place in water solution. These illustrate two important principles that apply in all cases. In writing a balanced equation for a reaction between species in solution:

1. *Only those species which take part in the formation of products are included in the equation.* In Reaction 4.7, the NO_3^- ions of the $AgNO_3$ and the Na^+ ions of the NaCl solution did not take part in the precipitation reaction. Similarly, in Reaction 4.8, the Cl^- ions of the HCl and the Na^+ ions of the NaOH were not involved in the acid-base reaction. Finally, in Reaction 4.9, the Cl^- ions of the HCl were neither oxidized nor reduced. Hence, none of these ions were included in the equations written to describe these reactions.

2. *The total charge must be the same on the two sides of the equation.* In Reactions 4.7 and 4.8, there is a total charge of zero on both sides.

$$+1 + (-1) = 0$$

In Reaction 4.9, there is a charge of $+2$ on the left (2 H^+ ions). The charge on the right is also $+2$ (1 Zn^{2+} ion).

For a reaction in water solution, as with all reactions, the coefficients in the balanced equation have a simple meaning. They tell us the relative numbers of moles of substances taking part in the reaction. Thus we have

In chemical equations the mole is king

Eq. 4.7: 1 mol of Ag^+ reacts with 1 mol of Cl^- to form 1 mol of AgCl
Eq. 4.8: 1 mol of H^+ reacts with 1 mol of OH^- to form 1 mol of H_2O
Eq. 4.9: 1 mol of Zn reacts with 2 mol of H^+ to form 1 mol of Zn^{2+} and 1 mol of H_2

We can use these relations to carry out calculations very similar to those discussed in Sections 4.2 and 4.3.

Example 4.8 In Reaction 4.9, determine
   a. the number of moles of H^+ required to form 1.22 mol of H_2.
   b. the mass in grams of Zn required to form 0.621 mol of H_2.

Solution
   a. From the balanced equation 2 mol $H^+ \simeq$ 1 mol H_2,

$$\text{moles } H^+ = 1.22 \text{ mol } H_2 \times \frac{2 \text{ mol } H^+}{1 \text{ mol } H_2} = 2.44 \text{ mol } H^+$$

   b. From the balanced equation: 1 mol Zn $\simeq$ 1 mol H_2
But the atomic mass of Zn is 65.38. So, 65.38 g Zn $\simeq$ 1 mol H_2.

$$\text{mass Zn} = 0.621 \text{ mol } H_2 \times \frac{65.38 \text{ g Zn}}{1 \text{ mol } H_2} = 40.6 \text{ g Zn}$$

Exercise How many moles of H_2 can be produced from 10.0 g of Zn? Answer: 0.153 mol H_2.

For reactions in water solution, a balanced equation serves another important purpose. Knowing the concentrations of two reactants, we can determine the volume of one solution required to react with a given volume of another solution. The calculations are illustrated in Example 4.9.

Example 4.9 What volume of 0.100 M HCl is required to react with 25.0 cm³ (0.0250 ℓ) of 0.200 M NaOH solution by Reaction 4.8?

Solution We see from the balanced equation that

$$1 \text{ mol } H^+ \simeq 1 \text{ mol } OH^-$$

From the information given, we can calculate the number of moles of OH^- in the NaOH solution. From the relation above, that must also be the number of moles of H^+ in the HCl solution. Finally, we can calculate the volume of 0.100 M HCl required.

(1) To find the number of moles of OH^-, we note that a 0.200 M NaOH solution is also 0.200 M in OH^-. Hence, by Equation 4.3a,

$$\text{moles } OH^- = 0.200 \, \frac{\text{mol } OH^-}{\ell} \times 0.0250 \, \ell = 5.00 \times 10^{-3} \text{ mol } OH^-$$

(2) Since 1 mol of HCl produces 1 mol of H^+, which in turn reacts with 1 mol of OH^-,

$$\text{moles HCl} = \text{moles } H^+ = \text{moles } OH^- = 5.00 \times 10^{-3} \text{ mol HCl}$$

(3) To obtain the volume of HCl required, we use Equation 4.3b:

$$\text{liters HCl} = \frac{\text{moles HCl}}{\text{molarity HCl}} = \frac{5.00 \times 10^{-3} \text{ HCl}}{1.00 \times 10^{-1} \text{ mol HCl}/\ell} = 5.00 \times 10^{-2} \, \ell \text{ (or 50.0 cm}^3\text{)}$$

Exercise What volume of 0.250 M NaOH is required to react with 40.0 cm³ of 0.100 M HCl? Answer: 16.0 cm³.

The principle illustrated by Example 4.9 is the basis for a very useful laboratory procedure called *titration*. This procedure involves measuring the volume of one solution required to react exactly with a known volume of a second solution. The experimental set-up for an acid-base titration is shown in Figure 4.6. By titration, it is possible to

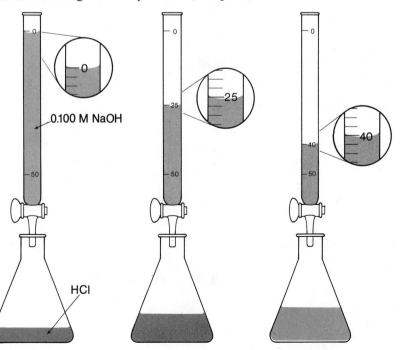

Figure 4.6 An acid-base titration. The desired volume of HCl (24.4 cm³) is measured into the flask. Sodium hydroxide of known concentration (0.100 M) is added from a buret; the volume used is carefully measured (40.0 cm³). A few drops of a species called an indicator is used to signal when to stop adding NaOH; it changes color when the reaction is complete. From the data obtained, the molarity of the HCl can be calculated (Example 4.10).

determine the molarity of an HCl solution of unknown concentration. To do this requires a standard solution (of known molarity) of the base, let us say NaOH. The calculations are illustrated in Example 4.10.

Example 4.10   A student determines the molarity of an HCl solution by titrating with a standard solution of NaOH. She finds that 40.0 cm³ of the 0.100 M NaOH solution is required to react exactly with 24.4 cm³ of the HCl. What is the molarity of the HCl?

Solution   We follow the same steps as in Example 4.9. The only difference is that in the final step we calculate the molarity of HCl, knowing its volume.

(1) moles $OH^- = 0.100 \frac{\text{mol } OH^-}{\ell} \times 0.0400 \ \ell = 4.00 \times 10^{-3} \text{ mol } OH^-$

(2) moles $HCl = 4.00 \times 10^{-3} \text{ mol } HCl$

(3) molarity $HCl = \frac{\text{moles } HCl}{\text{liters } HCl} = \frac{4.00 \times 10^{-3} \text{ mol}}{24.4 \times 10^{-3} \ \ell} = 0.164 \text{ mol}/\ell$

Exercise   A student finds that 16.8 cm³ of 0.100 M HCl is required to react with 14.5 cm³ of an NaOH solution. What is the molarity of the NaOH? Answer: 0.116 M.

SUMMARY

Chemical equations represent chemical reactions in which reactants are converted to products. Reactants are written on the left of the equation, products on the right. The symbols and formulas used must correspond to the actual reactants and products (Example 4.1). The coefficients of a balanced equation represent the relative numbers of moles. They can be used directly to relate moles of reactants to moles of products (Example 4.2). By converting moles to grams, we can use a balanced chemical equation to relate moles of one substance to grams of another (Example 4.3) or grams of one substance to grams of another (Example 4.4).

Many reactions occur in water solution. Among these are precipitate formation, acid-base reactions, and oxidation-reduction reactions. In writing an equation to describe such a reaction, we include only those species which actually take part in the reaction. The equation must balance with respect to both mass and charge. The coefficients in the equation can be used in the usual way to relate amounts of reactants and products (Example 4.8).

Concentrations of species in solution may be expressed in terms of molarity (moles per liter). The defining relation for molarity, Equation 4.3, is used to relate moles of solute to volume of solution (Examples 4.5 and 4.6). The molarity of an ion in solution is related in a simple way to that of the corresponding ionic compound (Example 4.7). The concentration unit molarity can also be used for a variety of practical calculations for reactions taking place in water solution (Examples 4.9 and 4.10).

KEY WORDS AND CONCEPTS

reactants	coefficient	precipitation	oxidation
products	mole	acid	reduction
chemical equation	molarity	base	titration

Catalog

4.1   Review and know the meanings of the key words and concepts in this chapter.

4.2   Balance the following equations:

    a. $Fe(s) + O_2(g) \rightarrow Fe_2O_3(s)$
    b. $Fe_2O_3(s) + CO(g) \rightarrow Fe_3O_4(s) + CO_2(g)$
    c. $NH_3(g) + O_2(g) \rightarrow NO(g) + H_2O(l)$
    d. $XeF_2(s) + H_2O(l) \rightarrow Xe(g) + HF(l) + O_2(g)$
    e. $Bi^{3+}(aq) + S^{2-}(aq) \rightarrow Bi_2S_3(s)$

4.3   Write balanced equations to represent

    a. the combustion of liquid heptane, C_7H_{16}, to carbon dioxide and water vapor.
    b. the reaction of liquid water with sodium metal to produce hydrogen gas and aqueous sodium hydroxide.
    c. the precipitation reaction that occurs when Mg^{2+} and OH^- ions are mixed in water.
    d. the decomposition of liquid hydrogen peroxide, H_2O_2, to oxygen gas and water.
    e. the reaction of solid aluminum sulfide with water to produce hydrogen sulfide gas and solid aluminum hydroxide.

4.4   For the reaction

$$Ca_3N_2(s) + 6\ H_2O(l) \rightarrow 3\ Ca(OH)_2(s) + 2\ NH_3(g)$$

complete the following:

    a. 2.40 mol Ca_3N_2 yields _____ mol NH_3.
    b. 1.35 mol Ca_3N_2 reacts with _____ mol H_2O.
    c. 0.450 mol H_2O produces _____ g $Ca(OH)_2$.
    d. _____ g Ca_3N_2 yields 2.94 g NH_3.

4.5   When acetylene gas, C_2H_2, burns in air, the products are $CO_2(g)$ and $H_2O(l)$.

    a. Write a balanced equation for this reaction.
    b. How many moles of CO_2 are produced from 0.624 mol of C_2H_2?
    c. How many grams of O_2 are required to react with 3.46 mol of C_2H_2?

4.6   Ethyl alcohol, C_2H_5OH, can be produced by the fermentation of grains, which contain glucose, $C_6H_{12}O_6$.

$$C_6H_{12}O_6(aq) \rightarrow 2\ C_2H_5OH(l) + 2\ CO_2(g)$$

    a. How many grams of ethyl alcohol are produced from 1.00 kg of glucose?
    b. Gasohol is a mixture of 10 cm³ of ethyl alcohol (d = 0.79 g/cm³) per 90 cm³ of gasoline. How many grams of glucose are required to produce the ethyl alcohol in one liter of gasohol?

4.18   Balance the following equations:

    a. $H_2O(g) + Cl_2O(g) \rightarrow HOCl(aq)$
    b. $NH_3(g) + O_2(g) \rightarrow N_2(g) + H_2O(l)$
    c. $(NH_4)_2Cr_2O_7(s) \rightarrow Cr_2O_3(s) + H_2O(l) + N_2(g)$
    d. $KClO_3(s) \rightarrow KClO_4(s) + KCl(s)$
    e. $Hg^{2+}(aq) + Cl^-(aq) \rightarrow HgCl_2(s)$

4.19   Write balanced equations to represent

    a. the reaction of liquid turpentine, $C_{10}H_{16}$, with chlorine gas to produce hydrogen chloride gas and carbon.
    b. the formation of liquid water by the reaction of gaseous hydrogen and oxygen.
    c. the precipitation reaction that occurs when Ca^{2+} and PO_4^{3-} ions are mixed in aqueous solution.
    d. the decomposition of solid potassium chlorate to solid potassium chloride and oxygen gas.
    e. the reaction of solid B_2O_3 with liquid hydrogen fluoride to form solid boron trifluoride and water.

4.20   For the reaction

$$2\ PbS(s) + 3\ O_2(g) \rightarrow 2\ PbO(s) + 2\ SO_2(g)$$

complete the following:

    a. 1.35 mol PbS yields _____ mol PbO.
    b. _____ mol O_2 reacts with 5.30 mol PbS.
    c. _____ g PbS produces 0.0252 mol SO_2.
    d. 11.3 g PbS yields _____ g SO_2.

4.21   The combustion of butane gas, C_4H_{10}, in air yields $CO_2(g)$ and $H_2O(l)$.

    a. Write a balanced equation for the reaction.
    b. How many moles of C_4H_{10} are required to form 12.6 mol of CO_2?
    c. How many grams of H_2O are formed from 1.69 mol of C_4H_{10}?

4.22   A commercial wine is about 9.7% ethyl alcohol by mass. Assume that 1.00 kg of wine is produced by the fermentation reaction given in Problem 4.6.

    a. How many grams of glucose are needed to produce the ethyl alcohol in the wine?
    b. What volume of $CO_2(g)$ is produced at the same time (d = 1.80 g/ℓ)?

4.7 In a blast furnace, iron ore, Fe_2O_3, is converted to iron by the reaction

$$Fe_2O_3(s) + 3\ CO(g) \rightarrow 2\ Fe(l) + 3\ CO_2(g).$$

a. How many moles of carbon monoxide are needed to produce 3.50 mol of iron?
b. How many grams of Fe_2O_3 are required to react with 0.500 mol CO?
c. Coke (essentially pure carbon) is burned to give the CO necessary for this reaction. How many kilograms of coke must be burned to form enough CO to react with 1.00 metric ton of Fe_2O_3?

4.8 Oxygen masks for producing O_2 in emergency situations contain potassium superoxide, KO_2. It reacts with CO_2 and H_2O in exhaled air to produce oxygen:

$$4\ KO_2(s) + 2\ H_2O(g) + 4\ CO_2(g) \rightarrow 4\ KHCO_3(s) + 3\ O_2(g)$$

If a person wearing such a mask exhales 0.667 g of CO_2 per minute, how many grams of KO_2 are consumed in five minutes?

4.9 A solution contains 12.5 g of glucose, $C_6H_{12}O_6$, in enough water to make 250 cm³ of solution.

a. What is the molarity of glucose?
b. What volume of this solution contains 0.012 mol of glucose?

4.10 Complete the table below for aqueous solutions:

Solute	g solute	V (liters)	Molarity
NH_4Cl	2.97	0.130	____
Na_2CO_3	3.42	____	0.129
$Sr(NO_3)_2$	4.50	0.750	____
$Sr(NO_3)_2$	____	2.25	0.325

4.11 What volume (cm³) of 0.500 M KOH has the same number of moles of solute as 150 cm³ of 0.200 M NaCl?

4.12 Calculate the number of moles of each ion in 2.00×10^2 cm³ of

a. 0.100 M NaCl.
b. 0.250 M K_2CrO_4.
c. 1.22 M $Al(NO_3)_3$.

4.13 Which member of the following pairs of solutions contains the greater number of grams of solute? Moles of ions?

a. 250 cm³ of 0.100 M NaCl or 100 cm³ of 0.300 M NaBr?
b. 2.5 ℓ of 0.050 M $CaCl_2$ or 1.2 ℓ of 0.075 M K_2SO_4?

4.14 Consider the reaction

$$Al(s) + 3\ H^+(aq) \rightarrow Al^{3+}(aq) + 3/2\ H_2(g)$$

How many grams of Al are required to produce 0.100 mol of H_2?

4.23 The air pollutant nitric oxide, NO, is produced by automobile engines. It forms another pollutant, NO_2, by reacting with O_2:

$$2\ NO(g) + O_2(g) \rightarrow 2\ NO_2(g)$$

a. How many moles of NO_2 are produced from 6.30 mol NO?
b. How many grams of NO are required to produce 21.4 g of NO_2?
c. How many grams of NO react to produce an average NO_2 concentration of 1.8×10^{-7} mol/ℓ in a 7.5-km column of air over a 100-km² metropolitan area?

4.24 A crude oil burned in electrical generating plants contains about 1.5% by mass of sulfur. When the oil burns the sulfur forms sulfur dioxide gas:

$$S(s) + O_2(g) \rightarrow SO_2(g)$$

How many liters of SO_2 (d = 2.60 g/ℓ) are produced when one kilogram of oil burns?

4.25 3.70 g of $CuSO_4$ is added to enough water to make 140 cm³ of solution.

a. What is the molarity of $CuSO_4$?
b. How many moles of $CuSO_4$ are there in 1.00 cm³ of this solution?

4.26 Complete the table below for aqueous solutions:

Solute	g solute	V (liters)	Molarity
LiBr	2.78	0.350	____
$Ba(OH)_2$	3.12	____	0.0900
$(NH_4)_2SO_4$	9.57	2.30	____
$(NH_4)_2SO_4$	____	6.32	0.434

4.27 How many cm³ of 0.150 M Li_2CO_3 contain the same number of moles of solute as 200 cm³ of 0.116 M K_2SO_4?

4.28 A solution is prepared by dissolving 12.0 g of $BaCl_2 \cdot 2\ H_2O$ to give 8.00×10^2 cm³ of solution.

a. What is the molarity of $BaCl_2$?
b. What is the molarity of Ba^{2+}? Cl^-?

4.29 Which member of the following pairs of solutions contains the lesser number of grams of solute? Moles of ions?

a. 1.25 cm³ of 0.150 M KCl or 150 cm³ of 0.230 M KBr?
b. 1.6 ℓ of 0.21 M $MgBr_2$ or 1.3 ℓ of 0.35 M K_2CrO_4?

4.30 For the reaction in Problem 4.14, calculate the mass in grams of Al required to form one liter of $H_2(g)$ (density = 0.0823 g/ℓ).

4.15 Solutions of K_2CrO_4 and $PbBr_2$ react to precipitate lead chromate.

    a. Write a balanced equation for the precipitation reaction.
    b. How many cm^3 of 0.100 M K_2CrO_4 are needed to form 0.225 mol of lead chromate?
    c. What volume of 0.120 M $PbBr_2$ is required for the reaction in (b)?

4.16 The titration of a 20.0-cm^3 sample of HBr requires 16.2 cm^3 of 0.110 M KOH. What is the molarity of the HBr?

4.17 A 1.35-g sample of a mixture is analyzed for Ba^{2+}. Enough Na_2SO_4 solution is added to precipitate all the Ba^{2+} as 0.730 g of $BaSO_4$.

    a. What is the mass % of Ba^{2+} in the mixture?
    b. How many cm^3 of 0.125 M Na_2SO_4 are required for the precipitation?

4.31 In solution, sodium sulfate and barium chloride react to precipitate barium sulfate.

    a. Write a balanced equation for the precipitation reaction.
    b. How many moles of barium sulfate are formed when 8.0 cm^3 of 0.20 M sodium sulfate reacts?
    c. How many cm^3 of 0.15 M barium chloride are required to form 0.040 g of $BaSO_4$?

4.32 What volume of 0.0500 M NaOH is required to react with 10.0 cm^3 of 0.250 M HI?

4.33 The percentage of Cl^- in a mixture was determined by adding 0.100 M $AgNO_3$ to a 0.350-g sample of the mixture dissolved in water. It was found that 21.5 cm^3 of the $AgNO_3$ solution was required for complete reaction.

    a. How many grams of AgCl precipitate?
    b. What is the mass % of Cl^- in the mixture?

*4.34 A 1.600-g sample of magnesium is burned in air to produce a mixture of two ionic solids, magnesium oxide and magnesium nitride, Mg_3N_2. Water is added to this mixture. It reacts with the magnesium oxide to form 3.454 g of magnesium hydroxide.

    a. Write balanced equations for the three reactions described above.
    b. How many grams of magnesium oxide are formed by the combustion of magnesium?
    c. How many grams of magnesium nitride are formed?

*4.35 One way to determine the percentage of sulfur in petroleum is to convert it to SO_4^{2-} ions and precipitate as $BaSO_4$. An 11.5-cm^3 sample of petroleum (d = 0.87 g/cm^3) forms 1.20 g of barium sulfate. From these data, determine the mass % of sulfur in the petroleum sample.

*4.36 Benzene, $C_6H_6(l)$, is reacted with nitric acid, $HNO_3(l)$. Two products are formed, one of which is water. The other product is an oily liquid which has a molecular mass of 213. Analysis shows that it contains 33.8% C, 1.42% H, 19.7% N and 45.1% O. From this information, write a balanced equation for the reaction of benzene with nitric acid.

*4.37 A mixture of $CaCl_2$ and KCl weighing 0.750 g was reacted with aqueous $AgNO_3$. The reaction converted all the Cl^- ions to 1.75 g of pure AgCl. What per cent by mass of the original mixture was KCl?

SOURCES OF THE ELEMENTS

As pointed out in Chapter 1, 91 different elements are found in nature. Their relative abundances differ greatly from one element to another. Figure 5.1 gives the percentages by mass of the elements in the world around us. This includes the atmosphere, the waters of the earth, and the earth's crust to a depth of 40 km. Notice that a single element, oxygen, makes up nearly half (49.5%) of the total mass. The ten most abundant elements, shown in color in Figure 5.1, account for about 99%. Some of the more familiar elements are really quite rare. For example, the ten metals chromium, cobalt, copper, gold, lead, mercury, nickel, silver, tin, and zinc together account for less than 0.1%. Clearly, abundance by itself is no measure of the importance of an element.

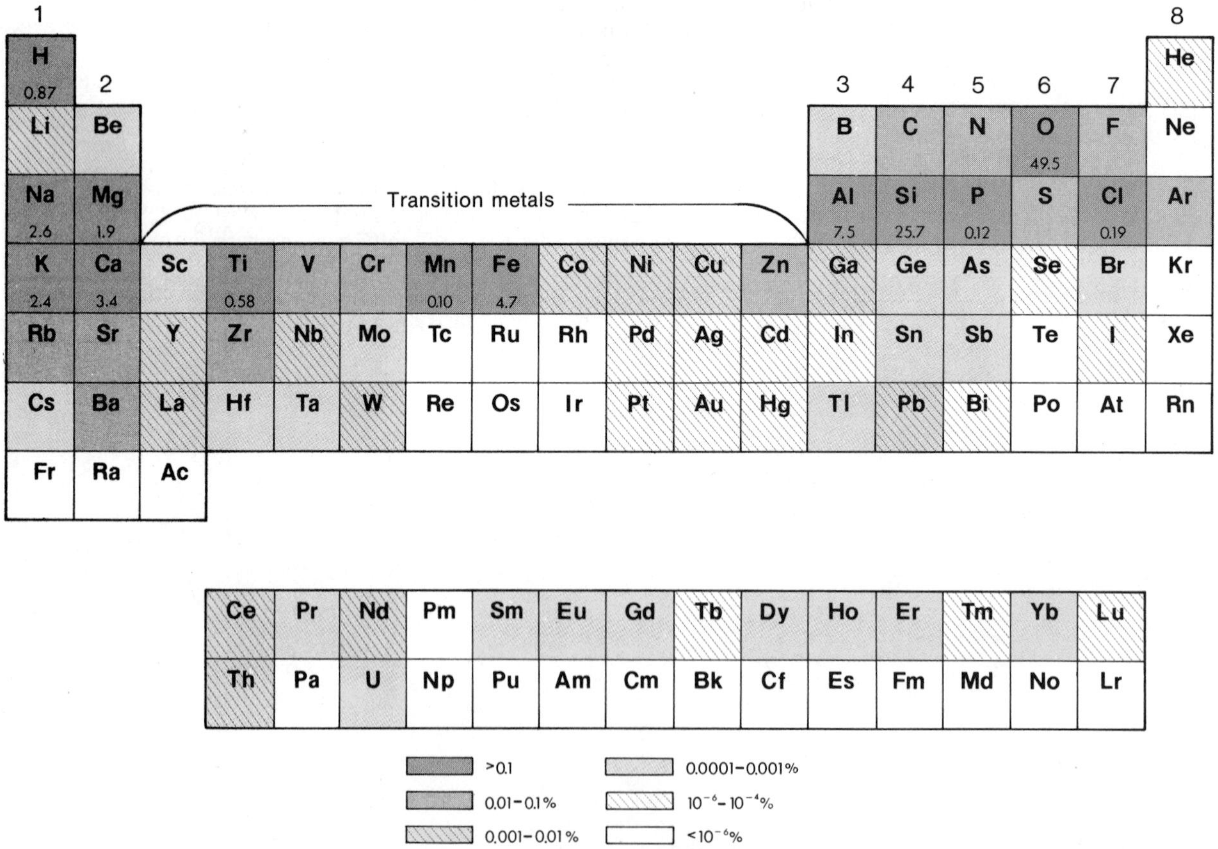

Figure 5.1   This Periodic Table is color-coded to show the relative abundances of elements. The ten most abundant elements are oxygen, silicon, aluminum, iron, calcium, sodium, potassium, magnesium, hydrogen, and titanium, in that order. Of these, only O_2, Al, Fe, Mg, and H_2 are used in large amounts. Many of the familiar heavy elements, such as silver, tin, iodine, gold, and mercury, are really quite rare.

In this chapter we will look at some of the processes used to obtain elements from natural sources. Industries which extract elements from the air, the oceans, or mineral deposits, called *ores,* employ several million people in the United States alone. The products of these industries may be used directly: iron and steel in automobiles, aluminum in airplanes, chlorine as a bleach. Other elements are converted to useful compounds: nitrogen and hydrogen to ammonia, NH_3; sulfur to sulfuric acid, H_2SO_4.

Many different processes are used to obtain elements in pure form. Sometimes a simple physical separation is all that is required. This is the case with the gaseous elements found in uncombined form in the atmosphere (Section 5.1). More often, elements in nature are chemically combined with one another. This is true of many of the more reactive nonmetals (Section 5.2) and virtually all of the metals (Sections 5.3, 5.4). These elements must be prepared from their compounds by chemical reactions.

Most elements are found in nature in compounds rather than in elementary form

Our discussion of the sources and preparation of the elements will be organized around the Periodic Table. We will make frequent use of the principles introduced in Chapters 1 through 4. A major purpose of this chapter is to review and extend such basic concepts as the mole (Chap. 2), formulas (Chap. 3), and the use of chemical equations (Chap. 4).

5.1 ELEMENTS FROM THE ATMOSPHERE

A total of seven different elements (Table 5.1) are found commonly in the atmosphere. These include nitrogen, oxygen, and the *noble gases* in Group 8 of the Periodic Table. All of these elements, with one exception, are obtained industrially from air. Helium is extracted from natural gas in certain wells found in the southwestern United States. It occurs there at concentrations up to 6%, far higher than in the atmosphere.

Although air is free for the taking, it has important commercial value. In 1978, 1.6×10^7 metric tons of oxygen were produced in the United States. Most of this oxygen was used to make steel, but it has many other applications: in hospitals, for welding, and for water treatment. Nitrogen is used mainly to make NH_3. Ammonia is an important component of fertilizers and is the starting material for making nearly all the other nitrogen-containing compounds.

TABLE 5.1   ELEMENTS IN THE ATMOSPHERE

	GROUP	bp (°C)	VOLUME % IN AIR	USES
N_2	5	−196	78.08	**Synthesis of NH_3.** Packaging of foods such as instant coffee to preserve flavor; liquid N_2 used as coolant (safer than liquid air)
O_2	6	−183	20.95	**Making steel** (Section 5.4). Life support systems, waste water treatment, high temperature flames
He	8	−269	0.000524	Balloons, dirigibles
Ne	8	−249	0.00182	Neon signs
Ar	8	−186	0.934	Provides inert atmosphere in light bulbs, arc welding of Al, Mg
Kr	8	−152	0.000114	High-speed flash bulbs
Xe	8	−109	0.000009	Experimental anesthetic; ^{133}Xe used as radioactive tracer in medical diagnostic studies

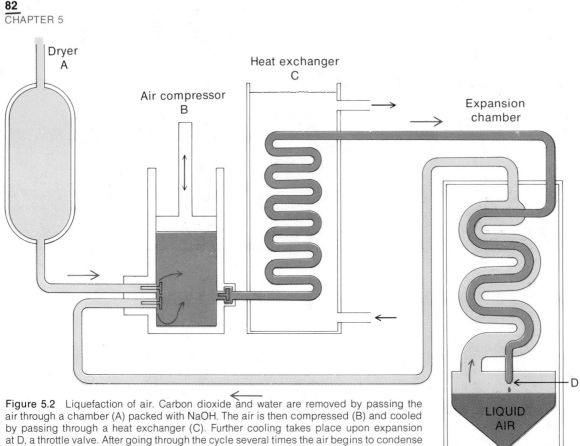

Figure 5.2 Liquefaction of air. Carbon dioxide and water are removed by passing the air through a chamber (A) packed with NaOH. The air is then compressed (B) and cooled by passing through a heat exchanger (C). Further cooling takes place upon expansion at D, a throttle valve. After going through the cycle several times the air begins to condense at D.

To separate the components of air, it is first liquefied by cooling (Fig. 5.2). The liquid is allowed to warm up and fractionally distilled. The first fraction that comes off is mostly N_2 (bp = $-196°C$), with small amounts of helium and neon. The residue is largely O_2 (bp = $-183°C$) and argon (bp = $-186°C$). It also contains small quantities of the higher boiling noble gases, krypton and xenon. By further fractionation, the noble gases Ne, Ar, Kr, and Xe are separated in pure form.

Kr and Xe are very expensive per gram

The first person to isolate a noble gas was the English scientist Sir Henry Cavendish (1731–1810). He subjected a sample of atmospheric nitrogen to repeated electrical discharges in the presence of oxygen. In this way, he formed oxides of nitrogen which were dissolved in water. Cavendish found that a bubble of gas, about 1% of the original volume, remained undissolved.

Nothing came of this observation for over a century. In 1892, Lord Rayleigh, at the Cavendish Laboratory in Cambridge, took the next step. He found that "atmospheric" nitrogen had a density about 0.5% greater than that of pure nitrogen. Rayleigh and a Scotsman, Sir William Ramsay, showed that this difference was due to the presence of a very unreactive element in the air. In 1894, they reported the discovery of argon ("the lazy one.") Over the next five years, Ramsay and William Travers isolated three more noble gases from liquid air. These were neon ("the new one"), krypton ("the hidden one"), and xenon ("the stranger").

Helium was first detected in the spectrum of the sun in 1868. Ramsay, in the same year he discovered argon, separated helium from a gas given off when a uranium mineral was heated. A year later, helium was found in trace amounts in the atmosphere. The last of the noble gases to be discovered was radon in 1898. It is produced by a natural nuclear reaction and is itself intensely radioactive. Radon is not found in detectable amounts in air.

Most elements must be extracted from the waters of the earth, its solid surface, or from underground mines. Included among these are 15 of the elements classified as nonmetals or metalloids. These are listed, with their sources and principal uses, in Table 5.2. In this section we will discuss the methods used to obtain six of these elements (S, H_2, F_2, Cl_2, Br_2, and I_2) from natural sources.

TABLE 5.2 NONMETALS AND METALLOIDS FOUND IN THE EARTH'S CRUST

	GROUP	mp (°C)	bp (°C)	PRINCIPAL SOURCE	USES
B	3	2300	2550	$Na_2B_4O_7 \cdot 10\,H_2O$ (borax)	Alloys, O_2 scavenger
C	4	3570		Coal, petroleum, natural gas	**Reduction of iron ore (coke); adsorbent (charcoal)**
Si	4	1414	2355	SiO_2 (sand, quartz)	Transistors, solar batteries
Ge	4	937	2830	Sulfides	Semiconductors
P	5	44	280	$Ca_3(PO_4)_2$ (phosphate rock)	Synthesis of P_4O_{10}, H_3PO_4
As	5	814		As_2S_3, other sulfides	Lead alloys (shot, batteries)
Sb	5	631	1380	Sb_2S_3	Bearings, battery plates
S	6	119	444	Free element	**Synthesis of H_2SO_4**, vulcanization of rubber
Se	6	217	685	PbSe, other selenides	Color treatment of glass
Te	6	450	990	PbTe, other tellurides	Alloys with metals
H_2		−259	−253	Natural gas, petroleum, H_2O	**Synthesis of NH_3**, reduction of metal oxides
F_2	7	−220	−188	CaF_2 (fluorite), Na_3AlF_6 (cryolite)	Synthesis of UF_6, SF_6, organic fluorine compounds
Cl_2	7	−101	−34	NaCl, Cl^- in ocean	**Bleach, water treatment**, organic chlorine compounds
Br_2	7	−7	59	Br^- in salt brines	Synthesis of $C_2H_4Br_2$ (antiknock additive)
I_2	7	114	184	I^- in salt brines	Antiseptic, synthesis of NaI, AgI

Example 5.1   The mineral known as borax is found in large quantities in dried-up lake beds in the Mojave Desert of California. What mass of borax (Table 5.2) is required to give 1.00 g of boron?

Solution   The formula of borax is $Na_2B_4O_7 \cdot 10\,H_2O$. In one mole of borax there are

$$
\begin{array}{rl}
2(22.99\ \text{g}) = & 45.98\ \text{g Na} \\
4(10.81\ \text{g}) = & 43.24\ \text{g B} \\
7(16.00\ \text{g}) = & 112.00\ \text{g O} \\
10(18.02\ \text{g}) = & \underline{180.2\ \ \text{g H}_2\text{O}} \\
& 381.4\ \ \text{g}
\end{array}
$$

Hence: 381.4 g borax ≏ 43.24 g B.

This gives us the conversion factor we need:

$$\text{mass borax} = 1.00 \text{ g B} \times \frac{381.4 \text{ g borax}}{43.24 \text{ g B}} = 8.82 \text{ g borax}$$

Exercise   What is the mass % of boron in borax? Answer: 11.37.

Sulfur

Sulfur is one of the very few elements found in pure form in nature. Huge deposits of sulfur are located in Texas and Louisiana, several hundred meters beneath the surface of the earth. In many places, these deposits are covered by water, quicksand, or marshes. The process used to mine the sulfur is named after its inventor, Herman Frasch, an American chemical engineer. A diagram of the Frasch process is shown in Figure 5.3.

The sulfur is heated to its melting point (119°C) by pumping superheated water at 165°C down one of three concentric pipes. Compressed air is used to bring the sulfur to the surface. The air and sulfur form a frothy mixture which rises through the middle pipe. Upon cooling, the sulfur solidifies, filling huge vats which may be half a kilometer long. The sulfur obtained this way has a purity approaching 99.9%.

Near the sulfur mine you may see a bright yellow mountain of pure sulfur

Hydrogen

Several different methods are used to produce hydrogen industrially. All of them are designed to supply the gas at the point where it is used. Because of its low density

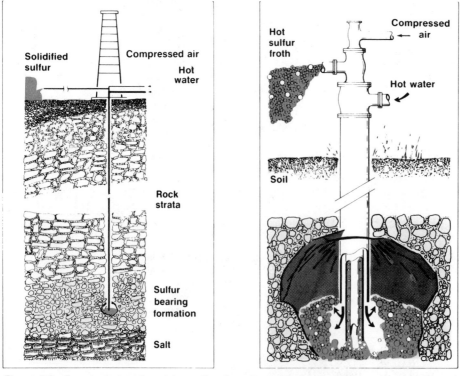

Figure 5.3 Frasch process for mining sulfur. Superheated water at 165°C is sent down through the outer pipe to form a pool of molten sulfur (mp = 119°C) at the base. Compressed air, pumped down the inner pipe, brings the sulfur to the surface. Sulfur deposits are often 100 m or more beneath the earth's surface, covered with quicksand and rock.

(0.0825 g/ℓ at 25°C and 1 atm), H_2 gas cannot be shipped economically over long distances. There is also a safety problem involved. Mixtures with air containing as little as 4% hydrogen are both flammable and explosive.

The main method used to prepare hydrogen in the United States today starts with natural gas. This gas, which consists largely of methane, is heated with steam at 600 to 1000°C. The principal reaction that takes place is

$$CH_4(g) + H_2O(g) \rightarrow CO(g) + 3\ H_2(g) \qquad (5.1)$$

A *catalyst* of finely divided nickel is used to make the reaction go faster. The hydrogen formed by Reaction 5.1 must be separated from the gaseous mixture with CO. One way to do this is to convert the carbon monoxide to CO_2. The carbon dioxide can then be removed by bubbling the H_2–CO_2 mixture through cold water. Carbon dioxide is much more soluble in water (0.034 mol/ℓ at 25°C, 1 atm) than is hydrogen (<0.001 mol/ℓ).

Example 5.2   In an industrial plant making H_2 from natural gas, 1.20 mol of CH_4 is heated with 2.62 mol of steam. How many moles of H_2 can be made from this mixture by Reaction 5.1?

Solution   We note from the balanced equation that CH_4 and H_2O react in a 1:1 mol ratio. In the mixture referred to here, steam is in excess. There is more than twice as much (2.62 mol) as is required to react with the methane (1.20 mol). After the reaction is over, we would expect to find 2.62 mol − 1.20 mol = 1.42 mol of H_2O unreacted.

To find out how much H_2 can be formed, our calculations must be based on the amount of CH_4 available. Since

$$1\ mol\ CH_4 \backsimeq 3\ mol\ H_2$$

$$no.\ moles\ H_2 = 1.20\ mol\ CH_4 \times \frac{3\ mol\ H_2}{1\ mol\ CH_4} = 3.60\ mol\ H_2$$

Exercise   How many moles of H_2 can be prepared from a mixture of 2.14 mol of CH_4 and 5.43 mol of H_2O? Answer: 6.42 mol H_2.

The situation described in Example 5.2 is common in many industrial and laboratory preparations. Frequently, an excess of the cheaper reactant, in this case H_2O, is used. Calculation of the amount of product obtained must be based on the reactant which is in short supply, in this case CH_4. This substance is often referred to as the **limiting reactant,** because its amount sets a limit on the yield of product.

The amount of product that can be obtained if the limiting reactant is completely converted is called the **theoretical yield.** In Example 5.2 we would say that the theoretical yield of H_2 is 3.60 mol. There is no way to get more than 3.60 mol of H_2 from this mixture of CH_4 and H_2O. Indeed, we are almost certain to get less. For one thing, some of the limiting reactant, CH_4, may be consumed by side reactions. Again, some of the product, H_2, may be lost in attempting to purify it.

For these and other reasons, the **actual yield** in a reaction is ordinarily less than the theoretical yield. Putting it another way, the **per cent yield** is expected to be less than 100.

The limiting reactant is the one that gets used up first

$$\%\ yield = \frac{actual\ yield}{theoretical\ yield} \times 100 \qquad (5.2)$$

Example 5.3 With the reaction mixture described in Example 5.2, 2.16 mol of H_2 is actually recovered, rather than the 3.60 mol calculated. What is the per cent yield of H_2?

Solution

$$\% \text{ yield} = \frac{2.16 \text{ mol}}{3.60 \text{ mol}} \times 100 = 60.0$$

Exercise Suppose the actual yield of H_2 in Example 5.3 were 1.81 mol. What would be the % yield? Answer: 50.3.

Very pure hydrogen can be prepared by the *electrolysis* of water. A simple type of apparatus used for this purpose is shown in Figure 5.4. A small amount of a compound such as NaOH is added to the water to provide ions to carry the current. Electrical energy is absorbed to bring about the reaction

$$2 \text{ H}_2\text{O(l)} \rightarrow 2 \text{ H}_2\text{(g)} + \text{O}_2\text{(g)} \tag{5.3}$$

The hydrogen produced this way is more expensive than that from natural gas. The major cost is for electrical energy. About 118 kJ of energy must be supplied to form one gram of hydrogen.

H_2 and CO from Reaction 5.1 are used to make industrial organic chemicals

Hydrogen is used mainly to make ammonia, NH_3. It is also used to remove sulfur from petroleum products, to produce certain metals from their oxides, and to make such compounds as sodium hydride, NaH.

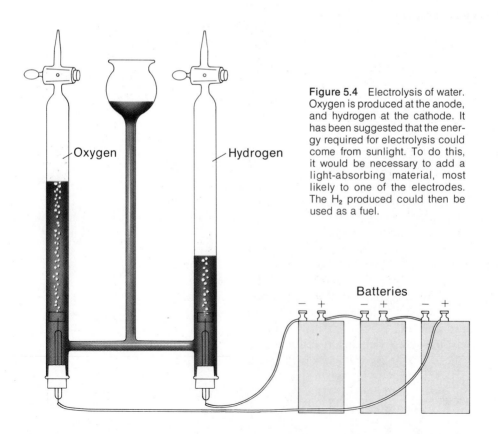

Figure 5.4 Electrolysis of water. Oxygen is produced at the anode, and hydrogen at the cathode. It has been suggested that the energy required for electrolysis could come from sunlight. To do this, it would be necessary to add a light-absorbing material, most likely to one of the electrodes. The H_2 produced could then be used as a fuel.

Oxygen

Hydrogen

Batteries

The elements in Group 7 of the Periodic Table are known as *halogens* ("salt-formers"). The first two members of the family, fluorine and chlorine, are gases at 25°C and 1 atm.* Bromine is a deep red, volatile liquid boiling at 59°C. Iodine is a shiny black solid. Upon heating, iodine is readily converted to a violet vapor (see color plate 4, center of book). Astatine, the last member of the family, is radioactive like its neighbor radon in Group 8.

The halogens are much too reactive to be found as the free elements in nature. Instead they occur as the *halide ions:* F^-, Cl^-, Br^-, and I^-. The principal source of fluorine is the water-insoluble mineral fluorite, CaF_2. Chlorine is present as the Cl^- ion in such ionic solids as NaCl, KCl, and $MgCl_2$. It is also the most abundant anion in seawater (0.53 M). Bromide and iodide ions are found in seawater, but at much lower concentrations. In certain natural brines (salt solutions), the concentrations of Br^- and I^- ions are higher.

The two heavier halogens, Br_2 and I_2, are the easiest to prepare. To extract bromine from brines, chlorine gas is bubbled through the solution. An oxidation-reduction reaction occurs. Chlorine, which has a strong attraction for electrons, is reduced to Cl^- ions. The electrons come from Br^- ions, which are oxidized to Br_2. The equation for the overall reaction is

$$Cl_2(g) + 2\ Br^-(aq) \rightarrow 2\ Cl^-(aq) + Br_2(l) \qquad (5.4)$$

In practice, an aqueous solution containing dissolved bromine and unreacted chlorine is the first product. Bromine can be separated from this solution in various ways. One method involves heating the solution to produce a mixture of $Br_2(g)$, $Cl_2(g)$, and water vapor. Upon cooling, liquid bromine separates out.

A reaction similar to 5.4 occurs when a brine containing I^- ions is treated with chlorine gas:

$$Cl_2(g) + 2\ I^-(aq) \rightarrow 2\ Cl^-(aq) + I_2(s) \qquad (5.5)$$

---

*Hydrogen is often included in Group 7. In certain of its chemical properties it resembles the halogens. For example, it forms the H^- ion in compounds such as NaH, analogous to the Cl^- ion in NaCl. On the other hand, in water solution, it forms the H^+ ion, analogous to the ions of the Group 1 metals (Li^+, Na^+, . . .). For this reason among others, hydrogen is often placed in two different groups of the Periodic Table, Groups 1 and 7.

Example 5.4   In a natural brine found in Arkansas, the concentration of Br^- ions is 5.00×10^{-3} M. For one cubic meter (1.00×10^3 ℓ) of this brine
   a. how many grams of Cl_2 are required to react with all the Br^- ions?
   b. if one kilogram of Cl_2 is used, what is the theoretical yield of Br_2 in moles?

Solution   In a cubic meter of the brine there are

$$5.00 \times 10^{-3}\ \frac{mol}{\ell} \times 1.00 \times 10^3\ \ell = 5.00\ mol\ Br^-$$

   a. From Equation 5.4, we see that 1 mol of Cl_2 reacts with 2 mol of Br^-. But, since the atomic mass of Cl is 35.45, 1 mol of Cl_2 weighs 70.90 g.

$$70.90\ g\ Cl_2 \doteq 2\ mol\ Br^-$$

$$mass\ Cl_2\ required = 5.00\ mol\ Br^- \times \frac{70.90\ g\ Cl_2}{2\ mol\ Br^-} = 177\ g\ Cl_2$$

b. Since only 177 g of Cl_2 is required, one kilogram of Cl_2 is clearly an excess. The limiting reactant is Br^-; at best, we can recover all of the Br^- ions from the brine. Since

$$2 \text{ mol } Br^- \rightleftharpoons 1 \text{ mol } Br_2$$

$$\text{theor. yield } Br_2 = 5.00 \text{ mol } Br^- \times \frac{1 \text{ mol } Br_2}{2 \text{ mol } Br^-} = 2.50 \text{ mol } Br_2$$

Exercise How many moles of Cl_2 are required to react with the Br^- ions in one liter of the brine referred to in Example 5.4? How many grams of Br_2 can be recovered? Answer: 2.50×10^{-3} mol Cl_2; 0.400 g Br_2.

Chloride ions hold on to their electrons more tightly than Br^- or I^- ions. Hence, they are more difficult to oxidize to the free element. Industrially, the oxidation of Cl^- to Cl_2 is carried out by electrolysis. The usual source of Cl^- is NaCl, either in the pure liquid state or in water solution. The electrolysis of molten sodium chloride is described in Section 5.4. In water solution, the reaction is

$H_2(g)$ is a useful by-product of this reaction, as is NaOH

$$2 \text{ Cl}^-(aq) + 2 \text{ H}_2O \rightarrow Cl_2(g) + H_2(g) + 2 \text{ OH}^-(aq) \tag{5.6}$$

This process will be discussed further in Chapter 9.

Of all the halide ions, F^- is the most difficult to oxidize. The industrial process for making F_2 starts with calcium fluoride, CaF_2. This is treated with sulfuric acid, H_2SO_4. The products are hydrogen fluoride, HF, and calcium sulfate, $CaSO_4$:

$$CaF_2(s) + H_2SO_4(l) \rightarrow 2 \text{ HF}(l) + CaSO_4(s) \tag{5.7}$$

The hydrogen fluoride is mixed with potassium fluoride, KF, and electrolyzed. The electrolysis reaction can be shown most simply as

$$2 \text{ HF}(l) \rightarrow H_2(g) + F_2(g) \tag{5.8}$$

F_2 is one of the most reactive of all chemical substances

The F_2 molecule has such a strong attraction for electrons that the element is very dangerous to work with. It reacts violently with water, asbestos, hot glass, and most metals. Fluorine is commonly stored in containers made of monel metal, an alloy of Ni, Cu, and Fe. This is one of the few materials that is not attacked by F_2.

5.3 THE LIGHT METALS (GROUPS 1, 2, AND Al)

The metals in the main groups toward the left of the Periodic Table have several properties in common. For one thing, as metals go, they have low densities, typically less than 3 g/cm³. They are all very reactive toward nonmetals and occur in nature in the form of ionic compounds. There, they are present as monatomic cations such as Na^+, Mg^{2+}, and Al^{3+}. The anion in the compound is most often Cl^-, CO_3^{2-}, SO_4^{2-}, or, in the case of Al^{3+}, the oxide ion, O^{2-} (Table 5.3). We will consider how three metals of this type are obtained from natural sources. These are sodium, magnesium, and aluminum in the third period of the Periodic Table.

TABLE 5.3 PROPERTIES, SOURCES, AND USES OF THE LIGHT METALS

	GROUP	mp (°C)	bp (°C)	DENSITY (g/cm³)	PRINCIPAL SOURCE	USES OF METAL
Li	1	186	1326	0.534	Complex silicates	Synthesis of organic Li compounds
Na	1	98	889	0.971	NaCl	Synthesis of antiknock additives, Na vapor lights
K	1	64	774	0.862	KCl	Synthesis of KO_2
Rb	1	39	688	1.53	RbCl (with K^+)	No major use
Cs	1	29	690	1.87	CsCl (with K^+)	Photocells, arc lamps
Be	2	1283	2970	1.85	Complex silicates	Alloys, nuclear shielding
Mg	2	650	1120	1.74	Mg^{2+} in seawater; $MgCO_3$	**Lightweight alloys,** used in aircraft, tools, etc.
Ca	2	845	1420	1.55	$CaCO_3$ (limestone) $CaSO_4 \cdot 2 H_2O$ (gypsum)	Reduction of metal oxides, hardening of alloys
Sr	2	770	1380	2.60	$SrCO_3$, $SrSO_4$	No major use
Ba	2	725	1640	3.51	$BaCO_3$, $BaSO_4$	Alloy with Ni, spark plugs
Al	3	660	2327	2.70	Al_2O_3 (bauxite)	**Structural metal** (aircraft, autos), foil, utensils

Example 5.5   A mineral called dolomite contains 13.2% Mg, 21.7% Ca, 13.0% C, and 52.1% O. What is the simplest formula of this ionic compound? What ions do you suppose are present?

Solution   Proceeding as in Example 3.3, p. 51, we find the number of moles of each element in 100 g of dolomite. A sample calculation is

$$\text{moles Mg} = 13.2 \text{ g} \times \frac{1 \text{ mol}}{24.3 \text{ g}} = 0.543 \text{ mol}$$

Making the same calculation for the other elements, we find

$$0.543 \text{ Mg} : 0.541 \text{ Ca} : 1.08 \text{ C} : 3.25 \text{ O}$$

Dividing by the smallest number, 0.541:

$$1 \text{ Mg} : 1 \text{ Ca} : 2 \text{ C} : 6 \text{ O}$$

The simplest formula could be written as $MgCaC_2O_6$. A moment's reflection should convince you that the ions present are Mg^{2+}, Ca^{2+}, and $2 CO_3^{2-}$. The formula of dolomite is often written as $MgCO_3 \cdot CaCO_3$. It is a 1:1 mixture of $MgCO_3$ and $CaCO_3$.

Exercise   The mineral called carnallite contains 23.0% K, 14.3% Mg, and 62.7% Cl. What is its simplest formula? What ions are present? Answer: $KMgCl_3$: K^+, Mg^{2+}, 3 Cl^-.

Sodium

Sodium metal is produced by the electrolysis of molten sodium chloride, NaCl. The electrical cell used is shown in Figure 5.5. Electrons, from a storage battery or other source of direct electric current, enter the cell through the iron bar at the left. At

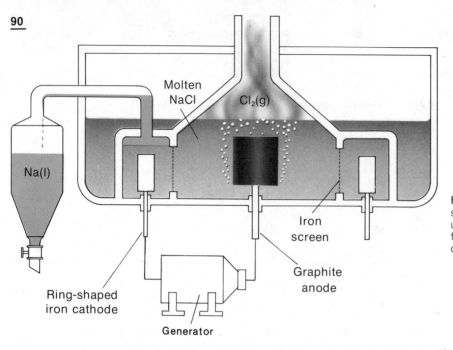

Figure 5.5 Electrolysis of molten sodium chloride. The iron screen is used to prevent sodium and chlorine from coming into contact with each other.

this *electrode,* called the **cathode, reduction** occurs. Sodium ions, Na^+, pick up electrons and are reduced to sodium atoms:

$$2\ Na^+ + 2\ e^- \rightarrow 2\ Na \qquad \text{(reduction at cathode)} \tag{5.9a}$$

Electric current is carried through the cell by the movement of ions. Cations (Na^+ ions) move to the cathode. Anions (Cl^- ions) move in the opposite direction, toward the graphite rod near the center of the cell. At this electrode, called the **anode, oxidation** occurs. Chloride ions, Cl^-, lose electrons to form Cl atoms. Pairs of Cl atoms immediately combine to form Cl_2 molecules:

$$2\ Cl^- \rightarrow Cl_2 + 2\ e^- \qquad \text{(oxidation at anode)} \tag{5.9b}$$

The overall cell reaction is obtained by summing 5.9a and 5.9b:

$$2\ NaCl(l) \rightarrow 2\ Na(l) + Cl_2(g) \tag{5.9}$$

> You can't make Na by electrolysis of an aqueous solution of NaCl

The cell must be operated at a high temperature, about 600°C, to keep the sodium chloride melted. (Solid NaCl does not conduct a current, since the ions are not free to move.) About 14 kJ of electrical energy must be supplied for every gram of sodium produced. The sodium metal (mp = 98°C) is drawn off as a liquid from the cathode compartment. The chlorine gas formed at the anode is a valuable by-product. The cell is designed so that the sodium and chlorine do not come into contact with each other. If that happened, they would react by the reverse of Reaction 5.9, and we would be back where we started.

Magnesium

The principal source of magnesium is seawater, where Mg^{2+} is the second most abundant cation after Na^+. The concentration of Mg^{2+} in the oceans is about 0.052 M. Magnesium metal is obtained by a three-step process.

1. The Mg^{2+} ions are removed from the seawater by adding a solution of calcium

hydroxide, $Ca(OH)_2$. The OH^- ions combine with Mg^{2+} to form a precipitate of magnesium hydroxide:

$$Mg^{2+}(aq) + 2\ OH^-(aq) \rightarrow Mg(OH)_2(s) \qquad (5.10)$$

The precipitate of $Mg(OH)_2$ is formed in large settling tanks. It is removed by filtration and dried.

2. The solid $Mg(OH)_2$ is treated with hydrochloric acid (H^+, Cl^- ions). An acid-base reaction occurs between the H^+ ions of the HCl and the OH^- ions of the solid:

$$Mg(OH)_2(s) + 2\ H^+(aq) \rightarrow Mg^{2+}(aq) + 2\ H_2O \qquad (5.11)$$

In this way, Mg^{2+} ions are brought into solution. In effect, a solution of $MgCl_2$ is formed; the Cl^- ions come from the HCl. Evaporation of this solution gives solid $MgCl_2$:

$$Mg^{2+}(aq) + 2\ Cl^-(aq) \rightarrow MgCl_2(s) \qquad (5.12)$$

3. The magnesium chloride is electrolyzed in the molten state. The process is very similar to that discussed with NaCl. The electrode reactions are

cathode: $Mg^{2+} + 2\ e^- \rightarrow Mg$ reduction
anode: $2\ Cl^- \rightarrow Cl_2 + 2\ e^-$ oxidation

To make a ton of Mg we have to process about 1000 tons of sea-water

The overall cell reaction is

$$MgCl_2(l) \rightarrow Mg(s) + Cl_2(g) \qquad (5.13)$$

Example 5.6   A saturated solution of $Ca(OH)_2$, about 0.10 M, is used to precipitate Mg^{2+} from seawater. Taking the concentration of Mg^{2+} to be 0.052 M, how many liters of 0.10 M $Ca(OH)_2$ are required to precipitate the Mg^{2+} in 1.0 ℓ of seawater?

Solution   This calculation is very similar to that called for in Example 4.9, p. 75. We first calculate the number of moles of Mg^{2+} in a liter of seawater. Then, using Equation 5.10, we determine the number of moles of $Ca(OH)_2$ required. Finally, we obtain the volume of 0.10 M $Ca(OH)_2$ needed:

(1) no. moles $Mg^{2+} = 0.052\ \dfrac{mol}{\ell}\ Mg^{2+} \times 1.0\ \ell = 0.052$ mol Mg^{2+}

(2) One mole of $Ca(OH)_2$ forms two moles of OH^-. According to Equation 5.10, two moles of OH^- are required to react with one mole of Mg^{2+}. In other words,

1 mol $Mg^{2+} \simeq 1$ mol $Ca(OH)_2$

The $Ca(OH)_2$ is made by heating seashells ($CaCO_3$) and driving off CO_2

So, to precipitate 0.052 mol of Mg^{2+}, we need an equal number, 0.052 mol of $Ca(OH)_2$.

(3) volume $Ca(OH)_2 = \dfrac{0.052\ mol}{0.10\ mol/\ell} = 0.52\ \ell$

In other words, for each liter of seawater we should add a little more than half a liter of 0.10 M $Ca(OH)_2$.

Exercise   From one liter of seawater, 0.052 mol of $Mg(OH)_2$ can be recovered. What volume of 6.0 M HCl is required to react with this by Reaction 5.11? Answer: 0.017 ℓ.

Aluminum

Aluminum is the third most abundant element in the earth's crust, after oxygen and silicon. In the form of Al^{3+} ions, it is a major component of such common materials as clay, feldspar, and granite. Unfortunately, aluminum cannot be extracted profitably from these materials. Instead, it is obtained from a relatively uncommon mineral called bauxite. This is a mixture of the oxides of aluminum (Al_2O_3), iron (Fe_2O_3), and silicon (SiO_2). The percentage of aluminum in bauxite is about 28.

The first step in preparing aluminum is the separation of aluminum oxide from bauxite. This is done by heating the ore with a concentrated solution of sodium hydroxide, NaOH. Of the three major components of bauxite, only Al_2O_3 dissolves. The impurities are then removed by filtration. When the solution is cooled and diluted with water, nearly pure aluminum oxide separates as a solid.

Aluminum metal is obtained from Al_2O_3 by electrolysis (Fig. 5.6). Cryolite, Na_3AlF_6, is added to the aluminum oxide to produce a mixture which melts at about 1000°C. (A mixture of AlF_3, NaF, and CaF_2 may be substituted for cryolite.) The cell is heated electrically to keep the mixture molten so that ions can move through it, carrying the electric current.

The iron wall of the cell serves as the cathode. Here Al^{3+} ions are reduced to form molten aluminum (mp = 660°C):

$$\text{cathode:} \qquad 4\ Al^{3+} + 12\ e^- \rightarrow 4\ Al \qquad \text{reduction}$$

The anodes are retractable carbon rods. There, O^{2-} ions lose electrons to form gaseous oxygen:

$$\text{anode:} \qquad 6\ O^{2-} \rightarrow 3\ O_2(g) + 12\ e^- \qquad \text{oxidation}$$

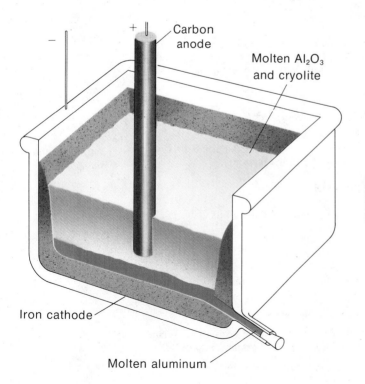

Carbon anode

Molten Al_2O_3 and cryolite

Iron cathode

Molten aluminum

Figure 5.6 Electrolytic preparation of aluminum. Aluminum, being more dense than cryolite, collects at the bottom of the cell and so is protected from oxidation by the air.

Figure 5.7 The first aluminum nuggets extracted by Charles Hall. (Courtesy of Alcoa.)

The oxygen formed attacks the carbon rods, slowly consuming them to give CO and CO_2.

The overall cell reaction can be written as

$$2 \ Al_2O_3(l) \rightarrow 4 \ Al(l) + 3 \ O_2(g) \tag{5.14}$$

This process consumes about 60 kJ of energy per gram of aluminum formed.

A century ago, aluminum sold for about $20 a kilogram, and the total production worldwide was about 2×10^3 kg a year. The metal was made by a costly process involving the reaction of sodium with aluminum chloride:

$$AlCl_3(s) + 3 \ Na(s) \rightarrow Al(s) + 3 \ NaCl(s) \tag{5.15}$$

The process for obtaining aluminum from bauxite was worked out in 1886 by Charles Hall, a graduate student at Oberlin College. The problem that Hall faced was to find a way to electrolyze Al_2O_3 at a temperature below its melting point of 2000°C. His general approach was to look for ionic compounds in which Al_2O_3 would dissolve at a reasonable temperature. After several unsuccessful attempts, Hall found that cryolite was the ideal "solvent." Curiously enough, the same electrolytic process was worked out by Heroult in France, also in 1886. Each young man (both were 22 years old) was entirely unaware of the other's work.

After the Hall-Heroult process began to be used, the price of aluminum fell rapidly. Within 10 years it was selling for $1 a kilogram, which is about what aluminum costs today. In the United States, the production of Al is now 2×10^6 metric tons per year, an amount greater than that of any other metal except copper and iron.

5.4 THE HEAVY METALS

The heavy metals (density > 3 g/cm³) include the elements in the center of the Periodic Table, the *transition metals*. The other metals in this category are the lower mem-

93

	Sc	Ti	V	Cr	Mn	Fe	Co	Ni	Cu	Zn	Ga		
mp(°C)	1541	1660	1890	1857	1244	1535	1495	1453	1083	420	30		
bp(°C)	2831	3287	3380	2672	1962	2750	2870	2732	2567	907	2403		
d(g/cm³)	3.0	4.51	6.11	7.19	7.43	7.87	8.92	9.91	8.94	7.13	5.91		
	Y	Zr	Nb	Mo	Tc	Ru	Rh	Pd	Ag	Cd	In	Sn	
mp(°C)	1509	1852	2468	2610	2200	2430	1966	1550	961	321	157	232	
bp(°C)	2930	3580	5127	5560	4700	3700	3700	3170	2210	767	2000	2270	
d(g/cm³)	4.48	6.49	8.57	10.22	11.48	12.45	12.41	12.01	10.49	8.65	7.31	7.30	
	La	Hf	Ta	W	Re	Os	Ir	Pt	Au	Hg	Tl	Pb	Bi
mp(°C)	920	2225	2980	3410	3180	2727	2448	1769	1063	−39	304	328	271
bp(°C)	3470	5200	5425	5930	5885	4100	4500	4530	2966	357	1457	1750	1560
d(g/cm³)	6.17	13.3	16.6	19.3	21.0	22.6	22.6	21.5	19.3	13.5	11.85	11.35	9.75

Oxide ores   Uncombined

Sulfide ores   Phosphates

Figure 5.8 Properties of the transition metals and the post-transition metals in Groups 3, 4, and 5. In general, these metals are higher melting, higher boiling, and more dense than those in Groups 1 and 2. Metals toward the left of the transition series tend to occur in nature as oxides; sulfide ores are more common among the metals toward the right. A few very unreactive metals, including platinum and gold, are found in elemental form.

bers of the main groups to the right of the transition metals. These elements (Ga, In, Tl; Sn, Pb, Bi) are often referred to as *post-transition metals*.

Some of the properties of the heavy metals are listed in Figure 5.8. The ores from which they are extracted are largely of three types:

1. *The free elements*. A few of the least reactive transition metals are found in the elementary state in nature. The most familiar of these is gold. It occurs as grains in river sand or, more commonly, as narrow veins in quartz rock. Copper and silver are also found in small amounts as the free metals.

2. *Sulfides*. This is the common type of ore among the transition metals toward the right of the Periodic Table. Typical sulfide ores include those of copper (Cu_2S), silver (Ag_2S), zinc (ZnS), and mercury (HgS). Most of the post-transition metals are also obtained from their sulfides (e.g., PbS, Bi_2S_3).

3. *Oxides*. The metals toward the left of the transition series occur largely as oxides (Cr_2O_3, MnO_2, Fe_2O_3). The principal ore of tin, a post-transition metal, is cassiterite, SnO_2.

We will consider the *metallurgy* (extraction process) for only two of these elements. These are the two most important heavy metals from an economic standpoint—copper and iron. About 2 million metric tons of copper are produced annually in the United States. The production of iron is even greater, about 100 million metric tons a year. Most of this iron is made into steel, an alloy containing Fe, C, and often other metals.

Most gold recovered these days is a by-product obtained during the refining of other metals

Copper

The metallurgy of copper depends upon the type of ore that is available. We will consider one of the more common ores called chalcocite. This contains copper(I) sulfide, Cu_2S, in highly impure form. Rocky materials lower the fraction of copper in the ore to less than 10%. The extraction of copper metal from this ore involves three separate steps:

1. The Cu_2S in the ore is concentrated by a process known as *flotation* (Fig. 5.9). The finely divided ore is mixed with oil and stirred with soapy water in a large tank. Compressed air is then blown through the mixture. The oil-coated particles of copper(I)

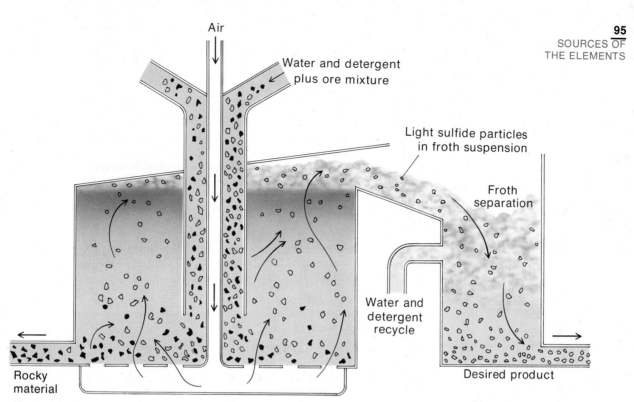

Figure 5.9 Low-grade sulfide ores, including Cu_2S, are often concentrated by flotation. The finely divided sulfide particles are trapped in soap bubbles, while the rocky material sinks to the bottom and is discarded.

sulfide are carried to the top of the tank. There they form a froth which can be skimmed off. Most of the rocky impurities, such as SiO_2, settle to the bottom of the tank.

2. The copper(I) sulfide is converted to the metal by blowing air through it at a high temperature. (Pure O_2 is often used instead of air.) The overall reaction that occurs is a simple one:

$$Cu_2S(s) + O_2(g) \rightarrow 2\ Cu(s) + SO_2(g) \qquad (5.16)$$

The solid produced is called "blister copper." It has an irregular appearance due to air bubbles that enter the copper while it is still molten. Blister copper is impure, containing small amounts of several other metals.

3. The copper is refined (purified) by electrolysis. A piece of impure blister copper is used as the anode. It is oxidized to Cu^{2+} ions, which enter the water solution. These ions are reduced to copper atoms at the cathode, a sheet of pure copper. As electrolysis proceeds, the copper is transferred from the impure anode to the pure cathode.

The solution contains copper sulfate, $CuSO_4$

$$
\begin{aligned}
\text{anode:} \quad & Cu(s, \text{impure}) \rightarrow Cu^{2+}(aq) + 2\ e^- \\
\text{cathode:} \quad & \underline{Cu^{2+}(aq) + 2\ e^- \rightarrow Cu(s, \text{pure})} \\
& Cu(s, \text{impure}) \rightarrow Cu(s, \text{pure}) \qquad (5.17)
\end{aligned}
$$

Less active metals including silver, gold, and platinum are not oxidized at the anode. Instead, they drop off and collect at the bottom of the cell. These metals are recovered, separated from each other, and purified. Their value exceeds the cost of the energy used in the electrolysis.

Metallurgy is often just
applied chemistry

Example 5.7 Zinc, like copper, occurs as a sulfide ore. The formula of the ore, known as zinc blende, is ZnS. When ZnS is heated with oxygen, a reaction similar to 5.16 occurs. However, the solid product is the oxide, ZnO, rather than the metal.
a. Write a balanced equation for the reaction.
b. How many grams of SO_2 are formed from 1.00 kg of ore containing 22% of ZnS?

Solution
a. The unbalanced equation is $ZnS(s) + O_2(g) \rightarrow ZnO(s) + SO_2(g)$. We could balance this equation by writing a coefficient of $^3/_2$ for O_2:

$$ZnS(s) + {}^3/_2\ O_2(g) \rightarrow ZnO(s) + SO_2(g)$$

Ordinarily, whole-number coefficients are preferred. Multiplying all the coefficients by 2 gives

$$2\ ZnS(s) + 3\ O_2(g) \rightarrow 2\ ZnO(s) + 2\ SO_2(g)$$

b. The mass of ZnS in the ore is

$$1.00 \times 10^3\ g \times 0.22 = 2.2 \times 10^2\ g\ ZnS$$

From the balanced equation

$$1\ mol\ ZnS \simeq 1\ mol\ SO_2$$

Changing to grams (at. mass Zn = 65.38, S = 32.06, O = 16.00)

$$97.44\ g\ ZnS \simeq 64.06\ g\ SO_2$$

$$mass\ SO_2 = 2.2 \times 10^2\ g\ ZnS \times \frac{64.06\ g\ SO_2}{97.44\ g\ ZnS} = 1.4 \times 10^2\ g\ SO_2$$

The SO_2 produced in the treatment of sulfide ores such as ZnS and Cu_2S is a major contributor to air polluton.

Exercise How many grams of SO_2 are formed from 1.00 kg of ore containing 8.0% of Cu_2S (Reaction 5.16)? Answer: 32 g.

Iron

The principal high-grade ore of iron is a mineral called hematite. This consists largely of iron(III) oxide, Fe_2O_3. The main impurity is silicon dioxide, SiO_2 (sand). Smaller amounts of compounds of manganese and phosphorus are also present. Reduction of Fe^{3+} ions in Fe_2O_3 is carried out in a blast furnace (Fig. 5.10). Typically, the furnace is about 30 m high and perhaps 10 m in diameter.

The solid "charge," admitted at the top of the blast furnace, consists of three materials. These are iron ore, limestone ($CaCO_3$), and coke, which is nearly pure carbon. To get the process started, a blast of compressed air or pure O_2 at 500°C is blown into the furnace through nozzles located near the bottom. Several different reactions occur, of which three are most important.

1. *Conversion of carbon to carbon monoxide.* In the lower part of the furnace coke burns to form carbon dioxide, CO_2. As the CO_2 rises through the solid mixture, it reacts further with the coke to form carbon monoxide, CO. The overall reaction is

$$2\ C(s) + O_2(g) \rightarrow 2\ CO(g) \tag{5.18}$$

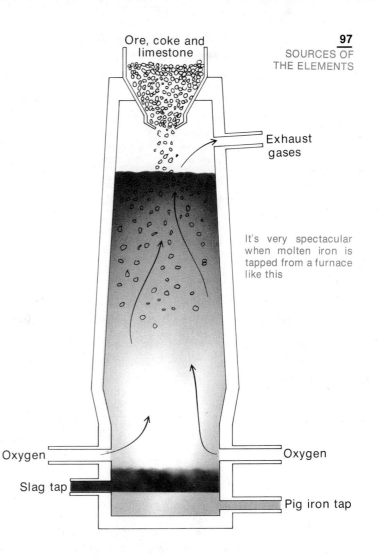

Ore, coke and limestone

Exhaust gases

It's very spectacular when molten iron is tapped from a furnace like this

Oxygen

Oxygen

Slag tap

Pig iron tap

Figure 5.10 Blast furnace. Oxygen reacting with coke furnishes the high temperatures needed for reduction of the iron oxide. The ore appears to be reduced by CO, formed by reaction of CO_2 with coke in the upper parts of the furnace. Using pure oxygen instead of air greatly increases efficiency of the process.

The heat given off by this reaction maintains a high temperature within the furnace.

   2. *Reduction of Fe^{3+} ions to Fe*. The CO produced by Reaction 5.18 reacts with the iron(III) oxide in the ore.

$$Fe_2O_3(s) + 3\ CO(g) \rightarrow 2\ Fe(l) + 3\ CO_2(g) \tag{5.19}$$

The C should react with Fe_2O_3, but CO is faster to react. Why?

Molten iron, formed at a temperature of 1600°C, collects at the bottom of the furnace. Four or five times a day, it is drawn off. The daily production of iron from a single blast furnace is about 1000 metric tons. This is enough to make 500 Cadillacs or 1000 Volkswagens.

   3. *Formation of slag*. The limestone added to the furnaces decomposes at about 800°C.

$$CaCO_3(s) \rightarrow CaO(s) + CO_2(g) \tag{5.20}$$

The calcium oxide formed reacts with impurities in the iron ore to form a glassy material called slag. The main reaction is with SiO_2 to form calcium silicate, $CaSiO_3$:

$$CaO(s) + SiO_2(s) \rightarrow CaSiO_3(l) \tag{5.21}$$

The slag, which is less dense than molten iron, forms a layer on the surface of the metal. This makes it possible to draw off the slag through an opening in the furnace above that used to remove the iron. The slag is used to make cement and as a base in road construction.

Example 5.8   In operating a blast furnace, enough limestone is used to give about two grams of CaO for every gram of SiO_2 impurity in the ore. In Reaction 5.21, what is the "theoretical yield" in grams of $CaSiO_3$, starting with 2.00 g of CaO and 1.00 g of SiO_2?

Solution   To determine the limiting reactant, we first convert grams of CaO and SiO_2 to moles. One mole of CaO weighs 56.08 g; one mole of SiO_2 weighs 60.09 g.

$$\text{moles CaO available} = 2.00 \text{ g} \times \frac{1 \text{ mol}}{56.08 \text{ g}} = 0.0357 \text{ mol}$$

$$\text{moles } SiO_2 \text{ present} = 1.00 \text{ g} \times \frac{1 \text{ mol}}{60.09 \text{ g}} = 0.0166 \text{ mol}$$

From Equation 5.21 we see that CaO and SiO_2 react in a 1:1 mol ratio. Hence, SiO_2 is the limiting reactant; CaO is in excess. After the reaction is over there will be almost 0.02 mol of CaO left.

To find the yield of $CaSiO_3$, we note that one mole of SiO_2 gives one mole of $CaSiO_3$. Hence we must form 0.0166 mol of $CaSiO_3$ (molar mass = 116.17 g).

$$\text{mass } CaSiO_3 = 0.0166 \text{ mol} \times \frac{116.17 \text{ g}}{1 \text{ mol}} = 1.93 \text{ g}$$

Exercise   How many grams of $CaSiO_3$ could be produced starting with a mixture of 1.00 g of CaO and 1.00 g of SiO_2? Answer: 1.93 g.

We use excess CaO to make sure all the SiO_2 is removed

Steel

The product that comes out of the blast furnace, called "pig iron," is highly impure. On the average, it contains about 4% of carbon. Other impurities, present in smaller amounts, include silicon, manganese, and phosphorus. To make steel from pig iron, two things must be done.

1. *Remove the Si, Mn, and P.* To do this, these three elements are oxidized:

$$Si(s) + O_2(g) \rightarrow SiO_2(s) \tag{5.22}$$

$$2 \text{ Mn}(s) + O_2(g) \rightarrow MnO(s) \tag{5.23}$$

$$P_4(s) + 5 \text{ } O_2(g) \rightarrow P_4O_{10}(s) \tag{5.24}$$

The oxides formed by these reactions are removed by adding $CaCO_3$. The limestone decomposes to CaO, which forms a slag as in the blast furnace.

2. *Lower the carbon content below 2%.* This is necessary to produce a strong ductile metal that can be heat-treated. Most of the carbon is burned to carbon dioxide:

The properties of steel vary greatly with carbon content

$$C(s) + O_2(g) \rightarrow CO_2(g) \tag{5.25}$$

By controlling the amount of oxygen used, it is possible to adjust the carbon content of the steel within very narrow limits.

Most of the steel produced in the world today is made by the "basic oxygen" pro-

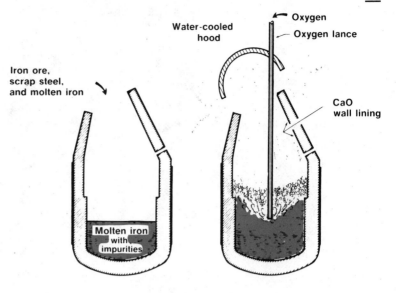

Figure 5.11 Converter used to make steel from pig iron by the basic oxygen process. Pure oxygen under pressure is blown into the molten metal to oxidize impurities such as Si, Mn, and P_4 to their oxides. The oxygen also reacts with carbon to form CO_2, lowering the carbon content to 2% or less.

cess. This is the only process used in Japan and Western Europe. It accounts for about 60% of steel production in the United States. A diagram of the "converter" used is shown in Figure 5.11. This is filled with a mixture of about 70% pig iron, 25% scrap iron or steel, and 5% limestone. The pig iron is brought over from the blast furnace while it is still molten. Most of the scrap comes from the steel plant itself. Some is obtained from recycling of steel products such as automobiles.

Pure oxygen under a pressure of about 10 atm is admitted to the converter through the water-cooled "lance" at the top. When it reaches the surface of the molten metal, it causes vigorous stirring. Reactions 5.22 through 5.25 occur rapidly. Their progress is followed by an automatic, computerized system of chemical analysis. When the carbon content drops to the desired level, the supply of oxygen is cut off. (Too much oxygen would oxidize the iron, a reaction we don't want.) At this stage, the steel is ready to be poured. The whole process takes from 30 min to an hour, and yields about 200,000 kg of steel in a single "blow." *That's enough to make about 30 Volkswagens*

5.5 RESOURCES OF METALS

In Sections 5.3 and 5.4, we saw how several metals are obtained from natural sources. You might well wonder about the extent of these sources. Are they limitless? Or will they, sometime in the future, dwindle away? Will we, in your lifetime, face a "metals crisis" similar to the "energy crisis" we live with today?

Table 5.4 suggests how long we can expect to continue extracting the common metals from ores now in use. The "static index" listed in column 3 is obtained by dividing known global reserves of ores by the annual consumption (in 1970). A more realistic estimate may be the "exponential index" in column 5. This takes into account the rate at which consumption of metal products is growing. On this basis we seem likely to run out of several of the heavy metals by the end of this century. Moreover, you will note from Table 5.4 that mineral deposits, like petroleum, are not distributed uniformly throughout the globe. The United States is dependent upon foreign ore deposits for many of its vital metals, including Hg, Sn, Mn, Al, and Cr.

TABLE 5.4 RESOURCES OF METALS (1970)*

1 METAL	2 KNOWN RE- SERVES (KG)	3 STATIC INDEX** (YR)**	4 ANNUAL % GROWTH*	5 EXPO- NENTIAL INDEX††	6 LOCATION OF RESERVES (% OF TOTAL)
Gold	1.10×10^7	11	4.1	9	Rep. So. Africa (40)
Mercury	1.15×10^8	13	2.6	13	Spain (30); Italy (21)
Silver	1.71×10^8	16	2.7	13	Communist coun- tries (36); U.S. (24)
Tin	4.4×10^9	17	1.1	15	Thailand (33); Malaysia (14)
Zinc	1.23×10^{11}	23	2.9	18	U.S. (27); Canada (20)
Lead	9.1×10^{10}	26	2.0	22	U.S. (39)
Copper	3.08×10^{11}	36	4.6	22	U.S. (28); Chile (19)
Tungsten	1.3×10^9	40	2.5	28	China (73)
Molybdenum	4.90×10^9	79	4.5	34	U.S. (58); U.S.S.R. (20)
Manganese	8×10^{11}	97	2.9	46	Rep. So. Africa (38); U.S.S.R. (25)
Aluminum	1.17×10^{12}	100	6.4	31	Australia (33); Guinea (20)
Cobalt	2.2×10^9	110	1.5	60	Rep. Congo (31); Zambia (20)
Pt metals	1.33×10^7	130	3.8	47	Rep. So. Africa (47); U.S.S.R. (47)
Nickel	6.67×10^{10}	190	3.4	53	Cuba (25); New Caledonia (22); Canada (14)
Iron	1×10^{14}	240	1.8	93	U.S.S.R. (33); Canada (14)
Chromium	7.75×10^{11}	420	2.6	95	Rep. So. Africa (75)

*From Meadows, D. H., et al., *Limits of Growth*. New York, Universe Books, 1972.
**Column 3: number of years reserves will last at 1970 rate of consumption.
†Column 4: average rate at which consumption is growing.
††Column 5: number of years reserves will last if growth continues at rate given in (4).

There are at least three ways in which the picture presented in Table 5.4 might change. We could:

Smaller cars will help to lower consumption

1. *Substitute* renewable or more abundant resources such as wood, glass, or stone for metals. This may be possible for at least a few metals such as iron and aluminum that are used mostly for construction.

2. *Recycle* metal products. At present, recycling projects make only a minor contribution to our supplies of most metals. Only about one tenth of the 10 billion aluminum and tin-plated steel cans thrown away each year are recycled. One diffi-culty is that waste metal products are often mixed with materials which are difficult to remove and adversely affect the properties of the metal. The copper used in wiring in automobiles is nearly impossible to remove from molten iron and forms a low strength alloy with it. One solution would be to replace copper with alumi-num, which can be separated easily from steel.

3. *Use lower-grade ores*. The reserves listed in Table 5.4 include only those ores from which metals can now be profitably extracted. In most cases, they repre-sent only a small fraction of the total amount of the metal in the earth's crust.

Bauxite, Al_2O_3, is an extreme example. Less than one billionth (10^{-9}) of the world's aluminum is in the form of that ore. In principle, there is no reason why aluminum could not be extracted from such sources as clay or feldspar. The difficulty is that to do this we need a cheap and essentially limitless source of energy. Here, as in so many other areas, the answer to the problem lies in finding new sources of energy, the ultimate resource.

SUMMARY

We obtain elements from a variety of sources: air, water, and minerals. Most of the elements are found in combined form, as compounds, in nature. A few, including sulfur, gold, and the atmospheric gases, exist naturally in elemental form. This chapter discussed the sources of selected elements and the processes used to extract them from these sources. Methods used to obtain elements from their compounds involve a variety of chemical reactions. At one stage or another, an oxidation-reduction reaction is involved. In many cases, the oxidation-reduction reaction occurs during electrolysis (Na, Mg, Al, Cl_2, F_2). Here, oxidation occurs at the anode, reduction at the cathode.

Throughout this chapter, we have applied principles discussed in Chapters 2 through 4. Several examples are used to review these principles. These include per cent composition (Example 5.1), mass relations in reactions (Examples 5.4, 5.6, and 5.7), and simplest formulas (Example 5.5).

The only new principle introduced in this chapter is that of limiting reactant. The amount(s) of product(s) formed in a reaction is governed by the amount of the least abundant reactant, the limiting reactant. To decide which reactant is limiting we compare the number of moles available to those required by the balanced equation (Example 5.2). The amount of product obtained if all the limiting reactant is consumed is called the theoretical yield (Example 5.3). The actual yield is ordinarily less than the theoretical yield (Example 5.8). In other words, the per cent yield is less than 100.

KEY WORDS AND CONCEPTS

ore	theoretical yield	electrolysis	transition metal
noble gas	actual yield	electrode	post-transition metal
mole	per cent yield	cathode	metallurgy
catalyst	molarity	anode	flotation
limiting reactant			

QUESTIONS AND PROBLEMS

Catalog

General: 5.1, 5.2–5.5, 5.24–5.27, 5.46–5.49
Writing and Balancing Equations: 5.6–5.8, 5.28–5.30
Per Cent Composition and Formulas: 5.9–5.11,
   5.31–5.33
Mole-Gram Relations: 5.12, 5.13, 5.34, 5.35

Molarity: 5.14, 5.15, 5.36, 5.37
Mass Relations in Equations: 5.16–5.18, 5.38–5.40
Limiting Reactant; Theoretical Yield: 5.19–5.23,
   5.41–5.45

5.1 Review and know the meaning of the key words and concepts in this chapter.

5.2 Which of the following elements are extracted from the air? the oceans? the earth's crust?

  a. Fe b. S c. Cl_2 d. N_2
  e. Ar f. Mg g. Cu h. Al

5.3 Describe, in your own words, the processes used to obtain the following elements from their natural sources:

  a. N_2 b. S c. Br_2 d. Cl_2 e. Fe

5.4 Explain why

  a. Al is obtained from bauxite, rather than feldspar or granite.
  b. sulfur is not obtained by strip mining.
  c. Br_2 but not F_2 can be obtained by oxidation with Cl_2.
  d. the sodium and chlorine produced by electrolysis of NaCl must be kept separated.

5.5 Indicate whether each of the following statements is true or false. If it is false, correct it.

  a. If, in a reaction between A and B, we have more moles of A than B, then B must be the limiting reactant.
  b. The percentage of iron in steel is greater than in pig iron.
  c. Pure oxygen is used in making steel because it oxidizes iron to Fe_2O_3.
  d. Oxidation occurs at the cathode, reduction at the anode.

5.6 Write balanced equations for

  a. the reaction of methane with steam.
  b. the electrolysis of water.
  c. the reaction of $Cl_2(g)$ with Br^- ions in water solution.
  d. the electrolysis of molten NaCl.

5.7 Write balanced equations for

  a. the reaction that occurs when Cu_2S is heated in air.
  b. the reaction between Fe_2O_3 and CO.
  c. the decomposition of limestone, $CaCO_3$.
  d. the electrolysis of molten Al_2O_3.

5.8 Write balanced equations for

  a. the oxidation of I^- ions in water solution by bromine.
  b. the reaction of CaO with the P_4O_{10} produced in steel manufacture. (The product is calcium phosphate.)

5.9 Feldspar, $KAlSi_3O_8$, is one of the most abundant minerals.

5.24 Which of the elements in Question 5.2 occur in the elementary form in nature? Which are commonly obtained by electrolysis?

5.25 Describe, in your own words, the processes used to obtain the following elements from their natural sources:

  a. O_2 b. I_2 c. Mg d. Cu e. Al

5.26 Explain why

  a. in the fractional distillation of liquid air, Ar comes off with O_2 rather than N_2.
  b. hydrogen is ordinarily obtained from natural gas rather than water.
  c. Ar occurs in nature in elementary form, but Na and Cl_2 do not.
  d. cryolite is used in the electrolysis of bauxite ore.

5.27 Criticize each of the following statements:

  a. The objective in steel-making is to form chemically pure iron.
  b. When copper(I) sulfide is heated in air, copper(II) oxide is formed.
  c. In the refining of copper, pure copper forms at the anode.
  d. Chloride ions oxidize bromine to bromide ions.

5.28 Write balanced equations for

  a. the reaction of chlorine gas with I^- ions in water solution.
  b. the electrolysis of a water solution containing Cl^- ions.
  c. the reaction of calcium fluoride with sulfuric acid.

5.29 Write balanced equations for

  a. the precipitation of Mg^{2+} ions from seawater.
  b. the reaction of $Mg(OH)_2$ with H^+ ions.
  c. the electrolysis of molten $MgCl_2$.
  d. the oxidation of phosphorus in the manufacture of steel.

5.30 Write balanced equations for

  a. the reaction that occurs upon heating dolomite, $MgCO_3 \cdot CaCO_3$, which is similar to the decomposition of limestone.
  b. the reaction that occurs when cinnabar, HgS, is heated in air (similar to Reaction 5.16).

5.31 Cryolite, Na_3AlF_6, is used in the electrolysis of bauxite.

a. What is the per cent composition of feldspar?
b. What mass of feldspar contains 5.00 g of aluminum?

5.10 Bismuth occurs in nature as a sulfide ore. The mass % of Bi and S are 81.3 and 18.7, in that order. What is the simplest formula of bismuth sulfide?

5.11 Azurite, a deep blue mineral used as a source of copper, contains 55.3% Cu, 6.97% C, 37.1% O, and 0.585% H.

a. What is the simplest formula of azurite?
b. Which of the following formulas corresponds to the composition of azurite?

$CuCO_3 \cdot 2 H_2O$ $CuCO_3 \cdot Cu(OH)_2$
$CuCO_3 \cdot 2 Cu(OH)_2$ $Cu(OH)_2 \cdot 2 CuCO_3$

5.12 Using Table 5.3, find the mass in grams of 1.61 mol of the principal ore of

a. Na   b. K   c. Rb   d. Cs

5.13 Complete the following table for Fe_3O_4, an ore of iron:

mol Fe_3O_4	g Fe_3O_4	g Fe
0.00250	——	——
——	81.2	——
——	——	16.6

5.14 In seawater, the concentration of Cl^- is about 0.53 M. What volume of seawater contains

a. 1.00 mol of Cl^-?   b. 1.00 g of Cl^-?

5.15 Reaction 5.7 is carried out with concentrated H_2SO_4, which is 18.0 M and has a density of 1.84 g/cm³. In 650 cm³ of this reagent, there is

a. —— mol H_2SO_4.   b. —— g H_2SO_4.
c. —— g H_2O.

5.16 Consider Reaction 5.5.

a. How many grams of Cl_2 are needed to make 1.26 mol of I_2?
b. How many grams of Cl_2 are needed to react with 1.00 g of I^- ions?
c. What volume of 0.10 M I^- reacts with 0.62 mol of Cl_2?

5.17 Consider the electrolysis of Al_2O_3 (Eq. 5.14).

a. How many grams of Al are produced from 1.00 kg of Al_2O_3?
b. How many grams of O_2 are formed in (a)?
c. About 15% of the O_2 formed in (a) reacts with the carbon anode to form CO_2. How many moles of CO_2 are produced?

a. What is the per cent composition of cryolite?
b. How many grams of cryolite are required to furnish 5.00 g of Al?

5.32 Manganese occurs in nature as the mineral pyrolusite (63.2% Mn, 36.8% O). What is the simplest formula of pyrolusite?

5.33 The mineral malachite contains 57.5% Cu, 5.43% C, 36.2% O, and 0.914% H.

a. What is the simplest formula of malachite?
b. Which of the formulas in Problem 5.11b corresponds to malachite?

5.34 Give the formulas of the carbonate ores of the following metals and find the number of moles of metal in 1.00 g of the carbonate.

a. Ca   b. Sr   c. Ba

5.35 Complete the following table for gypsum (see Table 5.3):

mol gypsum	mol H_2O	g Ca^{2+}
9.24	——	——
——	0.0750	——
——	——	1.26

5.36 In seawater, the concentration of Na^+ is about 0.46 M. In 1.24 ℓ of seawater, there are

a. —— mol Na^+.   b. —— g Na^+.

5.37 As a laboratory assistant, you are asked to prepare 500 cm³ of an aqueous solution 0.020 M in bromine. The density of liquid bromine is 2.96 g/cm³. How many cubic centimeters of liquid bromine should you use to prepare this solution?

5.38 Consider Reaction 5.4.

a. How many grams of Cl_2 are needed to form 0.0675 mol of Br_2?
b. How many grams of Br_2 are produced from 0.168 mol of Br^- ions?
c. How many grams of chlorine are needed to react with 2.00×10^2 ℓ of a solution 8.0×10^{-4} M in Br^-?

5.39 In the "roasting" of Cu_2S (Eq. 5.16), suppose one starts with 1.00 kg of Cu_2S.

a. How many moles of O_2 are required?
b. What volume of O_2 is required (1 mol O_2 occupies about 24.5 ℓ)?
c. What volume of air is required (refer to Table 5.1)?

5.18 Hydrofluoric acid, HF, cannot be stored in glass containers because it reacts with $CaSiO_3$ in the glass. The unbalanced equation for the reaction is

$$CaSiO_3(s) + HF(l) \rightarrow CaF_2(s) + H_2O(l) + SiF_4(g)$$

a. Balance the equation.
b. 2.40 mol of HF produces how many moles of water?
c. How many grams of HF react with 50.0 g of $CaSiO_3$?
d. How many grams of CaF_2 are formed in (c), assuming 100% yield?

5.19 The Super Widget Manufacturing Company makes widgets, an assembly of one bolt, three washers, and two nuts. Their current inventory is 12,000 washers, 4,000 bolts, and 7,000 nuts. What is the maximum number of widgets that could be produced with the stock on hand?

5.20 In the preparation of Br_2 from Br^- ions (Reaction 5.4), 2.4 mol of $Cl_2(g)$ is added to 6.0 ℓ of a solution 0.50 M in Br^-.

a. What is the limiting reactant?
b. What is the theoretical yield of Br_2, in moles?
c. If 1.24 mol of Br_2 is formed, what is the per cent yield?

5.21 Consider the reaction by which Fe_2O_3 is converted to iron (Eq. 5.19). Suppose 1.00×10^3 ℓ of CO is available to react with 1.00 kg of Fe_2O_3. Taking the density of CO to be 1.14 g/ℓ,

a. what is the limiting reactant?
b. what is the theoretical yield of iron, in grams?
c. if the actual yield of iron is 612 g, what is the per cent yield?

5.22 Hydrogen reacts with sodium to produce solid sodium hydride, NaH. A reaction mixture contains 5.75 g of Na and 3.03 g of hydrogen.

a. Write a balanced equation for NaH formation.
b. Which reactant is limiting?
c. What is the theoretical yield of NaH from the above reaction mixture?
d. What is the per cent yield if 4.00 g of NaH are formed?
e. How many grams of the excess reactant remain unreacted in (d)?

5.40 Small amounts of chlorine can be made by the reaction of MnO_2 with hydrochloric acid. The unbalanced equation for the reaction is

$$MnO_2(s) + Cl^-(aq) + H^+(aq) \rightarrow Mn^{2+}(aq) + Cl_2(g) + H_2O$$

a. Balance the equation.
b. Assuming 100% yield, how many moles of Cl^- must be used to make 5.68 g of chlorine?
c. How many grams of H^+ are required in (b)?
d. What volume of Cl_2 (d = 2.89 g/ℓ) is formed in (b)?

5.41 A 15-page booklet contains 7 green pages, 5 yellow pages, and 3 white pages. The paper stock at the printing company is 40,000 green sheets, 15,000 white sheets, and 20,000 yellow sheets. Can the company fill an order for 5000 booklets from its current stock of pages?

5.42 Consider the reaction between $Mg(OH)_2$ and H^+ ions (Eq. 5.11). If 2.4 ℓ of a solution 6.0 M in H^+ is added to 3.5 mol of solid $Mg(OH)_2$,

a. what is the limiting reactant?
b. what is the theoretical yield of Mg^{2+}, in moles?
c. assuming a 100% yield and a final volume of 2.5 ℓ, what is the concentration of Mg^{2+} in the final solution?

5.43 Consider the reactions by which $CaSiO_3$ slag is formed in the blast furnace (Eqs. 5.20 and 5.21). Suppose one starts with 2.00 g of $CaCO_3$ and 1.00 g of SiO_2.

a. What is the limiting reactant?
b. What is the theoretical yield of $CaSiO_3$, in grams?

5.44 Oxyacetylene torches are used for welding, reaching temperatures near 2000°C. These temperatures are due to the combustion of acetylene, C_2H_2, with oxygen.

$$C_2H_2(g) + O_2(g) \rightarrow CO_2(g) + H_2O(g) \quad \text{(unbalanced)}$$

a. Balance the equation.
b. Starting with 150 g of both C_2H_2 and O_2, which reactant is limiting?
c. What is the theoretical yield of H_2O from this reaction mixture?
d. If 30.5 g of water forms, what is the per cent yield?
e. How many grams of the excess reactant remain unreacted in (d)?

5.23 A student prepares HBr by reacting sodium bromide with phosphoric acid:

$$3 \text{ NaBr(s)} + \text{H}_3\text{PO}_4(l) \rightarrow 3 \text{ HBr(g)} + \text{Na}_3\text{PO}_4(s)$$

She needs 30.0 g of HBr. If NaBr is the limiting reagent and a 40% excess of H_3PO_4 is to be used, how much of each reagent should she weigh out, assuming a yield of

    a. 100%   b. 80%

5.45 A student in the inorganic laboratory wants to make 20 g of the compound $[\text{Co(NH}_3)_5\text{SCN]Cl}_2$ by the reaction

$$[\text{Co(NH}_3)_5\text{Cl]Cl}_2(s) + \text{KSCN(s)} \rightarrow [\text{Co(NH}_3)_5\text{SCN]Cl}_2(s) + \text{KCl(s)}$$

He is told to use a 50% excess of KSCN, and that he can expect to get an 85% yield in the reaction. How many grams of each reactant should he use?

5.46 A certain copper ore contains 6.1% by mass of Cu_2S. When the ore is concentrated by flotation, 20% of the Cu_2S is lost. Of the Cu_2S that remains, 15% is lost in the conversion to copper. How much ore is required to yield 1.00 kg of copper?

5.47 A sample of 1.00 g of Mg(OH)_2 is dissolved in 10.0 cm³ of 6.00 M HCl. Assuming a final volume of 10.5 cm³, what is the concentration of H^+ in the solution formed after reaction?

5.48 Hydrogen gas can be made by a reaction similar to 5.1 except that the starting material is octane, C_8H_{18}, rather than methane. Write a balanced equation for the reaction, and determine the mass % of H_2 in the gaseous H_2–CO mixture formed.

5.49
    The active ingredients of a certain biscuit mix are cream of tartar, $\text{KHC}_4\text{H}_4\text{O}_6$, and baking soda, NaHCO_3. The reaction that occurs when the biscuits "rise" is

$$\text{KHC}_4\text{H}_4\text{O}_6(s) + \text{NaHCO}_3(s) \rightarrow \text{KNaC}_4\text{H}_4\text{O}_6(s) + \text{CO}_2(g) + \text{H}_2\text{O}(l)$$

The mix contains 1.40% by mass of cream of tartar and 0.60% by mass of baking soda. What volume of CO_2 in cubic centimeters is produced per gram of mix? Assume the volume of one mole of CO_2 is about 30 ℓ.

6

THERMOCHEMISTRY _____

Chemical reactions such as those considered in Chapters 4 and 5 are accompanied by energy changes, which may take any of several forms. Consider, for example, the reactions

$$CH_4(g) + 2 O_2(g) \rightarrow CO_2(g) + 2 H_2O(l)$$

$$2 Al_2O_3(s) \rightarrow 4 Al(s) + 3 O_2(g)$$

The first reaction takes place when natural gas, which is mostly methane, burns in air. It supplies the heat required to cook a steak on a gas range or boil water with a Bunsen burner. In contrast, the decomposition of aluminum oxide to the elements requires the absorption of electrical energy (recall Fig. 5.6, p. 92).

In an exothermic reaction, the products are often hot when formed

Throughout most of the chapter, we will be interested in a particular type of energy change. This is the heat flow associated with a chemical reaction. Most familiar reactions are **exothermic;** that is, they evolve heat. An example is the combustion of methane mentioned above. In an exothermic reaction, heat flows from the reaction mixture into the surroundings. Usually, the effect is to increase the temperature of the surroundings.

Reactions which absorb heat are said to be **endothermic.** One example is the phase change observed when ice melts. In an endothermic reaction, heat flows into the reaction mixture from the surroundings. The usual effect, observed when an ice cube melts in a glass of tea, is to lower the temperature of the surroundings. In this case, the tea surrounding the ice becomes cold.

6.1   THE ENTHALPY CHANGE, ΔH

In a general way, we can relate the energy change in a reaction to the difference in energy between products and reactants. If the products have a higher energy than the reactants, we must supply energy to make the reaction proceed. Conversely, if the products are in a lower energy state than the reactants, we can get energy out of the reaction.

To make this argument quantitative, consider reactions taking place at a constant pressure. Most reactions that you carry out in general chemistry meet this condition. The "constant pressure" is that of the atmosphere. Such is the case, for example, when a fuel burns in an open flame. Again, any reaction taking place in a beaker, open test tube, or crucible is under a constant pressure. The heat flow for such reactions is related to an important property of the substances involved. This property is known as "heat content," or *enthalpy,* and is given the symbol H. For any reaction carried out directly at constant pressure, the heat flow is equal to the difference between the enthalpy of the products and that of the reactants. Using Q_p to represent the heat flow at constant pressure and ΔH for the enthalpy difference:

$$Q_p = H_{products} - H_{reactants} = \Delta H \tag{6.1}$$

Equation 6.1 can be regarded as a special case of the *Law of Conservation of Energy*. This law states that energy is neither created nor destroyed in an ordinary chemical reaction. Here we say that the heat flow observed is exactly compensated for by the change in enthalpy of the reaction mixture.

To understand what Equation 6.1 implies, consider a simple case. Suppose a lighted splint is inserted into a test tube containing hydrogen and oxygen. A vigorous exothermic reaction occurs:

$$H_2(g) + \tfrac{1}{2} O_2(g) \rightarrow H_2O(l) \tag{6.2}$$

When this reaction takes place at 25°C and 1 atm, 285.8 kJ of heat is evolved. At the same time, the enthalpy of the reaction system decreases by 285.8 kJ. This means that, under these conditions, the enthalpy of one mole of $H_2O(l)$ is 285.8 kJ less than that of the reactants (1 mol H_2 and $\tfrac{1}{2}$ mol O_2). Hence

$$Q_p = \Delta H = \text{H of 1 mol } H_2O(l) - (\text{H of 1 mol } H_2 + \text{H of } \tfrac{1}{2} \text{ mol } O_2)$$
$$= -285.8 \text{ kJ (at 25°C and 1 atm)}$$

This situation is shown in the enthalpy diagram at the left of Figure 6.1. Each division on the vertical scale corresponds to 40 kJ. Note that the product, $H_2O(l)$, lies 285.8 kJ below the reactants.

Exothermic reactions such as 6.2 are always associated with a *decrease* in enthalpy ($\Delta H < 0$). The enthalpy "lost" by the reaction mixture is the source of the heat evolved. In an *endothermic* reaction there is an *increase* in enthalpy ($\Delta H > 0$). Heat absorbed by the reaction mixture "raises" its enthalpy. An example of an endothermic reaction is that which occurs when mercury(II) oxide is heated in a test tube:

$Q_p < 0$ in exothermic reactions

$$HgO(s) \rightarrow Hg(l) + \tfrac{1}{2} O_2(g) \tag{6.3}$$

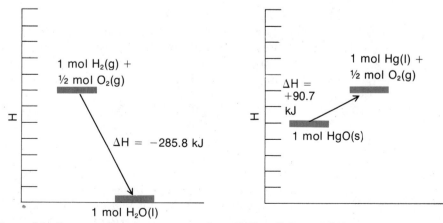

Figure 6.1   For an exothermic process, such as $H_2(g) + \tfrac{1}{2} O_2 \rightarrow H_2O(l)$, $H_{products} < H_{reactants}$, and $\Delta H < 0$. For an endothermic reaction, such as $HgO(s) \rightarrow Hg(l) + \tfrac{1}{2} O_2(g)$, $H_{products} > H_{reactants}$, and $\Delta H > 0$.

We find that 90.7 kJ of heat must be absorbed to decompose one mole of mercury(II) oxide. This means that the enthalpy of the products (1 mol Hg + ½ mol O_2) must be 90.7 kJ greater than that of the reactant, 1 mol of HgO.

$Q_p > 0$ in endothermic reactions

$$Q_p = \Delta H = (\text{H of 1 mol Hg} + \text{H of ½ mol } O_2) - \text{H of 1 mol HgO}$$
$$= +90.7 \text{ kJ (at 25°C and 1 atm)}$$

An enthalpy diagram for this reaction is shown at the right of Figure 6.1. Note that the products for this endothermic reaction lie above the reactant.

In summary we can say that

exothermic reaction: $H_{products} < H_{reactants}$; ΔH is a negative quantity

endothermic reaction: $H_{products} > H_{reactants}$; ΔH is a positive quantity

Nature of Enthalpy

Measurements of heat flow lead only to values of ΔH, the difference in enthalpy between products and reactants. We cannot measure the enthalpy of a single substance. Nevertheless, the enthalpy of a fixed amount of a pure substance, at a given temperature and pressure, is a characteristic property of that substance. In this sense, enthalpy, H, resembles volume, V. One gram of liquid water at 25°C and 1 atm has a definite volume. It also has a fixed, definite enthalpy under these conditions.

Quantities which behave in this way, such as H and V, are referred to as *state properties*. Their values are fixed when the temperature and pressure (the "state" of the substance) are specified. Their magnitude depends only upon the state of the substance and not upon its history. Consider, for example, one gram of $H_2O(l)$ at 25°C and 1 atm. It has the same enthalpy regardless of how the water was formed, whether by melting ice, by condensing steam, by Reaction 6.2, or by some other means.

If we carried some bottled water to the moon and back, its enthalpy would be the same before and after the trip The enthalpy of a substance changes at least slightly when its temperature changes (Fig. 6.2). One gram of liquid water has a somewhat higher enthalpy at 100°C than at 0°C. We have to add 418 J of heat to raise the temperature of one gram of $H_2O(l)$ from 0 to 100°C. This means that its enthalpy is 418 J greater at the higher temperature.

$$\text{H of 1 g } H_2O(l) \text{ at 100°C} - \text{H of 1 g } H_2O(l) \text{ at 0°C} = 418 \text{ J}$$

When a substance undergoes a phase change its enthalpy changes, even though the temperature remains constant. Heat must be absorbed to convert a solid to a liquid or a liquid to a gas. We have to add 333 J of heat to melt a gram of ice at 0°C. At 100°C, 2257 J must be absorbed to boil a gram of liquid water. Hence

$$\text{H of 1 g } H_2O(l) \text{ at 0°C} - \text{H of 1 g } H_2O(s) \text{ at 0°C} = 333 \text{ J}$$

$$\text{H of 1 g } H_2O(g) \text{ at 100°C} - \text{H of 1 g } H_2O(l) \text{ at 100°C} = 2257 \text{ J}$$

The fact that the enthalpy of steam is so much greater than that of liquid water at the same temperature explains why burns from steam are so painful.

Finally, we should point out that enthalpy, like volume, is directly proportional to mass. The enthalpy of 1 mol (18.0 g) of water at a given temperature is 18 times that of 1 g of water. Two moles of water have an enthalpy twice that of one mole, and so on.

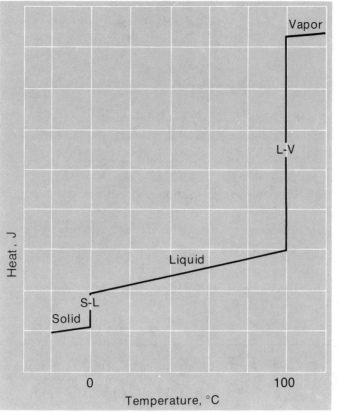

Figure 6.2 Variation with temperature of the enthalpy of one gram of water at 1 atm pressure. Enthalpy of a substance increases with increasing temperature. Note the large increases where phase changes occur; at 0°C, $\Delta H_{fusion} = 333$ J/g and at 100°C, $\Delta H_{vaporization} =$ J/g. (Each unit on the vertical scale is equivalent to 400 joules.)

6.2 THERMOCHEMICAL EQUATIONS

A thermochemical equation is simply a balanced chemical equation with one added piece of information. The heat flow for the reaction is specified. We will do this by listing, at the right of the equation, the value of ΔH in kilojoules. For the two reactions discussed in Section 6.1, we write the thermochemical equations

$$H_2(g) + \tfrac{1}{2} O_2(g) \rightarrow H_2O(l); \quad \Delta H = -285.8 \text{ kJ} \qquad (6.2)$$

$$HgO(s) \rightarrow Hg(l) + \tfrac{1}{2} O_2(g); \quad \Delta H = +90.7 \text{ kJ} \qquad (6.3)$$

There are several points to keep in mind in writing or interpreting equations such as these. In particular:

1. The coefficients always refer to number of moles. Thus -285.8 kJ is ΔH when 1 mol of $H_2O(l)$ is formed from 1 mol of $H_2(g)$ and $\tfrac{1}{2}$ mol of $O_2(g)$.

2. Since enthalpy changes in a phase change, the enthalpy of a substance depends upon whether it is a solid, liquid, or gas. This should be specified by using (s), (l), or (g) in writing the equation. If a species is in water solution, we indicate this by writing (aq) after its formula. The thermochemical equation for the precipitation of silver chloride is

$$Ag^+(aq) + Cl^-(aq) \rightarrow AgCl(s); \quad \Delta H = -65.5 \text{ kJ}$$

We can do thermochemistry on ions as well as on pure substances

3. The enthalpy of a substance depends at least slightly upon temperature.* Hence, we should really specify the temperature at which the reaction is carried out. Unless indicated otherwise, we will take the temperature to be 25°C.

Laws of Thermochemistry

To use thermochemical equations it is necessary to apply certain basic laws of thermochemistry. These laws are expressed most simply in terms of the enthalpy change, ΔH.

1. Since enthalpy is directly proportional to mass, *ΔH is directly proportional to the amount of substance that reacts or is produced in a reaction.* Thus, if we double the coefficients in an equation, the value of ΔH is multiplied by two. For example,

$$H_2(g) + \tfrac{1}{2} O_2(g) \rightarrow H_2O(l); \; \Delta H = -285.8 \text{ kJ}$$

$$2 H_2(g) + O_2(g) \rightarrow 2 H_2O(l); \; \Delta H = -571.6 \text{ kJ}$$

This law can be used to calculate ΔH when amounts of products or reactants are given in grams (Example 6.1).

Example 6.1   Hydrogen peroxide, a common bleaching agent, decomposes as follows:

$$H_2O_2(l) \rightarrow H_2O(l) + \tfrac{1}{2} O_2(g); \; \Delta H = -98.2 \text{ kJ}$$

How much heat is released when one gram of H_2O_2, stored in an open bottle, decomposes?

Solution   From the equation, ΔH for the decomposition of one mole of H_2O_2 is −98.2 kJ. Therefore

$$1 \text{ mol } H_2O_2 = 34.0 \text{ g } H_2O_2 \simeq -98.2 \text{ kJ}$$

Using the conversion factor approach:

$$1.00 \text{ g } H_2O_2 \times \frac{-98.2 \text{ kJ}}{34.0 \text{ g } H_2O_2} = -2.89 \text{ kJ}$$

Since ΔH is −2.89 kJ, we deduce that the decomposition of one gram of hydrogen peroxide gives off 2.89 kJ (2890 J) of heat.

Exercise   What is ΔH when 1.00 g of O_2 is formed by the decomposition of hydrogen peroxide? Answer: −6.14 kJ.

The heat flow is equivalent to, obtained from, 1 mol H_2O_2

2. *ΔH for a reaction is equal in magnitude but opposite in sign to ΔH for the reverse reaction.*

The validity of this law should be evident from Figure 6.1. In the diagram on the right, the enthalpy of HgO lies 90.7 kJ below the sum of the enthalpies of Hg and $\tfrac{1}{2} O_2$.

---

*The enthalpy change for a reaction, ΔH, is much less sensitive to temperature than the individual enthalpies of the substances involved. If the temperature increases, the enthalpies of both products and reactants increase. These effects tend to cancel, often leaving ΔH nearly unchanged over a rather wide temperature range.

Hence, 90.7 kJ must be absorbed to decompose one mole of HgO:

$$HgO(s) \rightarrow Hg(l) + \frac{1}{2} O_2(g); \Delta H = +90.7 \text{ kJ}$$

It's as far up from the lobby to the roof as it is down from the roof to the lobby

If we were to go in the opposite direction, forming one mole of HgO from the elements, 90.7 kJ would be evolved:

$$Hg(l) + \frac{1}{2} O_2(g) \rightarrow HgO(s); \Delta H = -90.7 \text{ kJ}$$

This law is readily applied to phase changes. Since 333 J of heat is absorbed when one gram of ice melts, it follows that 333 J of heat is evolved ($\Delta H = -333$ J) when one gram of water freezes. Again, since 2257 J is absorbed to vaporize one gram of water at 100°C, this same amount of heat is evolved when a gram of steam condenses. This is the source of heat in a steam heating system.

Example 6.2   For the reaction $H_2O_2(l) \rightarrow H_2O(l) + \frac{1}{2} O_2(g)$, $\Delta H = -98.2$ kJ. What is ΔH for the reaction $2 H_2O(l) + O_2(g) \rightarrow 2 H_2O_2(l)$?

Solution   Applying Rules 1 and 2 in succession:

$$2 H_2O_2(l) \rightarrow 2 H_2O(l) + O_2(g); \Delta H = 2(-98.2 \text{ kJ}) = -196.4 \text{ kJ}$$

$$2 H_2O(l) + O_2(g) \rightarrow 2 H_2O_2(l); \Delta H = -(-196.4 \text{ kJ}) = +196.4 \text{ kJ}$$

Exercise   Given that $2 Al_2O_3(s) \rightarrow 4 Al(s) + 3 O_2(g)$; $\Delta H = +3339.6$ kJ. Obtain ΔH for the formation of one mole of Al_2O_3 from the elements. Answer: -1669.8 kJ.

---

3. Since H is a state property, ΔH must be independent of the path followed in a reaction. In particular, *ΔH must be the same regardless of the number of steps involved.* Hence, if

$$\text{Reaction (1)} + \text{Reaction (2)} = \text{Reaction (3)}$$

then $\Delta H_3 = \Delta H_1 + \Delta H_2$.

To illustrate this situation, consider the reaction between tin and chlorine:

$$Sn(s) + 2 Cl_2(g) \rightarrow SnCl_4(l) \tag{6.4}$$

We can imagine that this reaction takes place in two steps. In the first step, one mole of Sn reacts with one mole of Cl_2 to form a mole of $SnCl_2$. Then, in a second step, the $SnCl_2$ reacts with another mole of Cl_2 to form $SnCl_4$:

$$Sn(s) + Cl_2(g) \rightarrow SnCl_2(s); \Delta H_1 = -349.8 \text{ kJ}$$
$$\underline{SnCl_2(s) + Cl_2(g) \rightarrow SnCl_4(l); \Delta H_2 = -195.4 \text{ kJ}}$$
$$Sn(s) + 2 Cl_2(g) \rightarrow SnCl_4(l)$$

Since these two equations add to give Equation 6.4:

$$\Delta H \text{ for Reaction } 6.4 = \Delta H_1 + \Delta H_2$$
$$= -349.8 \text{ kJ} - 195.4 \text{ kJ} = -545.2 \text{ kJ}$$

This situation is shown graphically in Figure 6.3, p. 112.

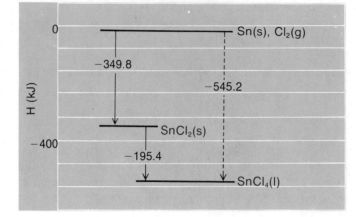

Figure 6.3 ΔH for the reaction $Sn(s) + 2\,Cl_2(g) \rightarrow SnCl_4(l)$ is the same whether it is carried out in one step *(dotted arrow)* or two steps *(solid arrows)*. Each division on the vertical axis represents 80 kJ.

This relationship between enthalpy changes in successive reactions is called **Hess's Law.** In essence, the law states that ΔH for a reaction is the same whether it occurs in a single step or a series of steps. It is useful for determining ΔH for a reaction which is difficult or impossible to carry out in a single step. We will illustrate its use later in this chapter and in succeeding chapters.

6.3   HEATS OF FORMATION

We have now written several thermochemical equations. In each case, we have cited the corresponding value of ΔH. Literally thousands of such equations would be needed to list the ΔH values for all the reactions that have been studied. Clearly we need some more concise way of recording data of this sort. These data should be in a form that can easily be used to calculate ΔH for a reaction in which we are interested. It turns out that there is a simple way to do this, using quantities known as heats of formation. Let us first consider how these quantities are defined. Then we will show how they can be used to calculate ΔH for reactions.

The molar heat of formation of a compound, ΔH_f, is equal to the enthalpy change, ΔH, when one mole of the compound is formed from the elements in their stable forms at 25°C and 1 atm. Thus, from the equations

$$Ag(s) + \tfrac{1}{2}\,Cl_2(g) \rightarrow AgCl(s); \quad \Delta H = -127.0 \text{ kJ} \tag{6.5}$$

$$\tfrac{1}{2}\,N_2(g) + O_2(g) \rightarrow NO_2(g); \quad \Delta H = +33.9 \text{ kJ} \tag{6.6}$$

we conclude that

Using the word heat, state what these equations mean

$$\Delta H_f\ AgCl(s) = -127.0 \text{ kJ}$$

$$\Delta H_f\ NO_2(g) = +33.9 \text{ kJ}$$

Heats of formation are listed for a variety of compounds in Table 6.1. Notice that, with a few exceptions, heats of formation are usually negative quantities. This means that the formation of a compound from the elements is ordinarily exothermic. Conversely, when a compound decomposes to the elements, heat must usually be absorbed.

TABLE 6.1 HEATS OF FORMATION (kJ/mol) AT 25°C AND 1 ATM

AgBr(s)	−99.5	$C_2H_2(g)$	+226.7	$H_2O(l)$	−285.8	$NH_4Cl(s)$	−315.4
AgCl(s)	−127.0	$C_2H_4(g)$	+52.3	$H_2O_2(l)$	−187.6	$NH_4NO_3(s)$	−365.1
AgI(s)	−62.4	$C_2H_6(g)$	−84.7	$H_2S(g)$	−20.1	NO(g)	+90.4
$Ag_2O(s)$	−30.6	$C_3H_8(g)$	−103.8	$H_2SO_4(l)$	−811.3	$NO_2(g)$	+33.9
$Ag_2S(s)$	−31.8	$n\text{-}C_4H_{10}(g)$	−124.7	HgO(s)	−90.7	NiO(s)	−244.3
$Al_2O_3(s)$	−1669.8	$n\text{-}C_5H_{12}(l)$	−173.1	HgS(s)	−58.2	$PbBr_2(s)$	−277.0
$BaCl_2(s)$	−860.1	$C_2H_5OH(l)$	−277.6	KBr(s)	−392.2	$PbCl_2(s)$	−359.2
$BaCO_3(s)$	−1218.8	CoO(s)	−239.3	KCl(s)	−435.9	PbO(s)	−217.9
BaO(s)	−558.1	$Cr_2O_3(s)$	−1128.4	$KClO_3(s)$	−391.4	$PbO_2(s)$	−276.6
$BaSO_4(s)$	−1465.2	CuO(s)	−155.2	KF(s)	−562.6	$Pb_3O_4(s)$	−734.7
$CaCl_2(s)$	−795.0	$Cu_2O(s)$	−166.7	$MgCl_2(s)$	−641.8	$PCl_3(g)$	−306.4
$CaCO_3(s)$	−1207.0	CuS(s)	−48.5	$MgCO_3(s)$	−1113	$PCl_5(g)$	−398.9
CaO(s)	−635.5	$CuSO_4(s)$	−769.9	MgO(s)	−601.8	$SiO_2(s)$	−859.4
$Ca(OH)_2(s)$	−986.6	$Fe_2O_3(s)$	−822.2	$Mg(OH)_2(s)$	−924.7	$SnCl_2(s)$	−349.8
$CaSO_4(s)$	−1432.7	$Fe_3O_4(s)$	−1120.9	$MgSO_4(s)$	−1278.2	$SnCl_4(l)$	−545.2
$CCl_4(l)$	−139.5	HBr(g)	−36.2	MnO(s)	−384.9	SnO(s)	−286.2
$CH_4(g)$	−74.8	HCl(g)	−92.3	$MnO_2(s)$	−519.7	$SnO_2(s)$	−580.7
$CHCl_3(l)$	−131.8	HF(g)	−268.6	NaCl(s)	−411.0	$SO_2(g)$	−296.1
$CH_3OH(l)$	−238.6	HI(g)	+25.9	NaF(s)	−569.0	$SO_3(g)$	−395.2
CO(g)	−110.5	$HNO_3(l)$	−173.2	NaOH(s)	−426.7	ZnO(s)	−348.0
$CO_2(g)$	−393.5	$H_2O(g)$	−241.8	$NH_3(g)$	−46.2	ZnS(s)	−202.9

To show how heats of formation can be used to calculate values of ΔH, consider the reaction

$$SnO_2(s) + 2\ H_2(g) \rightarrow Sn(s) + 2\ H_2O(l) \qquad (6.7)$$

We can imagine that this reaction takes place in two steps. The first of these is

$$SnO_2(s) \rightarrow Sn(s) + O_2(g); \quad \Delta H_a = -\Delta H_f\ SnO_2(s) \qquad (6.7a)$$

The oxygen produced then reacts with hydrogen to form two moles of water:

$$2\ H_2(g) + O_2(g) \rightarrow 2\ H_2O(l); \quad \Delta H_b = 2\ \Delta H_f\ H_2O(l) \qquad (6.7b)$$

Adding Equations 6.7a and 6.7b gives Equation 6.7:

$$SnO_2(s) \rightarrow Sn(s) + \cancel{O_2}(g)$$
$$\underline{2\ H_2(g) + \cancel{O_2}(g) \rightarrow 2\ H_2O(l)}$$
$$SnO_2(s) + 2\ H_2(g) \rightarrow Sn(s) + 2\ H_2O(l)$$

Hence, Hess's Law tells us that the enthalpy change for Reaction 6.7 must be

$$\Delta H = \Delta H_b + \Delta H_a = 2\ \Delta H_f\ H_2O(l) - \Delta H_f\ SnO_2(s)$$

From Table 6.1, ΔH = 2(−285.8) − (−580.7) = +9.1 kJ

We could go through this sort of analysis to relate ΔH for any reaction to the heats of formation of the compounds involved. In practice, it is unnecessary to do this. We need only apply the general rule illustrated by this example:

ΔH for any reaction is equal to the sum of the heats of formation of the product compounds minus the sum of the heats of formation of the reactant compounds. Using the symbol Σ to represent "the sum of" we have

$$\Delta H = \Sigma\ \Delta H_f\ \text{products} - \Sigma\ \Delta H_f\ \text{reactants} \qquad (6.8)$$

Any element in its stable form taking part in the reaction is omitted in taking these sums. Its heat of formation is zero because of the way in which ΔH_f is defined.

The use of Equation 6.8 to calculate ΔH of a reaction from heats of formation is shown in Example 6.3.

Example 6.3   Using Table 6.1, calculate ΔH for the reaction

$$8\ Al(s) + 3\ Fe_3O_4(s) \rightarrow 4\ Al_2O_3(s) + 9\ Fe(s)$$

Solution   Following Equation 6.8, and omitting terms for elements:

$$\Delta H = 4\ \Delta H_f\ Al_2O_3(s) - 3\ \Delta H_f\ Fe_3O_4(s)$$

From Table 6.1:

$$\Delta H = 4(-1669.8\ kJ) - 3(-1120.9\ kJ) = -3316.5\ kJ$$

Exercise   The reaction between aluminum and iron(III) oxide, Fe_2O_3, is strongly exothermic. Referred to as the thermite reaction, it is sometimes used to weld steel rails. Using Table 6.1, calculate ΔH for the reaction: $2\ Al(s) + Fe_2O_3(s) \rightarrow 2\ Fe(s) + Al_2O_3(s)$. Answer: $-847.6\ kJ$.

Heats of Formation of Ions in Solution

It is possible to set up a table, very much like Table 6.1, for heats of formation of ions in water solution. There is, however, one problem. We cannot measure the heat of formation of an individual ion. In any reaction involving ions, at least two of them are present. The principle of electrical neutrality requires that. Consider, for example, the reaction that occurs when HCl is added to water:

$$HCl(g) \rightarrow H^+(aq) + Cl^-(aq)$$

Here, two different ions are formed, H^+ and Cl^- ions. The same situation applies in all other reactions. We cannot carry out a reaction to form only the H^+ ion, with no other ions involved.

To get around this dilemma, the heat of formation of the H^+ ion is arbitrarily taken to be zero.

We need to fix ΔH_f for one ion and we choose H^+, for no really good reason

$$\Delta H_f\ H^+(aq) = 0 \tag{6.9}$$

Having done this, we can establish heats of formation for other ions. Take, for instance, the Cl^- ion. We can measure ΔH for the ionization of HCl in water; it is $-75.1\ kJ/mol$. That is,

$$HCl(g) \rightarrow H^+(aq) + Cl^-(aq);\ \Delta H = -75.1\ kJ \tag{6.10}$$

Applying Equation 6.8:

$$-75.1\ kJ = \Delta H_f\ H^+(aq) + \Delta H_f\ Cl^-(aq) - \Delta H_f\ HCl(g)$$

Taking the heat of formation of H^+ to be zero and that of HCl(g) to be $-92.3\ kJ$ (Table 6.1) we have

TABLE 6.2   HEATS OF FORMATION (kJ/mol) AT 25°C, 1 M

CATIONS					ANIONS			
$Ag^+(aq)$	+105.9	$K^+(aq)$	−251.2		$Br^-(aq)$	−120.9	$H_2PO_4^-(aq)$	−1302.5
$Al^{3+}(aq)$	−524.7	$Li^+(aq)$	−278.5		$Cl^-(aq)$	−167.4	$HPO_4^{2-}(aq)$	−1298.7
$Ba^{2+}(aq)$	−538.4	$Mg^{2+}(aq)$	−462.0		$ClO_3^-(aq)$	−98.3	$I^-(aq)$	−55.9
$Ca^{2+}(aq)$	−543.0	$Mn^{2+}(aq)$	−218.8		$ClO_4^-(aq)$	−131.4	$MnO_4^-(aq)$	−518.4
$Cd^{2+}(aq)$	−72.4	$Na^+(aq)$	−239.7		$CO_3^{2-}(aq)$	−676.3	$NO_3^-(aq)$	−206.6
$Cu^{2+}(aq)$	+64.4	$NH_4^+(aq)$	−132.8		$CrO_4^{2-}(aq)$	−863.2	$OH^-(aq)$	−229.9
$Fe^{2+}(aq)$	−87.9	$Ni^{2+}(aq)$	−64.0		$F^-(aq)$	−329.1	$PO_4^{3-}(aq)$	−1284.1
$Fe^{3+}(aq)$	−47.7	$Pb^{2+}(aq)$	+1.6		$HCO_3^-(aq)$	−691.1	$S^{2-}(aq)$	+41.8
$H^+(aq)$	0.0	$Sn^{2+}(aq)$	−10.0				$SO_4^{2-}(aq)$	−907.5
		$Zn^{2+}(aq)$	−152.4					

$$-75.1 \text{ kJ} = 0 + \Delta H_f \, Cl^-(aq) + 92.3 \text{ kJ}$$

or
$$\Delta H_f \, Cl^-(aq) = -75.1 \text{ kJ} - 92.3 \text{ kJ} = -167.4 \text{ kJ}$$

Heats of formation for other ions in water solution can be established in a similar way. Table 6.2 lists values for several common ions. This table can be used, much like Table 6.1, to calculate ΔH for reactions in solution. To do that, we use the relation

$$\Delta H = \Sigma \, \Delta H_f \text{ products} - \Sigma \, \Delta H_f \text{ reactants}$$

realizing that ΔH_f for $H^+(aq)$ is zero, as is ΔH_f for an element in its stable state.

Example 6.4   Calculate ΔH for the ionization of HBr

$$HBr(g) \rightarrow H^+(aq) + Br^-(aq)$$

using heats of formation.

Solution   For this reaction

$$\Delta H = \Delta H_f \, Br^-(aq) - \Delta H_f \, HBr(g)$$

Taking the heat of formation of HBr from Table 6.1 and that of Br^- from Table 6.2:

$$\Delta H = -120.9 \text{ kJ} - (-36.2 \text{ kJ}) = -84.7 \text{ kJ}$$

Exercise   Determine ΔH for the reaction $Zn(s) + 2 H^+(aq) \rightarrow Zn^{2+}(aq) + H_2(g)$. Answer: −152.4 kJ.

6.4   BOND ENERGIES (ENTHALPIES)

We have seen how tables of heats of formation can be used to obtain ΔH for a variety of reactions. Sometimes ΔH turns out to be a large negative number; the reaction is strongly exothermic. In other cases, ΔH is positive; heat must be absorbed for a reaction to occur. You might well wonder why ΔH should vary so widely from one reaction to another. More to the point, is there some basic property of the molecules involved in the reaction which determines the sign and magnitude of ΔH?

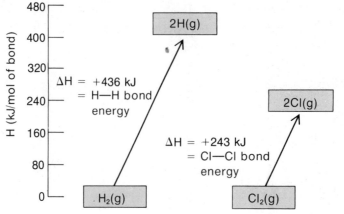

Figure 6.4 The H—H bond energy is greater than the Cl—Cl bond energy (+436 kJ/mol vs. +243 kJ/mol). This means that the bond in H_2 is stronger than that in Cl_2. Bond energies are always positive quantities; heat has to be absorbed to break a bond.

These questions can be answered at a molecular level in terms of a quantity known as bond energy. (More properly, but less commonly, it is called bond enthalpy.) **The bond energy is defined as ΔH when one mole of bonds is broken in the gaseous state.** From the equations (see also Figure 6.4):

Molecules don't fall apart unless they are forced to by addition of energy

$$H_2(g) \rightarrow 2\ H(g);\ \Delta H = +436\ kJ \qquad (6.11)$$

$$Cl_2(g) \rightarrow 2\ Cl(g);\ \Delta H = +243\ kJ \qquad (6.12)$$

we conclude that the H—H bond energy is +436 kJ/mol while that for Cl—Cl is +243 kJ/mol. In Reaction 6.11, 1 mol of H—H bonds in H_2 is broken; in 6.12, 1 mol of Cl—Cl bonds is broken. Notice that in both cases we are dealing with gaseous atoms and molecules. *Only if all reactants and products are gases can ΔH be obtained from bond energies.*

Bond energies for a variety of single bonds are listed in Table 6.3. You will note that bond energy is always a positive quantity. Energy is always absorbed when chemical bonds are broken. Conversely, heat is given off when bonds are formed from gaseous atoms. Thus we have

$$H(g) + Cl(g) \rightarrow HCl(g);\ \Delta H = -(B.E.\ H—Cl) = -431\ kJ$$

$$H(g) + F(g) \rightarrow HF(g);\ \Delta H = -(B.E.\ H—F) = -565\ kJ$$

Here the abbreviation B.E. is used to stand for bond energy (kJ/mol of bonds).

The concept of bond energy helps us to understand why some reactions are exothermic and others endothermic. If the bonds in the product molecules are stronger

TABLE 6.3   SINGLE BOND ENERGIES (kJ/mol) AT 25°C

	H	C	N	O	S	F	Cl	Br	I
H	436	414	389	464	339	565	431	368	297
C		347	293	351	259	485	331	276	238
N			159	222	—	272	201	243	—
O				138	—	184	205	201	201
S					226	285	255	213	—
F						153	255	255	—
Cl							243	218	209
Br								193	180
I									151

than those in the reactants, the products are more stable and have a lower energy than the reactants. Hence, the reaction is exothermic:

The bonds in reactants and products are all at 25°C

$$\text{``weak'' bonds} \rightarrow \text{``strong'' bonds}; \Delta H < 0$$

If the reverse is true, heat has to be absorbed to bring about reaction. The products have a higher energy than the reactants.

$$\text{``strong'' bonds} \rightarrow \text{``weak'' bonds}; \Delta H > 0$$

To cite a specific example, consider the reaction between hydrogen and fluorine:

$$H_2(g) + F_2(g) \rightarrow 2 \; HF(g); \Delta H = -541 \; kJ$$

The fact that this reaction evolves heat implies that the bonds in HF are "stronger" than those in the elements H_2 and F_2. More precisely, the heat evolved when 2 mol of H—F bonds is formed exceeds by 541 kJ that absorbed in breaking 1 mol of H—H and 1 mol of F—F bonds.

In principle, a table of bond energies can be used to calculate ΔH for reactions. To do this, we apply Hess's Law (Example 6.5).

Example 6.5   Using Table 6.3, estimate ΔH for the reaction

$$H_2(g) + Cl_2(g) \rightarrow 2 \; HCl(g)$$

Solution   Let us imagine this reaction as taking place in two steps:
(1) The reactant molecules, H_2 and Cl_2, break down to free atoms

$$H_2(g) + Cl_2(g) \rightarrow 2 \; H(g) + 2 \; Cl(g)$$

(2) The atoms formed in (1) combine to form HCl molecules

$$2 \; H(g) + 2 \; Cl(g) \rightarrow 2 \; HCl(g)$$

In the first step, we *break* a mole of H—H bonds and a mole of Cl—Cl bonds

$$\Delta H_1 = \text{B.E. H—H} + \text{B.E. Cl—Cl} = +436 \; kJ + 243 \; kJ = +679 \; kJ$$

In the second step, *two* moles of H—Cl bonds are *formed:*

$$\Delta H_2 = -2 \; \text{B.E. H—Cl} = -2(431 \; kJ) = -862 \; kJ$$

By Hess's Law: $\Delta H = \Delta H_1 + \Delta H_2 = +679 \; kJ - 862 \; kJ = -183 \; kJ$

Exercise   Using Table 6.3, estimate ΔH for $2 \; BrCl(g) \rightarrow Br_2(g) + Cl_2(g)$. Answer: 0 kJ.

Bond breaking requires energy

Bond making liberates energy

It may have occurred to you that the calculation called for in Example 6.5 could have been carried out very simply using heats of formation. That is, for the reaction

$$H_2(g) + Cl_2(g) \rightarrow 2 \; HCl(g)$$

$\Delta H = 2 \; \Delta H_f \; HCl = 2(-92.3 \; kJ) = -184.6 \; kJ$. Clearly, in this case it was unnecessary to work through bond energies. Consider, however, the reaction

$$CH_4(g) + OH(g) \rightarrow CH_3(g) + H_2O(g) \tag{6.13}$$

which is believed to be one of the steps involved in the combustion of methane. We cannot use Table 6.1 to calculate ΔH since heats of formation are not given for the unstable species OH and CH_3. However, if we look at the reaction from the standpoint of breaking and forming bonds, we see that what is really involved is

the breaking of a C—H bond:

$$
\begin{array}{ccc}
\quad\quad H & & H \\
\quad\quad | & & | \\
H—C—H & \rightarrow & H—C + H \\
\quad\quad | & & | \\
\quad\quad H & & H
\end{array}
\qquad (6.13a)
$$

followed by

the formation of an O—H bond: $H + O—H \rightarrow H—O—H$ (6.13b)

Hence ΔH for Reaction 6.13 must be the sum of the enthalpy changes for 6.13a and 6.13b, both of which can be related to bond energies:

$$\Delta H = \Delta H_a + \Delta H_b = \text{B.E. C—H} - \text{B.E. O—H} = 414\,\text{kJ} - 464\,\text{kJ} = -50\,\text{kJ}$$

In practice, we seldom calculate ΔH from bond energies if heat of formation data are available. The reason is that in most cases it is not possible to assign an exact value to a bond energy. The bond energy varies to some extent with the species in which the bond is found. Consider, for example, the enthalpy changes for the two reactions

$$\text{H—O—H(g)} \rightarrow \text{H(g)} + \text{O—H(g)}; \ \Delta H = +502\,\text{kJ} \qquad (6.14)$$

$$\text{O—H(g)} \rightarrow \text{H(g)} + \text{O(g)}; \ \Delta H = +426\,\text{kJ} \qquad (6.15)$$

ΔH as obtained from bond energies is usually an approximation

The ΔH values are quite different, even though both involve breaking an O—H bond. The bond energy quoted in Table 6.3, +464 kJ/mol, is an average, calculated from ΔH values for many different reactions in which O—H bonds are broken.

6.5 MEASUREMENT OF HEAT FLOW. CALORIMETRY

The amount of heat evolved or absorbed in a reaction is measured in a device called a calorimeter. A simple "coffee-cup" calorimeter commonly used in general chemistry is shown in Figure 6.5. It consists of a Styrofoam cup partially filled with water. A thermometer is lowered through the cover of the cup until its bulb is below the surface of the water.

To show how heat flow is measured with this calorimeter, let us consider a typical experiment. We start by weighing into the cup a known amount of water, let's say 200 g. The temperature of the water is read to be 25.0°C. Now, we carry out a reaction within the calorimeter. As a result of the reaction, the temperature of the water changes. We measure the final temperature to be 31.0°C.

Water "soaks up" the heat given off by the reaction

From these data, we can calculate the amount of heat flowing into the water. To do this, we use the general relation

$$Q = (\text{specific heat}) \times m \times \Delta t \qquad (6.16)$$

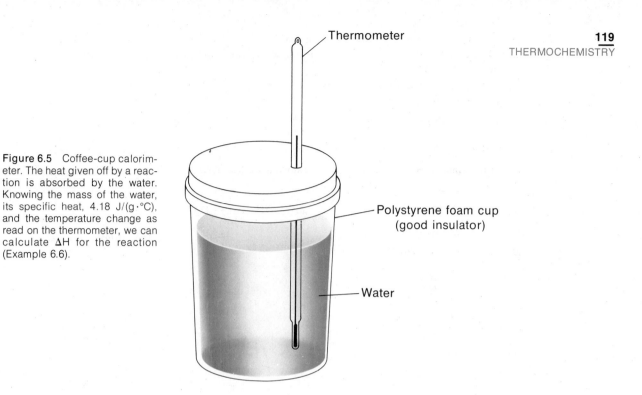

Thermometer

Polystyrene foam cup
(good insulator)

Water

Figure 6.5 Coffee-cup calorimeter. The heat given off by a reaction is absorbed by the water. Knowing the mass of the water, its specific heat, 4.18 J/(g·°C), and the temperature change as read on the thermometer, we can calculate ΔH for the reaction (Example 6.6).

Here, Q stands for heat flow, m is the mass in grams, and Δt is the change in temperature:

$$\Delta t = t_{final} - t_{initial} \tag{6.17}$$

The **specific heat** is the amount of heat required to raise the temperature of one gram of a substance by one degree Celsius. For water, the specific heat is 4.18 J/(g·°C).

Applying Equation 6.16 to the experiment described above:

$$Q_{water} = 4.18 \frac{J}{g \cdot °C} \times 200 \text{ g} \times (31.0°C - 25.0°C)$$

$$= +5.0 \times 10^3 \text{ J}$$

We conclude that in this experiment the water *absorbed* or "gained" 5000 J of heat. Since Styrofoam is a good insulator, there should be no heat flow through the walls of the calorimeter. Hence the 5000 J must have come from the reaction mixture. In other words, the reaction evolved or "lost" 5000 J of heat. Thus

$$\Delta H_{reaction} = -5.0 \times 10^3 \text{ J}$$

In general, for any reaction carried out in a coffee-cup calorimeter, ΔH for the reaction is equal in magnitude but opposite in sign to the heat flow for the water:

$$\Delta H_{reaction} = -(Q_{water}) \tag{6.18}$$

In a good calorimeter there is no heat "leak" to the surroundings

Equations 6.16 to 6.18 can be used to obtain ΔH for any reaction carried out in this simple calorimeter. Note that:

1. For an **exothermic** reaction ($\Delta H < 0$), Q_{water} is positive. The water absorbs the heat given off in the reaction. Hence Δt is positive; the temperature of the water **increases**—that is, $t_{final} > t_{initial}$.

2. For an **endothermic** reaction ($\Delta H > 0$), Q_{water} is negative. The water supplies the heat absorbed by the reaction. Hence Δt is negative; the temperature **decreases** ($t_{final} < t_{initial}$).

Example 6.6   In a coffee-cup calorimeter, the following acid-base reaction is carried out:

$$H^+(aq) + OH^-(aq) \rightarrow H_2O(l)$$

When 0.010 mol of H^+ reacts with 0.010 mol of OH^-, the temperature of 110 g of water rises from 25.0 to 26.2°C. Calculate

    a. Q_{water}.
    b. ΔH for the reaction.
    c. ΔH if 1.00 mol of H^+ reacts with 1.00 mol of OH^-.

Solution

Why not 3?

    a. Using Equation 6.16: $Q_{water} = 4.18 \dfrac{J}{g \cdot °C} \times 110 \text{ g} \times (26.2°C - 25.0°C)$

$$= 550 \text{ J (2 significant figures)}$$

    b. $\Delta H = -(Q_{water}) = -550 \text{ J}$

    c. When 0.010 mol of H^+ (or OH^-) reacts, ΔH is -550 J. Hence

$$0.010 \text{ mol } H^+ \simeq -550 \text{ J}$$

    For 1.00 mol of H^+ (or OH^-)

$$\Delta H = 1.00 \text{ mol } H^+ \times \frac{-550 \text{ J}}{0.010 \text{ mol } H^+} = -5.5 \times 10^4 \text{ J} = -55 \text{ kJ}$$

Exercise   Compare the result obtained in Example 6.6(c) to that calculated using heats of formation (Tables 6.1 and 6.2). Answer: -55.9 kJ.

Bomb Calorimeter

A coffee-cup calorimeter is suitable for measuring heat flows for reactions in solution. However, it cannot be used for reactions involving gases, which would escape from the cup. Neither would it be appropriate for reactions in which the products reach high temperatures. The bomb calorimeter, shown in Figure 6.6, is a more versatile instrument. This type of calorimeter was used to determine most of the enthalpy changes that we have referred to in this chapter.

Gaseous reactants may be pumped into the bomb

To use this instrument, we start by adding a weighed sample of the reactant(s) to the heavy-walled steel container, called a "bomb." This is then sealed and lowered into a metal vessel which fits snugly within the insulating walls of the calorimeter. A weighed amount of water sufficient to cover the bomb is added, and the entire apparatus is closed. The initial temperature is measured precisely. The reaction is then started, perhaps by electrical ignition. In an exothermic reaction, the hot products give off heat to the walls

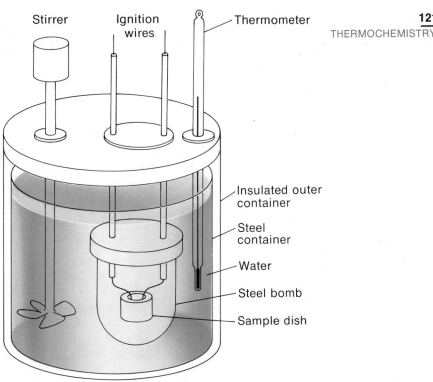

Figure 6.6 Bomb calorimeter. To determine Q for a reaction carried out in this apparatus, we must consider the heat absorbed by the metal parts of the calorimeter as well as by the water. To find Q_{bomb}, a reaction is carried out for which Q is known and Δt is measured.

of the bomb and to the water. The final temperature is taken to be the highest value read on the thermometer.

Analysis of the heat flow here is a bit more complex than in the case of the coffee cup. In particular, Equation 6.18 must be modified. We cannot ignore the heat flow into the metal parts of the bomb calorimeter. Our equation becomes

$$Q_{reaction} = -(Q_{water} + Q_{bomb}) \qquad (6.19)$$

All the heat from the reaction should go to the bomb or the water

Here, $Q_{reaction}$,* Q_{water}, and Q_{bomb} are the heat flows for the reaction, the water, and the bomb, in that order. Q_{water} is obtained as before:

$$Q_{water} = 4.18 \frac{J}{g \cdot °C} \times m_{water} \times \Delta t$$

The value of Q_{bomb} is obtained in a slightly different way. The bomb has a fixed mass and specific heat which remain constant for all the experiments carried out in the calorimeter. The product of the mass of the bomb times its specific heat is referred to as the "calorimeter constant" and is given the symbol C. This constant, which has the units of joules per degree Celsius, is determined by experiment (see Problem 6.14). In terms of C:

$$Q_{bomb} = C \times \Delta t$$

The use of these relations is shown in Example 6.7.

---

*In a bomb calorimeter, the pressure usually changes. Hence the heat flow measured may not be exactly equal to the enthalpy change (heat flow at constant pressure). For this reason, we write "$Q_{reaction}$" instead of $\Delta H_{reaction}$. This point is discussed in more detail in Section 6.7.

Example 6.7   A 1.000-g sample of the rocket fuel hydrazine, N_2H_4, is burned in a bomb calorimeter containing 1200 g of water. The temperature rises from 24.62 to 28.16°C. Taking C for the bomb to be 840 J/°C, calculate

    a. $Q_{reaction}$ for the combustion of the one-gram sample.

    b. $Q_{reaction}$ for the combustion of one mole of hydrazine in the bomb calorimeter.

Solution

    a. Using Equation 6.19:

$$Q_{reaction} = -(Q_{water} + Q_{bomb})$$

$$= -\left(4.18 \frac{J}{g \cdot °C} \times m_{water} \times \Delta t + C \times \Delta t\right)$$

$$= -\left(4.18 \frac{J}{g \cdot °C} \times m_{water} + C\right) \Delta t$$

Note that the mass of the water is 1200 g, C = 840 J/°C, Δt = 28.16°C − 24.62°C = 3.54°C. Hence:

$$Q_{reaction} = -\left(4.18 \frac{J}{g \cdot °C} \times 1200 \text{ g} + 840 \frac{J}{°C}\right)(3.54°C) = -20,700 \text{ J or } -20.7 \text{ kJ}$$

    b. From (a), we see that 20.7 kJ of heat is evolved per gram of hydrazine burned. One mole of hydrazine, N_2H_4, weighs 32.0 g. Hence, for the combustion of one mole of hydrazine,

$$Q_{reaction} = 32.0 \text{ g} \times -20.7 \frac{kJ}{g} = -662 \text{ kJ}$$

Exercise   Repeat the calculations of Example 6.7, ignoring the heat absorbed by the bomb. Answer: −17.8 kJ; −570 kJ.

6.6 SOURCES OF ENERGY

The United States today uses vast amounts of energy, about 8×10^{16} kJ/yr. With only about 6% of the world's population, it accounts for about 35% of worldwide energy usage. Each year, more energy is used than the year before. In the past 30 years, energy usage in the United States has more than doubled.

We can't create energy. We can only convert it from one form to another

Figure 6.7 shows where this energy comes from. As you can see, about 95% is obtained from the combustion of the fossil fuels petroleum, natural gas, and coal. Each of these materials consists largely of hydrocarbons, organic compounds containing the two elements carbon and hydrogen. Natural gas is mostly methane, CH_4. Petroleum is a mixture of liquid hydrocarbons with 5–20 carbon atoms per molecule. Coal is made up of solid hydrocarbons of high molecular mass and a low ratio of hydrogen to carbon atoms.

When hydrocarbons burn in air, they release energy as heat. Typical reactions include

$$CH_4(g) + 2 O_2(g) \rightarrow CO_2(g) + 2 H_2O(l); \Delta H = -890 \text{ kJ} \tag{6.20}$$

$$C_8H_{18}(l) + {}^{25}/_2 O_2(g) \rightarrow 8 CO_2(g) + 9 H_2O(l); \Delta H = -5440 \text{ kJ} \tag{6.21}$$

$$C_{10}H_8(s) + 12 O_2(g) \rightarrow 10 CO_2(g) + 4 H_2O(l); \Delta H = -5155 \text{ kJ} \tag{6.22}$$

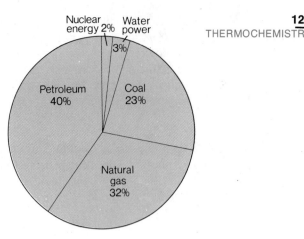

Figure 6.7 The major sources of energy in the United States are petroleum (40%), natural gas (32%), and coal (23%). Water power and nuclear energy make smaller contributions, although they account for about one fifth of electrical energy. Other sources, including solar energy, contribute less than 1% of total energy (as of 1981).

The heat evolved may be used directly to warm our homes and places of business. More commonly, it is converted to mechanical or electrical energy by machines.

Today we face an ''energy crisis'' whose magnitude we are just beginning to grasp. The steadily increasing price of all kinds of fuels is but one symptom of that crisis. Its origin is related to a simple fact: the United States is running out of petroleum and natural gas. We have already used more than half of our potential supplies of these fuels. The production of petroleum in the United States peaked in 1970 and has been decreasing ever since. *This is no small problem*

Until about 1950, all the petroleum used in the United States came from domestic wells. By the time of the oil embargo in 1974, this situation had changed drastically; about 35% of our oil was imported. Today this figure has increased to 45%. For economic and political reasons, our dependence upon foreign oil is, to say the least, disturbing. Moreover, world reserves of petroleum are limited. Their energy equivalent is about 600×10^{16} kJ. Each year, about 10×10^{16} kJ are used. Simple division gives

$$\frac{600 \times 10^{16} \text{ kJ}}{10 \times 10^{16} \text{ kJ/yr}} = 60 \text{ yr}$$

as a time limit for exhaustion of this fuel. Actually, this estimate is an optimistic one. Each year, consumption of petroleum is increasing by 4%. If that growth rate continues, the world will run out of oil in about 30 years.

Clearly, within your lifetime, the United States will have to make some drastic adjustments to meet its energy needs. Sources which now make only a minor contribution, if any at all, will have to be used. In this section, we will look at some of the prospects in this area.

Coal

Reserves of coal in the United States have an energy value 33 times that of petroleum and natural gas. Worldwide, the multiple is about 10. By shifting to coal, we could satisfy our energy needs for at least a century. For some applications, that would be quite easy to do. As late as 1950, nearly all of our electrical energy came from burning coal. Today, nearly half of it comes from petroleum or natural gas. Conversion back to coal could cut our consumption of oil and gas by perhaps 10%.

A solid fuel such as coal cannot be used directly as an energy source for transportation. However, it is possible to convert coal to a gaseous or liquid fuel. Heating coal with steam produces a gaseous mixture containing about 40 mol % H_2, 15% CO, 15% CH_4, and 30% CO_2. This product, called synthesis gas, can be used as a substitute for natural gas. However, it produces only about one third as much heat per unit volume. An alternative is to convert synthesis gas to a liquid fuel. If CO and H_2 are heated with a suitable catalyst, they form liquid hydrocarbons. A typical reaction is

$$8\ CO(g) + 17\ H_2(g) \rightarrow C_8H_{18}(l) + 8\ H_2O(l) \tag{6.23}$$

Good gasoline is now made by this process in South Africa

This process was used in Germany during World War II to make a low-grade gasoline. The product is more expensive than ordinary gasoline. However, as the price of petroleum rises, this process looks more attractive. It could be a major source of liquid fuels 20 years from now.

Nuclear Energy

At the present time, about 10% of our electrical energy comes from a process called *nuclear fission* (Chapter 24). Here, a heavy atom splits into smaller fragments when struck by a neutron. Energy is evolved as heat, which is then converted into electrical energy. In nuclear plants now in operation, the fuel used is the relatively rare isotope of uranium, U-235. A typical fission reaction is

$$^{235}_{92}U + ^{1}_{0}n \rightarrow ^{90}_{38}Sr + ^{144}_{54}Xe + 2\ ^{1}_{0}n; \quad \Delta H = -2 \times 10^{10}\ kJ \tag{6.24}$$

You can get an idea of the enormous amount of energy available from fission by calculating the amount of heat evolved per gram of uranium:

$$\frac{2 \times 10^{10}\ kJ}{235\ g} = 1 \times 10^8\ (100\ million)\ kJ/g$$

This compares to about 46 kJ/g for petroleum and natural gas.

Balanced against this advantage of nuclear fuels is the problem of storing the fission products. These products, such as strontium-90, are dangerously radioactive. They will remain so for many years to come. Moreover, there is always the possibility of a nuclear accident which could release radioactive material to the surroundings. For these reasons, among others, the development of nuclear energy has been much slower than predicted.

Three Mile Island witnessed such an accident in 1979

In many ways, the ideal energy source of the future would be *nuclear fusion* (Chapter 24). Here, light nuclei such as deuterium, $^{2}_{1}H$, and tritium, $^{3}_{1}H$, combine to form heavier nuclei. A typical fusion reaction is

$$^{2}_{1}H + ^{3}_{1}H \rightarrow ^{4}_{2}He + ^{1}_{0}n; \quad \Delta H = -1.7 \times 10^9\ kJ \tag{6.25}$$

This reaction, per gram of fuel, evolves about three to four times as much energy as fission.

$$\frac{1.7 \times 10^9\ kJ}{5.0\ g} = 3.4 \times 10^8\ kJ/g$$

The products of the fusion process are not radioactive. Hence, the safety hazards associated with fission reactors are greatly reduced. Most important, the light isotopes required for fusion are quite common. There are enough of them available to supply all our energy needs for hundreds of years.

Unfortunately, no one up to now has been able to use Reaction 6.25 to generate a sustained flow of energy. For this to happen, a basic problem must be solved. There is a huge electrical repulsion between two small, positively charged hydrogen nuclei (H$^+$ ions). To overcome this repulsion, the nuclei must be accelerated to enormous velocities, corresponding to temperatures of millions of degrees. Maintaining such temperatures requires the development of a whole new technology. It seems unlikely that fusion reactors will contribute to our energy needs before the year 2000, if then.

Solar Energy

The problem of trapping and using solar energy is one that has intrigued scientists for generations (and federal funding agencies since 1974). Each year the earth receives the enormous total of 2×10^{21} kJ of energy from the sun. A tiny fraction of this, through evaporation and condensation of water, supplies us with hydroelectric power. If this fraction could be increased to 0.0001 (1/100 of 1%), it would meet all the world's energy needs.

At the present time, the most practical application of solar energy is to heat water. Solar water heaters are used extensively in Israel, Japan, and Australia. At least half of the fuel used to heat water in the United States could be saved by using solar energy. This would cut fossil fuel consumption by 2%.

Solar energy is also being used on a small scale for space heating. The roof collector shown in Figure 6.8 consists of a shallow metal tray. The base of the tray is painted black to absorb as much sunlight as possible, and the glass cover exerts a "greenhouse" effect. It allows sunlight to penetrate but prevents radiant heat from escaping. Water, passing through the collector several times, is heated to about 65°C. This water passes into a storage tank, from which it is distributed throughout the house as in an ordinary hot water heating system.

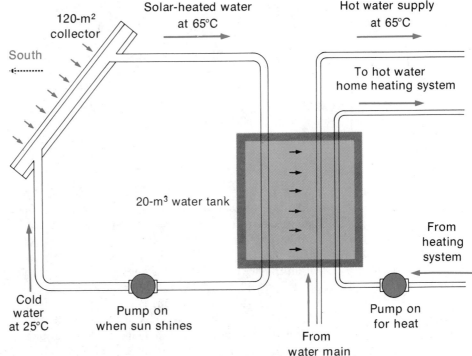

Figure 6.8 A simple solar heating system. In effect, the large insulated storage tank *(shaded box)* acts as a "heat reservoir" to transfer heat from the roof collector to hot water faucets and radiators in the house.

Unfortunately, there are several problems associated with solar heating systems. For one thing, the components tend to break down frequently. Leaks, freeze-ups, and failures due to corrosion are all too common. Many of these difficulties could be solved by circulating air through the collector instead of water. However, this requires a considerably larger storage tank, usually filled with rocks which have a specific heat only 40% that of water.

Example 6.8   In another type of solar heating system, the sun's energy is used to bring about the endothermic reaction

$$Na_2SO_4 \cdot 10\ H_2O(s) \rightarrow 2\ Na^+(aq) + SO_4^{2-}(aq) + 10\ H_2O(l); \Delta H = +78.7\ kJ$$

Upon cooling, the reverse reaction occurs, giving off heat which is used to raise the temperature of water in a storage tank.
   a. How much heat is absorbed when 1.00 g of $Na_2SO_4 \cdot 10\ H_2O$ decomposes?
   b. How many grams of water can be heated from 25.0 to 65.0°C by this amount of heat?

Solution
   a. One mole of this hydrate weighs

$$2(23\ g) + 32\ g + 4(16\ g) + 10(18\ g) = 322\ g$$

Hence: $\Delta H = 1.00\ g \times \dfrac{78.7\ kJ}{322\ g} = 0.244\ kJ = 244\ J$

   b. $Q_{water} = 4.18\ \dfrac{J}{g \cdot °C} \times m_{water} \times \Delta t$

Substituting $Q_{water} = 244\ J$, $\Delta t = 65.0°C - 25.0°C = 40.0°C$, and solving for m_{water},

$$m_{water} = \frac{244}{40.0 \times 4.18}\ g = 1.46\ g$$

One advantage of this system is that the hydrate decomposes at a rather low temperature, 32°C (90°F).

Exercise   In part (b) of Example 6.8, calculate the temperature increase for 1.00 g of water. Answer: 58.4°C.

What we really need is a practical way to store the summer heat for use in the winter. Any ideas?

All solar heating systems have one drawback. The supply of heat is cut off in periods of cold, cloudy weather and during long winter nights. These conditions prevail during the heating season in most of the United States. Hence, an auxiliary heating system of the ordinary type must be available. This factor makes the economics of solar heating marginal at the moment.

Within the next 20 years solar energy may be used in areas other than heating. If this happens, it will probably involve the use of solar cells. These devices use sunlight to produce an electric current. One type of solar cell is shown in Figure 6.9. It consists of two layers of specially treated silicon. When light strikes the cell, about 10% of the solar energy is converted to electrical energy. The major drawback is the high cost of the silicon, which must be very pure. Currently it costs about $60/kg. Other less expensive materials are available but they are less efficient. At present, the cost of electricity from solar cells is five to ten times that from conventional sources.

Figure 6.9 A solar cell converts sunlight directly into electrical energy. Research on solar cells is concentrated upon increasing their efficiency and reducing their cost. As of 1981, electrical energy from solar cells is too expensive for large scale use. (Department of Energy Photo.)

Conserving Petroleum and Natural Gas

Most of the alternative sources of energy that we have discussed are unlikely to have any major impact before the year 2000. There are, perhaps, two exceptions. We might be able to reduce the amount of petroleum and natural gas used to generate electricity by building more plants that use coal or nuclear fission. Otherwise, the only way we can conserve oil and gas over the next 20 years is to use less energy.

All of us are aware of several conservation measures. Turning down the thermostat at night saves fuel. So does setting the thermostat at 65°F (18°C) instead of 72°F (22°C). Driving at 55 mph instead of 65 cuts gasoline consumption by 15%. Walking to work instead of driving reduces it even more; car pooling falls somewhere in between.

Table 6.4 lists some methods of reducing energy usage which may be less obvious. All these are practical from an economic standpoint. None involves any reduction in our standard of living or any major disruption to our way of life. Taken together, they would reduce our energy needs by 24%. This is almost exactly the fraction of our total energy that comes from imported petroleum. By adopting these measures, the United States could become self-sufficient in energy.

TABLE 6.4 CONSERVATION MEASURES		
	SAVINGS (kJ/yr)	SAVINGS AS % OF TOTAL ENERGY USE
Reducing mass of automobiles to an average of one metric ton to decrease gasoline use	0.64×10^{16}	8
Insulation of buildings to meet minimal 1972 standards	0.48×10^{16}	6
Use of process steam in industry to generate electrical energy	0.32×10^{16}	4
Increased efficiency of air conditioning and refrigeration units	0.16×10^{16}	2
Regular maintenance of home heating systems	0.16×10^{16}	2
Replacement of incandescent by fluorescent lighting	0.08×10^{16}	1
Recycling of urban refuse or using it as fuel	$\underline{0.08 \times 10^{16}}$	$\underline{1}$
	1.92×10^{16}	24%

Throughout this chapter, we have mentioned many different kinds of energy changes. However, we have dealt quantitatively with only one type of energy change. This is the change in enthalpy, ΔH. You will recall that ΔH represents the heat flow in a process carried out at constant pressure. For our purposes, the "process" is usually a chemical reaction and the "constant pressure" is that of the atmosphere.

It is possible to develop equations relating all types of energy changes in all types of processes. To do this, we turn to the area of *thermodynamics*. Here, energy effects are arbitrarily divided into two types. Thermodynamics distinguishes between *heat*, given the symbol Q, and *work*, W. In principle, "work" refers to any type of energy other than heat. In chemistry, we ordinarily deal with only two kinds of work. One is mechanical work, such as that done by a gas in expanding against a restraining pressure. The other is electrical work, such as that supplied by a storage battery to electrolyze water.

In any thermodynamic treatment, the *system* which we are studying must be carefully defined. The system may be a reaction mixture whose enthalpy change is to be determined. In another case, it might be a gas undergoing an expansion. The system is separated from its *surroundings* by a boundary, which may be real or imaginary. For practical purposes, the surroundings may be a beaker of water or the air in the laboratory. In principle, however, they include all of the universe outside the system.

With this background, we can state the First Law of Thermodynamics. It expresses in a useful way the Law of Conservation of Energy. The First Law states that:

In any process, the total change in energy of the system, ΔE, is equal to the heat, Q, absorbed by the system minus the work, W, done by the system.

$$\Delta E = Q - W \qquad (6.26)$$

The quantity E in this equation is called the *internal energy*.

In an automobile, chemical energy is converted into work and heat

E and ΔE

Internal energy, like enthalpy, is a state property. That is, its value is fixed when the state of the system is specified. A sample of one mole of $H_2(g)$ at 25°C and 1 atm has a certain definite internal energy. This is the sum of all the kinds of energy possessed by the hydrogen molecules. It would include such things as:

—the bond energy holding the two atoms together in the H_2 molecule.

—the attractive energy between proton and electron in an H atom.

—the kinetic energy (energy of motion) of the H_2 molecule.

As you might guess, the internal energy of a system is not easy to evaluate. In thermodynamics, we make no attempt to do this. Instead, we concentrate upon the change in internal energy, ΔE. This can be determined by measuring Q and W for a process and using Equation 6.26. For example, suppose we find that when a gas expands it absorbs 100 J of heat and does 60 J of work. We conclude that in this process the internal energy of the gas has increased by 40 J:

$$\Delta E = 100 \text{ J} - 60 \text{ J} = 40 \text{ J}$$

For a reaction system, there is a simple way to determine the change in internal energy. This is to carry out the reaction in a bomb calorimeter. Under these conditions, there is no change in volume and no work is done. Hence, in Equation 6.26, $W = 0$, and ΔE is exactly equal to the heat flow, Q. To express this we write

$$\Delta E = Q_v \qquad (6.27)$$

where Q_v represents the heat flow for this constant volume process.

To show the usefulness of Equation 6.27, consider the combustion of methane. When one mole of methane burns in a bomb calorimeter, 885.3 kJ of heat are evolved.

$$CH_4(g) + 2\ O_2(g) \rightarrow CO_2(g) + 2\ H_2O(l); \ Q_v = -885.3\ kJ$$

It follows that ΔE for this reaction is -885.3 kJ. That is, the internal energy of the products (1 mol CO_2 + 2 mol H_2O) is 885.3 kJ less than that of the reactants (1 mol CH_4 + 2 mol O_2).

Internal Energy and Enthalpy

The enthalpy of a substance is related in a simple way to its internal energy. Thermodynamics defines enthalpy, H, to be

$$H = E + PV \qquad (6.28)$$

When H is defined this way, we can show that $\Delta H = Q_p$

Here, E is internal energy, P is pressure, and V is volume. Ordinarily, the PV term in this equation is a small quantity. One mole of a gas at 25°C and 1 atm has a volume of about 24.5 ℓ. This means that its PV product is 24.5 $\ell \cdot$ atm. Using the conversion factor: 1 $\ell \cdot$ atm = 0.1013 kJ, we see that PV for one mole of a gas under these conditions is only 2.5 kJ:

$$PV = 24.5\ \ell \cdot atm \times \frac{0.1013\ kJ}{1\ \ell \cdot atm} = 2.5\ kJ$$

For liquids and solids, PV is even smaller (Table 6.5).

TABLE 6.5 VALUES OF PV AT 25°C		
	$\dfrac{\ell \cdot atm}{mol}$	$\dfrac{kJ}{mol}$
$CH_4(g)$	24.5	2.5
$O_2(g)$	24.5	2.5
$CO_2(g)$	24.5	2.5
$H_2O(l)$	0.018	0.0018

This means that the enthalpy of a substance is very nearly (but not quite) equal to its internal energy. More important, it means that the difference between ΔH and ΔE in a chemical reaction is small. Here

$$\Delta H = \Delta E + \Delta(PV)$$
$$= \Delta E + (PV)_{products} - (PV)_{reactants} \qquad (6.29)$$

For the combustion of methane, using the data in Table 6.5:

$$\Delta(PV) = PV \ CO_2(g) + 2 \ PV \ H_2O(l) - (PV \ CH_4(g) + 2 \ PV \ O_2(g))$$
$$= -5.0 \text{ kJ}$$

Since ΔE is -885.3 kJ:

$$\Delta H = -885.3 \text{ kJ} - 5.0 \text{ kJ} = -890.3 \text{ kJ}$$

As you can see, the difference between ΔH and ΔE is small, often within experimental error. You might wonder why anyone would be concerned with such small differences. In self-defense, we can only say that thermodynamicists have a passion for precision, both in their theories and in their experiments. All things considered, this is a commendable virtue. Saying one thing when you mean something else can get you into all kinds of trouble, as you have probably already discovered.

SUMMARY

The energy change of a reaction can be related to the difference in enthalpy (heat content) between products and reactants. Exothermic reactions evolve heat because the enthalpy of the product(s) is less than that of the reactants; the reverse is true for an endothermic change (Fig. 6.1). In other words

exothermic reaction: $\Delta H < 0$ endothermic reaction: $\Delta H > 0$

The magnitude of ΔH for a reaction depends upon the amounts of substances reacting or formed (Example 6.1). The ΔH for a reaction in one direction is equal in magnitude but opposite in sign to ΔH for the reverse reaction (Example 6.2). Because ΔH is independent of reaction path, enthalpy changes for successive steps can be added to obtain ΔH for the overall reaction (Hess's Law).

The enthalpy change for a reaction can be calculated by using molar heats of formation of compounds or of ions in aqueous solution (Tables 6.1 and 6.2). The relation used is

$$\Delta H = \Sigma \ \Delta H_f \text{ products} - \Sigma \ \Delta H_f \text{ reactants}$$

Here it is understood that the heat of formation of an element in its stable state is zero; ΔH_f of $H^+(aq)$ is also taken to be zero (Examples 6.3 and 6.4).

Energy changes in reactions result from breaking bonds in reactants and making bonds in products. Bond breaking is endothermic; bond making is exothermic. The bond energy (B.E.) refers to the ΔH for the breaking of one mole of bonds in the gas state (Table 6.3). The enthalpy change for a reaction can be estimated from bond energies, using Hess's Law (Example 6.5).

Calorimetry is the study of heat flow. In a coffee-cup calorimeter, the heat evolved by a reaction is absorbed by water. The amount of heat absorbed can be calculated from the relation

$$Q_{water} = 4.18 \frac{J}{g \cdot °C} \times mass_{water} \times \Delta t_{water}$$

(Example 6.6). In a bomb calorimeter, part of the heat evolved in the reaction is absorbed by the metal parts of the calorimeter (Example 6.7).

Examination of our current energy sources (Fig. 6.7) indicates the need for conservation (Table 6.4). We will also need to make more use of energy sources not used on a large scale at this time (Example 6.8).

The First Law of Thermodynamics states that, in any process, the change in internal energy is equal to the difference between the heat absorbed by a system and the work done by the system ($\Delta E = Q - W$). The internal energy, E, like the enthalpy, H, is a state property. Ordinarily ΔH and ΔE for a reaction are nearly equal (Eq. 6.29).

KEY WORDS AND CONCEPTS

exothermic	mole	Hess's Law	calorimeter constant
endothermic	joule	heat of formation	First Law of Thermodynamics
enthalpy	kilojoule	bond energy	system
ΔH	state property	calorimeter	surroundings
Conservation of Energy	thermochemical equation	specific heat	internal energy

QUESTIONS AND PROBLEMS

Catalog

Laws of Thermochemistry: 6.2–6.5, 6.23–6.26
ΔH and Heats of Formation: 6.6–6.10, 6.27–6.31
ΔH and Bond Energies: 6.11, 6.12, 6.32, 6.33
Calorimetry: 6.13–6.15, 6.34–6.36

Sources of Energy: 6.16–6.18, 6.37–6.39
First Law: 6.19, 6.20, 6.40, 6.41
General: 6.1, 6.21, 6.22, 6.42–6.45

6.1   Review and know the meanings of the key words and concepts in this chapter.

6.2   For the reaction

$Al_2O_3(s) \rightarrow 2\ Al(s) + \frac{3}{2}\ O_2(g); \Delta H = +1669.8\ kJ$

   a. Draw a diagram similar to Figure 6.1 for this reaction.
   b. Calculate ΔH when 1.00 g of Al_2O_3 decomposes.
   c. Determine the number of grams of Al_2O_3 that decompose when 1.00×10^2 kJ is absorbed.

6.23   Upon heating, $CaCO_3$ decomposes:

$CaCO_3(s) \rightarrow CaO(s) + CO_2(g); \Delta H = +178.0\ kJ$

   a. Draw a diagram similar to Figure 6.1 for this reaction.
   b. Calculate ΔH when 1.00 g of $CaCO_3$ decomposes.
   c. Determine the number of grams of $CaCO_3$ that decompose when 1.00×10^2 kJ is absorbed.

6.3 Upon heating, 0.750 g of $KClO_3$ decomposes to KCl and $KClO_4$, evolving 262 J of heat.

  a. Write a balanced equation for the reaction.
  b. Calculate ΔH for the decomposition of one mole of $KClO_3$.
  c. What is ΔH for the formation of 0.150 mol of $KClO_4$?

6.4 When glucose combines with O_2, the following reaction occurs:

$$C_6H_{12}O_6(s) + 6\ O_2(g) \rightarrow 6\ CO_2(g) + 6\ H_2O(l);\ \Delta H = -2820\ kJ$$

How many grams of glucose would have to be burned to heat 1.00 kg of water from 20.00 to 30.00°C (specific heat = 4.18 J/g·°C)?

6.5 For the reaction

$$Mg^{2+}(aq) + SO_4^{2-}(aq) \rightarrow MgSO_4(s);\ \Delta H = +91.3\ kJ$$

Calculate ΔH when 1.00 g of $MgSO_4$ dissolves in water.

6.6 Using Table 6.1, calculate ΔH for the following reactions:

  a. $CH_4(g) + 2\ O_2(g) \rightarrow CO_2(g) + 2\ H_2O(l)$
  b. $2\ H_2O_2(l) \rightarrow 2\ H_2O(l) + O_2(g)$
  c. $3\ NO_2(g) + H_2O(l) \rightarrow 2\ HNO_3(l) + NO(g)$

6.7 Use Tables 6.1 and 6.2 to obtain ΔH for the reactions

  a. $Mg(s) + 2\ H^+(aq) \rightarrow Mg^{2+}(aq) + H_2(g)$
  b. $I_2(s) + 2\ Br^-(aq) \rightarrow Br_2(l) + 2\ I^-(aq)$
  c. $NaOH(s) \rightarrow Na^+(aq) + OH^-(aq)$

6.8 Use Table 6.1 to calculate

  a. ΔH for the combustion of one mole of propane, $C_3H_8(g)$, to form gaseous CO_2 and liquid water.
  b. ΔH for the combustion of one mole of propane to form liquid water and gaseous carbon monoxide.

6.9 For the reaction

$$SiO_2(s) + 4\ HF(g) \rightarrow 2\ H_2O(g) + SiF_4(g);\ \Delta H = -97.8\ kJ$$

Using Table 6.1, calculate the heat of formation of $SiF_4(g)$.

6.10 Given:

	ΔH (kJ)
$H^+(aq) + HCO_3^-(aq) \rightarrow CO_2(g) + H_2O(l)$	+11.8
$C(s) + O_2(g) \rightarrow CO_2(g)$	−393.5
$2\ H_2(g) + O_2(g) \rightarrow 2\ H_2O(l)$	−571.6

Calculate the heat of formation of the HCO_3^- ion.

6.24 When magnesium metal and carbon dioxide gas react, 18.4 kJ is evolved per gram of CO_2. The products are solid carbon and magnesium oxide.

  a. Write a balanced equation for the reaction.
  b. Calculate ΔH when 1.50 mol of Mg reacts.
  c. How much heat is evolved per gram of products formed in this reaction?

6.25 When glucose is oxidized in the body, about 40% of the energy evolved in the reaction referred to in Problem 6.4 is available for muscular activity. How much of this type of energy can be obtained from the oxidation of a spoonful of glucose, 3.8 g?

6.26 For the process

$$BaSO_4(s) \rightarrow Ba^{2+}(aq) + SO_4^{2-}(aq);\ \Delta H = +19.3\ kJ$$

Calculate ΔH when 1.00 g of $BaSO_4$ precipitates from water solution.

6.27 Using Table 6.1, calculate ΔH for the following reactions:

  a. $SiO_2(s) + 2\ C(s) \rightarrow Si(s) + 2\ CO(g)$
  b. $2\ C_2H_5OH(l) + 6\ O_2(g) \rightarrow 4\ CO_2(g) + 6\ H_2O(l)$
  c. $4\ NH_3(g) + 5\ O_2(g) \rightarrow 4\ NO(g) + 6\ H_2O(g)$

6.28 Use Tables 6.1 and 6.2 to obtain ΔH for the reactions

  a. $2\ H^+(aq) + CO_3^{2-}(aq) \rightarrow CO_2(g) + H_2O(l)$
  b. $Zn(s) + Cu^{2+}(aq) \rightarrow Zn^{2+}(aq) + Cu(s)$
  c. $Ca^{2+}(aq) + 2\ OH^-(aq) \rightarrow Ca(OH)_2(s)$

6.29 Use Table 6.1 to calculate ΔH for

  a. the combustion of one mole of acetylene gas, $C_2H_2(g)$, to form $CO_2(g)$ and $H_2O(l)$.
  b. the combustion of one mole of acetylene gas to form $CO_2(g)$ and $H_2O(g)$.

6.30 For the reaction

$$4\ PH_3(g) + 8\ O_2(g) \rightarrow P_4O_{10}(s) + 6\ H_2O(g);\ \Delta H = -4500\ kJ$$

The heat of formation of phosphine, PH_3, is +9.2 kJ/mol. Using Table 6.1, calculate the heat of formation of P_4O_{10}.

6.31 Given:

	ΔH (kJ)
$S(s) + {}^3/_2\ O_2(g) \rightarrow SO_3(g)$	−395.2
$2\ SO_2(g) + O_2(g) \rightarrow 2\ SO_3(g)$	−198.2

Calculate the heat of formation of $SO_2(g)$.

6.11 Using Table 6.3, estimate ΔH for the following processes:

a.

$$H—O—H(g) + Cl—O—Cl(g) \rightarrow 2\ H—O—Cl(g)$$

b. the first step in the decomposition of

assuming two NF_2 radicals are formed.

c.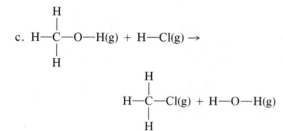

6.12 Using bond energies, estimate the heat of formation of $Cl(g)$.

6.13 When 1.00 g of $KClO_3$ is dissolved in 50.0 g of water in a coffee-cup calorimeter, the temperature drops from 25.00 to 23.36°C. Calculate ΔH for the process

$$KClO_3(s) \rightarrow K^+(aq) + ClO_3^-(aq)$$

Take the specific heat of water to be 4.18 J/g · °C.

6.14 To determine the calorimeter constant of a bomb calorimeter, a student adds 130 g of water at 65.0°C to the bomb which is initially at 25.0°C. The final temperature is 32.0°C. What is C, the calorimeter constant in J/°C?

6.15 A sample of sucrose, $C_{12}H_{22}O_{11}$, weighing 2.92 g is burned in a bomb calorimeter containing 6.00 kg of water (specific heat = 4.18 J/g·°C). The calorimeter constant is 3180 J/°C. The temperature rises from 23.40 to 25.10°C. Calculate Q for the combustion of one mole of sucrose.

6.16 Rank the following energy sources in the order you think they will contribute in the year 2000:

    a. solar energy   b. coal
    c. nuclear fusion   d. petroleum

6.17 About 16% of the total energy used in the United States is accounted for by automobiles. Using Table 6.4, calculate the percentage by which gasoline usage must be reduced to save 0.64×10^{16} kJ/yr.

6.32 Using Table 6.3, estimate ΔH for the following processes:

a. $H—O—O—H(g) \rightarrow 2\ O—H(g)$

b.

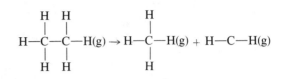

c.

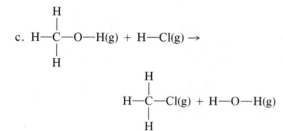

6.33 Using bond energies, estimate the heat of formation of $HF(g)$ and compare to the value given in Table 6.1.

6.34 Consider the reaction

$$Ni(s) + Cu^{2+}(aq) \rightarrow Ni^{2+}(aq) + Cu(s); \Delta H = -128.4\ kJ$$

This reaction is carried out in a coffee-cup calorimeter using 120 g of water. The copper formed weighs 1.00 g. Calculate the temperature change of the water (specific heat = 4.18 J/g · °C).

6.35 To determine the calorimeter constant of a bomb calorimeter, a one-gram sample of benzoic acid is burned in the calorimeter. The temperature of the 2.420 kg of water in the calorimeter rises from 23.10 to 25.09°C. Under these conditions Q for the combustion of benzoic acid is -26.42 kJ/g. What is the calorimeter constant of the bomb in J/°C?

6.36 When 4.20 g of ethyl alcohol, $C_2H_5OH(l)$, is burned in a bomb calorimeter containing 3.50 kg of water, the temperature rises 7.24°C. The calorimeter constant is 2550 J/°C; the specific heat of water is 4.18 J/g·°C. Calculate Q for the combustion of one mole of ethyl alcohol.

6.37 Discuss the advantages and disadvantages of concentrating research efforts on the development of the following energy sources as alternatives to petroleum:

    a. nuclear fusion   b. solar energy
    c. coal   d. nuclear fission

6.38 About 18% of the total energy used in the United States goes to heat buildings. Using Table 6.4, calculate the percentage of this heat which would be saved by insulating all buildings to meet 1972 standards.

6.18 To conserve energy, a home owner resets the hot water heater temperature control from 66°C to 49°C. The heater tank holds 150 ℓ of water (d = 1.00 g/cm³, specific heat = 4.18 J/g · °C).

    a. How much energy can be saved per tankful by this change?
    b. If the water is heated electrically at a rate of 6.0¢ per kilowatt hour, how much money is saved per tankful (1 kW·h = 3.60 × 10³ kJ)?

6.19 Find ΔE for a gas which

    a. absorbs 75 J of heat and does 15 J of work.
    b. evolves 40 J while doing 0.025 kJ of work.

6.20 Consider the reaction

$$C_2H_6(g) + {}^7/_2 O_2(g) \rightarrow 2 CO_2(g) + 3 H_2O(l)$$

    a. Using Table 6.1, calculate ΔH.
    b. For this reaction, $\Delta(PV)$ is −6.2 kJ. How much heat is evolved when this reaction is carried out in a bomb calorimeter?

6.21 Criticize each of the following statements:

    a. The heat of formation of Cl(g) is zero.
    b. ΔH for a reaction can be found by subtracting the bond energies of the reactant bonds from those of the products.
    c. The United States consumes about 8 × 10¹⁶ kJ of energy per year.

6.22 You are asked to calculate the final temperature reached when 22.0 g of CuSO₄ is dissolved in 650 cm³ of water, originally at 40.00°C. To do this, you would need data from three different tables, as well as two physical properties of water. Identify the tables and the physical properties.

6.39 An average home electric dishwasher uses 50 ℓ of hot water per load. If the temperature of the wash water is reduced from 70 to 50°C:

    a. How much energy is saved per month (30 days) if the dishwasher is run once a day?
    b. If the water is heated electrically at 6.0¢ per kilowatt hour, how much money is saved in a month? (1 kW·h = 3.60 × 10³ kJ)

6.40 Determine

    a. ΔE when a system absorbs 65 J of heat and does 60 J of work.
    b. Q when 165 J of work are done on a system and its internal energy increases by 212 J.

6.41 For the combustion of acetylene

$$C_2H_2(g) + {}^5/_2 O_2(g) \rightarrow 2 CO_2(g) + H_2O(l)$$

$\Delta(PV)$ is −3.8 kJ. When the reaction is carried out in a bomb calorimeter, 1296 kJ of heat is evolved. Calculate ΔH for the reaction.

6.42 Criticize each of the following statements:

    a. The most accurate way to calculate ΔH for a reaction is to use bond energies.
    b. Since the heats of formation of both H₂(g) and H⁺ are zero, there is no energy change for the reaction H₂(g) → 2 H⁺(g) + 2 e⁻.
    c. The heat of formation of Hg(g) is zero.

6.43 You are asked to calculate ΔH for the reaction

$$Cl_2(g) + 2 I^-(aq) \rightarrow 2 Cl^-(aq) + I_2(g)$$

To do this, you need one piece of information that is not given in this chapter. What is it?

*6.44 There are about 120 million automobiles in the United States, each driven about 2.0 × 10⁴ km/yr on the average. Fuel economy, on the average, is about 5.0 km/ℓ. The heat of combustion of gasoline is about 48 kJ/g and its density is 0.68 g/cm³.

    a. How much energy is used per year by automobiles?
    b. To reduce energy usage by automobiles by 0.64 × 10¹⁶ kJ/yr, what would have to be the average fuel economy, assuming the other factors stay constant?

*6.45 In walking a kilometer, you use about 100 kJ of energy. This energy comes from the oxidation of foods, which is about 30% efficient. How much energy do you "save" by walking a kilometer instead of driving a car that gets 8.0 km/ℓ of gasoline (density gasoline = 0.68 g/cm³, heat of combustion = 48 kJ/g)?

PHYSICAL BEHAVIOR
OF GASES

In the gas state, all pure substances show remarkably similar physical behavior. For example, consider their molar volumes. At 0°C and 1 atm, one mole of every gas occupies almost exactly the same volume, about 22.4 ℓ. This is true of $O_2(g)$, $N_2(g)$, $CH_4(g)$, and any other gas you care to mention. In contrast, molar volumes of liquids or solids vary widely from one substance to another. The volumes occupied by one mole of $H_2O(l)$, $Br_2(l)$, and $Hg(l)$ at 0°C and 1 atm are 18.0 cm³, 50.1 cm³, and 14.8 cm³, in that order.

The fact that different gases resemble each other so closely in their physical behavior has several interesting results. For one thing it allows us to write simple equations relating volume to pressure, temperature, and amount of gas (Section 7.3). Indeed, we can go one step further and write a single equation relating all four of these variables for all kinds of gases (Section 7.4). This equation, known as the **Ideal Gas Law,** is central to almost everything we will have to say about gases in this chapter. To understand the basis of this equation, it will be helpful to start by looking, in a qualitative way, at some of the common properties of gases.

7.1 GENERAL PROPERTIES OF GASES

When a liquid is heated sufficiently, it begins to evaporate or boil, forming a gas. During this change in phase the particles in the liquid are freed from one another and pass into space as particles of gas. This process is accompanied by a large increase in volume. From half a cupful of liquid water, we get a large enough volume of steam at 100°C and 1 atm to fill a 200-ℓ oil drum. Molecules in the gas phase have the same size as in the liquid. It follows then that the distances between molecules must be much greater in the gas than they are in the liquid.

Since molecules in the gas phase are far apart, it is rather easy to compress a gas, at least as compared to a liquid. In general we find that doubling the pressure reduces the volume of a gas to about half its previous value. Again, if we double the amount (mass) of gas in a container, we find that the pressure doubles. Increasing the temperature of a gas in a closed container increases the pressure of the gas.

Gases can be expanded indefinitely; they occupy their containers completely and uniformly. Suppose you were to release 1 cm³ of $H_2(g)$ at 1 atm into an evacuated 10,000-ℓ container. The gas molecules would diffuse almost instantly to give a constant density and pressure throughout the container. The end result would be the same if there were another gas at 1 atm in the container. However, the process would take much longer, perhaps hours rather than a fraction of a second. The diffusion of the hydrogen molecules would be hindered greatly by collisions with other molecules (see color plate 5).

In a gas at 1 atm, the molecules are about 10 diameters apart

All gases mix readily with one another to form completely uniform solutions. Ordinary air is one such solution. A mixture of gases never settles into two or more regions of different composition. This situation is quite different from that observed with liquids or solids. There, the solubility of one substance in another is usually limited.

7.2 ATMOSPHERIC PRESSURE AND THE BAROMETER

The most familiar gas, and the only one known until about 1750, is the air above us. This gas lies over the earth in a blanket about 80 km thick. Like all earthly matter, the air is subject to the pull of gravity. The air near the earth is compressed by the mass of the air above it. For this reason, the pressure of the atmosphere has its maximum value at sea level. In Denver, Colorado, 1.6 km above sea level, the atmospheric pressure has decreased by about 10%. Above 3 km, the low pressure makes breathing uncomfortable for people not used to such altitudes. For this reason, commercial airplanes have pressurized cabins. Spacecraft, which operate at much higher altitudes, are also pressurized. At a height of 15 km, atmospheric pressure is only about 10% of that at sea level. Uncomfortable is not the word to describe how an unpressurized person would feel at that altitude.

Although the facts of atmospheric pressure are really very simple, they were not clearly understood until about 1650. Torricelli, an Italian scientist, was the first person to measure the pressure of the atmosphere accurately. The device he built to do this is still used today in chemistry and physics laboratories. It is called the mercury barometer (Fig. 7.1). This consists of a closed glass tube filled with mercury and inverted over a pool of mercury. When the tube is first inverted, mercury flows into the reservoir, leaving

The tube needs to be longer than 760 mm

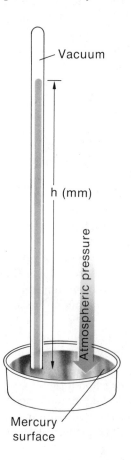

Vacuum

h (mm)

Atmospheric pressure

Mercury surface

Figure 7.1 The barometer. At the level of the lower mercury surface, the pressure both inside and outside the tube must be that of the atmosphere. Inside the tube the pressure is exerted by the mercury column h mm high. Hence, the atmospheric pressure must equal h mm Hg.

a nearly perfect vacuum above the mercury in the tube. The height of the liquid column remaining in the tube is a measure of the atmospheric pressure. It ranges from 740 to 760 mm, varying with both altitude and weather conditions.

Because of the way in which pressure is measured, it is often expressed in millimeters of mercury. Thus, we might say that the atmospheric pressure on a certain day is 752 mm Hg. Many other units are commonly used for pressure. These include:

—the **standard atmosphere,** or simply **atmosphere.** This is the pressure exerted by a column of mercury 760 mm high with the mercury at 0°C.

—pounds per square inch. A mass of 1 lb resting on a surface 1 in² in area exerts a pressure of 1 lb/in².

The force on your body due to atmospheric pressure is enormous

—the *kilopascal,* a metric pressure unit. A mass of 10 g resting on a surface 1 cm² in area exerts a pressure of approximately 1 kPa.

In these units:

$$1 \text{ atm} = 760 \text{ mm Hg} = 14.7 \text{ lb/in}^2 = 101.3 \text{ kPa} \qquad (7.1)$$

Example 7.1 An announcer on a radio station in Montreal reports the atmospheric pressure to be 99.6 kPa. What is the pressure in
  a. atmospheres? b. millimeters of mercury?

Solution Using the conversion factor approach:

a. $99.6 \text{ kPa} \times \dfrac{1 \text{ atm}}{101.3 \text{ kPa}} = 0.983 \text{ atm}$

b. $0.983 \text{ atm} \times \dfrac{760 \text{ mm Hg}}{1 \text{ atm}} = 747 \text{ mm Hg}$

Exercise Express a pressure of 729 mm Hg in atmospheres. Answer: 0.959 atm.

7.3 RELATIONS BETWEEN V AND P, T, OR n

The volume of a sample of gas is related in a simple way to
—the pressure of the gas, P.
—the absolute temperature, T (in K).
—the number of moles of gas, n.
In this section, we will look at each of these relations in turn.

V vs. P (Boyle's Law)

Let us suppose that we trap a sample of air and measure its volume at several different pressures, holding temperature constant. This allows us to determine the relation between volume and pressure, with temperature and amount fixed. When we carry out this experiment we find that when the pressure on the sample is increased, its volume decreases. The product P × V remains constant. In Table 7.1 we show typical data for the compression of a sample of air at 25°C.

We can express the results of this and similar experiments in the equation

$$P V = k_1 \qquad \text{or } V = k_1/P \qquad (\text{constant } n, T) \qquad (7.2)$$

where k_1 is a constant under these conditions.

TABLE 7.1   COMPRESSION OF A SAMPLE OF AIR AT 25°C

PRESSURE (mm Hg)	VOLUME (cm³)	PRESSURE × VOLUME (mm Hg × cm³)
400	100	4.0×10^4
670	60	4.0×10^4
800	50	4.0×10^4
1000	40	4.0×10^4
1200	33	4.0×10^4
1600	25	4.0×10^4

In Figure 7.2, the data in Table 7.1 are plotted. From the plot or from the data, it should be clear that *the volume of a gas sample at constant temperature is inversely proportional to its pressure.* For example, when the pressure is doubled (from 400 to 800 mm Hg), the volume decreases to one half its original value (100 cm³ to 50 cm³). This relationship is referred to as Boyle's Law, after Robert Boyle, who discovered it in 1660.

Robert Boyle was one of the first experimental scientists. In addition to his discovery of the law which bears his name, he made many contributions to scientific thought. One of the most important of these was a book entitled *The Sceptical Chemist* in which he challenged, on the basis of his experiments, the then prevailing notion that salt, sulfur, and mercury were the true principles of nature. In this regard he proposed that matter was ultimately composed of particles of various sorts which could arrange themselves into groups, and that groups of one kind constituted a chemical substance. Boyle thus used concepts of atomic and molecular theory similar to those we have today, and by some has been called one of the fathers of modern chemistry. Since his ideas on chemistry had of necessity to be extremely primitive and were not in any sense quantitative, we would prefer to consider the first modern chemists to be men like Dalton, Lavoisier, and Priestley, who actually performed experiments to establish fundamental chemical relationships.

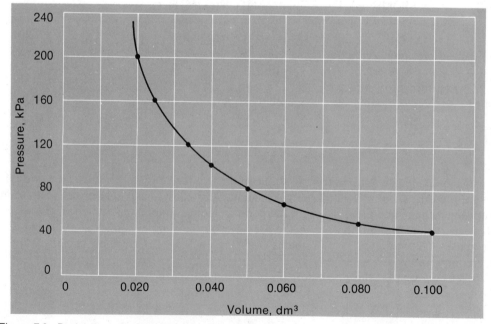

Figure 7.2   Boyle's Law. The pressure of a gas sample, maintained at constant temperature, is inversely related to its volume. Hence the product of P × V is constant.

V vs. T (Law of Charles and Gay-Lussac)

139

PHYSICAL
BEHAVIOR
OF GASES

We can readily extend the approach just described to find the relation between another pair of variables. This time let us see how the volume of a gas sample changes as it is heated at constant pressure. Here, V and T are the variables; n and P are fixed. If we study a sample of hydrogen under these conditions, we obtain data like that in Table 7.2.

TABLE 7.2   HEATING A SAMPLE OF HYDROGEN AT 1 ATM PRESSURE

VOLUME (cm³)	T (K)	t(°C)	V/T (cm³/K)
75	100	−173	0.75
150	200	−73	0.75
225	300	27	0.75
300	400	127	0.75
375	500	227	0.75

In this kind of experiment, we find that *the volume of a gas is directly proportional to the absolute temperature*. It doubles every time the Kelvin temperature doubles. This relation may be expressed as

$$V = k_2T \qquad \text{or } V/T = k_2 \qquad \text{(constant n, P)} \qquad (7.3)$$

where k_2 is a constant throughout the experiment. In Figure 7.3 we have plotted the data in Table 7.2. As you can see, the plot is a straight line passing through the origin (V = 0, T = 0). This is characteristic of a direct proportionality ($V = k_2T$).

The experiments upon which Equation 7.3 is based were first carried out by two French scientists, Charles and Gay-Lussac. Around 1800 they studied the effect of

Figure 7.3   Law of Charles and Gay-Lussac. The volume of a gas sample, maintained at constant pressure, is directly proportional to the absolute temperature.

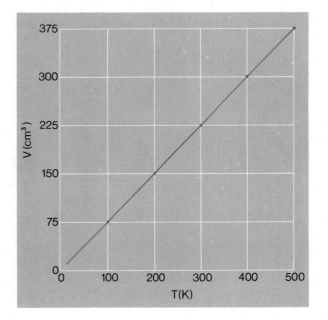

temperature upon the volume of a gas sample maintained at constant pressure. Actually, Charles and Gay-Lussac did not set out to discover this fundamental law that bears their names. Their interest was in the use of hot or light gases in lighter-than-air balloons. Charles was on the second balloon ever to lift a human being off the face of the earth. This happened on December 1, 1783. In 1804, Gay-Lussac made a solo flight in a balloon to a height of 7 km, which set an altitude record that lasted 50 years.

At the time of Charles and Gay-Lussac, the absolute temperature scale was unknown. Their results were expressed in terms of the Celsius scale. The volume of a gas sample at constant pressure is a linear (straight-line) function of the Celsius temperature, t:

$$V = a + bt \qquad \text{(at constant P, n)}$$

This is the relation that Charles and Gay-Lussac discovered

In this equation, a is the "y intercept," the value of V when t = 0. By the same token, b is the slope of the straight line obtained by plotting V vs. t.

As you can see from Figure 7.4, the "constants" a and b depend upon the pressure at which V and t are measured. (They also depend upon the amount of gas in the sample.) However, if the several straight lines in Figure 7.4 are extrapolated to V = 0, they intersect at a common point. From very accurate measurements with gases at low pressures, the temperature at this point is found to be -273.15 K. If then we define the absolute temperature, T, to be

$$T = t + 273.15°$$

Equation 7.3 results. That is, V is directly proportional to T; a plot of V vs. T passes through the origin. This argument was first presented by Lord Kelvin in 1848. In effect what he did was to use the Law of Charles and Gay-Lussac to establish the absolute scale of temperatures we use today.

One thing that might bother you, if you worry about matters of this sort, is the extrapolation to zero volume in Figures 7.3 and 7.4. Clearly this cannot go all the way; all gases condense to liquids before they reach 0 K. However, this is not a real objection to Kelvin's argument, since the absolute zero is determined from data at much higher temperatures, such as 0 and 100°C. Here, intermolecular attractions that cause condensation can be neglected. It is true, though, that one cannot attain exactly the absolute zero of temperature. So far, in laboratory experiments, the best we have been able to do is to get down to about 0.0001 K.

V vs. n

Thus far we have dealt with the behavior of samples containing a fixed amount of gas. We can readily go one step further, relating volume, V, to the number of moles of

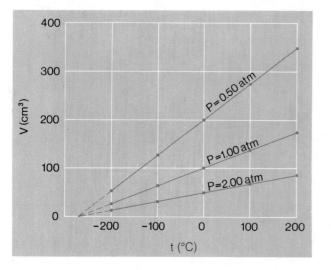

Figure 7.4 This graph shows a plot of V vs. t (°C) for three different gas samples. The value of V at a given temperature depends upon the pressure and amount of gas. However, all the straight lines extrapolate to give a volume of zero when t = -273°C = 0 K. Hence V must be directly proportional to the Kelvin temperature, T, where T = t + 273.

gas, n, at constant temperature and pressure. As you might expect, *volume is directly proportional to the number of moles of gas*. A sample containing two moles of gas has a volume twice that of one mole at the same T and P. The general relation is

$$V = k_3 n \qquad \text{(constant P, T)} \qquad (7.4)$$

where k_3 is a constant under these conditions.

The relationship between volume and amount is often expressed in a more fundamental way. We can say that:

Equal volumes of different gases at the same temperature and pressure contain the same number of moles.

In other words

$n_A = n_B$ for two gases A and B occupying the same volume at P and T

This relation was first suggested by Amadeus Avogadro in 1813. (He referred to the "number of molecules" rather than the "number of moles.") Equation 7.4 can be regarded as a modern form of Avogadro's Law.

Avogadro's Law, if you think about it for a while, offers a simple method to determine the relative masses of molecules. Suppose, for example, we compare equal volumes, let us say one liter, of O_2 and H_2 at 25°C and 1 atm. According to Avogadro's Law, these two samples should contain the same number of molecules. Therefore, the masses of the two samples must be in the same ratio as the molecular masses of O_2 and H_2. One liter of O_2 at 25°C and 1 atm weighs 1.31 g. Under the same conditions, one liter of H_2 weighs 0.0824 g. Hence, an O_2 molecule must weigh

$$\frac{1.31}{0.0824} = 15.9$$

or about 16 times as much as an H_2 molecule.

This argument, which seems so obvious today, was not accepted in Avogadro's time. Avogadro's ideas lay fallow for nearly 50 years. They were revived in 1860 by a fellow countryman, Stanislao Cannizzaro, professor of chemistry at Genoa. He showed that Avogadro's Law could be used to determine not only molecular masses but also, indirectly, atomic masses. Cannizzaro had more impact than Avogadro, perhaps because he presented his ideas more clearly. He made a major contribution to the development of the atomic mass scale discussed in Chapter 2.

Can you suggest how Cannizzaro found atomic masses by this approach?

7.4 THE IDEAL GAS LAW

In Section 7.3 we developed several equations involving the volumes of gases. These equations related volume to pressure, temperature, and number of moles of gas. They can be combined into a single equation relating the four variables P, V, n, and T. This equation is known as the Ideal Gas Law. It is usually written in the form

$$P V = n R T \qquad (7.5)$$

Here, P = pressure, V = volume, n = number of moles, and T = temperature. The quantity R appearing in the Ideal Gas Law is a true constant. It has the same value for all gases and is independent of P, V, n, or T.

We can readily show that the Ideal Gas Law combines the several equations written in Section 7.3. Suppose, for example, we hold n and T constant, as Boyle did. The entire right side of Equation 7.5 will then have a constant value (remember that R is constant). We might then set "nRT" equal to a constant, k_1, obtaining

$$P V = k_1 \text{ (n, T constant), which is Boyle's Law}$$

Similarly, if we hold n and P constant, we can rewrite Equation 7.5 as

If you know the Ideal
Gas Law, you can find
all of the related laws

$$V/T = nR/P = k_2 \text{ (at n, P constant), which is Charles' Law}$$

Thirdly, holding P and T constant, and again rewriting Equation 7.5, we get

$$V/n = RT/P = k_3 \text{ (at P, T constant), which is Equation 7.4}$$

Lastly, considering samples of gases A and B, each at the same volume, temperature, and pressure, we obtain for these samples,

$$P V = n_A R T \qquad \text{and} \qquad P V = n_B R T$$

Since P, V, R, and T have the same values in the two equations, it follows that

$$n_A = n_B, \text{ which is Avogadro's Law}$$

Evaluation of R, the Gas Constant

To determine the value of R, we need only establish by experiment one set of values for P, V, n, and T in Equation 7.5. This is readily done. Consider, for example gaseous oxygen at 0°C and 1.00 atm. These conditions are often referred to as *standard temperature and pressure* (**STP**). We find that at STP, 32.0 g (1.00 mol) of O_2 occupies a volume of 22.4 ℓ. Solving the Ideal Gas Law for R:

$$R = \frac{P V}{n T}$$

Could we find R using
$H_2(g)$?

Substituting P = 1.00 atm, V = 22.4 ℓ, n = 1.00 mol, T = 0 + 273 = 273 K

$$R = \frac{1.00 \text{ atm} \times 22.4 \ell}{1.00 \text{ mol} \times 273 \text{ K}} = 0.0821 \ \ell \cdot \text{atm}/(\text{mol} \cdot \text{K})$$

The value of R obtained under the most precise conditions, using oxygen at low pressures, is 0.082056 $\ell \cdot$atm/(mol$\cdot$K). Note that R involves the units of atmospheres, liters, moles, and K. These units must be used for pressure, volume, amount, and temperature in any problem where this value of R is employed.

In most of our work in this chapter, we will use 0.0821 $\ell \cdot$atm/(mol$\cdot$K) as the value of R. For certain purposes, however, we will need R in other units. Table 7.3 lists values of R in various sets of units.

TABLE 7.3 VALUES OF R IN DIFFERENT UNITS		
VALUE	**WHERE USED**	**HOW OBTAINED**
$0.0821 \dfrac{\ell \cdot \text{atm}}{\text{mol} \cdot \text{K}}$	Gas Law problems with V in liters, P in atm	From known values of P, V, T, n
$8.31 \dfrac{\ell \cdot \text{kPa}}{\text{mol} \cdot \text{K}}$	Gas Law problems with V in liters, P in kPa	1 atm = 101.3 kPa
$8.31 \dfrac{\text{J}}{\text{mol} \cdot \text{K}}$	Equations involving energy in joules	1 $\ell \cdot$atm = 101.3 J
$8.31 \times 10^7 \dfrac{\text{g} \cdot \text{cm}^2}{\text{s}^2 \cdot \text{mol} \cdot \text{K}}$	Calculation of average speed of molecules (Section 7.9)	$1 \text{ J} = 10^7 \dfrac{\text{g} \cdot \text{cm}^2}{\text{s}^2}$

The Ideal Gas Law can be used to solve a variety of problems dealing with the physical behavior of gases. It can, for example, be used to determine the effect of a change in conditions upon a particular variable (Examples 7.2 and 7.3).

Example 7.2   In a certain experiment a sample of helium was compressed at 25°C from a volume of 200 cm³ to a volume of 0.240 cm³, where its pressure was found to be 3.00 cm Hg. What was the original pressure of the helium?

Solution   For the gas in both states, PV = nRT

Initially:
$$P_1 = ?, \ V_1 = 200 \text{ cm}^3, \ n_1 = n, \ T_1 = T$$

$$P_1V_1 = nRT$$

Finally:
$$P_2 = 3.00 \text{ cm Hg}, \ V_2 = 0.240 \text{ cm}^3, \ n_2 = n, \ T_2 = T$$

$$P_2V_2 = nRT$$

Since both the temperature and amount of gas are held constant in this problem, the right sides of the two final equations are equal, and so

$$P_1V_1 = P_2V_2 \qquad P_1 = \frac{P_2V_2}{V_1}$$

$$P_1 = \frac{3.00 \text{ cm Hg} \times 0.240 \text{ cm}^3}{200 \text{ cm}^3} = 3.60 \times 10^{-3} \text{ cm Hg}$$

This problem is typical of many encountered with gases. The state of the gas is changed, with one or more of its variables remaining fixed; in this case, at constant n and T, we have essentially a Boyle's Law problem. By writing the Gas Law for the two states, one can usually quickly recognize which terms in the two equations are equal and use that information and the given data to find the unknown quantity. Note that in the problem here the units of volume (cm³) must be consistent if they are to cancel properly. The pressure is obtained in cm Hg, but can be readily converted to mm Hg or to atm if necessary:

$$3.60 \times 10^{-3} \text{ cm Hg} \times \frac{10 \text{ mm}}{1 \text{ cm}} = 3.60 \times 10^{-2} \text{ mm Hg}$$

$$3.60 \times 10^{-3} \text{ cm Hg} \times \frac{1 \text{ atm}}{76.0 \text{ cm Hg}} = 4.74 \times 10^{-5} \text{ atm}$$

Exercise   A gas with a volume of 20.0 cm³ at 1.00 atm expands to 50.0 cm³ at constant T. What is the final pressure? Answer: 0.400 atm.

> In working gas law problems it is helpful to write down the value of the known variables in the initial and final states

---

Example 7.3   A hydrogen gas volume thermometer has a volume of 100.0 cm³ when immersed in an ice-water bath at 0°C. When immersed in boiling liquid chlorine, the volume of the hydrogen at the same pressure is 87.2 cm³. Find the temperature of the boiling point of chlorine in K and °C.

We don't need to know
P, but do need to know
P is constant

Solution   For the hydrogen in the two states, PV = nRT

Initially: $P_1 = P$ $V_1 = 100.0 \text{ cm}^3$ $n_1 = n$ $T_1 = 0 + 273 = 273$ K

$$PV_1 = nRT_1$$

Finally: $P_2 = P$ $V_2 = 87.2 \text{ cm}^3$ $n_2 = n$ $T_2 = ?$

$$PV_2 = nRT_2$$

In this problem, P, n, and R are the same in the two states; we collect those variables on the left side of the two equations, obtaining

$$\frac{P}{nR} = \frac{T_1}{V_1} = \frac{T_2}{V_2}$$

Therefore $T_2 = \frac{V_2 T_1}{V_1}$

$$T_2 = \frac{87.2 \text{ cm}^3 \times 273 \text{ K}}{100.0 \text{ cm}^3} = 238 \text{ K}   t_2 = T_2 - 273 = -35°C$$

Exercise   A stoppered flask full of air at 20°C is heated until the pressure is doubled. What is the final temperature in °C? Answer: 313°C.

In Examples 7.2 and 7.3 the value of the gas constant R was not needed, since R was canceled from the calculations. In a different type of problem, we use R to calculate one of the four quantities, P, V, n, or T, knowing the values of the other three. Example 7.4 illustrates the kind of calculation involved. Later in this chapter and in succeeding chapters we will deal with other examples of this type.

Example 7.4   2.50 g of XeF_4 gas is introduced into an evacuated 3.00-ℓ container at 80°C. Find the pressure in atmospheres in the container.

Solution   In this problem only one state is involved and, for it, PV = nRT.

$$P = ?; V = 3.00 \; \ell; n = 2.50 \text{ g } XeF_4 \times \frac{1 \text{ mol}}{207.3 \text{ g } XeF_4} = 0.0121 \text{ mol}$$

$$R = 0.0821 \; \ell \cdot \text{atm}/(\text{mol} \cdot \text{K}); T = 273 + 80 = 353 \text{ K}$$

This example and those that follow show the real power of the Ideal Gas Law

Substituting: $P = \dfrac{nRT}{V} = \dfrac{0.0121 \text{ mol} \times 0.0821 \dfrac{\ell \cdot \text{atm}}{\text{mol} \cdot \text{K}} \times 353 \text{ K}}{3.00 \; \ell} = 0.117$ atm

Here all the elements in the Ideal Gas Law enter the calculation directly; if we use 0.0821 $\ell \cdot \text{atm}/(\text{mol} \cdot \text{K})$ for R, the units of all the terms are of necessity those that appear in R, and any quantities which do not have those units must be converted before substituting in the Gas Law. Here, for example, we converted the number of grams of XeF_4 to the number of moles by dividing by the molar mass, 207.3 g.

Exercise   What is the pressure exerted by 2.0 mol of O_2 in a 10.0-ℓ flask at 27°C? Answer: 4.9 atm.

For certain problems, it is convenient to rewrite the Ideal Gas Law in a different form. In particular, we need an equation relating P, V, and T to the mass of a gas rather than the number of moles. To obtain such a relation we note that

$$n = \frac{g}{GMM}$$

where g = mass of gas in grams. The quantity GMM appearing in this relation is the molar mass in grams per mole. Thus for O_2 the molar mass would be 32.0 g/mol; for N_2, GMM = 28.0 g/mol, and so on. Substituting for n in the Ideal Gas Law, we have

$$PV = \frac{gRT}{GMM} \qquad (7.6)$$

The Ideal Gas Law in this form is useful for calculating:
1. The molar mass (GMM) of a gas, knowing the mass (g) of a given volume (V) at a certain temperature (T) and pressure (P). Solving Equation 7.6 for GMM:

$$GMM = \frac{gRT}{PV} \qquad (7.7)$$

2. The density, d, of a gas of known molar mass (GMM) at a given temperature, T, and pressure, P. From Equation 7.6, we obtain, on solving for the density,

$$d = \frac{g}{V} = \frac{P \times GMM}{RT} \qquad (7.8)$$

Example 7.5   A sample of chloroform weighing 0.495 g is collected as a vapor (gas) in a flask with a volume of 127 cm³. At 98°C the pressure of the vapor in the flask is 754 mm Hg. Calculate the molar mass of chloroform.

Solution   We can substitute directly into Equation 7.7.

g = 0.495 g; R = 0.0821 $\ell \cdot atm/(mol \cdot K)$; T = 98 + 273 = 371 K

P = 754 mm Hg $\times \dfrac{1\ atm}{760\ mm\ Hg}$ = 0.992 atm; V = 127 cm³ $\times \dfrac{1\ \ell}{1000\ cm^3}$ = 0.127 ℓ

The data in this problem are easy to obtain experimentally

GMM = $\dfrac{0.495\ g \times 0.0821\ \ell \cdot atm/(mol \cdot K) \times 371\ K}{0.992\ atm \times 0.127\ \ell}$ = 120 g/mol

---

Exercise   What is the molar mass of a gas if 1.00 g occupies a volume of 2.00×10^2 cm³ at 25°C and 1.00 atm? Answer: 122 g/mol.

Example 7.6   Uranium hexafluoride, UF_6, is perhaps the most dense of all gases. What is the density of UF_6 at 100°C and 1.00 atm?

If we know P and T for
any given gas, we can
find its density

Solution   Substituting in Equation 7.8:

$$P = 1.00 \text{ atm; GMM UF}_6 = (238 + 6 \times 19) \text{ g/mol} = 352 \text{ g/mol}$$

$$R = 0.0821 \text{ } \ell \cdot \text{atm/(mol} \cdot \text{K); } T = 100 + 273 = 373 \text{ K}$$

$$d = \frac{352 \text{ g/mol} \times 1.00 \text{ atm}}{0.0821 \text{ } \ell \cdot \text{atm/(mol} \cdot \text{K)} \times 373 \text{ K}} = 11.5 \text{ g/}\ell$$

Exercise   What is the density of $O_2(g)$ at 1.00 atm and 27°C? Answer: 1.30 g/ℓ.

7.6   VOLUMES OF GASES INVOLVED IN REACTIONS

We saw in Chapter 4 that a balanced equation can be used to relate the masses of substances taking part in a reaction. Where gases are involved, it is possible to extend these relations to include volumes. The procedure is shown in Example 7.7.

Example 7.7   How many liters of pure oxygen, measured at 740 mm Hg and 24°C, would be required to burn 1.00 g of octane, $C_8H_{18}(l)$, to carbon dioxide and water?

Solution   We must first write the balanced equation for the reaction, using the approach described in Chapter 4:

$$C_8H_{18}(l) + {}^{25}\!/_2 \text{ } O_2(g) \rightarrow 8 \text{ } CO_2(g) + 9 \text{ } H_2O(l)$$

We now calculate the number of moles of O_2 required. From the coefficients of the balanced equation:

$$1 \text{ mol } C_8H_{18} \simeq {}^{25}\!/_2 \text{ mol } O_2$$

But, the molecular mass of C_8H_{18} is 8(12.0) + 18(1.0) = 114. So

$$114 \text{ g } C_8H_{18} \simeq {}^{25}\!/_2 \text{ mol } O_2$$

$$\text{moles } O_2 \text{ required} = 1.00 \text{ g } C_8H_{18} \times \frac{{}^{25}\!/_2 \text{ mol } O_2}{114 \text{ g } C_8H_{18}} = 0.110 \text{ mol } O_2$$

Now we can find the volume of oxygen required by substituting into the Ideal Gas Law:

$$P = 740 \text{ mm Hg} \times \frac{1 \text{ atm}}{760 \text{ mm Hg}} = 0.974 \text{ atm; } T = 273 + 24 = 297 \text{ K}$$

$$V = \frac{nRT}{P} = \frac{0.110 \text{ mol} \times 0.0821 \text{ } \ell \cdot \text{atm/(mol} \cdot \text{K)} \times 297 \text{ K}}{0.974 \text{ atm}} = 2.75 \text{ } \ell$$

This reaction is very similar to that occurring in an automobile engine. Clearly a large volume of air (which is only 21% O_2 by volume) must pass through the engine during combustion.

Exercise   How many liters of $O_2(g)$ at 1.00 atm and 27°C are required to burn 1.00 g of octane? Answer: 2.71 ℓ.

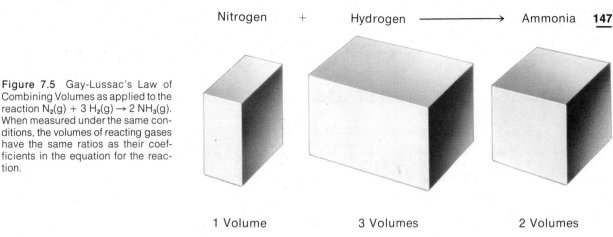

Figure 7.5 Gay-Lussac's Law of Combining Volumes as applied to the reaction $N_2(g) + 3 H_2(g) \rightarrow 2 NH_3(g)$. When measured under the same conditions, the volumes of reacting gases have the same ratios as their coefficients in the equation for the reaction.

1 Volume   3 Volumes   2 Volumes

Law of Combining Volumes

There is a very simple relationship between the volumes of different gases involved in a reaction. Consider the following reaction:

$$4 NH_3(g) + 5 O_2(g) \rightarrow 4 NO(g) + 6 H_2O(g)$$

According to the usual interpretation, we would say

$$4 \text{ moles } NH_3 + 5 \text{ moles } O_2 \rightarrow 4 \text{ moles } NO + 6 \text{ moles } H_2O$$

If all these gases are measured at the same temperature and pressure, their molar volumes will all be equal to some fixed value, V_m ℓ. Under such conditions it would be true that

$$4 V_m \text{ ℓ } NH_3 + 5 V_m \text{ ℓ } O_2 \rightarrow 4 V_m \text{ ℓ } NO + 6 V_m \text{ ℓ } H_2O$$

or, dividing through by V_m, simply

$$4 \text{ ℓ } NH_3 + 5 \text{ ℓ } O_2 \rightarrow 4 \text{ ℓ } NO + 6 \text{ ℓ } H_2O$$

We conclude that:

The volumes of different gases involved in a reaction, if measured at the same temperature and pressure, are in the same ratio as the coefficients in the balanced equation.

This relationship is known as the Law of Combining Volumes. It was first suggested, in a somewhat different form, by Gay-Lussac in 1808.

Example 7.8   In the reaction between ammonia and oxygen, what volume of $O_2(g)$ at 736 mm Hg and 22°C is required to react with 6.12 ℓ of $NH_3(g)$ at the same pressure and temperature?

Solution   Since 4 ℓ $NH_3 \backsimeq 5$ ℓ O_2

$$\text{Volume } O_2 = 6.12 \text{ ℓ } NH_3 \times \frac{5 \text{ ℓ } O_2}{4 \text{ ℓ } NH_3} = 7.65 \text{ ℓ } O_2$$

Note that it makes no difference what the temperature and pressure are, so long as they are the same for both gases.

Exercise   In this same reaction at the same temperature and pressure, what volume of water vapor is produced? Answer: 9.18 ℓ.

7.7   MIXTURES OF GASES; DALTON'S LAW

So far we have concentrated upon the behavior of pure gases. Frequently, however, we deal with gaseous mixtures where more than one substance is present. Here, a relation discovered by John Dalton in 1807 is very helpful. Dalton's Law states that:

The total pressure of a gas mixture is the sum of the partial pressures of the components of the mixture.

For a mixture of two gases A and B:

$$P_{tot} = P_A + P_B \tag{7.9}$$

Here, P_A is the partial pressure of gas A; P_B is the partial pressure of gas B. The partial pressure of a gas is the pressure it would exert if it were alone in the container at the same temperature as the mixture.

Dalton had more than one good idea, and some that weren't so good

Dalton's Law is perhaps applied most often in calculations involving gases collected over water (Fig. 7.6). Here, the gas being collected is mixed with water vapor. The only pressure that can be measured directly is the total pressure of the mixture. The partial pressure of the water vapor is readily obtained. It has a fixed value at a given temperature. This value can be obtained from a table of water vapor pressures (Appendix 1). By subtracting the partial pressure of $H_2O(g)$ from the total pressure of the mixture, we obtain the partial pressure of the gas under study. This in turn can be used to determine the number of moles of that gas present (Example 7.9).

Example 7.9   In a laboratory experiment, concentrated hydrochloric acid was reacted with aluminum. Hydrogen gas was evolved and collected over water at 25°C; it had a volume of 355 cm³ at a total pressure of 750 mm Hg. The vapor pressure of water at 25°C is known to be about 24 mm Hg.
  a. What was the partial pressure of hydrogen in the sample?
  b. How many moles of hydrogen were collected?

Solution
  a. The collected gas was a mixture of hydrogen and water vapor. The partial pressure of water in the sample would equal its vapor pressure, 24 mm Hg. By Dalton's Law: $P_{tot} = P_{H_2} + P_{H_2O}$. Therefore, $P_{H_2} = P_{tot} - P_{H_2O}$.

$$P_{H_2} = (750 - 24) \text{ mm Hg} = 726 \text{ mm Hg}$$

$$= 726 \text{ mm Hg} \times \frac{1 \text{ atm}}{760 \text{ mm Hg}} = 0.955 \text{ atm}$$

  b. By the Gas Law:

$$P_{H_2} = \frac{n_{H_2}RT}{V}; \qquad n_{H_2} = \frac{P_{H_2}V}{RT} = \frac{0.955 \text{ atm} \times 0.355 \, \ell}{0.0821 \, \frac{\ell \cdot atm}{mol \cdot K} \times 298 \text{ K}} = 0.0139 \text{ mol}$$

Exercise   Hydrogen gas is collected over water at a total pressure of 744 mm Hg at 20°C. Using the table of water vapor pressures on p. A.3, calculate the partial pressure of the $H_2(g)$. Answer: 726 mm Hg.

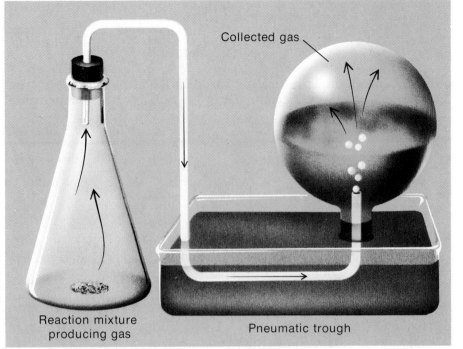

Figure 7.6 The collection of gases over water. With this apparatus the collected gas will contain water vapor at a partial pressure equal to the vapor pressure of water at the temperature of the system.

The validity of Dalton's Law is readily shown, using the Ideal Gas Law. For a mixture of two gases A and B, we can write

$$P_{tot} = n_{tot} \times \frac{RT}{V} = (n_A + n_B) \times \frac{RT}{V}$$

Does a molecule of A exert the same pressure as a molecule of B? Answer: Yes!

Separating terms on the right side of this equation:

$$P_{tot} = n_A\frac{RT}{V} + n_B\frac{RT}{V}$$

Applying the Ideal Gas Law again, this time to obtain the partial pressures of the individual gases:

$$P_A = n_A\frac{RT}{V}; \; P_B = n_B\frac{RT}{V}$$

Hence: $P_{tot} = P_A + P_B$, which is Equation 7.9.

Partial Pressure and Mole Fraction

As we have just seen, for a gas mixture,

$$P_A = n_A\frac{RT}{V} \qquad \text{and} \qquad P_{tot} = n_{tot}\frac{RT}{V}$$

If we divide P_A by P_{tot}, we obtain

$$\frac{P_A}{P_{tot}} = \frac{n_A}{n_{tot}} \qquad \text{or} \qquad P_A = \frac{n_A}{n_{tot}} \times P_{tot} \qquad (7.10)$$

The mole fraction is
often used to express
concentration
The fraction n_A/n_{tot} is referred to as the **mole fraction** of A in the mixture. It is the fraction of the total number of moles that is accounted for by gas A. Using X to represent mole fraction:

$$X_A = \frac{n_A}{n_{tot}} \qquad (7.11)$$

Substituting for n_A/n_{tot} in Equation 7.10, we have

$$P_A = X_A P_{tot} \qquad (7.12)$$

In other words, *the partial pressure of a gas in a mixture is equal to its mole fraction multiplied by the total pressure.*

Example 7.10   Air from the prairies of North Dakota in winter contains essentially only nitrogen, oxygen, and argon. The mole fractions of these three gases are 0.78, 0.21, and 0.01, in the order given. If the total pressure of the air is 742 mm Hg, what are the partial pressures of the three gases?

Solution   Applying Equation 7.12:

$$P_{N_2} = 0.78 \times 742 \text{ mm Hg} = 5.8 \times 10^2 \text{ mm Hg}$$
$$P_{O_2} = 0.21 \times 742 \text{ mm Hg} = 1.6 \times 10^2 \text{ mm Hg}$$
$$P_{Ar} = 0.01 \times 742 \text{ mm Hg} = 7 \text{ mm Hg}$$

Exercise   The partial pressure of O_2 in a different sample of air is 152 mm Hg. The total pressure is 750 mm Hg. What is the mole fraction of O_2? Answer: 0.203.

7.8   REAL GASES

In this chapter we have used the Ideal Gas Law in all our calculations, assuming it applies exactly. Under ordinary conditions (and in nearly all problems in this text), this assumption is a good one. However, all real gases deviate at least slightly from the Gas Law because it neglects two factors. These are:
—the finite volume of gas molecules.
—attractive forces between gas molecules.
The effect of these factors depends upon the temperature and pressure of the gas.

Temperature

As the temperature is lowered, the molecules in a gas move more slowly. This means that attractive forces become more important, making the gas easier to compress. The observed molar volume, V_{obs}, becomes less than that calculated from the Ideal Gas Law, V_{ideal}. To illustrate this effect, let us consider oxygen gas. At 25°C and 1 atm, the observed molar volume of O_2 is within 0.1% of that predicted by the Gas Law. We would say that under these conditions, O_2 behaves ideally for all practical purposes. When oxygen is cooled to $-150°C$, the situation changes. At $-150°C$ and 1 atm, V_{obs} for O_2 is about 1% less than V_{ideal}.

At 1 atm, the Ideal Gas Law works very well, all things considered

At the boiling point of a substance, attractive forces between molecules are large enough to make a gas condense to a liquid. The closer a gas is to its boiling point, the more it will deviate from the Ideal Gas Law. Consider, for example, SO_2. Sulfur dioxide boils at $-10°C$ at 1 atm. At 25°C and 1 atm, it is close enough to the boiling point for attractive forces to be significant. The molar volume of SO_2 under these conditions is about 1% less than that calculated from the Gas Law.

Pressure

In Figure 7.7, we have plotted the ratio V_{obs}/V_{ideal} for $CH_4(g)$ at 25°C as a function of pressure. As you might expect, this ratio starts off at 1.00 at zero pressure. As the pressure increases, V_{obs}/V_{ideal} drops below 1 and decreases steadily up to about 150 atm. This behavior is caused by the attraction between the molecules. The effect is in the same direction as that observed when the temperature is lowered.

The attractive forces go up as distance between molecules goes down

The behavior of $CH_4(g)$ at very high pressures at 25°C is shown at the right of Figure 7.7. Above about 150 atm, the ratio V_{obs}/V_{ideal} starts to increase. Here we are seeing the effect of the finite volume of the methane molecules. At high pressures their volume becomes significant compared to that of the container. Hence V_{obs}/V_{ideal} increases. By the time we reach 350 atm, the effect of molecular volume just about cancels that of the attractive forces. V_{obs}/V_{ideal} becomes 1.00 at this point. Above 350 atm, the molecular volume factor predominates and V_{obs}/V_{ideal} becomes greater than 1.

The graph that we have shown in Figure 7.7 for CH_4 is typical of most gases. Ordinarily:

—at low pressures attractive forces are most important, $V_{obs}/V_{ideal} < 1$.
—at high pressures the volume of the molecules is most important, $V_{obs}/V_{ideal} > 1$.
The exact shape of the curve will depend upon the gas studied and the temperature.

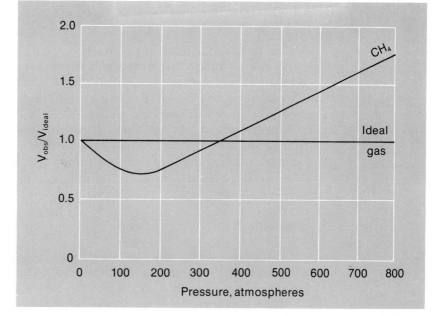

Figure 7.7 The pressure-volume behavior of $CH_4(g)$ at 25°C. Below about 350 atm, attractive forces between CH_4 molecules cause the observed volume to be less than that calculated from the Ideal Gas Law. At 350 atm, the effect of the attractive forces is just balanced by that of the finite volume of CH_4 molecules, and the gas appears to behave ideally. Above 350 atm, the effect of finite molecular volume predominates and $V_{obs} > V_{ideal}$.

van der Waals Equation

The van der Waals b is roughly equal to the molar volume of the liquefied gas. Example: For SO_2, V(l) = 0.045 ℓ/mol

It is possible to write equations involving P, V, and T for gases which take into account intermolecular attractions and finite molecular volumes. One of these is the van der Waals equation. For one mole of a gas it takes the form

$$P = \frac{RT}{V - b} - \frac{a}{V^2} \tag{7.13}$$

Here, a and b are independent of P, V, and T. They do, however, vary from one gas to another. Their values are selected to give the best possible agreement between the equation and the observed behavior of the gas. The term a/V^2 reflects the attractive forces between the molecules. The van der Waals constant b corrects for the effect of molecular volume. The van der Waals equation is considerably better than the Ideal Gas Law for predicting the properties of gases at high pressures.

TABLE 7.4   VAN DER WAALS CONSTANTS

	a $\left(\dfrac{\ell^2 \cdot atm}{mol^2}\right)$	b (ℓ/mol)
H_2	0.244	0.027
O_2	1.360	0.032
N_2	1.390	0.039
CH_4	2.253	0.043
CO_2	3.592	0.043
SO_2	6.714	0.056
Cl_2	6.493	0.056
H_2O	5.464	0.030

7.9   KINETIC THEORY OF GASES

The fact that the Ideal Gas Law applies to all gases indicates that the gaseous state is a relatively simple one to treat from a theoretical point of view. Gases must have certain properties in common which cause them to follow the same natural law. Between about 1850 and 1880, Maxwell, Boltzmann, Clausius, and others developed the kinetic theory of gases to explain these similarities in the behavior of gases. They based it on the idea that all gases behave similarly so far as molecular motion is concerned. Since that time, the kinetic theory has had to be modified only slightly. In its present form it is one of the most successful of scientific theories. It ranks in stature with the atomic theory of the nature of matter.

Postulates of the Kinetic Theory

1. **Gases consist of molecules in continuous, random motion.** The molecules undergo collisions with one another and with the container walls. The pressure of a gas arises from the forces associated with wall collisions.

2. **Molecular collisions are elastic.** During collisions there are no frictional losses

which result in loss of energy of motion. The temperature of a gas insulated from its surroundings does not change.

3. **The average energy of translational motion of a gas molecule is proportional only to the absolute temperature.** The energy associated with the motion of a molecule from one place to another depends on the temperature, but not on the pressure or the nature of the molecule. This postulate can be expressed in the form of a simple equation:

$$\epsilon_\tau = mu^2/2 = cT \tag{7.14}$$

Here, ϵ_τ is the average molecular energy of translation, m is the mass of the molecule, and u is its average speed. The quantity c is a constant which has the same value for all gases. We will say more later about its magnitude.

In addition to these postulates, it is assumed that the volumes of the molecules are negligible as compared to container volume. Moreover, attractive forces between molecules are neglected.

The postulates of the kinetic theory are easily stated. Their implications, however, are by no means obvious. The problem is that gas molecules move about in a completely random manner. Their velocities are constantly changing in both magnitude and direction as they collide with each other and with the walls of their container. To treat this kind of motion in a rigorous way requires mathematics of a high level of sophistication. In this text, we will not attempt to present such a mathematical development. Instead, we will concentrate upon some of the relationships which have resulted from the kinetic theory.

Gas Pressure

As noted in Sections 7.3 through 7.5, the pressure of a gas (P) is dependent upon container volume (V), number of moles (n), and temperature (T). We can readily understand the effect of these factors in terms of kinetic theory, which tells us that gas pressure is due to collisions with the walls of the container.

1. P increases as n increases (constant V, T) because, with more molecules present, there are more collisions per unit time.

2. P increases as V decreases (constant n, T) because, in a smaller volume, gas molecules strike the walls more often.

3. P increases as T increases (constant n, V) because, at a higher temperature, molecules move more rapidly. Collisions occur more frequently and with greater force.

Effusion of Gases. Graham's Law

The flow of gas molecules through a small opening or pinhole in a container is referred to as effusion. The rate of effusion is directly proportional to the average speed of gas molecules. Thus, the rates of effusion of two gases, A and B, from the same container at the same pressure are in the ratio of their average speeds:

$$\frac{\text{rate of effusion of A}}{\text{rate of effusion of B}} = \frac{\text{average speed of molecule A}}{\text{average speed of molecule B}} = \frac{u_A}{u_B} \tag{7.15}$$

By Equation 7.14, if the two gases are at the same temperature,

$$\frac{m_A u_A^2}{2} = \frac{m_B u_B^2}{2} = cT$$

Translational motion is motion through space

Experimentally, we let the gas effuse into a a vacuum and measure the drop in pressure in the container in a given time

or
$$\frac{u_A{}^2}{u_B{}^2} = \frac{m_B}{m_A} = \frac{MM_B}{MM_A} \tag{7.16}$$

where MM_A and MM_B are the molecular masses of A and B. If Equation 7.16 is solved for the ratio u_A/u_B, which is then substituted into 7.15, we obtain

$$\frac{\text{rate of effusion of A}}{\text{rate of effusion of B}} = \left(\frac{MM_B}{MM_A}\right)^{1/2} \tag{7.17}$$

This relation, in a slightly different form, was discovered experimentally by Thomas Graham in 1828. It is known as Graham's Law.

This law gives us a way of determining molecular masses of gases. All we need to do is compare the rate of effusion of the gas we are interested in to that of another gas of known molecular mass. Usually what is done is to measure the time required for equal amounts of the two gases to effuse under the same conditions of temperature and pressure. The measured times are *inversely* related to rates; the faster the gas effuses, the less the time required for a given amount to effuse. Hence,

How else can you obtain molecular masses experimentally?

$$\frac{\text{time}_B}{\text{time}_A} = \frac{\text{rate}_A}{\text{rate}_B} = \left(\frac{MM_B}{MM_A}\right)^{1/2} \tag{7.18}$$

In other words, the time required for effusion increases with molecular mass. Heavy molecules take longer to effuse.

Example 7.11   In an effusion experiment, it required 45 s for a certain number of moles of an unknown gas to pass through a small opening into a vacuum. Under the same conditions, it took 18 s for the same number of moles of O_2 to effuse. Find the molecular mass of the unknown gas.

Solution   Substituting into Equation 7.18:

$$\frac{\text{time}_{O_2}}{\text{time}_X} = \frac{18}{45} = \left(\frac{MM_{O_2}}{MM_X}\right)^{1/2} = \left(\frac{32}{MM_X}\right)^{1/2}$$

Squaring both sides: $\dfrac{32}{MM_X} = \left(\dfrac{18}{45}\right)^2 = 0.16$; $MM_X = \dfrac{32}{0.16} = 2.0 \times 10^2$

Exercise   A certain gas takes only one fourth as long to effuse as O_2. What is its molecular mass? Answer: 2.0.

During World War II, Graham's Law was applied to a rather complex chemical problem. It became necessary to separate U-235, which is fissionable (Chap. 6), from the more abundant isotope of uranium, U-238. Since the two isotopes have almost identical chemical properties, chemical separation was not feasible. Instead, an effusion process was worked out using uranium hexafluoride, UF_6. This compound is a gas at room temperature. Preliminary experiments indicated that $^{235}_{92}UF_6$ could indeed be separated from $^{238}_{92}UF_6$ by effusion. An enormous plant was built for this purpose in Oak Ridge, Tennessee. In this process, UF_6 effuses many thousands of times through porous barriers. The lighter fractions move on to the next stage while heavier fractions are recycled through earlier stages. Eventually, a nearly complete separation of the two isotopes is achieved.

The U-235 in the first atomic bomb was made this way

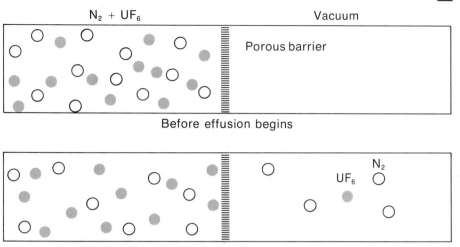

Figure 7.8 Effusion of gases. N_2 molecules move through the barrier faster than UF_6 molecules, since they are lighter (MM N_2 = 28, UF_6 = 352). After a short time, the gas to the right of the barrier is mostly N_2. The gas remaining in the compartment at the left is enriched in UF_6. This process can be used to separate a mixture of two gases.

Average Speeds of Gas Molecules

We have seen that Equation 7.14 can be used in a straightforward way to calculate relative speeds of two different kinds of molecules. For many purposes in chemistry, that kind of calculation is sufficient. It is possible, however, to go one step further and calculate the average speed of a particular kind of gas molecule. Here again we use Equation 7.14. Now, however, we need to evaluate the constant, c, in this equation. By a development which we will not attempt to go through here, it can be shown that

$$c = \frac{3R}{2N} \tag{7.19}$$

where R is the gas constant and N is Avogadro's number. Substituting in Equation 7.14, we obtain:

$$mu^2 = \frac{3RT}{N} \quad \text{or} \quad u^2 = \frac{3RT}{mN}$$

But the product of the mass of a molecule, m, times the number of molecules in a mole, N, is simply the molar mass, GMM. So

$$u^2 = \frac{3RT}{GMM}$$

or

$$u = \left(\frac{3RT}{GMM}\right)^{1/2} \tag{7.20}$$

We can calculate an average molecular speed by substituting into Equation 7.20. However, we must be careful to select the proper value of R (recall Table 7.3, p. 142). If we want to obtain u in centimeters per second, we should use

$$R = 8.31 \times 10^7 \frac{g \cdot cm^2}{s^2 \cdot mol \cdot K}$$

Example 7.12   Find the average speed of an oxygen molecule in air at room temperature (25°C).

Solution   In Equation 7.20: T = 25 + 273 = 298 K; GMM = 32.0 g/mol.

$$u = (3RT/GMM)^{1/2} = \left(\frac{3 \times 8.31 \times 10^7 \text{ g} \cdot \text{cm}^2/(\text{s}^2 \cdot \text{mol} \cdot \text{K}) \times 298 \text{ K}}{32.0 \text{ g/mol}}\right)^{1/2}$$

$$= 4.82 \times 10^4 \text{ cm/s} = 482 \text{ m/s}$$

By a simple conversion (1 m/s = 2.24 mile/h) we can show that, in more familiar units, the average speed is about 1000 miles/h!

Exercise   What is the average speed of an H_2 molecule (MM = 2.0) at 25°C? Answer: 1.9×10^3 m/s.

According to kinetic theory, molecular speeds are very high by ordinary standards. This prediction can be tested by experiments which show that measured speeds agree quite well with those calculated. Qualitatively, the average speed of O_2 and N_2 molecules agree with the observed speed of sound in air. Since sound is carried by molecular motion, one would expect its speed to be roughly the same as that of the molecules through which it passes. The speed of sound in air is about 800 miles/h, close to the average speeds of O_2 or N_2 molecules at 25°C.

Distribution of Molecular Speeds and Energies

Using Equation 7.20, it is possible to calculate the average speed of a gas molecule at a particular temperature. We must remember, however, that not all molecules in a gas sample will have that speed. The motion of molecules in a gas is utterly chaotic. In the course of a second, a molecule undergoes millions of collisions with other molecules. As a result, the speed and direction of motion of a molecule is constantly changing. Over a period of time, the speed of a molecule will vary from almost zero to some very high value, considerably above the average.

In 1860, James Clerk Maxwell showed that different possible speeds are distributed among molecules in a definite way. Indeed, he developed a mathematical expression for this distribution. His results are shown graphically in Figure 7.9 for O_2 at 25°C and 1000°C. On the graph, we plot the relative number of molecules having a certain speed, v, against that speed. We see that this number increases rapidly with the speed, up to a maximum of about 400 m/s at 25°C. This is the most probable speed of an oxygen molecule at 25°C. Above about 400 m/s, the number of molecules moving at any particular speed decreases. A molecule is only about one fifth as likely to be moving at 800 m/s as it is to be moving at 400 m/s. For speeds in excess of about 1200 m/s, the fraction of molecules drops off to nearly zero. In general, most molecules have speeds rather close to the average value.

As temperature increases, the speeds of molecules increase. The distribution curve for molecular speeds (Fig. 7.9) shifts to the right and becomes broader. The chance of a molecule having a very high speed is much greater at 1000°C than it is at 25°C. Note, for example, that large numbers of molecules have speeds greater than 1200 m/s at 1000°C.

Since translational energy is simply related to speed (Eq. 7.14), we can talk about a distribution of molecular energies as well as speeds. A plot of the relative number of

Sort of like people, most of them go along with the crowd

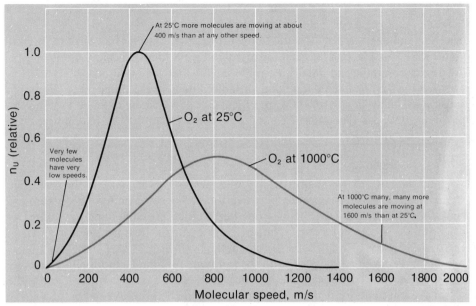

Figure 7.9 The graph shows the fraction of O_2 molecules having a given speed at 25°C and at 1000°C. At 25°C, most of the molecules are moving at a speed close to 400 m/s. At 1000°C, the average speed is about twice as great. More important, the fraction of molecules moving at very high speeds is much greater at the higher temperature.

molecules having a certain energy, E, against that energy would resemble Figure 7.9. Here again, *the fraction of molecules having very high energies increases sharply with temperature*. As we will see in Chapter 14, it is these high-energy molecules which take part in chemical reactions.

The mathematical relation obtained by Maxwell for the relative number of molecules having a speed u is one of the most formidable in all of physical science. It is

$$n_u \text{ (rel)} = 4\pi u^2 \left(\frac{GMM}{2\pi RT}\right)^{3/2} e^{-(GMM)u^2/2RT}$$

The fact that Maxwell was able to develop this equation from first principles is indicative of the intellectual ability of this remarkable man.

Maxwell was born in 1831 in Scotland. After education at the University of Edinburgh and at Cambridge he became a professor of natural philosophy, first in Scotland and later at Cambridge. His mathematical abilities became apparent at an early age and were applied in many areas. In addition to his accomplishments with gases, which laid the foundation of the science now known as statistical mechanics, Maxwell worked extensively in thermodynamics, developing several fundamental equations which bear his name. His greatest successes, however, were in connection with the theory of light and electricity, where he discovered and formulated the general equations of the electromagnetic field. He was first to recognize that light is a form of electromagnetic radiation and anticipated the development of what we now call radio waves. For several years during his life Maxwell was forced to be inactive because of illness: he was 48 years old when he died, having completed much of his work by the time he was 30.

The number of truly outstanding theoreticians the world has known is very small, certainly numbering less than one hundred. James Clerk Maxwell belongs among the elite of this group. He was truly an intellectual giant, to be ranked with Newton, Einstein, and J. Willard Gibbs (Chapter 23).

SUMMARY

The variables in gas behavior are pressure, volume, amount (number of moles), and temperature. These are related by the Ideal Gas Law:

Here, n is the number of moles, T is the temperature in K, and R is a constant [0.0821 $\ell \cdot$ atm/(mol$\cdot$K)]. This law can be used for many different kinds of calculations involving gases. It can be applied to determine relationships between two or more variables (Examples 7.2 and 7.3) or to calculate one variable, knowing the other three (Example 7.4). By using the relation n = g/GMM, the Ideal Gas Law can be extended to find the molecular mass of a gas (Example 7.5) or its density (Example 7.6). Going one step further, we can use the Law to relate the volumes of gases involved in reactions (Examples 7.7 and 7.8). Dalton's Law enables us to obtain the partial pressures of gases in a mixture (Examples 7.9 and 7.10).

Real gases deviate at least slightly from the Ideal Gas Law because of the finite size of gas molecules and attractive forces between them. These deviations become more significant when the molecules are close together, at high pressures and low temperatures. The van der Waals equation, although more complex than the Ideal Gas Law, represents the behavior of real gases more exactly.

The kinetic theory of gases deals with the random, continuous motion of gas molecules from place to place (translational motion). One of its basic postulates is that the average translational energy is directly proportional to the absolute temperature; the proportionality constant is the same for all gases. This relation can be used to compare the rates of effusion of different gases (Example 7.11) or to calculate average molecular speeds (Example 7.12). The distribution of molecular speeds at different temperatures is given in Figure 7.9.

KEY WORDS AND CONCEPTS

millimeter of Hg	Kelvin temperature scale	Law of Combining Volumes	kinetic theory
atmosphere	Avogadro's Law	Dalton's Law	translational energy
kilopascal	mole	partial pressure	effusion
Boyle's Law	Ideal Gas Law	mole fraction	Graham's Law
Law of Charles and Gay-Lussac	molar mass (GMM)	van der Waals equation	

QUESTIONS AND PROBLEMS

Catalog

Ideal Gas Law; Relation between Variables: 7.2–7.5, 7.23–7.29
Ideal Gas Law; Calculation of One Variable: 7.6, 7.7, 7.27, 7.28
Ideal Gas Law; Density, Molar Mass: 7.8–7.10, 7.29–7.31

Volumes of Gases in Reactions: 7.11–7.13, 7.32–7.34
Dalton's Law: 7.14–7.16, 7.35–7.37
Real Gases: 7.17, 7.38
Kinetic Theory: 7.18–7.20, 7.39–7.41
General: 7.1, 7.21, 7.22, 7.42–7.47

7.1 Review and know the meaning of the key words and concepts in this chapter.

7.2 A sample of oxygen gas occupies a volume of 2.30 ℓ at 740 mm Hg. What volume will it occupy if the pressure changes to 755 mm Hg at constant temperature?

7.23 Argon is used in light bulbs to reduce filament vaporization. What volume of Ar at 1.00 atm must be added to a bulb of volume 210 cm^3 to give a pressure of 1.30 mm Hg (constant T)?

7.3 The natural gas in a reservoir in Anchorage, Alaska has a volume of 3.00×10^6 m³ at 5°C and 1.00 atm. During the night, the temperature drops to -10°C but the pressure remains the same. What does the volume of the gas become?

7.4 Houses with well water systems have ballast tanks to hold a supply of water. Typically, water flows into the tank from the well until the gauge pressure (pressure in excess of 15 lb/in²) of the air in the tank reaches 50 lb/in². As water is drawn from the tank, the air above it expands and its pressure drops. When the gauge pressure reaches 20 lb/in², the pump delivers more water to the tank. Suppose a 1.20-m³ tank is 75% full of water when the gauge pressure is 50 lb/in². How much water can be withdrawn before the pump turns on at 20 lb/in²?

7.5 A 20.0-ℓ sample of steam at 100°C and 1.00 atm is cooled to 40°C and expanded until the pressure is 25 mm Hg. If no water condenses, what is the final volume of the water vapor?

7.6 A cylinder of xenon gas contains 50.0 ℓ of Xe at 18.4 atm and 25°C. How many moles of Xe are in the cylinder? How many grams?

7.7 A drum used to transport crude oil has a 159-ℓ volume. How many water molecules, as steam, are required to fill the drum at one atm and 100°C? What volume of liquid water is required to produce that amount of steam? (density of $H_2O(l) = 1.00$ g/cm³.)

7.8 Calculate the density of N_2 at 60°C and 1.00 atm. What is the density of O_2 under these conditions? Compare the number of molecules in one liter of each of these gases at the stated conditions.

7.9 A 0.939-g sample of a volatile liquid is completely vaporized in a 235-cm³ flask at 99.2°C and 754 mm Hg.

    a. Calculate the molecular mass of the gas.
    b. What is the formula of the liquid if it has a mass % composition of 29.3 % C, 5.7% H, and 65.0% Br?

7.10 A 1.59-g sample of $X_2H_6(g)$ has a volume of 603 cm³ at 1.00 atm and 15°C.

    a. What is the molecular mass of X_2H_6?
    b. Identify the element X.

7.11 Butane, C_4H_{10}, burns in air according to the equation

$$C_4H_{10}(g) + {}^{13}/_2\ O_2(g) \rightarrow 4\ CO_2(g) + 5\ H_2O(g)$$

If all the gases are measured under the same conditions, how many liters of carbon dioxide would be produced using 5.0 ℓ of oxygen? Assume 100% yield.

7.24 A balloonist puts 1.25×10^5 ℓ of air into her balloon at 25°C and atmospheric pressure. The air is heated to 200°C at constant pressure. What is the volume of air in the balloon at this temperature?

7.25 Frequently, ballast tanks of the type described in Problem 7.4 lose most of their air and become "waterlogged." When this happens, the pump operates more frequently. Suppose the 1.20-m³ tank referred to in Problem 7.4 has to be 95% full of water before the gauge pressure reaches 50 lb/in². How much water can be withdrawn from the tank before the pump turns on at 20 lb/in²?

7.26 Gauge pressure is the pressure in excess of atmospheric pressure, which we will take to be 100 kPa. The gauge pressure of an automobile tire at 21°C was 190 kPa. After an hour of driving, the tire heated up to 50°C and expanded by 1%. What should the new gauge pressure be?

7.27 The interior volume of a gas cylinder containing 3.3 kg of methane, CH_4, at 20°C is 45.0 ℓ. What is the pressure of the methane?

7.28 Standard temperature and pressure, STP, is 0°C and one atm. Find the volume occupied by one mole of an ideal gas at STP; determine the molar volume of the gas at 25°C and one atm.

7.29 One mole of air contains essentially 0.21 mol of O_2 and 0.79 mol of N_2.

    a. What is the "molecular mass" of air?
    b. Calculate the density of air at 755 mm Hg and 25°C.

7.30 A certain organic compound in the vapor state has a density of 3.27 g/ℓ at 95°C and 758 mm Hg.

    a. What is its molecular mass?
    b. If the compound contains 24.2% C, 4.1% H, and 71.7% Cl, what is its molecular formula?

7.31 A 0.0853-g sample of $X_4H_{10}(g)$ has a volume of 36.7 cm³ at 800 mm Hg and 20°C.

    a. What is the molecular mass of X_4H_{10}?
    b. Identify the element X.

7.32 Consider the reaction

$$4\ NH_3(g) + 5\ O_2(g) \rightarrow 4\ NO(g) + 6\ H_2O(g)$$

If all the gases are measured under the same conditions, what is the volume (liters) of products from the complete reaction of 2.5 ℓ of NH_3?

7.12 A student prepares oxygen by heating potassium chlorate:

$$KClO_3(s) \rightarrow KCl(s) + {}^3/_2 \, O_2(g)$$

What volume of O_2 at 20°C and 726 mm Hg is formed by the decomposition of 1.00 g of $KClO_3$?

7.13 Gasoline is a mixture of various hydrocarbons of which octane, C_8H_{18}, is typical.

    a. Write a balanced equation for the combustion of octane to give $CO_2(g)$ and $H_2O(l)$.
    b. What volume of O_2 at 50°C and 1.00 atm is required to react with 1.00 g of octane?
    c. Air is about 21% by volume of O_2. What volume of air is required in (b)?

7.14 The mole fractions of N_2, O_2 and Ar in dry air are 0.781, 0.210, and 0.009, in that order. What are the partial pressures of these three gases in dry air on a day when the barometric pressure is 732 mm Hg?

7.15 A mixture of 0.500 g of N_2 and 0.500 g of CO gases exerts a pressure of 740 mm Hg. What is the partial pressure of each gas?

7.16 A student collects 375 cm³ of O_2 gas saturated with water vapor at 23°C. The mixture exerts a total pressure of 768 mm Hg. At 23°C, the vapor pressure of $H_2O(l)$ = 21.0 mm Hg.

    a. What is the partial pressure of O_2 in the sample?
    b. How many grams of O_2 does the sample contain?

7.17 Calculate the pressure of one mole of H_2 in a 300-cm³ container at 0°C using the

    a. Ideal Gas Law.
    b. van der Waals equation (see Table 7.4).

7.18 It took 40.0 s for a sample of oxygen to effuse down a capillary. Another gas, which might be either N_2O or NO_2, took 48.0 s under the same T and P conditions. Which gas is it?

7.19 Consider SO_2 gas molecules at 50°C. At what temperature will oxygen molecules have the same

    a. average kinetic energy as SO_2 molecules?
    b. average velocity as SO_2 molecules at 50°C?

7.33 A student prepares hydrogen by the reaction of zinc with acid:

$$Zn(s) + 2 \, H^+(aq) \rightarrow Zn^{2+}(aq) + H_2(g)$$

What mass of zinc is required to form 1.00 ℓ of H_2 at 25°C and 760 mm Hg?

7.34 A Volvo engine has a cylinder volume of about 500 cm³. The cylinder is full of air at 60°C and 1.00 atm.

    a. How many moles of O_2 are in the cylinder? (mol % O_2 in air = 21)
    b. Assume that the hydrocarbons in gasoline have an average molecular mass of 100 and react with O_2 in a 1:12 mol ratio. How many grams of gasoline should be injected into the cylinder to react with the oxygen?

7.35 The mole fraction of CO in the air in a garage where a charcoal grill is operating can be dangerously high, about 5×10^{-5}.

    a. If the barometric pressure is 745 mm Hg, what is the partial pressure of CO in atmospheres?
    b. If the temperature is 30°C, what is the concentration of CO in moles per liter?

7.36 In a gaseous mixture of 0.750 g of N_2 and 0.750 g of O_2, the partial pressure of oxygen is 475 mm Hg. What is the total pressure of the mixture?

7.37 To prepare a sample of hydrogen gas, a student reacts zinc with hydrochloric acid. The overall reaction is

$$Zn(s) + 2 \, H^+(aq) \rightarrow Zn^{2+}(aq) + H_2(g)$$

The hydrogen is collected over water at 22°C and the total pressure is 750 mm Hg. (vp $H_2O(l)$ = 20 mm Hg)

    a. What is the partial pressure of H_2?
    b. How many grams of H_2 are there in a 3.00-ℓ sample of wet gas?

7.38 Using Figure 7.7, estimate the density of $CH_4(g)$ at 200 atm and 25°C and compare to the value calculated from the Ideal Gas Law.

7.39 A sample of a gas effuses 1.27 times as rapidly as Cl_2 under the same T and P conditions.

    a. Is the gas heavier or lighter than Cl_2?
    b. What is the ratio of the molecular mass of chlorine to that of the unknown gas?

7.40 Consider an O_2 molecule:

    a. At what temperature will its average kinetic energy be twice that at 20°C?
    b. At what temperature will its average speed be twice that at 20°C?

7.20 What is the average speed of an H_2 molecule at 0°C? an He atom?

7.21 2.00 mol of neon and 1.00 mol of hydrogen are in separate containers at the same T and P. Calculate each of the following ratios:

    a. volume Ne/volume H_2
    b. density Ne/density H_2
    c. average translational energy Ne/average translational energy H_2
    d. number of atoms Ne/number of atoms H

7.22 For a mole of an ideal gas, sketch graphs of

    a. V vs. P at constant T.
    b. P vs. T at constant V.
    c. u vs. T at constant P, V.
    d. E_{trans} vs. T at constant P, V.

7.41 At what temperature will an O_2 molecule have an average speed of 400 m/s? 600 m/s?

7.42 A mixture of 3.4 mol of helium and 6.1 mol of oxygen occupies a 4.75-ℓ container at 25°C. Which gas has the larger

    a. average molecular speed?
    b. average translational energy?
    c. partial pressure?
    d. mole fraction?

7.43 For an ideal gas, sketch graphs of

    a. V vs. T at constant P, n.
    b. P vs. n at constant V, T.
    c. n vs. T at constant P, V.
    d. E_{trans} vs. P at constant T, n.

*7.44 The Rankine temperature scale resembles the Kelvin scale in that 0° is taken to be the lowest attainable temperature (0°R = 0 K). However, the Rankine degree is the same size as the Fahrenheit degree, whereas the Kelvin degree is the same size as the Celsius degree. What is the value of the gas law constant in $\ell \cdot atm/(mol \cdot °R)$?

*7.45 A 0.150-g sample of an Al-Zn alloy reacts with HCl to form H_2:

$$Al(s) + 3\ H^+(aq) \rightarrow Al^{3+}(aq) + {}^3/_2\ H_2(g)$$

$$Zn(s) + 2\ H^+(aq) \rightarrow Zn^{2+}(aq) + H_2(g)$$

The hydrogen produced has a volume of 138 cm³ at 27°C and 1.00 atm. What is the mass % of aluminum in the alloy?

*7.46 An argon atom has a radius of about 0.094 nm.

    a. What is the volume (cm³) of a mole of Ar atoms? $\left(V = \dfrac{4\pi r^3}{3}\right)$
    b. In Ar(g) at 20°C and 50 atm, what fraction of the total volume is occupied by Ar atoms?

*7.47 The buoyant force on a balloon is equal to the mass of air it displaces. The gravitational force on the balloon is equal to the sum of the masses of the balloon, the gas it contains, and the balloonist. If the balloon and balloonist together weigh 200 kg, what would the diameter of a spherical hydrogen-filled balloon have to be in meters if the rig is to get off the ground at 25°C and one atm? (Take MM air = 28.9)

THE ELECTRONIC STRUCTURE OF ATOMS

In Chapter 2, we considered the structure of the atom briefly. You will recall that every atom has a tiny, positively charged nucleus, made up of protons and neutrons. The number of protons in the nucleus is characteristic of the element to which that atom belongs. It is referred to as the atomic number of the element.

The nucleus is surrounded by electrons which carry a negative charge. The charge of an electron is equal in magnitude but opposite in sign to that of a proton. In a neutral atom, the number of electrons is equal to the number of protons and, hence, to the atomic number of the element.

The chemical properties of atoms are determined by the way in which the electrons are arranged about the nucleus. Knowing the electronic structure of an atom, we can predict much of the chemical behavior of the corresponding element. In this chapter, we will look at electron arrangements in atoms. Before doing that, it will be helpful to examine the energies of these electrons. These are derived from the quantum theory, which was developed during the early part of this century.

8.1   THE QUANTUM THEORY

The quantum theory was first proposed by Max Planck in 1900 to explain the properties of the radiation given off by hot bodies. A few years later, in 1905, it was used by Albert Einstein to treat the emission of electrons by metals exposed to light. Still later, in 1913, Niels Bohr used quantum theory to develop a model of the hydrogen atom (Section 8.2). We now know that the quantum theory is a general one that applies to all the interactions of matter with energy. Here, we will discuss the postulates of the theory as they apply to electrons in atoms or molecules.

Postulates of the Quantum Theory

1. Atoms and molecules can only exist in certain states, characterized by definite amounts of energy. When an atom or molecule changes its state, it must absorb or emit an amount of energy just sufficient to bring it to another state.

Atoms and molecules can possess various kinds of energy. One form of energy of particular importance arises from the motion of electrons about the atomic nucleus and from the charge interactions among the electrons and between the electrons and the nucleus. This kind of energy is called *electronic energy*. Only certain values of electronic energy are allowed to an atom. When an atom goes from one allowed electronic state to another, it must absorb or emit just enough energy to bring its own energy to that of the final state.

The energy of systems that can exist only in discrete states is said to be *quantized*. A change in the energy level of such a system involves the absorption or emission of a definite amount ("quantum") of energy.

2. When atoms or molecules absorb or emit light in the process of changing their energies, the wavelength λ of the light is related to the magnitude of the energy change $|\Delta E|$ by the equation

$$|\Delta E| = hc/\lambda \qquad (8.1)$$

where h is a physical constant, called Planck's constant, and c is the speed of light.

A ray of light can be considered to consist of *photons* which appear to have some of the properties of particles. In particular, each photon of wavelength λ has an energy of hc/λ. An atom or molecule can move from one electronic energy state to another by absorbing or emitting a photon. If it absorbs a photon of energy hc/λ, it increases in energy by that amount. Conversely, if the atom or molecule gives off a photon, it decreases in energy by hc/λ.

3. The allowed energy states of atoms and molecules can be described by sets of numbers called quantum numbers.

It is possible to set up mathematical equations to describe the energies of electrons in atoms or molecules. Such an equation usually has several solutions. All of these solutions are of the same general form and contain one or more *quantum numbers*. These numbers distinguish a particular energy state from all the others. In the usual model, quantum numbers are associated with individual electrons in an atom. Each electron is assigned a set of quantum numbers according to a set of rules (Section 8.4).

Relation Between ΔE and λ

Figure 8.1 illustrates the meaning of Equation 8.1, the basic equation of simple quantum theory. Notice that the wavelength, λ, is inversely related to the energy difference, ΔE. If the energy states are widely separated (large ΔE), light of short wavelength, perhaps in the ultraviolet, is produced. Suppose, on the other hand, that the energy states are close together (ΔE small). In that case, light of longer wavelength, perhaps in the infrared, is emitted.

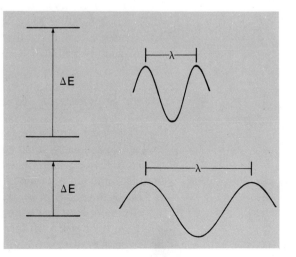

Figure 8.1 The wavelength of light given off when an electron returns to a lower energy state is inversely proportional to ΔE, the difference in energy between the two states. If ΔE is large, λ is small, perhaps in the ultraviolet. If ΔE is small, λ is large, perhaps in the infrared. By measuring λ, it is possible to determine ΔE, using Equation 8.2.

It is possible to determine the difference in energy, ΔE, between two electronic energy states. To do this, we measure the wavelength of the light absorbed or emitted when an electron moves from one state to the other. Equation 8.1 is used to obtain ΔE from λ. If we substitute

$$h = 6.626 \times 10^{-34}\ \frac{J \cdot s}{particle};\ c = 2.998 \times 10^8\ \frac{m}{s}$$

we obtain ΔE in *joules per particle* when λ is expressed in *meters:*

$$\Delta E = \frac{hc}{\lambda} = \frac{(6.626 \times 10^{-34})(2.998 \times 10^8)}{\lambda}\ \frac{J \cdot m}{particle}$$

$$= \frac{1.986 \times 10^{-25}}{\lambda}\ \frac{J \cdot m}{particle}$$

More frequently, we will need ΔE in *kilojoules per mole* with λ in *nanometers:*

$$\Delta E = \frac{1.986 \times 10^{-25}}{\lambda}\ \frac{J \cdot m}{particle} \times \frac{6.022 \times 10^{23}\ particles}{1\ mol} \times \frac{1\ kJ}{10^3\ J} \times \frac{10^9\ nm}{1\ m}$$

λ and ΔE are always related by this equation

$$\Delta E = \frac{1.196 \times 10^5}{\lambda}\ \frac{kJ \cdot nm}{mol} \tag{8.2}$$

We will find Equation 8.2 very useful in relating the spectrum of hydrogen to the Bohr model of the hydrogen atom.

8.2 THE ATOMIC SPECTRUM OF HYDROGEN AND THE BOHR MODEL

By 1900, scientists had begun to speculate on the way in which positive and negative charges are arranged in atoms. Part of the problem was solved in 1911 when Rutherford showed the existence of atomic nuclei (Chap. 2). Two years later, Niels Bohr, a young Danish physicist, took the next step. He developed a model for the behavior of the electron in the hydrogen atom.

Bohr based his approach on the Rutherford atom. He also had the quantum theory of Planck available. Most important, he had accurate information concerning energy differences between electronic states in the hydrogen atom. This came from a study of the atomic spectrum of hydrogen.

Atomic Spectrum of Hydrogen

If we heat a metal in a furnace or in a flame, it gives off visible light. At 1000°C, iron looks red; at 1500°C it appears white. The emitted light can be examined by a *spectroscope,* a device which breaks up light into its component colors. When this is done, the light is found to contain nearly all colors. We say that, over the region in which the iron radiates, its spectrum is *continuous,* containing light of all wavelengths.

If a sample of an element is excited electrically, as in a sodium vapor lamp or neon sign, the light given off looks quite different. We find that under these conditions the light consists of several individual lines at certain definite wavelengths. The lines ob-

TABLE 8.1 WAVELENGTHS (NM) OF LINES IN THE ATOMIC SPECTRUM OF HYDROGEN		
ULTRAVIOLET (LYMAN SERIES)	VISIBLE (BALMER SERIES)	INFRARED (PASCHEN SERIES)
121.53	656.28	1875.09
102.54	486.13	1281.80
97.23	434.05	1093.80
94.95	410.18	1004.93
93.75	397.01	
93.05		

Wavelengths of light can be measured very accurately

served are characteristic of a given element and can be used to identify it. Color plate 6 (center of book) shows the *atomic spectra* of some typical elements.

Some elements have very complex atomic spectra

The simplest atomic spectrum is that shown by hydrogen. In a hydrogen atom, we are dealing with a single electron moving between energy states. When the atom absorbs energy, the electron moves to a higher energy level. When the electron returns, it gives off energy as light at discrete wavelengths.

The atomic spectrum of hydrogen was first studied in the 1880's. At that time, a series of lines was discovered in the visible region. These lines comprise what is called the Balmer series. They were known to Bohr at the time he was developing his model of the hydrogen atom. Later, other series were found. One of these, called the Lyman series, is in the ultraviolet (<200 nm). Another, the Paschen series, lies in the infrared (>700 nm). The wavelengths (in nm) of some of the more prominent lines in each of these series are listed in Table 8.1.

Bohr Model for the Hydrogen Atom

Bohr assumed that a hydrogen atom consists of a central proton about which an electron moves in a circular orbit. He related the force of attraction of the proton for the electron to the centrifugal force due to the circular motion of the electron. In this way, Bohr was able to express the energy of the atom in terms of the radius of the electron's orbit. To this point, his analysis was purely classical, based on Coulomb's Law of electrostatic attraction and Newton's laws of motion. Bohr then introduced quantum theory into his model. Boldly and arbitrarily, he imposed a condition upon a property of the electron called its angular momentum. This is given by the product *mvr,* where m is the electron mass, v is its speed, and r is the radius of its orbit about the nucleus. Bohr proposed that the angular momentum be given by the equation

$$mvr = nh/2\pi \qquad (8.3)$$

where h is Planck's constant and **n** is a quantum number which can take on any positive integral value (1, 2, 3, . . .).

Bohr found that his quantum condition restricted the allowed energies of the electron in the hydrogen atom. It could have only those values given by the equation

$$E = -\frac{B}{n^2} \qquad (n = 1, 2, 3, \ldots) \qquad (8.4)$$

The value of the constant B in this equation could be calculated directly from theory; it is **1312 kJ/mol.**

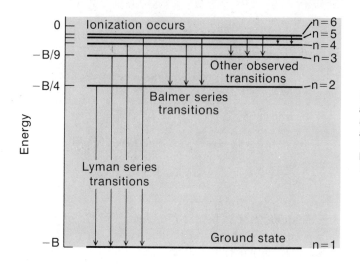

Figure 8.2 Some energy levels and transitions in the hydrogen atom. Lines in the Balmer series arise from transitions from upper levels (n > 2) to the n = 2 level. In the so-called Lyman series, the lower level is n = 1.

Before going further with the Bohr model, it may be well to make three points.

1. In setting up his model, Bohr took the zero of energy to be the point at which the proton and electron are completely separated. Energy has to be absorbed to reach that point. This means that the electron, in all its allowed energy states within the atom, must have an energy below zero. Hence, the minus sign in Equation 8.4.

2. In the normal hydrogen atom, the electron is in its **ground state,** for which **n** = 1. When an electron absorbs energy, it moves to a higher, **excited state.** These excited states correspond to **n** = 2, 3, 4, ...

3. When an excited electron gives off energy in the form of light, it drops back to a lower energy state. Some of these transitions are shown in Figure 8.2. Notice that the electron may return to

The energy is lowest in the ground state

—*the ground state* (**n** = 1). Electrons returning to this state produce the Lyman lines in the hydrogen spectrum. The transition between **n** = 2 and **n** = 1 yields one such line, that between **n** = 3 and **n** = 1 gives another, and so on.

—*an excited state.* Electrons returning to **n** = 2 from **n** = 3, 4, . . . are responsible for lines in the Balmer series. Transitions back to **n** = 3 give the Paschen series.

Bohr used his model to calculate the wavelengths of various lines in the spectrum of hydrogen. He found excellent agreement between theory and experiment. We can repeat his calculations by first using Equation 8.4 to obtain the energies of the individual states. Then, by subtraction, we get ΔE, the energy difference between two states. Finally, using Equation 8.2, we obtain a predicted value for the wavelength.

Example 8.1   Find the wavelength, in nanometers, of the line in the Balmer series that results from the transition from **n** = 2 to **n** = 3.

Solution   Using Equation 8.4 with B = 1312 kJ/mol:

$$E_3 = \frac{-1312}{9} \frac{kJ}{mol} = -145.8 \frac{kJ}{mol}$$

$$E_2 = \frac{-1312}{4} \frac{kJ}{mol} = -328.0 \frac{kJ}{mol}$$

$$\Delta E = E_3 - E_2 = -145.8 \frac{kJ}{mol} - \left(-328.0 \frac{kJ}{mol}\right) = 182.2 \frac{kJ}{mol}$$

Applying Equation 8.2:

$$182.2 \, \frac{kJ}{mol} = \frac{1.196 \times 10^5}{\lambda} \, \frac{kJ \cdot nm}{mol}$$

Solving for λ: $\lambda = \frac{1.196 \times 10^5}{182.2}$ nm $= 656.4$ nm

This is the first line in the Balmer series (Table 8.1).

Exercise   Repeat this calculation for the first line in the Lyman series ($n = 2$ to $n = 1$). Answer: 121.5 nm.

The calculation we have just gone through may seem routine. You can imagine, though, how Bohr must have felt when he did it for the first time. With his value of B, Bohr predicted the wavelengths of all the known lines in the Balmer series. He obtained agreement with experiment to within 0.1% or better. Bohr also predicted the existence of the Lyman series, unknown in 1913. When it was discovered some years later, the wavelengths agreed almost exactly with Bohr's predictions.

Bohr was 28 years old at that time

Another quantity which can be calculated from the Bohr theory is the ionization energy of the hydrogen atom. This is the energy which must be absorbed to remove an electron from the gaseous atom, starting from the ground state.

$$H(g) \rightarrow H^+(g) + e^-; \Delta E = \text{ionization energy}$$

From Equation 8.4, it is possible to calculate the ionization energy of hydrogen. To do this, we calculate ΔE when an electron moves from the ground state ($n = 1$, $E = -1312$ kJ/mol) to the state where it is completely removed from the atom ($n = \infty$, $E = 0$).

$$\Delta E = 0 - (-1312 \, kJ/mol) = 1312 \, kJ/mol$$

The measured value, 1318 kJ/mol, is in close agreement.

8.3   THE QUANTUM MECHANICAL ATOM

Bohr's theory for the structure of the hydrogen atom was highly successful. Scientists of the day must have thought they were on the verge of being able to predict the allowed energy levels of all atoms. However, the extension of Bohr's ideas to atoms with two or more electrons gave, at best, only qualitative agreement with experiment. Consider, for example, what happens when Bohr's theory is applied to the helium atom. Here, the errors in calculated energies and wavelengths are of the order of 5% instead of the 0.1% error with hydrogen. There appeared to be no way the theory could be modified to make it work well with helium or other atoms. Indeed, it soon became apparent that there was a fundamental problem with the Bohr model. The idea of an electron moving about the nucleus in a well-defined orbit at a fixed distance from the nucleus finally had to be abandoned.

Wave Nature of the Electron; the de Broglie Relation

Prior to 1900, it was supposed that light was wave-like in nature. However, the work of Planck and Einstein suggested that, in many processes, light behaves as if it consists

Waves have some particle properties, and particles have some wave properties

of particles, called photons. Within 20 years, the dualistic, wave-particle nature of light became generally accepted. Then, in 1924, a young French physicist, Louis de Broglie, came up with a revolutionary idea about the nature of matter. Reasoning as physicists do, often with striking success, de Broglie suggested that particles might well exhibit wave properties. He showed that the wavelength, λ, associated with a particle of mass m moving at speed v would be

$$\lambda = \frac{h}{mv} \tag{8.5}$$

where h is Planck's constant. Within a few years, Davisson and Germer, working at the Bell Telephone Laboratories, tested de Broglie's prediction. They showed that a beam of electrons does indeed have wave properties. Moreover, the observed wavelength of the electron was exactly that predicted by de Broglie.

Using Equation 8.5, it is possible to show how the Bohr quantum number **n** arises in a natural way. To do this, consider Figure 8.3. Here, we imagine an electron in the form of a wave moving about the nucleus along the circumference of a circle. Under these conditions, there is a restriction on the wavelengths that the electron can have. In successive revolutions, the waves must be exactly in phase with each other. That is, they must have exactly the same height (*amplitude*) at any given point. This means that a wave must fit into the circumference of the circle, 2πr, an integral number of times, **n.** In other words:

$$2\pi r = n\lambda \tag{8.6}$$

Quantization arises from relations like this one

where λ is the wavelength and **n** is a whole number—that is, 1, 2, 3, . . ., but *not* 1.5, 2.1, ... Combining Equations 8.5 and 8.6, we obtain

$$2\pi r = \frac{nh}{mv}$$

or
$$mvr = \frac{nh}{2\pi}$$

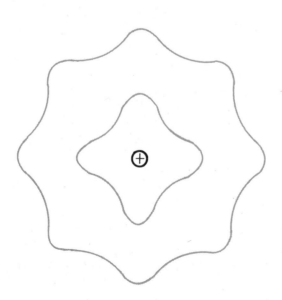

Figure 8.3 One way to rationalize the Bohr quantum condition is to assume that an electron wave moving in a circle about the nucleus must repeat itself an integral number of times in one revolution. This requires that λ = 2πr/n, where n is a whole number. Combining this relation with the de Broglie equation, λ = h/mv, gives the Bohr condition: mvr = nh/2π.

This is the condition that Bohr imposed arbitrarily on the momentum of the electron in the hydrogen atom (Eq. 8.3). Using the de Broglie relation, this condition becomes physically reasonable.

The Schrödinger Wave Equation

If an electron behaves like a wave, there is a fundamental problem with the Bohr atom. How does one specify the "position" of a wave at a particular instant? We can hope to determine its wavelength, its energy, and even its amplitude, but there is no obvious way to tell precisely where the electron is. Indeed, since a wave extends over space, the very idea of the position of the electron within an atom becomes nebulous, to say the least.

Scientists in the 1920's, speculating on this problem, became convinced that the Bohr model had to be abandoned. The idea of an electron revolving about the nucleus in a fixed orbit of definite radius simply did not correspond to reality. An entirely new approach was required to treat electrons in atoms and molecules. A new discipline called *quantum mechanics* was developed to describe the motion of small particles confined to tiny regions of space. The main concern of quantum mechanics is to find expressions for the energies of these particles.

In the quantum mechanical atom, no attempt is made to specify the position of an electron at a given instant. Neither does quantum mechanics concern itself with the path that an electron takes about the nucleus. (After all, if we can't say where the electron is, we certainly don't know how it got there.) Instead, quantum mechanics deals only with the probability of finding a particle within a given region of space. It tells us, for example, that there is a 90% chance of finding an electron within 0.14 nm of the hydrogen nucleus. Conversely, there is a 10% chance that, in the ground state, the hydrogen electron will be farther from the nucleus than this. You may find this way of describing electron distributions disturbing. At least Albert Einstein did; he is reputed to have said that "God does not play dice." Nevertheless, no one has ever been able to pin down an electron in an atom or molecule to a precise location. There really isn't much we can do about that.

Certainly no point in worrying about it

Erwin Schrödinger, in 1926, made a major contribution to quantum mechanics. Expanding upon the ideas of de Broglie, Schrödinger went one step further. He derived an equation from which one could calculate the amplitude, ψ, of the electron wave at various points in space. The equation is a complex one. We will not attempt to work any problems concerning it. However, we write down the Schrödinger wave equation to have something to talk about. For a single particle, such as an electron in a hydrogen atom, the equation has the form

$$\frac{\partial^2\psi}{\partial x^2} + \frac{\partial^2\psi}{\partial y^2} + \frac{\partial^2\psi}{\partial z^2} + \frac{8\pi^2 m}{h^2}(E - V)\psi = 0$$

Here, m is the mass of the particle, E is its total energy, V is its potential energy, and h is Planck's constant. The quantity ψ is, as we have mentioned, the amplitude of the wave associated with the motion of the particle. The notation $\partial^2\psi/\partial x^2$ comes from the calculus and is the symbol for the second derivative of ψ with respect to x. There are similar terms for the coordinates y and z.

For the electron in the hydrogen atom, the Schrödinger wave equation can be solved exactly for ψ. It turns out that there are several expressions for ψ that will satisfy this equation. Each of these solutions is associated with a set of quantum numbers (Section 8.4). Using these quantum numbers, it is possible to calculate the energies allowed to an electron in a hydrogen atom.

The energies are the same as Bohr obtained

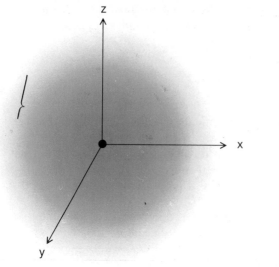

Figure 8.4 Electron cloud surrounding an atomic nucleus. The depth of color is proportional to the probability of finding an electron in a particular region. Notice that the probability decreases rapidly and smoothly as one moves out from the nucleus.

Once the Schrödinger equation has been solved for ψ, it becomes possible to determine the probability of finding a particle in a given region of space. It turns out that the square of the amplitude, ψ^2, is directly proportional to the probability of finding the particle at that point. For electrons, we can interpret the value of ψ^2 as being proportional to the electric charge density at that point. "Electron cloud" diagrams, showing charge densities in atoms and molecules, are often drawn on that basis. Figure 8.4 shows the "electron cloud" surrounding the nucleus of a typical atom.

Unlike the Bohr model, the Schrödinger wave equation can be applied to atoms other than hydrogen and to molecules as well. Unfortunately, this is easier said than done. The form of the equation for multi-electron systems is so complex as to defy exact solution. However, in the few cases where a satisfactory solution has been obtained, agreement with experiment is excellent. This convinces us that the approach is a correct although complex one.

Big computers are a big help on these problems

8.4 QUANTUM NUMBERS AND ALLOWED ENERGY STATES

Schrödinger found that the electron in the hydrogen atom could be described by three quantum numbers. These are now called **n**, ℓ, and $\mathbf{m}_\ell$. There are three such numbers because the electron requires three coordinates to describe its motion.

For reasons we will discuss later in this section, we find that it takes four, rather than three, quantum numbers to completely describe the state of an electron. The fourth quantum number is given the symbol $\mathbf{m}_s$. Each electron in an atom has a set of four quantum numbers, **n**, ℓ, $\mathbf{m}_\ell$, and $\mathbf{m}_s$, which fix its energy and the shape of its charge cloud. We will now discuss the quantum numbers of electrons as they are used with atoms in general. Here, we will be talking about *isolated, gaseous* atoms in their *ground states*.

*First Quantum Number, **n**. Principal Energy Levels*

The first quantum number, given the symbol **n,** is of primary importance in determining the energy of an electron. For the hydrogen atom, the energy depends only upon n (recall Equation 8.4). In other atoms, the energy of each electron depends mainly, but not completely, upon the value of **n**. As n increases, the energy of the electron increases and, on the average, it moves farther out from the nucleus. The quantum number n can take on only integral values, starting with 1:

$$n = 1, 2, 3, 4, \ldots \qquad (8.7)$$

In an atom, the value of **n** corresponds to what we call a **principal energy level.** Thus, an electron for which **n** = 1 is said to be in the first principal level. If **n** = 2, we are dealing with the second principal level, and so on.

Second Quantum Number, ℓ. Sublevels (s, p, d, f)

Each principal energy level includes one or more **sublevels.** The sublevels are denoted by the second quantum number, ℓ. As we shall see later, the general shape of the electron cloud associated with an electron is determined by ℓ.

The quantum numbers **n** and ℓ are related. We find that ℓ can take on any integral value starting with 0 and going up to a maximum of **(n − 1)**. That is:

$$\ell = 0, 1, 2, \ldots (n - 1) \qquad (8.8)$$

The conditions on n, ℓ, and m_ℓ come from the solution of the Shrödinger equation for the H atom

If **n** = 1, there is only one possible value of ℓ, namely 0. This means that, in the first principal level, there is only one sublevel, for which ℓ = 0. If **n** = 2, two values of ℓ are possible, 0 and 1. In other words, there are two sublevels (ℓ = 0 and ℓ = 1) within the second principal energy level. Similarly,

if **n** = 3: ℓ = 0, 1, or 2; three sublevels
if **n** = 4: ℓ = 0, 1, 2, or 3; four sublevels

In general, **in the nth principal level, there are n different sublevels.**

Another method is more commonly used to designate sublevels. Instead of giving the quantum number ℓ, we use a letter (s, p, d, or f*) to indicate the sublevel. A sublevel for which ℓ = 0 is referred to as an **s sublevel.** If ℓ = 1, we are dealing with a p sublevel. A d sublevel is one for which ℓ = 2; in an f sublevel, ℓ = 3. Usually, in designating a sublevel, we include a number to indicate the principal level as well. Thus we refer to a 1s sublevel (**n** = 1, ℓ = 0), a 2s sublevel (**n** = 2, ℓ = 0), and a 2p sublevel (**n** = 2, ℓ = 1).

---

*These letters come from the adjectives used by spectroscopists to describe spectral lines: *s*harp, *p*rincipal, *d*iffuse, and *f*undamental.

TABLE 8.2	SUBLEVEL DESIGNATIONS FOR THE FIRST FOUR PRINCIPAL LEVELS								
n	1	2		3			4		
ℓ	0	0	1	0	1	2	0	1	2	3
Sublevel	1s	2s	2p	3s	3p	3d	4s	4p	4d	4f

Within a given principal level (same value of **n**), sublevels always increase in energy in the order

$$\mathbf{ns} < \mathbf{np} < \mathbf{nd} < \mathbf{nf} \tag{8.9}$$

Thus a 2p sublevel has a slightly higher energy than a 2s sublevel. By the same token, when $\mathbf{n} = 3$, the 3s sublevel has the lowest energy, the 3p is intermediate, and the 3d has the highest energy.

Third Quantum Number, $\mathbf{m}_\ell$. Orbitals

The terminology seems complicated at first, but you need to learn it

Each sublevel contains one or more **orbitals,** designated by the third quantum number, $\mathbf{m}_\ell$. This quantum number tells us how the electron cloud surrounding the nucleus is directed in space. The value of $\mathbf{m}_\ell$ is related to that of ℓ. For a given value of ℓ, $\mathbf{m}_\ell$ can have any integral value, including 0, between ℓ and $-\ell$. That is,

$$\mathbf{m}_\ell = \ell, \ldots, +1, 0, -1, \ldots, -\ell \tag{8.10}$$

To illustrate how this rule works, consider an s sublevel ($\ell = 0$). Here, $\mathbf{m}_\ell$ can have only one value, 0. This means that an s sublevel contains only one orbital. In contrast, consider a p sublevel ($\ell = 1$). Here, $\mathbf{m}_\ell$ may be 1, 0, or -1. This means that within each p sublevel there are three different orbitals. In general, **for a sublevel of quantum number ℓ, there are a total of $2\ell + 1$ orbitals.** All these orbitals have essentially the same energy.

The shape of the electron cloud making up an orbital can be indicated in various ways. Consider, for example, the orbital within an s sublevel, called an s orbital. Two different ways of representing the 1s orbital are shown in Figure 8.5. The diagram at the left is similar to that shown in Figure 8.4. The density of shading is proportional to the probability of finding the electron at a given distance from the nucleus. The diagram at the right of Figure 8.5 is more commonly used, perhaps because it is easier to draw. The diameter of the sphere indicates the region in which there is a high probability, let us say 90%, of finding the electron.

The shapes of orbitals differ depending upon the sublevel in which they are located. Figure 8.6 shows the shape of the p orbitals. Notice that they are dumbbell-shaped, in contrast to the spherical shape of s orbitals. You will recall that in a p sublevel there are three different p orbitals. As you can see from Figure 8.6, these three orbitals are oriented at right angles to one another. We may consider them to be directed along the x, y, and z axes. They are sometimes referred to as p_x, p_y, and p_z orbitals.

It is possible to draw figures showing the shape and orientation of d orbitals. There

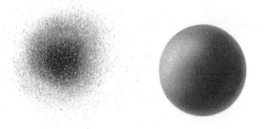

Figure 8.5 Two different ways of indicating the shape of the electron cloud of the 1s orbital. The diagram at the left shows how the probability of finding the electron decreases as one moves out from the nucleus. In the diagram at the right, the colored sphere encloses the region where the electron spends 90% of its time.

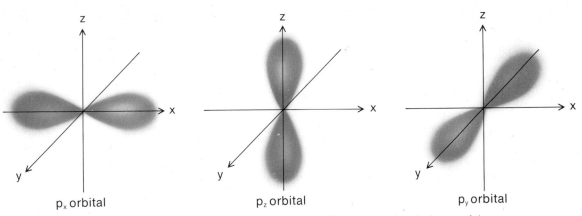

p_x orbital p_z orbital p_y orbital

Figure 8.6 Electron clouds corresponding to the three p orbitals. The electron density in one of these orbitals is symmetrical about the x axis (p_x orbital). In another orbital, it is symmetrical about the z axis (p_z orbital), and in the third it is symmetrical about the y axis (p_y orbital). We describe this situation by saying that the three orbitals are directed at 90° angles to each other.

are five such orbitals within a d sublevel ($\ell = 2$) with $m_\ell = 2, 1, 0, -1$, and -2. We will consider how d orbitals are represented in Chapter 19.

Fourth Quantum Number, m_s. Electron Spin

To describe an electron in an atom completely, we need to specify a fourth quantum number, m_s. This quantum number is associated with the spin of the electron about its axis. It was introduced to make theory consistent with experiment. In that sense it differs from the first three quantum numbers, which came from the solution to the Schrödinger wave equation for the hydrogen atom. The quantum number m_s is not related to n, ℓ, or m_ℓ. It can have either of two possible values:

$$m_s = +1/2 \text{ or } -1/2 \qquad (8.11)$$

Electrons which have the same value of m_s (i.e., both $+1/2$ or both $-1/2$) are said to have *parallel* spins. Electrons which have different m_s values (i.e., one $+1/2$ and the other $-1/2$) are said to have *opposed* spins.

Pauli Exclusion Principle

We have now considered the four quantum numbers which characterize an electron in an atom. There is an important rule called the Pauli exclusion principle which relates to these numbers. It tells us that *no two electrons in an atom can have the same set of four quantum numbers*. This principle was first stated by Wolfgang Pauli in 1925, again to make theory consistent with the properties of atoms.

The Pauli exclusion principle has an implication that may not be obvious at first glance. It requires that no more than two electrons can fit into an orbital. Moreover, if two electrons enter the same orbital they must have opposed spins. To see that this is the case, consider the 2s orbital. Any electron in this orbital must have

$$n = 2, \ell = 0, m_\ell = 0$$

To satisfy the Pauli exclusion principle, the electrons in this orbital must have different m_s values. But there are only two possible values of m_s. Hence only two electrons can enter that orbital. If the orbital is filled, one electron must have $m_s = +1/2$, and the other $m_s = -1/2$. In other words, the two electrons must have opposed spins. Their quantum numbers are

$$n = 2, \ell = 0, m_\ell = 0, m_s = +1/2$$
$$n = 2, \ell = 0, m_\ell = 0, m_s = -1/2$$

Electrons with op-
posed spins in the
same orbital are said
to be paired

The same argument can be applied equally well to any other orbital.

Capacities of Principal Levels, Sublevels, and Orbitals

The rules that we have given for quantum numbers fix the capacities of principal levels, sublevels, and orbitals. In summary:

1. Each principal level of quantum number **n** contains a total of **n** sublevels.
2. Each sublevel of quantum number ℓ contains a total of $2\ell + 1$ orbitals. That is,

> an s sublevel ($\ell = 0$) contains 1 orbital
> a p sublevel ($\ell = 1$) contains 3 orbitals
> a d sublevel ($\ell = 2$) contains 5 orbitals
> an f sublevel ($\ell = 3$) contains 7 orbitals

3. Each orbital can hold two electrons, which must have opposed spins.

Applying these rules to the first three principal energy levels, we obtain Table 8.3. On the bottom line of the table, each arrow indicates an electron. In each orbital, there are two electrons with opposed spins. The number of electrons in a sublevel is found by

Since orbital capacity
is fixed, the capacities
of sublevels and levels
are also fixed

adding up the electrons in the orbitals within that sublevel. For example, in a p sublevel ($\ell = 1$), there are six electrons, two in each of three orbitals. To find the total number of electrons in a principal level, we add the electrons in the sublevels within that principal level.

TABLE 8.3	ALLOWED SETS OF QUANTUM NUMBERS FOR ELECTRONS IN ATOMS												
Level **n**	1	2			3								
Sublevel ℓ	0	0	1			0	1			2				
Orbital m_ℓ	0	0	1	0	-1	0	1	0	-1	2	1	0	-1	-2
Spin m_s ↑ = +1/2 ↓ = $-1/2$	⇅	⇅	⇅	⇅	⇅	⇅	⇅	⇅	⇅	⇅	⇅	⇅	⇅	⇅

Example 8.2
    a. How many electrons can fit into the principal level for which **n** = 2?
    b. What is the capacity for electrons of the 3d sublevel?

Solution
    a. In the **n** = 2 level, there are two sublevels, the 2s and the 2p. Of these, the 2s contains one orbital with a capacity of 2 e^-. The 2p contains three orbitals, each of which can hold 2 e^-.

$$\text{total capacity} = 1(2) + 3(2) = 8$$

Note that under the **n** = 2 level in Table 8.3 there are eight electrons, in four pairs.

b. Like any d sublevel, the 3d contains five orbitals. Each orbital can hold 2 e⁻. This result is confirmed in Table 8.3, where we have five electron pairs (10 e⁻) in the ℓ = 2 sublevel.

Exercise   What is the total capacity for electrons of the third principal energy level?
Answer: 18.

Using the approach in Example 8.2, we can readily obtain the total electron capacities of any principal level or sublevel. In Table 8.4, we list these capacities through **n** = 4. Notice that the capacity of a principal level is $2n^2$, where **n** is the quantum number.

TABLE 8.4   CAPACITIES OF ELECTRONIC LEVELS AND SUBLEVELS
IN ATOMS

LEVEL n	TOTAL NUMBER OF ELECTRONS IN LEVEL	NUMBER OF ELECTRONS IN SUBLEVELS s p d f			
1	2	2	–	–	–
2	8	2	6	–	–
3	18	2	6	10	–
4	32	2	6	10	14

8.5   ELECTRON ARRANGEMENTS IN ATOMS

There are several ways to show how electrons in an atom are distributed among principal levels, sublevels, and orbitals. The **electron configuration** gives the number of electrons in each principal level and sublevel. With an **orbital diagram,** we go one step further and indicate the arrangement of electrons within orbitals. Finally, we can specify the set of four quantum numbers for each electron.

Table 8.5 gives the electron configuration, orbital diagram, and quantum numbers of the electrons in hydrogen (1 e⁻) and helium (2 e⁻). Note that

—in electron configurations, a superscript is used to indicate the number of electrons in a given sublevel.

—in an orbital diagram, arrows are used to indicate electron spins. Thus ↑ indicates an electron with m_s = +1/2; ↓ shows an electron with the opposite spin, m_s = −1/2.

—in specifying quantum numbers, they are arranged in the order **n**, ℓ, m_ℓ, m_s.

TABLE 8.5   ELECTRON ARRANGEMENTS IN THE HYDROGEN AND
HELIUM ATOMS

	₁H (one electron)	₂He (two electrons)
Electron configuration	$1s^1$	$1s^2$
Orbital diagram	1s (↑)	1s (↑↓)
Quantum numbers	1, 0, 0, +1/2	↑: 1, 0, 0, +1/2 ↓: 1, 0, 0, −1/2

To arrive at the electron configurations of atoms, we must know the order in which different sublevels are filled. Electrons enter the available sublevels in order of their increasing energy. Usually, a sublevel is filled to capacity before the next one is entered. The relative energies of different sublevels can be obtained from experiment. Figure 8.7 is a plot of these energies through the $n = 4$ principal level.

From Figure 8.7, it is possible to predict the electron configurations of atoms of elements with atomic numbers 1 through 30. Since an s sublevel can hold only two electrons, the 1s is filled at helium ($1s^2$). With lithium (at. no. = 3), the third electron has to enter a new sublevel. This is the 2s, the lowest sublevel of the second principal energy level. Lithium puts one electron in this sublevel ($1s^2 2s^1$). With beryllium (at. no. = 4), the 2s sublevel is filled ($1s^2 2s^2$). The next six elements (at. nos. 5 through 10) fill the 2p sublevel. Their electron configurations are

$_5$B	$1s^2 2s^2 2p^1$	$_8$O	$1s^2 2s^2 2p^4$
$_6$C	$1s^2 2s^2 2p^2$	$_9$F	$1s^2 2s^2 2p^5$
$_7$N	$1s^2 2s^2 2p^3$	$_{10}$Ne	$1s^2 2s^2 2p^6$

Beyond neon, we enter the third principal level. The 3s sublevel is filled at magnesium:

$$_{12}\text{Mg} \qquad 1s^2 2s^2 2p^6 3s^2$$

Six more electrons are required to fill the 3p sublevel. This is filled to capacity at argon:

$$_{18}\text{Ar} \qquad 1s^2 2s^2 2p^6 3s^2 3p^6$$

After argon, we observe an "overlap" of principal energy levels. The next electron enters the *lowest* sublevel of the fourth principal level (4s) instead of the *highest* sublevel of the third principal level (3d). Potassium (at. no. = 19) puts one electron in the 4s sublevel; calcium (at. no. = 20) fills it with two electrons:

$$_{20}\text{Ca} \qquad 1s^2 2s^2 2p^6 3s^2 3p^6 4s^2$$

Now, the 3d sublevel starts to fill with scandium (at. no. = 21). Recall that a d sublevel has a capacity of ten electrons. Hence the 3d sublevel becomes filled at zinc (at. no. = 30):

$$_{30}\text{Zn} \qquad 1s^2 2s^2 2p^6 3s^2 3p^6 4s^2 3d^{10}$$

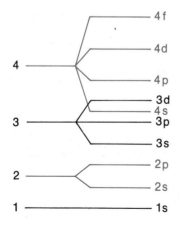

Figure 8.7   In general, energy increases with the principal quantum number, **n**. However, it is possible for the lowest sublevel of **n** = 4 (i.e., 4s) to be below the highest sublevel of **n** = 3 (i.e., 3d). This appears to be the situation in the potassium and calcium atoms, where successive electrons enter the 4s rather than the 3d sublevel.

> **Example 8.3** Find the electron configuration of the sulfur atom and the nickel atom.
>
> **Solution** S atom: at. no. 16; 16 electrons. We fill the sublevels to capacity in order of increasing energy. There are two 1s electrons, two 2s, six 2p, two 3s, and four 3p electrons, making a total of sixteen. The configuration of the S atom is therefore $1s^2 2s^2 2p^6 3s^2 3p^4$.
>
> Ni atom: at. no. 28; 28 electrons. Proceeding as before, this time until we have added 28 electrons, we obtain as the electron configuration of the Ni atom: $1s^2 2s^2 2p^6 3s^2 3p^6 4s^2 3d^8$. (Note that the 4s sublevel fills before the 3d.)
>
> ---
>
> **Exercise** Give the electron configuration of Mn (at. no. = 25). Answer: $1s^2 2s^2 2p^6 3s^2 3p^6 4s^2 3d^5$.

With a little practice, problems like this become easy

We can use this general approach to find the electron configuration of any atom. The method, as we have seen, involves adding electrons one by one as atomic number increases. This is sometimes referred to as the Aufbau (building-up) process. To follow it, all we need to know is the order in which sublevels are filled. Figure 8.7 gives this information for the first 30 electrons. Beyond that, it is probably simplest to deduce the order of filling from the Periodic Table (Section 8.6).

In Table 8.6 we list the electron configurations of elements of atomic numbers 1 through 101. They generally follow the rules we have illustrated. The major exceptions involve elements where a sublevel is close to being filled or half-filled. It appears that having a full or half-full sublevel leads to increased stability. Consider, for example, chromium (at. no. = 24). Here, a 4s electron is "promoted" to the 3d sublevel, giving chromium five 3d electrons. In copper, a 4s electron is also promoted to the 3d sublevel, thereby filling that sublevel.

All things considered, electron configurations follow the rules very well

You will note that in Table 8.6 we have saved space by using the phrases "neon core" to indicate the first ten electrons, "argon core" for the first 18 electrons, and so on. This method is often used in writing out what we call *abbreviated electron configurations*. Thus for nickel (at. no. = 28) we write

$$[_{18}Ar]4s^2 3d^8$$

Here the symbol $[_{18}Ar]$ indicates that the first 18 electrons have the argon configuration $1s^2 2s^2 2p^6 3s^2 3p^6$. Usually, in an abbreviated electron configuration, we start from the preceding noble gas.

Orbital Diagrams. Hund's Rule

For many purposes, electron configurations are sufficient to describe the arrangement of electrons in atoms. The energy of an electron is determined primarily by the principal level and sublevel in which it is located. Sometimes, however, we want to indicate how electrons are distributed within orbitals. To do this, we use orbital diagrams such as those shown in Table 8.5, p. 175, for hydrogen and helium.

To show how orbital diagrams are obtained from electron configurations, consider the boron atom (at. no. = 5). Its electron configuration is $1s^2 2s^2 2p^1$. We know that the pair of electrons in the 1s orbital must have opposed spins. The same is true for the two electrons in the 2s orbital. There are three orbitals in the 2p sublevel. The single 2p electron in boron could be in any one of these orbitals. Its spin could be either "up" ($m_s = +1/2$) or "down" ($m_s = -1/2$). The orbital diagram is ordinarily written

TABLE 8.6   GROUND STATE ELECTRON CONFIGURATIONS OF GASEOUS ATOMS

ELEMENT	ATOMIC NUMBER	POPULATIONS OF SUBSHELLS										
		1s	2s	2p	3s	3p	3d	4s	4p	4d	4f	5s
H	1	1										
He	2	2										
Li	3	2	1									
Be	4	2	2									
B	5	2	2	1								
C	6	2	2	2								
N	7	2	2	3								
O	8	2	2	4								
F	9	2	2	5								
Ne	10	2	2	6								
Na	11				1							
Mg	12				2							
Al	13	Neon core			2	1						
Si	14				2	2						
P	15				2	3						
S	16				2	4						
Cl	17				2	5						
Ar	18	2	2	6	2	6						
K	19							1				
Ca	20							2				
Sc	21						1	2				
Ti	22						2	2				
V	23						3	2				
Cr	24						5	1				
Mn	25						5	2				
Fe	26	Argon core					6	2				
Co	27						7	2				
Ni	28						8	2				
Cu	29						10	1				
Zn	30						10	2				
Ga	31						10	2	1			
Ge	32						10	2	2			
As	33						10	2	3			
Se	34						10	2	4			
Br	35						10	2	5			
Kr	36	2	2	6	2	6	10	2	6			
Rb	37											1
Sr	38											2
Y	39									1		2
Zr	40									2		2
Nb	41									4		1
Mo	42			Krypton core						5		1
Tc	43									6		1
Ru	44									7		1
Rh	45									8		1
Pd	46									10		
Ag	47									10		1
Cd	48									10		2

TABLE 8.6 GROUND STATE ELECTRON CONFIGURATIONS OF GASEOUS ATOMS (*Continued*)

ELEMENT	ATOMIC NUMBER		4d	4f	5s	5p	5d	5f	6s	6p	6d	7s
In	49		10		2	1						
Sn	50		10		2	2						
Sb	51		10		2	3						
Te	52		10		2	4						
I	53		10		2	5						
Xe	54		10		2	6						
Cs	55		10		2	6			1			
Ba	56		10		2	6			2			
La	57		10		2	6	1		2			
Ce	58		10	2	2	6			2			
Pr	59		10	3	2	6			2			
Nd	60		10	4	2	6			2			
Pm	61		10	5	2	6			2			
Sm	62		10	6	2	6			2			
Eu	63		10	7	2	6			2			
Gd	64		10	7	2	6	1		2			
Tb	65		10	9	2	6			2			
Dy	66		10	10	2	6			2			
Ho	67		10	11	2	6			2			
Er	68		10	12	2	6			2			
Tm	69		10	13	2	6			2			
Yb	70		10	14	2	6			2			
Lu	71		10	14	2	6	1		2			
Hf	72		10	14	2	6	2		2			
Ta	73		10	14	2	6	3		2			
W	74		10	14	2	6	4		2			
Re	75	Krypton	10	14	2	6	5		2			
Os	76	core	10	14	2	6	6		2			
Ir	77		10	14	2	6	9					
Pt	78		10	14	2	6	9		1			
Au	79		10	14	2	6	10		1			
Hg	80		10	14	2	6	10		2			
Tl	81		10	14	2	6	10		2	1		
Pb	82		10	14	2	6	10		2	2		
Bi	83		10	14	2	6	10		2	3		
Po	84		10	14	2	6	10		2	4		
At	85		10	14	2	6	10		2	5		
Rn	86		10	14	2	6	10		2	6		
Fr	87		10	14	2	6	10		2	6		1
Ra	88		10	14	2	6	10		2	6		2
Ac	89		10	14	2	6	10		2	6	1	2
Th	90		10	14	2	6	10		2	6	2	2
Pa	91		10	14	2	6	10	2	2	6	1	2
U	92		10	14	2	6	10	3	2	6	1	2
Np	93		10	14	2	6	10	5	2	6		2
Pu	94		10	14	2	6	10	6	2	6		2
Am	95		10	14	2	6	10	7	2	6		2
Cm	96		10	14	2	6	10	7	2	6	1	2
Bk	97		10	14	2	6	10	9	2	6		2
Cf	98		10	14	2	6	10	10	2	6		2
Es	99		10	14	2	6	10	11	2	6		2
Fm	100		10	14	2	6	10	12	2	6		2
Md	101		10	14	2	6	10	13	2	6		2

Atom	Orbital diagram			Electron configuration
B	$(\uparrow\downarrow)$	$(\uparrow\downarrow)$	$(\uparrow\ )(\ \ )(\ \ )$	$1s^2 2s^2 2p$
C	$(\uparrow\downarrow)$	$(\uparrow\downarrow)$	$(\uparrow\ )(\uparrow\ )(\ \ )$	$1s^2 2s^2 2p^2$
N	$(\uparrow\downarrow)$	$(\uparrow\downarrow)$	$(\uparrow\ )(\uparrow\ )(\uparrow\ )$	$1s^2 2s^2 2p^3$
O	$(\uparrow\downarrow)$	$(\uparrow\downarrow)$	$(\uparrow\downarrow)(\uparrow\ )(\uparrow\ )$	$1s^2 2s^2 2p^4$
F	$(\uparrow\downarrow)$	$(\uparrow\downarrow)$	$(\uparrow\downarrow)(\uparrow\downarrow)(\uparrow\ )$	$1s^2 2s^2 2p^5$
Ne	$(\uparrow\downarrow)$	$(\uparrow\downarrow)$	$(\uparrow\downarrow)(\uparrow\downarrow)(\uparrow\downarrow)$	$1s^2 2s^2 2p^6$
	1s	2s	2p	

Figure 8.8 Orbital diagrams showing electron arrangements for atoms with five to ten electrons. Electrons entering orbitals of equal energy remain unpaired as long as possible.

$$\qquad\qquad\qquad 1s\qquad 2s\qquad\qquad 2p$$
$$_5B\qquad\qquad (\uparrow\!\downarrow)\quad (\uparrow\!\downarrow)\quad (\uparrow)(\ \ )(\ \ )$$

With the next element, carbon, a complication arises. Where should we put the sixth electron? We could put it in the same orbital as the other 2p electron, in which case it would have to have the opposite spin, $\downarrow$. It could go into one of the other two orbitals, either with a parallel spin, $\uparrow$, or an opposed spin, $\downarrow$. Experiment shows that there is an energy difference between these arrangements. The most stable is the one in which the two electrons are in different orbitals with parallel spins. The orbital diagram of the carbon atom is

$$\qquad\qquad\qquad 1s\qquad 2s\qquad\qquad 2p$$
$$_6C\qquad\qquad (\uparrow\!\downarrow)\quad (\uparrow\!\downarrow)\quad (\uparrow)(\uparrow)(\ \ )$$

This situation arises frequently with other atoms. There is a general principle that applies in all such cases. Known as Hund's rule, it can be stated as follows:

Within a given sublevel, the order of filling is such that there is the maximum number of half-filled orbitals. The single electrons in these half-filled orbitals have parallel spins.

Following this principle, we show the orbital diagrams for the elements boron through neon in Figure 8.8. Note that the effect of Hund's rule is to give, in each case, the maximum number of electrons with parallel spins (two for C, three for N, two for O).

Example 8.4 Construct the orbital diagrams of the sulfur and nickel atoms.

Solution We first need to know the electron configurations of these atoms. In Example 8.3 we found them to be: S atom, $1s^2 2s^2 2p^6 3s^2 3p^4$; Ni atom, $1s^2 2s^2 2p^6 3s^2 3p^6 4s^2 3d^8$. In the orbital diagrams we list all orbitals, pairing all electrons in orbitals in filled sublevels. With unfilled sublevels, electrons are put one by one in the available orbitals, keeping spins parallel, unpaired, as much as possible.

S atom: $(\uparrow\!\downarrow)$ $(\uparrow\!\downarrow)$ $(\uparrow\!\downarrow)(\uparrow\!\downarrow)(\uparrow\!\downarrow)$ $(\uparrow\!\downarrow)$ $(\uparrow\!\downarrow)(\uparrow)(\uparrow)$
 1s   2s   2p   3s   3p

Ni atom: (↿⇂)   (↿⇂) (↿⇂)(↿⇂)(↿⇂)   (↿⇂)   (↿⇂)(↿⇂)(↿⇂)   (↿⇂)   (↿⇂)(↿⇂)(↿⇂)(↑)(↑)
　　　　　 1s　　2s　　　2p　　　　　3s　　　3p　　　4s　　　　　3d

We would expect that there would be two electrons left unpaired in the S atom. In Ni there would also be two unpaired electrons, in the 3d sublevel.

Exercise   Show the orbital diagram for the 3d sublevel for Mn (at. no. = 25). Answer: (↑)(↑)(↑)(↑)(↑).

Hund's rule, like the Pauli exclusion principle, is based upon experiment. It is possible to determine the number of unpaired electrons in an atom. With solids, this is done by studying their behavior in a magnetic field. If there are unpaired electrons present, the solid will be attracted into the field. Such a substance is said to be *para-magnetic*. If the atoms in the solid contain only paired electrons, it is slightly repelled by the field. Substances of this type are called *diamagnetic*. With gaseous atoms, the atomic spectrum can also be used to establish the presence and number of unpaired electrons.

Experimentally, we find that certain gaseous atoms are paramagnetic. Among these are H(g) and Li(g). Both of these have a single electron in a half-filled orbital. In contrast, He(g) and Be(g) are diamagnetic. In these two atoms, each orbital is completely full. Since we find that C(g) and N(g) are paramagnetic, we infer that they, like H(g) and Li(g), have half-filled orbitals. Indeed, we can go one step further and show that C(g) has two "unpaired" electrons, in different orbitals with parallel spins. Similarly, N(g) can be shown to have three unpaired electrons.

Unlike atoms, most molecules are diamagnetic

Quantum Numbers

Within limits, it is possible to state a complete set of quantum numbers for electrons in atoms. This was done in Table 8.5 for hydrogen and helium. Ordinarily, to specify the quantum numbers, it is simplest to start with the orbital diagram. Consider, for example, the boron atom, whose orbital diagram is given on p. 180. For the first four electrons, we readily arrive at the sets of quantum numbers:

$$1, 0, 0, +1/2; \quad 1, 0, 0, -1/2; \quad 2, 0, 0, +1/2; \quad 2, 0, 0, -1/2$$

We note that the fifth electron is in a 2p orbital. Its m_ℓ value could be 1, 0, or −1. If we agree to use the highest m_ℓ value for the first orbital to be filled, the quantum numbers for the fifth electron in boron are

$$2, 1, 1, +1/2$$

Following the same procedure, the set of quantum numbers for the sixth electron in carbon would be

2, 1, 0, +1/2　　(notice that the fifth and sixth electrons must have parallel spins)

For the seventh electron in nitrogen, we have

2, 1, −1, +1/2　　(all three 2p electrons have parallel spins)

This is one set, but there are others. Can you see why?

> **Example 8.5** Write a complete set of quantum numbers for the four 3p electrons in the sulfur atom. Its orbital diagram is shown in Example 8.4.
>
> **Solution** For all of these electrons, $\mathbf{n} = 3$, $\ell = 1$. Let us agree to assign $\mathbf{m}_\ell$ values of 1, 0, and −1 from left to right in the diagram. Within the first 2p orbital, the electrons have opposed spins ($\mathbf{m}_s = +1/2$ and −1/2). The two electrons in the half-filled 2p orbitals have parallel spins. Since they are drawn as ↑, we will assign them $\mathbf{m}_s$ values of +1/2. Hence we have
>
> $$3, 1, 1, +1/2; \qquad 3, 1, 1, -1/2; \qquad 3, 1, 0, +1/2; \qquad 3, 1, -1, +1/2$$
>
> _____
>
> **Exercise** Give the complete set of four quantum numbers for the three 2p electrons in N (Fig. 8.8). Answer: 2, 1, 1, +1/2; 2, 1, 0, +1/2; 2, 1, −1, +1/2.

There is one difficulty with the model just presented that should be mentioned. The theory is based on the assumption that the electrons in an atom can be described by assigning them quantum numbers. This means that in the theory we consider the electrons to have properties as individuals. (Indeed the quantum numbers we use are based on the wave mechanical solution of the one-electron problem and, hence, involve only the electron-nucleus interactions.) In such an atomic model the electronic properties of the atom will simply be equal to the sum of the properties of all the electrons in the atom. Such a model is clearly incorrect. In an atom there are important electron-electron charge interactions and electron-electron spin interactions, as well as the electron-nucleus charge interaction covered by the model. So far, the theory of atomic structure has been unable to properly treat electron-electron interactions in atoms. Such interactions are usually considered to be "averaged in" when one interprets electron configurations. The qualitative structure of atoms, involving the existence of shells and their populations, does not seem to be appreciably altered by such an averaging. This explains the usefulness of electron configurations. Clearly, however, the necessity for such an averaging renders impossible the use of electron configurations in quantitative calculations of atomic energy levels and atomic dimensions.

8.6 ELECTRON ARRANGEMENTS AND THE PERIODIC TABLE

In Section 8.5, we discussed electron arrangements in atoms. In essence, what we did was to follow a set of arbitrary rules to assign electrons to principal levels, sublevels, and orbitals. These assignments are based, in part, upon theory, particularly the solution to the wave equation for the hydrogen atom. Ultimately, however, they must be consistent with fact. Some types of experimental evidence related to this problem have been discussed in this chapter. Atomic spectra, ionization energies, and magnetic behavior are all very helpful in deciding upon the electronic structures of atoms.

When the elements with the same outermost electron configurations are put in columns, the result is the Periodic Table

By far the most important evidence for electron arrangements comes from the Periodic Table. This Table was first proposed by Mendeleev in 1869, before the electron itself was discovered. We now know, however, that there is a direct correlation between the position of an element in the Periodic Table and its electron configuration. **Atoms of elements in the same vertical group of the Periodic Table have the same outer electron configuration.** To see that this is the case, consider the configurations listed in Table 8.7. Notice that all

—atoms of Group 1 elements have one s electron in the outermost principal level.

—atoms of Group 2 elements have two s electrons in the outermost principal level.

This relation explains why elements in a given group closely resemble one another in their chemical properties. When an element takes part in a chemical reaction, it is ordinarily the electrons in the outermost energy levels that are involved. Consider, for

TABLE 8.7   ELECTRON CONFIGURATIONS OF THE GROUP 1
AND 2 ELEMENTS

GROUP 1		GROUP 2	
$_3$Li	$[_2$He$]2s^1$	$_4$Be	$[_2$He$]2s^2$
$_{11}$Na	$[_{10}$Ne$]3s^1$	$_{12}$Mg	$[_{10}$Ne$]3s^2$
$_{19}$K	$[_{18}$Ar$]4s^1$	$_{20}$Ca	$[_{18}$Ar$]4s^2$
$_{37}$Rb	$[_{36}$Kr$]5s^1$	$_{38}$Sr	$[_{36}$Kr$]5s^2$
$_{55}$Cs	$[_{54}$Xe$]6s^1$	$_{56}$Ba	$[_{54}$Xe$]6s^2$

example, the Group 1 metals. When atoms of these elements react with nonmetals or with water, they lose their outer s electrons. Since all the Group 1 metals have a single outer s electron, they all form +1 ions (Li^+, Na^+, K^+, Rb^+, Cs^+). This explains why the formulas of the compounds they form resemble each other so closely—that is,

chlorides (Cl^- ion): LiCl, NaCl, KCl, RbCl, CsCl
hydroxides (OH^- ion): LiOH, NaOH, KOH, RbOH, CsOH
sulfates (SO_4^{2-} ion): Li_2SO_4, Na_2SO_4, K_2SO_4, Rb_2SO_4, Cs_2SO_4

The electrons in the outermost shell fix the chemical properties of an atom

Filling of Sublevels in the Periodic Table

Figure 8.9, p. 184, indicates the regions of the Periodic Table where various sublevels are filling. Notice that:

1. The elements in Groups 1 and 2 are filling an s sublevel. Thus, Li and Be in the second period fill the 2s sublevel. Na and Mg in the third period fill the 3s sublevel, and so on.

2. The elements in Groups 3 through 8 (six elements in each period) fill p sublevels, which have a capacity of six electrons. In the second period, the 2p sublevel starts to fill with B (at. no. = 5) and is completed with Ne (at. no. = 10). In the third period, the elements Al (at. no. = 13) through Ar (at. no. = 18) fill the 3p sublevel.

3. The ten transition metals, in the center of the Periodic Table, fill d sublevels. Remember that a d sublevel has a capacity of ten electrons. The elements Sc (at. no. = 21) through Zn (at. no. = 30) fill the 3d sublevel. In the fifth period, the elements Y (at. no. = 39) through Cd (at. no. = 48) fill the 4d sublevel. The ten transition metals in the sixth period fill the 5d sublevel.

4. There are two sets of 14 elements each listed separately (to save space) at the bottom of the Table. These elements are filling f sublevels. The 14 elements following lanthanum, which have atomic numbers 58 through 71, are filling the 4f sublevel. They are frequently referred to as *lanthanides*, or sometimes as *rare earths*. The 14 elements following actinium are called *actinides*. These elements, of atomic number 90 through 103, are filling the 5f sublevel.

Because they have similar properties, compounds of the rare earths are difficult to separate from one another. Until quite recently, samples of pure compounds of these elements were not available except for those of cerium, the most abundant member of the series. Chromatographic processes are now used to separate compounds of the lanthanide metals. The availability of these compounds in highly pure form has led to several commercial applications. A brilliant red phosphor now used in color TV receivers contains a small amount of europium oxide, Eu_2O_3. This is added to a base of yttrium oxide, Y_2O_3, or gadolinium oxide, Gd_2O_3. Another rare earth oxide, Nd_2O_3, is being used as part of a liquid laser system.

The actinide metals are all radioactive. Only two of these elements, uranium and thorium, are found in appreciable amounts in nature. The other elements were first observed in the products

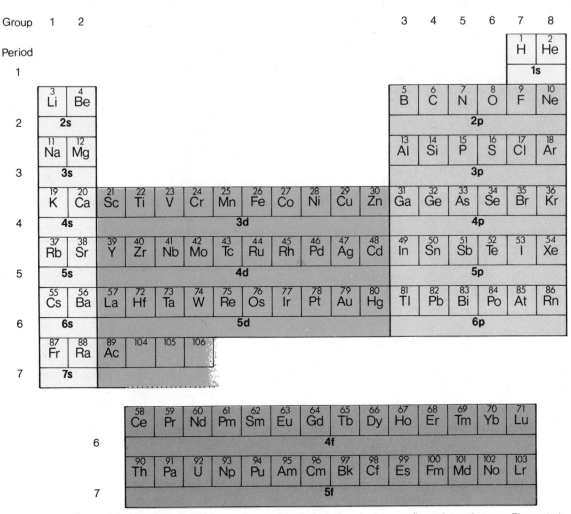

Figure 8.9 The Periodic Table can be used to deduce the electron configurations of atoms. Elements in Groups 1 and 2 are filling an ns sublevel, where n is the number of the period. Elements in Groups 3 through 8 (except for H and He) fill an np sublevel. The transition metals fill an (n − 1)d sublevel. For example, the elements Sc through Zn in the fourth period fill the 3d sublevel. The lanthanides fill the 4f sublevel, while the actinides fill the 5f.

of controlled nuclear reactions. In most cases, they have been produced in only very small amounts. Uranium and plutonium are used as the fuel elements in nuclear reactors and nuclear weapons (Chap. 24).

Outer Electron Configurations from the Periodic Table

We could, if we had to, use Figure 8.9 to deduce the complete electron configuration of any element. More commonly, we use the Periodic Table itself for a less ambitious purpose. Specifically, we use it to obtain the outer electron configurations of the main-group elements. These are:

Group	1	2	3	4	5	6	7	8
Outer Configuration	ns^1	ns^2	ns^2np^1	ns^2np^2	ns^2np^3	ns^2np^4	ns^2np^5	ns^2np^6

where *n is the number of the period in which the element is located*. That is, n = 1 for the first period (H, He), n = 2 for the second period (Li → Ne), and so on.

Example 8.6   Using the Periodic Table, give the outer electron configuration of
 a. Ba (at. no. = 56)   b. Te (at. no. = 52)

Solution
 a. Barium is in the sixth period, in Group 2. Outer configuration: $6s^2$.
 b. Tellurium is in the fifth period, in Group 6. Outer configuration: $5s^25p^4$.

Exercise   Give the outer electron configuration of lead. Answer: $6s^26p^2$.

8.7 ELECTRON CONFIGURATIONS AND THE PHYSICAL PROPERTIES OF ATOMS

Many of the trends in the properties of atoms are readily explained in terms of their electron configurations. In this section, we will consider two such properties, atomic radius and ionization energy. We will be interested in seeing how these properties relate to electron configurations and how they change as we move across or down in the Periodic Table.

Atomic Radius

The size of an atom is a rather nebulous concept. The electron cloud surrounding the nucleus does not have a sharp boundary. We can, however, define and measure a quantity known as the atomic radius, assuming spherical atoms. Quite simply, the atomic radius is taken to be one half the distance between centers of touching atoms (Fig. 8.10). One way of measuring this quantity will be discussed in Chapter 9.

The atoms are studied in the element, not in a compound

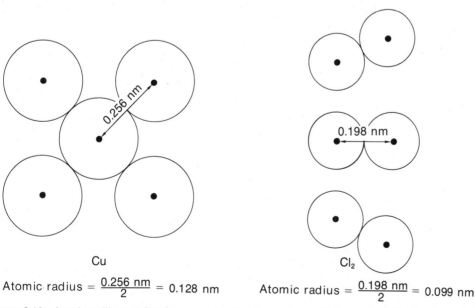

Cu

Atomic radius = $\dfrac{0.256 \text{ nm}}{2}$ = 0.128 nm

Cl₂

Atomic radius = $\dfrac{0.198 \text{ nm}}{2}$ = 0.099 nm

Figure 8.10   Atomic radii are defined by assuming that those atoms which are closest in the element are touching. The atomic radius is taken to be one half of the closest internuclear distance.

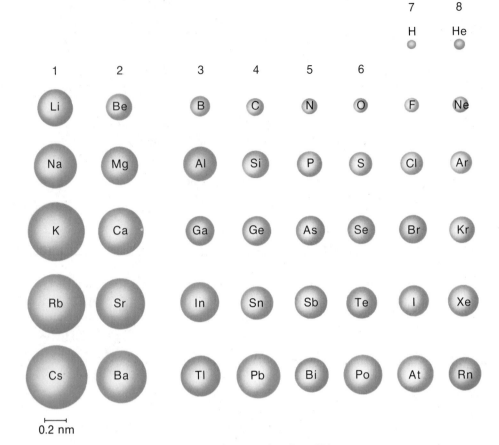

Figure 8.11 Atomic radii of the main-group elements. Atomic radii increase as one goes down a group and in general decrease going across a row in the Periodic Table. Hydrogen has the smallest atom, cesium the largest.

The atomic radii of the main-group elements are shown in Figure 8.11. Notice that, in general, **atomic radius decreases as we move across the Periodic Table from left to right and increases as we move down within a given group.**

These trends can be rationalized in terms of the electron configurations of the corresponding atoms. Consider first the increase in radius as we move down the Table, let us say among the Group 1 metals. All of these elements have a single s electron outside a filled level or filled p sublevel. Electrons in these inner levels are much closer to the nucleus than the outer s electron. Hence, they "shield" it from the positive charge of the nucleus. Each inner electron just about cancels the charge of one proton in the nucleus. Thus, the outer s electron is attracted toward the nucleus by a net charge of +1. In this sense, it behaves like the electron in the hydrogen atom. But the average distance of the electron from the hydrogen nucleus increases with the first quantum number, **n**. So, we would expect the radius to increase as we move from Li (2s electron) to Na (3s electron), and so on down the group.

The decrease in atomic radius as we move across the Periodic Table can be explained in a similar way. Consider the third period, where electrons are being added to the third principal energy level. The added electrons should be rather poor shields for one another, since they are all at about the same distance from the nucleus. Only the ten inner electrons in filled levels should shield the outer electrons from the nuclear charge. Hence, the "effective nuclear charge" increases steadily as we move across the period. It should be about +1 for Na (at. no. = 11), +2 for Mg (at. no. = 12), and so on. As the effective nuclear charge increases, outer electrons are pulled in closer to the nucleus and atomic radius decreases.

Seems reasonable, once you think about it

1	2		3	4	5	6	7	8
							H 1318	He 2377
Li 527	Be 904		B 808	C 1092	N 1410	O 1322	F 1686	Ne 2088
Na 502	Mg 745		Al 586	Si 791	P 1021	S 1004	Cl 1264	Ar 1527
K 427	Ca 594		Ga 586	Ge 770	As 954	Se 946	Br 1146	Kr 1356
Rb 410	Sr 556		In 565	Sn 715	Sb 841	Te 879	I 1017	Xe 1176
Cs 377	Ba 510		Tl 594	Pb 724	Bi 711	Po 820	At	Rn 1042

Figure 8.12 First ionization energies of the main-group elements (in kilojoules per mole). In general, ionization energy decreases as one moves down in the Periodic Table and increases as one moves across from left to right. Cesium has the smallest ionization energy, and helium the largest.

Ionization Energy

The first ionization energy of an element is the energy change, ΔE, for the removal of the outermost electron from a gaseous atom to form a $+1$ ion.

$$M(g) \rightarrow M^+(g) + e^-; \quad \Delta E = \text{ionization energy} \qquad (8.12)$$

Ionization energy is always a positive quantity. Energy must be absorbed to remove an electron. The magnitude of ΔE is a measure of the ease with which an atom gives up an electron. *The smaller the ionization energy, the more readily the electron is removed.*

Ionization energies of the main-group elements are listed in Figure 8.12. Notice that **the general trend is for ionization energy to increase as we move across the Periodic Table and decrease as we move down.** Comparing Figure 8.12 with 8.11, we see an inverse correlation between ionization energy and atomic radius. The smaller the atom, the more tightly its electrons are held to the nucleus and the more difficult they are to remove. Large atoms such as those of the heavier Group 1 metals have rather low ionization energies. The outer electrons are far away from the nucleus and hence are relatively easy to remove.

Example 8.7   Consider the three elements B, C, and Al. Using only the Periodic Table, predict which of the three elements has
   a. the largest atomic radius; the smallest atomic radius.
   b. the largest ionization energy; the smallest ionization energy.

Solution   The three elements form a block in the Table:

$$\text{B} \qquad \text{C}$$
$$\text{Al}$$

a. Since atomic radius increases as we go down in the Table and decreases as we go across, Al should be the largest and C the smallest atom. The observed atomic radii are: B = 0.088 nm; C = 0.077 nm; Al = 0.143 nm.

b. Since ionization energy increases as we go across in the Table and decreases as we go down, C should have the highest value and Al the smallest. The observed values are: B = 808 kJ/mol; C = 1092 kJ/mol; Al = 586 kJ/mol.

Exercise   Compare the two elements Ca and Rb in regard to atomic radius and ionization energy. Answer: Rb has the larger radius and the smaller ionization energy.

If you look carefully at Figure 8.12, you will note a few exceptions to the general trends referred to above and illustrated in Example 8.7. For example, ionization energy actually decreases as one moves across from Be in Group 2 (904 kJ/mol) to B in Group 3 (808 kJ/mol). This is readily explained when we realize that the first electron removed from boron comes from the 2p sublevel as opposed to 2s for beryllium. Since 2p is higher in energy than 2s, it is not too surprising that less energy is required to remove an electron from that sublevel.

8.8   HISTORICAL PERSPECTIVE ON THE PERIODIC TABLE: NEWLANDS, MENDELEEV, AND MEYER

In 1867, Dmitri Mendeleev, professor of chemistry at the University of St. Petersburg, started writing a textbook of general chemistry. One problem he faced was that of finding a logical way to organize the descriptive chemistry of the more than 60 elements known at that time. After several false starts, he hit upon the idea of devoting separate chapters to families of elements with similar properties. Early in 1869, he was working with the chapters on the elements of Groups 7, 1, and 2. Listing successive members of these families one after the other in order of increasing atomic mass,

ΔAM →								
16.5	44.5	47	F = 19	Cl = 35.5	Br = 80	I = 127		
14	46	48	Na = 23	K = 39	Rb = 85	Cs = 133		
16	48	49	Mg = 24	Ca = 40	Sr = 88	Ba = 137		

Mendeleev was struck by the nearly constant increase in atomic mass between corresponding members of the three families. Further speculation along these lines led him to the discovery of the Periodic Table, which first appeared in a paper presented before a meeting of the Russian Chemical Society on March 6, 1869. The acclaim that greeted this and successive papers lifted Mendeleev from obscurity to fame. One result was that he had no trouble finding a publisher for his textbook, *Principles of Chemistry*. It appeared in 1870, went through eight editions, and was translated into English, French, and German.

Although Mendeleev was unaware of it at the time, he was not the first to come up with the idea of a Periodic Table. That distinction belongs to J. A. R. Newlands, professor of chemistry at the City of London College. In 1864 he arranged the elements in order of increasing atomic mass in rows of seven and found that such a sequence tended to put elements with similar properties into the same group. Unfortunately, Newlands' Periodic Table, based on what he called the "Law of Octaves," met with only ridicule and scorn from members of the English Chemical Society. Indeed, they refused to publish his paper.

Looking at Figure 8.13, you can readily see why these two versions of the Periodic Table met such different receptions. Newlands' arrangement worked very well for the elements through calcium, but beyond that he was in trouble. One can

Newlands (1864)

						H	
Li	Be	B	C	N	O	F	
Na	Mg	Al	Si	P	S	Cl	
K	Ca	Cr	Ti	Mn	Fe	Co, Ni	
Cu	Zn	Y	In	As	Se	Br	
Rb	Sr	La, Ce	Zr	Nb, Mo	Ru, Rh	Pd	
Ag	Cd	U	Sn	Sb	Te	I	
Cs	Ba, V						

Figure 8.13 Two early versions of the Periodic Table. Both are in the condensed form used by early chemists, with transition metals placed with main-group elements. The elements in color were out of place in Newland's table.

Mendeleev (as revised, 1871)

I	II	III	IV	V	VI	VII	VIII
R_2O	RO	R_2O_3	RO_2	R_2O_5	RO_3	R_2O_7	RO_4
H							
Li	Be	B	C	N	O	F	
Na	Mg	Al	Si	P	S	Cl	
K	Ca	—	Ti	V	Cr	Mn	Fe, Co, Ni
Cu	Zn	—	—	As	Se	Br	Ru, Rh, Pd
Ag	Cd	In	Sn	Sb	Te	I	
Cs	Ba						

TABLE 8.8 PREDICTED AND OBSERVED PROPERTIES OF GERMANIUM (EKASILICON)		
	PREDICTED BY MENDELEEV (1871)	**OBSERVED (1886)**
Atomic mass	72	72.3
Density	5.5 g/cm³	5.47 g/cm³
Specific heat	0.31 J/(g·°C)	0.32 J/(g·°C)
Melting point	Very high	960°C
Formula of oxide	RO_2	GeO_2
Formula of chloride	RCl_4	$GeCl_4$
Density of oxide	4.7 g/cm³	4.70 g/cm³
Boiling point of chloride	100°C	86°C

make a logical case for locating chromium below aluminum (they both form oxides of formula R_2O_3), but this requires that chromium (AM = 52) precede titanium (AM = 48). Things get worse as we move across this period. Clearly, iron should not be in the same group with oxygen. Neither cobalt nor nickel (let alone both of them!) belong with the halogens.

Mendeleev, facing the same problem that Newlands "solved" so unsatisfactorily, got around it in a novel way. He was shrewd enough to realize that the elements beyond calcium would fall in the proper sequence only if he left gaps in certain places in the Table. This he justified by arguing that the elements which should fill these slots had not yet been discovered. Mendeleev went out on a very long limb by predicting in detail the chemical and physical properties of three such elements: "ekaboron" (scandium), "ekaaluminum" (gallium), and "ekasilicon" (germanium). By 1886, all of these elements had been isolated and shown to have properties virtually identical with those predicted (Table 8.8). The spectacular agreement between prediction and experiment removed any doubts about the validity and value of Mendeleev's Periodic Table.

This story would not be complete without mentioning the contribution of a man whom history seems almost to have forgotten, the German chemist Lothar Meyer. In July of 1868, in the process of revising his highly successful text *Modern Theories of Chemistry,* Meyer compiled a Periodic Table containing 56 elements. He also prepared extensive graphs showing that such properties as molar volume were a periodic function of atomic mass. Unaware of Mendeleev's work, Meyer published his results in 1870. In 1882, the two men jointly were awarded the Davy Medal, the highest honor of the Royal Society. Five years later, the Society belatedly awarded the same medal to its own member, J. A. R. Newlands.

SUMMARY

An electron in an atom can have only certain discrete (quantized) energies. When it moves from a higher to a lower level, energy may be given off in the form of light. The energy change in this transition is related to the wavelength of the light by the equation $\Delta E = hc/\lambda$. The Bohr model for the allowed energies of the hydrogen electron agreed with the observed wavelengths of lines in the hydrogen spectrum (Example 8.1).

The quantum mechanical model of the atom describes electrons in terms of quantum numbers. There are four such numbers:

n, which designates the principal level; $n = 1, 2, 3, \ldots$

ℓ, which designates the sublevel (s, p, d, or f); $\ell = 0, 1, \ldots (n - 1)$

m_ℓ, which designates the orbital; $m_\ell = +\ell, \ldots, +1, 0, -1, \ldots, -\ell$

m_s, which designates the electron spin; $m_s = +1/2, -1/2$

No two electrons in an atom can have the same set of four quantum numbers. This means, for example, that two electrons in the same orbital must have opposed spins.

These rules fix the capacities of principal levels, sublevels, and orbitals (Example 8.2). With one more piece of information, the order in which sublevels fill, they allow us to derive electron configurations of atoms (Example 8.3). Going one step further, we can write orbital diagrams (Example 8.4), taking into account Hund's rule (p. 180). Finally, we can write a set of four quantum numbers for each electron in an atom (Example 8.5).

The positions of elements in the Periodic Table can be correlated with electron configurations (Fig. 8.9). Elements in the same group have atoms with the same outer electron configuration (Example 8.6). This makes them similar chemically (Table 8.7). Trends in ionization energy (Fig. 8.12) can be correlated with those for atomic radii (Fig. 8.11); the two quantities are inversely related (Example 8.7).

KEY WORDS AND CONCEPTS

quantum theory	amplitude (ψ)	electron spin	unpaired electron
photon	quantum mechanics	Pauli exclusion principle	main-group element
Planck's constant	de Broglie relation	electron configuration	transition metal
wavelength	Schrödinger equation	orbital diagram	lanthanide (rare earth)
atomic spectrum	quantum number	Hund's rule	actinide
Bohr model	principal energy level	paramagnetic	atomic radius
ground state	sublevel	diamagnetic	ionization energy
excited state	orbital		

QUESTIONS AND PROBLEMS

Catalog

ΔE, λ, and the Bohr Model: 8.2–8.6, 8.24–8.28
Electron Configurations: 8.7–8.10, 8.29–8.32
Orbital Diagrams: 8.11–8.13, 8.33–8.35

Quantum Numbers: 8.14–8.17, 8.36–8.39
Periodic Table: 8.18–8.21, 8.40–8.43
General: 8.1, 8.22, 8.23, 8.44–8.48

8.1   Review and know the meaning of the key words and concepts in this chapter.

8.2   The meter is defined as 1.65076373×10^6 wavelengths of a certain line in the krypton spectrum.

   a. What is the wavelength of this line, in nanometers (four sig. fig.)?
   b. What is ΔE (kJ/mol) for the transition that gives this line?

8.3   Calculate the wavelength (nm) of the Paschen series line that results from the $n = 5$ to $n = 3$ transition.

8.4   There is a line in the hydrogen spectrum at 434.05 nm.

   a. What is ΔE (kJ/mol) for the transition associated with this line?
   b. The lower level in this transition is $n = 2$. What is the quantum number of the upper level?

8.5   Calculate the ionization energy (kJ/mol) for the hydrogen electron in the first excited state ($n = 2$).

8.6   The energy of any one-electron species in its **n**th state is $-Z^2B/n^2$. Z is the charge on the nucleus and B is 2.179×10^{-18} J. Find ΔE for the He$^+$ ion (J/atom) when the electron moves from $n = 2$ to $n = 1$.

8.7   Write electron configurations for

   a. S   b. K$^+$ c. V

8.24   Two energy states in the hydrogen atom have a ΔE of 9.331×10^4 J/mol. What is the wavelength of the corresponding line in the hydrogen spectrum?

8.25   Find the wavelength (nm) for the Lyman series line that results from an $n = 4$ to $n = 1$ transition.

8.26   Another line in the hydrogen spectrum is at 1004.9 nm.

   a. What is ΔE (kJ/mol) for the transition that produces this line?
   b. The higher level in this transition is $n = 7$. What is the quantum number of the lower level?

8.27   Calculate the ionization energy (kJ/mol) for the hydrogen electron in the $n = 6$ state.

8.28   Using the equation given in Problem 8.6, find the ionization energy of the He$^+$ ion in its ground state, first in J/atom and then in kJ/mol.

8.29   Write electron configurations for:

   a. Cl   b. Cl$^-$ c. Co

8.8 Give the symbol of the element of lowest atomic number which has

a. two 2p electrons.
b. three 3d electrons.
c. a completed 3p sublevel.

8.9 Which of the following electron configurations are for atoms in ground states? excited states? Which are not possible?

a. $1s^2 2s^2$
b. $1s^2 2s^2 3s^1$
c. $[_{10}Ne]3s^2 3p^8 4s^1$
d. $[_2He]2s^2 2p^6 2d^2$
e. $[_{18}Ar]4s^2 3d^3$
f. $[_{10}Ne]3s^2 3p^5 4s^1$

8.10 What fraction of the total number of electrons is in s sublevels in

a. He? b. Ne? c. Zn?

8.11 Give the orbital diagram of

a. N b. Cl c. Fe

8.12 Give the symbols of the atoms which have the following orbital diagrams.

```
        1s      2s        2p          3s         3p
a.    ( ↑↓ )  ( ↑↓ )  ( ↑↓ )( ↑↓ )( ↑↓ )  ( ↑ )
b.    ( ↑↓ )  ( ↑↓ )  ( ↑↓ )( ↑↓ )( ↑↓ )  ( ↑↓ )  ( ↑ )( ↑ )(  )
c.    ( ↑↓ )  ( ↑↓ )  ( ↑↓ )( ↑ )( ↑ )
```

8.13 Give the number of unpaired electrons in

a. O b. Al c. V

8.14 When **n** = 4, what are the possible values of ℓ? Identify each ℓ value with an s, p, d, or f sublevel.

8.15 Consider the sulfur atom. For how many electrons in this atom does

a. **n** = 3? b. ℓ = 1?
c. **n** = 3 and ℓ = 1?

8.16 Give a set of four quantum numbers for each electron in Mg.

8.17 Arrange the following sets of quantum numbers of electrons in order of increasing energy. If they have the same energy, place them together.

a. 3, 2, −1, +1/2   b. 1, 0, 0, +1/2
c. 2, 1, 1, −1/2   d. 3, 2, 1, +1/2
e. 3, 1, 0, +1/2   f. 2, 0, 0, +1/2

8.18 What would you expect to be the atomic number of the next noble gas beyond radon, Rn?

8.30 Give the symbol of the element of lowest atomic number which has

a. a d electron.
b. a completed s sublevel.
c. three 3p electrons.

8.31 Which of the following electron configurations are for atoms in the ground state? in excited states? Which are impossible?

a. $1s^1 2s^1$
b. $1s^2 2s^2 2p^3$
c. $[_{10}Ne]3s^2 3p^3 4s^1$
d. $[_{10}Ne]3s^2 3p^6 4s^3 3d^2$
e. $[_{10}Ne]3s^2 3p^6 4f^4$
f. $1s^2 2s^2 2p^4 3s^2$

8.32 What fraction of the total number of electrons is in p sublevels in

a. Be? b. Mg? c. Fe?

8.33 Give the orbital diagram of

a. Ne b. Mg^{2+} c. Ti

8.34 A certain +3 ion has the orbital diagram

```
       1s      2s         2p
     ( ↑↓ )  ( ↑↓ )  ( ↑↓ )( ↑↓ )( ↑↓ )
```

Identify the atom from which this ion is derived.

8.35 Give the number of unpaired electrons in

a. F b. F^- c. Mn

8.36 What letter is used to designate an ℓ = 3 sublevel? In what principal level does this sublevel first appear? How many orbitals does it contain?

8.37 Consider the neon atom. For how many electrons in this atom does

a. m_ℓ = 1? b. m_s = +1/2?
c. m_ℓ = 1 and m_s = +1/2?

8.38 Cr has the electron configuration $[_{18}Ar]4s^1 3d^5$. Give a set of four quantum numbers for each unpaired electron in Cr.

8.39 Given the following sets of electron quantum numbers, indicate those which could not occur and explain your reasoning.

a. 2, 2, 1, +1/2   b. 3, 2, 0, −1/2
c. 3, 3, 2, +1/2   d. 1, 0, 0, 1
e. 4, 0, 2, +1/2

8.40 What outer electron configuration would you expect for the proposed superheavy element of atomic number 120?

8.19 In the fourth period of the Periodic Table, how many elements have one or more

    a. 4s electrons?   b. 3d electrons?
    c. 4p electrons?   d. 4d electrons?

8.20 Use the Periodic Table to compare the elements O, F, and Na with respect to

    a. atomic radius   b. ionization energy

8.21 Give the symbol of the element which

    a. is in Group 8 but has no p electrons.
    b. has a single electron in the 3d sublevel.
    c. forms a +1 ion with a $1s^2 2s^2 2p^6$ electron configuration.

8.22 Explain, in terms of electron configurations or orbital diagrams, why

    a. the atomic radius of lithium is smaller than that of sodium.
    b. selenium and tellurium have similar chemical properties.
    c. oxygen atoms are paramagnetic.

8.23 Indicate whether each of the following statements is true or false. If false, correct the statement.

    a. An electronic transition from $n = 3$ to $n = 1$ absorbs energy.
    b. Light emitted by an $n = 4$ to $n = 2$ transition will have a shorter wavelength than that from an $n = 5$ to $n = 2$ transition.
    c. A sublevel with $\ell = 4$ has a capacity of 18 electrons.
    d. An atom of a Group 6 element has two unpaired electrons.

8.41 Of the first 100 elements in the Periodic Table, how many have one or more

    a. 1s electrons?   b. 2p electrons?
    c. 3d electrons?   d. 4d electrons?

8.42 Follow the directions of Problem 8.20 for the elements Ca, Sr, and Rb.

8.43 Give the symbols of all the elements which

    a. have the outer electron configuration ns^2np^3.
    b. just fill a p sublevel.

8.44 Explain, in terms of orbital diagrams or electron configurations, why

    a. in the Periodic Table, H can be placed in either Group 1 or 7.
    b. the ionization energy of Mg^+ is greater than that of Na even though they both have 11 electrons.
    c. K has a relatively simple atomic spectrum while Cr has a very complex one.

8.45 Criticize the following statements:

    a. Atomic radius increases across a period because "effective nuclear charge" increases steadily.
    b. Atoms with high ionization energy readily form +1 ions.
    c. The energy of a d orbital electron is greater than that of a p orbital electron.
    d. Doubling the energy of a photon doubles its wavelength.

*8.46 In 1885, Johann Balmer, a numerologist, derived the following relation for the wavelengths of lines in the visible spectrum of hydrogen:

$$\lambda = 364.6 \; n^2/(n^2 - 4)$$

where λ is in nanometers and **n** is an integer which can be 3, 4, 5, ... Show that this relation follows from the Bohr equation and Equation 8.2. Note that, for the Balmer series, the electron is returning to the $n = 2$ level.

*8.47 Suppose the rules for assigning quantum numbers were as follows:

$$n = 1, 2, 3, \ldots$$
$$\ell = 0, 1, 2, \ldots, n$$
$$m_\ell = 0, 1, 2, \ldots, \ell$$
$$m_s = +1/2 \text{ or } -1/2$$

Prepare a table similar to Table 8.3, based on these rules, for $n = 1$ and $n = 2$. Give the electron configuration for an atom with eight electrons.

*8.48 Suppose that the spin quantum number could have the values 1/2, 0, and −1/2. Assuming that the rules governing the values of the other quantum numbers and the order of filling sublevels were unchanged,

    a. what would be the electron capacity of an s sublevel? a p sublevel?
    b. how many electrons could fit in the $n = 2$ level?
    c. what would be the electron configuration of the element with atomic number 8? 17?

GROUPS 1 AND 2; THE METALS AND THEIR COMPOUNDS

In Chapter 8 we considered how the electronic structures of atoms relate to their positions in the Periodic Table. We also saw how certain properties of atoms vary as we move across or down the Table. In this chapter, we will review and expand upon these and other ideas presented in Chapters 6 through 8. To do that, we focus upon two groups in the Periodic Table, 1 and 2. The elements in these groups are often referred to as the *alkali metals* (Group 1) and *alkaline earth metals* (Group 2).

The sources, uses, and methods of preparation of these metals were described in Chapter 5 (recall Table 5.3 and Section 5.3). Here, we will concentrate upon their physical and chemical properties. To start with, we should point out that the alkali and alkaline earth metals have two common characteristics.

1. They show, to an unusual extent, the properties we associate with metals. The elements in both groups are excellent conductors of electricity and have low ionization energies (lose electrons readily). They are among the most reactive of all metals.

2. Their compounds are ionic. In these compounds, the alkali metals are present as $+1$ ions (Li^+, Na^+, K^+, Rb^+, Cs^+), and the alkaline earth metals as $+2$ ions (Mg^{2+}, Ca^{2+}, Sr^{2+}, Ba^{2+}).

The properties of the first element in Group 2, beryllium, differ considerably from those of the other elements in these groups. Due to its position in the Periodic Table, Be is the least metallic of these elements. It is less reactive toward nonmetals than are the other alkaline earths. The compounds of beryllium tend to have a molecular rather than an ionic structure. For these reasons, among others, we will have relatively little to say about the chemistry of beryllium in this chapter.

9.1 PHYSICAL PROPERTIES OF THE METALS

Table 9.1 lists several physical properties of the metals in Groups 1 and 2. (Some of these were cited in previous chapters but are repeated here for comparative purposes.) Several trends are evident from the data in the table. We will examine a few of these in this section.

Atomic and Ionic Radii

Looking at Table 9.1, we note the trends in atomic radii referred to in Chapter 8. Atomic radius decreases as we move across in the Periodic Table from Group 1 to 2

TABLE 9.1 PHYSICAL PROPERTIES OF THE GROUP 1 AND GROUP 2 METALS

	$_3$Li	$_{11}$Na	$_{19}$K	$_{37}$Rb	$_{55}$Cs
Atomic mass	6.94	22.99	39.10	85.47	132.91
Melting point (°C)	186	98	64	39	28
Boiling point (°C)	1326	889	774	688	690
Density (g/cm³)	0.534	0.971	0.862	1.53	1.87
Atomic radius (nm)	0.152	0.186	0.231	0.244	0.262
Ionic radius (nm)	0.060	0.095	0.133	0.148	0.169
Ionization energy (kJ/mol)	527	502	427	410	377
Flame color	Red	Yellow	Violet	Purple	Blue
Strongest lines in atomic	670.8	589.6	404.7	780.0	459.3
spectrum (nm)	610.4	589.0	404.4	420.2	455.5

	$_4$Be	$_{12}$Mg	$_{20}$Ca	$_{38}$Sr	$_{56}$Ba
Atomic mass	9.01	24.30	40.08	87.62	137.34
Melting point (°C)	1283	650	845	770	725
Boiling point (°C)	2970	1120	1420	1380	1640
Density (g/cm³)	1.85	1.74	1.55	2.60	3.51
Atomic radius (nm)	0.111	0.160	0.197	0.215	0.217
Ionic radius (nm)	0.031	0.065	0.099	0.113	0.135
Ionization energy (kJ/mol)	904	745	594	556	510
Flame color	—	—	Red	Crimson	Green

(compare Na, 0.186 nm, to Mg, 0.160 nm). As we move down a group, atomic radius increases (compare Mg, 0.160 nm, to Ca, 0.197 nm).

The horizontal trend in atomic radius explains in part why the Group 2 metals have higher densities than do the Group 1 metals. An alkaline earth atom is smaller than that of an alkali metal in the same period. It is also slightly heavier (at. mass Na = 22.99, Mg = 24.30). This means that atoms of the Group 2 metals have the larger density, as do the metals themselves (d Na = 0.971 g/cm³, Mg = 1.74 g/cm³).

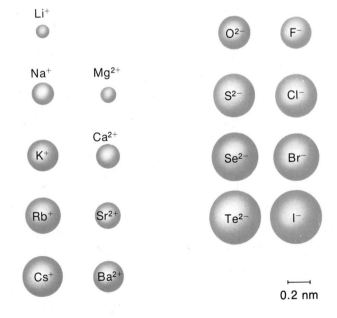

Figure 9.1 Sizes of ions of elements in Groups 1, 2, 6, and 7. In general, negative ions are larger than positive ions. (See discussion on p. 196.)

0.2 nm

Table 9.1 also lists the *ionic radii* of these metals (+1 ions for Group 1, +2 ions for Group 2). These are obtained from measurements of the distances between ions in compounds (Section 9.5). The horizontal and vertical trends are the same as those cited with atomic radii. Notice, however, that the positive ions are considerably smaller than the atoms from which they are formed (Na = 0.186 nm; Na⁺ = 0.095 nm). The situation is quite different with negative ions. Typically, they are larger than the corresponding atoms (Cl = 0.099 nm; Cl⁻ = 0.181 nm). As a result, negative ions are usually larger than positive ions (Fig. 9.1).

Ionization Energies

The ionization energies listed in Table 9.1 are for the removal of the first electron from the gaseous atom. They represent ΔE for the process

$$M(g) \rightarrow M^+(g) + e^-$$

These ionization energies follow the trends referred to in Chapter 8. As we move across from Group 1 to 2, ΔE increases; as we move down a given group, it decreases. Of all these metals, beryllium, at the top of Group 2, has the largest ionization energy, 904 kJ/mol. This explains in part why it is the least metallic element in these two groups.

A given element has several ionization energies, one for each electron in its atom. For sodium:

$$Na(g) \rightarrow Na^+(g) + e^-; \Delta E_1 = \text{first ion. energy} = 502 \text{ kJ/mol}$$
$$Na^+(g) \rightarrow Na^{2+}(g) + e^-; \Delta E_2 = \text{second ion. energy} = 4560 \text{ kJ/mol}$$
$$Na^{2+}(g) \rightarrow Na^{3+}(g) + e^-; \Delta E_3 = \text{third ion. energy} = 6950 \text{ kJ/mol}$$

The first electron always is the easiest to remove

Successive ionization energies always increase in magnitude. As the positive charge on the ion increases, it becomes more and more difficult to remove electrons.

Notice, however, that ionization energy does not increase smoothly as we remove successive electrons. For sodium, there is a big jump between the first and second ionization energies; ΔE_2 is nine times ΔE_1. In contrast, ΔE_3 is less than twice ΔE_2. This is readily explained when we recall the electron configuration of Na, $1s^2 2s^2 2p^6 3s^1$. The first electron to come out is the 3s electron, which is far from the nucleus. It should leave rather easily ($\Delta E_1 = 502$ kJ/mol). To go beyond that, we have to break into a complete 2p sublevel, which is much more difficult ($\Delta E_2 = 4560$ kJ/mol). Successive electrons come from that same sublevel, so we would expect ΔE to increase steadily but rather slowly ($\Delta E_3 = 6950$ kJ/mol, . . .).

Similar effects are observed with other atoms. In Figure 9.2, we show successive ionization energies of magnesium, which has the electron configuration $1s^2 2s^2 2p^6 3s^2$. Here, as we might expect, the large "jump" comes after the second electron is removed from the 3s sublevel.

Atomic Spectra

In Chapter 8, we discussed the atomic spectrum of hydrogen. Atoms of other elements show similar, although more complex, spectra. Those of the alkali metals—and, to a lesser extent, the alkaline earth metals—are rather easy to produce. Heating compounds of these metals in a Bunsen flame excites an electron to a higher energy state. When the electron returns to the ground state or to a lower excited state, energy is given off in the form of light.

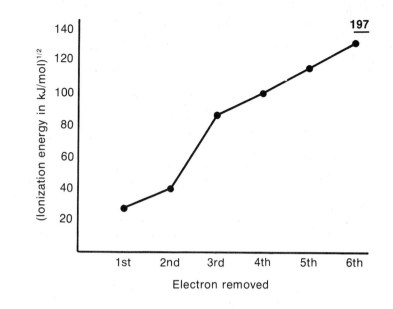

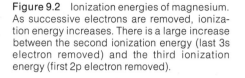

Figure 9.2 Ionization energies of magnesium. As successive electrons are removed, ionization energy increases. There is a large increase between the second ionization energy (last 3s electron removed) and the third ionization energy (first 2p electron removed).

The flame colors listed in Table 9.1 and shown in color plate 7 (center of book) can be used to identify the alkali metals. Lithium gives an orange-red color because it emits strongly at two wavelengths in this region, 671 nm (red) and 610 nm (orange). Sodium yields two lines close together at 589 nm, in the yellow region. Potassium emits violet light at the lower edge of the visible range, around 404 nm.

From these wavelengths, it is possible to calculate the energy differences between two electronic energy states in the alkali metal atoms. To do this, we use an equation from Chapter 8:

$$\Delta E = \frac{1.196 \times 10^5}{\lambda} \frac{kJ \cdot nm}{mol} \tag{9.1}$$

Example 9.1   The line at 670.8 nm in the atomic spectrum of Li is produced when an electron, after being excited to the 2p state, returns to the 2s ground state. What is the difference in energy between these two states?

Solution

$$\Delta E = \frac{1.196 \times 10^5}{670.8} \frac{kJ}{mol} = 178.3 \text{ kJ/mol}$$

Exercise   Calculate ΔE for the other line listed for Li in Table 9.1 (this comes from a 3d → 2p transition). Answer: 195.9 kJ/mol.

9.2   CHEMICAL PROPERTIES

Anyone who has worked with the alkali and alkaline earth metals is aware that they are very reactive chemically. They react spontaneously with air, with water, and with most of the nonmetals. In their compounds, the metals (except for Be) exist as simple cations (+1 for Group 1, +2 for Group 2). These cations have the electron configuration of the preceding noble gas:

TABLE 9.2   REACTIONS OF ALKALI METALS AND ALKALINE EARTH METALS

COMBINING SUBSTANCE	REACTION	METAL
	Group 1	
Hydrogen	$2 \text{ M(s)} + \text{H}_2\text{(g)} \rightarrow 2 \text{ MH(s)}$	All
Halogens	$2 \text{ M(s)} + \text{X}_2 \rightarrow 2 \text{ MX(s)}$	All
Nitrogen	$6 \text{ M(s)} + \text{N}_2\text{(g)} \rightarrow 2 \text{ M}_3\text{N(s)}$	Li
Sulfur	$2 \text{ M(s)} + \text{S(s)} \rightarrow \text{M}_2\text{S(s)}$	All
Oxygen	$4 \text{ M(s)} + \text{O}_2\text{(g)} \rightarrow 2 \text{ M}_2\text{O(s)}$	Li
	$2 \text{ M(s)} + \text{O}_2\text{(g)} \rightarrow \text{M}_2\text{O}_2\text{(s)}$	Na
	$\text{M(s)} + \text{O}_2\text{(g)} \rightarrow \text{MO}_2\text{(s)}$	K, Rb, Cs
Water	$2 \text{ M(s)} + 2 \text{ H}_2\text{O(l)} \rightarrow 2 \text{ M}^+\text{(aq)} + 2 \text{ OH}^-\text{(aq)} + \text{H}_2\text{(g)}$	All
	Group 2	
Hydrogen	$\text{M(s)} + \text{H}_2\text{(g)} \rightarrow \text{MH}_2\text{(s)}$	Ca, Sr, Ba
Halogens	$\text{M(s)} + \text{X}_2 \rightarrow \text{MX}_2\text{(s)}$	All
Nitrogen	$3 \text{ M(s)} + \text{N}_2\text{(g)} \rightarrow \text{M}_3\text{N}_2\text{(s)}$	Mg, Ca, Sr, Ba
Sulfur	$\text{M(s)} + \text{S(s)} \rightarrow \text{MS(s)}$	Mg, Ca, Sr, Ba
Oxygen	$2 \text{ M(s)} + \text{O}_2\text{(g)} \rightarrow 2 \text{ MO(s)}$	Be, Mg, Ca, Sr, Ba
	$\text{M(s)} + \text{O}_2\text{(g)} \rightarrow \text{MO}_2\text{(s)}$	Ba
Water	$\text{M(s)} + 2 \text{ H}_2\text{O(l)} \rightarrow \text{M}^{2+}\text{(aq)} + 2 \text{ OH}^-\text{(aq)} + \text{H}_2\text{(g)}$	Ca, Sr, Ba
	$\text{M(s)} + \text{H}_2\text{O(g)} \rightarrow \text{MO(s)} + \text{H}_2\text{(g)}$	Mg

Noble Gas Atom	Group 1 Cation	Group 2 Cation	Electron Configuration
$_2\text{He}$	$_3\text{Li}^+$		$1s^2$
$_{10}\text{Ne}$	$_{11}\text{Na}^+$	$_{12}\text{Mg}^{2+}$	$[_2\text{He}]2s^22p^6$
$_{18}\text{Ar}$	$_{19}\text{K}^+$	$_{20}\text{Ca}^{2+}$	$[_{10}\text{Ne}]3s^23p^6$
$_{36}\text{Kr}$	$_{37}\text{Rb}^+$	$_{38}\text{Sr}^{2+}$	$[_{18}\text{Ar}]3d^{10}4s^24p^6$
$_{54}\text{Xe}$	$_{55}\text{Cs}^+$	$_{56}\text{Ba}^{2+}$	$[_{36}\text{Kr}]4d^{10}5s^25p^6$

Table 9.2 summarizes some of the more important reactions of the alkali and alkaline earth metals. As you can perhaps judge from the table, the Group 1 metals are somewhat more reactive than those in Group 2.

Example 9.2   Write balanced equations for the reaction of calcium with
  a. chlorine   b. iodine   c. nitrogen   d. oxygen

Solution
  a. $\text{Ca(s)} + \text{Cl}_2\text{(g)} \rightarrow \text{CaCl}_2\text{(s)}$
  b. $\text{Ca(s)} + \text{I}_2\text{(s)} \rightarrow \text{CaI}_2\text{(s)}$
  c. $3 \text{ Ca(s)} + \text{N}_2\text{(g)} \rightarrow \text{Ca}_3\text{N}_2\text{(s)}$
  d. $2 \text{ Ca(s)} + \text{O}_2\text{(g)} \rightarrow 2 \text{ CaO(s)}$

Exercise   Identify the ions present in each of the compounds formed in Example 9.3.
Answer: Ca^{2+}; Cl^-; I^-; N^{3-}; O^{2-}.

As you can see from Table 9.2, all the metals in Groups 1 and 2 react with the halogens to form stable compounds. Except for the halides of beryllium, all these compounds are ionic. They are stable in the sense that they have a negative heat of formation. Consider, for example, NaCl:

$$Na(s) + \frac{1}{2} Cl_2(g) \rightarrow NaCl(s); \Delta H = -411 \text{ kJ}$$

Clearly this reaction is exothermic; 411 kJ of heat is evolved per mole of NaCl formed. To understand where this heat comes from, let us imagine a possible three-step path for the reaction:

1. *The elementary substances in their stable states at 25°C and 1 atm are converted to gaseous atoms.*

$$Na(s) + \frac{1}{2} Cl_2(g) \rightarrow Na(g) + Cl(g); \Delta H_1 = +230 \text{ kJ}$$

As you might expect, this step is endothermic. It requires that enough energy be absorbed to break the bonds holding sodium atoms together in sodium metal and those joining chlorine atoms in Cl_2 molecules.

2. *Electron transfer occurs between the gaseous atoms to form ions.*

$$Na(g) + Cl(g) \rightarrow Na^+(g) + Cl^-(g); \Delta H_2 = +130 \text{ kJ}$$

Interestingly enough, this step is also endothermic. The energy which must be absorbed to remove the electrons from a mole of sodium atoms ($\Delta H = +502$ kJ/mol) exceeds that which is evolved when these electrons are acquired by chlorine atoms ($\Delta H = -372$ kJ/mol).

3. *The gaseous ions combine to form an ionic solid.*

$$Na^+(g) + Cl^-(g) \rightarrow NaCl(s); \Delta H_3 = -771 \text{ kJ}$$

This is a highly exothermic process because of the strong electrostatic attraction between oppositely charged ions.

The value of ΔH for the overall reaction must be, by Hess's Law, the sum of the enthalpy changes for the individual steps:

$$\Delta H = \Delta H_1 + \Delta H_2 + \Delta H_3$$

$$= +230 \text{ kJ} + 130 \text{ kJ} - 771 \text{ kJ} = -411 \text{ kJ}$$

From this analysis we see that the evolution of energy in the overall reaction is due to the exothermic nature of the final step. This enthalpy change, in which the crystal lattice is formed from gaseous ions, is referred to as the **lattice energy.**

When gaseous ions combine to form a lattice, a lot of energy is released

It is possible to go through an analysis of this type for the formation of any alkali halide. The magnitudes of the several enthalpy changes vary with the particular compound. However, the signs are always the same. Steps 1 and 2 are endothermic. The lattice energy, ΔH_3, is always a large negative quantity. It makes the overall reaction exothermic and the ionic compound stable.

Example 9.3 Going through a similar analysis for KCl, it is found that $\Delta H_1 = +209$ kJ and $\Delta H_2 = +63$ kJ. The heat of formation of KCl is -436 kJ. Calculate the lattice energy of KCl (ΔH_3).

Solution Applying Hess's Law with $\Delta H = -436$ kJ:

$$-436 \text{ kJ} = +209 \text{ kJ} + 63 \text{ kJ} + \Delta H_3$$

$$\Delta H_3 = -708 \text{ kJ}$$

Exercise Why is ΔH_2 a smaller positive quantity for KCl than for NaCl? How would you expect ΔH_1 for NaBr to compare to that for NaCl (recall Table 6.3, p. 116)? Answer: Step (2) involves the ionization of a metal atom; the ionization energy of K(g) is less than that of Na(g). Step (1) involves breaking the covalent bond in a halogen molecule; the bond energy of Br_2, 193 kJ/mol, is less than that of Cl_2, 243 kJ/mol. Hence ΔH_1 should be smaller for NaBr.

Reactions with Hydrogen

The metals of Groups 1 and 2 react directly with hydrogen at high temperatures to form ionic hydrides. For example:

$$2 \text{ K(s)} + H_2(g) \rightarrow 2 \text{ KH(s)}$$

The cation in solid KH is K^+ These white, crystalline solids contain H^- ions. They are often referred to as saline hydrides because of their physical resemblance to NaCl.

The saline hydrides react with water to produce hydrogen gas. A typical reaction is

$$CaH_2(s) + 2 H_2O(l) \rightarrow 2 H_2(g) + Ca^{2+}(aq) + 2 OH^-(aq)$$

In this way, saline hydrides can serve as compact, portable sources of hydrogen gas for inflating life rafts and balloons.

Reactions with Oxygen

As you can see from Table 9.2, several different products can be formed when a Group 1 or 2 metal reacts with oxygen. To be specific, three products are possible.
1. The *normal oxide,* containing the oxide anion, O^{2-}:
    Group 1 oxides: general formula M_2O (2 M^+ ions, 1 O^{2-} ion).
    Group 2 oxides: general formula MO (M^{2+} ion, O^{2-} ion).
2. The *peroxide,* containing the peroxide ion, O_2^{2-}:
    Group 1 peroxides: general formula M_2O_2 (2 M^+ ions, 1 O_2^{2-} ion).
    Group 2 peroxides: general formula MO_2 (M^{2+} ion, O_2^{2-} ion).
3. The *superoxide,* containing the superoxide anion, O_2^-:
    Group 1 superoxides: general formula MO_2 (M^+ ion, O_2^- ion).

OXIDES. Lithium is the only Group 1 metal that forms the normal oxide in good yield by direct reaction with oxygen. The other Group 1 oxides (Na_2O, K_2O, Rb_2O, Cs_2O) can be prepared by other means. In contrast, the Group 2 metals usually react with oxygen to give the normal oxide. Beryllium and magnesium must be heated strongly to give BeO and MgO. Calcium and strontium react more readily to give CaO and SrO. Barium, the most reactive of the Group 2 metals, catches fire when exposed to moist air. The product is a mixture of the normal oxide, BaO, and the peroxide, BaO_2.

The oxides of the Group 1 and Group 2 metals are high-melting, white solids. Indeed, the melting point of MgO is so high, 2800°C, that it is used as a lining for high-temperature furnaces. All the Group 1 oxides react with water to form a basic solution containing OH^- ions. For this reason, such oxides are called *basic anhydrides* (bases without water). The reaction of Li_2O is typical:

Not many substances are still solid at 2800°C

$$Li_2O(s) + H_2O(l) \rightarrow 2\ Li^+(aq) + 2\ OH^-(aq) \qquad (9.2)$$

CaO, SrO, and BaO are Group 2 oxides which are basic anhydrides. Typically,

$$CaO(s) + H_2O(l) \rightarrow Ca^{2+}(aq) + 2\ OH^-(aq) \qquad (9.3)$$

Reaction 9.3 is often referred to as the "slaking" of lime. It is exothermic, giving off enough heat to ignite paper or even wood.

Example 9.4 Using Tables 6.1 and 6.2, Chapter 6, calculate ΔH for Reaction 9.3.

Solution Recall from Chapter 6 that

$$\Delta H = \Sigma\ \Delta H_f \text{ products} - \Sigma\ \Delta H_f \text{ reactants}$$

Here:

$$\Delta H = \Delta H_f\ Ca^{2+}(aq) + 2\ \Delta H_f\ OH^-(aq) - \Delta H_f\ CaO(s) - \Delta H_f\ H_2O(l)$$

$$= -543.0 \text{ kJ} - 459.8 \text{ kJ} + 635.5 \text{ kJ} + 285.8 \text{ kJ} = -81.5 \text{ kJ}$$

Exercise How many grams of water can be heated from 20°C to the boiling point, 100°C, by the heat given off in this reaction [specific heat of water = 4.18 J/(g·°C)]? Answer: 240 g. (Note that only 18 g of water is involved in the reaction. Hence, unless a large excess is used, some of it will be converted to steam.)

PEROXIDES. The most important peroxides are those of sodium and barium, Na_2O_2 and BaO_2. Upon addition to water, they react vigorously to form hydrogen peroxide, H_2O_2.

$$Na_2O_2(s) + 2\ H_2O(l) \rightarrow 2\ Na^+(aq) + 2\ OH^-(aq) + H_2O_2(aq) \qquad (9.4)$$

$$BaO_2(s) + 2\ H_2O(l) \rightarrow Ba^{2+}(aq) + 2\ OH^-(aq) + H_2O_2(aq) \qquad (9.5)$$

These reactions can be used to produce small amounts of hydrogen peroxide in the laboratory. Commercially, both Na_2O_2 and BaO_2 are used as bleaching agents.

SUPEROXIDES. Virtually all the potassium metal produced today is used to make potassium superoxide, KO_2. The reaction

$$K(s) + O_2(g) \rightarrow KO_2(s) \qquad (9.6)$$

is carried out by spraying the metal into air. This serves not only to furnish the oxygen required but also to cool the product. Potassium superoxide is used in self-contained

Peroxides and superoxides are to be used with caution

breathing apparatus for fire fighters and miners. It reacts with the moisture in exhaled air to generate oxygen:

$$4 \ KO_2(s) + 2 \ H_2O(g) \rightarrow 3 \ O_2(g) + 4 \ KOH(s) \tag{9.7}$$

The carbon dioxide in the exhaled air is removed by reaction with the KOH formed in Reaction 9.7.

$$KOH(s) + CO_2(g) \rightarrow KHCO_3(s) \tag{9.8}$$

A person using a mask charged with KO_2 can rebreathe the same air for an extended period of time. This allows that person to enter an area where there are poisonous gases or oxygen-deficient air.

Example 9.5   It is possible to prepare potassium peroxide, K_2O_2, by heating potassium carefully with exactly the equivalent amount of oxygen. If 1.00 g of K is to be converted to K_2O_2, what volume of $O_2(g)$ at 720 mm Hg and 20°C should be used?

Solution   We start by writing the balanced equation for the reaction

$$2 \ K(s) + O_2(g) \rightarrow K_2O_2(s)$$

Noting that the atomic mass of potassium is 39.1, we have

$$78.2 \ g \ K \simeq 1 \ mol \ O_2$$

$$\text{no. moles } O_2 \text{ required} = 1.00 \ g \ K \times \frac{1 \ mol \ O_2}{78.2 \ g \ K} = 0.0128 \ mol \ O_2$$

Now, using the Ideal Gas Law (Chapter 7), we calculate the volume of oxygen:

$$V = \frac{nRT}{P} = \frac{(0.0128 \ mol)\left(0.0821 \ \frac{\ell \cdot atm}{mol \cdot K}\right) \times 293 \ K}{(720/760 \ atm)} = 0.325 \ \ell$$

Exercise   Suppose air (21.0% O_2) is used instead of O_2. What volume of air, at the same P and T, should be used? Answer: 1.55 ℓ.

Reactions with Water

The alkali metals react vigorously with water to liberate hydrogen. At the same time, a water solution of the alkali hydroxide is formed. The reaction of sodium is typical:

$$Na(s) + 2 \ H_2O(l) \rightarrow 2 \ Na^+(aq) + 2 \ OH^-(aq) + H_2(g); \ \Delta H = -367.6 \ kJ \tag{9.9}$$

As you can see from the thermochemical equation, the reaction is strongly exothermic. Enough heat is evolved for the hydrogen to catch fire.

Among the Group 2 metals, Ca, Sr, and Ba react with water in much the same way as the alkali metals. The reaction with calcium is

$$Ca(s) + 2 \ H_2O(l) \rightarrow Ca^{2+}(aq) + 2 \ OH^-(aq) + H_2(g); \ \Delta H = -431.2 \ kJ \tag{9.10}$$

Beryllium does not react with water at all. Magnesium reacts very slowly with boiling water, but reacts more readily with steam at high temperatures:

$$Mg(s) + H_2O(g) \rightarrow MgO(s) + H_2(g); \Delta H = -360.0 \text{ kJ} \qquad (9.11)$$

This reaction, like 9.9 or 9.10, produces enough heat to ignite the hydrogen. Fire fighters who try to put out a magnesium fire by spraying water on it have discovered this reaction, often with tragic results. The best way to extinguish burning magnesium is to dump dry sand on it.

Mg was used in fire bombs in World War II

9.3 COMPOUNDS OF THE GROUP 1 METALS

Behavior of Li and Its Compounds

Nearly all the common compounds of the Group 1 metals are soluble in water. The only major exceptions are a few lithium compounds. The fluoride, LiF, carbonate, Li_2CO_3, and phosphate, Li_3PO_4, all have low water solubilities. In this sense, lithium resembles magnesium and other Group 2 elements, which form several insoluble compounds (see Table 9.3, p. 207).

The solubility behavior of lithium compounds is an example of a *diagonal relationship* in the Periodic Table. Elements which lie along a diagonal line (upper left to lower right) often have somewhat similar properties. Thus, lithium resembles magnesium in some ways, and beryllium has many properties in common with aluminum.

$$
\begin{array}{ccc}
\text{Li} & \text{Be} & \text{B} \\
\text{Na} & \text{Mg} & \text{Al}
\end{array}
$$

Looking back at Table 9.2, you can see some similarities in the chemical properties of lithium and magnesium. Lithium is the only Group 1 element that reacts directly with nitrogen, forming an ionic nitride, Li_3N. Magnesium reacts similarly, forming Mg_3N_2. Again, lithium is the only Group 1 element that reacts with oxygen to form a normal oxide, Li_2O. Magnesium and the other Group 2 elements commonly react like this, forming oxides such as MgO.

Relationships such as the one between Li and Mg have a simple explanation. Elements lying on this type of diagonal line in the Periodic Table have similar atomic properties. Notice, for example, that the atomic radius of Li (0.152 nm) is closer to that of Mg (0.160 nm) than it is to Na (0.186 nm). The same is true with ionic radii (Li^+ = 0.060 nm; Mg^{2+} = 0.065 nm; Na^+ = 0.095 nm).

Sodium Chloride

As pointed out in Chapter 5, NaCl is the major source of sodium metal and the compounds of that element. It occurs in impure form ("rock salt") in large underground deposits. The largest of these, more than 100 m thick, underlies parts of Oklahoma, Texas, and Kansas. The salt may be obtained by forming a water solution, which is pumped to the surface. Upon evaporation, crystals of NaCl form.

One chemical we won't run out of is NaCl

The sodium chloride produced this way is contaminated with small amounts of $MgCl_2$ and $CaCl_2$. These compounds are present in ordinary table salt. They tend to pick up water from the air, causing the salt to "cake" and refuse to pour. Small amounts of chemical drying agents are added to prevent this. About 0.01% of KI is also added to table salt. This prevents a condition called goiter, an enlargement of the thyroid gland caused by iodine deficiency.

Sodium Hydroxide

About 10^7 metric tons of NaOH are produced annually in the United States. You may be familiar with sodium hydroxide in the form of lye. Larger amounts of NaOH are used to make soap, textiles, and paper. It is also used in the metallurgy of aluminum (Chap. 5). Most of our NaOH comes from an electrolytic process which was mentioned in Chapter 5. When a concentrated water solution of NaCl is electrolyzed, Cl^- ions are oxidized to Cl_2 at the anode:

$$\text{anode:} \qquad 2\ Cl^-(aq) \rightarrow Cl_2(g) + 2\ e^- \qquad (9.12a)$$

At the cathode, H_2O molecules rather than Na^+ ions are reduced:

$$\text{cathode:} \qquad 2\ H_2O + 2\ e^- \rightarrow H_2(g) + 2\ OH^-(aq) \qquad (9.12b)$$

The overall cell reaction is obtained by summing these two half-equations.

$$2\ Cl^-(aq) + 2\ H_2O \rightarrow Cl_2(g) + H_2(g) + 2\ OH^-(aq) \qquad (9.12)$$

In effect, Cl^- ions in the NaCl solution are replaced by OH^- ions. Evaporation gives NaOH, contaminated by some NaCl. The sodium hydroxide can be purified by fractional crystallization (Chap. 1). Since the solubility of NaCl is nearly independent of temperature, it remains in solution when the NaOH crystallizes.

Sodium hydroxide is very soluble in water, about 20 M at 25°C. The solution process is exothermic:

$$NaOH(s) \rightarrow Na^+(aq) + OH^-(aq); \Delta H = -42.9\ kJ \qquad (9.13)$$

The heat can be used to speed up the solution process

Often, enough heat is released to raise the temperature to near the boiling point. The solution is strongly basic because of the high concentration of OH^- ions. Over a period of time, concentrated NaOH solutions attack glass, becoming cloudy in the process. They also pick up CO_2 from the air:

$$2\ OH^-(aq) + CO_2(g) \rightarrow H_2O + CO_3{}^{2-}(aq) \qquad (9.14)$$

This contaminates the NaOH solution with sodium carbonate, Na_2CO_3.

Sodium hydroxide is known commonly as lye or caustic soda. As a solid or concentrated solution, it is used in commercial oven and drain cleaners. These products should be used with caution because concentrated NaOH is corrosive to skin and clothing.

Sodium Carbonate and Sodium Hydrogen Carbonate

These two compounds have been made for over 100 years by a process worked out by Albert and Ernest Solvay in Belgium. The key step in the *Solvay process* can be carried out in the general chemistry laboratory (Fig. 9.3). A water solution of sodium chloride and ammonia is saturated with carbon dioxide at 0°C. The CO_2 and NH_3 react to form $NH_4{}^+$ and $HCO_3{}^-$ ions:

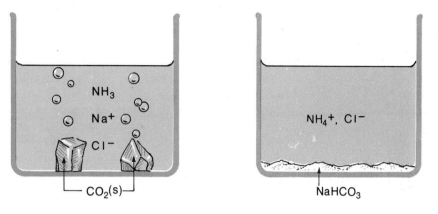

Figure 9.3 The Solvay process. A water solution containing ammonia and sodium chloride is saturated with carbon dioxide at 0°C. A precipitate of sodium hydrogen carbonate forms. The equation for the overall reaction is

$$Na^+(aq) + CO_2(g) + NH_3(aq) + H_2O \rightarrow$$
$$NaHCO_3(s) + NH_4^+(aq)$$

Much of the $NaHCO_3$ obtained is later converted to Na_2CO_3 by heating to 300°C.

$$CO_2(g) + NH_3(aq) + H_2O \rightarrow NH_4^+(aq) + HCO_3^-(aq) \qquad (9.15)$$

The concentration of HCO_3^- is about 0.1 M. Under these conditions, the Na^+ ions of the NaCl solution react with HCO_3^- ions to give a precipitate of sodium hydrogen carbonate, $NaHCO_3$:

$$Na^+(aq) + HCO_3^-(aq) \rightarrow NaHCO_3(s) \qquad (9.16)$$

In a separate step, the $NaHCO_3$ is heated to about 300°C to convert it to sodium carbonate:

$$2\ NaHCO_3(s) \rightarrow Na_2CO_3(s) + CO_2(g) + H_2O(g) \qquad (9.17)$$

Na_2CO_3 was one of the first chemicals to be produced industrially by synthesis from other substances

In order for any industrial process such as this to be profitable, several factors must be considered. The starting materials must be inexpensive. Any species which takes part in the reaction but does not appear in the products must be recovered. Any by-products must be used or sold. To see how the Solvay process meets these requirements, let us consider some of its features.

1. The CO_2 required for Reaction 9.15 is produced by heating limestone, $CaCO_3$. This is a common, inexpensive mineral.

$$CaCO_3(s) \rightarrow CaO(s) + CO_2(g)$$

The other starting materials, H_2O and NaCl, are both cheap.

2. The CO_2 produced by Reaction 9.17 is recycled to form more $NaHCO_3$ by Reaction 9.15.

3. The NH_3 required to carry out Reaction 9.15 is recovered. The solution remaining after the precipitation of $NaHCO_3$, which contains NH_4^+ ions, is saved. It is heated with the CaO that is formed as a by-product of the decomposition of limestone. The following reaction occurs:

$$CaO(s) + 2\ NH_4^+(aq) \rightarrow Ca^{2+}(aq) + 2\ NH_3(g) + H_2O$$

In this way, the ammonia can be used over and over again.

4. The final by-product of the Solvay process is calcium chloride, $CaCl_2$. Indeed, the net result of the whole process can be represented by the equation

$$2 \; NaCl(s) + CaCO_3(s) \rightarrow Na_2CO_3(s) + CaCl_2(s) \tag{9.18}$$

The calcium chloride finds some use as a drying agent and for melting ice on highways.

Despite these economies, the Solvay process has decreased in importance in recent years. The market for $CaCl_2$ has decreased and its disposal causes pollution problems. More and more Na_2CO_3 and $NaHCO_3$ are being produced from a mineral called trona, $Na_2CO_3 \cdot NaHCO_3 \cdot 2 \; H_2O$. Large deposits of trona were found near Green River, Wyoming in 1938. Recently, they have been extensively mined and now supply more than half of our sodium carbonate and sodium hydrogen carbonate.

About 7×10^6 metric tons of Na_2CO_3 are produced annually in the United States. Commonly called *washing soda,* it is used in the manufacture of glass, soap, paper, and many chemicals. Nowadays, it is also used in detergents as a substitute for phosphates. Washing soda is not a serious pollutant, since carbonates occur naturally in surface waters. However, it forms a precipitate of $CaCO_3$ with Ca^{2+} ions in hard water. This, as we will see in Section 9.4, creates a few problems.

It is also known as soda ash

Sodium hydrogen carbonate, $NaHCO_3$, is commonly called bicarbonate of soda or baking soda. It is an ingredient of many commercial products used to relieve indigestion. The HCO_3^- ions react with H^+ ions, thus relieving "excess stomach acidity."

$$HCO_3^-(aq) + H^+(aq) \rightarrow CO_2(g) + H_2O \tag{9.19}$$

A similar reaction accounts for the use of $NaHCO_3$ in baking powders. An acidic component of the baking powder provides the H^+ ions required for Reaction 9.19. The carbon dioxide is formed as tiny bubbles which expand upon warming, causing bread or pastries to "rise."

Example 9.6   A certain baking powder contains 20% by mass of $NaHCO_3$. Assume that $NaHCO_3$ is the limiting reagent and that 2 teaspoons (about 10 g) of the baking powder are used to make a batch of biscuits. What volume of CO_2 is produced when the biscuits rise in an oven at 475°F (246°C) and one atmosphere pressure?

Solution   The mass of $NaHCO_3$ in the baking powder is $0.20 \times 10 \; g = 2.0 \; g$. The molar mass of $NaHCO_3$ is $23 \; g + 1 \; g + 12 \; g + 48 \; g = 84 \; g$. According to Equation 9.19

$$1 \; mol \; NaHCO_3 \simeq 1 \; mol \; CO_2$$

$$84 \; g \; NaHCO_3 \simeq 1 \; mol \; CO_2$$

$$\text{no. moles } CO_2 = 2.0 \; g \; NaHCO_3 \times \frac{1 \; mol \; CO_2}{84 \; g \; NaHCO_3} = 0.024 \; mol \; CO_2$$

The volume of CO_2 can be calculated from the Ideal Gas Law with $n = 0.024 \; mol$, $R = 0.0821 \; \frac{\ell \cdot atm}{mol \cdot K}$, $T = 519 \; K$, and $P = 1.00 \; atm$.

$$V = \frac{(0.024)(0.0821)(519)}{1.00} \; \ell = 1.0 \; \ell$$

Exercise   Suppose the acidic component of the baking powder is $KHC_4H_4O_6$. How many grams of this compound are required to react with 2.0 g of $NaHCO_3$? (The two compounds react in a 1:1 mole ratio.) Answer: 4.5 g.

TABLE 9.3 SOLUBILITIES OF COMPOUNDS OF THE GROUP 1 AND GROUP 2 METALS									
	Li^+	Na^+	K^+	Rb^+	Cs^+	Be^{2+}	Mg^{2+}	Ca^{2+}	Sr^{2+}	Ba^{2+}
F^-	ss	S	S	S	S	S	I	I	I	I
Cl^-	S	S	S	S	S	S	S	S	S	S
Br^-	S	S	S	S	S	S	S	S	S	S
I^-	S	S	S	S	S	S	S	S	S	S
NO_3^-	S	S	S	S	S	S	S	S	S	S
SO_4^{2-}	S	S	S	S	S	S	S	ss	I	I
OH^-	S	S	S	S	S	I	I	ss	ss	S
CO_3^{2-}	ss	S	S	S	S	ss	I	I	I	I
PO_4^{3-}	ss	S	S	S	S	S	I	I	I	I

S = soluble (>0.1 M); ss = slightly soluble (0.1–0.01 M); I = insoluble (<0.01 M).

9.4 COMPOUNDS OF THE GROUP 2 METALS

Solubilities

As you can see from Table 9.3, there are many more insoluble compounds of the Group 2 metals than of the Group 1 metals. This has several interesting consequences. For one thing, it accounts for the presence of large deposits of such minerals as limestone ($CaCO_3$), $CaCO_3 \cdot MgCO_3$ (dolomite), CaF_2 (fluorite), and $BaSO_4$ (barite) at or near the earth's surface. It also explains why seawater contains so much more Na^+ than Mg^{2+} (0.46 M vs. 0.052 M), even though the natural abundances of the two elements are close to one another (2.6% Na vs. 1.9% Mg).

Water Softening. Deionization

In many areas, water supplies contain appreciable amounts of various cations. Of these, Ca^{2+} and Mg^{2+} are particularly common. This is because wells, rivers, and lakes often come in contact with such minerals as limestone, $CaCO_3$, or dolomite, $CaCO_3 \cdot MgCO_3$. The presence of Ca^{2+} or Mg^{2+} ions in water causes "hardness." Hard water has several undesirable properties. For one thing, these ions form precipitates with soap. This wastes soap and leaves greasy deposits such as bathtub rings. From an industrial standpoint, a more serious problem is the tendency of calcium or magnesium compounds to precipitate when hard water is heated or evaporated. Among these compounds are $CaCO_3$, $CaSO_4$, and $Mg(OH)_2$. They deposit, often as a tightly adherent scale, on steam irons, teakettles, and boiler and steam pipes. This lowers the heat conductivity and, with pipes, may even block the flow of water.

Ca^{2+} and Mg^{2+} do not affect the taste of drinking water

The most common method of softening water today uses a process called *cation exchange*. The purpose is to replace Ca^{2+} or Mg^{2+} ions in hard water with Na^+ ions, which are much less objectionable. This exchange takes place when hard water is passed through a column containing a "water softener," usually a mineral called a *zeolite*. In appearance, zeolite crystals look very much like grains of sand. The structure of a portion of such a crystal is shown in Figure 9.4. This zeolite differs from simple ionic compounds in one important respect. The negative ion is huge, consisting of Al, Si, and O atoms bonded together to form the backbone of the crystal. Trapped in "holes" within this anionic lattice are Na^+ ions. There are just enough Na^+ ions to compensate for the negative charge of the network-like anion.

When you "soften" water, you take out the "hardness"

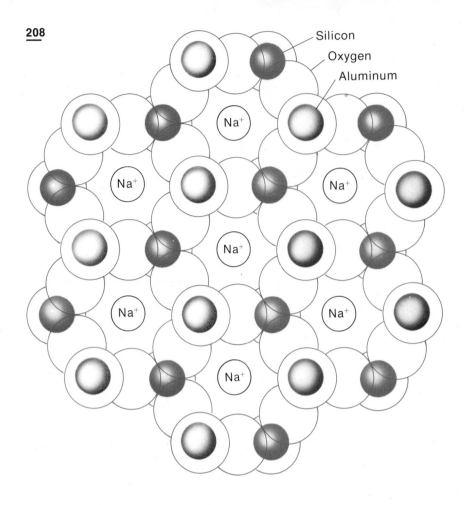

Silicon
Oxygen
Aluminum

Figure 9.4 Structure of a natural zeolite with the simplest formula NaAlSiO₄. Na⁺ ions are held loosely in the anionic lattice, which consists of Al, Si, and O atoms. When water containing Ca²⁺ ions is passed through the zeolite, Ca²⁺ ions enter the lattice while Na⁺ ions take their place in solution.

When pure water passes through the zeolite, nothing much happens. The Na^+ ions cannot leave the crystal; that would create an imbalance of charge. However, if hard water comes in contact with the zeolite, the situation changes. Now, Na^+ ions migrate out of the lattice, to be replaced by Ca^{2+} or Mg^{2+} ions. The process can be represented by the equations

Home water softeners operate via these reactions

$$Ca^{2+}(aq) + 2\ NaZ(s) \rightarrow CaZ_2(s) + 2\ Na^+(aq) \qquad (9.20)$$

$$Mg^{2+}(aq) + 2\ NaZ(s) \rightarrow MgZ_2(s) + 2\ Na^+(aq) \qquad (9.21)$$

Here Z represents a small portion of the zeolite anion. The result of these reactions is to replace a Ca^{2+} or Mg^{2+} ion with two Na^+ ions.

After a zeolite has been used for some time, most of the holes in the lattice become filled with alkaline earth cations. At that point, exchange of cations with hard water stops. However, the zeolite is readily returned to its original state. To do this, the column is flushed with a concentrated solution of NaCl. Reactions 9.20 and 9.21 are reversed and the zeolite water softener is ready for reuse.

Most cation exchangers nowadays use synthetic organic resins. These have structures similar in many ways to zeolites. However, they have a much greater capacity for picking up Ca^{2+} or Mg^{2+} ions. Moreover, they can be used to form part of a unit that will remove all the ions from water. Such a system is shown in Figure 9.5. The column at the left contains a synthetic cation exchange resin which we will represent as HR.

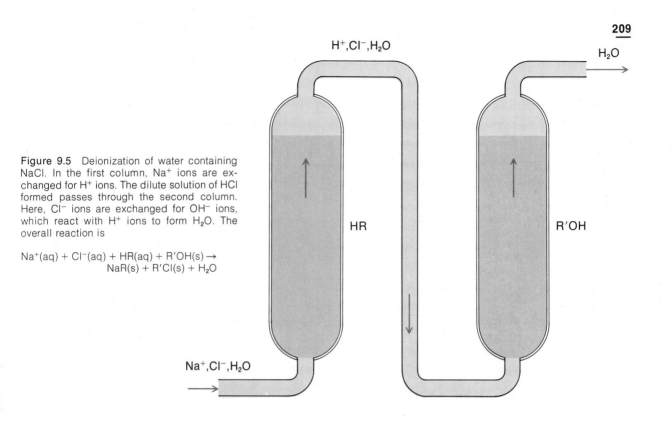

Figure 9.5 Deionization of water containing NaCl. In the first column, Na^+ ions are exchanged for H^+ ions. The dilute solution of HCl formed passes through the second column. Here, Cl^- ions are exchanged for OH^- ions, which react with H^+ ions to form H_2O. The overall reaction is

$$Na^+(aq) + Cl^-(aq) + HR(aq) + R'OH(s) \rightarrow$$
$$NaR(s) + R'Cl(s) + H_2O$$

It is similar to a zeolite except that the exchangeable cation is H^+ rather than Na^+. The second column is packed with an anion exchange resin. Here, replaceable OH^- ions are located in "holes" in a cationic network. We represent this resin by the formula $R'OH$, with the understanding that R' is a complex cation.

When water containing a dissolved salt MX is passed through the two columns in succession, the following reactions occur:

first column: $$M^+(aq) + HR(s) \rightarrow MR(s) + H^+(aq)$$

second column: $$X^-(aq) + R'OH(s) \rightarrow R'X(s) + OH^-(aq)$$

The H^+ and OH^- ions produced react with each other to form water:

$$H^+(aq) + OH^-(aq) \rightarrow H_2O$$

Adding the three equations just written, we arrive at the overall reaction for the removal of MX:

$$M^+(aq) + X^-(aq) + HR(s) + R'OH(s) \rightarrow MR(s) + R'X(s) + H_2O$$

We see from these equations that there are no ions in solution at the end of the process, which is referred to as *deionization*. In principle at least, any salt present in the water will be removed; M^+ can be any cation (e.g., Na^+) and X^- any anion (e.g., Cl^-). To regenerate the first column, it can be flushed with a strong solution of hydrochloric acid, which acts as a source of H^+ ions. The second column is returned to its original state by flushing with sodium hydroxide solution.

Example 9.7   Write balanced equations for the removal by ion exchange of dissolved $Ca(NO_3)_2$.

Solution

cation exchange: $Ca^{2+}(aq) + 2 HR(s) \rightarrow CaR_2(s) + 2 H^+(aq)$

anion exchange: $2 NO_3^-(aq) + 2 R'OH(s) \rightarrow 2 R'NO_3(s) + 2 OH^-(aq)$

neutralization: $2 H^+(aq) + 2 OH^-(aq) \rightarrow 2 H_2O$

overall: $Ca^{2+}(aq) + 2 NO_3^-(aq) + 2 HR(s) + 2 R'OH(s) \rightarrow CaR_2(s) + 2 R'NO_3(s) + 2 H_2O$

Exercise   Write a balanced equation for the removal of Na_2CO_3 by ion exchange.
Answer:

$2 Na^+(aq) + CO_3^{2-}(aq) + 2 HR(s) + 2 R'OH(s) \rightarrow 2 NaR(s) + R'_2CO_3(s) + 2 H_2O$

The major drawback of this process is the cost of the resins and the chemicals used to regenerate them. These expenses are great enough to rule out deionization of seawater for public use. For that purpose, ordinary distillation is cheaper. However, ion exchange columns are widely used to obtain high-purity water in the laboratory. The chances are that the "distilled" water you use in the lab is really deionized.

Deionized water is not necessarily sterile

Calcium Compounds

By far the most abundant and widely used compound of calcium is the carbonate, $CaCO_3$. Calcium carbonate is found in many different forms in nature. The purest of these is the transparent mineral calcite. In less pure form, $CaCO_3$ is found as marble, limestone, and dolomite ($CaCO_3 \cdot MgCO_3$). Huge quantities of $CaCO_3$ are used in the metallurgy of iron (Chap. 5), in making other chemicals such as Na_2CO_3 (Section 9.3), and in the manufacture of glass.

Limestone is a sedimentary rock

Calcium carbonate is very insoluble in pure water. However, it dissolves to an

Figure 9.6   Stalactites (upper) and stalagmites (lower) consist of calcium carbonate. They are formed when a water solution containing Ca^{2+} and HCO_3^- ions enters a cave. Carbon dioxide is released and calcium carbonate precipitates:

$$Ca^{2+}(aq) + 2 HCO_3^-(aq) \rightarrow CaCO_3(s) + H_2O(l) + CO_2(g)$$

(Photograph courtesy of the National Park Service, U.S. Department of the Interior.)

appreciable extent in ground water which contains dissolved carbon dioxide from the atmosphere:

$$CaCO_3(s) + H_2O(l) + CO_2(g) \rightarrow Ca^{2+}(aq) + 2\ HCO_3^-(aq) \qquad (9.22)$$

This reaction is responsible for the formation of limestone caves. Such caves often contain icicle-like formations of calcium carbonate, called stalactites and stalagmites (Fig. 9.6). These are formed when Reaction 9.22 reverses within the cave.

When limestone is heated to about 800°C, it decomposes:

$$CaCO_3(s) \rightarrow CaO(s) + CO_2(g)$$

The solid product, calcium oxide, is often referred to as *quicklime*. As pointed out earlier, it reacts with water to form calcium hydroxide:

$$CaO(s) + H_2O(l) \rightarrow Ca(OH)_2(s) \qquad (9.23)$$

Calcium hydroxide is sometimes called slaked lime or, more simply, *lime*. It is used to "sweeten" acidic soils and in many processes where a cheap, strong base is required. Mortar is made by mixing $Ca(OH)_2$ with sand and water. The mortar "sets" by picking up CO_2 from the air:

$$Ca(OH)_2(s) + CO_2(g) \rightarrow CaCO_3(s) + H_2O(l) \qquad (9.24)$$

The reaction occurs slowly

The calcium carbonate formed binds the particles of sand together.

Another important calcium compound is the mineral called gypsum. This is a hydrate of calcium sulfate, $CaSO_4 \cdot 2\ H_2O$. Blackboard "chalk" is nearly pure gypsum. On a larger scale, gypsum is used in making cement, wallboard, and pottery. Other uses depend upon a reaction that takes place when gypsum is heated, losing three fourths of its water of crystallization:

$$CaSO_4 \cdot 2\ H_2O(s) \rightarrow CaSO_4 \cdot 1/2\ H_2O(s) + 3/2\ H_2O(g)$$

The product is called plaster of Paris, or sometimes simply plaster. When it is ground to a fine powder and mixed with water, plaster is converted back to gypsum.

$$CaSO_4 \cdot 1/2\ H_2O(s) + 3/2\ H_2O(l) \rightarrow CaSO_4 \cdot 2\ H_2O(s)$$

This process takes place with an increase in volume. Hence, the material expands to fill completely any space to which it is confined. This explains its use in making models of statues, patching holes in walls, and forming casts for broken bones.

9.5 CRYSTAL STRUCTURES

Solids tend to crystallize in definite geometric forms which can often be seen by the naked eye. In ordinary table salt, we can distinguish small, cubic crystals of NaCl. Large, beautifully formed crystals of such minerals as fluorite, CaF_2, are found in nature. With many metals, it is possible to observe distinct crystal forms under a microscope.

Crystals have distinct geometric forms because the particles that form them are arranged in a definite three-dimensional pattern. These particles may all be of the same type, as is the case with atoms in a metal. In an ionic compound, two different kinds of particles are involved, cations and anions. In this section, we will look at some of the

The lattice in a crystal persists over millions upon millions of atoms or ions

simpler ways in which atoms or ions can be arranged in a crystal. Our discussion will focus on the metals of Groups 1 and 2 and their ionic compounds.

Metals

The crystal structures of metals are most simply described in terms of what is known as a **unit cell.** The unit cell is the smallest unit which, repeated over and over again, generates the crystal. Three types of unit cells are shown in Figure 9.7. These are:

1. The **simple cubic cell,** which consists of eight atoms located at the corners of the cell. Although this cell is the easiest to visualize, it is seldom found in nature. It is unstable because it contains a relatively large amount of empty space. Only about one half of the total volume (52.36%) is occupied by the atoms themselves. The rest of the space (47.64%) is accounted for by the ''holes'' between the atoms when they are in contact with each other.

2. The **body-centered cubic cell.** This is a cube with atoms at each corner and one in the center of the cube. Here, the atoms are packed more efficiently than in the simple cubic cell. About two thirds of the total volume (68.02%) is occupied by the atoms. All the Group 1 metals and barium in Group 2 crystallize in this pattern.

Face-centered cubic is a "close packed" structure

3. The **face-centered cubic cell.** This is a cube with an atom at each corner and in the center of each face of the cube. In a face-centered cubic structure, the atoms are packed as efficiently as possible. Nearly three fourths of the total volume (74.04%) is occupied by the atoms. Calcium and strontium in Group 2 have this type of unit cell, as do many of the transition metals. Magnesium crystallizes in a closely related structure (hexagonal closest packing) in which the packing is equally efficient.

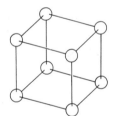

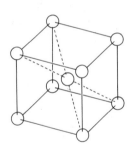

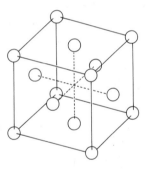

Simple cubic   Body-centered cubic   Face-centered cubic

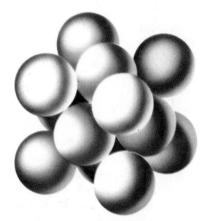

Figure 9.7 Three types of cubic lattices. In the simple cubic lattice, there are 8 atoms at the corners of the cube. In the body-centered lattice, in addition to the eight atoms, there is an atom at the center of the cube. In the face-centered lattice, in addition to the eight atoms, there is an atom at the center of each of the six faces of the cube.

In considering the unit cells shown in Figure 9.7, it is important to keep one point in mind. *Only in the simple cubic cell do atoms at the corners of the cube* touch each other. In the *body-centered cubic* (BCC) cell, *atom contact is along a body diagonal.* That is, the atom in the center of the cube touches atoms at opposite corners of the cube. In the *face-centered cubic* (FCC) cell, *atoms touch along a face diagonal.* The atom at the center of each face is in contact with atoms at opposite corners of the face.

By experiment, it is possible to determine the nature and dimensions of the unit cell in a metal crystal. This information can be used to obtain the atomic radius of the metal. The approach used is indicated in the diagram below, for a simple cubic cell. Here, the atoms at the corners of the cube are touching. The centers of the four atoms (assumed to be spherical) locate the corners of one face of the cube. The atomic radius, r, is related in a very simple way to the length of one side of the cube, s:

The experiment involves diffraction of x-rays

simple cubic cell: $r = s/2$

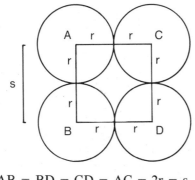

$$AB = BD = CD = AC = 2r = s$$

For a face-centered or body-centered cubic cell, the relation between r and s is more complex:

face-centered cubic cell: $\mathbf{r = s\sqrt{2}/4}$ (Example 9.8)

body-centered cubic cell: $\mathbf{r = s\sqrt{3}/4}$ (Problem 9.40)

Example 9.8   Calcium crystallizes in a structure with a FCC unit cell 0.559 nm on an edge—that is, s = 0.559 nm.
   a. Show that, for a face-centered cubic cell, $r = s\sqrt{2}/4$.
   b. Calculate the atomic radius of calcium.

Solution
   a. It will be helpful to draw a face of the unit cell. Points A, B, C, and D locate the centers of the four atoms at the corners of this face.

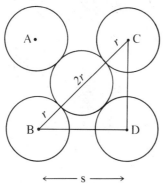

Here is one place where geometry is applied to chemistry

We need a relation between the atomic radius, r, and the distance along an edge of the cube, s. The atoms touch along the face diagonal. Clearly, the length of the diagonal, BC, is 4r; BD and CD are both equal to s. Applying the Pythagorean theorem:

$$(4r)^2 = s^2 + s^2 = 2s^2$$

Taking the square root of both sides:

$$4r = \sqrt{2}s, \text{ or } r = s\sqrt{2}/4$$

b. Substituting s = 0.559 nm, $\sqrt{2}$ = 1.41:

$$r = \frac{(0.559 \text{ nm})(1.41)}{4} = 0.197 \text{ nm}$$

Exercise   For a BCC unit cell, r = s$\sqrt{3}$/4. Potassium crystallizes in a BCC cell 0.533 nm on an edge. What is the atomic radius of K? Answer: 0.231 nm.

Ionic Compounds

The geometry of ionic crystals, where there are two different kinds of ions, is more difficult to describe than that of metals. However, in many cases we can visualize the packing in terms of the unit cells discussed above. Lithium chloride, LiCl, is a case in point. Here, the larger Cl^- ions form a face-centered cubic lattice (Fig. 9.8). The smaller Li^+ ions fit into "holes" between the Cl^- ions. This puts a Li^+ ion at the center of each edge of the cube.

The structure of the sodium chloride crystal is shown at the right of Figure 9.8. The Na^+ ion is slightly too large to fit into holes in a face-centered lattice of Cl^- ions. As a result, the Cl^- ions are pushed slightly apart so that they are no longer touching. Instead, Na^+ ions are in contact with Cl^- ions. However, the relative positions of positive and negative ions remain the same as in LiCl. In particular, each anion is surrounded by six cations and each cation by six anions.

From data on crystals such as those shown in Figure 9.8, it is possible to calculate the radii of anions and cations. Consider, for example, the LiCl cell, where Cl^- ions are

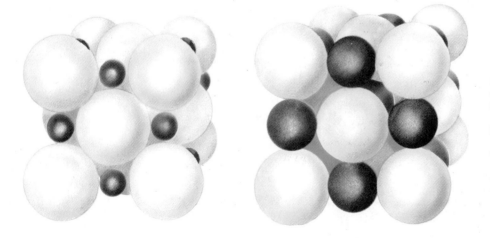

Figure 9.8   In the LiCl cell the Cl^- ions are in contact, while Li^+ ions "rattle around" in the holes between Cl^- ions. In NaCl, the Cl^- ions are forced slightly apart by the Na^+ ions, which are in contact with Cl^- ions. If the ratio r cation/r anion were 0.414, there would be both anion-anion and anion-cation contact. In LiCl this ratio is 0.33; in NaCl, it is 0.52.

LiCl   NaCl

in contact along a face diagonal. The length, s, of this cell is found to be 0.513 nm. Applying the argument used in Example 9.8:

$$r\ Cl^- = s\sqrt{2}/4 = \frac{(0.513\ nm) \times 1.41}{4} = 0.181\ nm$$

Having established the radius of the Cl^- ion, it now becomes possible to obtain the radii
of alkali metal ions. To do this, measurements must be made on crystals where there is
cation-anion contact, such as NaCl. In this case, we find that the length, s, of the cell
is 0.552 nm. From Figure 9.8, we see that

$$2r\ Na^+ + 2r\ Cl^- = 0.552\ nm$$

$$r\ Na^+ = \frac{0.552\ nm - 2r\ Cl^-}{2} = \frac{0.552\ nm - 0.362\ nm}{2} = 0.095\ nm$$

Example 9.9   LiBr has a structure similar to that of LiCl (anions touching along a face
diagonal). KBr has a structure similar to that of NaCl (cation-anion contact along the edge
of the cube). The length, s, of the cube is 0.552 nm in LiBr and 0.656 nm in KBr. What is the
ionic radius of
    a. Br^-?   b. K^+?

Solution

a. $r\ Br^- = s\sqrt{2}/4 = \dfrac{(0.552\ nm) \times 1.41}{4} = 0.195\ nm$

b. $2r\ K^+ + 2r\ Br^- = 0.656\ nm$

$$r\ K^+ = \frac{0.656\ nm - 2r\ Br^-}{2} = \frac{0.656\ nm - 0.390\ nm}{2} = 0.133\ nm$$

Exercise   NaBr has a structure similar to that of NaCl. Taking the radii of the Na^+ and
Br^- ions to be 0.095 nm and 0.195 nm, determine the length of the cubic cell in NaBr. Answer:
0.580 nm.

The structures shown in Figure 9.8 are typical of all the alkali halides (Group 1
cation, Group 7 anion) except those of cesium. In CsCl, the Cs^+ ion is much too large
to fit into a face-centered cubic array of Cl^- ions. We find that CsCl crystallizes in a quite
different structure, shown at the right of Figure 9.9, p. 216. Here each Cs^+ ion is located
at the center of a cube outlined by Cl^- ions, located at the corners of the cube. The unit
cell is a body-centered cube. The Cs^+ ion at the center touches all the Cl^- ions at the
corners. The Cl^- ions do not touch each other. As you can see, each Cs^+ ion is surrounded
by eight Cl^- ions. It is also true that each Cl^- ion is surrounded by eight Cs^+ ions.

SUMMARY

The alkali and alkaline earth metals (Groups 1 and 2) show trends in physical properties (Table 9.1) related to their positions in the Periodic Table. They have relatively
large atomic radii (Group 1 > Group 2) and small ionization energies (Group 1 < Group
2). The positive ions they form (+1 for Group 1, +2 for Group 2) are smaller than the
corresponding atoms or such negative ions as Cl^-. The atomic spectra of these elements

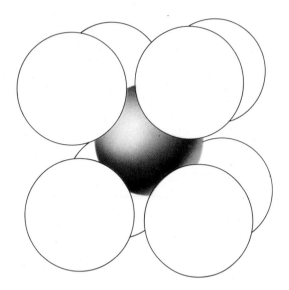

Figure 9.9 In CsCl, a large Cs^+ ion at the center of a cube touches eight Cl^- ions at the corners.

are readily excited. Characteristic colors are produced in a flame due to electronic transitions.

All the metals in these two groups are reactive toward nonmetals (Table 9.2). The compounds formed, except for those of beryllium, are ionic. With oxygen, the product may be a normal oxide (Li_2O, MgO), a peroxide (Na_2O_2, BaO_2), or a superoxide (KO_2). The preparation and properties of some of the more important compounds of sodium and calcium ($NaCl$, $NaOH$, Na_2CO_3, $NaHCO_3$, $CaCO_3$, CaO, $Ca(OH)_2$, $CaSO_4$) were discussed in Sections 9.3 and 9.4. Compounds of the Group 2 metals tend to be less water soluble than those of Group 1. For this reason, Ca^{2+} and Mg^{2+} ions are often removed from water supplies by "softening" or deionization.

The metals in Groups 1 and 2 tend to crystallize in body-centered or face-centered cubic structures (Fig. 9.7 and Example 9.8). Many of their ionic compounds have similar crystal structures (Fig. 9.8 and 9.9; Example 9.9).

In discussing the physical and chemical properties of these metals, we reviewed many of the principles introduced in Chapters 6 to 8. These are emphasized in several examples, including 9.1 (electronic transitions), 9.3 and 9.4 (thermochemistry), and 9.5 and 9.6 (Ideal Gas Law).

KEY WORDS AND CONCEPTS

alkali metal	atomic spectra	Solvay process	unit cell
alkaline earth metal	oxide	hard water	simple cubic cell
atomic radius	base anhydride	cation exchange	body-centered cubic cell
ionic radius	peroxide	zeolite	face-centered cubic cell
ionization energy	superoxide	deionization	

QUESTIONS AND PROBLEMS

Catalog

General: 9.1, 9.2–9.7, 9.21–9.26, 9.40–9.43
Atomic Spectra: 9.8, 9.9, 9.27, 9.28
Thermochemistry: 9.10–9.12, 9.29–9.31
Gas Laws: 9.13, 9.14, 9.32, 9.33

Electronic Structure, Periodic Table: 9.15–9.17,
  9.34–9.36
Crystal Structures: 9.18–9.20, 9.37–9.39

9.1 Review and know the meanings of the key words and concepts in this chapter.

9.2 Complete the following table.

Name	Formula
barium peroxide	_____
_____	$KMnO_4$
strontium nitrate	_____
_____	Li_2CO_3

9.3 Give the symbol of

a. the Group 2 metal with the largest atomic radius.
b. the Group 1 metal with the smallest ionic radius (+1 ion).
c. the alkali metal that reacts with nitrogen.
d. the alkali metal that resembles Mg most closely.

9.4 Give the formula of the compound formed when lithium reacts with

a. I_2 b. H_2 c. H_2O d. S   e. N_2

9.5 Write a balanced equation for the reaction that occurs when

a. sodium reacts with oxygen.
b. barium reacts with water.
c. a water solution of NaCl is electrolyzed.
d. CO_2 and NH_3 react in water solution.
e. Ca^{2+} ions come in contact with a zeolite (NaZ).

9.6 Show by means of balanced equations how NaCl is removed from water by deionization.

9.7 Describe in words how you would prepare

a. CaO from $CaCO_3$.
b. NaOH from NaCl.
c. $CaSO_4 \cdot {}^1/_2 \, H_2O$ from $CaSO_4 \cdot 2 \, H_2O$.

9.8 Barium atoms impart a green color to a flame due to an electronic transition that yields a 554-nm photon. What is the energy of this radiation (kJ/mol)?

9.21 Complete the following table.

Name	Formula
lithium hydroxide	_____
_____	Na_2O_2
calcium phosphate	_____
_____	$Mg(HCO_3)_2$

9.22 Name

a. the alkaline earth metal which most resembles aluminum.
b. the alkali metal compound called "washing soda."
c. the only alkali metal that reacts directly with O_2 to form a "normal" oxide.
d. the least metallic metal in Group 2.

9.23 Give the formula of the compound formed when magnesium reacts with

a. Br_2 b. H_2O c. N_2 d. O_2

9.24 Write a balanced equation for the reaction that occurs when

a. potassium reacts with oxygen.
b. lithium reacts with water.
c. a water solution of NaOH comes in contact with CO_2.
d. sodium hydrogen carbonate is heated.
e. Mg^{2+} ions are removed from water by a zeolite (NaZ).

9.25 Show by means of balanced equations how $BaCl_2$ is removed from water by deionization.

9.26 Describe in words how you would prepare

a. baking soda from table salt.
b. slaked lime from limestone.
c. H_2O_2 from sodium peroxide.

9.27 Cesium metal emits electrons upon exposure to radiation with an energy greater than 181 kJ/mol. What is the maximum wavelength of radiation that can cause this emission?

9.9 Magnesium does not give a flame test. The strongest line in the Mg spectrum arises from a transition between two states which differ in energy by 419.3 kJ/mol. What is the wavelength of this line? In what region of the spectrum does it fall?

9.10 Use data from Tables 6.1 and 6.2 to calculate ΔH for the slaking of lime, which can be represented as either (a) or (b) below.

a. $CaO(s) + H_2O(l) \rightarrow Ca(OH)_2(s)$
b. $CaO(s) + H_2O(l) \rightarrow Ca^{2+}(aq) + 2\ OH^-(aq)$

9.11 During the making of a cast to set a broken bone, the following exothermic reaction occurs:

$$CaSO_4 \cdot {}^1\!/_2\ H_2O(s) + {}^3\!/_2\ H_2O(l) \rightarrow CaSO_4 \cdot 2\ H_2O(s)$$
$$\text{plaster} \hspace{4.5cm} \text{cast}$$

Calculate ΔH for the fabrication of a 2.00-kg cast. ΔH_f: $CaSO_4 \cdot {}^1\!/_2\ H_2O = -1574$ kJ/mol; $H_2O = -286$ kJ/mol; $CaSO_4 \cdot 2\ H_2O = -2021$ kJ/mol.

9.12 The formation of NaF from the elements can be analyzed by a process similar to that described for NaCl on p. 199. For NaF, $\Delta H_1 = 184$ kJ, $\Delta H_2 = 163$ kJ, and $\Delta H = -569$ kJ. What is the lattice energy (ΔH_3) for NaF?

9.13 Above 700°C, barium peroxide decomposes:

$$2\ BaO_2(s) \rightarrow 2\ BaO(s) + O_2(g)$$

If the oxygen liberated by heating 10.0 g of barium peroxide is collected in a 500-cm³ flask at 20°C, what is the pressure in the flask?

9.14 3.0 g of sodium metal react completely with water.

a. At 20°C and 1.00 atm, what volume of hydrogen gas is liberated?
b. What volume of 0.10 M HCl is required to react completely with the NaOH produced by the reaction?

9.15 Which of the following are isoelectronic—that is, have the same electron configuration?

a. H^+, He, Li, Li^+, Be
b. F^-, Ne, Na, Mg^{2+}

9.16 Give the outer electron configuration of

a. K b. Sr c. Na d. Mg

9.28 Sodium vapor lamps provide high-intensity lighting. One of the strongest lines emitted is at 589.6 nm. What is the energy difference (kJ/mol) for the electronic transition that produces this line?

9.29 Barium peroxide decomposes upon heating to barium oxide and oxygen gas. When 10.0 g of barium peroxide decomposes, ΔH is +5.03 kJ. Calculate ΔH_f for barium peroxide (ΔH_f BaO = -558 kJ/mol).

9.30 Use data from Tables 6.1 and 6.2 to

a. calculate ΔH for the reaction given by Equation 9.22.
b. Calculate ΔH for the formation of a 10.0-kg stalactite (reverse of Reaction 9.22).

9.31 Given:

	ΔH (kJ)
$Li(g) + Br(g) \rightarrow LiBr(s)$	-607
$Li(g) \rightarrow Li^+(g) + e^-$	$+527$
$Li^+(g) + Br^-(g) \rightarrow LiBr(s)$	-787

Calculate ΔH for $Br(g) + e^- \rightarrow Br^-(g)$. This is known as the "electron affinity" of bromine.

9.32 1.00 g of solid sodium is to be converted to NaCl by reaction with chlorine gas. What volume of Cl_2 at 25°C and 780 mm Hg is required?

9.33 24.8 g of molten KCl are electrolyzed to produce potassium metal and chlorine gas.

a. What volume of chlorine is produced at 25°C and 1.00 atm (assume 100% yield)?
b. How many moles of KOH are produced when the amount of potassium produced in (a) above reacts completely with water?

9.34 Give the formulas of four Group 2 halides in which the cation and anion have the same electron configuration.

9.35 Give the symbols of all the atoms in Groups 1 and 2 that have electrons in

a. the 3d sublevel b. 2p c. 4f

9.17 Consider the following species: Na, Mg, K. List them in order of

a. increasing first ionization energy.
b. increasing atomic radius.
c. decreasing ionic radius.

9.18 Rubidium crystallizes in a body-centered cubic cell 0.564 nm on an edge. Calculate the atomic radius of Rb.

9.19 Consider the BCC cell of CsCl (Fig. 9.9). The ionic radii of Cs^+ and Cl^- are 0.169 and 0.181 nm, in that order. What is the length of

a. the body diagonal?
b. the side of the cell?

9.20 Potassium iodide has a unit cell similar to that of NaCl (Fig. 9.8). The ionic radii of K^+ and I^- are 0.133 and 0.216 nm, in that order. What is the length of

a. one side of the cube?
b. the face diagonal of the cube?

9.36 List Be, Na, and Mg in order of

a. decreasing first ionization energy.
b. decreasing atomic radius.
c. decreasing chemical reactivity.

9.37 The crystal structure of strontium consists of face-centered cubic unit cells. The atomic radius of strontium is 0.215 nm. Calculate the volume of its unit cell: $V = (cell\ edge)^3$.

9.38 Barium (at. rad. = 0.217 nm) crystallizes with a body-centered cubic unit cell. What is the length of the side of the cell?

9.39 In the LiCl structure shown in Figure 9.8, the chloride ions form a FCC unit cell 0.513 nm on an edge. The ionic radius of Cl^- is 0.181 nm.

a. Along a cell edge, how much space is there between the Cl^- ions?
b. Would a Na^+ ion (r = 0.095 nm) fit into this space? a K^+ ion (r = 0.133 nm)?

*9.40 Show, using principles of geometry, that for a body-centered cubic cell, the relation between the atomic radius, r, and the length of a side of the cell, s, is $r = s\sqrt{3}/4$.

*9.41 Sodium hydrogen carbonate is a home remedy for "excess stomach acid." The relieving reaction is

$$HCO_3^-(aq) + H^+(aq) \rightarrow CO_2(g) + H_2O(l)$$

To relieve 100 cm³ of excess stomach acid (0.12 M HCl), how many teaspoonfuls (7.8 g) of $NaHCO_3$ should you take? What volume of CO_2 will be produced at body temperature, 98.6°F, and 1.00 atm?

*9.42 As shown in Figure 9.8, Li^+ ions fit into a face-centered array of Cl^- ions but Na^+ ions do not. What is the value of the r cation/r anion ratio so that a cation will just fit into a structure of this type?

*9.43 Barium crystallizes in a BCC unit cell 0.501 nm on an edge. In effect, two Ba atoms occupy a unit cell. Calculate the density of barium metal and compare to the experimental value of 3.5 g/cm³.

COVALENT BONDING; MOLECULAR SUBSTANCES

In Chapter 2, we discussed briefly two different kinds of "building blocks" of matter: ions and molecules. In later chapters, we considered many of the properties of ions and the forces that hold them together, called ionic bonds. You will recall that monatomic ions, such as Na^+ and Cl^-, are formed by a transfer of electrons, in this case from a Na to a Cl atom.

In this chapter we will look at the structures and properties of molecular substances. The atoms within a molecule are held together by strong forces called **covalent bonds.** A covalent bond consists of an electron pair shared between two bonded atoms. Perhaps the simplest bond of this type is that which joins H atoms in the H_2 molecule. When two hydrogen atoms, each with one electron, come together, they form a bond. This may be shown as

$$H : H$$

where the dots represent electrons. More commonly, we show the structure of the H_2 molecule as

$$H—H$$

The understanding is that the dash represents a covalent bond, an electron pair shared by the two atoms.

10.1 NATURE OF THE COVALENT BOND

We can interpret ionic bonds found in compounds such as NaCl in a quite simple way. Oppositely charged ions (Na^+ and Cl^-) are held together by electrostatic forces. The nature of the covalent bond in H_2 is more difficult to visualize. The structures

$$H : H \qquad or \qquad H—H$$

can be misleading if they are taken to mean that the two electrons are fixed in position between the two nuclei. A more accurate picture of the electron density in H_2 is shown

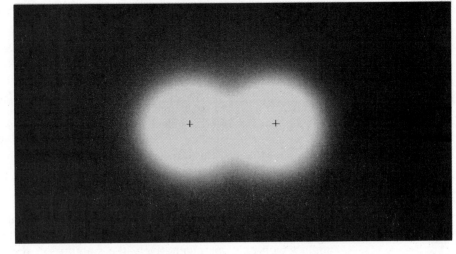

Figure 10.1 Electron density in H_2. The depth of shading is proportional to the probability of finding an electron in a particular region. In chemical bonds there tends to be a concentration of electronic charge between the nuclei.

in Figure 10.1. At a given instant, the two electrons may be located at any of various points about the two nuclei. However, they are more likely to be between the nuclei than at the far ends of the molecule.

A question that has long intrigued chemists is: why should the sharing of two electrons between two nuclei result in increased stability? Why, for example, should the H_2 molecule be more stable, by about 400 kJ/mol, than two isolated H atoms? The first plausible answer to this question was put forth in 1927 by two physicists, W. A. Heitler and T. London. They used quantum mechanics to calculate the interaction energy of two hydrogen atoms as a function of the distance between them.

At large distances of separation (far right, Fig. 10.2), there is no interaction between two hydrogen atoms. As they come closer together (moving to the left in Fig. 10.2), the two atoms experience an attraction. This leads gradually to an energy minimum, at an internuclear distance of 0.074 nm. The attractive energy at this point is 436 kJ/mol, the H—H bond energy. At this distance of separation the molecule is in its most stable state. If we attempt to bring the atoms closer together, repulsive forces become important and the energy curve rises steeply.

The existence of the energy minimum shown in Figure 10.2 accounts for the stability

For the reaction:

$$H_2(g) \rightarrow 2\ H(g)$$

$$\Delta H = 436\ kJ$$

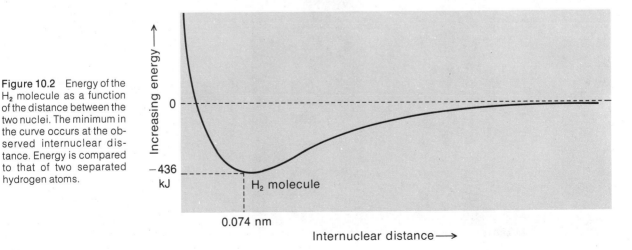

Figure 10.2 Energy of the H_2 molecule as a function of the distance between the two nuclei. The minimum in the curve occurs at the observed internuclear distance. Energy is compared to that of two separated hydrogen atoms.

of the H_2 molecule. The question remains, however: what causes this energy minimum? There are two major factors involved.

1. As two hydrogen atoms approach each other, the electron of one atom is attracted to the nucleus of the other. The electrostatic attraction increases steadily as the distance shortens. However, repulsion between particles of like charge (electron-electron and nucleus-nucleus) also becomes important. The attractive and repulsive forces are balanced at 0.074 nm. There the system has its lowest energy and so is most stable.

2. In the H_2 molecule, the electrons interact with both nuclei equally. In the electrostatic analysis, we treat the electrons as distinguishable particles. It turns out that we really don't know which electron is which. When we take this into account, the electrons are given a little more freedom, which leads to a lower energy. In the jargon of quantum mechanics, this effect is described in terms of "orbital overlap." We say that the H_2 molecule is stabilized by the overlap of the two 1s orbitals. Each electron, in effect, is spread over both orbitals.

The Heitler-London model explains the stability of the H_2 molecule reasonably well. Unfortunately, the equations upon which it is based cannot be solved for any but the simplest of molecules. Later, in Section 10.5, we will describe an extension of this model which explains qualitatively the stability of more complex molecules. In the remainder of this section, our goal is more modest. We will look at how electron pairs are distributed between two atoms joined by one or more covalent bonds.

Computers are getting better and better in dealing with this kind of problem

Polar and Nonpolar Covalent Bonds

As we might expect, the two electrons in the H_2 molecule are shared equally by the two nuclei. Stated another way, a bonding electron is as likely to be found in the vicinity of one nucleus as another. Bonds of this type are described as **nonpolar.** We find nonpolar bonds whenever the two atoms joined are identical, as in H_2 or F_2.

In the HF molecule, the distribution of the bonding electrons is somewhat different from that found in H_2 or F_2. Here, the density of the electron cloud is greatest about the fluorine atom. The bonding electrons, on the average, are shifted toward fluorine and away from hydrogen (Fig. 10.3). Bonds in which the electron density is unsymmetrical are referred to as **polar bonds.**

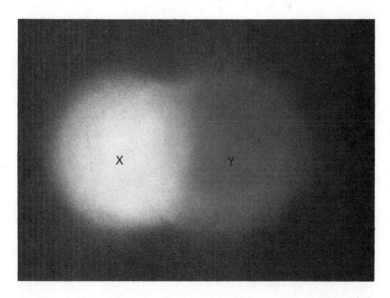

Figure 10.3 If two atoms, X and Y, differ in electronegativity, the bond between them is polar. The electron cloud associated with the bonding electrons is concentrated around the more electronegative atom, in this case X.

TABLE 10.1   ELECTRONEGATIVITY VALUES

H 2.1						
Li 1.0	Be 1.5	B 2.0	C 2.5	N 3.0	O 3.5	F 4.0
Na 0.9	Mg 1.2	Al 1.5	Si 1.8	P 2.1	S 2.5	Cl 3.0
K 0.8	Ca 1.0	Sc 1.3	Ge 1.8	As 2.0	Se 2.4	Br 2.8
Rb 0.8	Sr 1.0	Y 1.2	Sn 1.8	Sb 1.9	Te 2.1	I 2.5
Cs 0.7	Ba 0.9	La–Lu 1.0–1.2	Pb 1.9	Bi 1.9	Po 2.0	At 2.2

Atoms of two different elements always differ at least slightly in their affinity for electrons. Hence covalent bonds between unlike atoms are always polar. Consider, for example, the H—F bond. Since fluorine has a stronger attraction for electrons than does hydrogen, the bonding electrons are displaced toward the fluorine atom. The H—F bond is polar, with a partial negative charge at the fluorine atom and a partial positive charge at the hydrogen atom.

The ability of an atom to attract to itself the electrons in a covalent bond is described in terms of **electronegativity.** The greater the electronegativity of an atom, the greater its affinity for bonding electrons. Electronegativities can be estimated in various ways. One method, based on bond energies, leads to the scale listed in Table 10.1. Here each element is assigned a number ranging from 4.0 for the most electronegative element, fluorine, to 0.7 for cesium, the least electronegative. Among the main-group elements, electronegativity increases moving from left to right in the Periodic Table. Ordinarily, it decreases as we move down a given group.

The electronegativity is always a positive number

The extent of polarity of a covalent bond is related to the difference in electronegativities of the bonded atoms. If this difference is large, as in HF (E.N. H = 2.1, F = 4.0), the bond will be strongly polar. Where the difference is small, as in H—C (E.N. H = 2.1, C = 2.5), the bond will be only slightly polar. Thus, in the carbon-hydrogen bond, the bonding electrons are only slightly displaced toward the carbon atom.

In a pure (nonpolar) covalent bond, the electrons are equally shared. In a pure ionic bond, there has been a complete transfer of electrons from one atom to another. We can think of a polar covalent bond as being intermediate between these two extremes. In this sense, we can relate bond polarity to *partial ionic character*. The greater the difference in electronegativity between two elements, the more ionic will be the bond between them. The relation between these two variables is shown in Figure 10.4. Notice that a difference of 1.7 units corresponds to a bond with 50% ionic character. Such a bond might be described as being halfway between a pure covalent and a pure ionic bond.

It is clearly an oversimplification to refer to a bond between two elements as being "ionic" or "covalent." Consider, for example, the bonding in compounds formed by a Group 1 or 2 metal with a nonmetal in Group 6 or 7. The difference in electronegativity ranges from a minimum of 0.6 for Be—Te to a maximum of 3.3 for Cs—F. The percentage of ionic character varies similarly, from about 10% for BeTe to 95% for CsF.

The difference in electronegativity between oxygen or fluorine on the one hand and

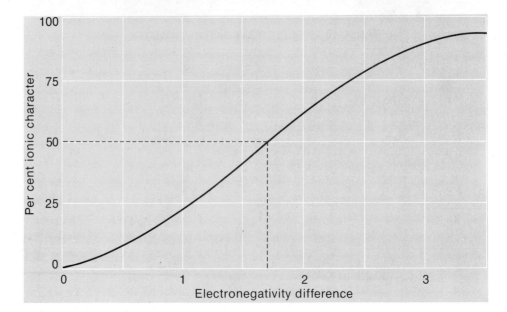

Figure 10.4 Relation between ionic character of a bond and the difference in electronegativity of the bonded atoms.

a Group 1 or 2 metal on the other always exceeds 1.7 units. In this sense, the bonding in the oxides and fluorides of these metals is mainly ionic. The same is true for the oxide and fluoride of aluminum in Group 3. Here, electronegativity differences of 2.0 and 2.5 units correspond to about 65% and 80% ionic character in Al_2O_3 and AlF_3, respectively. The situation is quite different with the chloride, bromide, and iodide of aluminum. In each of these compounds, electronegativity differences less than 1.7 units imply that the bonding is mainly covalent.

Example 10.1   Using only the Periodic Table, arrange the following bonds in order of increasing polarity (ionic character): O—Cl, O—O, O—Na.

Solution   The O—O bond must come first; since the atoms are identical, the bond is nonpolar. To decide which bond comes next, recall that electronegativity increases as we move across in the Table and decreases as we move down. Chlorine lies below but to the right of oxygen. We might expect the two elements to have about equal electronegativities. On the other hand, sodium lies far to the left of oxygen and is below it. Hence, sodium should have a much lower electronegativity than oxygen. The proper sequence is:

$$O—O < O—Cl < O—Na$$

We progress from a nonpolar to a slightly polar to a strongly polar, essentially ionic, bond.

Exercise   Which of the three bonds, B—C, C—N, or B—Si would you expect to be least polar? Answer: B—Si.

Single, Double, and Triple Bonds

A bond which consists of a single electron pair is commonly referred to as a **single bond.** This is the type of bond found in the H_2 molecule:

H—H

TABLE 10.2 CARBON-CARBON BOND DISTANCES AND ENERGIES		
	BOND DISTANCE (nm)	**BOND ENERGY** (kJ/mol)
C_2H_6	0.154	347
C_3H_8	0.154	347
C_4H_{10}	0.154	347
C_2H_4	0.133	598
C_2H_2	0.120	820

and in many other molecules as well. Ordinarily, the distance between like atoms joined by a single bond is the same in different molecules. The same is true of *bond energy*, referred to in Chapter 6. Consider, for example, the C—C bond. This is present in molecules of ethane, C_2H_6, propane, C_3H_8, and butane, C_4H_{10}.

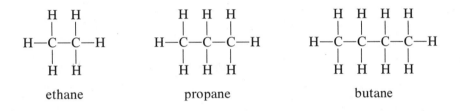

ethane   propane   butane

As you can see from Table 10.2, the distance between C atoms in these molecules is the same, 0.154 nm. Again, the C—C bond energy is the same, 347 kJ/mol.

When we find that either bond distance or bond energy changes sharply, we suspect a change in bond type. This is the case with the last two compounds listed in Table 10.2 (ethylene, C_2H_4, and acetylene, C_2H_2). The distance between the carbon atoms in ethylene, 0.133 nm, is significantly shorter than the C—C bond distance, 0.154 nm. At the same time, the bond holding the carbon atoms together is stronger than the C—C bond, 598 vs 347 kJ/mol. These effects are even more pronounced in acetylene.

This evidence is taken to mean that there are *multiple bonds* between the carbon atoms in ethylene and acetylene. We say that there is a **double bond** in ethylene, consisting of two pairs of electrons joining the two carbon atoms:

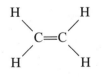

There is a **triple bond** in acetylene (three pairs of electrons):

$$H—C\equiv C—H$$

We always find that double or triple bonds are stronger than single bonds between the same two atoms. This comparison is made in Table 10.3, where energies are listed for the more common types of multiple bonds. *Notice that the atoms joined by multiple bonds are most commonly C, O, N, or S.*

Only a few elements form multiple bonds

TABLE 10.3 COMPARISON OF BOND ENERGIES (kJ/mol)					
SINGLE BOND	BOND ENERGY	DOUBLE BOND	BOND ENERGY	TRIPLE BOND	BOND ENERGY
C—C	347	C=C	598	C≡C	820
C—N	293	C=N	615	C≡N	890
C—O	351	C=O	715	C≡O	1075
C—S	259	C=S	477	—	—
N—N	159	N=N	418	N≡N	941
N—O	222	N=O	607	—	—
O—O	138	O=O	402	—	—
S—O	347	S=O	665	—	—

The C≡O bond is the strongest in all of chemistry

10.2 LEWIS STRUCTURES; THE OCTET RULE

The idea of the covalent bond was first suggested by the American physical chemist G. N. Lewis in 1916. He pointed out that the electron configuration of the noble gases appears to be a particularly stable one. Noble-gas atoms are themselves extremely unreactive. Moreover, the monatomic ions formed from elements in Groups 1, 2, 6, and 7 have noble-gas structures (recall Fig. 3.3, p. 56). Lewis suggested that *atoms by sharing electrons to form an electron-pair bond can acquire a stable, noble-gas structure.* Consider, for example, two hydrogen atoms, each with one electron. The process by which they combine to form an H_2 molecule can be shown as

$$H\cdot + H\cdot \rightarrow H:H$$

using dots for electrons. In H_2, each hydrogen atom has a share in two electrons. In that sense, it has the electronic structure of the noble gas, helium (at. no. = 2).

The idea is readily extended to other simple molecules. An example is the F_2 molecule, formed from two fluorine atoms. You will recall that a fluorine atom has the electron configuration $1s^2 2s^2 2p^5$. It has seven electrons in its outermost principal energy level ($n = 2$). These electrons are referred to as **valence electrons.** Showing them as dots about the symbol of the element, we can represent the fluorine atom as

$$:\ddot{F}\cdot$$

When two of these atoms combine, we have

$$:\ddot{F}\cdot \;+\; \cdot\ddot{F}: \rightarrow \;:\ddot{F}:\ddot{F}:$$

From our drawing of the F_2 molecule, you can see that each atom owns six valence electrons outright and shares two others. Putting it another way, each F atom is surrounded by eight valence electrons. By this model, both F atoms achieve the electron configuration $1s^2 2s^2 2p^6$, which is that of the noble gas neon. This, according to Lewis, explains why the F_2 molecule is stable and why F atoms combine to form F_2 rather than F_3, F_4, ...

The structures just written are referred to as Lewis structures, or sometimes as "flyspeck formulas." In writing the Lewis structure for a species, we include only those electrons in the outermost energy level. These so-called *valence electrons* are the ones that take part in covalent bonding. For the main-group elements, the only ones that we will be concerned with here, **the number of valence electrons is equal to the group number in the Periodic Table** (Table 10.4).

In a sense, we count the bonding electrons twice

TABLE 10.4   LEWIS STRUCTURES OF ATOMS

Group	Outer Electron Configuration	Number of Valence Electrons	Example
1	ns^1	1	Li·
2	ns^2	2	·Be·
3	ns^2np^1	3	·Ḃ·
4	ns^2np^2	4	·Ċ·
5	ns^2np^3	5	·N̈·
6	ns^2np^4	6	·Ö·
7	ns^2np^5	7	:F̈·
8	ns^2np^6	8	:N̈e:

In drawing Lewis structures we don't distinguish between s and p electrons

In the Lewis structure of a molecule, a covalent bond between atoms is ordinarily shown as a straight line between bonded atoms. Unshared electron pairs, belonging entirely to one atom, are shown as dots. The Lewis structures for the molecules formed by hydrogen with C, N, O, and F are

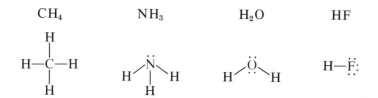

CH_4 NH_3 H_2O HF

Notice that in each case the central atom (C, N, O, F) is surrounded by eight valence electrons.

Many polyatomic ions can be assigned simple Lewis structures. For example, the OH^- and NH_4^+ ions can be shown as

$$(:\ddot{O}-H)^- \text{ and } \left[H-\underset{H}{\overset{H}{N}}-H \right]^+$$

In both these ions, hydrogen atoms are joined by covalent bonds to nonmetal atoms (O, N). In both ions there are eight valence electrons. With the OH^- ion, this is one more than the number contributed by the neutral atoms ($6 + 1 = 7$). The extra electron is accounted for by the -1 charge of the ion. With the NH_4^+ ion, four hydrogen atoms and a nitrogen atom supply nine valence electrons ($4 + 5 = 9$). One of these is missing in the NH_4^+ ion, accounting for its $+1$ charge.

These examples illustrate the principle that atoms in covalently bonded species tend to have noble-gas structures. This rule is often referred to as the **octet rule.** Nonmetals, except for hydrogen, achieve a noble-gas structure by acquiring an "octet" of electrons (eight). Hydrogen atoms, in stable molecules or polyatomic ions, are surrounded by two electrons.

Most molecular substances obey the octet rule

Writing Lewis Structures

For very simple molecules, Lewis structures can often be written by inspection. Usually, though, you will save time and avoid confusion by following these rules:

1. *Draw a skeleton structure for the molecule or ion, joining atoms by single bonds.* In some cases, only one arrangement of atoms is possible; in others, experimental evidence must be used to decide between two or more alternative structures.

2. *Count the number of valence electrons.* For a molecule, we simply sum up the valence electrons of the atoms present. For a *polyatomic anion*, electrons are *added* to take into account the negative charge. For a polyatomic *cation*, a number of electrons equal to the positive charge must be *subtracted*.

3. *Deduct two valence electrons for each single bond written in step 1. Distribute the remaining electrons as unshared pairs so as to give each atom eight electrons, if possible.*

The application of these rules and some further guiding principles are shown in Example 10.2.

Example 10.2   Draw Lewis structures for
   a. ClO^- b. SO_4^{2-} c. CH_4O

Solution

a. Only one skeleton structure is possible for the hypochlorite ion

$$(Cl-O)^-$$

To obtain the total number of valence electrons we add one (the charge of the ion) to the number contributed by chlorine in Group 7 (7) and oxygen in Group 6 (6).

$$\text{no. of valence electrons} = 7 + 6 + 1 = 14$$

Deducting the two electrons used to make one covalent bond leaves 12. Putting six of these electrons around each atom, we arrive at a reasonable structure for the ClO^- ion which satisfies the octet rule, since both the Cl and O atoms have a share in eight electrons. (The bonding electrons are counted for both atoms.)

$$(:\overset{\cdot\cdot}{\underset{\cdot\cdot}{Cl}}-\overset{\cdot\cdot}{\underset{\cdot\cdot}{O}}:)^-$$

b. Various skeletons could be written for the sulfate ion. However, in *ions* such as SO_4^{2-}, NO_3^-, and CO_3^{2-} we ordinarily find that *each oxygen atom is bonded to the central, nonmetal atom.* Following this general rule, we write

$$\left[\begin{array}{c} O \\ | \\ O-S-O \\ | \\ O \end{array} \right]^{2-}$$

The number of valence electrons is found by adding the charge of the ion to the total contributed by the sulfur and oxygen atoms:

$$\text{no. of valence electrons} = 6 + 4(6) + 2 = 32$$

Deducting eight electrons for the four covalent bonds in the skeleton structure leaves 24. Putting six electrons around each oxygen atom gives us a plausible Lewis structure for the sulfate ion:

c. In principle, one could write several different skeletons for a molecule of methyl alcohol, CH_4O. The correct skeleton is easily arrived at if you realize that

—*a hydrogen atom can form only one covalent bond.* (It needs only two electrons to have a noble-gas structure.)

—*a carbon atom typically forms four bonds, with no unshared pairs.*

The only skeleton consistent with these rules is

There are almost no
exceptions to these
two rules

$$
\begin{array}{c}
H \\
| \\
H-C-O-H \\
| \\
H
\end{array}
$$

The number of valence electrons in the molecule is simply the sum of those from the carbon, hydrogen, and oxygen atoms:

$$\text{no. of valence electrons} = 4 + 4(1) + 6 = 14$$

Deducting ten electrons for the five covalent bonds in the skeleton leaves four. This is just enough to complete the octet of oxygen. The Lewis structure is

---

Exercise   Draw the Lewis structure of the phosphate ion, PO_4^{3-}. Answer:

$$
\left[
\begin{array}{c}
\ddot{O}: \\
| \\
:\ddot{O}-P-\ddot{O}: \\
| \\
:\ddot{O}:
\end{array}
\right]^{3-}
$$

Sometimes, when you follow these rules, you find in the last step that there are "too few electrons to go around." That is, there are not enough electrons left to give each atom an octet. This dilemma can be avoided by making electron pairs do "double duty." By moving an unshared pair to a position between bonded atoms, it is counted in the octet of both atoms. The result of such a shift is to form a multiple bond from a single bond. Forming a double bond by moving one electron pair repairs a deficiency of two electrons. If you are four electrons shy, you must form a triple bond (or two double bonds). The process involved is shown in Example 10.3.

Example 10.3   Draw Lewis structures for
   a. SO_2 b. N_2

Solution

   a. Here, as in SO_4^{2-}, a central sulfur atom is bonded to oxygens. The skeleton is

The number of valence electrons is $6 + 2(6) = 18$. Subtracting four electrons for the two covalent bonds leaves 14. These electrons could be spent by filling out the oxygen octets, which accounts for 12, and putting two around the sulfur:

This leaves sulfur with only six valence electrons, two less than an octet. To correct this, shift an unshared pair from one of the oxygen atoms to form a double bond with sulfur. The final Lewis structure is*

We could have equally well put the double bond on the right side. See footnote

b. The skeleton is N—N. Since nitrogen is in Group 5, the total number of valence electrons is $2(5) = 10$. The single bond in the skeleton consumes two electrons, so there are eight left. Distributing these as unshared pairs gives the structure

$$:\ddot{N}—\ddot{N}:$$

Note that both N atoms are two electrons shy of an octet. In other words, there is a deficiency of four electrons. Move two unshared pairs, one from each nitrogen, to give a triple bond:

$$:N\equiv N:$$

Exercise   Derive the Lewis structure of the nitrate ion, NO_3^-. Answer:

Exceptions to the Octet Rule

Most of the molecules and polyatomic ions that we work with in general chemistry follow the octet rule. However, there are some important species that do not. Several of these will be discussed in Chapter 12. Here, we will consider two general types of molecules in which at least one atom does not have a noble-gas structure.

1. *Odd-electron species* (free radicals), which have an odd number of valence electrons. One such species is the nitric oxide molecule, NO.

$$\text{no. of electrons in NO} = \text{no. of valence electrons} = 5 + 6 = 11$$

11 is not an even number

There is no way that we can write a Lewis structure for the NO molecule to give each atom an octet. The best we can do is to write a structure such as

$$·\overset{··}{N}=\overset{··}{O}:$$

---

*This structure implies that there are two different kinds of bonds in the SO_2 molecule. Experimentally, the two bonds are identical. This apparent inconsistency can be explained by invoking the concept of resonance, discussed in Chapter 12.

2. *Species in which there are two or three rather than four pairs of electrons* around the central atom. Examples include the fluorides of beryllium and boron. In the gas state, these exist as molecules of BeF_2 and BF_3, which seem to have the structures

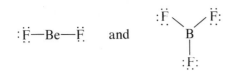

Here, the central atom is surrounded by four valence electrons (Be) or six (B) rather than eight.

10.3 MOLECULAR GEOMETRY

A molecule such as Cl_2 or HF must be linear; the two atoms define a straight line. With molecules containing more than two atoms, the geometry is not so obvious. Here we must be concerned with the angles between bonded atoms. Consider, for example, a molecule of the type XY_2, where X is the central atom. Two geometries are possible. The atoms might be in a straight line, giving a linear molecule with a bond angle of 180°. On the other hand, they could be arranged in a triangular pattern. In that case, we would have a bent molecule with a bond angle less than 180° (Fig. 10.5).

The major features of molecular geometry can be predicted on the basis of a quite simple principle. It was first suggested by Sidgwick and Powell in 1940, and was later developed and expanded by R. J. Gillespie of McMaster University in Canada:

The electron pairs surrounding an atom repel each other and are directed to be as far apart as possible.

This is called the electron-pair repulsion rule

In this section, we will apply this rule to predict the geometry of some rather simple molecules and polyatomic ions. In all of these species, *a central atom is surrounded by two, three, or four electron pairs*. We will be interested in
—the bond angles in the molecule.
—the position of other atoms relative to the central atom.
—the general shape of the molecule, as described by such words as ''linear,'' ''bent,'' and so on.
In later chapters, we will consider the geometry of more complex species.

Species with Single Bonds, No Unshared Electron Pairs Around the Central Atom

Figure 10.6 shows the geometries to be expected when two, three, or four electron pairs around a central atom are used to form single bonds. In each case, the electron pairs are directed to be as far apart as possible. With two electron pairs (Be in BeF_2),

Figure 10.5 Two possible geometries for a molecule of general formula XY_2. Certain molecules of this type, including BeF_2 and CO_2, are linear (180° bond angle). Others, including H_2O and SO_2, are bent (bond angle < 180°). Knowing the Lewis structure, it is possible to predict whether a molecule will be linear or bent (see text discussion).

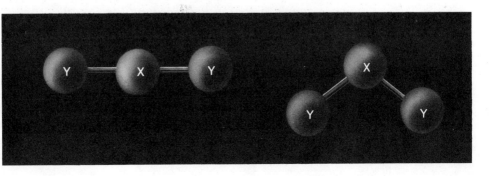

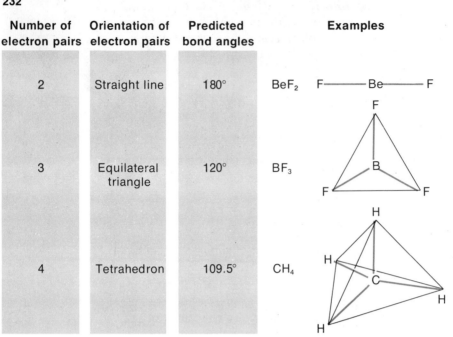

Number of electron pairs	Orientation of electron pairs	Predicted bond angles	Examples
2	Straight line	180°	BeF_2
3	Equilateral triangle	120°	BF_3
4	Tetrahedron	109.5°	CH_4

Figure 10.6 Geometry of electron pairs around a central atom (Be, B, C). The electron pairs orient themselves so as to be as far apart as possible. The most common situation involves four electron pairs directed toward the corners of a regular tetrahedron, as in CH_4.

These geometries allow the electrons to get as far apart as they can

this gives a 180° bond angle and, hence, a **linear** molecule. Three electron pairs around a central atom (B in BF_3) are directed toward the corners of an **equilateral triangle.** The bond angles are 120°. The molecule is planar; that is, all four atoms are in the same plane.

In the most common case, with molecules that follow the octet rule, the central atom is surrounded by four electron pairs. Suppose now that all of these pairs are used to form a total of four single bonds. Here, electron pair repulsion predicts a three-dimensional structure. The four electron pairs are directed toward the corners of a regular **tetrahedron.** This is a three-dimensional structure with four faces and four corners. All the bond angles are 109.5°, the tetrahedral angle. Methane, CH_4, is a classic example of a molecule with this structure. The carbon atom is at the center of the tetrahedron. There is a hydrogen atom at each corner.

The tetrahedron is a basic unit in the structure of many organic molecules. Indeed, whenever carbon forms four single bonds, it shows tetrahedral geometry. Many polyatomic ions also show this structure. Consider the Lewis structures of NH_4^+ (p. 227) and SO_4^{2-} (p. 229). Note that in both cases the central atom is surrounded by four electron pairs, all of which are used to form single bonds. In both ions, the central atom is in the center of a regular tetrahedron, symmetrically surrounded by four other atoms.

Species with Unshared Pairs Around the Central Atom

In many molecules and polyatomic ions, one or more of the electron pairs around the central atom is unshared. The electron pair repulsion principle is readily extended to predict the geometries of these species. Here, we should stress one point. In describing the geometries of these species, *we refer only to the positions of the atoms.* Although unshared pairs affect the geometry, their positions are not included in the description.

The geometry indicates the positions of atomic nuclei, not the electrons

Figure 10.7 shows the geometry of the NH_3 molecule. You will recall that in ammonia the central nitrogen atom forms three bonds and has one unshared pair. The dia-

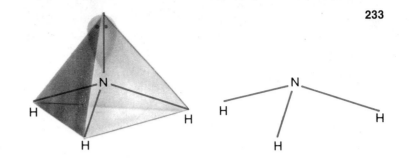

Figure 10.7 Two ways of showing the geometry of the NH_3 molecule. The drawing at the left includes the unshared pair *(shown in color)*. The drawing at the right shows the orientation of the atoms. The nitrogen atom is located directly above the center of the equilateral triangle formed by the three hydrogen atoms. The NH_3 molecule is commonly described as being pyramidal.

gram at the left of Figure 10.7 shows the orientation of the four electron pairs about the N atom. Notice that, as in CH_4, these electron pairs are directed toward the corners of a regular tetrahedron. At the right of Figure 10.7, we show the positions of the atoms in NH_3. The nitrogen atom is located above the center of an equilateral triangle formed by the three hydrogen atoms. We describe this molecule as a **pyramid.** Any molecule XY_3 in which a central atom is surrounded by three bonds and one unshared electron pair will have this geometry. Such molecules are three-dimensional, as the word "pyramid" implies.

In Figure 10.8, we show the geometry of the H_2O molecule. Here, the O atom is surrounded by two bonds and two unshared pairs. The diagram at the left emphasizes that the four electron pairs are, as always, oriented at the tetrahedral angle. At the right, the geometry of the atoms is shown. We describe the water molecule as being **bent.** Any molecule XY_2 in which a central atom is surrounded by two bonds and two unshared electron pairs will have this geometry. The three atoms are located in a triangular pattern, within a plane.

This same principle is readily extended to species in which there are three electron pairs about a central atom, one of which is an unshared pair (Example 10.4).

Example 10.4   Predict the geometry of a molecule with the Lewis structure:

$$:\ddot{Y}—\ddot{X}—\ddot{Y}:$$

Solution   The three pairs of electrons around the central atom should be directed toward the corners of an equilateral triangle. This means that the molecule will be bent, with a bond angle of about 120°. That is,

$$120°$$

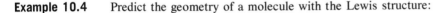

Exercise   Suppose, instead of one unshared pair on X, there were zero or two. What would you predict the bond angle to be in these cases? Answer: 180° with no unshared pairs (compare BeF_2); 109° with two unshared pairs (compare H_2O).

Figure 10.8 Two ways of showing the geometry of the H_2O molecule. At the left, the two unshared pairs are shown. As you can see from the drawing at the right, H_2O is a bent molecule. The bond angle, 105°, is a little smaller than the predicted value, 109°; the unshared pairs spread out over a larger volume than that occupied by the shared pairs.

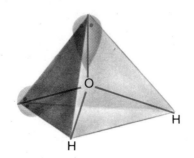

Experimentally, we find that bond angles in molecules where the central atom has at least one unshared pair of electrons tend to be somewhat smaller than those listed in Figure 10.6, p. 232. For example, in NH_3 the measured bond angle is about 107°, a little less than the predicted tetrahedral angle of 109.5°. The effect is slightly greater in H_2O, where the bond angle is 105°. These discrepancies have been attributed to the influence of the unshared pair(s). We might expect the electron cloud formed by the unshared pair in NH_3 to spread out over a greater volume than that of the three pairs connected to hydrogen atoms. This would tend to force the bonding pairs closer to one another and thereby reduce the bond angle. Where there are two pairs of unshared electrons, as in H_2O, this effect is more pronounced.

Species Containing Multiple Bonds

The argument we have just gone through is readily extended to cover species in which double or triple bonds are present.

So far as molecular geometry is concerned, a multiple bond behaves as if it were a single electron pair.

To illustrate this point, consider the CO_2 molecule. Its Lewis structure can be written as

$$:\ddot{O}\!=\!C\!=\!\ddot{O}:$$

The carbon atom is surrounded by two double bonds. So far as molecular geometry is concerned, we can pretend that these are single bonds, ignoring the "extra" electron pairs. The bonds are directed to be as far apart as possible, giving a 180° bond angle. The CO_2 molecule, like BeF_2, is linear.

$$F\!-\!Be\!-\!F \qquad O\!=\!C\!=\!O$$
$$180° \qquad\qquad 180°$$

Using the rules in this section, you can properly predict the geometry of most molecules

We can extend this idea to the acetylene molecule, C_2H_2. You will recall that here the two carbon atoms are joined by a triple bond:

$$H\!-\!C\!\equiv\!C\!-\!H$$

Each carbon atom behaves as if it were surrounded by two electron pairs. Both of the bond angles ($H\!-\!C\!\equiv\!C$ and $C\!\equiv\!C\!-\!H$) are 180°. The molecule is linear; the four atoms are in a straight line. The two "extra" electron pairs in the triple bond do not affect the geometry of the molecule.

In ethylene, C_2H_4, there is a double bond between the two carbon atoms. The molecule has the geometry to be expected if each carbon atom had only three pairs of electrons around it.

The six atoms are located in a plane, with bond angles of 120°.

Example 10.5   Predict the geometry of the SO_2 molecule.

Solution   Consider the Lewis structure of SO_2, shown on p. 230. Around the central sulfur atom there is one double bond, one single bond, and one unshared pair of electrons. Pretending that the double bond is a single electron pair, we predict a bent molecule with a 120° bond angle. The observed angle is 119.5°.

Exercise   Predict the geometry of the NO_3^- ion, whose Lewis structure is shown on p. 230. Answer: The structure is like that of BF_3. The nitrogen atom is at the center of an equilateral triangle with the three oxygen atoms at the corners.

All of the statements we have made concerning molecular geometry are summarized in a slightly different form in Table 10.5. We should emphasize one important point concerning molecular geometries. **You must know or be able to derive the Lewis structure of a species before you can predict its geometry.**

With a little practice, Lewis structures are easily obtained

TABLE 10.5   GEOMETRIES OF SPECIES IN WHICH A CENTRAL ATOM, X, IS SURROUNDED BY FOUR, THREE, OR TWO ELECTRON PAIRS

NUMBER OF ELECTRON PAIRS AROUND X	NUMBER OF ATOMS BONDED TO X	NUMBER OF UNSHARED PAIRS	PREDICTED BOND ANGLE	GEOMETRY OF SPECIES
4	4	0	109°	Tetrahedral
	3	1	109°	Pyramidal
	2	2	109°	Bent
3	3	0	120°	Equilateral triangle
	2	1	120°	Bent
2	2	0	180°	Linear

10.4   POLARITY OF MOLECULES

A polar molecule is one in which there is a separation of charge. We refer to such a species as a **dipole**. There is a positive pole (partial + charge) at one point in the molecule and a negative pole (partial − charge) of equal magnitude at another point. As shown in Figure 10.9, polar molecules line up in the electrical field. In contrast, a nonpolar molecule is one in which there is no separation of charge. Nonpolar molecules are oriented randomly in an electrical field.

If a molecule is diatomic, we can readily decide whether it is polar or nonpolar. Here we need only be concerned with bond polarity. If the two atoms are the same, as in H_2 or F_2, the bond is nonpolar; so is the molecule. Suppose, on the other hand, that the two atoms differ, as in HF. Here the bond is polar; the bonding electrons are shifted toward the more electronegative F atom. There is a negative pole at the F atom and a positive pole at the H atom. The HF molecule is a dipole; it lines up in an electrical field as shown

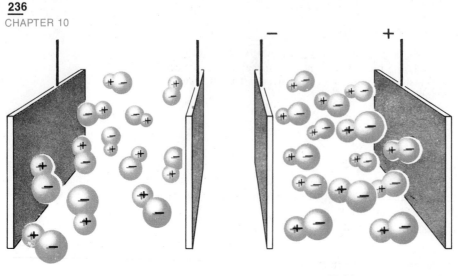

Figure 10.9 In an electric field, polar molecules line up as shown. Since the molecule has a net charge of zero, it does not move in the field. Ions such as Na^+ and Cl^- would move; nonpolar molecules such as H_2 would neither line up nor move.

Field off   Field on

in Figure 10.9. Any diatomic molecule composed of two different atoms behaves this way; it is a dipole.

In molecules containing more than two atoms, we must know the bond angles to decide whether the molecule is polar. Consider, for example, the two molecules shown at the top of Figure 10.10, BeF_2 and H_2O. The linear BeF_2 molecule is nonpolar; the two

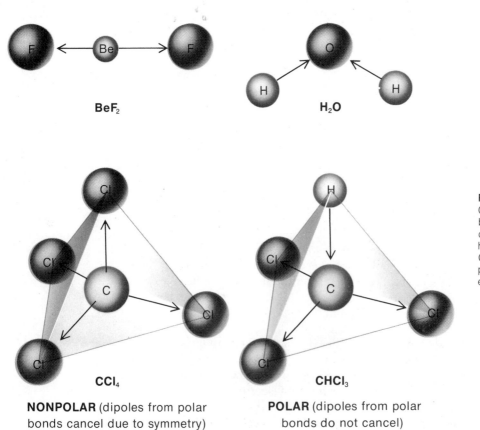

BeF₂

H₂O

CCl₄

NONPOLAR (dipoles from polar bonds cancel due to symmetry)

CHCl₃

POLAR (dipoles from polar bonds do not cancel)

Figure 10.10 The BeF_2 and CCl_4 molecules are nonpolar because the polar bonds cancel each other. This does not happen in the polar H_2O and $CHCl_3$ molecules. (The arrows point to the atom of higher electronegativity.)

polar bonds cancel each other. Putting it another way, the centers of positive and negative charge coincide in this molecule. In contrast, the bent water molecule is polar. The centers of positive and negative charge in H_2O do not coincide. There is a negative pole located at the oxygen atom. Midway between the two hydrogen atoms, there is a compensating positive pole. Thus H_2O, like HF, is polar.

Another molecule which is nonpolar despite the presence of polar bonds is CCl_4. The four C—Cl bonds are polar; the bonding electrons are slightly displaced toward chlorine. However, the four chlorines are symmetrically arranged about the carbon atom. As a result, the polar bonds cancel. If one of the Cl atoms in CCl_4 is replaced by hydrogen, the situation changes. The symmetry is destroyed and we obtain a polar species, $CHCl_3$.

Nonpolar species:
X_2
linear XY_2
planar XY_3
tetrahedral XY_4

Example 10.6   Would you expect the SO_2 molecule to be a dipole?

Solution   In Example 10.5, we decided that the SO_2 molecule is bent, with a bond angle of 120°. We would expect it to be slightly polar. There should be a positive pole (partial + charge) at the less electronegative sulfur atom and a negative pole (partial − charge) midway between the two oxygens.

Exercise   Is the CO_2 molecule, whose geometry is shown on p. 234, a dipole? Answer: No, symmetrical.

10.5   ATOMIC ORBITALS; HYBRIDIZATION

Throughout this chapter, we have used Lewis structures to describe the arrangement of atoms in molecules. These can be very useful. Lewis structures allow us to predict molecular geometries and polarities. However, they do not give us information about the energies of electrons in molecules. Neither do they tell us anything about the orbitals in which the bonding electrons are located.

In the 1930's a theoretical treatment of the covalent bond was developed by Linus Pauling and J. C. Slater, among others. It is referred to as the *atomic orbital* or **valence bond** model. According to this model, a covalent bond consists of a pair of electrons of opposed spin within an atomic orbital. For example, a hydrogen atom forms a covalent bond by accepting an electron from another atom to complete its 1s orbital. Using orbital diagrams, we could write

	1s
isolated H atom	(↑)
H atom in a stable molecule	(↑↓)

The second electron, shown in color, is contributed by another atom. This could be another H atom in H_2, a F atom in HF, a C atom in CH_4, and so on.

This simple model is readily extended to other atoms. The fluorine atom (electron configuration $1s^2 2s^2 2p^5$) has a half-filled p orbital:

	1s	2s	2p
isolated F atom	(↑↓)	(↑↓)	(↑↓) (↑↓) (↑)

By accepting an electron from another atom, F can complete this 2p orbital:

	1s	2s	2p
F atom in HF, F_2, ...	(⇅)	(⇅)	(⇅) (⇅) (↑)

According to this model, it would seem that in order for an atom to form a covalent bond, it must have an unpaired electron. Indeed, the number of bonds formed by an atom should be determined by the number of unpaired electrons. Since hydrogen has an unpaired electron, a H atom should form one covalent bond. The same holds for the F atom. Noble gas atoms, such as He and Ne, which have no unpaired electrons, should not form bonds at all. This is, of course, the case for He and Ne.

When we try to extend this simple idea beyond hydrogen, the halogens and the noble gases, problems arise. Consider, for example, the three atoms Be (at. no. = 4), B (at. no. = 5), and C (at. no. = 6).

	1s	2s	2p
Be atom	(⇅)	(⇅)	( ) ( ) ( )
B atom	(⇅)	(⇅)	(↑) ( ) ( )
C atom	(⇅)	(⇅)	(↑) (↑) ( )

Notice that the beryllium atom has no unpaired electrons, the boron atom has one, and the carbon atom two. Simple valence bond theory would predict that Be, like He, should not form covalent bonds. A boron atom should form one bond, carbon two. Experience tells us that these predictions are wrong. Beryllium forms two bonds in BeF_2; B forms three bonds in BF_3. Carbon, in all its stable compounds, forms four bonds, not two.

To explain these and other discrepancies, simple valence bond theory must be modified. It is necessary to invoke a new kind of atomic orbital, called a hybrid orbital.

Hybrid Orbitals: sp, sp², sp³

The formation of two bonds by beryllium can be explained if we assume that, prior to reaction, one of the 2s electrons is promoted to the 2p level. Thus we could write

	1s	2s	2p
excited Be atom	(⇅)	(↑)	(↑) ( ) ( )

Now that the Be atom has two unpaired electrons, it can form two covalent bonds, as in BeF_2:

	1s	2s	2p
Be atom in BeF_2	(⇅)	(↑⇂)	(↑⇂) ( ) ( )

(The colored arrows indicate electrons supplied by fluorine; the horizontal lines indicate the orbitals involved in bond formation.) The formation of two Be—F bonds releases enough energy to more than compensate for that absorbed in promoting the 2s electron.

There is one basic objection to this model. It implies that two different kinds of bonds are formed. One would be an "s" bond, since the two electrons are filling an s orbital. The other would be a "p" bond, with a pair of electrons in a p orbital. Experimentally, we find that the properties of the two bonds in BeF_2 are identical. This suggests that the two orbitals used for bond formation by beryllium must be equivalent. In atomic orbital terminology, we say that an s and a p orbital have been mixed, or *hybridized*, to give two new orbitals. These orbitals are referred to as **sp hybrid** orbitals.

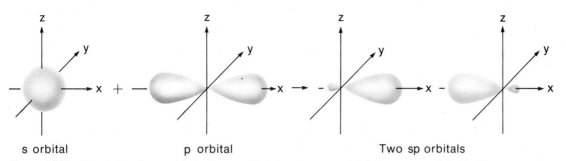

s orbital   p orbital   Two sp orbitals

Figure 10.11 The mixing of an s orbital with a p orbital gives two new orbitals known as sp hybrids. These two orbitals, oriented at a 180° angle, are capable of holding two electron pairs. sp hybrid orbitals are found in BeF_2 and C_2H_2.

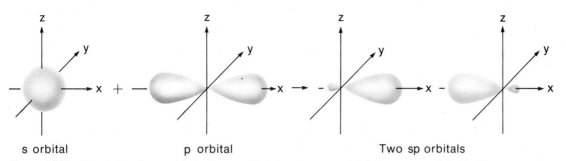

one s orbital + one p orbital → two sp hybrid orbitals

A similar kind of argument can be used with boron. To explain why it forms three bonds in BF_3, we assume that an s orbital is mixed with two p orbitals to give three new orbitals. These are called **sp² hybrid** orbitals.

one s orbital + two p orbitals → three sp² hybrid orbitals

Thus we would have

	1s	2s	2p
B atom in BF_3	(↑↓)	(↑↓)	(↑↓) (↑↓) ( )

The excited B atom has 3 unpaired electrons

Here, as before, the black arrows represent electrons contributed by boron. The blue arrows refer to electrons from F atoms. The three hybrid orbitals used in bonding are enclosed by horizontal lines.

To explain why carbon forms four bonds, we again invoke hybridization. This time we hybridize one s orbital with three p orbitals. Four new orbitals, known as **sp³ hybrid** orbitals, are formed.

one s orbital + three p orbitals → four sp³ hybrid orbitals

In CH_4 or any other molecule *in which carbon forms four single bonds,* the carbon atom would have the orbital diagram

	1s	2s	2p
C atom in CH_4	(↑↓)	(↑↓)	(↑↓) (↑↓) (↑↓)

The excited C atom has 4 unpaired electrons

You will recall that the bond angles in ammonia and water are very nearly tetrahedral. This suggests that the four electron pairs surrounding the central atom in these molecules occupy sp³ hybrid orbitals. If the bonds in NH_3 or H_2O were pure "p" bonds, we would expect them to be oriented at right angles to each other.

We should emphasize that hybrid orbitals have their own unique properties, quite different from those of the orbitals from which they are formed. Table 10.6 gives the orientation in space of hybrid orbitals. The geometries are exactly what we would predict on the basis of electron pair repulsion. In each case, the several hybrid orbitals are directed so as to be as far apart as possible.

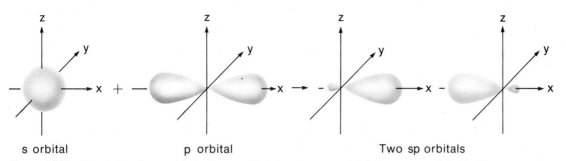

TABLE 10.6   HYBRID ORBITALS AND THEIR GEOMETRIES

NUMBER OF BONDS	ATOMIC ORBITALS	HYBRID ORBITALS	ORIENTATION	EXAMPLE
2	s, p	sp	linear	BeF_2
3	s, two p	sp^2	equilateral triangle	BF_3
4	s, three p	sp^3	tetrahedron	CH_4

The energy required to make hybrid orbitals is furnished by the bonds that are formed when those orbitals are filled

Hybridization in Molecules Containing Multiple Bonds

In Section 10.4, we pointed out that, so far as geometry is concerned, a multiple bond acts as if it were a single bond. In other words, the "extra" electron pairs in a double or triple bond have no effect upon the geometry of the molecule. This behavior is explained in terms of hybridization.

The extra electron pairs in a multiple bond (one pair in a double bond, two in a triple bond) are not hybridized.

From this point of view, the geometry of a molecule is fixed by the electron pairs in hybrid orbitals about a central atom. These orbitals are directed to be as far apart as possible. The hybrid orbitals contain

—all unshared electron pairs.
—electron pairs forming single bonds.
—one and only one of the electron pairs in a double or triple bond.

To illustrate this rule, consider the C_2H_4 and C_2H_2 molecules. You will recall that the bond angles in these molecules are 120° for ethylene and 180° for acetylene. This implies sp^2 hybridization in C_2H_4 and sp hybridization in C_2H_2 (Table 10.6). Using colored lines to represent hybridized electron pairs:

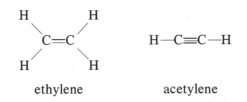

ethylene   acetylene

In both cases, only one of the electron pairs in the multiple bond is hybridized.

Example 10.7   Referring to the Lewis structure of SO_2

describe the hybridization about the sulfur atom.

Solution   In SO_2, we can hybridize three electron pairs around sulfur (the unshared pair, the single bond, and one of the electron pairs in the double bond). Hence we have sp^2 hybridization.

> **Exercise**   What is the hybridization around the carbon atom in CO_2, whose Lewis structure is given on p. 234? Answer: sp.

Sigma and Pi Bonds

We have seen that the extra electron pairs in a multiple bond are not hybridized and have no effect upon molecular geometry. At this point you might well wonder what happened to these electrons. Where are they in the molecule? In Figure 10.12, we attempt to answer that question for the ethylene molecule. The sp^2 hybrid orbitals, three extending from each carbon atom, are shown in color. Above and below the C—C axis are two lobes shown in gray. These two lobes form an orbital in which the extra electron pair of the double bond is located.

Looking at Figure 10.12, we see a basic difference between the two bonds joining the carbon atoms in C_2H_4. The sp^2 hybrid bond, shown in color, occupies an orbital that is symmetrical about the C—C axis. If we move out a given distance in any direction from a point on that axis, the electron density is the same. Bonds formed from orbitals of this type, where the electron density is symmetrical about the bond axis, are called **sigma (σ)** bonds.

The orbital occupied by the extra electron pair of the double bond in C_2H_4 has a quite different shape. The electron density in this orbital is not symmetrical about the C—C axis. Indeed, the electron density is high "north" or "south" of the axis, within the plane of the page. In other areas around the axis, it is very low. Bonds of this type, in which the electron density is not symmetrical about the bond axis, are called **pi (π)** bonds. In general, the extra electron pairs in a multiple bond form pi bonds. Thus:

—in C_2H_4 the C=C bond consists of one sigma bond + one pi bond

—in C_2H_2 the C≡C bond consists of one sigma bond + two pi bonds

The C—H bonds in these molecules, like all single bonds, are sigma bonds.

Single bond:
$1\ \sigma$
Double bond:
$1\ \sigma + 1\ \pi$
Triple bond:
$1\ \sigma + 2\ \pi$

> **Example 10.8**   Give the number of sigma and pi bonds in SO_2, referring to the Lewis structure shown in Example 10.7.
>
> **Solution**   The extra electron pair in the double bond, which is not hybridized, is a pi bond. The other two bonds in the molecule are sigma bonds. We could describe the bonds in SO_2 as being:
>
> > one single bond + one double bond
> >
> > or: two sigma bonds + one pi bond
>
> ---
>
> **Exercise**   Give the number of sigma and pi bonds in CO_2. Answer: Two of each.

Figure 10.12 In C_2H_4, three of the bonds from each carbon atom are sp^2 hybrids. The other bond (one of the electron pairs in the double bond) has a quite different electron cloud associated with it. The cloud is concentrated in two regions, one above the carbon-carbon axis and the other below. This bond is referred to as a π bond. Bonds such as sp^2 hybrids where the electron cloud is symmetrical about the axis joining the bonded atoms are called σ bonds.

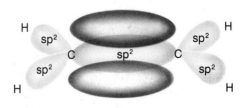

SUMMARY

Covalent bonds (shared electron pairs) join atoms within a molecule or polyatomic ion. Two atoms may share one, two, or three electron pairs to form a single, double, or triple bond, respectively. As the number of bonds between two atoms increases, bond distance decreases and bond energy increases (Tables 10.2 and 10.3).

By sharing valence electrons, atoms ordinarily achieve an octet structure; H atoms in molecules are surrounded by only one shared pair. Lewis structures are drawn to locate shared and unshared pairs. They can be derived for molecules and polyatomic ions. The rules are given on p. 228; see also Examples 10.2 and 10.3. Species with an odd number of valence electrons (NO) or fewer than four pairs around the central atom (BeF_2, BF_3) do not follow the octet rule.

The shape of a polyatomic ion or simple molecule can be predicted from the Lewis structure by considering repulsion among electron pairs about a central atom (Examples 10.4 and 10.5). The geometry depends upon the number of bonds and unshared pairs about the central atom (Table 10.5). So far as geometry is concerned, a multiple bond behaves as if it were a single bond.

A covalent bond between two unlike atoms is polar. The extent of bond polarity depends upon the difference in electronegativity between the atoms. A diatomic molecule containing two unlike atoms (HF) is polar. With more complex molecules, one must consider molecular geometry as well as bond polarity to decide whether the species is a dipole (Example 10.6).

By mixing (hybridizing) orbitals, atoms can make maximum use of their valence electrons to form two bonds (Be), three bonds (B), or four bonds (C). The hybrid orbitals considered in this chapter were sp, sp^2, and sp^3. Their spatial orientation is that predicted by electron pair repulsion (Table 10.6). The extra electron pairs in a multiple bond are not hybridized (Example 10.7); instead they form pi bonds (Example 10.8). Pi and sigma bonds differ in the way the electron cloud is oriented around the axis of the bond (Fig. 10.12).

KEY WORDS AND CONCEPTS

covalent bond	triple bond	linear molecule	nonpolar molecule
nonpolar bond	valence electron	tetrahedral molecule	dipole
polar covalent bond	Lewis structure	pyramidal molecule	hybrid orbital
electronegativity	octet rule	bent molecule	sigma bond
single bond	free radical	valence bond model	pi bond
double bond	electron pair repulsion	polar molecule	

QUESTIONS AND PROBLEMS

Catalog

Bond Polarity: 10.2, 10.3, 10.23, 10.24
Lewis Structures: 10.4–10.9, 10.25–10.30
Molecular Geometry: 10.10–10.13, 10.31–10.34

Molecular Polarity: 10.14–10.16, 10.35–10.37
Hybridization: 10.17–10.19, 10.38–10.40
General: 10.1, 10.20–10.22, 10.41–10.46

10.1   Review and know the meaning of the key words and concepts in this chapter.

10.2   Using trends in the Periodic Table only (not Table 10.1) arrange the following bonds in order of increasing polarity: O—Te, O—S, O—Se.

10.3   Which one of the following bonds would you expect to be least polar: P—S, S—Cl, As—Te, or Sb—Se?

10.4   Write the Lewis structure of

    a. $CHCl_3$ b. NF_3
    c. ClO_3^- d. CS_2

10.5   Draw the Lewis structure of each of the following polyatomic ions:

    a. CN^- b. ClO_4^-
    c. PCl_4^+ d. NO_2^-

10.6   Draw Lewis structures for the following species (the skeleton structure is indicated by the way the molecule is written):

    a. H_2N—NCl_2 b. F_2P—CN
    c. ClS—SCl   d. HO—NO_2

10.7   Two different molecules have the formula $C_2H_4Cl_2$. Draw Lewis structures for both.

10.8   Give the formula of a polyatomic ion which you would expect to have the same Lewis structure as

    a. HCl   b. F_2
    c. N_2 d. CCl_4

10.9   Write reasonable Lewis structures for the following species, none of which obey the octet rule:

    a. NO   b. ClF^+ c. $BeCl_2$ d. OH

10.10   Describe the geometries of the species in Question 10.4.

10.11   Draw Lewis structures and describe the geometry of

    a. Cl_2O b. ONCl   c. PF_3 d. ICl_2^+

10.12   One of the most objectionable compounds in photochemical smog is peroxyacetyl nitrate, PAN. Its skeleton is

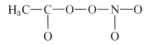

Write the Lewis structure for this molecule and give all the bond angles.

10.23   Arrange the following bonds in order of decreasing polarity by using Periodic Table trends (not Table 10.1): S—Cl, Si—Cl, Al—Cl, P—Cl.

10.24   Which one of the following bonds would you expect to be most polar: O—F, O—S, I—F, or N—O?

10.25   Draw Lewis structures for

    a. $SiCl_4$ b. PH_3
    c. NO^+ d. CO

10.26   Follow the directions for Problem 10.5 for

    a. SeO_4^{2-} b. N_3^-
    c. ClF_2^+ d. SCN^-

10.27   Follow the directions for Problem 10.6 for the following species:

    a. FN—NF   b. O—N—N—O
    c. H—O—N—O   d. F_3C—N—SF_2

10.28   There are two compounds with the molecular formula C_2H_6O. Draw a Lewis structure for each compound.

10.29   Give the formula of a molecule which you would expect to have the same Lewis structure as

    a. ClO^- b. $H_2PO_4^-$
    c. PH_4^+ d. SiO_4^{4-}

10.30   Write reasonable Lewis structures for the following, none of which obey the octet rule:

    a. NO_2 b. SO_2^- c. BCl_3 d. CH_3

10.31   Describe the geometries of the species in Question 10.25.

10.32   Draw Lewis structures and describe the geometry of

    a. SF_2 b. NCl_3 c. BCl_4^- d. $OCCl_2$

10.33   A major eye irritant in smog is acrolein, which has the skeleton structure

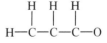

Draw the Lewis structure for this molecule and indicate all the bond angles.

10.13 Give all of the bond angles (109°, 120°, or 180°) in the following ions:

a. $(:\ddot{O}-\ddot{N}=\ddot{O})^-$

b. $(:\ddot{O}-C-\ddot{O}:)^{2-}$
$\quad\quad\quad\underset{\displaystyle :O:}{\overset{\displaystyle \|}{}}$

c. $(:\ddot{O}-\underset{\displaystyle :\ddot{O}:}{\overset{\displaystyle :\ddot{O}:}{Si}}-\ddot{O}:)^{4-}$

10.14 Which of the species in Question 10.4 are dipoles?

10.15 Which of the species in Question 10.13 are dipoles?

10.16 Which of the following molecules would be expected to be polar: CO_2, CS_2, or SCO? Explain.

10.17 What is the hybridization of nitrogen in

a. $(O-N-O)^-$
$\quad\quad\underset{\displaystyle O}{\overset{\displaystyle \|}{}}$
b. $H-\underset{\displaystyle Cl}{\overset{\displaystyle |}{\ddot{N}}}-H$

c. $:N{\equiv}N:$

d. $:N{\equiv}N-O$

10.18 Give the hybridization of each C, N, and O atom in nitrobenzene:

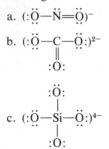

10.19 Give the formula of a molecule or ion in which an atom of

a. B forms three bonds using sp^2 hybrid orbitals.
b. B forms four bonds using sp^3 hybrid orbitals.
c. oxygen forms a pi bond.
d. carbon forms two pi bonds.

10.34 Give all the bond angles in the following molecules:

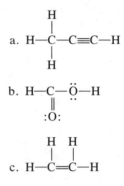

a. $H-\underset{\displaystyle H}{\overset{\displaystyle H}{C}}-C{\equiv}C-H$

b. $H-C-\ddot{O}-H$
$\quad\quad\underset{\displaystyle :O:}{\overset{\displaystyle \|}{}}$

c. $H-\underset{\displaystyle H}{\overset{\displaystyle H}{C}}{=}\underset{\displaystyle H}{\overset{\displaystyle }{C}}-H$

10.35 Which of the species in Question 10.25 are dipoles?

10.36 Which of the species in Question 10.34 are dipoles?

10.37 Given

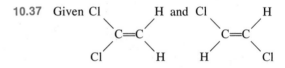

Is either a polar molecule? Explain.

10.38 What is the hybridization of carbon in

a. CH_3Cl
b. $(O-C-O)^{2-}$
$\quad\quad\quad\underset{\displaystyle O}{\overset{\displaystyle \|}{}}$

c. $O{=}C{=}O$
d. $H-C-OH$
$\quad\quad\underset{\displaystyle O}{\overset{\displaystyle \|}{}}$

10.39 Give the hybridization of each C and N atom in pyridoxamine, a form of vitamin B_6. (Some of the unshared pairs are not shown.)

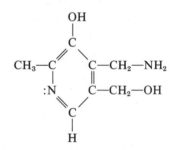

10.40 Give the formula of a molecule or ion in which an atom of

a. N forms four bonds using sp^3 hybrid orbitals.
b. N forms a pi and two sigma bonds.
c. carbon forms four bonds, three in which it uses sp^2 hybrid orbitals.

10.20 List the steps you take to decide whether or not a

a. bond is polar.
b. molecule is polar.

10.21 Consider the following Lewis structures. For each, indicate all bond angles and indicate whether the species is a polar molecule.

a.
$$\begin{array}{c} H \\ | \\ H-C-H \\ | \\ :\ddot{C}l: \end{array}$$

b. $:\ddot{F}-\ddot{O}-\ddot{O}-\ddot{F}:$

c. $H-\ddot{O}-\ddot{B}r:$

10.22 Explain in your own words

a. how sigma and pi bonds differ.
b. why Cl_2 is nonpolar but ICl is polar.
c. why an AX_2 type molecule is not always linear.
d. the differences between sp^2 and sp^3 hybridization.

10.41 List the steps you take to decide the hybridization of a carbon atom in a particular compound.

10.42 Draw Lewis structures for the following molecules and arrange them in order of increasing carbon-to-oxygen bond length:

a. CO_2 b. CO   c. H_3COH

10.43 Criticize the following statements:

a. Covalent bonds are not present in ionic compounds.
b. The bond energy of a single bond is one half that of a double bond between the same atoms.
c. The linear molecule X—Y—Z is nonpolar.
d. The number of covalent bonds formed by an atom equals the number of unpaired electrons in the isolated, gaseous atom.

*10.44 Consider the hydrazine molecule, N_2H_4. Draw its Lewis structure and describe its geometry. Would you expect this molecule to be polar?

*10.45 A certain compound contains 48.6% C, 8.18% H, and 43.2% O by mass. The vapor at 150°C and 1.00 atm has a density of 2.13 g/ℓ. Suggest two possible Lewis structures for the compound.

*10.46 The percentage of ionic character in a polar covalent bond can be calculated by the equation

$$\% \text{ ion. char.} = 100[1 - e^{-(\Delta EN)^2/4}]$$

where e is the base of natural logarithms and ΔEN is the difference in electronegativity between the two atoms joined by the bond. Use this equation to calculate the percentage of ionic character when $\Delta EN = 1.7$.

11

LIQUIDS AND SOLIDS _____

In Chapter 7, we discussed the physical behavior of gases. You will recall that we were able to describe the behavior of all gases by a single equation, the Ideal Gas Law. This equation applies to all substances in the gas state, regardless of their chemical composition. It can be used with elements such as $H_2(g)$, $N_2(g)$, $O_2(g)$, and $Cl_2(g)$, and with compounds such as $CH_4(g)$, $NH_3(g)$, and $H_2O(g)$.

In this chapter, we turn our attention to liquids and solids. We cannot write a single equation, analogous to the Ideal Gas Law, to describe their behavior. The properties of each solid and each liquid are unique. This reflects two important differences at the molecular level between gases and the condensed states of matter.

1. At ordinary temperatures and pressures, the particles in liquids or solids are much closer together than they are in a gas. In a liquid or solid, the particles are in contact with one another. A gas, on the other hand, is mostly empty space. Only a small fraction of the total volume is occupied by the gas molecules themselves.

This difference in behavior is perhaps shown most clearly by a comparison of molar volumes. All gases, at a given temperature and pressure, have the same molar volume. At 25°C and one atmosphere, the molar volume, V_m, of a gas, as calculated from the Ideal Gas Law, is 24.5 ℓ:

$$V_m = \frac{nRT}{P} = \frac{(1.00 \text{ mol})\left(0.0821 \dfrac{\ell \cdot \text{atm}}{\text{mol} \cdot \text{K}}\right)(298 \text{ K})}{1.00 \text{ atm}} = 24.5 \, \ell$$

In contrast, each liquid and solid has its own molar volume, which is much smaller than that of a gas. Thus, for water and benzene at 25°C we have

$$V_m \, H_2O(l) = \frac{\text{molar mass}}{\text{density}} = \frac{18.0 \text{ g}}{1.00 \text{ g/cm}^3} = 18.0 \text{ cm}^3$$

$$V_m \, C_6H_6(l) = \frac{78.1 \text{ g}}{0.879 \text{ g/cm}^3} = 88.9 \text{ cm}^3$$

2. In a liquid or solid, attractive forces between particles are much more important than in gases. With gases, where the molecules are far apart, these forces can usually be neglected. With molecules in close contact, as in liquids and solids, this is no longer true. As we will see later, the strength of these forces depends upon the nature of the liquid or solid. This factor, more than any other, explains why each substance has its own characteristic properties in the liquid or solid state.

In liquids and solids, the particles are essentially touching each other

11.1 PROPERTIES OF LIQUIDS

Most of the properties that we associate with liquids reflect, directly or indirectly, the strength of the forces *between* molecules. In this section, we will look at some of these properties. Later, in Section 11.2, we will seek to explain them in terms of intermolecular forces.

Vapor Pressure

All of us are familiar with the process of vaporization, in which a liquid is converted to a vapor. In an open container, this process continues until all the liquid is gone. Water in a beaker or a flower pot evaporates into the atmosphere, as you may have noted if you forgot to have someone water your plants when you were on vacation. In evaporation, molecular movement is mainly in one direction. Molecules steadily leave the liquid and diffuse away into the air.

In a closed container, such as the flask shown in Figure 11.1, the situation is quite different. Suppose we put a small amount of a volatile liquid into the flask and stopper it. At first, the movement of molecules is primarily in one direction, from liquid to vapor. Here, however, the vapor molecules cannot escape from the container. Some of them collide with the surface and re-enter the liquid. As time passes, and the concentration of molecules in the vapor increases, so does the rate of condensation. Eventually, it becomes equal to the rate of vaporization. When this happens, we say that the liquid and vapor are in a state of *dynamic equilibrium*.

$$\text{liquid} \rightleftharpoons \text{vapor}$$

The double arrow implies that the forward process (vaporization) and the reverse process (condensation) are occurring at the same rate.

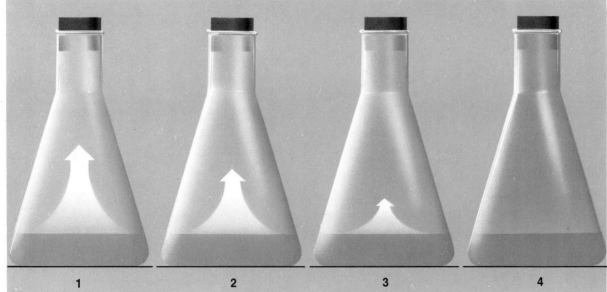

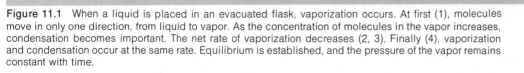

Figure 11.1 When a liquid is placed in an evacuated flask, vaporization occurs. At first (1), molecules move in only one direction, from liquid to vapor. As the concentration of molecules in the vapor increases, condensation becomes important. The net rate of vaporization decreases (2, 3). Finally (4), vaporization and condensation occur at the same rate. Equilibrium is established, and the pressure of the vapor remains constant with time.

Once equilibrium is reached, the concentration of molecules in the gas does not change with time. Putting it another way, *the pressure exerted by the vapor over the liquid remains constant.* The pressure of vapor in equilibrium with a liquid is referred to as the **vapor pressure.** This quantity is a characteristic property of a particular liquid. It varies from one liquid to another, depending upon the strength of the intermolecular forces. At 25°C, the vapor pressure of water is 24 mm Hg. That of benzene at the same temperature is 92 mm Hg.

It is important to realize that, *so long as both liquid and vapor are present, the pressure exerted by the vapor is independent of the volume of the container.* To see what this statement means, consider Figure 11.2. This shows the vaporization of benzene at 25°C in a container whose volume can be varied by raising the piston. We start by placing a small amount of liquid benzene in the container with the piston at the level shown in A. Vaporization occurs until a pressure of 92 mm Hg is established. The equilibrium is then disturbed by raising the piston to the level shown in B. The pressure drops for a moment. However, liquid quickly evaporates to establish the original pressure, 92 mm Hg. This process is repeated each time we raise the piston. Eventually, all the liquid is converted to vapor (Fig. 11.2C). If we continue to raise the piston, no more molecules can vaporize to compensate for the increase in volume. Hence (Fig. 11.2D) the pressure drops off, as predicted by Boyle's Law.

If it can, Bz(l) will evaporate until $P_{Bz(g)}$ equals 92 mm Hg

The process just described can be reversed. If the vapor is compressed to the point where its pressure becomes equal to the vapor pressure of the liquid (point C in Fig. 11.2), liquid begins to condense. Further decrease in volume does not cause an increase in pressure. Instead, condensation continues, keeping the pressure of the vapor constant (points A and B in Fig. 11.2).

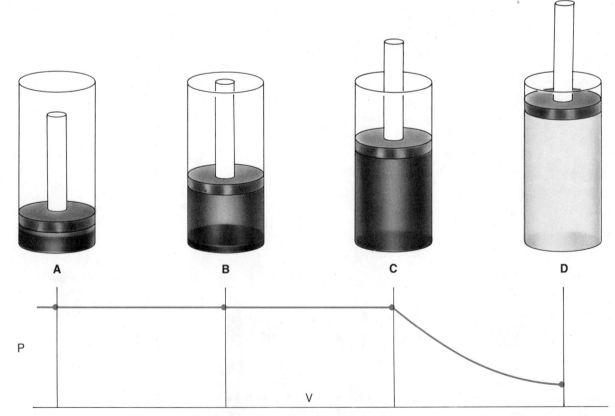

Figure 11.2 The pressure of a vapor in equilibrium with a liquid is independent of the volume of the container as long as there is liquid present (A and B). When all the liquid is vaporized (C), a further increase in volume (D) decreases the pressure in accordance with Boyle's Law. Since System C consists only of gas, its volume can be calculated from the Ideal Gas Law if the temperature and amount of sample are known.

Example 11.1   Suppose that in the experiment just described we start with 0.100 mol of benzene at 25°C (vp = 92 mm Hg). As the piston is slowly raised, the pressure stays constant at 92 mm Hg until all the liquid has vaporized. At what volume does this occur?

Solution   When all the liquid has vaporized, we have 0.100 mol of benzene vapor at 25°C and 92 mm Hg. Its volume can be estimated from the Ideal Gas Law:

$$V = \frac{nRT}{P} = \frac{(0.100 \text{ mol})\left(0.0821 \frac{\ell \cdot \text{atm}}{\text{mol} \cdot \text{K}}\right)(298 \text{ K})}{(92/760 \text{ atm})} = 20 \ \ell$$

We conclude that when the volume reaches 20 ℓ all the liquid is vaporized. If we continue to raise the piston, the pressure will drop below 92 mm Hg. If, on the other hand, we lower the piston, liquid will condense and the pressure will stay constant at 92 mm Hg.

Exercise   Suppose we started with 0.200 mol of benzene. At what volume would we first observe a pressure drop? Answer: 40 ℓ.

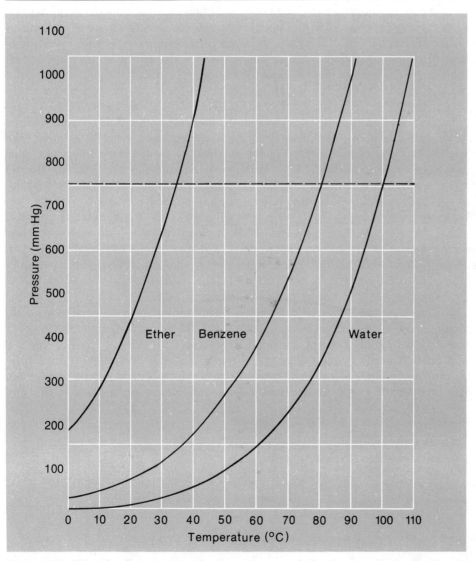

Figure 11.3   Effect of temperature upon the vapor pressures of ether, benzene, and water. Note that vapor pressure increases ever more rapidly as temperature rises. At a given temperature, vp ether > vp benzene > vp water. The normal boiling points of these three liquids (vp = 1 atm) are 35, 80, and 100°C.

At high T, molecules
have more energy and
so escape more easily
from the liquid surface

The vapor pressure of a liquid always increases as temperature rises. Water evaporates more readily on a hot, dry day. Stoppers in bottles of volatile liquids such as ether or gasoline may pop out when the temperature rises. Unpleasant odors such as those of pyridine or butyric acid become more noticeable at higher temperatures.

We can study the effect of temperature upon vapor pressure quite simply. We confine a liquid-vapor mixture in a container of fixed volume and steadily increase the temperature. At higher temperatures, more molecules enter the vapor and their concentration increases rapidly. The pressure of vapor in equilibrium with the liquid goes up. With benzene, the vapor pressure, which is 92 mm Hg at 25°C, becomes 181 mm Hg at 40°C. At 80°C, the vapor pressure reaches one atmosphere, 760 mm Hg. The data for benzene are plotted in Figure 11.3, p. 249. On the same graph, we show the effect of temperature on the vapor pressures of two other liquids, water and diethyl ether.

Boiling Point

When we heat a liquid in an open container bubbles form, usually at the bottom where heat is being applied. The first bubbles we see are air, driven out of solution by the increase in temperature. Eventually, however, at a certain temperature, vapor bubbles form throughout the liquid. These vapor bubbles rise to the surface and break. When this happens, we say that the liquid is boiling.

The temperature at which a liquid boils depends upon the pressure above it. To understand why this is the case, consider Figure 11.4. This shows vapor bubbles rising in a boiling liquid. For a vapor bubble to form, the pressure within it, P_1, must be at least equal to the pressure above it, P_2. But P_1 is simply the vapor pressure of the liquid. We conclude that **a liquid boils at a temperature at which its vapor pressure becomes equal to the pressure above its surface.** If this pressure is one atmosphere, 760 mm Hg, we refer to the temperature as the **normal boiling point.** (When the phrase "boiling point" is used without qualification, normal boiling point is implied.)

At the normal boiling
point the vapor pres-
sure is one atmosphere

Referring back to Figure 11.3, we can readily obtain the normal boiling points of ether, 35°C, benzene, 80°C, and water, 100°C. Notice that there is an inverse relation between normal boiling point and vapor pressure at room temperature. Diethyl ether, with a vapor pressure of 537 mm Hg at 25°C, need be heated only to 35°C to bring its vapor pressure to 760 mm Hg. Water, with a vapor pressure of only 24 mm Hg at 25°C, must be heated to 100°C to boil at one atmosphere pressure.

As you might perhaps expect, the boiling point of a liquid can be reduced by lowering the pressure above it. Water can be made to boil at 25°C by evacuating the space above it. When a pressure of 24 mm Hg, the equilibrium vapor pressure at 25°C, is reached, the water starts to boil. Chemists often take advantage of this effect in purifying a high-boiling compound which might decompose or oxidize at its normal boiling point. They boil it at a reduced temperature under vacuum and condense the vapor.

Example 11.2   Suppose you want to boil benzene at 20°C. Referring to Figure 11.3, at what pressure will this occur?

Solution   The vapor pressure of benzene at 20°C appears to be about 75 mm Hg. When a flask containing benzene at 20°C is evacuated to 75 mm Hg, it boils.

Exercise   At what temperature would benzene boil under a pressure of 1000 mm Hg? Answer: 90°C.

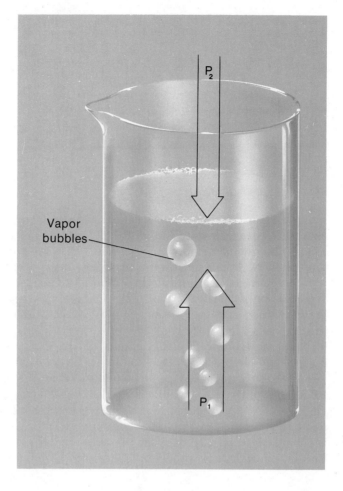

Figure 11.4 A liquid boils when its vapor pressure (P_1) exceeds the pressure above it (P_2).

Vapor bubbles

If you have been fortunate enough to camp in the high Sierras or the Rockies, you may have noticed that it takes longer to boil foods at high altitudes. The reduced pressure lowers the temperature at which water boils in an open container. This slows down the physical and chemical changes that take place when foods like potatoes or eggs are cooked. In principle this problem can be solved by using a pressure cooker. In that device, the pressure developed is high enough to raise the boiling point of water above 100°C. Pressure cookers are indeed used in places like Cheyenne, Wyoming (elevation 1900 m), but not by mountain climbers, who have to carry all their equipment on their backs.

They have enough trouble as it is

Critical Temperature and Pressure

Consider an experiment in which a sample of liquid benzene is placed in an evacuated, heavy-walled glass tube. The tube is then sealed and heated to higher and higher temperatures. The pressure of the vapor rises steadily. It is 1 atm at 80°C, 14 atm at 200°C, and 43 atm at 280°C. Nothing spectacular happens (unless there happens to be a weak spot in the tube) until we reach 289°C at a vapor pressure of 48 atm. Suddenly, as we pass this temperature, the meniscus between the liquid and vapor disappears! The tube now contains only one phase, benzene vapor.

We find that it is impossible to have liquid benzene at temperatures above 289°C, regardless of how much pressure is applied. Even at pressures as high as 1000 atm, benzene vapor refuses to liquefy at 290 or 300°C. This behavior is typical of all sub-

TABLE 11.1   CRITICAL TEMPERATURES

"Permanent Gases"		"Condensable Gases"		"Liquids"	
Helium	−268°C	Carbon dioxide	31°C	Ether	194°C
Hydrogen	−240	Ethane	32	Ethyl alcohol	243
Nitrogen	−147	Propane	97	Benzene	289
Argon	−122	Ammonia	132	Bromine	302
Oxygen	−119	Chlorine	146	Water	374
Methane	−82	Sulfur dioxide	158	Mercury	1460

The vapor pressure of a liquid cannot be greater than the critical pressure. True or false?

stances. There is a temperature, called the **critical temperature,** above which the liquid phase of a pure substance cannot exist. The pressure which must be applied to cause condensation at that temperature is called the *critical pressure*. Quite simply, the critical pressure is the vapor pressure of the liquid at the critical temperature.

Table 11.1 lists the critical temperatures of several common substances. The species in the column at the left all have critical temperatures below 25°C. They are often referred to as "permanent gases." Applying pressure at room temperature will not condense a permanent gas. It must be cooled as well. The permanent gases are stored and sold under high pressures, often 150 atm or greater. When the valve on the cylinder is opened, gas escapes and the pressure drops, in accordance with the Ideal Gas Law.

The gases listed in the center column of Table 11.1 have critical temperatures above 25°C. They are available commercially as liquids in high pressure cylinders. When the valve on a cylinder of propane is opened, the gas that escapes is replaced by vaporization of liquid. The pressure quickly returns to its original value. Only when the liquid is completely vaporized does the pressure drop as gas is withdrawn (recall Fig. 11.2). This indicates that almost all of the propane is gone, and it's time to order a new tank.

11.2   INTERMOLECULAR FORCES

Nearly all the substances that are liquids or gases at 25°C are composed of molecules. The forces between these molecules are weak, at least in comparison to chemical bonds between atoms. Consider, for example, the element hydrogen, which consists of H_2 molecules. To boil liquid hydrogen, it is only necessary to overcome the attractive forces between molecules. The low boiling point of hydrogen, −253°C at 1 atm, reflects the weakness of these forces. The covalent bond holding the hydrogen molecule together is much, much stronger. At a temperature of 2400°C, only about 1% of the H_2 molecules are broken down to atoms.

Forces between molecules are much smaller than the forces within molecules

In this section we will look at three different types of intermolecular forces. We will consider the origin of these forces and the factors that influence their strength. Finally, we will see how the magnitude of intermolecular forces affects such properties as boiling point.

Dipole Forces

Polar molecules in a crystal tend to line up in a regular pattern. The most stable arrangement is one in which the positive pole of one molecule is as close as possible to the negative pole of its neighbor. Such a pattern is shown in Figure 11.5 for iodine chloride, ICl. Note that the positive pole (I atom) of one molecule is next to the negative

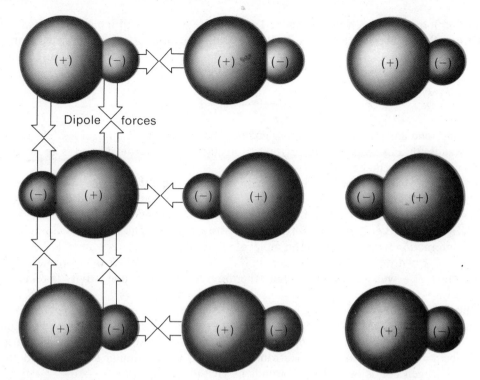

Figure 11.5 Dipole forces in the ICl crystal. The (+) and (−) indicate relatively small charges on the I and Cl atoms in the polar molecules. The existence of these partial charges causes the molecules to line up in the pattern shown. Adjacent molecules are attracted to each other by dipole forces between the (+) pole of one molecule and the (−) pole of the other.

pole (Cl atom) of an adjacent molecule. Under these conditions, there is an electrostatic attraction, called a *dipole force,* between neighboring molecules. These forces persist in the liquid, where the ICl molecules are still close to each other. Only in the gas, where the molecules are far apart, do dipole forces become negligible.

The effect of dipole forces on normal boiling point is shown in Table 11.2. Here we compare the boiling points of polar substances to nonpolar substances of comparable molecular mass. In each pair, the polar species boils at a higher temperature.

Hydrogen Bonds

In certain molecules, there is an unusually strong type of dipole force known as a *hydrogen bond*. This is found in molecules in which a hydrogen atom is bonded to an

TABLE 11.2   BOILING POINTS OF POLAR VS. NONPOLAR SUBSTANCES

	NONPOLAR			POLAR	
FORMULA	MOLECULAR MASS	BOILING POINT (°C)	FORMULA	MOLECULAR MASS	BOILING POINT (°C)
N_2	28	−196	CO	28	−192
SiH_4	32	−112	PH_3	34	−85
GeH_4	77	−90	AsH_3	78	−55
Br_2	160	59	ICl	162	97

atom of fluorine, oxygen, or nitrogen. The simplest examples of molecules containing hydrogen bonds are HF, H_2O, and NH_3. The hydrogen atom of one molecule is strongly attracted to the F, O, or N atom of an adjacent molecule. We might represent the situation in HF as

$$H\text{—}F\text{------}H\text{—}F$$

Here the solid lines represent the polar covalent bonds within HF molecules. The dotted line stands for a hydrogen bond between molecules. The hydrogen bond is about 10 to 15% as strong as the covalent bond.

There are two reasons why hydrogen bonds are stronger than ordinary dipole forces.

1. The difference in electronegativity between hydrogen (2.1) and fluorine (4.0), oxygen (3.5), or nitrogen (3.0) is quite large. It causes the bonding electrons in HF, H_2O, or NH_3 to be markedly displaced from the hydrogen atom. The hydrogen atom, so far as its interaction with a neighboring molecule is concerned, behaves almost like a bare proton. The hydrogen bond is strongest in HF, where the difference in electronegativity is greatest. It is weakest in NH_3, where the difference in electronegativity is relatively small.

2. The small size of hydrogen allows the F, O, or N atom of one molecule to approach the hydrogen atom in another very closely. It is significant that hydrogen bonding occurs only with these three elements. All of them have small atomic radii. The larger chlorine and sulfur atoms, with electronegativities (3.0, 2.8) similar to nitrogen, do not form hydrogen bonds in such compounds as HCl and H_2S.

Example 11.3   Would you expect to find hydrogen bonds in

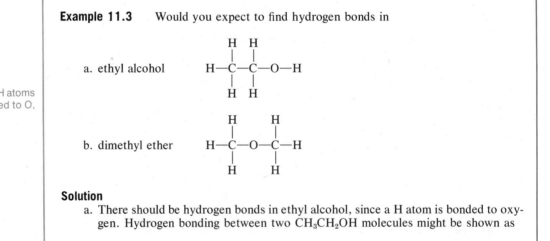

a. ethyl alcohol

b. dimethyl ether

For H bonding, H atoms must be attached to O, F, or N atoms

Solution
   a. There should be hydrogen bonds in ethyl alcohol, since a H atom is bonded to oxygen. Hydrogen bonding between two CH_3CH_2OH molecules might be shown as

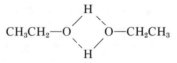

$$CH_3CH_2\text{—}O \quad \quad O\text{—}CH_2CH_3$$

   b. Since all the hydrogen atoms are bonded to carbon in dimethyl ether, there should be no hydrogen bonds. Reflecting this difference in behavior, dimethyl ether boils at −25°C, as compared to 78°C for ethyl alcohol.

Exercise   Would you predict hydrogen bonding in acetone, $CH_3\text{—}\overset{\parallel}{\underset{O}{C}}\text{—}CH_3$; in acetic

acid, $CH_3\text{—}\overset{\parallel}{\underset{O}{C}}\text{—}OH$? Answer: In acetic acid, but not in acetone.

TABLE 11.3 EFFECT OF MOLECULAR MASS UPON BOILING POINT		
	MM	**bp (°C)**
CH_4	16.0	−161
CF_4	88.0	−128
CCl_4	153.8	77
CBr_4	331.6	190

The presence of hydrogen bonding has a pronounced effect upon physical properties. Substances which show hydrogen bonding tend to have unusually high boiling points. Consider, for example, H_2O (MM = 18.0), which has a normal boiling point of 100°C. This is very much higher than that of methane, CH_4 (bp = −161°C), a compound of comparable molecular mass (16.0) but with no hydrogen bonding. In order to boil water, hydrogen bonds must be broken, so a high temperature is required. The heat of vaporization of water is also very high for a substance of molecular mass 18. The amount of heat which must be absorbed to vaporize one gram of water, 2257 J, is greater than that for any other liquid.

Dispersion Forces*

We ordinarily find that, in the absence of hydrogen bonding (and dipole forces), **boiling point increases with molecular mass among substances with similar molecular structures.** Compare, for example, the compounds shown in Table 11.3. All of these consist of nonpolar molecules; boiling point increases steadily from CH_4 to CBr_4.

This implies that there is a type of intermolecular force, common to all substances, whose magnitude increases with molecular mass. This is known as a *dispersion force*. Its origin is more difficult to visualize than that of a dipole force or hydrogen bond. Like them, it is basically electrical in nature. However, dispersion forces are due to what we might call temporary rather than permanent dipoles.

On the average, electrons in a nonpolar molecule such as H_2 are as close to one nucleus as the other. However, at a given instant, the electron cloud may be concen-

---

*These forces are sometimes referred to as van der Waals forces. Properly speaking, van der Waals forces include all the intermolecular forces referred to in this section, as well as one other, the force between a dipole in one molecule and an induced dipole in an adjacent molecule. It is the sum of all these forces which determines the magnitude of the deviation of real gases from ideal behavior (cf. van der Waals equation, Chapter 7).

Figure 11.6   Temporary dipoles in H_2 molecules create an attractive force between adjacent molecules. This dispersion force is present in all molecules, but is the only intermolecular force with nonpolar molecules such as H_2.

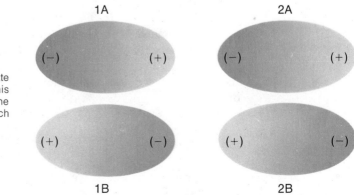

trated at one end of the molecule (position 1A in Fig. 11.6). A fraction of a second later it may be at the opposite end of the molecule (position 1B). The situation is similar to that of a person watching a tennis match from a position directly in line with the net. At one instant his eyes are focused on the player to his left. A moment later, they shift to the player on his right. Over a period of time, he looks to one side as often as the other. The "average" position of focus of his eyes is straight ahead.

The momentary concentration of the electron cloud on one side or the other sets up a temporary dipole in the H_2 molecule. This in turn induces a similar dipole in an adjacent molecule. When the electron cloud in the first molecule is at 1A, the electrons in the second molecule are attracted to 2A. As the first electron cloud shifts to 1B, the electrons of the second molecule are pulled back to 2B. These temporary dipoles, both in the same direction, lead to an attractive force between the molecules. This is the dispersion force.

The strength of dispersion forces depends upon how readily electrons in a molecule can be moved about, or "polarized." As one might expect, the ease of polarization depends upon molecular size. Large molecules where the electrons are far removed from the nuclei are relatively easy to polarize. Small, compact molecules are difficult to polarize. In general, molecular size and molecular mass parallel each other. This explains why dispersion forces increase in strength with molecular mass.

Example 11.4   What types of intermolecular forces are present in H_2? CCl_4? $CHCl_3$? H_2O?

Solution   The H_2 and CCl_4 molecules are nonpolar (Chap. 10). Hence they are attracted to each other only by dispersion forces. The $CHCl_3$ and H_2O molecules are both polar (Chap. 10). In $CHCl_3$ there are dipole forces as well as dispersion forces. In H_2O, hydrogen bonds as well as dispersion forces are present.

Exercise   In which of these substances would you expect dispersion forces to be strongest? weakest? Answer: CCl_4; H_2.

11.3   TYPES OF SOLIDS

The properties of solids depend upon the nature of the particles (atoms, ions, or molecules) present and upon the forces between these particles. It is convenient to distinguish, on the basis of particle structure, between four different kinds of solids:

1. **Ionic** (NaCl, MgO, $CaCO_3$, . . .)
2. **Molecular** (I_2, H_2O, . . .)
3. **Macromolecular** (C, SiO_2, . . .)
4. **Metallic** (Na, Mg, Fe, . . .)

In this section, we will look at the structures and general properties of these four types of solids.

Ionic Solids

In previous chapters, we discussed the particle structures of typical solids of this type. You will recall that an ionic solid consists of oppositely charged ions (e.g., Na^+,

Cl⁻). These are held together by strong electrical forces called ionic bonds. In terms of this structure, we can readily see why:

1. Ionic solids are nonvolatile and high-melting (typically 600–2000°C). Ionic bonds must be broken to melt the solid, separating oppositely charged ions from each other. Only at high temperatures do the ions acquire enough kinetic energy for this to happen.

2. Ionic solids do not conduct electricity, because the charged ions are fixed in position. However, they become good conductors when melted or dissolved in water. There the ions (e.g., Na^+, Cl^-) are free to move through the liquid, carrying an electric current.

3. Many, but by no means all, ionic compounds (e.g., NaCl but not $CaCO_3$) are soluble in the polar solvent water. In contrast, ionic solids are insoluble in nonpolar solvents such as CCl_4 or benzene, C_6H_6. We will have more to say about the water solubilities of ionic compounds in Chapter 16.

Simple ionic solids such as NaCl and MgO melt without any change in chemical composition. However, if a polyatomic ion is present, heating may result in decomposition rather than melting. This behavior is typical of compounds containing OH^- or CO_3^{2-} ions. In either case, one of the decomposition products is a solid metal oxide, containing the O^{2-} ion. The other product, with hydroxides, is $H_2O(g)$. With carbonates, it is $CO_2(g)$. Thus slaked lime, $Ca(OH)_2$, at 500°C breaks down to quicklime, CaO, and water vapor:

$$Ca(OH)_2(s) \rightarrow CaO(s) + H_2O(g) \qquad (11.1)$$

When limestone, $CaCO_3$, is heated to 800°C it decomposes as follows:

$$CaCO_3(s) \rightarrow CaO(s) + CO_2(g) \qquad (11.2)$$

Reactions 11.1 and 11.2 are typical of the hydroxides and carbonates of
—all the Group 2 metals.
—most of the transition metals (e.g., $Zn(OH)_2$ and $ZnCO_3$).
—Li in Group 1 (hydroxides and carbonates of the other Group 1 metals melt without decomposition).

Example 11.5   Write balanced equations for the thermal decomposition of
    a. Li_2CO_3 b. $Fe(OH)_3$

Solution
    a. $Li_2CO_3(s) \rightarrow Li_2O(s) + CO_2(g)$
    b. The products are $H_2O(g)$ and iron(III) oxide, Fe_2O_3. The balanced equation is
       $2 Fe(OH)_3(s) \rightarrow Fe_2O_3(s) + 3 H_2O(g)$

Exercise   Silver hydroxide, AgOH, is very unstable. It decomposes when formed by mixing solutions containing Ag^+ and OH^- ions. Write a balanced equation for the decomposition of AgOH. Answer: $2 AgOH(s) \rightarrow Ag_2O(s) + H_2O(l)$

You will recall from Chapter 3 that many ionic compounds separate from water solution as *hydrates*. In these compounds, H_2O molecules are incorporated into the crystal lattice. Upon heating, water is driven off as a vapor. Hydrated barium chloride, $BaCl_2 \cdot 2 H_2O$, is typical. It loses water at about 100°C to form anhydrous barium chloride:

$$BaCl_2 \cdot 2 H_2O(s) \rightarrow BaCl_2(s) + 2 H_2O(g) \qquad (11.3)$$

Certain ionic hydrates lose water upon exposure to dry air at room temperature. This process is referred to as *efflorescence*. The hydrated crystals crumble to form a fine powder of the anhydrous substance. Often a color change is also involved (see color plate 8, center of book). A familiar example of such a change involves the hydrates of cobalt(II) chloride, $CoCl_2$. These are often used in devices to indicate humidity. The pink compound $CoCl_2 \cdot 6\ H_2O$ is stable in very humid air. When the pressure of water vapor in the air drops, a lower hydrate, blue $CoCl_2 \cdot 4\ H_2O$, is formed.

$$CoCl_2 \cdot 6\ H_2O(s) \rightarrow CoCl_2 \cdot 4\ H_2O(s) + 2\ H_2O(g) \qquad (11.4)$$

$$\text{pink} \qquad\qquad\qquad \text{blue}$$

Molecular Solids

Molecular substances which are solids at room temperature have relatively strong intermolecular forces. They include nonpolar substances of high molecular mass such as iodine, I_2 (MM = 254) and naphthalene, $C_{10}H_8$ (MM = 128). Other molecular solids contain hydrogen bonds. Substances of this type include ice (mp = 0°C), and urea, which has the structure

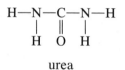

urea

Sugars such as glucose and sucrose are other examples of molecular solids. Here, as in water, there are —OH groups in the molecule which form hydrogen bonds.

As a class, molecular solids tend to be:

1. *Low melting, with melting points usually below 300°C.*

2. *Volatile, with appreciable vapor pressures at room temperature.* This accounts for the odor of naphthalene (mothballs). It also explains why certain molecular substances can *sublime,* passing directly from the solid to the vapor upon heating. Iodine and naphthalene are commonly purified by sublimation (see color plate 9, center of book). Solid carbon dioxide (dry ice) is another molecular solid that sublimes readily. At room temperature, dry ice "disappears" by vaporizing.

3. *Insoluble in water but soluble in nonpolar solvents such as CCl_4 or benzene.* Iodine is only slightly soluble in water (0.0013 mol/ℓ at 25°C); the water solubility of naphthalene is even lower (0.0002 mol/ℓ at 25°C). Both of these molecular solids are much more soluble in benzene (0.48 mol/ℓ for iodine, 4.6 mol/ℓ for naphthalene). A few molecular solids dissolve in water because they readily form hydrogen bonds with H_2O. Examples include urea, glucose, and sucrose.

4. *Nonconductors of electricity, either as solids or pure liquids.* Since molecules are uncharged, they cannot carry an electric current.

Summer in Minnesota is very pleasant

By far the most common molecular solid, at least in Minnesota in winter, is ice. In one sense, ice is very unusual. It is almost unique in having a density less than that of the liquid from which it is formed (d ice = 0.917 g/cm³; d water at 0°C = 1.000 g/cm³). This behavior is an indirect result of hydrogen bonding. When water freezes to ice, an open hexagonal pattern of molecules results (Fig. 11.7). Each oxygen atom in an ice crystal is bonded to four hydrogens. Two of these are attached by ordinary covalent bonds at a distance of 0.099 nm. The other two involve hydrogen bonds 0.177 nm in length. The large proportion of "empty space" in the ice structure explains why ice is less dense than water. Indeed, water starts to decrease in density if cooled below 4°C.

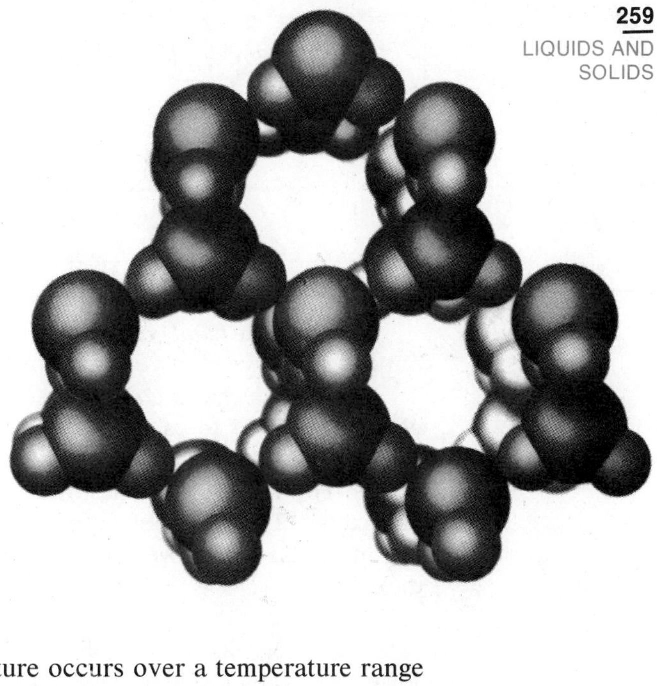

Figure 11.7 Crystal structure of ice. The large spheres represent oxygen atoms, and the small ones hydrogen. Each oxygen atom of an H_2O molecule forms hydrogen bonds with H atoms in two adjacent molecules. This leads to a hexagonal pattern with a large amount of empty space, which accounts for the low density of ice.

This implies that the formation of an open structure occurs over a temperature range rather than taking place solely at the freezing point. It appears that even in water at room temperature some of the molecules are in an open, ice-like pattern. More and more molecules assume this pattern as the temperature drops. Below 4°C, the transition to the open structure outweighs the normal contraction upon cooling. So, water expands as its temperature drops below 4°C.

The fact that water has its greatest density at 4°C has profound effects upon our environment. In winter, at the bottoms of lakes and rivers, there is a layer of water at this temperature, in which fish and other marine life can find food and oxygen. Ice collects at the surface. If ice were more dense than water, it would sink to the bottom. This would make it easier for lakes or rivers to freeze solidly, thus destroying marine life.

This would make ice skating rather difficult

Macromolecular Solids

Covalent bond formation need not, and often does not, lead to small, discrete molecules. Instead, it can lead to structures of the type shown in Figure 11.8. Here, all the atoms are held together by a network of electron-pair bonds. Substances with this type of structure are referred to as *macromolecular*. The entire crystal in effect consists of one huge molecule.

Macromolecular solids have certain properties in common. They are:

1. *High melting, often with melting points above 1000°C.* To melt a macromolecular solid, covalent bonds between atoms in the crystal must be broken. In this respect, macromolecular solids differ markedly from ones containing small, discrete molecules. There, only weak intermolecular forces need be overcome to melt the solid.

Figure 11.8 Two-dimensional representation of macromolecular crystals of an element X or a compound XY. Because of the continuous network of covalent bonds, there are no small, discrete molecules.

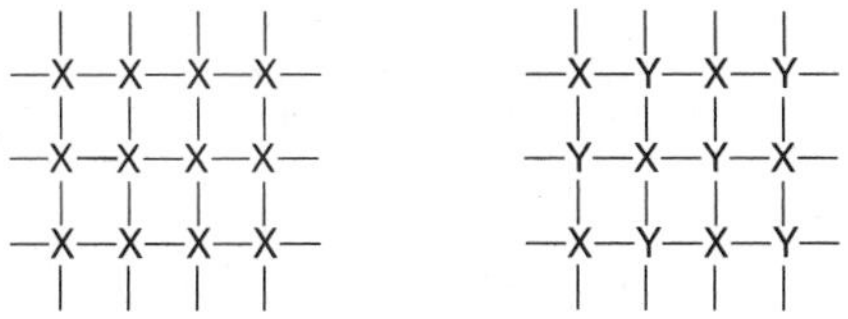

TABLE 11.4   MELTING POINTS OF MACROMOLECULAR SUBSTANCES

	mp (°C)		mp (°C)
C	3570	SiC	>2700
Si	1414	BN	3500
Ge	937	SiO₂	1700

2. *Insoluble in all common solvents*. For solution to occur, covalent bonds throughout the solid would have to be broken.

3. *Poor electrical conductors*. In a typical macromolecular substance, there are no charged particles to carry a current.

Table 11.4 lists some typical macromolecular substances, along with their melting points. You will note that they include elements (C, Si, Ge) as well as compounds. Later, in Chapter 12, we will look at the structures of the macromolecular crystals of the Group 4 nonmetals. Here, we will consider a typical macromolecular compound, silicon dioxide.

These materials are often very hard

SILICON DIOXIDE.   Silicon forms a strong bond with oxygen (B.E. Si—O = 368 kJ/mol). Oxygen in turn can form a bond to another silicon, giving chains of the type

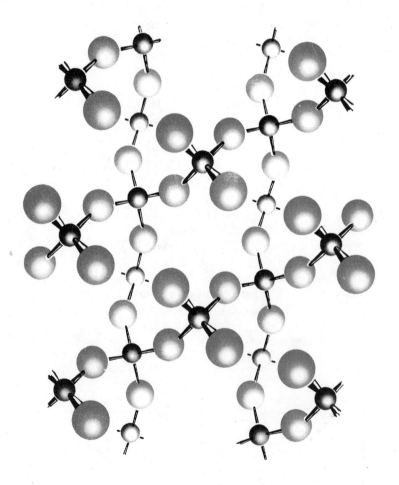

Figure 11.9 Crystal structure of quartz, SiO₂. In SiO₂ every silicon atom *(small)* is linked tetrahedrally to four oxygen atoms *(large)*.

$$-\overset{\displaystyle |}{\underset{\displaystyle |}{Si}}-O-\overset{\displaystyle |}{\underset{\displaystyle |}{Si}}-O-\overset{\displaystyle |}{\underset{\displaystyle |}{Si}}-O-$$

These chains link together to form a macromolecular structure such as the one shown in Figure 11.9. In the crystal, each silicon atom is at the center of a tetrahedron, bonded to four oxygen atoms. Each oxygen atom is bonded to two silicons.

Quartz is the most common form of SiO_2 and is the main component of sea sand. Like graphite and diamond, quartz has a high melting point, about 1700°C. Unlike most solids, it does not melt sharply to a liquid. Instead quartz turns to a viscous mass over a wide temperature range, first softening at about 1400°C. The viscous fluid probably contains long —Si—O—Si—O— chains, with enough bonds broken so the material can flow.

Ordinary glass is made by heating a mixture of sand, limestone ($CaCO_3$), and soda ash (Na_2CO_3) to the melting point. Carbon dioxide is given off from the hot mass, which contains the oxides of silicon, sodium, and calcium in a mole ratio of about 7:1:1. The "soft" glass produced this way softens at about 600°C and has a wider range of high viscosity than quartz. Hard glass, called Pyrex or Kimax, is made from a melt of silicon, boron, aluminum, sodium, and potassium oxides. It is much superior to soft glass in withstanding chemical attack and thermal shock. Hard glass softens at about 800°C and is readily worked in the flame of a gas-oxygen torch. Nearly all the glassware used in chemical equipment nowadays is made from hard glass.

Pyrex was invented in about 1920 at the Corning Glass Co.

Small amounts of various substances are often added to glass to produce different colors. Chromium(III) oxide, Cr_2O_3, gives a green glass, CoO a blue glass, and MnO_2 a violet glass. Milky white glass contains calcium fluoride, CaF_2; SnO_2 produces an opaque glass. Eyeglasses that darken in the sunlight contain small amounts of white, finely dispersed silver chloride, AgCl. Exposure to sunlight converts some of the Ag^+ ions to metallic silver, which is black. The reaction is reversible; in a dark room, AgCl is re-formed and the glass becomes clear again.

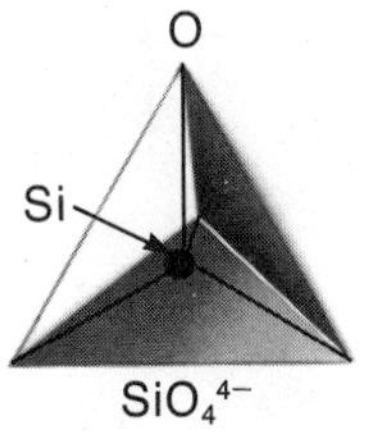

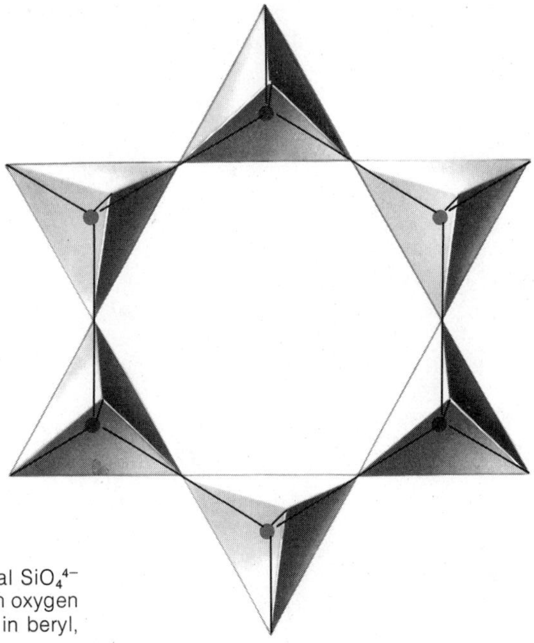

Figure 11.10 Two discrete silicate anions. At the left is the tetrahedral SiO_4^{4-} anion found in zircon. At the right, six such tetrahedra are linked through oxygen atoms at the corners to give a hexagonal pattern. In this anion, found in beryl, the Si:O ratio is 1:3 (see discussion, p. 262).

SILICATES.   The bulk of the rocks and minerals found in the earth's crust are classified as silicates. These minerals have a structure that resembles SiO_2 in that each silicon atom is bonded tetrahedrally to four oxygens. However, silicates are ionic. The negative ion is an oxyanion containing silicon as a central atom. The positive ion is monatomic (e.g., Na^+, Mg^{2+}, Al^{3+}).

The simplest silicates are those containing discrete anions (Fig. 11.10). The SiO_4^{4-} anion is found in the semiprecious stone zircon, $ZrSiO_4$. In the anion shown at the right of Figure 11.10, six tetrahedra are linked through oxygen atoms at two corners. This gives a hexagonal ring structure found in the mineral beryl, $Be_3Al_2Si_6O_{18}$.

In most silicates, the anion forms a network solid similar in structure to a macro-molecule. Here, SiO_4^{4-} tetrahedra are linked together through oxygen atoms to form networks in one, two, or three dimensions. Two such structures are shown in Figure 11.11. In the structure at the left, tetrahedra are linked to form an infinite chain. This is typical of the *fibrous* minerals such as diopside, empirical formula $CaMg(SiO_3)_2$. Silicates of this type are dangerous air pollutants. The long fibers are readily adsorbed on the surface of lung tissue. Over time, they attack and destroy the tissue, causing the disease known as silicosis. Asbestos, a fibrous silicate with a somewhat more complex structure than that shown in Figure 11.11, is one of the worst offenders. Indeed, asbestos, which is used in building construction and in brake linings, is known to cause lung cancer.

The structure shown at the right of Figure 11.11 is typical of *layer* silicates such as talc, $Mg_3(OH)_2Si_4O_{10}$. Here, SiO_4^{4-} tetrahedra are linked together in two dimensions to give an infinite layer. The layers are held together only by weak dispersion forces. Talcum powder, made from talc, has a slippery feeling, caused by layers slipping past one another. Mica is another layer silicate with a somewhat more complex structure. In mica, cations are located between the layers. This makes it more difficult to peel off sheets of mica.

The simplest three-dimensional silicate network is that of SiO_2 (Fig. 11.9). Here, all the oxygen atoms in a SiO_4^{4-} tetrahedron are shared. Many common minerals are derivatives of SiO_2 in which some of the silicon atoms are replaced by aluminum. Feldspar is one example of such a silicate. Here an Al atom is substituted for every fourth Si atom. This gives feldspar the empirical formula $KAlSi_3O_8$. The zeolite minerals used in water softening also have this type of structure.

The cations are held in position by electro-static forces

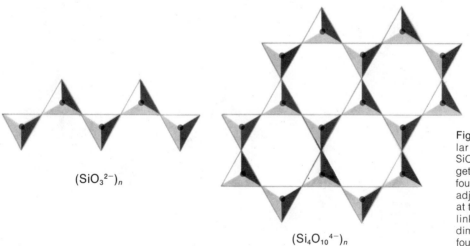

$(SiO_3{}^{2-})_n$

$(Si_4O_{10}{}^{4-})_n$

Figure 11.11 Two macromolecu-lar silicate anions. At the left, SiO_4^{4-} tetrahedra are linked to-gether to form a chain; two of the four oxygen atoms are shared with adjacent tetrahedra. in the anion at the right. SiO_4^{4-} tetrahedra are linked together to give a two-dimensional sheet; three of the four oxygen atoms are shared.

Of the 106 known elements, about 81 can be classified as metals. These elements have several properties in common. Metals as a class are:

1. *Nonvolatile*. With one exception (mercury), all metals are solids at 25°C. They show a wide variation in melting point (mp Cs = 29°C, W = 3380°C).

2. *Insoluble in water and other common solvents*. No metals "dissolve" in water in the true sense. As we saw in Chapter 9, a few very active metals react chemically with water to form hydrogen gas. Liquid mercury dissolves many metals, forming solutions called amalgams. Perhaps the most familiar amalgams are those of silver and gold. These are formed when ores of these metals are extracted with mercury. A Ag-Sn-Hg amalgam is used in filling teeth.

3. *Excellent conductors of electricity*. Of the four types of substances we have discussed, only metals are good conductors in the solid state. Silver is the best metallic conductor but is too expensive for general use. Copper and aluminum are commonly used in electrical wiring. Mercury is one of the poorest metallic conductors, but is used in many electrical devices where a liquid conductor is needed. In all metals, the current is carried by electrons moving through the metal under the influence of an external electrical field.

Metals have certain other properties which distinguish them from other types of solids. They are good conductors of heat. Metals are *ductile* (capable of being drawn out into wire) and *malleable* (capable of being hammered into thin sheets). Finally, metal surfaces are good reflectors of light. Most metals have a silvery white color, indicating that light of all wavelengths is being reflected. Gold and copper absorb some light in the blue region and so appear yellow (gold) or red (copper).

The properties of metals suggest a structure in which electrons are relatively mobile. Only if this is true can we explain why metals are good electrical conductors. A simple approach to metallic bonding, known as the **electron-sea** model, is consistent with this idea. The metallic lattice is pictured as a regular array of positive ions (metal atoms minus their valence electrons). These are anchored in position like bell buoys in a mobile "sea" of electrons. The valence electrons can wander throughout the lattice somewhat like gas molecules in a closed container. They are not held down to a particular

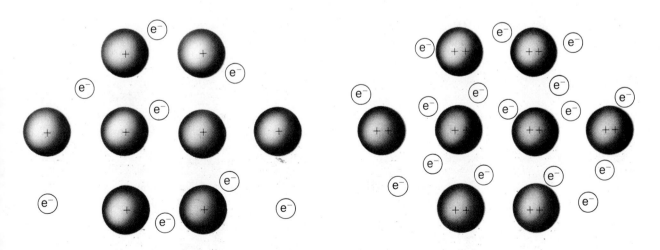

Figure 11.12 Electron-sea model for metallic bonding in sodium (Na^+, e^-) and magnesium (Mg^{2+}, $2e^-$). The hardness of metals increases with the number of electrons available for metallic bonding.

positive ion. In Figure 11.12, p. 263, we show what a tiny portion of a metal crystal might look like according to the electron-sea model.

This simple picture of metallic bonding offers an obvious explanation of the high electrical conductivity of metals. It can also be used to explain many of the other properties of metals. High thermal conductivity is explained by assuming that heat is carried through the metal by collisions between electrons, which occur frequently. Since electrons are not tied down to a particular bond, they can absorb and re-emit light over a wide wavelength range. This explains why metal surfaces are excellent reflectors.

According to the electron-sea model, the strength of the metallic bond is directly related to the charge of the positive ions in the lattice. Looking at Figure 11.12, we would expect the "lattice energy" of magnesium to be greater than that of sodium. With magnesium, we are dealing with +2 ions as compared to +1 ions for sodium. This agrees with the fact that magnesium has a higher melting point than sodium (650°C vs. 98°C).

As we might expect, Mg is also much harder than Na

Classification of Solids

It is possible to summarize many of the ideas expressed in this section by Table 11.5. Here the properties of different kinds of solids are related to their structures.

TABLE 11.5	TYPES OF SUBSTANCES; PROPERTIES AS RELATED TO STRUCTURE			
Structural Units	**Forces Within Units**	**Forces Between Units**	**Properties**	**Examples**
1. Ions	———	Ionic bond	High *mp*. Conductors in molten state or water solution. Usually soluble in water, insoluble in organic solvents.	NaCl MgO
2. Molecules a. Non-polar	Covalent bond	Dispersion	Low *mp, bp*; often gas or liquid at 25°C. Non-conductors. Insoluble in water, soluble in organic solvents.	H_2 CCl_4
b. Polar	Covalent bond	Dispersion, dipole, H bond	Similar to nonpolar but generally higher *mp* and *bp*, more likely to be water-soluble.	HCl NH_3
3. Macromolecules	Covalent bond	———	Hard, very high-melting solids. Nonconductors. Insoluble in common solvents.	C SiO_2
4. Cations, mobile electrons	———	Metallic bond	Variable *mp*. Good conductors in solid. Insoluble in common solvents.	Na Fe

Example 11.6   A certain substance is a liquid at room temperature and is insoluble in water. Suggest which of the four categories in Table 11.5 it probably fits into, and list additional experiments which could be carried out to confirm your prediction.

Solution   The fact that it is a liquid suggests that it is probably molecular in nature. The fact that it is insoluble in water agrees with this classification. To confirm that it is indeed molecular, we might measure the conductivity of the liquid, which should be essentially zero. It should also be soluble in most organic solvents.

Exercise   A certain solid is high-melting (above 1000°C) and is a nonconductor. Which type of solid is it likely to be? How could you make a final decision as to type of solid? Answer: It could be ionic or macromolecular. Test conductivity of melt.

11.4 PHASE CHANGES

Earlier in this chapter, we discussed the equilibrium between a liquid and its vapor. For a pure substance, at least two other types of phase equilibria need be considered: solid-liquid and solid-vapor. Many of the important relations in all these equilibria can be summarized in a **phase diagram.** Figure 11.13 shows a portion of the phase diagram for the substance water. Pressure is plotted along the vertical axis, and temperature along the horizontal axis.

To understand what this diagram implies, consider first the three lines AB, AC, and AD. Each of these lines tells us the pressures and temperatures at which two phases are in equilibrium with each other. To be specific:

1. Line AB is a portion of the vapor pressure-temperature curve of liquid water. At any temperature and pressure along this line, liquid water is in equilibrium with water vapor. From the curve we see that at point A, these two phases are in equilibrium at 0°C and about 5 mm Hg (more exactly 0.01°C and 4.56 mm Hg). At B, corresponding to 25°C, the pressure exerted by the vapor in equilibrium with liquid water is about 24 mm Hg. If line AB were extended, we would read an equilibrium pressure of 760 mm Hg at 100°C, the normal boiling point of water. The line would end at 374°C, the critical temperature of water, where the pressure would be 218 atm.

So, at 25°C the vapor pressure of water is 24 mm Hg

2. Line AC represents the vapor pressure curve of ice. At any point along this line,

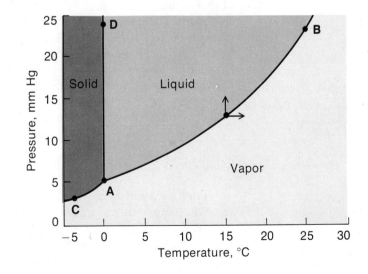

Figure 11.13 Phase diagram for water in the region from −5 to 30°C. The triple point is at A. AC represents the vapor pressure curve of ice, and AB that of liquid water. The line AD, which is inclined very slightly toward the left, gives the temperatures and pressures at which ice and water are in equilibrium.

such as −3°C and 3 mm Hg (point C) or 0°C and 5 mm Hg (point A), ice and vapor are in equilibrium with each other.

3. Line AD gives the temperatures and pressures at which liquid water is in equilibrium with ice.

Point A on the phase diagram is the only one at which all three phases, liquid, solid, and vapor, are in equilibrium with each other. It is called, simply, the **triple point.** For water, the triple point temperature is 0.01°C. At this temperature liquid water and ice have the same vapor pressure, 4.56 mm Hg.

In the three areas of the phase diagram labeled "solid," "liquid," and "vapor," only one phase is present. To show this, consider what happens to an equilibrium mixture of two phases when the pressure or temperature is changed. Suppose we start at the point on AB indicated by a filled circle. Here liquid water and vapor are in equilibrium at 15°C and 13 mm Hg. If we increase the pressure on such a mixture, condensation should occur. The phase diagram confirms this. By increasing the pressure at 15°C (*vertical arrow*), we move up into the liquid region. In another experiment, we might hold the pressure constant but increase the temperature. This change should cause the liquid to vaporize. The phase diagram tells us that this is indeed what happens. An increase in temperature (*horizontal arrow*) shifts us into the vapor region.

The phase diagrams of other substances resemble that of water. However, they differ in such features as the triple point temperature and the orientation of the solid-liquid line AD. Differences of this type cause substances such as iodine, carbon dioxide, and benzene to behave quite differently from water when they sublime or melt.

Sublimation

As pointed out earlier, the process by which a solid changes directly to vapor without passing through the liquid state is called sublimation. From Figure 11.13 we see that *a solid will sublime at any temperature below the triple point when the pressure above it is reduced below the equilibrium vapor pressure.* To illustrate what this means, consider the conditions under which ice sublimes. This happens on a cold, dry winter day when the temperature is below 0°C and the pressure of water vapor in the air is less than the equilibrium value (4.56 mm Hg at 0°C). The rate of sublimation can be increased by evacuating the space above the ice. This is how foods are freeze-dried. The food is frozen, put into a vacuum chamber, and evacuated to a pressure of 1 mm Hg or less. The ice crystals formed upon freezing sublime to give a product whose mass is only a fraction of that of the original food.

Iodine sublimes more readily than ice because its triple point pressure, 90 mm Hg, is much higher. Iodine sublimes upon heating in an open test tube, provided we stay below the triple point temperature, 115°C. If we exceed the triple point, the solid melts. No such problem arises with solid carbon dioxide (dry ice). It has a triple point pressure above one atmosphere (5.2 atm at −57°C). Liquid carbon dioxide cannot be made by heating dry ice in an open container. No matter what we do, solid CO_2 passes directly to the vapor at one atmosphere pressure.

How could you make liquid CO_2?

Fusion

For a pure substance, the **melting point** is identical with the **freezing point.** It represents the temperature at which solid and liquid phases are in equilibrium. Melting points are usually measured in an open container at atmospheric pressure. For most substances, the melting point under these conditions is virtually identical with the triple point. For water, the difference is only 0.01°C.

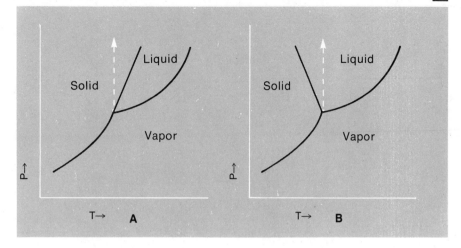

Figure 11.14 Effect of pressure on the melting point of a solid. When the solid is the more dense phase, the behavior will be as in A. An increase in pressure converts liquid to solid; the melting point increases. If the liquid is the more dense phase (B) an increase in pressure converts solid to liquid, and the melting point decreases.

Although the effect of pressure upon melting point is very small, we are often interested in its direction. To decide whether the melting point will be increased or decreased by compression, we apply a simple principle. **An increase in pressure favors the formation of the more dense phase.** We can distinguish between two types of behavior (Fig. 11.14).

1. *The solid is the more dense phase* (Fig. 11.14A). The solid-liquid equilibrium line is inclined to the right, shifting away from the y axis as it rises. At higher pressures, the solid becomes stable at temperatures above the normal melting point. In other words, the melting point is raised by an increase in pressure. This behavior is shown by most substances.

2. *The liquid is the more dense phase* (Fig. 11.14B). The liquid-solid line is inclined to the left, toward the y axis. An increase in pressure favors the formation of liquid. That is, the melting point is decreased by raising the pressure. Water is one of the few substances that behaves this way; ice is more dense than liquid water. The effect is very small. An increase in pressure of 134 atm is required to lower the melting point of ice by 1°C.

ΔH for Phase Changes

Experience tells us that melting a solid and vaporizing a liquid are endothermic processes. An ice cube melting in a glass of water absorbs heat from the water, thereby lowering its temperature. Water in a saucepan on a gas range stops boiling if we turn off the burner, thereby removing the source of heat. The reverse processes, freezing of a liquid or condensation of a vapor, are exothermic. The amounts of heat evolved for a given amount of substance are exactly equal to those absorbed when the phase change occurs in the opposite direction.

The heat flow for a phase change at constant pressure is equal to the enthalpy difference between the two phases. Fig. 11.15, p. 268, shows the enthalpy of one mole of H_2O as a function of temperature. The large vertical jumps at 0°C and 100°C correspond to the molar heat of fusion and the molar heat of vaporization, in that order. We can also use Figure 11.15 to obtain three other quantities. The slopes of lines AB, CD, and EF give us the molar heat capacities $\left(\dfrac{J}{mol \cdot °C}\right)$ of ice, liquid water, and water vapor,

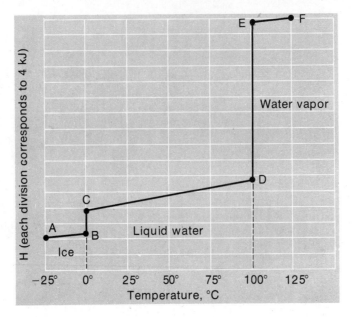

Figure 11.15 Molar enthalpy of water between −25 and 125°C. It takes a lot more heat to vaporize the liquid (D → E) than to melt the solid (B → C). Heat must also be absorbed to raise the temperature of the solid (A → B), the liquid (C → D), or the vapor (E → F).

in that order. (The molar heat capacity is the amount of heat required to raise the temperature of one mole of a substance by 1°C.)

The enthalpy changes just described for H_2O are listed in Table 11.6. On the left, we give the values per mole, and on the right, per gram (18.0 g H_2O = 1 mol). The information given in Table 11.6 is used in many practical calculations (Example 11.7).

TABLE 11.6   ENTHALPY CHANGES INVOLVING H_2O

PER MOLE		PER GRAM	
ΔH_{fus} (at 0°C)	5.99 kJ/mol	ΔH_{fus} (at 0°C)	333 J/g
*ΔH_{vap} (at 100°C)	40.63 kJ/mol	*ΔH_{vap} (at 100°C)	2257 J/g
Molar heat cap. (s)	37 J/(mol·°C)	Specific heat (s)	2.05 J/(g·°C)
Molar heat cap. (l)	75 J/(mol·°C)	Specific heat (l)	4.18 J/(g·°C)
Molar heat cap. (g)	33 J/(mol·°C)	Specific heat (g)	1.84 J/(g·°C)

*The heat of vaporization of water depends somewhat upon temperature. It increases from 40.63 kJ/mol (2257 J/g) at 100°C to 44.88 kJ/mol (2491 J/g) at 0°C.

Example 11.7   An ice cube weighing 30.0 g melts in water originally at 20.0°C.
  a. How much heat is absorbed from the water when the ice cube melts?
  b. How many grams of water can be cooled to 0°C by the melting of the ice cube?

Solution
  a. This calculation can be made from the molar heat of fusion or, perhaps more simply, from the heat of fusion per gram.

$$Q_{ice} = 30.0 \text{ g} \times 333 \frac{J}{g} = 9990 \text{ J}$$

  b. Since heat is transferred from the water to the ice, Q_{water} is equal in magnitude but opposite in sign to Q_{ice}:

$$Q_{water} = -Q_{ice} = -9990 \text{ J}$$

But, Q_{water} = (mass water in g)(specific heat)(temperature change)

Substituting numbers: $-9990 \text{ J} = (\text{no. g water})\left(\dfrac{4.18 \text{ J}}{g\cdot°C}\right)(0.0 - 20.0)°C$

Solving: no. g water $= \dfrac{-9990 \text{ J}}{\left(\dfrac{4.18 \text{ J}}{g\cdot°C}\right)(-20.0°C)} = 120 \text{ g water}$

Notice that the amount of water cooled from 20°C to 0°C is four times the amount of ice melted (120 g vs. 30 g). This means that if you want to prepare a cold drink starting with water at room temperature, and make sure there is some ice left at equilibrium, you had better add somewhat more than one part of ice per four parts of water.

Exercise What is ΔH when one gram of ice at 0°C is converted to steam at 100°C?
Answer: 3008 J.

The amount of heat absorbed when one mole of a solid sublimes is called the **molar heat of sublimation.** Applying Hess's Law, we have, at a given temperature,

$$\Delta H_{subl} = \Delta H_{fus} + \Delta H_{vap} \qquad (11.5)$$
$$(s \rightarrow g) \quad (s \rightarrow l) \quad (l \rightarrow g)$$

For water at 0°C, these quantities are 50.9, 6.0 and 44.9 kJ/mol, in that order.

In Table 11.7, we list the molar enthalpy changes for fusion, vaporization, and sublimation for several different substances. Notice that in each case, $\Delta H_{vap} \gg \Delta H_{fus}$. The heat of sublimation, since it is the sum of the other two enthalpy changes, is the largest of the three.

TABLE 11.7 ΔH (kJ/mol) FOR PHASE CHANGES

SUBSTANCE		mp(°C)	ΔH_{fus}*	ΔH_{vap}*	ΔH_{subl}*
Mercury	Hg	−39	2.3	56.5	58.8
Water	H_2O	0	6.0	44.9	50.9
Benzene	C_6H_6	5	10.7	34.6	45.3
Naphthalene	$C_{10}H_8$	80	19.3	40.6	59.9
Sodium chloride	NaCl	800	29	180	209

*Values given at the melting point.

SUMMARY

In solids and liquids, the distances between particles are shorter and the inter-molecular forces more important than in gases. As intermolecular forces increase from one liquid to another, vapor pressure decreases and boiling point increases. The vapor pressure of a liquid is independent of volume (Example 11.1) but increases exponentially with temperature. The boiling point is reached when the vapor pressure equals the external pressure (Example 11.2). The maximum vapor pressure of a liquid (its critical pressure) occurs at the critical temperature. Above that temperature, the liquid cannot exist.

There are several types of intermolecular forces:

Type of Force	Occurrence (Example 11.3, 11.4)	Effect
Dispersion	All substances; only type in nonpolar substances	Boiling point increases with increase in molecular mass (Table 11.3)
Dipole	Polar substances	Small effect on boiling point (Table 11.2)
Hydrogen bond	Between H atom in one molecule and N, O, or F atom in neighboring molecule	Large increase in boiling point

Solids can be classified into several types (Table 11.5 and Example 11.6). The strong electrostatic attractions between cations and anions give ionic solids high melting points. Ionic solids containing water of hydration or polyatomic anions often decompose on heating (Example 11.5). Molecular solids contain discrete molecules held in the lattice by rather weak intermolecular forces; they tend to be low melting. In a macromolecular solid, there is a continuous network of covalent bonds. Typically, these solids are water-insoluble and high melting (Table 11.4). The properties of metals can be explained, at least qualitatively, in terms of the electron-sea model.

The temperature/pressure conditions for phase changes are summarized in a phase diagram (Fig. 11.13). An increase in pressure favors the more dense phase (Fig. 11.14). Fusion, evaporation, and sublimation are endothermic phase changes; condensation and solidification are exothermic (Table 11.7 and Example 11.7).

KEY WORDS AND CONCEPTS

dynamic equilibrium	dipole force	macromolecular solid	triple point
vapor pressure	hydrogen bond	ductile	sublimation
normal boiling point	dispersion force	malleable	heat of vaporization
critical temperature	polarization	electron-sea model	heat of fusion
critical pressure	efflorescence	phase diagram	heat of sublimation

QUESTIONS AND PROBLEMS

Catalog

Vapor Pressure: 11.2–11.4, 11.23–11.25
Boiling Point, Critical Point: 11.5–11.7, 11.26–11.28
Intermolecular Forces: 11.8–11.10, 11.29–11.31

Types of Solids: 11.11–11.13, 11.32–11.34
Phase Changes: 11.14–11.17, 11.35–11.38
General: 11.1, 11.18–11.22, 11.39–11.46

11.1   Review and know the meanings of the key words and concepts in this chapter.

11.2   A sample of liquid water is placed in a 200-cm³ container at 30°C. Using Figure 11.3 and assuming that some liquid remains in each case, determine the pressure of water vapor at equilibrium

    a. in the 200-cm³ container at 30°C.
    b. if the volume of the container is increased to 400 cm³.
    c. if the temperature is raised to 50°C.

11.23 Repeat the calculations in Problem 11.2 for benzene.

11.3 A mother uses a water vaporizer to raise the humidity in a child's bedroom which is at 20°C and has a volume of $2.7 \times 10^4 \ \ell$. Assume the air is originally dry and no moisture leaves the room when the vaporizer is operating.

    a. How many grams of water must be put into the vaporizer to ensure that the air becomes saturated with water vapor (vp water at 20°C = 18 mm Hg)?

    b. What will be the final pressure of water vapor in the room if 800 g of water are initially in the vaporizer?

    c. What will be the final pressure of the water vapor in the room if she puts 400 g of water into the vaporizer?

11.4 A sample of water vapor in a flask of constant volume exerts a pressure of 400 mm Hg at 100°C. The flask is slowly cooled.

    a. Assuming no condensation, use the Ideal Gas Law to calculate the pressure of the vapor at 90°C; at 80°C.

    b. Compare your answers in (a) to the equilibrium vapor pressures of water: 526 mm Hg at 90°C; 355 mm Hg at 80°C. Will condensation occur at 90°C; 80°C?

    c. On the basis of your answers to (a) and (b), predict the pressure exerted by the water vapor at 90°C; at 80°C.

11.5 Referring to Figure 11.3, give the temperature at which water boils when the external pressure is

    a. 1000 mm Hg   b. 500 mm Hg
    c. 100 mm Hg

11.6 Consider the following data for the variation of atmospheric pressure with elevation:

Elevation (km)	Atm. press. (mm Hg)
0.0	760
2.0	596
4.0	462
6.0	355
8.0	268
10.0	200

Plot these data and use the curve obtained, with Figure 11.3, to estimate the boiling point of water on Mount Everest, elevation 8.85 km.

11.7 The normal boiling point of SO_2 is −10°C. At 32°C its vapor pressure is 5 atm. Which of the following statements concerning sulfur dioxide must be true?

    a. A tank of SO_2 at 32°C which has a pressure of 2 atm must contain liquid SO_2.

    b. A tank of SO_2 at 32°C which has a pressure of 1 atm cannot contain liquid SO_2.

    c. The critical temperature of SO_2 must be greater than 20°C.

11.24 Carbon disulfide, CS_2, has a vapor pressure of 300 mm Hg at 20°C. A sample of 5.00 g of CS_2 is put into a stoppered flask at that temperature.

    a. What is the maximum volume the flask can have if equilibrium is to be established between liquid and vapor?

    b. If the flask has a volume of 3.0 ℓ, what will be the pressure of $CS_2(g)$?

    c. If the flask has a volume of 5.0 ℓ, what will be the pressure of $CS_2(g)$?

11.25 A sample of benzene vapor in a flask of constant volume exerts a pressure of 300 mm Hg at 80°C.

    a. Assuming no condensation, use the Ideal Gas Law to construct a graph of P (mm Hg) vs. t (°C) between 80 and 50°C.

    b. Comparing the graph in (a) to the vapor pressure curve for benzene in Figure 11.3, estimate the temperature at which condensation first occurs upon cooling.

11.26 Answer Problem 11.5, substituting benzene for water.

11.27 At the top of Mount Whitney, in California, water boils at about 85°C. Using Figure 11.3 and the graph made in Problem 11.6, estimate the elevation of Mount Whitney (the recorded elevation is 4.4 km).

11.28 The critical point of CO is −139°C, 35 atm. Liquid CO has a vapor pressure of 5 atm at −171°C. Which of the following statements must be true?

    a. CO is a gas at −171°C and 1 atm.

    b. A tank of CO at 20°C can have a pressure of 5 atm.

    c. CO gas cooled to −145°C and 60 atm pressure will condense.

    d. The normal boiling point of CO lies between −171°C and −139°C.

11.8 Which of the following would you expect to show dipole forces?

    a. CO   b. CO_2 (linear)   c. F_2
    d. H_2S (bent)

11.9 Which of the following would show hydrogen bonding?

    a. CH_4 b. NH_3 c. H—N—N—H
                                | |
                                H H
    d. $(H—F—H)^+$

11.10 Arrange the following in order of increasing boiling point:

    a. Ar   b. He   c. Ne   d. Xe

11.11 Classify each of the following solids as molecular, macromolecular, ionic, or metallic.

    a. Na   b. SiO_2 c. $C_{10}H_8$ d. NaCl
    e. $MgSO_4$

11.12 Write balanced equations for any reactions that occur on heating:

    a. $MgCO_3$ b. $Ba(OH)_2$ c. NaCl
    d. $CaCl_2 \cdot 6 H_2O$

11.13 Classify as metallic, macromolecular, ionic, or molecular a solid which

    a. melts below 100°C to give a nonconducting liquid.
    b. conducts electricity as a solid.
    c. dissolves in water to give a conducting solution.

11.14 Referring to Figure 11.13, state what phase(s) is (are) present at

    a. −3°C, 15 mm Hg.   b. 3°C, 1 mm Hg.
    c. 15°C, 13 mm Hg.

11.15 The triple point of iodine is 115°C, 90 mm Hg. What phase(s) is (are) present at

    a. 115°C, 90 mm Hg?
    b. 115°C, 80 mm Hg?
    c. 120°C, 90 mm Hg?

11.16 Using data from Tables 11.6 and 11.7, calculate ΔH for

    a. freezing 1.00 kg of benzene.
    b. subliming 50.0 g of naphthalene at 80°C.
    c. converting 10.0 g $H_2O(s)$ at −10°C to $H_2O(l)$ at 60°C.

11.17 An ice cube at 0°C weighing 40.0 g is added to water originally at 30.0°C. How many grams of water can be cooled to 0.0°C by the melting of the ice?

11.29 Which of the following would show dispersion forces? dipole forces?

    a. CH_4 b. CH_3Cl c. CH_2Cl_2
    d. $CHCl_3$ e. CCl_4

11.30 Which of the following would show hydrogen bonding?

    a. CH_3F b. $CH_3—OH$
    c. $CH_3—O—CH_3$ d. HO—OH

11.31 Arrange the following in order of decreasing boiling point:

    a. I_2 b. F_2 c. Cl_2 d. Br_2

11.32 Give the formula of a solid compound containing nitrogen which is

    a. molecular   b. ionic
    c. macromolecular

11.33 Give the formula of a solid compound of calcium which, upon heating,

    a. melts without decomposition.
    b. gives off water.
    c. gives off CO_2.

11.34 Classify as metallic, molecular, ionic, or macromolecular a solid which

    a. is a nonconductor but conducts when melted.
    b. dissolves in water to give a nonconducting solution.
    c. melts below 100°C and reacts violently with water.

11.35 Referring to Figure 11.13, indicate the point, line, or area where the following phase(s) is (are) present:

    a. liquid and vapor   b. solid
    c. solid, liquid, vapor

11.36 The density of solid iodine is greater than that of the liquid. The melting point of I_2 at 1 atm will be

    a. the triple point temperature, 115°C.
    b. slightly less than 115°C.
    c. slightly greater than 115°C.

11.37 Use data from Tables 11.6 and 11.7 to obtain ΔH for

    a. melting 100 g NaCl.
    b. subliming 10.0 g benzene at 5°C.
    c. condensing 250 g of $H_2O(g)$ at 100°C to $H_2O(l)$ at 60°C.

11.38 An ice cube at 0°C weighing 40.0 g is added to 200 g of water originally at 30.0°C. The ice cube melts completely. What is the final temperature?

11.18 Explain in your own words the difference between

    a. polar bonds and dipole forces.
    b. dipole forces and hydrogen bonding
    c. sublimation and vaporization.

11.19 How would you explain to a high school chemistry student why

    a. water cools upon standing in slightly porous clay pots?
    b. a sealed tin can filled with steam at 100°C collapses when it is cooled?
    c. a pressure cooker is used to prepare foods at high altitudes?
    d. on a hot summer day, road tar softens but beach sand (SiO_2) does not?

11.20 What are the strongest attractive forces that must be overcome to

    a. melt benzene, C_6H_6?
    b. dissolve iodine in CCl_4?
    c. melt NaF?
    d. decompose Li_2CO_3 to Li_2O?

11.21 For each of the following pairs, choose the higher boiling member. Explain your reason in each case.

    a. H_2 or O_2 b. NaCl or C_4H_{10}
    c. SiO_2 or CO_2 d. H_2S or H_2O

11.22 The density of iron at 25°C is 7.87 g/cm³.

    a. What is the volume of one mole of solid iron at 25°C?
    b. Iron has an atomic radius of 0.126 nm. Calculate the volume of one mole of iron atoms. ($V = 4\pi r^3/3$)
    c. From your answers to (a) and (b), calculate the fraction of "empty space" in an iron crystal at 25°C.

11.39 In your own words, distinguish between

    a. a molecular and a macromolecular substance.
    b. intermolecular forces and covalent bonds.
    c. triple point and critical temperature.

11.40 How would you explain to your parents

    a. how freeze-dried foods are made?
    b. why a cook does not "cry" as much when cutting a cold onion as when cutting a warm one?
    c. atmospheric pressure decreases with altitude?
    d. why mothballs in a clothes closet "disappear" without forming a puddle?

11.41 Follow the directions for Problem 11.20.

    a. sublime dry ice (CO_2)?
    b. melt HF?
    c. vaporize $CaCl_2$?
    d. vaporize liquid chlorine?

11.42 Follow directions for Problem 11.21 for the following substances.

    a. SO_2 or SiO_2 b. Mg or Na
    c. HF or HCl   d. SiH_4 or CH_4

11.43 The density of liquid mercury at 20°C is 13.6 g/cm³; its vapor pressure is 1.2×10^{-3} mm Hg.

    a. What volume (cm³) is occupied by one mole of Hg(l) at 20°C?
    b. What volume (cm³) is occupied by one mole of Hg(g) at 20°C and the equilibrium vapor pressure?
    c. The atomic radius of Hg is 0.155 nm. Calculate the volume (cm³) of one mole of Hg atoms. ($V = 4\pi r^3/3$)
    d. From your answers to (a), (b), and (c), calculate the percentage of the total volume occupied by the atoms in Hg(l) and Hg(g) at 20°C and 1.2×10^{-3} mm Hg.

*11.44 In a fibrous silicate (Fig. 11.11), how many of the oxygen atoms in a SiO_4^{4-} tetrahedron are shared by another tetrahedron? Show that your answer leads to a Si:O atom ratio of 1:3 in such a mineral. Applying the same reasoning to the layer silicate shown at the right of the figure, find the simplest atom ratio of Si:O in that structure.

*11.45 It has been suggested that the pressure exerted on a skate blade is sufficient to melt the ice beneath it and form a thin film of water which makes it easier for the blade to slide over the ice. Assume that a skater weighs 120 lb and the blade has an area of 0.10 in². Calculate the pressure exerted on the blade (1 atm = 15 lb/in²). From information in the text, calculate the decrease in melting point at this pressure. Comment on the plausibility of this explanation, and suggest another mechanism by which the water film might be formed.

*11.46 The relationship between vapor pressure and temperature is given by the Clausius-Clapeyron equation

$$\log_{10} \frac{P_2}{P_1} = \frac{\Delta H_{vap}}{2.30\ R} \left[\frac{T_2 - T_1}{T_2 T_1} \right]$$

where P_2 and P_1 are the vapor pressures at two different temperatures, T_2 and T_1, expressed in K; R is the gas law constant, 8.31 J/(mol·K); and the heat of vaporization is expressed in joules per mole. What temperature must be reached for the vapor pressure of water to be twice that at 37°C? Take ΔH_{vap} of water to be 42.7 kJ/mol.

12

THE NONMETALS AND
THEIR COMPOUNDS

In Chapter 9 we looked at the chemical properties of the metals in Groups 1 and 2 of the Periodic Table. You will recall that these elements form ionic compounds. In this chapter we will consider the elements at the right of the Periodic Table. These are the nonmetals and metalloids in Groups 4, 5, 6, 7, and 8. In these elements themselves, and in the compounds that they form with each other, the bonding is primarily covalent. Electron-pair bonds join nonmetal atoms, forming either discrete molecules or macro-molecular structures.

We start by examining the electronic structures and properties of the nonmetallic elements. Later we will look at the compounds of these elements with hydrogen (Section 12.2), oxygen (Section 12.3), and the halogens (Section 12.4). We will not attempt to cover all the properties of these substances. Instead, we will concentrate upon a few of the more important compounds of the nonmetals. In doing so, we will review and expand upon the principles of covalent bonding and intermolecular forces developed in Chapters 10 and 11.

12.1 THE NONMETALLIC ELEMENTS

Table 12.1 lists some of the physical properties of the nonmetals and indicates their molecular structures. Some of these structures were presented in Chapter 10. You will recall that the noble gases exist as individual atoms. The halogens (Group 7 elements) form diatomic molecules in which there is a single bond; each atom has an octet of electrons.

$$:\ddot{X}-\ddot{X}: \qquad (X = F, Cl, Br, I)$$

In the N_2 molecule, there is a triple bond between the nitrogen atoms. That is,

$$:N\equiv N:$$

The entries in Table 12.1 show many of the trends referred to in previous chapters. Note in particular that:

—the boiling points of the Group 8 and Group 7 elements increase with atomic or molecular mass (He < Ne < Ar < Kr < Xe; $F_2 < Cl_2 < Br_2 < I_2$). This reflects the fact that dispersion forces become stronger as the size of the atom or molecule increases.

—elements with macromolecular structures (C, Si, and Ge in Group 4) tend to

TABLE 12.1   PROPERTIES AND STRUCTURES OF THE NONMETALS

GROUP 8	Helium	Neon	Argon	Krypton	Xenon
Mp, bp (°C)	−272, −269	−249, −246	−189, −186	−157, −152	−112, −107
At. rad. (nm)	0.05	0.070	0.094	0.109	0.130
Structure	He	Ne	Ar	Kr	Xe
GROUP 7		Fluorine	Chlorine	Bromine	Iodine
Mp, bp (°C)		−220, −188	−101, −34	−7, 59	114, 184
At. rad. (nm)		0.064	0.099	0.114	0.133
Radius X⁻ (nm)		0.136	0.181	0.195	0.216
Structure		F_2	Cl_2	Br_2	I_2
GROUP 6		Oxygen	Sulfur	Selenium	Tellurium
Mp, bp (°C)		−218, −183	119, 444	217, 685	450, 990
At. rad. (nm)		0.066	0.104	0.117	0.137
Radius X²⁻ (nm)		0.140	0.184	0.198	0.221
Structure		O_2, O_3	S_8	Se_8*	Te_8*
GROUP 5		Nitrogen	Phosphorus	Arsenic	Antimony
Mp, bp (°C)		−210, −196	44, 280	814, subl.	631, 1380
At. rad. (nm)		0.070	0.110	0.121	0.141
Structure		N_2	P_4*	As_4*	Sb_4*
GROUP 4		Carbon	Silicon	Germanium	
Mp, bp (°C)		3570, subl.	1414, 2355	937, 2830	
At. rad. (nm)		0.077	0.117	0.122	
Structure		Macro.	Macro.	Macro.	

*These elements also form macromolecular solids.

have very high melting points and boiling points. Contrast carbon (mp = 3570°C) to the other nonmetals in the second period. Nitrogen, oxygen, fluorine, and neon, all of which exist as atoms or as small, discrete molecules (N_2, O_2, F_2, Ne), melt below −200°C.

Carbon (graphite) makes an excellent container, even at 2500°C, if no O_2 is around

Oxygen

In its most common and most stable form, oxygen exists as the diatomic molecule O_2. It is possible to write a Lewis structure for O_2 which follows the octet rule:

$$:\overset{..}{O}=\overset{..}{O}:$$

However, experimental evidence suggests that this structure is not correct. Oxygen is paramagnetic (see color plate 10, center of book). This could be explained by assuming a structure

$$:\overset{.}{\underset{.}{O}}-\overset{.}{\underset{.}{O}}:$$

The octet rule fails here

in which there are two unpaired electrons. Again, there are problems with this structure. The distance between the two oxygen atoms in O_2 (0.121 nm) is considerably smaller than that expected for an O—O single bond (0.148 nm). The simple picture of covalent bonding presented in Chapter 10 cannot adequately explain the properties of O_2. In Section 12.5, we will describe a more sophisticated model which better explains the electron distribution in O_2.

Gaseous oxygen can also exist in another form called ozone, molecular formula O_3. Ozone can be prepared in the laboratory by passing $O_2(g)$ through an electric discharge. It is extremely reactive. Among other things, ozone combines vigorously with rubber, accelerating automobile tire deterioration. It has been used to kill bacteria in water and as a bleach for fabrics and paper.

The two species O_2 and O_3 are referred to as **allotropes** of the element oxygen. The term allotropy is used to describe the situation where an element exists in two different forms in the same physical state. We will encounter other examples of allotropy of a somewhat different type later in this chapter.

We can write a simple Lewis structure for O_3 in which there is one single bond and one double bond:

In agreement with this structure, the observed bond angle is about 120°. However, there is a problem. From experiment, the two bonds in O_3 are identical. In particular, the two bond distances are the same, 0.128 nm.

We can rationalize this behavior by assuming that each of the bonds in O_3 is a "hybrid," intermediate between a single and double bond. To express this idea we write two Lewis structures, with a double arrow between them:

The understanding is that the true structure of O_3 is intermediate between these. The two structures are referred to as **resonance forms.** The idea of resonance is introduced when a single Lewis structure does not describe the properties of a substance adequately.

Several of the molecules and polyatomic ions discussed in Chapter 10 have resonance forms. Among these is SO_2 (Example 12.1). Later in this chapter we will see other examples. Here it is well to keep two points in mind:

1. Resonance forms do not imply different kinds of molecules. There is only one O_3 molecule whose structure is believed to lie between those of the two resonance forms.

2. Resonance can be expected when it is possible to write two or more Lewis structures which are about equally reasonable. In these structures, the nuclei are in the same position. That is, the atoms must be bonded in the same order. They differ only in the distribution of valence electrons between the atoms.

In drawing resonance forms, you can only move the electrons

Example 12.1   Write two resonance forms for SO_2.

Solution SO_2 has the same number of atoms (3) and valence electrons (18) as O_3. We would expect the Lewis structures to be very similar. The resonance forms are

In SO_2, as in O_3, the two bond distances are equal.

Exercise   In Chapter 10, we showed the Lewis structure of CO_2 as

$$\ddot{O}{=}C{=}\ddot{O}{:}$$

Can you suggest other resonance forms for CO_2? Answer:

$$:O{\equiv}C-\ddot{\underset{..}{O}}:\quad\text{and}\quad:\ddot{\underset{..}{O}}-C{\equiv}O:$$

Sulfur

Solid sulfur exists in several allotropic forms, of which the most common are called rhombic and monoclinic. Both of these are built up of S_8 molecules, with the structure shown in Figure 12.1. The two allotropes are in equilibrium with each other at 96°C. Above this temperature, the monoclinic form is stable; below 96°C, rhombic sulfur is the stable form. When liquid sulfur freezes at 119°C, monoclinic crystals separate. These crystals can be kept for some time if the melt is cooled quickly to room temperature. At 25°C, the change to the more stable rhombic form is quite slow.

The free-flowing, pale yellow liquid formed when sulfur melts contains S_8 molecules. However, upon heating to 160°C, a striking change occurs. The liquid becomes so viscous that it cannot be poured readily. At the same time its color changes to a deep reddish-brown. These effects reflect a change in molecular structure. The S_8 rings break apart and then link to one another to form long chains such as

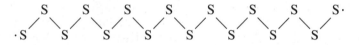

Liquid sulfur between 160 and 250°C contains a high proportion of such chains. They vary in length from eight to several thousand atoms. The chains become tangled, producing a highly viscous liquid. The deep color is due to the absorption of light by the unpaired electrons at the ends of the chain.

If liquid sulfur at 200°C is quickly poured into water, a rubbery mass results (see color plate 11, center of book). This is referred to as "plastic sulfur." It consists of long-chain molecules which did not have time to rearrange to the S_8 molecules stable at room temperature. Within a few hours, the plastic sulfur loses its elasticity as it converts to rhombic crystals.

At 200°C sulfur is a polymer

Rhombic sulfur

a

Monoclinic sulfur

b

c

Figure 12.1 The two allotropes of solid sulfur, rhombic *(a)* and monoclinic *(b)*, differ only in the way in which S_8 molecules are packed in the crystal. These molecules consist of eight-membered, puckered rings *(c)*.

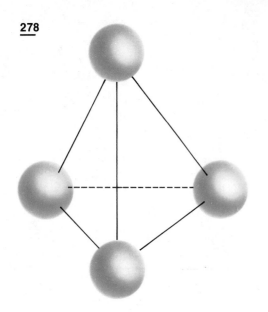

Figure 12.2 In the P_4 molecule, each phosphorus atom is at the corner of a tetrahedron, bonded to three other phosphorus atoms.

Phosphorus

Phosphorus, unlike nitrogen, does not form multiple bonds readily. Solid phosphorus has several allotropes. The two most common are:

1. *White phosphorus,* which consists of P_4 molecules with the structure shown in Figure 12.2. It is a soft, waxy substance with a low melting point (44°C) and boiling point (280°C). Like most molecular substances, white phosphorus is readily soluble in such nonpolar solvents as CCl_4. The chemical reactivity of white phosphorus is so great that it is stored under water to protect it from O_2. A piece of P_4 exposed to air in a dark room glows because of the light given off upon oxidation. White phosphorus is extremely toxic. As little as 0.1 g taken internally can be fatal. Direct contact with the skin produces painful burns.

Like many things, P_4 is both fascinating and dangerous

Before the undesirable properties of white phosphorus were known, it was used in matches. Today, two different kinds of matches are available, neither of which contains white phosphorus. The head of "strike-anywhere" matches contains a mixture of P_4S_3, sulfur, and an oxidizing agent such as PbO_2 or $KClO_3$. When struck against a rough surface, the mixture ignites. Safety matches contain Sb_2S_3 and an oxidizing agent; the special surface against which they are struck contains red phosphorus and powdered glass. Friction sets off a reaction between red phosphorus and the oxidizing agent. This in turn ignites the antimony sulfide.

Example 12.2 With the aid of Figure 12.2, and following the rules given in Chapter 10, write the Lewis structure of P_4.

Solution Figure 12.2 gives the skeleton of P_4. Note that there are six single bonds in the structure, accounting for 12 valence electrons. Since phosphorus is in Group 5, each atom has 5 valence electrons. The total number of valence electrons available is 4(5) = 20. Subtracting 12 leaves 8 valence electrons to distribute. Putting two of these as an unshared pair on each P atom gives a Lewis structure in which each atom has an octet:

This is the only common molecule with this kind of bonding

Exercise   Referring to Figure 12.2, what is the bond angle in P_4? Answer: 60°.

2. *Red phosphorus,* which is the form in which the element is usually found in the laboratory. This allotrope has properties quite different from those of white phosphorus. It is much higher melting (mp = 590°C at 43 atm) and is insoluble in common solvents. The low volatility of red phosphorus makes it much less toxic than the white form. It is also less reactive and must be heated to 250°C to burn in air. These properties are consistent with the structure of red phosphorus, which is known to be macromolecular.

Carbon

Carbon exists in two different crystalline forms, diamond and graphite. The two allotropes differ in the way the atoms are bonded to one another. Both are macromolecular. This explains why both diamond and graphite have very high melting points, above 3500°C. However, as you can see from Figure 12.3, the bonding patterns in the two crystals are quite different.

The graphite crystal is planar, with the carbon atoms arranged in a hexagonal pattern. Each carbon atom is bonded to three others, forming one double bond and two single bonds. The forces between the layers in graphite are of the dispersion type and are quite weak. Thus, the layers can readily slide past one another so that graphite is soft and slippery to the touch.* When you write with a "lead" pencil, which is really made of graphite, thin layers of graphite rub off onto the paper.

In diamond, each carbon atom forms single bonds with four other carbon atoms

---

*Studies show that the lubricating properties of graphite disappear under high vacuum. Apparently these properties require the presence of adsorbed H_2O or O_2 molecules, which act much like a film of oil on a metal surface, allowing graphite layers to slide readily past one another.

Graphite layer   Diamond crystal

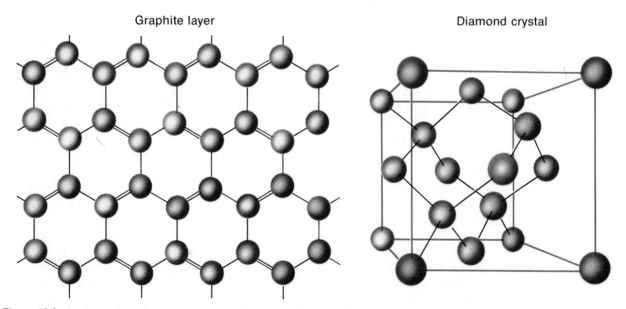

Figure 12.3 In diamond, each carbon atom is at the center of a tetrahedron, on each corner of which is another carbon atom. In graphite, the carbon atoms are linked together in planes of hexagons. Within a layer, a carbon atom is bonded to three other carbon atoms; one third of the bonds are double bonds. The layers are held to one another by weak dispersion forces.

arranged tetrahedrally around it. The bonds are strong enough (bond energy C—C = 347 kJ/mol) to produce a rugged, three-dimensional lattice. Diamond is one of the very hardest of substances, and is used in industry in cutting tools and quality grindstones.

At room temperature and atmospheric pressure, graphite is the stable form of carbon. Diamond, in principle, should slowly transform to graphite under ordinary conditions. Fortunately for the owners of diamond rings this transition occurs at zero rate unless the diamond is heated to about 1500°C, where the conversion occurs rapidly. For understandable reasons, no one has ever become very excited over the commercial possibilities of this process. The more difficult task of converting graphite to diamond has aroused much greater enthusiasm.

Since diamond has a higher density than graphite (3.51 vs. 2.26 g/cm^3), its formation should be favored by high pressures. Theoretically, at 25°C and 8000 atm graphite should turn to diamond. However, under those conditions the reaction has a negligible rate. At higher temperatures it goes faster, but the required pressure goes up too; at 2000°C, a pressure of about 100,000 atm is needed. In 1954 scientists at the General Electric laboratories were able to achieve these high temperatures and pressures and converted graphitic carbon to diamond for the first time. The synthetic diamonds produced by this process were at first quite small, but their size and quality have been improved so that one-carat diamonds are now occasionally formed. At present, most of our industrial diamonds are synthetic.

Research directors like projects like this one

Several amorphous (noncrystalline) forms of carbon are also known. Two of these are charcoal and carbon black. Charcoal can be made by heating wood or other high-carbon materials to a high temperature in the absence of air. This drives off water and other volatile substances. Carbon black is made by burning natural gas in limited air and collecting the soot on cold metal plates. This material is blended with rubber for automobile tires and the soles of running shoes, where it increases wear resistance. A form of charcoal with a very large surface area, called activated carbon, is produced in much the same way as charcoal. This material is very porous and absorbent. Activated carbon is used in industry to decolorize sugar solutions. It has also been applied to remove odors in inner soles, refrigerators and, on a larger scale, in public water supply systems. In all the amorphous forms of carbon, the atoms are arranged in irregular hexagonal patterns. These are similar in many ways to the structure of graphite.

Silicon and Germanium

Both silicon and germanium have only one crystalline form, that of diamond. Extremely pure crystals of the two elements are nonconductors. The valence electrons are tied down in the four covalent bonds that each atom forms with its neighbors. The conductivity increases dramatically when small amounts of arsenic (Group 5) or boron (Group 3) are introduced into the crystal. As little as 0.0001 mol % of these elements present in Si or Ge helps to produce the semiconductor devices used in transistors or solar cells.

To understand the effect of impurities on the conductivity of silicon or germanium, consider Figure 12.4. An atom of arsenic, with five valence electrons, can fit into the crystal lattice of Si or Ge. Its atomic radius, 0.121 nm, is close to that of Si (0.117 nm) or Ge (0.122 nm). To do so, however, an arsenic atom must give up its fifth valence electron. This electron can move through the crystal under the influence of an electrical field. This gives an **n-type semiconductor** (current carried by the flow of negative charge). If an atom of boron or another element with three valence electrons is introduced into the lattice, a different situation arises. An electron deficiency is created at the site occupied by the foreign atom. It is surrounded by seven valence electrons rather than eight. In this sense, there is a "positive hole" in the lattice. In an electrical field, an electron

The first transistors were made at the Bell Telephone Laboratories

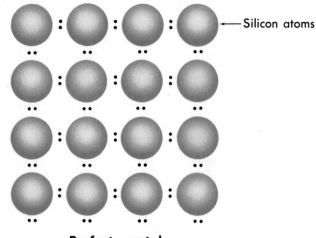

Silicon atoms

Perfect crystal

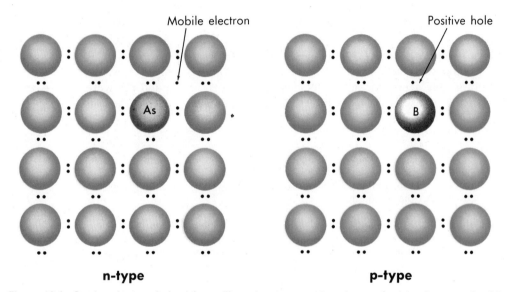

Mobile electron

Positive hole

n-type   **p-type**

Figure 12.4   Semiconductors derived from silicon. In n-type semiconductors the impurity atoms furnish mobile electrons to the crystal. In p-type semiconductors there is a deficiency of electrons, since the impurity atoms have three rather than four valence electrons.

moves from a neighboring atom to fill that hole. By so doing, it creates an electron deficiency around the atom which it leaves. The result is a **p-type semiconductor.** In effect positive holes move through the lattice.

The two heavier members of Group 4, tin and lead, are not included in this chapter. They are metals rather than nonmetals or metalloids. However, tin does have an allotropic form (grey tin) which has the diamond structure. When tin metal is kept at temperatures below 13°C for long periods of time, grey tin forms as a powder. Hence articles made of tin, notably organ pipes, sometimes crumble in very cold weather. The formation of grey tin seems to spread, like an infection, from a single point. This problem was common in the cold cathedrals of northern Europe in the nineteenth century. The organ pipes were said to suffer from "tin disease," for which at that time there was no known cure.

TABLE 12.2 HYDROGEN COMPOUNDS OF THE NONMETALS		
GROUP 5	**GROUP 6**	**GROUP 7**
NH_3, N_2H_4, HN_3	H_2O, H_2O_2	HF
PH_3, P_2H_4	H_2S	HCl
AsH_3	H_2Se	HBr
SbH_3	H_2Te	HI

12.2 HYDROGEN COMPOUNDS

Table 12.2 gives the molecular formulas of the hydrogen compounds of the elements in Groups 5, 6, and 7. The noble gases (Group 8) do not form compounds with hydrogen. In contrast, the Group 4 elements form many hydrogen compounds. Silicon and germanium form series of compounds with the general formulas Si_nH_{2n+2} and Ge_nH_{2n+2}. Here, n ranges from 1 to 10. Because these compounds catch fire spontaneously in air, they must be stored and handled in an oxygen-free environment. Carbon forms literally thousands of stable compounds with hydrogen, called hydrocarbons. These will be discussed separately at the end of this section.

All the compounds listed in Table 12.2 have rather simple structures. The Lewis structures and geometries of molecules such as NH_3, H_2O, and HF were discussed in Chapter 10. Those of H_2O_2 (hydrogen peroxide), N_2H_4 (hydrazine), and HN_3 (hydrazoic acid) are readily derived.

Example 12.3   Consider hydrogen peroxide, H_2O_2.
   a. Write a reasonable Lewis structure for H_2O_2.
   b. Predict the bond angle.

Solution
   a. Hydrogen atoms can form only one bond. The most reasonable skeleton would be a symmetrical one in which a hydrogen is bonded to each oxygen atom:

We start with a total of $2(1) + 2(6) = 14$ valence electrons. With three single bonds in the skeleton (6 valence electrons), that leaves 8 valence electrons to be distributed. Putting two unshared pairs on each oxygen, we arrive at the correct Lewis structure:

There is free rotation around the O—O bond

   b. The four electron pairs around each oxygen should be directed toward the corners of a regular tetrahedron. The predicted bond angle is 109°. Experimentally, it is found to be 105°.

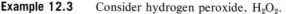

Exercise   The skeleton of hydrazine, N_2H_4, is similar to that of hydrogen peroxide. Write a reasonable Lewis structure for N_2H_4. Answer: H—N̈—N̈—H
                                                                            | |
                                                                            H   H

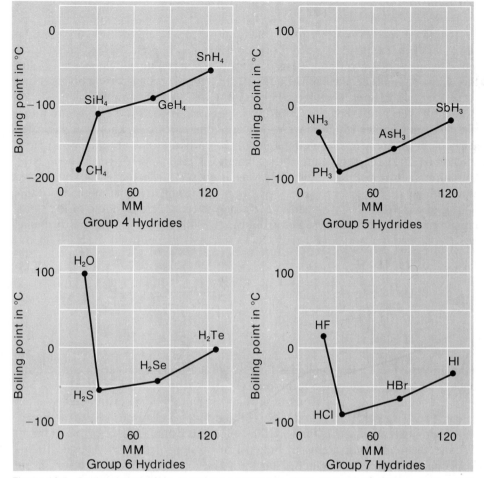

Figure 12.5 Boiling points of the hydrogen compounds of the elements in Groups 4 through 7. Among the Group 4 hydrides, boiling point increases steadily with molecular mass. In Groups 5, 6, and 7 the first member (NH_3, H_2O, HF) has an abnormally high boiling point because of hydrogen bonding.

Boiling Points of Hydrogen Compounds

Figure 12.5 shows the normal boiling points of hydrogen compounds containing one nonmetal atom (Groups 4, 5, 6, 7) per molecule. Notice that:

1. In the absence of hydrogen bonding, boiling point increases steadily with molecular mass. Compare, for example, the boiling points of the hydrogen compounds of the Group 4 elements. They increase steadily from CH_4 to SnH_4. These molecules are all nonpolar; the increase in boiling point reflects an increase in the strength of dispersion forces.

2. The three compounds NH_3, H_2O, and HF have unusually high boiling points, much higher than one would predict by extrapolation of the curves in Figure 12.5. This is because of the strong hydrogen bonds formed by these molecules which make boiling more difficult.

Hydrogen peroxide, H_2O_2, and hydrazine, N_2H_4, also have unusually high boiling points, 151 and 114°C, in that order. Compare these values to that for O_2 (bp = −183°C), a nonpolar molecule of comparable molecular mass. Hydrogen peroxide and hydrazine, like H_2O, NH_3, and HF, show hydrogen bonding.

Hydrocarbons

Organic compounds
contain C, H, and pos-
sibly other nonmetal
atoms

As pointed out earlier, carbon forms many more compounds with hydrogen than does any other element. In all of the hydrocarbons except methane, CH_4, carbon atoms are bonded to one another to form chains or rings. Carbon is almost unique in its ability to do this. In part, this is because of the unusual strength of carbon-carbon bonds (bond energy C—C = 347 kJ/mol, C=C = 598 kJ/mol, C≡C = 820 kJ/mol).

There are several different classes of hydrocarbons. These will be considered in a systematic way in Chapter 25. Here we will discuss only one class, known as saturated hydrocarbons or **alkanes.** In these compounds, the carbon atoms are linked in chains by single bonds. The general formula of the alkanes is C_nH_{2n+2}, where n can be as large as 100. The first member of the series, CH_4, was discussed in Chapter 10. The next two alkanes are ethane, C_2H_6, and propane, C_3H_8. In propane and in higher members of this series, the carbon atoms are arranged in a zigzag pattern, corresponding to the tetrahedral angle of 109°. This is not evident from the two-dimensional structural formulas

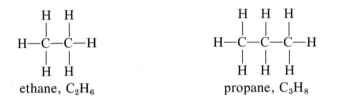

ethane, C_2H_6 propane, C_3H_8

Notice that in these molecules, and indeed in all hydrocarbons, each carbon atom forms four bonds and each hydrogen atom forms one bond (to carbon). In this way, both carbon and hydrogen acquire noble-gas structures. Each carbon atom is surrounded by eight valence electrons (Ne structure), and each H atom by two electrons (He structure).

Two different alkanes are known with the molecular formula C_4H_{10}. In one of these, called butane, the four carbon atoms are linked in a "straight chain." In the other, called 2-methylpropane, there is a "branched chain." The longest continuous chain in the molecule contains three carbon atoms; there is a CH_3 branch from the central carbon atom. The two-dimensional structural formulas are

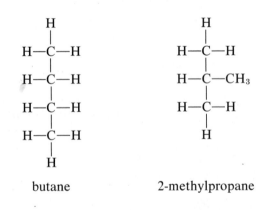

butane   2-methylpropane

Compounds having the same molecular formula but different molecular structures are called **structural isomers.** Butane and 2-methylpropane are referred to as structural isomers of C_4H_{10}. They are two distinct compounds with their own characteristic physical and chemical properties. Isomerism of this and other types is common among hydrocarbons and indeed among organic compounds in general. For small molecules, it is quite easy to identify the various isomers.

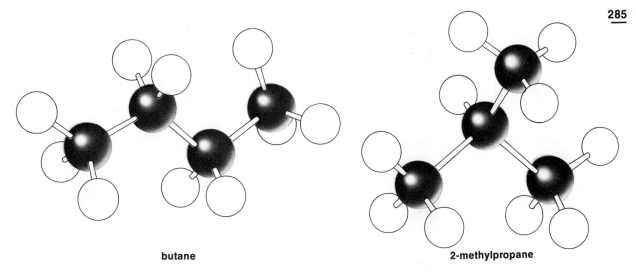

butane   **2-methylpropane**

Figure 12.6 There are two isomers of C_4H_{10}. The isomer shown at the right (2-methylpropane) has a more compact molecular structure and a lower boiling point.

Essentially all organic molecules obey the octet rule

Example 12.4   Draw structural formulas for the isomers of C_5H_{12}.

Solution   First sketch the straight-chain structure

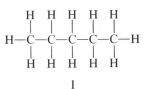

I

Counting hydrogens, we find there are 12, as required. Since each carbon atom has four bonds and each hydrogen one, this is a correct Lewis structure.

Having found one correct structure, we need to determine all the *nonequivalent* alternate structures. Two such structures are shown below (only the carbon skeletons are shown).

II   III

In structure II the longest carbon chain consists of only four atoms, so that II and I are clearly different. Similarly, in structure III, there are only three carbon atoms in the longest carbon chain, so it differs from both I and II. Looking at structures I, II, and III you will notice another important difference which distinguishes one from another. In I, no carbon atom is attached to more than two other carbon atoms. In II there is one carbon atom that is bonded to three other carbons. Finally, in III, the carbon atom in the center is bonded to four other carbon atoms.

At this point, working only with pencil and paper, you might be tempted to draw other structures, such as

However, a few moments' reflection (or access to a molecular model kit) should convince you that these are in fact equivalent to structures written previously. In particular, the first one, like I, has a five-carbon chain in which no carbon atom is attached to more than two other carbons. The second structure, like II, has a four-carbon chain with one carbon atom bonded to three other carbons. Structures I, II, and III represent the three possible isomers of C_5H_{12}; there are no others.

Exercise   How many isomers are there of C_6H_{14}? Answer: Five (one has a six-carbon chain, two have five-carbon chains, and two have four-carbon chains).

The intermolecular forces in alkanes are primarily of the dispersion type; the polarity of these molecules is very small and there is no hydrogen bonding. As we might expect on this basis, the boiling points of alkanes increase with molecular mass.

	MM	bp (°C)
methane, CH_4	16	-161
ethane, C_2H_6	30	-88
propane, C_3H_8	44	-42
butane, C_4H_{10}	58	-1

Ordinarily, branched-chain alkanes have lower boiling points than straight-chain isomers of the same molecular mass. Thus 2-methylpropane boils at $-10°C$ as compared to $-1°C$ for butane. This reflects the fact that dispersion forces are weaker for compact molecules. Presumably, this is because the electrons in such molecules have less room to move about and so are more difficult to polarize.

A spherical molecule has a lower bp than its linear isomer

Example 12.5   Of the three isomers of C_5H_{12} shown in Example 12.4, which would you expect to have the lowest boiling point? the highest?

Solution   Isomer III is the most compact molecule and should have the lowest boiling point. Isomer I is the least compact and should have the highest boiling point. The observed values are: I (36°C), II (28°C), III (10°C).

Exercise   Of the five isomers of C_6H_{14}, which one would you expect to be highest boiling? Answer: The straight-chain isomer, hexane (bp = 69°C).

12.3 OXYGEN COMPOUNDS

Table 12.3 lists the nonmetal oxides that are stable enough to be isolated in more or less pure form. Several other oxygen compounds (e.g., SO) have been reported,

TABLE 12.3   OXYGEN COMPOUNDS OF THE NONMETALS*

GROUP 4	GROUP 5	GROUP 6	GROUP 7	GROUP 8
$CO_2(g)$, $CO(g)$	$N_2O_5(s)$, $N_2O_4(g)$, $N_2O_3(d)$, $N_2O_2(g)$, $N_2O(g)$, $NO_2(g)$, $NO(g)$		$F_2O(g)$, $F_2O_2(l)$	
$SiO_2(s)$	$P_4O_{10}(s)$, $P_4O_6(l)$	$SO_3(l)$, $SO_2(g)$	$Cl_2O_7(l)$, $Cl_2O_6(l)$, $Cl_2O(g)$, $ClO_2(g)$	
$GeO_2(s)$	$As_2O_5(s)$, $As_4O_6(s)$	$SeO_3(s)$, $SeO_2(s)$	$Br_3O_8(d)$, $BrO_2(d)$, $Br_2O(d)$	
	$Sb_2O_5(s)$, $Sb_4O_6(s)$	$TeO_3(s)$, $TeO_2(s)$	$I_4O_9(s)$, $I_2O_5(s)$, $I_2O_4(s)$	$XeO_4(d)$, $XeO_3(d)$

*Physical states at 25°C and 1 atm are indicated by (s), (l), and (g). Compounds that decompose below 25°C are listed as (d).

usually as unstable intermediates in reactions. With some exceptions, the formulas given in the table represent molecules. The oxides of selenium (SeO_2, SeO_3), tellurium (TeO_2, TeO_3), silicon (SiO_2), and germanium (GeO_2) are known to be macromolecular. The structures of As_2O_5 and Sb_2O_5 are unknown; these are simplest formulas.

As you can see from Table 12.3, most nonmetals form more than one compound with oxygen. Typically, the lower oxide (lesser amount of oxygen) is formed at high temperatures in a limited supply of oxygen or air. Cooling with excess O_2 or air leads to further reaction. The final product is a higher oxide, containing more oxygen. As an example, white phosphorus burns in a limited supply of oxygen to form P_4O_6:

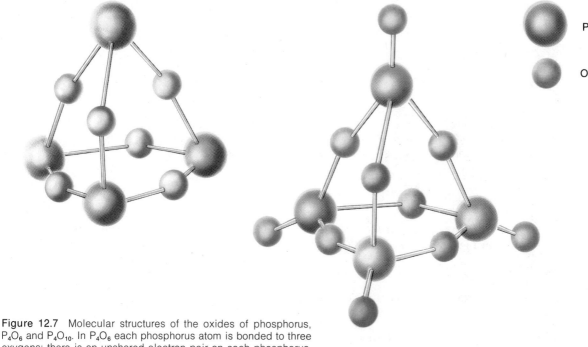

P

O

Figure 12.7 Molecular structures of the oxides of phosphorus, P_4O_6 and P_4O_{10}. In P_4O_6 each phosphorus atom is bonded to three oxygens; there is an unshared electron pair on each phosphorus. In P_4O_{10}, each phosphorus atom is at the center of a tetrahedron, bonded to four oxygen atoms.

$$P_4(s) + 3 \; O_2(g) \rightarrow P_4O_6(l) \qquad (12.1)$$

In excess oxygen, the higher oxide, P_4O_{10}, is produced:

$$P_4O_6(l) + 2 \; O_2(g) \rightarrow P_4O_{10}(s) \qquad (12.2)$$

We will see further examples of this principle when we examine the structures and methods of preparation of the oxides of sulfur, nitrogen, and carbon.

Oxides of Sulfur

When the element sulfur or a metal sulfide reacts with oxygen, sulfur dioxide is formed:

$$S(s) + O_2(g) \rightarrow SO_2(g) \qquad (12.3)$$

$$2 \; ZnS(s) + 3 \; O_2(g) \rightarrow 2 \; ZnO(s) + 2 \; SO_2(g) \qquad (12.4)$$

Equation 12.4 is typical of the reactions that occur when a sulfide ore is roasted as the first step in the extraction of a metal. The Lewis structure of SO_2 was derived in Chapter 10; its resonance forms are shown on p. 276.

Sulfur dioxide in the air is slowly oxidized to the higher oxide, SO_3. This reaction is catalyzed by various solids, as will be discussed in Chapter 15.

$$2 \; SO_2(g) + O_2(g) \rightarrow 2 \; SO_3(g) \qquad (12.5)$$

In the gaseous state, sulfur trioxide consists of discrete molecules with three resonance forms:

3 σ bonds
1 π bond

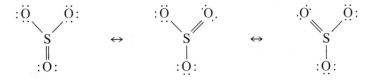

As predicted by electron-pair repulsion, SO_3 is a planar molecule. The sulfur atom is at the center of an equilateral triangle formed by the three oxygen atoms. The bond angles are 120°, giving a symmetrical, nonpolar molecule.

Oxides of Nitrogen

The Lewis structures of the various oxides of nitrogen, seven in all, are shown in Figure 12.8. For each oxide, it is possible to draw resonance forms (Example 12.6).

Example 12.6   Consider the N_2O molecule shown in Figure 12.8.
   a. Draw another resonance form of N_2O.
   b. What is the bond angle in N_2O?
   c. Is the N_2O molecule polar or nonpolar?

Solution

   a. :Ö=N=N̈: or :O≡N−N̈:

b. In any of the resonance forms, the central nitrogen atom, so far as geometry is concerned, would behave as if it were surrounded by two electron pairs. The bond angle is 180°; the molecule is linear, like BeF_2.

c. Polar (unsymmetrical).

Exercise What is the hybridization about the central nitrogen atom in N_2O? Answer: sp.

Of the several oxides of nitrogen, NO, NO_2, and N_2O are the most common. Nitrogen monoxide (nitric oxide) forms when the elements combine at high temperatures, above 1000°C.

$$N_2(g) + O_2(g) \rightarrow 2\ NO(g) \qquad\qquad (12.6)$$

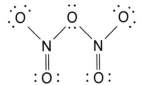

N_2O_5 (dinitrogen pentoxide)

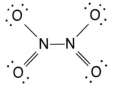

N_2O_4 (dinitrogen tetroxide)

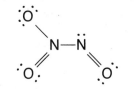

N_2O_3 (dinitrogen trioxide)

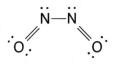

N_2O_2 (dinitrogen dioxide)

NO_2 (nitrogen dioxide)

·N=Ö:

NO (nitrogen monoxide
or nitric oxide)

:Ö—N≡N:

N_2O (dinitrogen monoxide or nitrous oxide)

Figure 12.8 Lewis structures of the oxides of nitrogen. Many other resonance forms are possible.

This reaction occurs in an automobile engine; the exhaust gases contain small but significant amounts of NO. Upon cooling in air, NO is converted to nitrogen dioxide, NO_2:

$$2 \text{ NO(g)} + O_2(g) \rightarrow 2 \text{ NO}_2(g) \tag{12.7}$$

NO$_2$ is a brown poisonous gas

Nitrogen dioxide is a major culprit in smog formation. These two oxides can also be formed in the laboratory from nitric acid, HNO_3, by processes to be discussed in Chapter 22.

Dinitrogen oxide (nitrous oxide) is often referred to as "laughing gas." It was one of the first anesthetics and is still used for that purpose, particularly in dentistry. It was also the first aerosol propellant, used in whipped cream dispensers. Dinitrogen oxide can be made by heating ammonium nitrate *carefully:*

$$NH_4NO_3(s) \rightarrow N_2O(g) + 2 \text{ H}_2O(g) \tag{12.8}$$

At high temperatures, in an enclosed space, NH_4NO_3 can detonate, so this is not a reaction to be carried out casually.

Oxides of Carbon

When carbon or any of its compounds is heated to a high temperature in a limited amount of air, carbon monoxide, CO, is formed:

$$2 \text{ C(s)} + O_2(g) \rightarrow 2 \text{ CO(g)} \tag{12.9}$$

$$2 \text{ C}_8H_{18}(l) + 17 \text{ O}_2(g) \rightarrow 16 \text{ CO(g)} + 18 \text{ H}_2O(l) \tag{12.10}$$

Reaction 12.9 occurs just below the surface of a bed of burning charcoal or coal. Reaction 12.10 takes place when an automobile engine is left running in a closed garage (octane, C_8H_{18}, is typical of the hydrocarbons found in gasoline). Carbon monoxide is extremely poisonous. As little as 0.02 mol % can cause unconsciousness. At the 0.1% level, CO is fatal if inhaled for only a few minutes.

In principle, CO should react with oxygen of the air to form carbon dioxide, CO_2:

CO_2 is much more stable than CO

$$2 \text{ CO(g)} + O_2(g) \rightarrow 2 \text{ CO}_2(g) \tag{12.11}$$

In practice, this reaction occurs very slowly under ordinary conditions. Carbon dioxide is formed directly when hydrocarbons such as octane are burned in an excess of air:

$$2 \text{ C}_8H_{18}(l) + 25 \text{ O}_2(g) \rightarrow 16 \text{ CO}_2(g) + 18 \text{ H}_2O(l) \tag{12.12}$$

(Compare the coefficients of O_2 in Reactions 12.12 and 12.10.)

The Lewis structures of CO and CO_2 are quite simple. Carbon dioxide is a resonance hybrid of three structures:

$$:O\equiv C-\ddot{O}: \quad \leftrightarrow \quad :\ddot{O}=C=\ddot{O}: \quad \leftrightarrow \quad :\ddot{O}-C\equiv O:$$

It is a linear, nonpolar molecule. Carbon monoxide has the structure

$$:C\equiv O:$$

The molecule is of necessity linear. It is also slightly polar, since oxygen has a stronger attraction for electrons than carbon.

Example 12.7   How many pi bonds are there in the CO molecule? in CO_2?

Solution   In CO, the triple bond consists of one sigma bond and two pi bonds. With CO_2, whichever resonance form you look at, there are two pi bonds.

Exercise   Referring to the N_2O_4 molecule in Figure 12.8, how many pi bonds are there? how many sigma bonds? Answer: Two; five.

Reactions of Nonmetal Oxides with Water

Many nonmetal oxides react with water to form oxygen-containing acids such as:

$$SO_2(g) + H_2O \rightarrow H_2SO_3(aq), \text{ sulfurous acid}$$

$$SO_3(g) + H_2O \rightarrow H_2SO_4(aq), \text{ sulfuric acid}$$

$$N_2O_5(g) + H_2O \rightarrow 2\ HNO_3(aq), \text{ nitric acid}$$

$$CO_2(g) + H_2O \rightarrow H_2CO_3(aq), \text{ carbonic acid}$$

$$P_4O_{10}(s) + 6\ H_2O \rightarrow 4\ H_3PO_4(aq), \text{ phosphoric acid}$$

Nonmetal oxides of this type are known as *acid anhydrides* (anhydride: without water). The acids formed ionize to furnish H^+ ions in the solution.

12.4 HALOGEN COMPOUNDS

The halogens (F, Cl, Br, I) form more than a hundred different compounds with nonmetals. We will not attempt to discuss the molecular structures, let alone the properties, of all these compounds. Many of them have Lewis structures which follow the octet rule. Their geometries can be predicted from the principles discussed in Chapter 10. Examples of such compounds include

Many more than 100 if you include all organic compounds

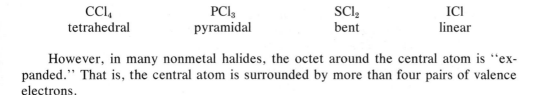

CCl₄	PCl₃	SCl₂	ICl
tetrahedral	pyramidal	bent	linear

However, in many nonmetal halides, the octet around the central atom is "expanded." That is, the central atom is surrounded by more than four pairs of valence electrons.

Expanded Octets

The number of valence electrons surrounding the central atom in a nonmetal halide is readily calculated. Almost without exception, a halogen atom contributes one valence electron to the central atom, forming a single covalent bond. To find the number of valence electrons around an atom such as phosphorus in PF_5 we:

Add one electron for each halogen atom to the number contributed by the central atom itself (its group number).

Thus, for the fluorides of elements in Groups 5, 6, 7, and 8:

PF_5	x = 5 + 5 = 10	SF_6	x = 6 + 6 = 12
SF_4	x = 4 + 6 = 10	ClF_5	x = 5 + 7 = 12
ClF_3	x = 3 + 7 = 10	XeF_4	x = 4 + 8 = 12
XeF_2	x = 2 + 8 = 10		

In these molecules, as with most, all electrons are paired

where x is the number of valence electrons around the central atom in the molecule (shown in bold type). We conclude that in PF_5, SF_4, ClF_3, and XeF_2 there are 10 valence electrons (five pairs) around the central atom. In SF_6, ClF_5, and XeF_4, there are 12 valence electrons (six pairs) around the central atom. In all these molecules, the octet around the central atom is "expanded" to include five or six electron pairs.

None of the nonmetal atoms in the second period of the Periodic Table form expanded octets. This phenomenon first occurs in the third period (Si, P, S, Cl), where the principal quantum number of the valence electrons is 3. Expanded octets are also observed with atoms in successive periods with $n = 4$ or $n = 5$. The "extra" electron pairs enter **d orbitals** which are available in these levels (3d, 4d, 5d). Invoking the idea of hybridization, we would say that:

1. Five electron pairs around a central atom are located in **sp³d hybrid orbitals.**

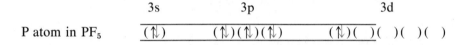

2. Six electron pairs around a central atom are located in **sp³d² hybrid orbitals.**

	3s	3p	3d
S atom in SF_6	(↑↓)	(↑↓)(↑↓)(↑↓)	(↑↓)(↑↓)( )( )( )

Example 12.8   Consider the IF_3 molecule, where three fluorine atoms are bonded to a central iodine atom.
   a. How many pairs of valence electrons are there around the I atom?
   b. What is the hybridization of I in IF_3?

Solution
   a. I is in Group 7. Hence x = 3 + 7 = 10. There are five electron pairs (three shared, two unshared).
   b. sp³d. (Since I is in the fifth period, one 5s orbital, three 5p orbitals, and one 5d orbital are hybridized to accommodate the five electron pairs.)

Exercise   How many pairs of valence electrons are there around the Xe atom in XeF_4? What is its hybridization? Answer: Six; sp³d².

Geometry of Expanded Octets

We can apply the electron pair repulsion rule

In Figure 12.9, we show the geometries of the PF_5 and SF_6 molecules. The electron pairs involved in bonding (five for PF_5, six for SF_6) are directed so as to be as far apart as possible. Note that:

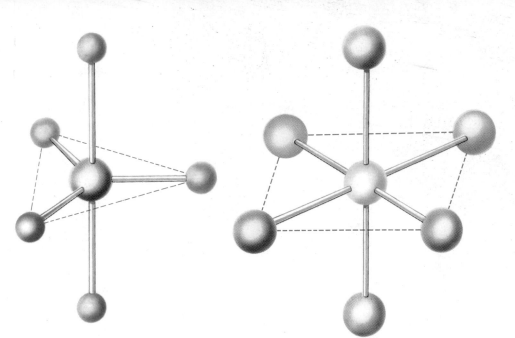

Figure 12.9 In PF$_5$, the phosphorus atom is at the center of a triangular bipyramid (two pyramids sharing a triangular base). In SF$_6$, the sulfur atom is at the center of a regular octahedron. This is a figure with eight sides, all of which are equilateral triangles, and six vertices, each equidistant from the center.

1. The PF$_5$ molecule is a **triangular bipyramid** (two pyramids with triangular faces joined through the base). The P atom is at the center of an equilateral triangle formed by three F atoms. The other two fluorine atoms are located above and below the phosphorus, in a straight line with it. The five F atoms are equidistant from the P atom.

2. The SF$_6$ molecule is an **octahedron.** We can think of an octahedron as being formed by fusing two square pyramids through the base. The S atom is at the center of a square formed by four F atoms. The other two fluorine atoms are located above and below the sulfur, in a straight line with it. The six F atoms are equidistant from the S atom.

In molecules such as XeF$_4$ or IF$_3$, one or more of the electron pairs around the central atom will be unshared. To deduce the geometries of these molecules, we need to know what happens when successive corners are removed from an octahedron or triangular bipyramid. In several cases, more than one geometry is possible. The structures ordinarily observed are shown in Figure 12.10, p. 294.

Example 12.9 What is the geometry of IF$_3$?

Solution In Example 12.8, we found that there are five electron pairs around the I atom. Of these, three are used to form bonds with F atoms; two are unshared. Using Figure 12.10, we predict a T-shaped molecule.

Exercise How many unshared pairs are there around the Xe atom in XeF$_4$? What is the geometry of the molecule? Answer: Two unshared pairs; square planar.

Among the compounds listed in Figure 12.10 are two involving the noble gas xenon, XeF$_2$ and XeF$_4$. Several other compounds of xenon are known, including XeF$_6$, XeO$_3$, XeO$_4$, and XeOF$_4$. In addition, a fluoride of krypton, KrF$_2$, has been prepared. Of these, perhaps the most

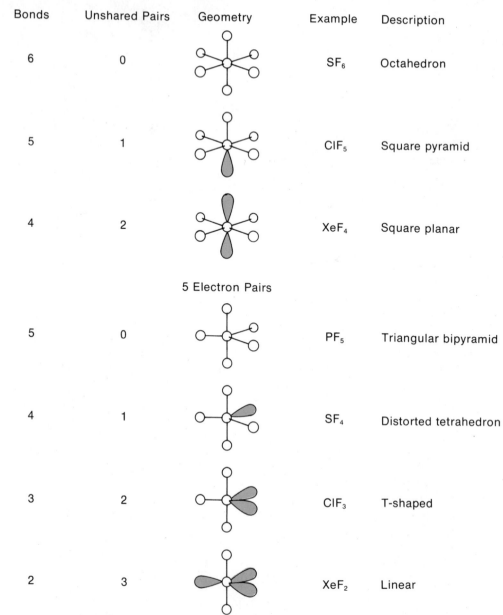

Figure 12.10 Geometries of molecules with expanded octets; the colored ovals represent unshared electron pairs. The structures of ClF_5 and XeF_4 are derived from that of SF_6 by removing successive corners of an octahedron.

thoroughly studied compound is xenon tetrafluoride, XeF_4. It is made by heating the elements together at 400°C:

$$Xe(g) + 2 F_2(g) \rightarrow XeF_4(s)$$

The product is a pure crystalline substance with a melting point of 140°C. It is stable in air and can be stored indefinitely without decomposition.

Until about 20 years ago, the noble gases were generally referred to as "inert" gases. They were believed to be completely unreactive toward other elements and compounds. The first noble gas compound was discovered at the University of British Columbia by Neil Bartlett, a 29-year-old chemist. In the course of his research on platinum-fluorine compounds, he isolated a reddish solid which he showed to be $O_2^+(PtF_6)^-$. Bartlett realized that the ionization energy of Xe (1130 kJ/mol) is virtually identical with that of O_2 (1110 kJ/mol). This encouraged him to attempt to make the analogous compound, $Xe^+(PtF_6)^-$. His success opened up a new era in noble gas chemistry.

12.5 MOLECULAR ORBITALS

In Chapter 10, we used the valence bond approach to explain bonding in molecules. It accounts, at least qualitatively, for the stability of the covalent bond in terms of the overlap of atomic orbitals. By invoking hybridization, valence bond theory can account for the molecular geometries predicted by electron-pair repulsion. Where Lewis structures are inadequate, as in O_3 or SO_2, the concept of resonance allows us to explain the observed properties.

A major weakness of valence bond theory has been its inability to predict the magnetic properties of molecules. An example cited in this chapter is O_2. The same problem arises with the B_2 molecule found in boron vapor at high temperatures. This molecule, like O_2, is paramagnetic but has an even number, six, of valence electrons. The octet rule, or valence bond theory, would predict that all electrons are paired in these molecules. This is inconsistent with their paramagnetism, which requires two unpaired electrons.

The deficiencies of valence bond theory arise from an inherent weakness. It assumes that the electrons in a molecule occupy atomic orbitals of the individual atoms. In the CH_4 molecule, the bonding is described in terms of the 1s orbitals of the H atom and the four sp^3 orbitals of the C atom. Clearly, this is an approximation. Each bonding electron in CH_4 must really be in an orbital characteristic of the molecule as a whole.

Following this idea, molecular orbital theory tries to treat bonds in terms of orbitals involving an entire molecule. The molecular orbital approach involves three basic operations:

1. The atomic orbitals of atoms are combined to give a new set of molecular orbitals characteristic of the molecule as a whole. Here, *the number of molecular orbitals formed is equal to the number of atomic orbitals combined*. When two H atoms combine to form H_2, two s orbitals, one from each atom, yield two molecular orbitals. In another case, six p orbitals, three from each atom, give a total of six molecular orbitals.

2. The molecular orbitals are arranged in order of increasing energy. In principle these energies can be obtained by solving the Schrödinger equation. In practice, we cannot make precise calculations for any but the simplest of molecules. Instead, the relative energies of molecular orbitals are deduced from experimental observations. Spectra and magnetic properties of molecules are used.

3. The valence electrons in a molecule are distributed among the available molecular orbitals. The process followed is much like that used with electrons in atoms. In particular, we find that:

   a. *Each molecular orbital can hold a maximum of two electrons.*
   b. *Electrons go into the lowest energy molecular orbital available*. A higher orbital starts to fill only when each orbital below it has its quota of two electrons.
   c. *Hund's rule is obeyed*. When two orbitals of equal energy are available to two electrons, one electron goes into each, giving two half-filled orbitals.

To illustrate molecular orbital theory, we will apply it to the diatomic molecules of the elements in the first two periods of the Periodic Table.

2 AO → 2 MO
6 AO → 6 MO

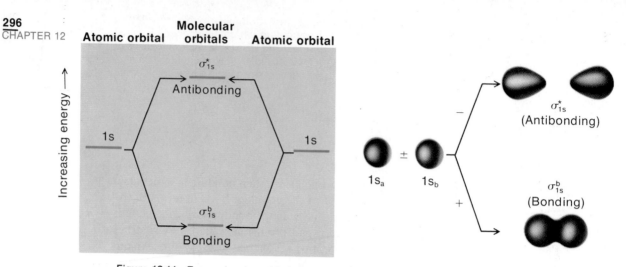

Figure 12.11 Two molecular orbitals are formed by combining two 1s atomic orbitals. In the bonding molecular orbital there is a high electron density between the atoms, which leads to a low energy. In the antibonding orbital the electron density is low between the atoms, giving a higher energy.

Hydrogen and Helium; Combination of 1s Orbitals

Molecular orbital (MO) theory predicts that two 1s orbitals will combine to give two molecular orbitals. One of these has an energy lower than that of the atomic orbitals from which it is formed (Fig. 12.11). Placing electrons in this orbital gives a species which is more stable than the isolated atoms. For that reason the lower molecular orbital in Figure 12.11 is called a **bonding orbital.** The other molecular orbital has a higher energy than the corresponding atomic orbitals. Electrons entering it are in an unstable, higher energy state. It is referred to as an **antibonding orbital.** In energy, this orbital lies about as far above the individual atomic orbitals as the bonding orbital lies below them.

The electron density in these two molecular orbitals is shown at the right of Figure 12.11. Notice that the bonding orbital has a high density in the region between the nuclei. This accounts for its stability. In the antibonding orbital, the chance of finding the electron between the nuclei is very small. The electron density is concentrated at the far ends of the "molecule." This means that the nuclei are less shielded from each other than they are in the isolated atoms. Small wonder that the antibonding orbital is higher in energy than the atomic orbitals from which it is formed.

The electron density in both molecular orbitals is symmetrical about the axis between the two nuclei. This means that both of these are sigma orbitals. In MO notation, the 1s bonding orbital is designated as σ_{1s}^b. The antibonding orbital is given the symbol σ_{1s}^*. In general, the superscript "b" refers to a bonding orbital. An asterisk is used to designate an antibonding orbital.

In the H_2 molecule, there are two 1s electrons. They fill the σ_{1s}^b orbital, giving a single bond. In the He_2 molecule, there would be four electrons to distribute, two from each atom. These would fill both the bonding and antibonding orbitals. As a result, the net number of bonds in He_2 would be zero. The general relation is

$$\text{number of bonds} = \frac{B - NB}{2}$$

where B is the number of electrons in bonding orbitals and NB the number of electrons in antibonding orbitals. Here, $B = NB = 2$, so the number of bonds is zero. The stability of He_2 would be no greater than that of two isolated He atoms. We would predict that the He_2 molecule should not exist, and indeed it does not.

When 2 AO's of equal energy combine, the resulting MO's lie above and below the energy of the AO's

Among the diatomic molecules of the second period elements are three familiar ones, N_2, O_2, and F_2. The molecules Li_2, B_2, and C_2 are less common but have been observed and studied in the gas phase. In contrast, the molecules Be_2 and Ne_2 are either highly unstable or nonexistent. Let us see what molecular orbital theory predicts as to the structure and stability of these molecules. We start by considering how the atomic orbitals containing the valence electrons (2s and 2p) are used to form molecular orbitals.

Combining two 2s atomic orbitals, one from each atom, gives two molecular orbitals. These are very similar to the ones discussed on p. 296. They are designated as σ_{2s}^b (sigma, bonding, 2s) and σ_{2s}^* (sigma, antibonding, 2s).

Consider now what happens to the 2p orbitals. In an isolated atom, there are three such orbitals, oriented at right angles to each other. We designate these atomic orbitals as p_x, p_y, and p_z. When two p_x atomic orbitals, one from each atom, combine, they form the two molecular orbitals shown at the upper right of Figure 12.12. These are both sigma orbitals, symmetrical about the x axis. One of them, called σ_{2p}^b, is a bonding orbital. The other, σ_{2p}^*, is antibonding. These orbitals have electron cloud densities that resemble those of the sigma molecular orbitals shown in Figure 12.11.

The p_y or p_z atomic orbitals combine with one another in a different way. As you can see from Figure 12.12, two such atomic orbitals, one from each atom, combine to form two pi molecular orbitals. In these orbitals, the electron density is not symmetric about the axis. The higher energy orbital, π_{2p}^*, where the electron density is concentrated away from the nuclei, is antibonding. The lower energy orbital, π_{2p}^b, with a high electron density between the nuclei, is a bonding orbital. There are two such sets of orbitals. One is formed by p_y orbitals. The other, not shown in the cloud density diagram at the right of Figure 12.12, is formed by p_z orbitals.

Summarizing the molecular orbitals available to the valence electrons of the second period elements, we have

one σ_{2s}^b and one σ_{2s}^* orbital
one σ_{2p}^b and one σ_{2p}^* orbital
two π_{2p}^b and two π_{2p}^* orbitals

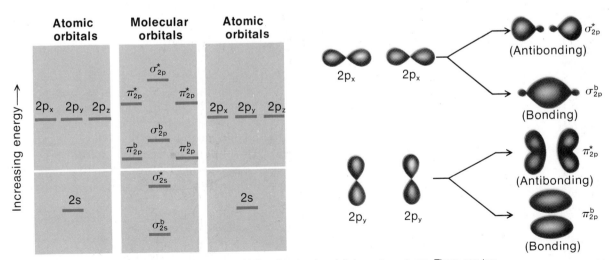

Figure 12.12 Molecular orbitals obtained by combining 2p atomic orbitals on two atoms. There are two π_{2p}^b orbitals and two π_{2p}^* orbitals because the $2p_y$ and $2p_z$ orbitals can participate in π orbital formation. The molecular orbitals from the 2s electrons have the structure shown in Figure 12.11.

TABLE 12.4 PREDICTED AND OBSERVED PROPERTIES OF DIATOMIC MOLECULES OF SECOND PERIOD ELEMENTS

OCCUPANCY OF ORBITALS

	σ_{2s}^b	σ_{2s}^*	π_{2p}^b	π_{2p}^b	σ_{2p}^b	π_{2p}^*	π_{2p}^*	σ_{2p}^*
Li_2	(↿⇂)	( )	( )	( )	( )	( )	( )	( )
Be_2	(↿⇂)	(↿⇂)	( )	( )	( )	( )	( )	( )
B_2	(↿⇂)	(↿⇂)	(↿)	(↿)	( )	( )	( )	( )
C_2	(↿⇂)	(↿⇂)	(↿⇂)	(↿⇂)	( )	( )	( )	( )
N_2	(↿⇂)	(↿⇂)	(↿⇂)	(↿⇂)	(↿⇂)	( )	( )	( )
O_2	(↿⇂)	(↿⇂)	(↿⇂)	(↿⇂)	(↿⇂)	(↿)	(↿)	( )
F_2	(↿⇂)	(↿⇂)	(↿⇂)	(↿⇂)	(↿⇂)	(↿⇂)	(↿⇂)	( )
Ne_2	(↿⇂)	(↿⇂)	(↿⇂)	(↿⇂)	(↿⇂)	(↿⇂)	(↿⇂)	(↿⇂)

	PREDICTED PROPERTIES		**OBSERVED PROPERTIES**	
	No. of Un-paired e^-	No. of Bonds	No. of Un-paired e^-	Bond Energy (kJ/mol)
Li_2	0	1	0	105
Be_2	0	0	0	unstable
B_2	2	1	2	289
C_2	0	2	0	628
N_2	0	3	0	941
O_2	2	2	2	494
F_2	0	1	0	153
Ne_2	0	0	0	nonexistent

The relative energies of these orbitals are shown at the left of Figure 12.12. This order applies in the molecules of the second period elements, at least through N_2.

To obtain the MO structure of the diatomic molecules of the elements in the second period, we fill the available molecular orbitals in order of increasing energy. The results are shown in Table 12.4. Note the excellent agreement between MO theory and the properties of these molecules. In particular, the number of unpaired electrons predicted agrees with experiment. There is also a general correlation between the predicted number of bonds

$$\text{number of bonds} = \frac{B - NB}{2}$$

and the bond energy. We would expect a double bond (C_2 or O_2) to be stronger than a single bond (Li_2, B_2, F_2). A triple bond, as in N_2, should be still stronger.

A major triumph of MO theory is its ability to explain the properties of O_2. It explains how this molecule can have a double bond and, at the same time, have two unpaired electrons. You will recall that simple valence bond theory could not do this. Again, molecular orbital theory succeeds where valence bond theory fails in explaining paramagnetism in B_2. For N_2 and F_2, the two theories predict the same number of bonds (triple, single), with no unpaired electrons.

SUMMARY

Nonmetal atoms combine with one another to form molecules or, in some cases, macromolecules. In this chapter we considered the structures and properties of several nonmetallic elements (Table 12.1). Among these were oxygen, sulfur, phosphorus, and carbon, all of which show allotropy. Later in the chapter, we discussed saturated hydrocarbons (Section 12.2) and the oxides of sulfur, nitrogen, and carbon (Section 12.3).

Throughout this chapter, we used principles of covalent bonding (Chapter 10) and intermolecular forces (Chap. 11). Several of the examples (12.2, 12.3, 12.5, 12.6, and 12.7) reviewed these principles. A few new concepts were introduced in this chapter. These include:

1. *Resonance* (Examples 12.1 and 12.6). A species such as O_3 for which a single Lewis structure is inadequate is said to show resonance. Its structure is intermediate between those of resonance forms, which are written by shifting electron pairs from one atom to another.

2. *Structural isomerism* (Example 12.4), in which atoms are bonded in different patterns. Structural isomers have the same molecular formula but are distinctly different compounds.

3. *Expanded octets* (Examples 12.8 and 12.9), which are common among nonmetal atoms in the third, fourth, and fifth periods. These atoms may be surrounded by five electron pairs (sp^3d hybridization) or six electron pairs (sp^3d^2 hybridization). The geometry of molecules in which the central atom has an expanded octet can be predicted from electron-pair repulsion (Fig. 12.10).

4. *Molecular orbitals* (Section 12.5). Here, atomic orbitals are combined to form an equal number of molecular orbitals. A given "MO" may be bonding or antibonding (Figs. 12.11 and 12.12). The relative energies of different molecular orbitals govern the order in which they are filled by electron pairs (Table 12.4).

KEY WORDS AND CONCEPTS

macromolecular	n-type semiconductor	sigma bond	triangular bipyramid
octet	p-type semiconductor	pi bond	octahedron
allotropy	structural isomer	acid anhydride	molecular orbital
resonance	saturated hydrocarbons (alkanes)	expanded octet	bonding orbital
Lewis structure	dispersion force	hybrid orbitals	antibonding orbital

QUESTIONS AND PROBLEMS

Catalog

General: 12.1–12.6, 12.23–12.27, 12.44–12.47
Lewis Structures, Geometry, Polarity: 12.7–12.9, 12.28–12.30
Hybridization: Pi and Sigma Bonds: 12.10, 12.31
Intermolecular Forces: 12.11–12.13, 12.32–12.34

Resonance: 12.14, 12.15, 12.35, 12.36
Isomerism: 12.16, 12.17, 12.37, 12.38
Expanded Octets: 12.18–12.20, 12.39–12.41
Molecular Orbitals: 12.21, 12.22, 12.42, 12.43

12.1   Review and know the meaning of the key words and concepts in this chapter.

12.2   Give the formula of the atom or molecule present in the following gaseous elements at 25°C:

    a. chlorine   b. neon   c. nitrogen
    d. oxygen   e. argon

12.23   Which of the following elements exist as molecules in the solid? Which are macromolecular?

    a. iodine   b. sulfur   c. phosphorus
    d. carbon   e. silicon

12.3 Explain how the allotropic forms of each of the following elements differ from each other:

    a. oxygen and ozone
    b. rhombic and monoclinic sulfur
    c. diamond and graphite
    d. white and red phosphorus

12.4 Write balanced equations for the reaction of oxygen with

    a. P_4O_6 b. SO_2 c. ZnS   d. N_2
    e. CO

12.5 Give the symbol of the element that

    a. has two gaseous allotropes.
    b. is used in jewelry and automobile tires.
    c. forms an oxide that serves up giggles and whipped cream.

12.6 Criticize each of the following statements:

    a. $Br_2(l)$ and $Br_2(g)$ are allotropes.
    b. Nitrogen and phosphorus in Group 5 form oxides of similar formulas.
    c. A high reaction temperature tends to favor the formation of a higher oxide (e.g., NO_2 as opposed to NO).
    d. The resonance forms of O_3 have different properties.
    e. The xenon atom in XeF_4 is surrounded by four electron pairs.

12.7 Give the Lewis structures, predict the geometry, and describe the polarity of

    a. H_2O_2 b. F_2O c. $SeBr_2$ d. $AsCl_3$

12.8 Consider the N_2O_3 molecule shown in Figure 12.8. Give the bond angles and predict whether the molecule is polar.

12.9 At one time, it was believed that the O_3 molecule was a three-membered ring with the Lewis structure

How would the properties of such a molecule differ from those of the O_3 molecule shown on p. 276?

12.10 Consider the oxides of nitrogen shown in Figure 12.8. Give the hybridization of each N atom in each molecule.

12.11 The boiling points of the elements silicon, phosphorus, sulfur, chlorine, and argon are 2355, 280, 444, −34, and −186°C, in that order. Can you explain why silicon has such a high boiling point? Why does chlorine boil below sulfur but above argon?

12.24 Describe the structure of:

    a. liquid sulfur between 160 and 250°C
    b. plastic sulfur
    c. amorphous carbon
    d. a p-type semiconductor derived from silicon

12.25 Describe how you would prepare

    a. NO_2 from NO   b. N_2O c. SO_2
    d. P_4O_6

12.26 Write the symbol of

    a. a nonmetal that forms a molecule containing eight atoms.
    b. the only nonmetal that is a liquid at room temperature and 1 atm.
    c. a solid nonmetal that reacts vigorously with O_2 at room temperature.

12.27 Classify the following statements as true or false. Correct any false statements to make them true.

    a. Allotropic forms of an element have different properties.
    b. Oxygen gas is diamagnetic.
    c. Diamond is converted to graphite at high pressures.
    d. An n-type semiconductor can be made by adding a trace of Ge to Si.
    e. Two substances with the same molecular formula have the same boiling point.

12.28 Follow the directions of Problem 12.7 for

    a. SO_3 b. SbH_3 c. F_2O_2 d. $SiCl_4$

12.29 Follow the directions of Problem 12.8 for the N_2O_2 molecule shown in Figure 12.8.

12.30 A plausible Lewis structure for P_4 is

$$:P{=\!=}P:$$
$$| \quad |$$
$$:P{=\!=}P:$$

How would the properties of this molecule differ from those of the P_4 molecule shown on p. 278?

12.31 Give the number of pi and sigma bonds in each of the molecules shown in Figure 12.8.

12.32 The compounds SiH_4, Si_2H_6, Si_3H_8, and Si_4H_{10} are silicon analogs of hydrocarbons.

    a. Which of these compounds has the highest boiling point? lowest boiling point?
    b. Explain your reasoning.

12.12 Which one of the following compounds would you expect to have the higher boiling point? Explain your reasoning.

$$CH_3-CH_2-CH_2-CH_2Cl \quad \text{or} \quad CH_3-\overset{\overset{\displaystyle H}{|}}{\underset{\underset{\displaystyle CH_3}{|}}{C}}-CH_2Cl$$

12.13 What is the major type of attractive force that must be overcome to

    a. sublime CO_2?
    b. melt Si?
    c. convert S_8 rings to S_8 chains?
    d. vaporize $NH_3(l)$?

12.14 Draw resonance forms for

    a. SO_2 b. NO_2 c. CO_2 d. CO_3^{2-}

12.15 The oxalate ion, $C_2O_4^{2-}$, has the skeleton structure

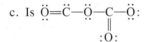

    a. Complete the Lewis structure of this ion.
    b. Draw resonance forms for $C_2O_4^{2-}$, equivalent to the Lewis structure drawn in (a).

    c. Is $\ddot{O}=\ddot{C}-\ddot{O}-C-\ddot{O}:$
                             $\overset{\displaystyle \|}{\underset{\displaystyle :O:}{}}$

another resonance form of the oxalate ion? Explain.

12.16 Draw the structural isomers of C_6H_{14}.

12.17 Draw Lewis structures for the structural isomers of $C_3H_6Cl_2$.

12.18 Consider the species

    a. RnF_4 b. $SeCl_4$ c. PCl_5 d. SF_5Cl

How many electron pairs are there (five or six) around the central atom in each species? What is the hybridization of the central atom?

12.19 Describe the geometry of each species in Problem 12.18.

12.20 Give the formula of a species in which sulfur shows the following types of hybridization:

    a. sp^2 b. sp^3 c. sp^3d d. sp^3d^2

12.33 Sulfur hexafluoride, SF_6, and decane, $CH_3-(CH_2)_8-CH_3$, have nearly the same molecular mass. Yet, decane boils at 174°C while SF_6 sublimes at −64°C. Suggest an explanation for this difference.

12.34 What is the major type of attractive force that must be overcome to

    a. vaporize Br_2?
    b. melt ice?
    c. melt diamond?
    d. vaporize C_5H_{12}?

12.35 Draw resonance forms for

    a. SO_3 b. O_3 c. CS_2 d. NO_3^-

12.36 The Lewis structure for hydrazoic acid may be written as

    a. Draw two other resonance forms for this molecule
    b. Is

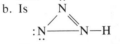

another resonance form of hydrazoic acid? Explain.

12.37 Draw the structural isomers of Si_5H_{12}.

12.38 Draw the structural isomers of C_4H_9Cl.

12.39 Consider the species

    a. SeF_6 b. KrF_2 c. $TeBr_4$ d. PCl_4^-

How many electron pairs are there (five or six) around the central atom in each species? What is the hybridization of the central atom?

12.40 Describe the geometry of each species in Problem 12.39.

12.41 Give the formula of a species in which a Group 5 element shows the following types of hybridization:

    a. sp b. sp^2 c. sp^3 d. sp^3d
    e. sp^3d^2

12.21 Using MO theory, list the number of electrons in each of the 2s and 2p MO's, the number of bonds, and the number of unpaired electrons in

a. O_2 b. O_2^- c. O_2^{2-}

12.22 Suppose in building up molecular orbitals the σ_{2p}^b were placed below the π_{2p}^b. Prepare a diagram similar to Table 12.4 based on this assignment. For which species would this change in relative energies affect the prediction of number of bonds? number of unpaired electrons?

12.42 Follow the directions of Problem 12.21 for

a. CO   b. C_2^- c. F_2^-

12.43 Consider the +1 ions formed by removing an electron from each of the diatomic molecules of the elements of atomic numbers 3 through 9. Using the MO approach, determine the number of bonds and unpaired electrons in each of these ions.

* 12.44 In principle, there are several structural isomers of N_2O_3 in addition to the one shown in Figure 12.8. Draw Lewis structures for six of these isomers.

* 12.45 The density of hydrogen fluoride vapor at 28°C and 1.00 atm is 2.30 g/ℓ. What information does this yield concerning intermolecular forces in gaseous hydrogen fluoride?

* 12.46 The geometries shown in Figure 12.10 are not the only ones possible for species in which the central atom has an expanded octet and one or more pairs of unshared electrons. Suggest other possible geometries for XeF_4 and ClF_3.

* 12.47 Consider the polyatomic ion IO_6^{5-}. How many pairs of electrons are there around the central iodine atom? What is its hybridization? Describe the geometry of the ion.

CHEMICAL EQUILIBRIA
IN GASEOUS SYSTEMS

In Chapter 11, we described several different equilibrium systems. All of these involved a single substance distributed between two phases. Recall our discussion of what happens when a sample of liquid water is placed in a closed container at constant temperature. Part of the liquid vaporizes. After a few minutes, equilibrium is established between liquid water and water vapor. To show this, we write the equation

$$H_2O(l) \rightleftharpoons H_2O(g)$$

The double arrow implies that the equilibrium is *dynamic*. That is, the two processes, vaporization and condensation, are occurring at the same rate. As a result, the equilibrium concentration of water vapor does not change with time. At 100°C, it has a fixed value of about 0.0327 mol/ℓ.

Chemical reactions carried out in closed containers show some resemblance to phase changes. Ordinarily, the reactants are not completely consumed. Instead, we obtain an equilibrium mixture containing both products and reactants. At equilibrium the two processes, forward and reverse reactions, are taking place at the same rate. Unless we disturb the equilibrium in some way, the concentrations of all species will remain constant as time passes.

In this chapter we will examine the properties of chemical systems which reach equilibrium in the gas state. Our main interest will be the position of such equilibria. We will develop principles to predict the relative concentrations of products and reactants at equilibrium. We will also look at ways in which the position of an equilibrium can be shifted to favor the formation of either products or reactants. To start with, let us consider how chemical equilibrium is established in a typical gaseous system.

This concentration is equivalent to a vapor pressure of 1 atm

13.1 THE N₂O₄–NO₂ EQUILIBRIUM

Consider what happens when a sample of N_2O_4, a colorless gas, is placed in a closed, evacuated container at 100°C. Instantly, a reddish-brown color develops. This color is due to nitrogen dioxide, NO_2, formed by decomposition of part of the N_2O_4:

$$N_2O_4(g) \rightarrow 2\ NO_2(g) \tag{13.1a}$$

At first, this is the only reaction taking place. However, as soon as some NO_2 is formed, the reverse reaction can occur:

$$2\ NO_2(g) \rightarrow N_2O_4(g) \tag{13.1b}$$

TABLE 13.1	ESTABLISHMENT OF EQUILIBRIUM AT 100°C IN THE SYSTEM $N_2O_4(g) \rightleftharpoons 2 NO_2(g)$					
Time (seconds)	0	20	40	60	80	100
Conc. N_2O_4 (mol/ℓ)	0.100	0.070	0.050	**0.040***	**0.040**	**0.040**
Conc. NO_2 (mol/ℓ)	0.000	0.060	0.100	**0.120***	**0.120**	**0.120**

*Equilibrium concentrations are in bold type.

As time passes, Reaction 13.1a slows down and 13.1b speeds up. Soon, their rates become equal. A dynamic equilibrium has been established.

$$N_2O_4(g) \rightleftharpoons 2 NO_2(g) \tag{13.1}$$

At equilibrium, appreciable amounts of both gases are present. From that point on their concentrations are constant, so long as the volume of the container and the temperature remain unchanged.

The approach to equilibrium in this system is illustrated by the data in Table 13.1, plotted in Figure 13.1. We start with 0.100 mol of N_2O_4 in a 1.00-ℓ container at 100°C. The original concentration of N_2O_4 is thus 0.100 mol/ℓ. Since there is no NO_2 present at the beginning of the experiment, its original concentration is zero. As equilibrium is approached, the *net reaction* taking place is that in the forward direction (13.1a). The

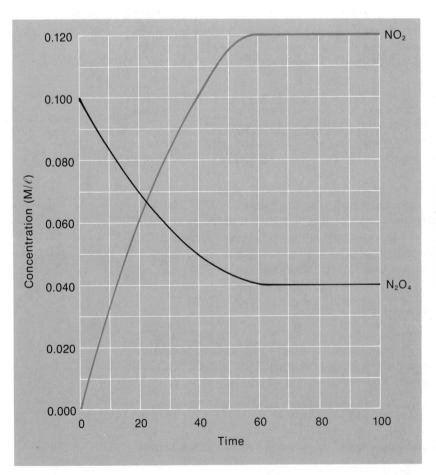

Figure 13.1 Approach to equilibrium in the N_2O_4–NO_2 system. The concentration of N_2O_4 starts off at 0.100 M, drops sharply at first, and finally levels off at its equilibrium value, 0.040 M. Meanwhile, the concentration of NO_2 rises rapidly at first, then increases more slowly, and finally becomes constant at the equilibrium value, 0.120 M.

concentration of N_2O_4 drops, rapidly at first and then more slowly. The concentration of NO_2 increases. Finally, both concentrations level off and become constant. This tells us that we are at equilibrium. From Table 13.1 or Figure 13.1 we see that

$$\text{equilibrium conc. } N_2O_4 = 0.040 \text{ mol/} \ell$$
$$\text{equilibrium conc. } NO_2 = 0.120 \text{ mol/} \ell$$

One feature of the data in Table 13.1 deserves further comment. Notice that, over any time interval, the increase in concentration of NO_2 is exactly *twice* the decrease in concentration of N_2O_4. Consider, for example, what happens in the 60 s required to reach equilibrium. The concentration of N_2O_4 decreases by 0.060 mol/ℓ:

$$\Delta \text{ conc. } N_2O_4 = 0.040 \text{ mol/} \ell - 0.100 \text{ mol/} \ell = -0.060 \text{ mol/} \ell$$

The concentration of NO_2 increases by twice this amount, 0.120 mol/ℓ:

$$\Delta \text{ conc. } NO_2 = 0.120 \text{ mol/} \ell - 0.000 \text{ mol/} \ell = 0.120 \text{ mol/} \ell$$

This relationship is explained quite simply. As equilibrium is approached, N_2O_4 decomposes:

$$N_2O_4(g) \rightarrow 2 NO_2(g)$$

From the coefficients of the balanced equation, we see that two moles of NO_2 are formed for every mole of N_2O_4 that decomposes. Hence if x moles of N_2O_4 break down, 2x moles of NO_2 must form. If the concentration of N_2O_4 decreases by x mol/ℓ, that of NO_2 must increase by twice that amount, 2x mol/ℓ.

Equilibrium is reached in other chemical systems in much the way we have described for the N_2O_4–NO_2 system. However, as you might expect, the concentrations of products and reactants at equilibrium are quite different from those in Table 13.1. Moreover, the ratio of the changes in concentration as equilibrium is approached depends upon the system involved (Example 13.1).

Example 13.1   Consider the HI–H_2–I_2 system. To establish equilibrium, we start with 0.100 mol of HI in a 1.00-ℓ container at 520°C. Its original concentration is 0.100 mol/ℓ; those of H_2 and I_2 are 0.000 mol/ℓ. As time passes, some of the HI decomposes by the reaction

$$2 HI(g) \rightarrow H_2(g) + I_2(g)$$

At equilibrium, the concentration of H_2 is found to be 0.010 mol/ℓ. What is the equilibrium concentration of
 a. I_2?   b. HI?

Solution
 a. From the balanced equation, we see that one mole of I_2 is formed for every mole of H_2. The concentrations of H_2 and I_2 must therefore increase by the same amount. Since they both start at zero, the two concentrations must be equal at equilibrium. The equilibrium concentration of I_2, like that of H_2, is 0.010 mol/ℓ.
 b. Two moles of HI are required to form one mole of H_2. The decrease in HI concentration must be twice the increase in the concentration of H_2.

$$\text{decrease in HI conc.} = 2(0.010 \text{ mol/} \ell) = 0.020 \text{ mol/} \ell$$

You need to think about this for awhile, but it pays off

The original concentration of HI is 0.100 mol/ℓ. Hence,

equilibrium conc. HI = 0.100 mol/ℓ − 0.020 mol/ℓ = 0.080 mol/ℓ

Exercise   When 0.100 mol of PCl_5 is placed in a 1.00-ℓ container at 250°C, part of it decomposes by the reaction $PCl_5(g) \rightarrow PCl_3(g) + Cl_2(g)$. The equilibrium concentration of PCl_5 is 0.060 mol/ℓ. What is the equilibrium concentration of PCl_3? Cl_2? Answer: 0.040 mol/ℓ.

13.2 THE EQUILIBRIUM CONSTANT FOR THE N_2O_4–NO_2 SYSTEM

In Section 13.1, we described one way of establishing equilibrium at 100°C in the N_2O_4–NO_2 system. There, we started with pure N_2O_4 at a concentration of 0.100 mol/ℓ. We found that

$$[N_2O_4] = 0.040 \text{ mol}/\ell \qquad [NO_2] = 0.120 \text{ mol}/\ell$$

The square brackets, here and elsewhere throughout this text, represent equilibrium concentrations in moles per liter. It is important that these be distinguished from original or other nonequilibrium concentrations.

There are many ways in which we can approach equilibrium in this system. Table 13.2 gives data for three experiments, in which the original concentrations are quite different. Experiment 1 is that just described, in which we start with pure N_2O_4. In Experiment 2, we start from the other side of the equilibrium system, with pure NO_2 at a concentration of 0.100 mol/ℓ. As we approach equilibrium, some of the NO_2 combines to form N_2O_4:

$$2 \, NO_2(g) \rightarrow N_2O_4(g)$$

Finally, in Experiment 3, we start with a mixture of N_2O_4 and NO_2, both at a concentration of 0.100 mol/ℓ. Some of the N_2O_4 decomposes in this experiment. At equilibrium, there is less N_2O_4 and more NO_2 than there was originally.

Looking at the data in Table 13.2, you might wonder whether these three experiments have anything in common. Specifically, is there any relationship between the equilibrium concentrations of NO_2 and N_2O_4 which is valid for all the experiments? It turns out that there is, although it is not an obvious one. The quotient $[NO_2]^2/[N_2O_4]$ is the same, about 0.36, in each case:

TABLE 13.2 EQUILIBRIUM MEASUREMENTS IN THE N_2O_4–NO_2 SYSTEM AT 100°C

		ORIGINAL CONC. (mol/ℓ)	EQUILIBRIUM CONC. (mol/ℓ)
Expt. 1	N_2O_4	0.100	0.040
	NO_2	0.000	0.120
Expt. 2	N_2O_4	0.000	0.014
	NO_2	0.100	0.072
Expt. 3	N_2O_4	0.100	0.070
	NO_2	0.100	0.160

Expt. 1 $\quad \dfrac{[NO_2]^2}{[N_2O_4]} = \dfrac{(0.120)^2}{0.040} = 0.36$

Expt. 2 $\quad \dfrac{[NO_2]^2}{[N_2O_4]} = \dfrac{(0.072)^2}{0.014} = 0.37$

Expt. 3 $\quad \dfrac{[NO_2]^2}{[N_2O_4]} = \dfrac{(0.160)^2}{0.070} = 0.36$

This relationship holds for any equilibrium mixture containing NO_2 and N_2O_4 at 100°C. Regardless of where we start, we find that at equilibrium,

$$\frac{[NO_2]^2}{[N_2O_4]} = 0.36 \qquad \text{(at 100°C)}$$

Reaction proceeds until this condition is satisfied

Further experiments with this system at various temperatures lead to the general conclusion:

At any given temperature, the quantity

$$\frac{[NO_2]^2}{[N_2O_4]}$$

is a constant, independent of the amounts of N_2O_4 and NO_2 that we start with, the volume of the container, or the total pressure. This constant is referred to as the **equilibrium constant, K_c,** for the reaction

$$N_2O_4(g) \rightleftharpoons 2\ NO_2(g)$$

13.3 GENERAL EXPRESSION FOR K_c

We have seen that for the system

$$N_2O_4(g) \rightleftharpoons 2\ NO_2(g);\ K_c = \frac{[NO_2]^2}{[N_2O_4]}$$

For every gaseous system, a similar expression can be written. The form of that expression can be deduced from the Law of Chemical Equilibrium. For the general gas phase system

$$a\ A(g) + b\ B(g) \rightleftharpoons c\ C(g) + d\ D(g)$$

where A, B, C, D represent different substances and a, b, c, d are their coefficients in the balanced equation

$$K_c = \frac{[C]^c \times [D]^d}{[A]^a \times [B]^b} \tag{13.2}$$

Notice that in the expression for K_c:

—the equilibrium concentrations of **products** (right side of equation) appear in the **numerator.**

—the equilibrium concentrations of **reactants** (left side of equation) appear in the **denominator.**

To illustrate the application of this law, consider the systems

$$2 \text{ HI(g)} \rightleftharpoons \text{H}_2(g) + \text{I}_2(g) \qquad\qquad (13.3)$$

$$\text{N}_2(g) + 3 \text{ H}_2(g) \rightleftharpoons 2 \text{ NH}_3(g) \qquad\qquad (13.4)$$

For these reactions we have

$$K_c = \frac{[\text{H}_2] \times [\text{I}_2]}{[\text{HI}]^2} \qquad \text{(for Reaction 13.3)}$$

$$K_c = \frac{[\text{NH}_3]^2}{[\text{N}_2] \times [\text{H}_2]^3} \qquad \text{(for Reaction 13.4)}$$

Determination of K_c

The values of K_c for reactions such as 13.3 and 13.4 are found in much the same way as we described for the N_2O_4–NO_2 system. A typical calculation is shown in Example 13.2.

Example 13.2   An equilibrium mixture of N_2, H_2, and NH_3 at 300°C is analyzed, and it is found that:

$$[\text{N}_2] = 0.25 \text{ mol}/\ell \qquad [\text{H}_2] = 0.15 \text{ mol}/\ell \qquad [\text{NH}_3] = 0.090 \text{ mol}/\ell$$

Find K_c at 300°C for the reaction $\text{N}_2(g) + 3 \text{ H}_2(g) \rightleftharpoons 2 \text{ NH}_3(g)$.

Solution K_c is found by simply substituting into the equilibrium equation:

$$K_c = \frac{[\text{NH}_3]^2}{[\text{N}_2] \times [\text{H}_2]^3} = \frac{(0.090)^2}{(0.25) \times (0.15)^3} = 9.6$$

Exercise   Using the data in Example 13.1, calculate K_c at 520°C for the system $2 \text{ HI(g)} \rightleftharpoons \text{H}_2(g) + \text{I}_2(g)$. Answer: 0.016.

Reactions Involving Liquids or Solids As Well As Gases

In certain gas-phase reactions, one or more of the substances involved is present as a pure liquid or solid. Examples of such systems include

$$\text{CaCO}_3(s) \rightleftharpoons \text{CaO}(s) + \text{CO}_2(g) \qquad\qquad (13.5)$$

$$\text{CO}_2(g) + \text{H}_2(g) \rightleftharpoons \text{CO}(g) + \text{H}_2\text{O}(l) \qquad\qquad (13.6)$$

In such cases we find that **there is no term for the liquid or solid in the expression for K_c.** Thus we have

$$K_c = [CO_2] \qquad \text{(for Reaction 13.5)}$$

$$K_c = \frac{[CO]}{[CO_2] \times [H_2]} \qquad \text{(for Reaction 13.6)}$$

The terms for solids ($CaCO_3$, CaO) and liquids (H_2O) are omitted.

To see why this simplification is possible, consider Reaction 13.6 taking place at, say, 100°C. The equation tells us that liquid water is present. There must also be water vapor present, in equilibrium with the liquid. Indeed, it is actually $H_2O(g)$ which takes part in the gas phase reaction. The pressure of the vapor is 1.00 atm, the vapor pressure of liquid water at 100°C. The concentration of water vapor, calculated from the Ideal Gas Law, is 0.0327 mol/ℓ. This concentration remains constant so long as there is any liquid water present, regardless of what [CO], [H_2], or [CO_2] may be. Hence, it can be incorporated into the expression for K_c, and we do not need a term for the concentration of H_2O.

In K_c the concentration terms must all be able to vary

Example 13.3   Write expressions for K_c for
   a. $2\ SO_3(g) \rightleftharpoons 2\ SO_2(g) + O_2(g)$
   b. $CuO(s) + H_2(g) \rightleftharpoons Cu(s) + H_2O(g)$

Solution

   a. $K_c = \dfrac{[SO_2]^2 \times [O_2]}{[SO_3]^2}$ b. $K_c = \dfrac{[H_2O]}{[H_2]}$

Exercise   Suppose the reaction in (b) is carried out at a temperature low enough so that liquid water is present. What is the expression for K_c? Answer: $1/[H_2]$.

13.4 INTERPRETATION OF K_c

A basic goal of chemistry is to predict the extent to which a reaction can be expected to occur in the laboratory or in nature. A knowledge of the equilibrium constant for the system allows us to make such predictions. As we shall see, the extent of reaction depends upon the magnitude of K_c.

K_c Very Small

Consider the equilibrium system

$$N_2(g) + O_2(g) \rightleftharpoons 2\ NO(g); \ K_c = \frac{[NO]^2}{[N_2] \times [O_2]} = 1 \times 10^{-30} \text{ at } 25°C \qquad (13.7)$$

We note that K_c for this system is very small. An equilibrium mixture of these three gases at 25°C will contain very little NO as compared to N_2 or O_2. Suppose, for example, that

$$[N_2] = [O_2] = 1 \text{ mol}/\ell$$

Then: $\qquad [NO]^2 = 1 \times 10^{-30}; [NO] = 1 \times 10^{-15} \text{ mol}/\ell$

We see that the concentration of NO in this mixture is very small indeed.

This analysis implies that at 25°C the reaction

$$N_2(g) + O_2(g) \rightarrow 2 \ NO(g)$$

If $K_c \ll 1$, the system consists mainly of reactants will not occur to any appreciable extent. Only a trace of NO will be formed. The equilibrium mixture will consist almost entirely of unreacted N_2 and O_2. Conversely, if we start with pure NO, the reverse reaction

$$2 \ NO(g) \rightarrow N_2(g) + O_2(g) \quad .$$

will go nearly to completion. Virtually all the NO will be converted to N_2 and O_2 when equilibrium is reached. In this way [NO] will be very small, as will the ratio $\dfrac{[NO]^2}{[N_2] \times [O_2]}$.

In general we can say that, if K_c is very small,
—**the forward reaction** (left to right) **will not occur to an appreciable extent.**
—**the reverse reaction** (right to left) **will go nearly to completion.**

K_c Very Large

Consider the equilibrium system:

$$2 \ Cl(g) \rightleftharpoons Cl_2(g); \ K_c = \frac{[Cl_2]}{[Cl]^2} = 1 \times 10^{38} \text{ at } 25°C \qquad (13.8)$$

Since K_c is so large, the equilibrium mixture will consist almost entirely of Cl_2 molecules. There will be virtually no free Cl atoms. Suppose, for example, that

$$[Cl_2] = 1 \ mol/\ell$$

Then:
$$[Cl]^2 = \frac{1}{1 \times 10^{38}} = 1 \times 10^{-38}; \ [Cl] = 1 \times 10^{-19} \ mol/\ell$$

The concentration of Cl atoms in this mixture is very, very small compared to that of Cl_2 molecules.

This argument tells us that at 25°C the reaction

If $K_c \gg 1$, the system consists mainly of products
$$2 \ Cl(g) \rightarrow Cl_2(g)$$

goes virtually to completion. Nearly all the Cl atoms are converted to Cl_2 molecules at equilibrium. Conversely, the reaction

$$Cl_2(g) \rightarrow 2 \ Cl(g)$$

will not occur to an appreciable extent. Very few Cl atoms are formed at equilibrium by the dissociation of Cl_2 molecules. Hence [Cl] is very small compared to $[Cl_2]$, as required by the relation $[Cl_2]/[Cl]^2 = 1 \times 10^{38}$.

In general we can say that, if K_c is very large,
—**the forward reaction** (left to right) **will go nearly to completion.**
—**the reverse reaction** (right to left) **will not occur to an appreciable extent.**

K_c Neither Very Large nor Very Small

Recall from Section 13.2 the system

$$N_2O_4(g) \rightleftharpoons 2\ NO_2(g);$$

$$K_c = \frac{[NO_2]^2}{[N_2O_4]} = 0.36 \text{ at } 100°C$$

Here K_c is close to 1. We expect to find significant amounts of both NO_2 and N_2O_4 at equilibrium. From the data in Table 13.2, we see that this is the case. Starting with 0.100 mol/ℓ of N_2O_4, about 60% of it dissociates to NO_2; 40% remains unreacted. In Experiment 2, starting with 0.100 mol/ℓ of NO_2, 28% of it reacts to form N_2O_4; 72% is left at equilibrium.

In general, if K_c is neither very large nor very small, we expect to obtain an equilibrium mixture containing appreciable amounts of both reactants and products. This is true regardless of which side of the equilibrium system we start from. Neither the forward nor the reverse reaction goes to completion.

This is the type of reaction we will be talking about in the remainder of this chapter. If K_c is neither very large nor very small, we cannot use qualitative terms such as "far to the right" or "far to the left" to describe the position of an equilibrium. Instead we need to know, quite precisely, the concentrations of products and reactants at equilibrium. In Section 13.5, we will consider how such calculations can be made.

13.5 CALCULATION OF EQUILIBRIUM CONCENTRATIONS FROM K_c

The equilibrium constant for a chemical system can be used to calculate the concentrations of the species present at equilibrium. In the simplest case, we can use K_c to obtain the equilibrium concentration of one species, knowing those of all the other species (Example 13.4).

Example 13.4 We have seen that at room temperature the equilibrium constant for

$$N_2(g) + O_2(g) \rightleftharpoons 2\ NO(g)$$

is very small. However, the equilibrium constant increases with temperature, becoming 0.10 at 2000°C. In an equilibrium mixture at 2000°C, $[N_2] = [O_2] = 0.10$ mol/ℓ. Calculate the equilibrium concentration of NO.

Solution The expression for K_c is

$$K_c = \frac{[NO]^2}{[N_2] \times [O_2]} = 0.10$$

Substituting for $[N_2]$ and $[O_2]$, we have

$$\frac{[NO]^2}{(0.10) \times (0.10)} = 0.10$$

$$[NO]^2 = 1.0 \times 10^{-3} = 10 \times 10^{-4}$$

$$[NO] = (10 \times 10^{-4})^{1/2} = 3.2 \times 10^{-2} = 0.032 \text{ mol}/\ell$$

The equilibrium concentrations must satisfy the equation for K_c

---

Exercise In a different equilibrium mixture at 2000°C, $[NO] = [N_2] = 0.10$ mol/ℓ. What is $[O_2]$? Answer: 1.0 mol/ℓ.

More often, we use K_c to determine the extent of a particular reaction. Here we are given the *original* concentrations of reactants, let us say N_2 and O_2. We are asked to calculate the *equilibrium* concentrations of both reactants and products (NO). The reasoning here is somewhat more complex than that in Example 13.4. In general, we follow a three-step process:

1. Express the equilibrium concentrations of all species in terms of a single unknown, x.

2. Substitute these concentration terms into the expression for K_c. This gives an algebraic equation which can be solved for x.

3. Having found x, calculate the equilibrium concentrations of all species.
   This approach is illustrated in Example 13.5.

Example 13.5   Consider again the system

$$N_2(g) + O_2(g) \rightleftharpoons 2\ NO(g);\ K_c = 0.10 \text{ at } 2000°C$$

Suppose now that the original concentrations of N_2, O_2, and NO are 0.100 mol/ℓ, 0.100 mol/ℓ, and 0.000 mol/ℓ, in that order. Some of the N_2 and O_2 combine to form NO. At equilibrium, what are the concentrations of NO, N_2, and O_2?

Solution

1. We must first express the equilibrium concentrations of N_2, O_2, and NO in terms of a single unknown, x. We will choose x to be the number of moles per liter of N_2 that react to achieve equilibrium. The balanced equation tells us that 1 mol of O_2 reacts with 1 mol of N_2. Hence, an equal number of moles per liter, x, of O_2 must be consumed. Again, 2 mol of NO are formed for every mole of N_2 that reacts. So, we must form 2x moles per liter of NO. Putting our reasoning in the form of a table:

	Orig. Conc. (mol/ℓ)	Change in Conc. (mol/ℓ)	Equil. Conc. (mol/ℓ)
N_2	0.100	−x	0.100 − x
O_2	0.100	−x	0.100 − x
NO	0.000	+2x	2x

Do you see how the entries in this table are found from the equation for the reaction?

2. We are now ready to substitute into the expression for K_c:

$$K_c = 0.10 = \frac{[NO]^2}{[N_2] \times [O_2]} = \frac{(2x)^2}{(0.100 - x)^2}$$

This is a second order equation in x. Such equations can always be solved by use of the quadratic formula (p. 313). Here, however, we can simplify the arithmetic by noting that the right side of the equation is a perfect square. Taking the square root of both sides we have

$$(0.10)^{1/2} = 0.32 = \frac{2x}{0.100 - x}$$

Solving for x: $$2x = 0.32(0.100 - x) = 0.032 - 0.32x$$

$$2.32x = 0.032;\ x = 0.032/2.32 = 0.014$$

3. Referring back to the equilibrium table:

$$[NO] = 2x = 0.028 \text{ mol/}\ell$$

$$[N_2] = [O_2] = 0.100 - x = 0.100 - 0.014 = 0.086 \text{ mol/}\ell$$

Exercise   Suppose the original concentrations of N_2, O_2, and NO were 0.20 M, 0.30 M, and 0.00 M, in that order. In that case, what would be the algebraic equation obtained in Step 2? Answer: $0.10 = (2x)^2/(0.20 - x)(0.30 - x)$.

The calculations in Example 13.5 suggest three general comments:

1. There are many different ways in which the unknown can be chosen in a problem of this type. It doesn't matter what choice we make, provided we are consistent in relating equilibrium concentrations. In Example 13.5, we could have taken y, the unknown, to be the number of moles per liter of NO formed. In that case [NO], [N_2], and [O_2] would have been y, $0.100 - (y/2)$, and $0.100 - (y/2)$, in that order. With this choice, our equations would have looked slightly different. However, the final answers would have been the same.

2. The arithmetic involved in equilibrium calculations can be relatively simple or somewhat more complex (see Example 13.6). The reasoning involved is the same. It is always helpful to set up an equilibrium table to summarize the analysis of the problem.

3. At 2000°C, appreciable quantities of nitric oxide are formed from nitrogen and oxygen. This explains why NO resulting from high-temperature combustion processes such as in an automobile engine is a serious air pollutant.

Example 13.6   Consider the system

$$N_2O_4(g) \rightleftharpoons 2\ NO_2(g);\ K_c = 0.36 \text{ at } 100°C$$

Suppose we start with pure N_2O_4 at a concentration of 0.100 mol/ℓ. What are the equilibrium concentrations of NO_2 and N_2O_4?

Solution
1. Some of the N_2O_4 will decompose to achieve equilibrium. Let x be the number of moles per liter of N_2O_4 that decomposes. The balanced equation tells us that two moles of NO_2 are formed for every mole of N_2O_4 that decomposes. Hence, if the concentration of N_2O_4 decreases by x, that of NO_2 must increase by 2x. Setting up an equilibrium table:

	Orig. Conc. (mol/ℓ)	Change in Conc. (mol/ℓ)	Equil. Conc. (mol/ℓ)
N_2O_4	0.100	$-x$	$0.100 - x$
NO_2	0.000	$+2x$	$2x$

2. $K_c = 0.36 = \dfrac{(2x)^2}{0.100 - x} = \dfrac{4x^2}{0.100 - x}$

This time, we cannot solve for x as simply as in Example 13.5. The denominator of the right side is not a perfect square. Instead, we use the general method of solving a quadratic equation. This involves rearranging to the form

$$ax^2 + bx + c = 0$$

and applying the quadratic formula

$$x = \frac{-b \pm \sqrt{b^2 - 4ac}}{2a}$$

To convert our expression to the desired form, we proceed as follows:

$$4x^2 = 0.36(0.100 - x) = 0.036 - 0.36x$$

$$4x^2 + 0.36x - 0.036 = 0$$

Another way to solve is by trial and error, using your calculator. Try it

We can simplify a bit by dividing both sides of the equation by 4:

$$x^2 + 0.090x - 0.0090 = 0$$

Thus, a = 1, b = 0.090, and c = −0.0090.

$$x = \frac{-0.090 \pm \sqrt{(0.090)^2 + 4(0.0090)}}{2} = \frac{-0.090 \pm \sqrt{0.0441}}{2}$$

$$= \frac{-0.090 \pm 0.21}{2} = \frac{0.12}{2} \text{ or } \frac{-0.30}{2} \text{ (i.e., 0.060 or } -0.15)$$

Of the two answers, only 0.060 is plausible. A value of −0.15 for x would give a negative concentration for NO_2, which is impossible.

3. $NO_2 = 2x = 2(0.060 \text{ mol}/\ell) = 0.120 \text{ mol}/\ell$

$N_2O_4 = 0.100 - x = 0.040 \text{ mol}/\ell$

Compare these values with those listed in Experiment 1, Table 13.2. The calculations we have just carried out refer to that experiment.

Exercise   Suppose the original concentrations of N_2O_4 and NO_2 are 0.000 and 0.100 mol/ℓ, in that order. Taking x to be the equilibrium concentration of N_2O_4, what would be the algebraic equation obtained in Step 2? Answer: $0.36 = (0.100 - 2x)^2/x$. (The solution to this equation is given in Table 13.2, under Expt. 2.)

13.6   EFFECT OF CHANGES IN CONDITIONS UPON AN EQUILIBRIUM SYSTEM

Once a system has attained equilibrium, it is possible to change the ratio of products to reactants by changing the external conditions. We will consider three ways in which a chemical equilibrium can be disturbed:
1. Adding (or removing) a reactant or product.
2. Changing the volume of the system.
3. Changing the temperature.

We can deduce the direction in which an equilibrium will shift when one of these changes is made by applying Le Châtelier's Principle. This states that:

There is a natural tendency, in people and reactions, to resist change

If a system at equilibrium is disturbed by some change, the system will shift so as to partially counteract the effect of the change.

We can determine the extent of the shift by working with the equilibrium constant, K_c. In the discussion that follows, we will illustrate both of these approaches.

Adding or Removing a Species

According to Le Châtelier's Principle, **if we disturb a chemical system at equilibrium by adding a species (reactant or product), the reaction will proceed in such a direction as to consume part of the added species. Conversely, if we remove a species, the system will shift so as to restore part of that species.**

Let us apply this general rule to the equilibrium system

$$N_2O_4(g) \rightleftharpoons 2 \ NO_2(g)$$

Suppose this system has reached equilibrium at a certain temperature. We might disturb the equilibrium by:

—*adding* N_2O_4. If we do this, reaction will occur in the forward direction (left to right). In this way, part of the N_2O_4 will be consumed.

—*adding* NO_2, which causes the reverse reaction (right to left) to occur, using up part of the NO_2 added.

—*removing* N_2O_4. Here, reaction occurs in the reverse direction to restore part of the N_2O_4.

—*removing* NO_2, which causes the forward reaction to occur, restoring part of the NO_2 removed.

It is possible to use K_c to calculate the extent to which reaction occurs when an equilibrium is disturbed by adding or removing a product or reactant. To show how this is done, we consider the effect of adding hydrogen iodide to the HI–H_2–I_2 system (Example 13.7).

$$2\ HI(g) \rightleftharpoons H_2(g) + I_2(g); \qquad K_c = \frac{[H_2] \times [I_2]}{[HI]} = 0.016 \text{ at } 520°C$$

Example 13.7   In Example 13.1, we saw that this system is at equilibrium at 520°C when $[HI] = 0.080$ mol/ℓ, and $[H_2] = [I_2] = 0.010$ mol/ℓ. Suppose that to this mixture we add enough HI to raise its concentration temporarily to 0.096 mol/ℓ. When equilibrium is restored, what will $[HI]$, $[H_2]$, and $[I_2]$ be?

Solution

1. Here, the "original concentrations" are those that prevail immediately after the equilibrium is disturbed:

orig. conc. HI = 0.096 mol/ℓ; orig. conc. H_2 = orig. conc. I_2 = 0.010 mol/ℓ

Since we added HI, some of it must decompose to bring the system back to equilibrium. When that happens, some H_2 and some I_2 are formed. Let x be the number of moles per liter of H_2 that is produced by this process. Since H_2 and I_2 are formed in a 1:1 mole ratio, the concentration of I_2 must also increase by x. Again, since two moles of HI are required to form one mole of H_2, the concentration of HI must decrease by 2x. The equilibrium table is:

	Orig. Conc. (mol/ℓ)	Change in Conc. (mol/ℓ)	Equil. Conc. (mol/ℓ)
H_2	0.010	+x	0.010 + x
I_2	0.010	+x	0.010 + x
HI	0.096	−2x	0.096 − 2x

2. Substituting into the expression for K_c:

$$0.016 = \frac{(0.010 + x)(0.010 + x)}{(0.096 - 2x)^2}$$

To solve this equation, we first take the square root of both sides. Since $(0.016)^{1/2} = 0.13$, we have

$$0.13 = \frac{0.010 + x}{0.096 - 2x}$$

which solves to give x = 0.002.

3. Thus,

$$[H_2] = 0.010 + 0.002 = 0.012 \text{ mol}/\ell$$
$$[I_2] = 0.010 + 0.002 = 0.012 \text{ mol}/\ell$$
$$[HI] = 0.096 - 0.004 = 0.092 \text{ mol}/\ell$$

Note that the term on the right is a perfect square

Notice that, in Example 13.7,

1. Reaction occurs in the forward direction, as Le Châtelier's Principle predicts. Some HI (0.004 mol/ℓ) is converted to H_2 and I_2.

2. Only a portion of the added HI reacts. Its final concentration, 0.092 mol/ℓ, is intermediate between the original value, 0.080 mol/ℓ, and that immediately after equilibrium is disturbed, 0.096 mol/ℓ.

Changes in Volume

To understand how a change in container volume can change the position of an equilibrium, consider again the N_2O_4–NO_2 system

$$N_2O_4(g) \rightleftharpoons 2 \, NO_2(g)$$

Suppose we decrease the volume of this system, as shown in the center drawing of Figure 13.2. The immediate effect is to increase the number of molecules per unit volume. According to Le Châtelier's Principle, the system will shift so as to partially counteract this change. There is a simple way that this can happen. Some of the NO_2 molecules combine with each other to form N_2O_4 (diagram at right of Fig. 13.2). That is, reaction occurs in the reverse direction. Since two molecules of NO_2 form only one molecule of N_2O_4, this reduces the number of molecules.

It is possible, using the value of K_c, to calculate the extent to which NO_2 is converted to N_2O_4 when the volume is decreased. The results of such calculations are given in Table 13.3. As the volume is decreased from 10.0 ℓ to 1.0 ℓ, more and more of the NO_2 is converted to N_2O_4. Notice that the total number of moles of gas decreases steadily as a result of this conversion.

N$_2$O$_4$

NO$_2$

Original equilibrium

Equilibrium disturbed

Equilibrium re-established

Figure 13.2 Effect of a decrease in volume upon the $N_2O_4(g) \rightleftharpoons 2\,NO_2(g)$ system. The immediate effect *(middle cylinder)* is to crowd the same number of moles of gas into a smaller volume and so increase the total pressure. This is partially compensated for by the conversion of some of the NO_2 to N_2O_4, thereby reducing the total number of moles of gas present.

TABLE 13.3   EFFECT OF CHANGE IN VOLUME ON THE EQUILIBRIUM
SYSTEM: $N_2O_4(g) \rightleftharpoons 2 NO_2(g)$; $K_c = 0.36$ at 100°C

V (ℓ)	n NO_2	n N_2O_4	n TOTAL
10.0	1.20	0.40	1.60
5.0	0.96	0.52	1.48
2.0	0.68	0.66	1.34
1.0	0.52	0.74	1.26

The analysis we have gone through for the N_2O_4–NO_2 system can be applied to any chemical equilibrium involving gases. **When the volume of an equilibrium system is decreased, reaction takes place in the direction that decreases the total number of moles of gas. When the volume is increased, the reaction which increases the total number of moles of gas takes place.**

The application of this principle to several different systems is shown in Table 13.4. In System 2, the number of moles of gas decreases, from $3/2$ to 1, as the reaction goes to the right. Hence, decreasing the volume causes the forward reaction to occur; an increase in volume has the reverse effect. Notice that it is the change in the number of moles of *gas* which determines which way the equilibrium shifts (System 4). When there is no change in the number of moles of gas (System 5), a change in volume has no effect on the position of the equilibrium.

Changes in volume of gaseous systems at equilibrium ordinarily result in changes in pressure. For example, if we cut the volume of the N_2O_4–NO_2 system in half, we momentarily increase the pressure by a factor of two. Instead of saying that the shift in the position of the equilibrium comes about because of a change in volume, we might ascribe the shift to the pressure change that accompanies the volume decrease. Specifically, we could say that an *increase in pressure shifts the position of the equilibrium in such a way as to decrease the number of moles of gas* ($2 NO_2 \rightarrow N_2O_4$).

In discussing the effect of pressure on an equilibrium system, we must be careful to specify how the pressure change comes about. There are many different ways in which we could change the total pressure in the N_2O_4–NO_2 system without changing the volume. We might, for example, add helium at constant volume. This increases the total number of moles and hence the total pressure. However, it has no effect upon the position of the equilibrium. In general, we do not shift the position of an equilibrium by adding an unreactive gas.

Changes in Temperature

Let us suppose that we increase the temperature of an equilibrium mixture of N_2O_4 and NO_2. According to Le Châtelier's Principle, the system will shift so as to partially

TABLE 13.4   EFFECT OF A CHANGE IN VOLUME UPON THE
POSITION OF GASEOUS EQUILIBRIA

SYSTEM	V INCREASES	V DECREASES
1. $N_2O_4(g) \rightleftharpoons 2 NO_2(g)$	→	←
2. $SO_2(g) + ½ O_2(g) \rightleftharpoons SO_3(g)$	←	→
3. $N_2(g) + 3 H_2(g) \rightleftharpoons 2 NH_3(g)$	←	→
4. $C(s) + H_2O(g) \rightleftharpoons CO(g) + H_2(g)$	→	←
5. $N_2(g) + O_2(g) \rightleftharpoons 2 NO(g)$	0	0

counteract the temperature increase. This can be achieved if the forward reaction occurs:

$$N_2O_4(g) \rightarrow 2\ NO_2(g); \Delta H = +58.2\ kJ$$

Since the reaction is endothermic, it absorbs some of the heat used to raise the temperature. The result is to increase the concentration of NO_2 at the expense of N_2O_4. This effect is shown in color plate 12 (center of book). Note that the reddish-brown color of NO_2 is much deeper at 50°C than at 0°C.

In general, we can say that, for a system at equilibrium,

—**an increase in temperature causes the endothermic reaction to occur;** e.g.,

$$N_2O_4(g) \rightarrow 2\ NO_2(g); \Delta H = +58.2\ kJ$$

—**a decrease in temperature causes the exothermic reaction to occur;** e.g.,

$$2\ NO_2(g) \rightarrow N_2O_4(g); \Delta H = -58.2\ kJ$$

The system shifts as T changes because K_c changes with T

You will recall from our earlier discussion that the equilibrium constant of a system changes with temperature. The effect of a change in temperature upon an equilibrium system is often expressed in terms of its effect upon K_c.

If the forward reaction (left to right) is endothermic, K_c becomes larger as the temperature increases. If the forward reaction is exothermic, K_c becomes smaller as the temperature increases.

The effect of temperature upon the magnitude of K_c is often very large. For the synthesis of ammonia from the elements, K_c changes by a factor of 50,000 when the temperature is raised from 200 to 600°C (Table 13.5).

Example 13.8 Consider the system $I_2(g) \rightleftharpoons 2\ I(g)$; $\Delta H = +151\ kJ$. Suppose the system is at equilibrium at 1000°C. In which direction will reaction occur if
  a. I atoms are added?
  b. the system is compressed?
  c. the temperature is increased?

Solution
  a. $2\ I(g) \rightarrow I_2(g)$
  b. $2\ I(g) \rightarrow I_2(g)$ This decreases the number of particles per unit volume.
  c. $I_2(g) \rightarrow 2\ I(g)$ This reaction absorbs heat.

Exercise What effect, if any, will each of these changes (a, b, c) have on the magnitude of K_c? Answer: Only (c) changes K_c, making it larger.

TABLE 13.5   EFFECT OF TEMPERATURE UPON K_c

1. $N_2O_4(g) \rightleftharpoons 2\ NO_2(g)$; $\Delta H = +58.2\ kJ$

t (°C)	0	50	100
K_c	0.0005	0.022	0.36

2. $N_2(g) + 3\ H_2(g) \rightleftharpoons 2\ NH_3(g)$; $\Delta H = -92.4\ kJ$

t (°C)	200	400	600
K_c	650	0.50	0.014

SUMMARY

When a reaction system is confined, it reaches a dynamic equilibrium where forward and reverse reactions occur at the same rate. From that point on, there is no net change in the concentrations of products or reactants. The equilibrium system can be described in terms of a ratio known as the equilibrium constant, K_c. The expression for K_c follows directly from the equation written to represent the equilibrium system (Section 13.2; Example 13.3). The magnitude of K_c governs the extent to which forward and reverse reactions will take place. If K_c is very large, the forward reaction goes nearly to completion while the reverse reaction does not occur to an appreciable extent. If K_c is very small, the opposite is true; that is, the reverse reaction dominates. Finally, if K_c is neither very large nor very small, there will be appreciable quantities of both products and reactants at equilibrium.

Several calculations involving equilibrium systems were considered in this chapter. Generally, they are of three types, depending upon what is known and what is to be calculated.

Known	To be calculated	Example no.
1. Equilibrium concentrations of all species	K_c	13.2
2. K_c and all but one equilibrium concentration	Equilibrium concentration of a species	13.4
3. K_c and initial concentrations	Equilibrium concentrations of all species	13.5 and 13.6

If a system at equilibrium is disturbed in some way, either the forward or the reverse reaction will occur to restore equilibrium. We can use Le Châtelier's Principle (p. 314) to predict which way the equilibrium will shift under stress (Example 13.8). In general, we find that:

—a decrease in volume (compression) causes the reaction to occur which decreases the number of moles of gas (Tables 13.3, 13.4).

—an increase in temperature causes the reaction which is endothermic to occur; in this case, the magnitude of K_c changes (Table 13.5).

The equilibrium constant expression can be used in the usual way to determine the extent to which a system shifts when a substance is added or removed (Example 13.7).

KEY WORDS AND CONCEPTS

mole
[ ] equilibrium concentration (mol/ℓ)

Law of Chemical Equilibrium
equilibrium constant, K_c

quadratic formula
Le Châtelier's Principle

QUESTIONS AND PROBLEMS

Catalog

Expression for K_c: 13.2, 13.3, 13.22, 13.23
K_c from Equilibrium Concentrations: 13.4–13.6, 13.24–13.26
Relation between Equilibrium Concentrations: 13.7, 13.8, 13.27, 13.28

Equilibrium from Original Concentrations: 13.9–13.14, 13.29–13.34
Magnitude of K_c: 13.15, 13.16, 13.35, 13.36
Shifts in Equilibria (Qualitative): 13.17–13.19, 13.37–13.39
General: 13.1, 13.20, 13.21, 13.40–13.45

13.1   Review and know the meanings of the key words and concepts in this chapter.

13.2   Write equilibrium constant expressions for each of the following systems:

  a. $SO_2(g) + \frac{1}{2} O_2(g) \rightleftharpoons SO_3(g)$
  b. $3 Fe(s) + 4 H_2O(g) \rightleftharpoons Fe_3O_4(s) + 4 H_2(g)$
  c. $2 N_2O_5(g) \rightleftharpoons 4 NO_2(g) + O_2(g)$
  d. $H_2(g) + CO_2(g) \rightleftharpoons CO(g) + H_2O(l)$

13.3   Write an equation for an equilibrium system that would lead to each of the following expressions for K_c:

  a. $\dfrac{[H_2]^2 \times [O_2]}{[H_2O]^2}$ b. $\dfrac{[H_2] \times [O_2]^{1/2}}{[H_2O]}$

  c. $\dfrac{[H_2O]^2}{[H_2]^2 \times [O_2]}$ d. $[H_2]^2 \times [O_2]$

13.4   Calculate K_c at 1500°C for

$$2 NO(g) \rightleftharpoons N_2(g) + O_2(g)$$

given that the equilibrium concentrations of N_2, O_2, and NO are 0.050 M, 0.050 M, and 0.00055 M, in that order.

13.5   For the reaction in Problem 13.4 at a different temperature, if one starts with pure NO at 4.00 mol/ℓ, the equilibrium concentrations of N_2 and O_2 are each 1.82 mol/ℓ. Calculate K_c at this temperature.

13.6   When 0.250 mol of CO_2 is placed in a one-liter container, some of it decomposes by the reaction

$$2 CO_2(g) \rightleftharpoons 2 CO(g) + O_2(g)$$

The equilibrium concentration of CO is 0.040 mol/ℓ. Calculate

  a. $[O_2]$ b. $[CO_2]$ c. K_c

13.7   At 800 K, K_c is 0.016 for

$$2 HI(g) \rightleftharpoons H_2(g) + I_2(g)$$

In an equilibrium mixture at 800 K, $[H_2] = [I_2] = 0.030$ mol/ℓ. What is the equilibrium concentration of HI?

13.8   For the system

$$PCl_5(g) \rightleftharpoons PCl_3(g) + Cl_2(g)$$

$K_c = 0.050$ at 250°C. In a certain equilibrium mixture at 250°C, $[PCl_3] = 2 \times [PCl_5]$. What is the equilibrium concentration of Cl_2?

13.22   Write expressions for K_c for each of the following systems:

  a. $CS_2(g) + 3 Cl_2(g) \rightleftharpoons S_2Cl_2(g) + CCl_4(g)$
  b. $2 NOCl(g) \rightleftharpoons 2 NO(g) + Cl_2(g)$
  c. $Ni(s) + 4 CO(g) \rightleftharpoons Ni(CO)_4(g)$
  d. $SiCl_4(l) + 2 H_2O(g) \rightleftharpoons SiO_2(s) + 4 HCl(g)$

13.23   Write an equation for an equilibrium system that would lead to each of the following expressions for K_c:

  a. $\dfrac{[NO]}{[N_2]^{1/2} \times [O_2]^{1/2}}$ b. $\dfrac{[HI]^2}{[H_2] \times [I_2]}$

  c. $\dfrac{[HI]^2}{[H_2]}$ d. $\dfrac{[H_2O]}{[H_2]}$

13.24   For the system

$$2 NO(g) + Br_2(g) \rightleftharpoons 2 NOBr(g)$$

analysis shows that at equilibrium at 350°C, the concentrations of NO, Br_2, and NOBr are 0.30 M, 0.11 M, and 0.046 M, in that order. Calculate K_c.

13.25   The equilibrium in Problem 13.24 is studied at a different temperature. Starting with NO and Br_2 each at 0.50 mol/ℓ, the equilibrium concentration of NOBr is found to be 0.16 mol/ℓ. Calculate K_c at this temperature.

13.26   Starting with 0.0100 mol/ℓ of NH_3, the reaction

$$2 NH_3(g) \rightleftharpoons N_2(g) + 3 H_2(g)$$

produces 0.0020 mol/ℓ of N_2 at equilibrium. Calculate

  a. $[H_2]$ b. $[NH_3]$ c. K_c

13.27 K_c is 55 at 1500°C for

$$2 NOCl(g) \rightleftharpoons 2 NO(g) + Cl_2(g)$$

What is the concentration of NOCl at equilibrium if those of NO and Cl_2 are 0.040 M and 0.020 M, in that order?

13.28   For the system

$$2 HI(g) \rightleftharpoons H_2(g) + I_2(g)$$

$K_c = 0.016$ at 800 K. If, at 800 K, $[HI] = 0.10$ M and $[H_2] = [I_2]$, calculate the equilibrium concentration of H_2.

13.9 K_c for the reaction

$$2 \; ICl(g) \rightleftharpoons I_2(g) + Cl_2(g)$$

is 0.11 at a certain temperature. Suppose the initial concentrations (mol/ℓ) of ICl, I_2, and Cl_2 are 0.10, 0.00, and 0.00, respectively. Some of the ICl decomposes and the system reaches equilibrium. What is the equilibrium concentration of each species?

13.10 The reaction

$$CO(g) + H_2O(g) \rightleftharpoons CO_2(g) + H_2(g)$$

has a K_c value of 4.0 at a certain temperature. Calculate the concentration of all species at equilibrium starting with

    a. conc. CO = conc. H_2O = 0.20 mol/ℓ.
    b. conc. CO = conc. H_2O = conc. CO_2 = conc. H_2 = 0.020 mol/ℓ.

13.11 For the system

$$PCl_5(g) \rightleftharpoons PCl_3(g) + Cl_2(g)$$

K_c = 0.050 at 250°C. If 0.40 mol of PCl_5 is placed in a 1.0-ℓ container at this temperature, calculate the equilibrium concentrations of all species.

13.12 Consider the data in Table 13.3. Show by calculation that, starting with 1.00 mol N_2O_4 in a 5.0-ℓ container, the number of moles of NO_2 at equilibrium is indeed 0.96.

13.13 When chlorine gas is heated, it decomposes:

$$Cl_2(g) \rightleftharpoons 2 \; Cl(g)$$

K_c for this reaction is 1.2×10^{-6} at 1000°C and 3.6×10^{-2} at 2000°C. Starting with a concentration of Cl_2 of 0.20 M, what will be the equilibrium concentration of Cl at

    a. 2000°C?
    b. 1000°C? Note that K_c at 1000°C is so small that $[Cl_2]$ will be very nearly equal to its original concentration.

13.14 For the system

$$SnO_2(s) + 2 \; H_2(g) \rightleftharpoons Sn(s) + 2 \; H_2O(g)$$

at 500°C, $[H_2O] = [H_2]$ = 0.10 M.

    a. Calculate K_c at this temperature.
    b. Enough H_2 is added to raise its concentration temporarily to 0.20 M. When equilibrium is restored, what are the concentrations of H_2O and H_2?

13.29 For the equilibrium in Problem 13.28, 1.00 mol of HI is placed in a 10.0-ℓ flask at 800 K. What are the equilibrium concentrations of H_2, I_2, and HI?

13.30 For the reaction

$$2 \; IBr(g) \rightleftharpoons I_2(g) + Br_2(g)$$

K_c is 2.5×10^{-3} at a certain temperature. Calculate the equilibrium concentration of each species in a 4.0-ℓ vessel starting with

    a. 0.40 mol IBr.
    b. 0.20 mol I_2, 0.20 mol Br_2.
    c. 0.20 mol I_2, 0.20 mol Br_2, 0.20 mol IBr.

13.31 For the system

$$CO(g) + Cl_2(g) \rightleftharpoons COCl_2(g)$$

K_c = 3.0. 2.0 mol of CO and 1.0 mol of Cl_2 are put in a 5.0-ℓ container. What are the equilibrium concentrations of all species?

13.32 Proceeding as in Problem 13.12, check by calculation the values given for moles of NO_2 and N_2O_4 in 2.0 ℓ (Table 13.3). Start with 1.00 mol of N_2O_4.

13.33 For the system

$$N_2(g) + O_2(g) \rightleftharpoons 2 \; NO(g)$$

K_c = 1×10^{-30} at 25°C and 0.10 at 2000°C. Starting with 0.040 mol/ℓ of N_2 and 0.010 mol/ℓ of O_2, calculate the equilibrium concentration of NO

    a. at 2000°C.
    b. at 25°C, making the simplification indicated in Problem 13.13b.

13.34 For the system

$$CO(g) + H_2O(g) \rightleftharpoons CO_2(g) + H_2(g)$$

equilibrium is established at a certain temperature when the concentrations of CO, H_2O, CO_2, and H_2 are 1.0×10^{-2}, 2.0×10^{-2}, 1.2×10^{-2}, and 1.2×10^{-2} M, respectively.

    a. Calculate K_c.
    b. If enough CO is added to raise its concentration temporarily to 2.0×10^{-2} M, what will be the final concentrations of all species?

13.15 For the reaction

$$Br_2(g) \rightleftharpoons 2\ Br(g)$$

$K_c = 4 \times 10^{-18}$ at 200°C. If one starts with 1.0 mol of Br_2 in a 1.0-ℓ container at 25°C, will the equilibrium system contain:

      a. mostly Br_2?   b. mostly Br atoms?
      c. about equal amounts of Br_2 and Br?

13.16 Suppose you want to remove water vapor from a container. Using the information below, suggest the best reagent for this purpose.

	K_c
$Cu(s) + H_2O(g) \rightleftharpoons CuO(s) + H_2(g)$	2×10^{-18}
$CO(g) + H_2O(g) \rightleftharpoons CO_2(g) + H_2(g)$	1×10^2
$CO(g) + H_2(g) \rightleftharpoons C(s) + H_2O(g)$	2×10^{17}
$2\ H^+(aq) + SO_4^{2-}(aq) \rightleftharpoons SO_3(g) + H_2O(g)$	1×10^{-28}

13.17 For the equilibrium

$$SO_2(g) + \tfrac{1}{2}\ O_2(g) \rightleftharpoons SO_3(g)$$

$\Delta H = -99.1$ kJ. Predict whether the forward or reverse reaction will occur when the equilibrium is disturbed by

      a. expanding the container.
      b. removing SO_3.
      c. raising the temperature.
      d. adding O_2.

13.18 For the system

$$H_2(g) + I_2(g) \rightleftharpoons 2\ HI(g)$$

K_c is 64 at 800 K and 80 at 700 K. Is the forward reaction exothermic or endothermic? Explain.

13.19 The system

$$A(g) + 2\ B(g) \rightleftharpoons C(g)$$

is at equilibrium when the concentration of A is 1.0 M. Enough A is added to raise its concentration temporarily to 1.4 M. The final equilibrium concentration of A could be

      a. 0.8 M   b. 1.0 M   c. 1.3 M
      d. 1.4 M   e. 1.6 M

13.35 At 1500°C, K_c for the reaction in Problem 13.15 is 3×10^{-3}. At this temperature, will the equilibrium system contain mostly Br_2, mostly Br, or appreciable amounts of both species?

13.36 Given the following data at 25°C:

	K_c
$N_2(g) + O_2(g) \rightleftharpoons 2\ NO(g)$	1×10^{-30}
$2\ H_2(g) + O_2(g) \rightleftharpoons 2\ H_2O(g)$	2×10^{81}
$2\ CO_2(g) \rightleftharpoons 2\ CO(g) + O_2(g)$	4×10^{-92}

Which of the three compounds, H_2O, CO_2, or NO, would be most likely to dissociate to give O_2 at 25°C?

13.37 For the system

$$Xe(g) + 2\ F_2(g) \rightleftharpoons XeF_4(g)$$

$\Delta H = -218$ kJ. Predict what effect each of the following changes would have upon the per cent conversion of Xe to XeF_4 at equilibrium:

      a. increasing container size.
      b. adding F_2.
      c. lowering the temperature.
      d. compressing the system.

13.38 K_c is 3×10^{-3} at 900 K and 0.2 at 1200 K for

$$H_2O(g) + C(s) \rightleftharpoons CO(g) + H_2(g)$$

Is the forward reaction exothermic or endothermic? Explain.

13.39 Based on your answer to Problem 13.19, calculate the change in concentration of both B and C produced by adding A.

13.20 On an exam, students are told that for the system

$$2 NH_3(g) \rightleftharpoons N_2(g) + 3 H_2(g)$$

1.00 mol of NH_3, 0.50 mol of N_2, and 1.20 mol of H_2 are present at equilibrium in a 10.0-ℓ container. They are asked to calculate K_c. Three students give the following answers, all of which are wrong. Can you deduce the error(s) in their reasoning?

a. $K_c = \dfrac{(1.20)^3(0.50)}{(1.00)^2} = 0.86$

b. $K_c = \dfrac{(0.360)^3(0.050)}{(0.200)^2} = 0.058$

c. $K_c = \dfrac{(0.100)^2}{(0.050)(0.120)^3} = 120$

13.21 Consider the equilibrium in Problem 13.20. It is found in a particular system that $[N_2] = [H_2] = [NH_3] = 1.0$ M. Which of the following are possible values for the *original* concentrations?

	Orig. conc. NH_3	Orig. conc. N_2	Orig. conc. H_2
a.	1.0	0.0	0.0
b.	2.0	0.0	3.0
c.	1.2	0.9	0.7
d.	0.0	1.0	3.0
e.	0.0	1.5	2.5

13.40 In another question dealing with the equilibrium in Problem 13.20, students are asked to set up an expression to solve for $[N_2]$, starting with 1.00 mol of NH_3 in a 1.00-ℓ container. Again, analyze the following incorrect set-ups to find the error in each ($x = [N_2]$).

a. $K_c = \dfrac{(x)(x)^3}{(1 - 2x)^2}$

b. $K_c = \dfrac{(x)(x)^3}{1}$

c. $K_c = \dfrac{(x)(3x)^3}{(1 - x)^2}$

d. $K_c = \dfrac{(x)(x)}{1 - x}$

13.41 Consider again the equilibrium

$$2 NH_3(g) \rightleftharpoons N_2(g) + 3 H_2(g)$$

Suppose the original concentrations of NH_3, N_2, and H_2 are all 1.0 M. Which of the following are possible values for the concentrations at equilibrium?

	$[NH_3]$	$[N_2]$	$[H_2]$
a.	0.8	1.1	1.3
b.	1.2	0.9	0.7
c.	1.2	1.2	1.2
d.	1.2	0.8	0.8
e.	0.6	1.2	1.6

*13.42 For the system $SO_3(g) \rightleftharpoons SO_2(g) + {}^1/_2 O_2(g)$ at 1000 K, $K_c = 0.050$. Sulfur trioxide, originally at 1.00 atm pressure, partially dissociates to SO_2 and O_2 at 1000 K. What is the original concentration of SO_3 in moles per liter? What is its final concentration?

*13.43 At a certain temperature the reaction $Xe(g) + 2 F_2(g) \rightleftharpoons XeF_4(g)$ gives a 60% yield of XeF_4, starting with 0.40 mol Xe and 0.80 mol F_2 in a 2.0-ℓ vessel. Calculate K_c at this temperature. How many additional moles of F_2 must be added to increase the yield of XeF_4 from Xe to 75%?

*13.44 A student studying the equilibrium $I_2(g) \rightleftharpoons 2 I(g)$, at a high temperature puts 0.10 mol of I_2 in a 1.0-ℓ container. He finds that the total pressure at equilibrium is 50% greater than it was originally. What is K_c for this reaction?

*13.45 The relation between K_c and temperature is approximately

$$\log_{10} \frac{K_c \text{ at } T_2}{K_c \text{ at } T_1} = \frac{\Delta H(T_2 - T_1)}{2.30 R T_2 T_1}$$

where ΔH is the enthalpy change for the forward reaction in joules, T_2 and T_1 are the absolute temperatures, and R is the gas law constant, 8.31 J/(mol·K). For the system $2 HI(g) \rightleftharpoons H_2(g) + I_2(g)$, $K_c = 0.016$ at 520°C and $\Delta H = +10.4$ kJ. Calculate K_c at 600°C.

RATES OF REACTION _____

We saw in Chapter 13 that reactions for which the equilibrium constant is large tend to go virtually to completion. Consider, for example,

$$CO(g) + NO(g) \rightarrow CO_2(g) + \frac{1}{2} N_2(g) \tag{14.1}$$

At 25°C, K_c for this reaction is a huge number, about 10^{60}. On this basis, it would seem that Reaction 14.1 offers a way to remove the toxic gases CO and NO from automobile exhausts. Unfortunately, there is a practical difficulty. Under ordinary conditions, Reaction 14.1 occurs at what amounts to an extremely slow rate. We would have to wait essentially forever for CO and NO to be converted completely to CO_2 and N_2.

The situation just described is quite common. Many reactions which should in principle go to completion occur very, very slowly. Often, this is to our benefit. A case in point involves the fossil fuels—coal, petroleum, and natural gas. From equilibrium considerations, these fuels should be converted to CO_2 and H_2O upon exposure to air.

Actually, so should we mortals

We conclude that there is no correlation between the rate of a reaction and its equilibrium constant. To predict how rapidly a reaction will occur, we must become familiar with the principles of *chemical kinetics*. These principles are the subject of this chapter. Here, we will apply them mainly to reactions involving gases. All the reactions we will consider have very large equilibrium constants. Hence, we need only be concerned with the rate of the forward reaction and can ignore the reverse reaction.

14.1 MEANING OF REACTION RATE

To discuss reaction rate in a meaningful way, we must first define precisely what this term means. *The rate of reaction is a positive quantity that tells us how the concentration of a reactant or product changes with time.* Most often, reaction rate is expressed in terms of reactant concentrations. Consider, for example, the decomposition of dinitrogen pentoxide, N_2O_5:

$$2 N_2O_5(g) \rightarrow 4 NO_2(g) + O_2(g) \tag{14.2}$$

We ordinarily take the rate of this reaction to be

$$rate = \frac{-\Delta \text{ conc. } N_2O_5}{\Delta t} \tag{14.3}$$

The minus sign here is required to make the rate a positive quantity; the concentration of N_2O_5 decreases with time. Let us suppose that the concentration of N_2O_5 changes from 0.16 to 0.11 mol/ℓ in one minute. Then

$$\Delta \text{ conc. } N_2O_5 = 0.11 \text{ mol}/\ell - 0.16 \text{ mol}/\ell = -0.05 \text{ mol}/\ell$$

and, $$\text{rate} = \frac{-\Delta \text{ conc. } N_2O_5}{\Delta t} = \frac{-(-0.05 \text{ mol}/\ell)}{1 \text{ min}} = +0.05 \frac{\text{mol}}{\ell \cdot \text{min}}$$

Reaction rate has the units of concentration divided by time. We will always express concentration in moles per liter. On the other hand, time may be given in seconds, minutes, hours, days, or years. Thus, the units of reaction rate may be

$$\frac{\text{mol}}{\ell \cdot \text{s}}, \quad \frac{\text{mol}}{\ell \cdot \text{min}}, \quad \frac{\text{mol}}{\ell \cdot \text{h}}, \ldots$$

The magnitude of the reaction rate will depend upon the time unit used. The rate with time in hours will be 60 times that with time in minutes. Thus, a rate of 0.05 mol/($\ell \cdot$ min) would be 3 mol/($\ell \cdot$ h):

$$0.05 \frac{\text{mol}}{\ell \cdot \text{min}} \times \frac{60 \text{ min}}{1 \text{ h}} = 3 \frac{\text{mol}}{\ell \cdot \text{h}}$$

Measurement of Rate

Let us consider how the rate of Reaction 14.2 might be measured. We start by putting 0.160 mol of N_2O_5 in a one-liter container at 67°C. At one-minute intervals, we withdraw small samples of the reaction mixture and analyze them for N_2O_5. In this way, we obtain the concentration-time data listed in Table 14.1, p. 326, and plotted in Figure 14.1.

From Table 14.1 or Figure 14.1, it should be clear that the concentration of N_2O_5 decreases rapidly in the early stages of the reaction. Later, it drops more and more slowly. In other words, the rate of reaction is a function of time.

To find the rate of reaction from the concentration-time data, we use Figure 14.1.

Figure 14.1 The rate of reaction can be determined by measuring concentration of reactant as a function of time and graphing the data. To determine the rate at a particular point, we draw a tangent to the curve and find its slope. This gives us the ratio Δconc./Δt, which is numerically equal to the reaction rate.

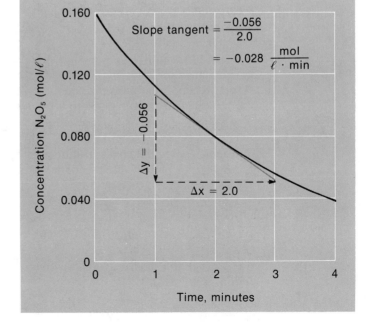

TABLE 14.1 RATE OF DECOMPOSITION OF N_2O_5 AT 67°C					
Time (min)	0	1	2	3	4
Conc. N_2O_5 (mol/ℓ)	0.160	0.113	0.080	0.056	0.040
Rate (mol/($\ell \cdot$ min))	0.056	0.039	0.028	0.020	0.014

If we draw a tangent to the curve at any point, its slope will equal $\dfrac{\Delta \text{ conc. } N_2O_5}{\Delta t}$. From Equation 14.3, we see that the rate of reaction is equal to $\dfrac{-\Delta \text{ conc. } N_2O_5}{\Delta t}$. Putting these relations together:

$$\text{rate} = -(\text{slope of tangent to conc. vs. time curve})$$

From Figure 14.1, we see that the slope of the tangent at t = 2 min is -0.028 mol/($\ell \cdot$min). Hence, the rate at that point is

$$\text{rate} = -\left(-0.028 \, \frac{\text{mol}}{\ell \cdot \text{min}}\right) = 0.028 \, \frac{\text{mol}}{\ell \cdot \text{min}}$$

Rates at other times are obtained in a similar way. These are listed at the bottom of Table 14.1.

14.2 DEPENDENCE OF REACTION RATE UPON CONCENTRATION

In discussing the decomposition of N_2O_5, we pointed out that the rate of reaction decreases with time. From a slightly different point of view, the rate decreases as the concentration of N_2O_5 decreases. This behavior is typical of most reactions. We ordinarily find that reactions proceed more slowly as the concentration of reactant decreases. To increase the rate, we start with a higher concentration of reactant (see color plate 13, center of book).

Note that if rate vs. conc. gives a straight line, it does *not* follow that conc. vs. time gives a straight line (Fig. 14.1)

It turns out that there is a simple relation between concentration and rate for the decomposition of N_2O_5. You may be able to deduce this relation from the data in Table 14.1. Notice what happens when the concentration decreases by a factor of two (from 0.160 to 0.080 or from 0.080 to 0.040 mol/ℓ). The rate is cut in half (from 0.056 to 0.028 or from 0.028 to 0.014 $\dfrac{\text{mol}}{\ell \cdot \text{min}}$). This suggests that the rate of this reaction is directly proportional to the concentration of N_2O_5. Indeed, this is true, as you can see from Figure 14.2. A plot of rate vs. concentration is a straight line. If extrapolated, it would pass through the origin (rate = 0 when conc. N_2O_5 = 0).

Rate Expression and Rate Constant

Since the rate of decomposition of N_2O_5 is directly proportional to its concentration, it follows that

$$\text{rate} = k(\text{conc. } N_2O_5) \tag{14.4}$$

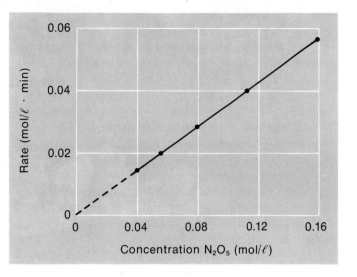

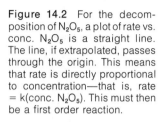

Figure 14.2 For the decomposition of N_2O_5, a plot of rate vs. conc. N_2O_5 is a straight line. The line, if extrapolated, passes through the origin. This means that rate is directly proportional to concentration—that is, rate = k(conc. N_2O_5). This must then be a first order reaction.

Equation 14.4 is referred to as the **rate expression** for the decomposition of N_2O_5. It tells us how the rate of the reaction

$$2\ N_2O_5(g) \rightarrow 4\ NO_2(g) + O_2(g)$$

depends upon the concentration of reactant. The proportionality constant k in Equation 14.4 is called a **rate constant.** It is independent of the other quantities in the equation. However, k does depend upon:

—*the nature of the reaction.* "Fast" reactions typically have large rate constants. If k is very small, reaction occurs slowly at ordinary concentrations.

—*the temperature.* Ordinarily, k increases with temperature.

The value of k in Equation 14.4 can be calculated knowing the rate of reaction at any given concentration of N_2O_5. Solving this equation for k, we have

$$k = \frac{\text{rate}}{\text{conc. } N_2O_5}$$

Note from Table 14.1 that rate = 0.056 mol/(ℓ · min) when conc. N_2O_5 = 0.160 mol/ℓ. Hence,

$$k = \frac{0.056\ \text{mol}/(\ell \cdot \text{min})}{0.160\ \text{mol}/\ell} = 0.35/\text{min}$$

The same result would be obtained, within experimental error, by using any of the other data points in Table 14.1.

Knowing k, we can calculate the reaction rate at any concentration of N_2O_5. Suppose, for example, we wish to know the rate when conc. N_2O_5 = 0.100 mol/ℓ. Substituting in the rate expression, Equation 14.4:

$$\text{rate} = k(\text{conc. } N_2O_5) = \frac{0.35}{\text{min}} \times 0.100\ \frac{\text{mol}}{\ell} = 0.035\ \frac{\text{mol}}{\ell \cdot \text{min}}$$

The rate constant is related to, but not equal to, the rate

Rate expressions have been established by experiment for a large number of reactions. For a process involving a single reactant

$$a\ A(g) \rightarrow products$$

the rate expression has the general form

$$rate = k(conc.\ A)^m \qquad (14.5)$$

The power to which the concentration of reactant A is raised in the rate expression describes the **order** of the reaction. If m in Equation 14.5 is 0, we say that the reaction is "zero order." If m = 1, the reaction is "first order"; if m = 2, it is "second order"; and so on. Note that:

— For a zero order reaction (m = 0), the rate is independent of the concentration of reactant. Doubling the concentration leaves the rate unchanged.

— For a first order reaction (m = 1), the rate is directly proportional to the concentration of reactant. Doubling the concentration increases the rate by a factor of two.

— For a second order reaction (m = 2), the rate is proportional to the *square* of the concentration of reactant. Doubling the concentration increases the rate by a factor of four (Example 14.1).

Example 14.1   Consider the decomposition of acetaldehyde, CH_3CHO:

$$CH_3CHO(g) \rightarrow CH_4(g) + CO(g)$$

Experimentally, this reaction is found to be second order. The rate at a certain temperature is 0.18 mol/($\ell \cdot$ s) when conc. CH_3CHO = 0.10 mol/ℓ.
  a. Write the rate expression for this reaction.
  b. Determine the rate constant, k.
  c. Calculate the rate of reaction when conc. CH_3CHO = 0.20 mol/ℓ.

Solution
  a. rate = $k(conc.\ CH_3CHO)^2$
  b. Solving for k and substituting:

$$k = \frac{rate}{(conc.\ CH_3CHO)^2} = \frac{0.18\ mol/(\ell \cdot s)}{(0.10\ mol/\ell)^2} = 18\ \frac{\ell}{mol \cdot s}$$

  c. rate = $18\ \dfrac{\ell}{mol \cdot s} \times (0.20\ mol/\ell)^2 = 0.72\ \dfrac{mol}{\ell \cdot s}$

Exercise   At what concentration of CH_3CHO is the rate of reaction 0.36 mol/($\ell \cdot$ s)?
Answer: 0.14 mol/ℓ.

Our discussion to this point illustrates the fact that *the order of a reaction must be determined experimentally. It cannot, in general, be deduced from the coefficients of the balanced equation*. The decomposition of acetaldehyde is second order even though the coefficient of CH_3CHO in the balanced equation is 1:

$$CH_3CHO(g) \rightarrow CH_4(g) + CO(g); rate = k(conc.\ CH_3CHO)^2$$

Again, the decomposition of N_2O_5 is first order, even though the coefficient of N_2O_5 in the balanced equation is 2:

$$2 N_2O_5(g) \rightarrow 4 NO_2(g) + O_2(g); \text{ rate} = k(\text{conc. } N_2O_5)^1$$

Many reactions (indeed, most reactions) involve more than one reactant. For a reaction between A and B:

$$a A(g) + b B(g) \rightarrow \text{products}$$

the general form of the rate expression is

$$\text{rate} = k(\text{conc. A})^m (\text{conc. B})^n \qquad (14.6)$$

Here we refer to m as "the order of the reaction with respect to A." Similarly, n is the order of the reaction with respect to B. The **overall order** of the reaction is the sum of the exponents $m + n$. Thus, for the reaction

$$CO(g) + NO_2(g) \rightarrow CO_2(g) + NO(g) \qquad (14.7)$$

the experimentally determined rate expression is

$$\text{rate} = k(\text{conc. CO}) \times (\text{conc. } NO_2)$$

Hence we say that this reaction is
 —first order with respect to CO (m = 1).
 —first order with respect to NO_2 (n = 1).
 —second order overall (m + n = 2).

Frequently it is found that the exponents m and n in Equation 14.6 are integers (0, 1, 2). This is the case for the $CO-NO_2$ reaction and also for the reaction between NO and H_2:

$$2 NO(g) + 2 H_2(g) \rightarrow N_2(g) + 2 H_2O(g); \text{ rate} = k(\text{conc. NO})^2 \times (\text{conc. } H_2)^1$$

Sometimes, however, m or n may be a fraction, as in

$$CHCl_3(g) + Cl_2(g) \rightarrow CCl_4(g) + HCl(g); \text{ rate} = k(\text{conc. } CHCl_3)^1 \times (\text{conc. } Cl_2)^{1/2}$$

14.3   RELATION BETWEEN REACTANT CONCENTRATION AND TIME

A rate expression such as Equation 14.4 tells us how the rate of a reaction changes with concentration. However, from a practical standpoint, we are usually more interested in the relation between concentration and time. Suppose, for example, you are studying the decomposition of N_2O_5. Most likely, you would want to know how much N_2O_5 is left after 5 min, 1 h, or several days. Equation 14.4 is not of much help for that purpose.

It is possible to develop algebraic equations relating reactant concentration to time. The form of the equation may be very simple or quite complex, depending upon the order of the reaction and the number of reactants. Here, we will discuss in detail only one type of concentration-time relation. This is for a first order reaction involving a single reactant:

$$a \; A(g) \rightarrow \text{products; rate} = k(\text{conc. A})$$

First Order Reactions

As we have seen, the reaction

$$2 \; N_2O_5(g) \rightarrow 4 \; NO_2(g) + O_2(g)$$

is first order; rate $= k(\text{conc. } N_2O_5)$.

We would like to obtain an equation relating the concentration of N_2O_5 to time. This is most readily done by using calculus (see small print section, p. 331). The equation obtained is

$$\log_{10} (\text{conc. } N_2O_5) = \log_{10} (\text{conc. } N_2O_5)_0 - kt/2.30$$

Here, k is the rate constant and t is the time. The subscript $_0$ is used to indicate the original concentration of N_2O_5, at t = 0.

Even if you don't know anything about calculus, you can show that this equation is valid by plotting the base 10 logarithm of conc. N_2O_5 vs. time. This is done in Figure 14.3, using the concentration-time data in Table 14.1. As you can see, we obtain a straight line. The general equation of a straight line is

$$y = a + bx$$

In this case, "y" is $\log_{10} (\text{conc. } N_2O_5)$ and "x" is time, t. The y intercept, a, is $\log_{10} (\text{conc. } N_2O_5)_0$. The slope, b, is $-k/2.30$. From Figure 14.3, we see that the slope is $-0.15/\text{min}$, so

$$\frac{-k}{2.30} = \frac{-0.15}{\text{min}}; \; k = \frac{(2.30)(0.15)}{\text{min}} = 0.35/\text{min}$$

This is the same value of k found earlier (p. 327).

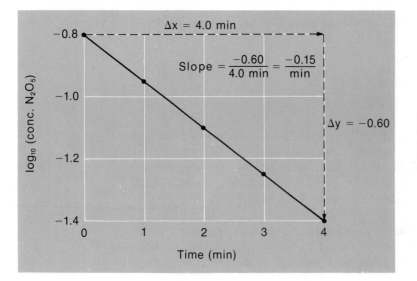

Figure 14.3 For a first order reaction, a plot of log conc. vs. time is a straight line. The rate constant can be determined from the slope of the line, k = −2.30 (slope). In this case, k = −2.30(−0.15/min) = 0.35/min.

The relation between concentration and time that we have "discovered" for the decomposition of N_2O_5 applies to any first order reaction. It is usually written in the form

$$\log_{10} \frac{X_0}{X} = \frac{kt}{2.30} \qquad (14.8)$$

Here, X_0 is the original concentration of reactant, X is the concentration of reactant at time t, and k is the first order rate constant.

To obtain Equation 14.8 by using calculus, we start by writing the relation between concentration of reactant, X, and time, t, in derivative notation:

$$\text{rate} = \frac{-dX}{dt}$$

The calculus allows one to work with systems which are changing

For a first order reaction, rate is directly proportional to concentration (rate $= kX$), so

$$\frac{-dX}{dt} = kX; \frac{dX}{X} = -kdt$$

Integrating from $t = 0$ to t and from X_0 to X (remember, X_0 is the concentration at $t = 0$):

$$\int_{X_0}^{X} \frac{dX}{X} = -k \int_0^t dt$$

$$\ln X - \ln X_0 = -k(t - 0) = -kt$$

where "ln" stands for a natural logarithm (base e). To convert natural to base 10 logarithms, we divide by 2.30 and obtain

$$\log_{10} X - \log_{10} X_0 = \frac{-kt}{2.30}$$

and, finally,

$$\log_{10} \frac{X_0}{X} = \frac{kt}{2.30}$$

which is the first order rate law.

Equation 14.8 is a very useful one. For first order reactions, we can use it to calculate:

—the concentration of reactant remaining after a given time (Example 14.2a).

—the time required for reactant concentration to drop to a certain level (Example 14.2b and c).

Example 14.2   The decomposition of hydrogen peroxide to water and oxygen

$$2\ H_2O_2(l) \rightarrow 2\ H_2O(l) + O_2(g)$$

is a first order reaction with a rate constant of 0.0410/min.

a. If we start with a 0.500 M H_2O_2 solution, what will be its concentration after 10.0 min?

b. How long will it take for the concentration of H_2O_2 to drop from 0.500 M to 0.100 M?
c. How long will it take for one half of a sample of H_2O_2 to decompose?

Solution

a. Substituting numbers into Equation 14.8, letting X = conc. H_2O_2:

$$\log_{10} \frac{0.500}{X} = \frac{0.0410 \times 10.0}{2.30} = 0.178$$

The antilog of 0.178 is 1.51, so $\frac{0.500}{X} = 1.51$; solving, X = 0.331 mol/ℓ.

b. Solving Equation 14.8 for t, we obtain

$$t = \frac{2.30}{k} \log_{10} \frac{X_0}{X} = \frac{2.30}{0.0410} \log_{10} \frac{0.500}{0.100} = 56.1 \log_{10} 5.00$$

But, $\log_{10} 5.00 = 0.699$, so t = 56.1(0.699) = 39.2 min.

c. When half of the sample has decomposed, X = X_0/2; X_0 = 2X; X_0/X = 2. Using the equation from part b:

$$t = \frac{2.30}{k} \log_{10} 2 = \frac{2.30(0.301)}{k} = \frac{0.693}{k}$$

With k = 0.0410/min, we have t = $\frac{0.693}{0.0410}$ min = 16.9 min.

Exercise   Looking back at the data in Table 14.1, how long does it take for one half of a sample of N_2O_5 to decompose? Answer: 2 min.

The analysis of Example 14.2c and the exercise that follows reveals an important feature of a first order reaction. *The time required for one half of a reactant to decompose via a first order reaction has a fixed value, independent of concentration.* This quantity, called the half-life, is given by the expression:

$$t_{1/2} = \frac{0.693}{k} \tag{14.9}$$

where k is the rate constant for the first order reaction. For the reaction discussed in Example 14.2, where

$$t_{1/2} = 17 \text{ min; orig. conc. } H_2O_2 = 0.500 \text{ mol}/\ell$$

we have

| | | Conc. H_2O_2 | Fraction of H_2O_2 | |
t (min)	No. half-lives	(mol/ℓ)	unreacted	reacted
17	1	0.250	1/2	1/2
34	2	0.125	1/4	3/4
51	3	0.0625	1/8	7/8

How long would it take for all the H_2O_2 to react?

Notice from Equation 14.9 that the half-life, $t_{1/2}$, is *inversely* proportional to the rate constant, k. A "fast" reaction, for which k is large, will have a short half-life. Conversely, a "slow" reaction (small value of k) will have a relatively long half-life.

TABLE 14.2   CHARACTERISTICS OF ZERO, FIRST, AND SECOND ORDER
REACTIONS

$$aA(g) \rightarrow \text{products}$$
$$(X, X_0 = \text{conc. A at t and t} = 0, \text{respectively})$$

ORDER	RATE EXPRESSION	CONC.-TIME RELATION	HALF-LIFE	LINEAR PLOT
0	rate $= k$	$X_0 - X = kt$	$X_0/2k$	X vs. t
1	rate $= kX$	$\log_{10} \dfrac{X_0}{X} = \dfrac{kt}{2.30}$	$0.693/k$	$\log_{10} X$ vs. t
2	rate $= kX^2$	$\dfrac{1}{X} - \dfrac{1}{X_0} = kt$	$1/kX_0$	$\dfrac{1}{X}$ vs. t

Reactions of Other Integral Orders

Among gas phase reactions, second order processes are perhaps more common than those of first order. Zero order reactions, in which the rate is independent of reactant concentrations, are much less common. One such reaction is the thermal decomposition of hydrogen iodide on a gold surface:

$$2 \text{ HI(g)} \xrightarrow{\text{Au}} \text{H}_2(g) + \text{I}_2(g); \text{ rate} = k(\text{conc. HI})^0 = k$$

This reaction occurs at a constant rate independent of the concentration of HI. Third order reactions are very rare. One of the few known examples is the reaction between nitric oxide and hydrogen:

$$2 \text{ NO(g)} + 2 \text{ H}_2(g) \rightarrow \text{N}_2(g) + 2 \text{ H}_2\text{O(g)}; \text{ rate} = k(\text{conc. NO})^2 \times (\text{conc. H}_2)^1$$

Rather than discuss each of these reaction types in detail, we list in Table 14.2 some of the properties of zero, first, and second order reactions. All the relations apply to the decomposition of a single reactant:

$$a \text{ A(g)} \rightarrow \text{products}$$

In the last column of Table 14.2, we list the functions (X, $\log_{10} X$, and $1/X$) which, when plotted against t, give a straight line for zero, first, and second order reactions, respectively. These relations offer a simple method of deciding upon the order of a reaction which has been studied in the laboratory.

Example 14.3   The following data were obtained for the gas phase decomposition of hydrogen iodide:

t (h)	0	2	4	6
conc. HI	1.00	0.50	0.33	0.25

Is this reaction zero, first, or second order in HI?

Solution   It will be useful to prepare a table in which we list X, $\log X$, and $1/X$ as a function of time, letting $X = $ conc. HI.

Given X as a function of time, we can find the order by finding the plot which yields a straight line

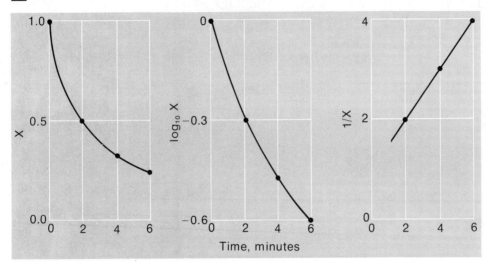

Figure 14.4 Decomposition of HI (Example 14.3). Here, X is the concentration of HI at time t. If the reaction were zero order, a plot of X vs. t would be linear. If it were first order, a plot of log X vs. t would be linear. For a second order reaction such as this one, a plot of 1/X vs. t is a straight line.

t	X	$\log_{10} X$	1/X
0	1.00	0.00	1.0
2	0.50	−0.30	2.0
4	0.33	−0.48	3.0
6	0.25	−0.60	4.0

If it is not obvious from the table above that the only linear plot will be that of 1/X vs. t, that point should be clear from Figure 14.4. We conclude that we are dealing with a second order reaction.

Exercise The half-life of a certain reaction is directly proportional to the original concentration of reactant. Using Table 14.2, determine the order of the reaction. Answer: 0.

14.4 ACTIVATION ENERGY

Consider the reaction

$$CO(g) + NO_2(g) \rightarrow CO_2(g) + NO(g)$$

We believe that this reaction takes place as the result of collisions between CO and NO_2 molecules. This is consistent with the rate expression

$$\text{rate} = k(\text{conc. CO}) \times (\text{conc. } NO_2)$$

If we double the concentration of CO, holding that of NO_2 constant, the number of collisions in a given time doubles (Fig. 14.5). Doubling the concentration of NO_2 (holding that of CO constant) has the same effect. In general, the number of collisions per unit time is directly proportional to the concentration of CO or NO_2. The fact that the rate is also directly proportional to these concentrations implies that reaction occurs as a direct result of collisions between CO and NO_2 molecules.

There is one restriction on this simple model for the CO–NO_2 reaction. We can easily show that *not every collision leads to reaction*. From the kinetic theory of gases, it is possible to calculate the rate at which molecules collide with each other. For a mix-

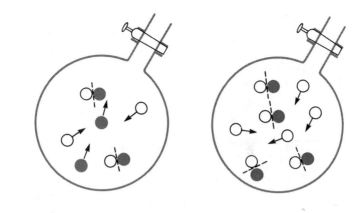

Figure 14.5 Effect of concentration upon reaction rate for the process $CO(g) + NO_2(g) \rightarrow CO_2(g) + NO(g)$. When the concentration of CO molecules *(open circles)* is doubled, collisions occur twice as often. More generally, the number of collisions per unit time and hence the reaction rate is directly proportional to the concentration of CO or NO_2.

ture of CO and NO_2 at 400°C and concentrations of 0.10 mol/ℓ, it turns out that every molecule should collide with about a billion other molecules in one second. If every collision were effective, the reaction between CO and NO_2 should be over in a fraction of a second. By experiment, we find that this is not the case. Under the conditions specified, the half-life of the reaction is about 20 s.

It would, in fact, be explosive

If you think about it for a moment, you can see why only a small fraction of molecular collisions should be effective. The reactant molecules are held together by strong chemical bonds. For reaction to occur, these bonds must be weakened to the point of breaking. This can happen only if the molecules collide with considerable force. Slow-moving molecules do not have enough kinetic energy to react when they collide. They bounce off one another and retain their identity. Only molecules moving at very high speeds have enough kinetic energy for collision to result in reaction.

For every reaction, there is a certain minimum energy required to bring about reaction. This is referred to as the **activation energy.** It has the symbol E_a and is expressed in kilojoules. For the reaction between 1 mol of CO and 1 mol of NO_2, E_a is 134 kJ. The colliding molecules (CO and NO_2) must have a total kinetic energy of at least 134 kJ/mol if they are to react.

We find that the activation energy for a reaction:

—*is a positive quantity* ($E_a > 0$).

—*depends upon the nature of the reaction.* Other factors being equal, we expect "fast" reactions to have a small activation energy. A reaction with a large activation energy takes place slowly under ordinary conditions. The larger the value of E_a, the smaller will be the fraction of molecules having enough kinetic energy to react when they collide.

—*is independent of temperature or concentration.*

Activation Energy Diagrams

Figure 14.6, p. 336, is an energy diagram for the CO–NO_2 reaction. Reactants, CO and NO_2, are shown at the left. Products, CO_2 and NO, are at the right. They have an energy 234 kJ less than that of the reactants; ΔH for the reaction is −234 kJ. In the center of the figure is an intermediate called an *activated complex*. This is an unstable, high-energy species which must be formed before the reaction can occur. It has an energy 134 kJ greater than the reactants and 368 kJ greater than the products. The activation energy, 134 kJ, is absorbed in converting the reactants to the activated complex. The exact nature of this species is difficult to determine. For this reaction, the activated complex might be a "pseudomolecule" made up of CO and NO_2 molecules in close contact. The path of the reaction might be more or less as follows:

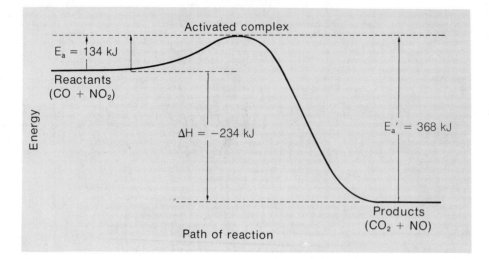

Figure 14.6 Concept of activation energy. During the reaction step, about 134 kJ, the activation energy E_a, must be furnished to the reactants for each mole of CO that reacts. This energy activates each CO–NO_2 complex to the point where reaction can proceed.

The collision must shake up the electrons pretty well if it is to be effective

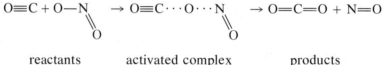

reactants   activated complex   products

The dotted lines stand for "partial bonds" in the activated complex. The N—O bond in the NO_2 molecule has been partially broken. A new bond between carbon and oxygen has started to form.

From Figure 14.6, we see that

$$CO(g) + NO_2(g) \rightarrow CO_2(g) + NO(g); E_a = 134 \text{ kJ}$$

$$CO_2(g) + NO(g) \rightarrow CO(g) + NO_2(g); E_a' = 368 \text{ kJ}$$

where E_a and E_a' are the activation energies for the forward and reverse reactions! The difference between these two quantities is equal to the enthalpy change for the forward reaction

$$CO(g) + NO_2(g) \rightarrow CO_2(g) + NO(g); \Delta H = -234 \text{ kJ}$$

In general, for any process,

$$\Delta H = E_a - E_a' \qquad (14.10)$$

If the forward reaction is exothermic ($\Delta H < 0$), E_a is smaller than E_a'. This is the case with the CO–NO_2 reaction, where

$$\Delta H = -234 \text{ kJ}; E_a = E_a' - 234 \text{ kJ}$$

Suppose, on the other hand, that the forward reaction is endothermic ($\Delta H > 0$). In that case, the activation energy for the forward reaction, E_a, is larger than that for the reverse reaction. For example, if ΔH were +200 kJ, then

$$\Delta H = +200 \text{ kJ}; E_a = E_a' + 200 \text{ kJ}$$

Certain substances called *catalysts* can increase the rate of a reaction without being consumed by it. One reaction which is subject to catalysis is the decomposition of hydrogen peroxide:

$$H_2O_2(aq) \rightarrow H_2O(l) + \tfrac{1}{2} O_2(g)$$

This occurs rather slowly under ordinary conditions (recall Example 14.2). However, it takes place almost instantly if a pinch of manganese dioxide, MnO_2, is added. All the MnO_2 can be recovered when the reaction is over. This indicates that it is indeed a catalyst. Another reaction of this type is the decomposition of nitrous oxide:

$$N_2O(g) \rightarrow N_2(g) + \tfrac{1}{2} O_2(g) \qquad (14.11)$$

This reaction can be speeded up by bringing the N_2O into contact with a metal such as gold.

A catalyst operates by lowering the activation energy required for reaction (Fig. 14.7). Consider, for example, Reaction 14.11. For the uncatalyzed reaction, E_a is 250 kJ. For the catalyzed reaction on a gold surface, E_a is only 120 kJ. The reduction in activation energy comes about because the catalyst provides an alternate pathway of lower energy for the reaction. In the decomposition of N_2O on gold, the gas is chemically adsorbed on the metal surface. A bond is formed between the oxygen of the N_2O molecule and a gold atom. This weakens the bond joining nitrogen to oxygen, making it easier for the N_2O molecule to break apart.

This lowering of E_a speeds up the reaction fantastically

We see from Figure 14.7 that a catalyst does not affect the relative energies of reactants and products. Neither does it change the equilibrium constant, K_c. Adding a catalyst does not affect the position of an equilibrium. It neither increases nor decreases the yield of product. However, by speeding up the reaction, a catalyst allows us to reach equilibrium more quickly.

Many reactions which take place slowly under ordinary conditions occur readily in the body in the presence of catalysts called *enzymes*. Enzymes are protein molecules of high molecular mass. An example of an enzyme-catalyzed reaction is that of sugar (sucrose) with oxygen:

$$C_{12}H_{22}O_{11}(s) + 12 O_2(g) \rightarrow 12 CO_2(g) + 11 H_2O(l)$$

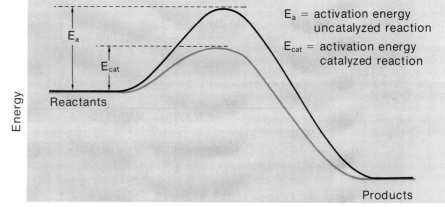

Figure 14.7 By changing the path by which a reaction occurs, a catalyst can lower the activation energy that is required, and so speed up the reaction.

E_a = activation energy uncatalyzed reaction

E_{cat} = activation energy catalyzed reaction

Energy

Reactants

E_a

E_{cat}

Products

Path of reaction

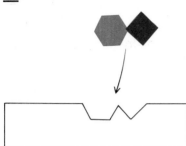 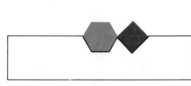 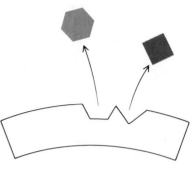

Substrate diffuses to active site   Enzyme-substrate complex   Products diffuse from active site

Figure 14.8 In enzyme catalysis, the substrate appears to fit, in a "lock and key" arrangement, on the enzyme. After absorption, the enzyme configuration often changes, which assists in cleaving the crucial substrate bond, and thereby increases the rate of reaction.

This reaction is difficult to bring about directly, as by heating a sample of sugar in a test tube. In the body, sugar is metabolized at 37°C (98.6°F) in a series of biochemical reactions. The end products are carbon dioxide and water. Each step in the sequence is catalyzed by a particular enzyme adapted for that purpose.

Catalysis by enzymes can be interpreted in terms of the "lock and key" analogy shown in Figure 14.8. The reactant ("substrate") fits into a specific site on an enzyme surface. There it is held in position by intermolecular forces. The substrate-enzyme complex can then react with another species such as a water molecule.

Enzyme activity is diminished in the presence of certain substances known as inhibitors. One way in which an inhibitor can operate is to occupy sites on an enzyme molecule which are supposed to be reserved for the substrate. Frequently inhibitors have geometries closely resembling those of the substrates they replace. For example, the metabolism of citric acid is inhibited by its close relative, fluorocitric acid, which can presumably fit into the same slot in an enzyme.

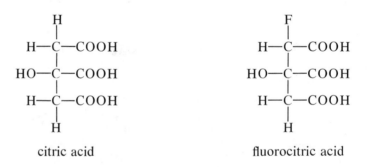

citric acid   fluorocitric acid

Until quite recently we knew almost nothing about the molecular structure of enzymes. The lock and key model was little more than a convenient way to rationalize the kinetics of enzyme reactions. Within the past two decades the structures of some of the simpler enzymes have been established by x-ray crystallography. In some cases, the structures of substrate-enzyme complexes have been determined. Research in this area led in 1969 to the first laboratory synthesis of an enzyme, ribonuclease. For this work, Drs. Stein and Moore of Rockefeller University and Dr. Anfinsen of NIH won the 1972 Nobel Prize in chemistry.

These structures are detailed and may involve over 1000 atoms

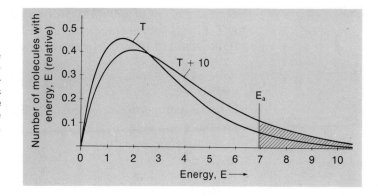

Figure 14.9 When the temperature is increased, the fraction of molecules with very high energies increases sharply. Hence, many more molecules possess the activation energy, E_a, and reaction occurs more rapidly. If E_a is of the order of 50 kJ, an increase in temperature of 10°C approximately doubles the number of molecules having energy E_a or greater, and thus doubles the reaction rate.

14.6 DEPENDENCE OF REACTION RATE UPON TEMPERATURE

The rates of most reactions increase as the temperature rises. A person in a hurry to prepare dinner applies this principle in using a pressure cooker to raise the temperature for cooking potatoes, apples, or a pot roast (not all at the same time, we trust). By storing the leftovers in a refrigerator, we slow down the chemical reactions responsible for food spoilage. As a general and very approximate rule, it is often stated that an increase in temperature of 10°C doubles the reaction rate. If this rule holds, foods should cook twice as fast in a pressure cooker at 110°C as in an open saucepan, and deteriorate four times as rapidly at room temperature (25°C) as they do in a refrigerator at 5°C.

The effect of temperature on reaction rate can be explained in terms of the kinetic theory of gases. Recall from Chapter 7 that raising the temperature greatly increases the fraction of molecules having very high kinetic energies. These are the molecules that are most likely to react when they collide. The higher the temperature, the larger the fraction of molecules that can provide the activation energy required for reaction. This effect is seen in Figure 14.9, where we show the distribution of kinetic energies among gas molecules at two different temperatures. The shaded areas include those molecules having a kinetic energy equal to or greater than E_a. Note that this area is considerably larger at the higher temperature. This means that, at the higher temperature, a larger fraction of the molecules will have sufficient energy to react when they collide. Hence, the reaction will go faster. Putting it another way, the rate constant, k, becomes larger as the temperature increases. Table 14.3 shows this effect for the CO–NO_2 reaction. Notice that the rate constant increases by a factor of nearly a thousand when the temperature rises from 600 to 800 K.

Relation between k and T

The argument we have just gone through can be made quantitative. Kinetic theory tells us that the fraction, f, of molecules having an energy equal to or greater than E_a is

$$f = e^{-E_a/RT} \qquad (14.12)$$

TABLE 14.3	TEMPERATURE DEPENDENCE OF THE RATE CONSTANT FOR THE REACTION $CO(g) + NO_2(g) \rightarrow CO_2(g) + NO(g)$				
T (K)	600	650	700	750	800
k (ℓ/(mol · s))	0.028	0.22	1.3	6.0	23

Here, e is the base of natural logarithms, R is the gas constant, E_a is the activation energy, and T is the absolute temperature in K. If we assume that the rate constant, k, is directly proportional to f (which is approximately true),

$$k = cf = ce^{-E_a/RT}$$

where c is a proportionality constant. Taking the natural logarithm of both sides of this equation:

$$\ln k = \ln c - E_a/RT$$

Converting to base 10 logarithms:

$$\log_{10} k = \log_{10} c - \frac{E_a}{2.30 \, RT}$$

Substituting R = 8.31 J/(mol · K) and setting $\log_{10} c = A$, we obtain

$$\log_{10} k = A - \frac{E_a}{19.14 \, T} \tag{14.13}$$

Here k is the rate constant for the reaction, A is a constant independent of the other quantities in the equation, E_a is the activation energy for the reaction in *joules,* and T is the temperature in K. This equation was first shown to be valid by the Swedish physical chemist Svandte Arrhenius in 1887. It is referred to as the Arrhenius equation.

Equation 14.13 is of the form

$$y = a + bx$$

where "y" is $\log_{10} k$ and "x" is 1/T. A plot of $\log_{10} k$ vs. 1/T should be a straight line.

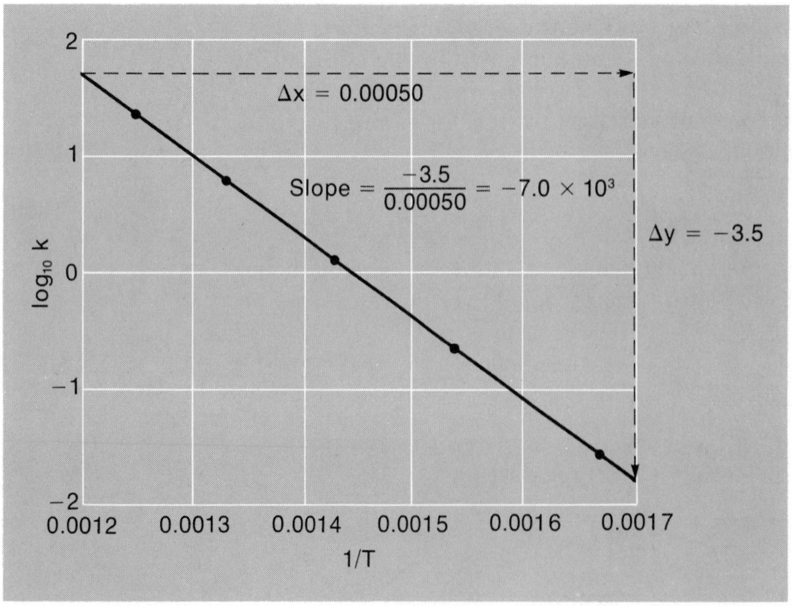

Figure 14.10 A plot of $\log_{10}k$ vs. 1/T (k = rate constant, T = Kelvin temperature) is ordinarily a straight line. From the slope of this line, the activation energy can be determined: $E_a = -2.30(8.31)(\text{slope})$. For the CO–$NO_2$ reaction shown here, $E_a = -2.30(8.31)(-7.0 \times 10^3) \, J = 1.34 \times 10^5 \, J = 134$ kJ.

In Figure 14.10, we have made this plot for the CO–NO$_2$ reaction, taking the data from Table 14.3. As you can see, it is indeed a straight line. The activation energy for the reaction can be obtained from the slope of the line:

$$\text{slope} = -E_a/19.14 \tag{14.14}$$

For the CO–NO$_2$ reaction, we see from Figure 14.10 that the slope is about -7.0×10^3. Hence

$$E_a = -19.14 \times (-7.0 \times 10^3) = 1.34 \times 10^5 \text{ J} = 134 \text{ kJ}$$

This tells us that the activation energy for this reaction is 134 kJ.

"Two-Point" Equation Relating k and T

We can use Equation 14.13 to obtain a relation between rate constants, k_2 and k_1, at two different temperatures, T_2 and T_1. At the two temperatures, we have

$$\log_{10} k_2 = A - \frac{E_a}{19.14 \, T_2}$$

$$\log_{10} k_1 = A - \frac{E_a}{19.14 \, T_1}$$

Subtracting the second equation from the first:

$$\log_{10} k_2 - \log_{10} k_1 = \log_{10} \frac{k_2}{k_1} = \frac{E_a}{19.14}\left(\frac{1}{T_1} - \frac{1}{T_2}\right)$$

or

$$\log_{10} \frac{k_2}{k_1} = \frac{E_a}{19.14}\left(\frac{T_2 - T_1}{T_2 T_1}\right) \tag{14.15}$$

In using Equation 14.15, we ordinarily take T_2 to be the higher temperature. This avoids the use of negative logarithms. Remember that E_a must be expressed in joules; T_2 and T_1 are in K.

Example 14.4

  a. The activation energy for a certain reaction is 1.14×10^5 J. At 600 K, k = 0.75 ℓ/(mol·s). Calculate k at 700 K.
  b. What must be the value of E_a if the rate constant for a reaction is to double when the temperature increases from 27 to 37°C?

Solution

  a. Applying Equation 14.15 with $T_2 = 700$ K, $T_1 = 600$ K.

$$\log_{10} \frac{k_2}{k_1} = \frac{1.14 \times 10^5}{19.14}\left(\frac{700 - 600}{700 \times 600}\right) = 1.42$$

Working with this
equation on your cal-
culator is easy, once
you have done it a few
times

Taking antilogs, $\dfrac{k_2}{k_1}$ = antilog 1.42 = 26.

$$k_2 = 26k_1 = 26 \times 0.75 \; \ell/(mol \cdot s) = 20 \; \ell/(mol \cdot s)$$

b. If $k_2/k_1 = 2.00$, then $\log_{10} k_2/k_1 = \log_{10} 2 = 0.301$.
Substituting in Equation 14.15

$$0.301 = \frac{E_a(310 - 300)}{(19.14)(310)(300)}$$

Solving, $\qquad\qquad\qquad E_a = 5.36 \times 10^4 \, J = 53.6 \; kJ$

Note that if E_a were appreciably greater than 53.6 kJ, k would more than double for a 10° rise in temperature; if E_a were smaller than 53.6 kJ, k would increase by less than a factor of two. Clearly the empirical rule that a temperature increase of 10°C doubles the reaction rate is at best a crude approximation.

Exercise For what value of E_a would k be independent of temperature (i.e., k_2 at T_2 = k_1 at T_1)? Answer: $E_a = 0$.

14.7 REACTION MECHANISMS

A reaction mechanism is a description of the path, or sequence of steps, by which a reaction occurs. In the simplest case, only a single step is involved. This is a collision between two reactant molecules. We believe this to be the mechanism for the reaction of CO with NO_2.

$$CO(g) + NO_2(g) \rightarrow CO_2(g) + NO(g)$$

Reactions which occur in a single step involving a collision between reactant molecules have a simple rate expression. They are first order in both reactants, second order overall. Thus for the CO–NO_2 reaction:

$$rate = k(conc. \; CO) \times (conc. \; NO_2)$$

In practice, most reactions occur in more than one step. The first step may involve the formation of an intermediate which then undergoes further reaction. Several steps may be required before product molecules are formed. In this section, we will look at the kinetics of reactions that occur by a mechanism which involves more than one step. We start by considering an important problem. *Given a mechanism for a several-step reaction, how do we obtain the rate expression for the reaction?*

Rate Expression from Mechanisms

Consider the reaction between hydrogen and iodine:

$$H_2(g) + I_2(g) \rightarrow 2 \; HI(g) \tag{14.16}$$

It has been suggested that this reaction occurs in two steps:

Many mechanisms
have steps that go to
equilibrium

Step A: $I_2 \rightleftharpoons 2 \; I$
Step B: $H_2 + 2 \; I \rightarrow 2 \; HI$

PLATE 1 *Color of Potassium Permanganate.* The purple color of the solution is caused by strong absorption in the green region of the spectrum, as shown at the bottom of this plate. (See Table 1.6, p. 14.)

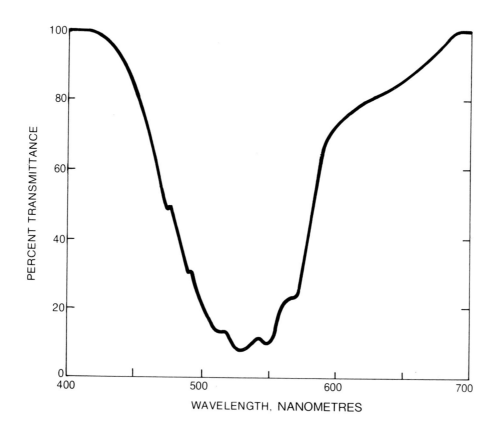

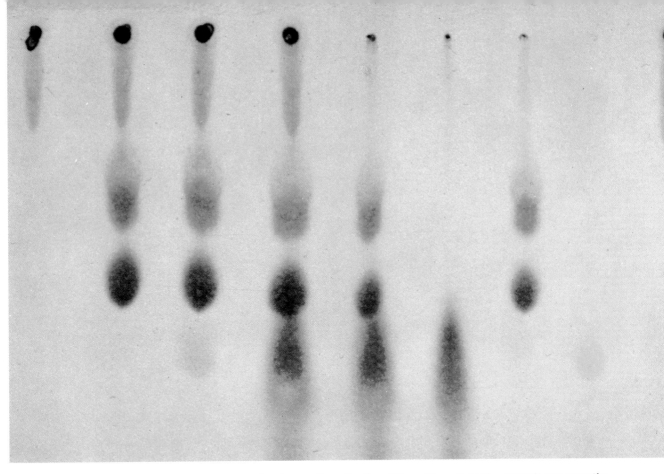

PLATE 2 *Paper Chromatography.* Separation of colored dyes. The original spot is shown near the top of the plate; the solvent has advanced to the bottom of the plate. (See Chapter 1.)

PLATE 3 One mole of carbon (12.01 g), sulfur (32.06 g), and copper (63.55 g). Each sample contains the same number of atoms, 6.022×10^{23}. (See Chapter 2.)

PLATE 4   *The Halogens.* Flasks containing Cl_2, Br_2, and I_2 show a gradation in color from greenish-yellow through deep red to violet. The colors shown for bromine and iodine are those of the vapors in equilibrium with $Br_2(l)$ and $I_2(s)$. (See Table 5.2, p. 153.)

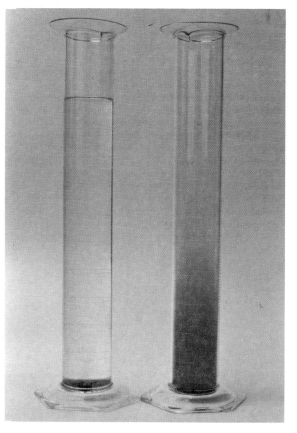

PLATE 5   At the left, liquid bromine slowly diffuses upward into water. At right, gaseous bromine rapidly diffuses into air. (See Chapter 7.)

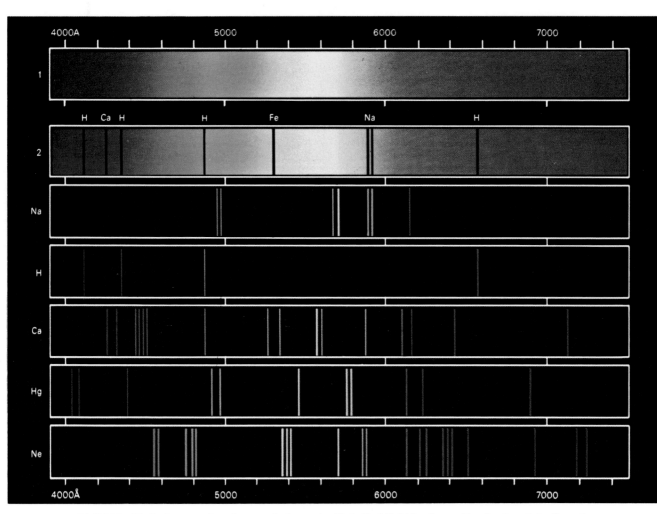

PLATE 6 Emission spectra of several elements. Note that light is given off only at certain discrete wavelengths. (See Chapter 8.) *From Keenan, C.W., Wood, J.H., and Kleinfelter, D.C.,* General College Chemistry, *5th ed., New York, Harper & Row.*

PLATE 7 Flame tests for the alkali metals: Li (red), Na (yellow), K (violet). (See Table 9.1.)

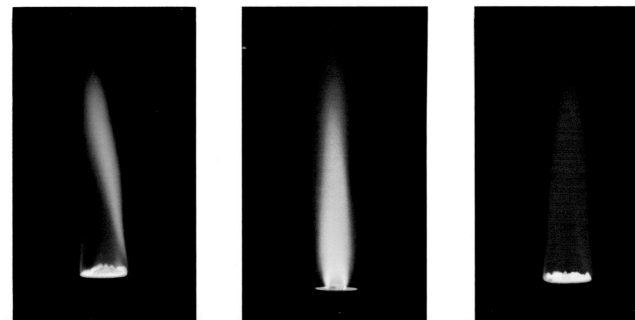

PLATE 8 CuSO₄•5H₂O (upper left) is blue; the anhydrous salt (lower left) is white. CoCl₂•6H₂O (upper right) is pink; lower hydrates such as CoCl₂•4H₂O are purple (lower right) or blue. (See Chapter 11.)

PLATE 9 *Sublimation of Iodine.* Solid iodine, heated in a beaker of water at the bottom of the plate, vaporizes up the tube (violet color) and condenses at the top when it comes in contact with a layer of Dry Ice. (See Chapter 11.)

PLATE 10 *Physical Properties of Liquid Oxygen.* Oxygen is attracted into a magnetic field; one consequence is that liquid oxygen can be suspended between the poles of an electromagnet. Both the paramagnetism and the blue color are due to the unpaired electrons in the O_2 molecule. (See Chapter 12.)

PLATE 11 When liquid sulfur at 200°C is poured into cold water, a rubbery material called "plastic sulfur" forms. Here, S. Ruven Smith shows some of its (and his) properties. (See Chapter 12.)

PLATE 12 *Effect of Temperature on the N_2O_4-NO_2 Equilibrium.* At 0°C (tube in ice bath at left), N_2O_4, which is colorless, predominates. At 50°C (tube in water bath at right), some of the N_2O_4 has dissociated to give the deep brown color of NO_2. (See Chapter 13.)

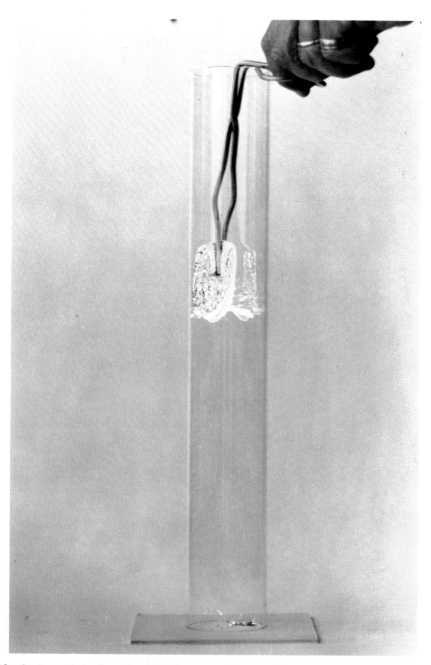

PLATE 13 A piece of steel wool, which has a very large surface area, burns spectacularly when heated to redness and immersed in pure oxygen. (See Chapter 14.)

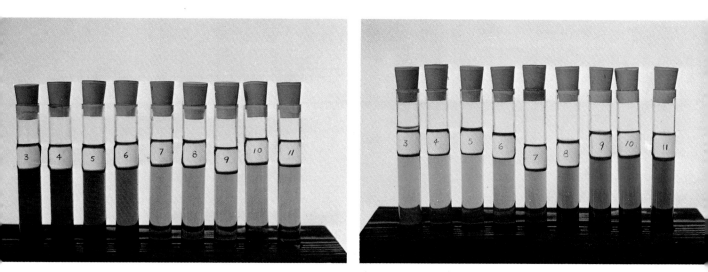

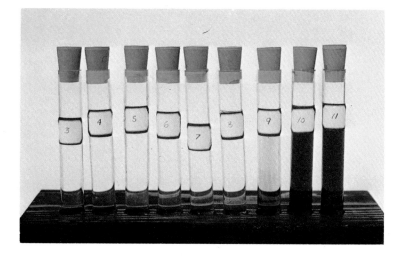

PLATE 14 *Color of Indicators As a Function of pH.* The photographs show the colors of methyl red (upper left), bromthymol blue (upper right), and phenolphthalein (below) in solutions ranging in pH from 3 to 11. (The labels were attached by W.L.M. just before an 8 A.M. class.) Note that each indicator is useful only in a narrow pH range of 1 to 2 units. (See Chapter 17.)

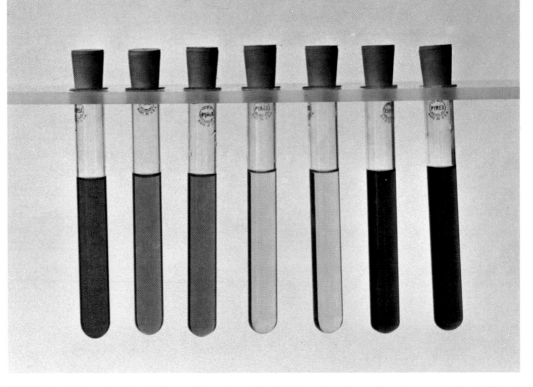

PLATE 15 Color of salt solutions with universal indicator, which is red in strong acid (far left) and purple in strong base (far right). NH_4I and $Zn(NO_3)_2$ are weakly acidic at about pH 5 (orange); $KClO_4$, like pure water, has a pH of 7 (yellow-green); Na_3PO_4 is basic at about pH 12 (purple). (See Chapter 17.)

PLATE 16 The three tubes at the left show the effect of adding a few drops of strong acid or strong base to pure water. The pH changes drastically, giving a pronounced color change with universal indicator. In the three tubes at the right, this experiment is repeated, using a buffer of pH 7 instead of water. This time the pH changes only very slightly, and there is no change in the color of the indicator. (See Chapter 18.)

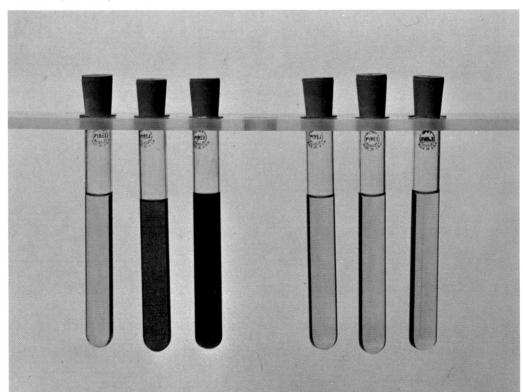

PLATE 17 *Nitrates of Metals of the 4th Period.* Ions having no d electrons (Ca^{2+} at far left) or a full d sublevel (Zn^{2+} at far right) are colorless. In contrast, salts of transition metals with a partially filled d sublevel (Co^{2+}, Ni^{2+}, Cu^{2+}) are deeply colored. (See Chapter 19.)

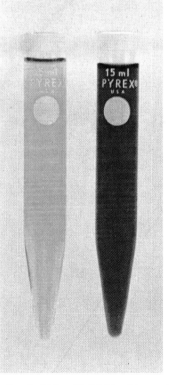

PLATE 18 A solution containing the Cu^{2+} ion (light blue) turns to a deep blue when ammonia is added because of the formation of the $Cu(NH_3)_4^{2+}$ complex. (See Chapter 19.)

PLATE 19   *Color of Complex Ions of Co³⁺.* The five compounds at the top form a spectrochemical series (Table 19.5), in which NH_3 molecules are replaced by ligands of lower field strength ($NH_3 >$ $NCS^- > H_2O > Cl^-$). The absorption shifts to higher wavelengths with a corresponding change in color.

   The two compounds at the bottom illustrate the difference in color frequently observed between geometrical isomers. The compound at the left contains the *cis*-$Co(en)_2Cl_2^+$ ion; the one at the right is the *trans* isomer.

PLATE 20 *Rate of Ligand Substitution* (Chapter 19). These solutions were prepared by dissolving *trans*-[Co(en)$_2$Cl$_2$]Cl in water at approximately 10-minute intervals. The tube at the far left shows the characteristic green color of the *trans*-Co(en)$_2$Cl$_2$$^+$ cation. As time passes, this is replaced by the red color of aquo complexes such as Co(en)$_2$H$_2$OCl^{2+} and Co(en)$_2$(H$_2$O)$_2$$^{3+}$. The overall reaction may be represented as:

$$Co(en)_2Cl_2^+ (aq) + 2\ H_2O \rightarrow Co(en)_2(H_2O)_2^{3+} (aq) + 2\ Cl^- (aq)$$
$$\text{green} \qquad\qquad\qquad \text{red}$$

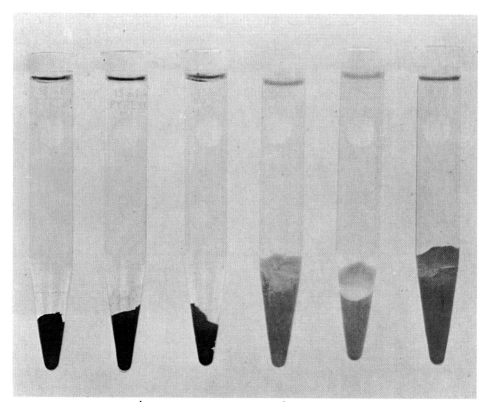

PLATE 21 *Sulfides of Group II Cations.* (See Chapter 20.)

a. CuS   b. Bi_2S_3 c. HgS   d. CdS   e. SnS_2 f. Sb_2S_3

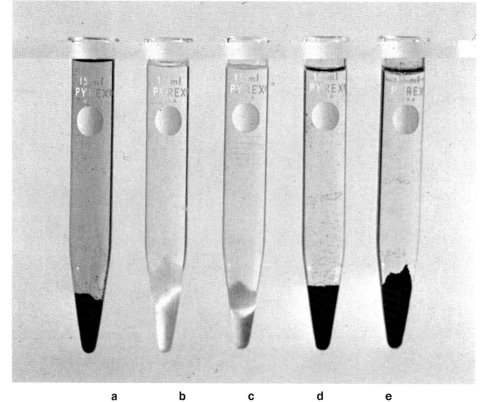

PLATE 22 *Sulfides of Group III Cations.* (See Chapter 20.)

a. FeS   b. ZnS   c. MnS   d. CoS   e. NiS

PLATE 23 When a strip of zinc is placed **in a solution containing** Cu^{2+} ions (left), an oxidation-reduction reaction occurs. The final result **is shown at the right. Copper metal plates out and the** blue color due to Cu^{2+} **fades (Chapter 21).**

PLATE 24 A copper **wire reacts with nitric acid by oxidation-reduction. The brown gas formed is** nitrogen dioxide, NO_2 (Chapter 21).

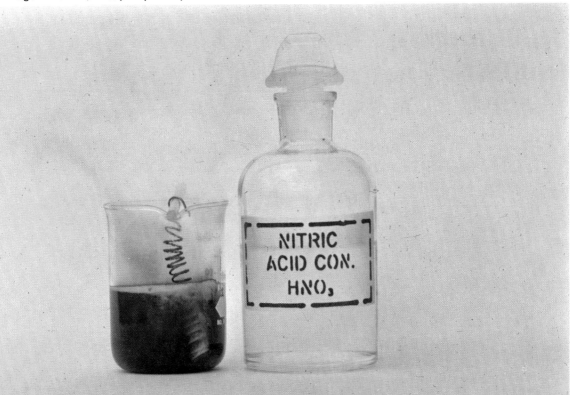

PLATE 25 *Oxidation States of Chromium.* In the +6 state, chromium forms two oxyanions, $Cr_2O_7^{2-}$ (red) and CrO_4^{2-} (yellow). Compounds of Cr^{3+} are most often purple due to the presence of ions such as $Cr(H_2O)_6^{3+}$. Solutions of $CrCl_3$ are an exception; species such as $Cr(H_2O)_5Cl^{2+}$ and $Cr(H_2O)_4Cl_2^+$ shift the color to green. Compounds of Cr^{2+} (far right) are blue when freshly prepared but are rapidly oxidized to Cr^{3+} by contact with air. (See Chapter 22.)

PLATE 26 *Oxidation of Br^- and I^- by Cl_2.* The tube at the far left contains a colorless solution of KBr in water. Addition of chlorine brings about the oxidation-reduction reaction:

$$Cl_2(aq) + 2\ Br^-(aq) \rightarrow Br_2(aq) + 2\ Cl^-(aq)$$

giving rise to the light color of dissolved bromine seen in the second tube. This color is intensified (third tube) by shaking with a little CCl_4, in which Br_2 is more soluble.

Tubes 4 to 6 show a similar sequence, starting with KI solution (colorless), oxidizing with Cl_2:

$$Cl_2(aq) + 2\ I^-(aq) \rightarrow I_2(aq) + 2\ Cl^-(aq)$$

(pale red) and concentrating the color by adding CCl_4 (deep violet). (See Chapter 22.)

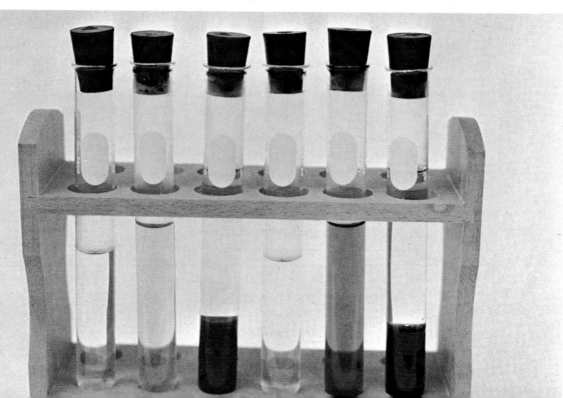

In Step A an equilibrium is established between I_2 molecules and iodine atoms. Step B involves a simultaneous collision of two I atoms with a H_2 molecule. The product, HI, is formed in this step.

Our goal is to obtain the rate expression for Reaction 14.16, using this two-step mechanism. In other words, we want to determine m and n in the equation

$$\text{rate} = k(\text{conc. } H_2)^m \times (\text{conc. } I_2)^n$$

To do this, we follow three rules.

1. *For any step, the order with respect to a reactant species is its coefficient in that step.* Thus, for the H_2–I_2 reaction we have

$$\text{rate of step A (forward)} = k_a(\text{conc. } I_2)$$
$$\text{rate of step B} = k_b(\text{conc. } H_2) \times (\text{conc. } I)^2$$

where k_a and k_b are the rate constants for the individual steps.

2. Often, one step in a mechanism is much slower than any other. In such cases, **the slow step is rate-determining.** That is, the rate of the overall reaction is determined by that of the slow step. The situation is analogous to that in a relay race where there is one slow runner (WLM) and two fast runners (EJS and CLS). The time required to reach the finish line will be determined mainly by WLM, provided he is much slower than EJS or CLS (as indeed he is).

WLM is a better trout fisherman than we are
EJS
CLS

In Reaction 14.16, the first step is almost instantaneous. Equilibrium is established very quickly. In contrast, Step B is quite slow. Three-body collisions occur very rarely. (How often have you seen three cars on a highway collide simultaneously?) Hence

$$\text{rate of Reaction 14.16} = \text{rate of Step B} = k_b(\text{conc. } H_2) \times (\text{conc. } I)^2$$

3. Frequently, reactive intermediates are present as reactants in the rate-determining step. Their concentrations are not known experimentally. Hence they must be eliminated from the rate expression if it is to be compared with experiment. *The final rate expression must include only species which appear in the balanced equation for the overall reaction.*

We saw that I atoms are involved in the rate-determining (slow) step for Reaction 14.16:

$$\text{rate} = k_b(\text{conc. } H_2) \times (\text{conc. } I)^2$$

The term $(\text{conc. } I)^2$ must be eliminated to obtain the final rate expression. To do this, we note that in Step A, I atoms are in equilibrium with I_2 molecules. Writing the equilibrium constant expression for this step:

$$K_c = \frac{(\text{conc. } I)^2}{(\text{conc. } I_2)}$$

Solving for $(\text{conc. } I)^2$ and substituting in the rate expression, we have

$$(\text{conc. } I)^2 = K_c \times (\text{conc. } I_2)$$

$$\text{rate} = k_b \times K_c \times (\text{conc. } H_2) \times (\text{conc. } I_2)$$

The product of the two constants, k_b and K_c, is simply the observed rate constant for the reaction. Calling that k, the predicted rate expression is

$$\text{rate} = k(\text{conc. } H_2) \times (\text{conc. } I_2) \qquad (14.17)$$

Clearly, this mechanism predicts that the reaction should be first order in H_2 and first order in I_2. These are the orders observed experimentally. In other words, this mechanism is consistent with the observed rate expression.

Example 14.5   Another possible mechanism for the H_2–I_2 reaction is:

Step A: $I_2(g) \rightleftharpoons 2\ I(g)$	fast
Step B: $H_2(g) + I(g) \rightarrow HI(g) + H(g)$	slow, rate determining
Step C: $H(g) + I(g) \rightarrow HI(g)$	fast

For this mechanism, derive a rate expression for the overall reaction and determine whether it is consistent with the observed expression (Eq. 14.17).

Solution   The rate expression for the overall reaction must be the same as that for Step B:

$$\text{rate B} = k_b \times (\text{conc. } H_2) \times (\text{conc. } I)$$

To eliminate the term (conc. I), we again consider the equilibrium in Step A:

$$K_c = \frac{(\text{conc. } I)^2}{(\text{conc. } I_2)}$$

$$(\text{conc. } I)^2 = K_c \times (\text{conc. } I_2)$$

$$(\text{conc. } I) = K_c^{1/2} \times (\text{conc. } I_2)^{1/2}$$

Substituting in the rate expression:

$$\text{rate B} = k_b \times K_c^{1/2} \times (\text{conc. } H_2) \times (\text{conc. } I_2)^{1/2}$$

The rate expression for the overall reaction would then be

$$\text{rate} = k \times (\text{conc. } H_2) \times (\text{conc. } I_2)^{1/2}$$

where k is the observed rate constant. In other words, this mechanism predicts that the reaction should be first order in H_2 and one-half order in I_2. From experiment, we know that the reaction is actually first order in both H_2 and I_2. This tells us that the mechanism proposed here is *not* the correct one.

Exercise   Suppose the reaction $H_2(g) + I_2(g) \rightarrow 2\ HI(g)$ occurs in a single step, via a collision between H_2 and I_2 molecules. What would the rate expression be? Answer: rate $= k \times (\text{conc. } H_2) \times (\text{conc. } I_2)$.

Comparison of the exercise just worked out with the text discussion preceding it points to a limitation of mechanism studies. Often, two or more mechanisms lead to the same rate expression. When this happens, we cannot be sure which one is correct. Other evidence must be considered to make a choice. For many years, it was assumed that the H_2–I_2 reaction occurs in a single step. More recent studies suggest that the two-step mechanism discussed in the text is the correct one. Of one thing we can be sure. The three-step mechanism discussed in Example 14.5 is *not* correct, since it does not lead to the observed rate expression.

It is not easy to prove that a given mechanism is actually the correct one

Chain Reactions

Many gas phase reactions are started by the formation, at very low concentrations, of an extremely reactive species. This species sets off a series of reactions leading to the formation of products. Such processes are referred to as **chain reactions.** Typically, they occur very rapidly after a short waiting period during which the reactive species is formed. An example of a chain reaction is the formation of hydrogen chloride from the elements. A mixture of hydrogen and chlorine stored at room temperature in the dark shows no evidence of reaction over long periods of time. However, if the mixture is exposed to ultraviolet light, a vigorous reaction occurs. The first step in this reaction is referred to as **chain initiation.** It involves the dissociation of a Cl_2 molecule into atoms:

$$Cl_2 \rightarrow 2\ Cl \qquad\qquad (14.18a)$$

The very reactive chlorine atoms react with hydrogen molecules:

$$Cl + H_2 \rightarrow HCl + H \qquad\qquad (14.18b)$$

This step forms another highly reactive species, a hydrogen atom. It attacks a chlorine molecule:

$$H + Cl_2 \rightarrow HCl + Cl \qquad\qquad (14.18c)$$

In this way, the chlorine atoms are quickly re-formed and can react with more H_2 molecules. The **chain propagation,** represented by Equations 14.18b and 14.18c, occurs over and over again until the H_2 and Cl_2 are almost completely converted to HCl.

For every Cl atom made via 14.18a, we may get 10^6 HCl molecules by 14.18b and 14.18c

The hydrogen and chlorine atoms, which act as **chain carriers,** can be consumed by reaction with each other:

$$H + Cl \rightarrow HCl;\ Cl + Cl \rightarrow Cl_2;\ H + H \rightarrow H_2 \qquad\qquad (14.18d)$$

These processes represent **chain termination,** since they break the chain mechanism.

As you can well imagine, kinetic studies of chain reactions are not easy to carry out in the laboratory. Frequently a chemist trying to establish the order of a chain reaction finds that it is over before he has had time to make any measurements. In extreme cases the only result of his efforts may be shattered glassware and broken or badly bent apparatus. The H_2–Cl_2 reaction is particularly notorious in this respect.

SUMMARY

The rate of a reaction indicates how rapidly a reactant is converted into a product. The change of concentration with time can be treated graphically to establish the reaction rate (Fig. 14.1 and Table 14.1). A rate expression (law) gives the relation between concentration and rate of reaction. The rate expression includes a rate constant, k, and concentration terms. The exponent of a concentration term indicates the order of the reaction with respect to that species (Example 14.1). The order of a reaction must be determined experimentally.

As a reaction proceeds, the reactant concentration decreases with time. The relationships between concentration and time are summarized for zero, first, and second order reactions in Table 14.2 and Figure 14.4. These relationships can be used to determine a concentration at a given time (Example 14.2) or to find the order of a reaction (Example 14.3).

A certain minimum energy, the activation energy, is necessary for a reaction to occur when two molecules collide (Fig. 14.6). In general, reaction rate is inversely related to activation energy; a fast reaction implies a low activation energy. A catalyst increases reaction rate by lowering the activation energy.

As temperature increases, a greater fraction of molecules have energies which exceed the activation energy (Fig. 14.9). Thus, we expect an increase in temperature to increase the rate of reaction (Table 14.3). Equation 14.15 relates temperatures, rate constants, and activation energy (Example 14.4).

A reaction mechanism shows the individual steps by which a reaction occurs. Using certain rules (p. 343), a rate expression may be derived from the mechanism (Example 14.5).

KEY WORDS AND CONCEPTS

kinetics	order of reaction	activated complex	reaction mechanism
reaction rate	overall order	catalyst	rate-determining step
rate expression	half-life	enzyme	chain reaction
rate constant, k	activation energy	Arrhenius equation	

QUESTIONS AND PROBLEMS

Catalog

Rate Expressions: 14.2–14.4, 14.21–14.23
Determination of Reaction Order: 14.5, 14.6, 14.24, 14.25
First Order Reactions: 14.7–14.9, 14.26–14.28
Activation Energy and ΔH: 14.10, 14.11, 14.29, 14.30

Effect of Temperature upon Reaction Rate: 14.12–14.15, 14.31–14.34
Mechanism: 14.16, 14.17, 14.35, 14.36
General: 14.1, 14.18–14.20, 14.37–14.43

14.1   Review and know the meanings of the key words and concepts in this chapter.

14.2   The reaction $Q(g) \rightarrow R(g)$ is found to be second order. At a certain temperature, the rate is 0.040 mol/($\ell \cdot$ s) when the concentration of Q is 0.10 mol/ℓ.

   a. Write the rate expression for this reaction.
   b. Determine the rate constant.
   c. Calculate the rate of reaction when conc. Q is 0.25 mol/ℓ.

14.3   The gas phase reaction

$$2\ ClO_2(g) + F_2(g) \rightarrow 2\ FClO_2(g)$$

is first order with respect to each reactant. At 250 K, the rate constant is 1.2 ℓ/(mol $\cdot$ min).

   a. What is the overall reaction order?
   b. What will be the reaction rate at 250 K when conc. ClO_2 = 0.020 mol/ℓ and conc. F_2 = 0.035 mol/ℓ?

14.21   The rate of the second order reaction $A(g) \rightarrow D(g)$ is 1.2 mol/($\ell \cdot$ min) when the concentration of A is 0.050 mol/ℓ.

   a. Write the rate expression for this reaction.
   b. Determine the rate constant.
   c. At this temperature, what would be conc. A for the reaction rate to be twice that given?

14.22   The reaction

$$Br_2(g) + 2\ NO(g) \rightarrow 2\ NOBr(g)$$

is determined experimentally to be first order with respect to Br_2 and second order with respect to NO. At a certain temperature the rate constant is 0.050 ℓ^2/(mol$^2 \cdot$ s).

   a. What is the overall order of the reaction?
   b. What will be the reaction rate at this temperature when conc. Br_2 = 0.10 mol/ℓ and conc. NO = 0.050 mol/ℓ?

14.4 The reaction $X(g) \rightarrow$ products(g) has a rate of 0.014 mol/($\ell \cdot$ s) when conc. X is 0.50 mol/ℓ. What will be the rate when conc. X is 1.0 mol/ℓ if the reaction is

    a. zero order in X?
    b. first order in X?
    c. second order in X?

14.5 In a certain reaction, $A(g) \rightarrow B(g)$, the rate is measured when the concentration of A is 0.10 M and again when it is 0.050 M. What is the order of the reaction if the ratio of the new rate to the original rate is

    a. 0.50?   b. 1.0?   c. 0.25?

14.6 The following data were obtained for the concentration of HBr in the reaction

$$4 HBr(g) + O_2(g) \rightarrow 2 Br_2(g) + 2 H_2O(g)$$

holding the concentration of O_2 constant:

Conc. HBr (M)	1.00	0.79	0.63	0.50
Time (min)	0	1	2	3

Following the procedure of Example 14.3, find the order of reaction with respect to HBr.

14.7 The following data are for the gas phase decomposition of CH_3NO_2 at 473 K.

t (min)	Conc. (M)	t (min)	Conc. (M)
0	0.200	4	0.187
1	0.197	8	0.175
2	0.193	16	0.153
3	0.190		

    a. By plotting the data, show that the reaction is first order.
    b. From the graph, determine k.
    c. Using k, find the time for the concentration to drop to one fourth of the original concentration.

14.8 The first order decomposition of nitromethane, CH_3NO_2 has a rate constant of 0.0170/min at 200°C. Starting with a concentration of 0.400 mol/ℓ of CH_3NO_2, calculate

    a. the concentration of CH_3NO_2 after 15.0 min.
    b. how long it will take for the concentration to drop from 0.400 mol/ℓ to 0.100 mol/ℓ.
    c. the half-life.

14.9 In the first order decomposition of CH_3NO_2 at 400 K, it is found that 20.0% of a sample has decomposed in 82.4 min. How long will it take 50.0% of the sample to decompose?

14.23 The rate of the reaction $D(g) \rightarrow$ products is 0.030 mol/($\ell \cdot$ min) when conc. D is 0.150 mol/ℓ. What is the rate constant, k, if the reaction is

    a. zero order in D?
    b. first order in D?
    c. second order in D?

14.24 For the reaction referred to in Problem 14.5, the rate is measured again when the concentration of A is 0.020 M. What is the order of the reaction if the ratio of this rate to the original rate is

    a. 0.040?   b. 0.20?   c. 0.0080?

14.25 For a certain reaction $A(g) \rightarrow$ products, what is the order of the reaction (Table 14.2) if the half-life is

    a. independent of the original concentration.
    b. directly proportional to the original concentration.
    c. inversely proportional to the original concentration.

14.26 The following data applies to the first order decomposition of dimethyl ether, $(CH_3)_2O$.

t (s)	Conc. (M)
0	0.01000
200	0.00916
400	0.00839
600	0.00768
800	0.00703

    a. From a plot of conc. vs. time, estimate the rate at t = 800 s.
    b. From a plot of log conc. vs. time, determine k, the rate constant. Use k to obtain the rate at t = 800 s.
    c. Which of the rates just calculated in (a) or (b) do you think is more accurate? Explain.

14.27 The half-life for the first order decomposition of acetone, $(CH_3)_2CO$, is 5.8 s at 650°C. Calculate

    a. the rate constant, k.
    b. how long it will take for the acetone concentration to change from 0.020 mol/ℓ to 0.0020 mol/ℓ.
    c. the acetone concentration 8.0 min after an initial concentration of 6.0 mol/ℓ.

14.28 The first order decomposition of diazomethane, CH_2N_2, has a half-life of 17.3 min at 873 K. The concentration of CH_2N_2 is 0.076 mol/ℓ after 5.0 min.

    a. What was the original concentration?
    b. How long will it take for 40% of the original sample to decompose?

14.10 Consider the data for several systems for the conversion of reactants (R) to products (P) at the same temperature.

System	E_a (kJ)	E_a' (kJ)
1	30	55
2	70	20
3	16	35

a. Which system has the highest forward rate?
b. What is ΔH of reaction for System 1?
c. For which system(s) is the forward reaction endothermic?

14.11 For the reaction

$$H_2(g) + Cl_2(g) \rightarrow 2\ HCl(g)$$

E_a is 155 kJ. Use the data in Table 6.1 to calculate the activation energy for the reverse reaction. Prepare a diagram similar to Figure 14.6 for this reaction.

14.12 The following data were obtained for the reaction

$$SiH_4(g) \rightarrow Si(s) + 2\ H_2(g)$$

k (s^{-1})	0.048	2.3	49	590
t (°C)	500	600	700	800

Plot these data (log k vs. $1/T$) and find the activation energy for the reaction.

14.13 The rate constants $\dfrac{\ell}{\text{mol} \cdot \text{s}}$ are 1.1 and 3.0 at 823 and 873 K, respectively, for the gas phase reaction

$$CH_4(g) + 2\ S_2(g) \rightarrow CS_2(g) + 2\ H_2S(g)$$

Calculate the activation energy of the reaction.

14.14 For a certain reaction, $E_a = 82$ kJ. The rate constant k is 1.2×10^{-2} ℓ/(mol · s) at 300 K. What is k at 400 K?

14.15 Cold-blooded animals decrease their body temperature in cold weather to match that of their environment. The activation energy of a certain enzyme-catalyzed reaction in a cold-blooded animal is 65 kJ. By what percentage is the rate of this reaction decreased if the body temperature of the animal drops from 30°C to 20°C?

14.29 The activation energy of a certain reaction is 50 kJ. In the presence of a catalyst, E_a is reduced to 30 kJ. For the reaction, ΔH is −70 kJ. Draw a diagram similar to Figure 14.7 for this reaction.

14.30 The activation energy for the decomposition of ammonia to the elements is about 300 kJ.

$$NH_3(g) \rightarrow \tfrac{1}{2}\ N_2(g) + \tfrac{3}{2}\ H_2(g)$$

Using Table 6.1, obtain the activation energy for the reverse reaction and draw a diagram for this reaction similar to Figure 14.6.

14.31 The following data are for the gas phase decomposition of acetaldehyde.

$k\left(\dfrac{\ell}{\text{mol} \cdot \text{s}}\right)$	0.0105	0.101	0.60	2.92
T (K)	700	750	800	850

Plot these data and determine the activation energy for the reaction.

14.32 For the gas phase decomposition of N_2O_5, k is 1.4/s at 400 K and 43/s at 450 K. What is the activation energy for this reaction?

14.33 For the reaction in Problem 14.14, at what temperature does $k = 1.2 \times 10^{-4}$ ℓ/(mol · s)?

14.34 The chirping rate of a cricket, X, in chirps per minute, near room temperature is given by

$$X = 7.2\ t - 32$$

where t is the temperature in °C. Calculate the chirping rates at 20 and 30°C, and use them to estimate the activation energy for this reaction.

14.16 For the reaction

$$H_2(g) + Br_2(g) \rightarrow 2\ HBr(g)$$

a proposed mechanism is

Step 1: $Br_2 \rightleftharpoons 2\ Br$ (fast)
Step 2: $Br + H_2 \rightarrow HBr + H$ (slow)
Step 3: $H + Br_2 \rightarrow HBr + Br$ (fast)

Using this mechanism, determine the rate expression in terms of the concentrations of H_2 and Br_2.

14.17 At low temperatures, the rate law for the reaction $CO(g) + NO_2(g) \rightarrow CO_2(g) + NO(g)$ is rate = constant $\times$ (conc. $NO_2)^2$. Which of the following mechanisms is consistent with this rate law?

a. $CO + NO_2 \rightarrow CO_2 + NO$
b. $2\ NO_2 \rightleftharpoons N_2O_4$ (fast)
   $N_2O_4 + 2\ CO \rightarrow 2\ CO_2 + 2\ NO$ (slow)
c. $2\ NO_2 \rightarrow NO_3 + NO$ (slow)
   $NO_3 + CO \rightarrow NO_2 + CO_2$ (fast)
d. $2\ NO_2 \rightarrow 2\ NO + O_2$ (slow)
   $2\ CO + O_2 \rightarrow 2\ CO_2$ (fast)

14.18 In your own words explain why

a. a decrease in temperature slows the rate of a reaction.
b. doubling the concentration of a reactant does not always double the rate of the reaction.
c. a flame lights a cigarette but the cigarette continues to burn after the flame is removed.
d. a catalyst does not change the equilibrium constant, K_c.

14.19 The decomposition of acetaldehyde occurs at a rate of 0.060 mol/($\ell \cdot$ min). Express this rate in

a. mol/($\ell \cdot$ s)   b. mol/($\ell \cdot$ h)

14.20 For the reaction

$$4\ HBr(g) + O_2(g) \rightarrow 2\ Br_2(g) + 2\ H_2O(g)$$

the rate expression is most often written as rate = $-\Delta$ conc. $O_2/\Delta t$. Write expressions equivalent to this in terms of

a. conc. HBr   b. conc. Br_2 c. conc. H_2O

*14.40 Using calculus, derive the equation for

14.35 Two mechanisms proposed for the reaction

$$2\ NO(g) + O_2(g) \rightarrow 2\ NO_2(g)$$

are   Mechanism 1: $NO + O_2 \rightleftharpoons NO_3$ (fast)
           $NO_3 + NO \rightarrow 2\ NO_2$ (slow)
       Mechanism 2: $NO + NO \rightleftharpoons N_2O_2$ (fast)
           $N_2O_2 + O_2 \rightarrow 2\ NO_2$ (slow)

Show that each of these mechanisms is consistent with the observed rate law, rate = k(conc. $NO)^2 \times$ (conc. O_2).

14.36 Suggest a mechanism for the CO–NO_2 reaction, other than a simple collision, which would lead to the observed rate expression rate = k(conc. CO) $\times$ (conc. NO_2). (Recall the discussion of the H_2–I_2 reaction.)

14.37 In terms of the concepts of this chapter, explain in your own words why

a. diamond is not converted to graphite even though the process has a large K_c value.
b. the rate constant for a zero order reaction equals the reaction rate.
c. an inhibitor reduces the rate of an enzyme-catalyzed reaction.
d. the slow step in a reaction determines the overall rate of the reaction.

14.38 The reaction

$$2\ ICl(g) + H_2(g) \rightarrow I_2(g) + 2\ HCl(g)$$

is second order overall. Write three rate expressions that are consistent with this.

14.39 Express the rate of the reaction

$$2\ HI(g) \rightarrow H_2(g) + I_2(g)$$

a. in terms of Δ conc. H_2.
b. in terms of Δ conc. HI, if you want the rate to be the same as in (a).

a. the concentration-time relation for a second order reaction (see Table 14.2).
b. the concentration-time relation for a third order reaction, 3 A $\rightarrow$ products.

*14.41 The following data apply to the reaction

$$A(g) + 3 B(g) + 2 C(g) \rightarrow products$$

Concentrations are given in mol/ℓ.

Conc. A	Conc. B	Conc. C	Rate
0.20	0.40	0.10	X
0.40	0.40	0.20	8X
0.20	0.20	0.20	X
0.40	0.40	0.10	4X

Determine the rate law for the reaction.

*14.42 For the reaction $CO(g) + Cl_2(g) \rightarrow COCl_2(g)$, the rate expression is rate = k × (conc. CO) × (conc. Cl_2)$^{3/2}$. It is believed that the slow step in the mechanism is $COCl(g) + Cl_2(g) \rightarrow COCl_2(g) + Cl(g)$. Write a mechanism consistent with these facts.

*14.43 The decomposition of N_2O_5 is believed to occur by the following mechanism:

$$2 N_2O_5(g) \rightleftharpoons N_2O_5^*(g) + N_2O_5(g)$$
$$N_2O_5^*(g) \rightleftharpoons NO_2(g) + NO_3(g)$$
$$NO_2(g) + NO_3(g) \rightarrow NO_2(g) + O_2(g) + NO(g)$$
$$NO(g) + NO_3(g) \rightarrow 2 NO_2(g)$$

where $N_2O_5^*$ represents an activated molecule. Show that the rate of formation of O_2 is directly proportional to the concentration of N_2O_5.

THE ATMOSPHERE

Life on this planet depends upon the relatively thin layer of air that surrounds it. The atmosphere accounts for only about 0.0001% of the total mass of the earth. Yet it is the reservoir from which we draw oxygen for metabolism, carbon dioxide for photosynthesis, and nitrogen, whose compounds are essential to plant growth. Our climate is governed by the movement of water vapor from the earth's surface into the atmosphere and back again.

Even trace components of the atmosphere can affect the delicate balance of life. Small amounts of ozone at a height of about 30 km absorb most of the harmful ultraviolet radiation of the sun. On the other hand, as little as 0.2 part per million of ozone near the earth's surface promotes smog formation.

Table 15.1 gives the mole fractions of gases in the atmosphere. Two species are omitted. One is water vapor, whose mole fraction may vary from 0.02 in the tropics to 0.0005 in polar regions. The other comprises suspended particles (e.g., dust, smoke) which vary in both concentration and chemical composition. You will note that there are five major components of air (N_2, O_2, Ar, H_2O, and CO_2). Together, their mole fractions total about 0.99997.

The concentrations of minor components of the atmosphere are often expressed in *parts per million* (ppm) or parts per billion (ppb). These units are simply related to mole fraction. If a species is present at a concentration of 1 ppm, there is 1 mol of that species in 10^6 mol of air (or one molecule per million molecules of air). In other words,

$$\text{parts per million (ppm)} = 10^6 \, X$$

where X is mole fraction. Similarly,

$$\text{parts per billion (ppb)} = 10^9 \, X$$

TABLE 15.1 COMPOSITION OF CLEAN, DRY AIR AT SEA LEVEL					
COMPONENT	**MOLE FRACTION**	**COMPONENT**	**MOLE FRACTION**	**COMPONENT**	**MOLE FRACTION**
N_2	0.7808	Ne	1.82×10^{-5}	SO_2	$<1 \times 10^{-6}$
O_2	0.2095	He	5.24×10^{-6}	O_3	$<1 \times 10^{-7}$
Ar	0.00934	CH_4	2×10^{-6}	NO_2	$<2 \times 10^{-8}$
CO_2	0.00033	Kr	1.14×10^{-6}	I_2	$<1 \times 10^{-8}$
		H_2	5×10^{-7}	NH_3	$<1 \times 10^{-8}$
		N_2O	5×10^{-7}	CO	$<1 \times 10^{-8}$
		Xe	8.7×10^{-8}	NO	$<1 \times 10^{-8}$

Thus, from the data in Table 15.1

$$\text{ppm He} = 5.24 \times 10^{-6} \times 10^6 = 5.24$$

$$\text{ppb Xe} = 8.7 \times 10^{-8} \times 10^9 = 87$$

In previous chapters (5, 8, and 12) we looked at the physical and chemical properties of several components of the atmosphere (N_2, O_2, and the noble gases). With that background, we will consider in this chapter what might be called "selected topics" in atmospheric chemistry. These include:

—the "fixation" of atmospheric nitrogen—that is, its conversion to useful nitrogen compounds such as ammonia and nitric acid (Section 15.1).

—the effect of water vapor and carbon dioxide on our weather and climate (Section 15.2).

—the chemistry of the upper atmosphere (Section 15.3).

—air pollution (Section 15.4).

Throughout this chapter, we will review and expand upon the principles of chemical equilibrium and reaction rates, introduced in Chapters 13 and 14.

15.1 NITROGEN FIXATION

Combined nitrogen in the form of protein is essential to both plant and animal life. There is more than enough elementary nitrogen in the air, about 5×10^{18} kg, to meet all our needs. The problem is to convert the element to compounds which can be used by plants to make proteins. At room temperature and atmospheric pressure, N_2 does not react with any other element. Its inertness is due to the strength of the triple bond holding the N_2 molecule together:

$$:N\equiv N:(g) \rightarrow 2 \cdot \ddot{N}\cdot(g); \; \Delta H = 941 \text{ kJ}$$

The high stability of the bond implies that the activation energy for any reaction of N_2 is likely to be high. Consequently, we expect the rate of reaction to be slow.

Certain bacteria found on the roots of plants such as peas, beans, clover, and alfalfa have the ability to "fix" nitrogen. That is, they can convert N_2 to compounds such as NH_3. These bacteria contain enzymes which catalyze the conversion. The mechanism of the process is poorly understood. However, it seems that two different enzymes are involved. Both contain iron atoms; one of the enzymes contains molybdenum as well. It appears that molecular nitrogen forms a weak complex with the metal atoms of the enzymes. This is then converted to a more stable ammonia complex. Recently, this process has been simulated in the laboratory, using certain compounds of another transition metal, titanium.

For thousands of years, nitrogen compounds have been added to the soil to increase the yield of food crops. Until about 100 years ago, the only way to do this was to add "organic nitrogen" (i.e., manure). Late in the nineteenth century, it became common practice in the United States and Western Europe to use sodium nitrate, $NaNO_3$, imported from Chile. Then, in 1908, Fritz Haber in Germany showed that atmospheric nitrogen could be fixed by reacting it with hydrogen to form ammonia.

The Haber Process for Making Ammonia

The reaction which Haber used was

$$N_2(g) + 3 \; H_2(g) \rightleftharpoons 2 \; NH_3(g); \; \Delta H = -92.4 \text{ kJ} \tag{15.1}$$

TABLE 15.2 EFFECT OF TEMPERATURE AND PRESSURE UPON THE YIELD OF AMMONIA IN THE HABER PROCESS ($[H_2] = 3[N_2]$)

		MOL % NH₃ IN EQUILIBRIUM MIXTURE				
°C	K_c	10 atm	50 atm	100 atm	300 atm	1000 atm
200	650	51	74	82	90	98
300	9.5	15	39	52	71	93
400	0.5	4	15	25	47	80
500	0.08	1	6	11	26	57
600	0.014	0.5	2	5	14	13

His research was supported by German industrialists, who wanted to convert the ammonia to nitric acid, a starting material for making explosives. In 1913, the first large scale ammonia plant went into production. During World War I, the Haber process produced enough ammonia and nitric acid to make Germany independent of foreign supplies of sodium nitrate, which were cut off by the British blockade. As has happened so many times, a scientific development was used first for military purposes and only later to meet social needs.

The Haber process is now the main synthetic source of fixed nitrogen in the world. Its feasibility depends upon choosing conditions under which nitrogen and hydrogen will react rapidly to give a high yield of ammonia. At room temperature and atmospheric pressure, the position of the equilibrium favors the formation of ammonia ($K_c = 5 \times 10^8$). However, the rate of reaction is virtually zero. Equilibrium can be reached more rapidly by increasing the temperature. But, since Reaction 15.1 is exothermic, high temperatures reduce K_c and hence the yield of ammonia. High pressures, on the other hand, have a favorable effect on both the rate of reaction and the position of the equilibrium. An increase in pressure brings the gas molecules closer together. They collide more frequently, so equilibrium is reached more rapidly. High pressure also increases the relative amount of ammonia at equilibrium, since Reaction 15.1 results in a decrease in the number of moles of gas (4 mol → 2 mol).

Note how changing the conditions influences yield

Much of Haber's research involved finding a catalyst to make Reaction 15.1 take place at a reasonable rate without going to very high temperatures. Nowadays, the catalyst used is a special mixture of iron, potassium oxide, and aluminum oxide. The reaction is carried out at 400 to 450°C and pressures of 200 to 600 atm. Ammonia (bp = −33°C) is condensed out as a liquid from the gaseous mixture. Unreacted hydrogen and nitrogen are recycled to raise the yield of ammonia.

Example 15.1   According to Table 15.2, the mol % of NH₃ at equilibrium at 300°C and 100 atm is 52. The equilibrium concentration of H_2 is three times that of N_2. What are the mol % of N_2 and H_2?

Solution   The total mol % (N_2, H_2, NH_3) must add up to 100:

$$\text{mol } \% \ N_2 + \text{mol } \% \ H_2 = 100 - 52 = 48$$

But, if $[H_2] = 3[N_2]$, then

$$\text{mol } \% \ H_2 = 3 \times \text{mol } \% \ N_2$$

Hence $4 \times \text{mol } \% \ N_2 = 48$; $\text{mol } \% \ N_2 = 12$; $\text{mol } \% \ H_2 = 36$

Under these extreme conditions the gases are not ideal

Exercise   What are the mol % of N_2 and H_2 at 400°C and 1000 atm? Answer: 5, 15.

Most of the ammonia produced today is used to make fertilizers. The pure liquid under pressure can be used directly. More commonly, it is converted to compounds containing the NH_4^+ ion. This is done by adding the ammonia to an acidic water solution:

$$NH_3(aq) + H^+(aq) \rightarrow NH_4^+(aq) \tag{15.2}$$

If the acid used is nitric acid, the final product is ammonium nitrate, NH_4NO_3. When sulfuric acid is used, ammonium sulfate, $(NH_4)_2SO_4$, is formed. Still another fertilizer made from ammonia is urea, an organic compound with the structure

$$
\begin{array}{c}
\text{H---N---C---N---H} \\
\; | \quad\; \| \quad\; | \\
\text{H} \quad \text{O} \quad \text{H}
\end{array}
$$

urea

Urea is made from ammonia by reaction with carbon dioxide:

$$2 NH_3(g) + CO_2(g) \rightarrow (NH_2)_2CO(s) + H_2O(l) \tag{15.3}$$

The hydrogen used as a reactant in the Haber process accounts for virtually the entire cost of the ammonia formed. As pointed out in Chapter 5, most of our hydrogen comes from the reaction of natural gas with steam. Until quite recently, the price of hydrogen had been stable for decades at about 2 cents a kilogram. A sharp rise in the price of natural gas changed this situation in the early 1970's. It raised the price of hydrogen, the ammonia made from hydrogen, and the fertilizers made from ammonia. The effect has been particularly severe for many underdeveloped countries. They must have an abundant supply of cheap fertilizer to avert mass starvation. Two approaches are possible. One is to find a cheaper way to make hydrogen. The other is to discover a process for fixing nitrogen that does not involve hydrogen as a reactant.

The N_2 is obtained by distilling liquid air

The Ostwald Process for Making Nitric Acid

About 15% of the ammonia made by the Haber process is converted to nitric acid. The process by which this is done was developed by a German chemist, Wilhelm Ostwald. The reaction takes place in three steps. All of these are carried out at a relatively low pressure, 1 to 10 atm.

1. Ammonia is burned in air at about 1000°C, using a platinum catalyst. Under these conditions, more than 95% of the ammonia is converted to nitric oxide, NO:

With no catalyst you get $N_2(g)$

$$4 NH_3(g) + 5 O_2(g) \rightarrow 4 NO(g) + 6 H_2O(g) \tag{15.4}$$

2. The gaseous mixture produced is mixed with more air. This lowers the temperature and brings about the reaction

$$2 NO(g) + O_2(g) \rightleftharpoons 2 NO_2(g) \tag{15.5}$$

3. The nitrogen dioxide produced in the second step is passed through water. In this way a solution of nitric acid is formed:

$$3 NO_2(g) + H_2O(l) \rightarrow NO(g) + 2 HNO_3(aq) \tag{15.6}$$

The NO formed as a by-product is recycled in Reaction 15.5. The aqueous solution formed by Reaction 15.6 contains about 60 mass % HNO_3. To obtain the anhydrous acid, H_2SO_4 is added and the mixture is distilled. Nearly pure nitric acid (bp = 86°C) boils off and is condensed.

So all the NO is finally converted to HNO_3

Example 15.2   Consider Reaction 15.5: $2 NO(g) + O_2(g) \rightleftharpoons 2 NO_2(g)$; $\Delta H = -113.0$ kJ. Applying the principles discussed in Chapters 13 and 14, what would you expect to happen to the rate and the yield of NO_2 at equilibrium if
   a. the pressure were increased?
   b. the temperature were increased?
   c. a catalyst were used?

Solution
   a. An increase in pressure should increase the rate (higher concentration of reactants). It should also increase the yield of NO_2 (3 mol gas → 2 mol gas).
   b. One would expect an increase in temperature to increase the rate; k usually increases with T (see, however, Problems 15.21 and 15.42). An increase in temperature would decrease the yield of NO_2 since the forward reaction is exothermic.
   c. A suitable catalyst would speed up the reaction but would have no effect upon the yield of NO_2.

Exercise   Suppose pure O_2 were used in this reaction instead of air (same total pressure). What effect would you expect this to have on the rate and yield of NO_2? Answer: Should increase both.

The major use of nitric acid today is in the manufacture of ammonium nitrate, NH_4NO_3, a fertilizer component and industrial explosive. Nitric acid is also used to make other explosives, such as nitroglycerine and trinitrotoluene (TNT).

Could we make NH_4NO_3 from just air and water?

15.2 WATER VAPOR AND CARBON DIOXIDE; WEATHER AND CLIMATE

The properties and composition of the atmosphere determine our weather over the short term and our climate over the years. Many factors affect both weather and climate. Among these are the presence in air of two gases: water vapor and carbon dioxide.

Relative Humidity

The concentration of water vapor in the air is often expressed in terms of the *relative humidity:*

$$\text{R. H.} = \frac{P_{H_2O}}{P^0_{H_2O}} \times 100\% \qquad (15.7)$$

Here P_{H_2O} is the partial pressure of water vapor in the air while $P^0_{H_2O}$ is the equilibrium vapor pressure of water at the same temperature. On a day when the temperature is 25°C ($P^0_{H_2O} = 23.8$ mm Hg) and the partial pressure of water vapor in the air is 20.0 mm Hg,

$$\text{R. H.} = \frac{20.0}{23.8} \times 100\% = 84.0\%$$

At 84% R.H., the air contains 84% of the water that it could hold at that temperature

Our comfort depends upon relative humidity as well as temperature. At relative humidities below about 30%, evaporation of water from body surfaces is extensive and rapid. Mouth and nasal membranes dry out, allowing viruses to enter the lungs more readily. This may be why colds are so common in the winter months when the relative humidity indoors is often quite low. Most of us are familiar with the uncomfortable effects of high relative humidities, above 80%. Perspiration fails to evaporate, leaving us feeling clammy and hot.

Changes in temperature can bring about marked variations in relative humidity. From Equation 15.7, we note that relative humidity is inversely related to the equilibrium vapor pressure of water, $P^0_{H_2O}$. Since $P^0_{H_2O}$ increases with temperature, this means that relative humidity tends to drop when the temperature rises (Example 15.3).

Example 15.3   On a cold winter day, when the outside temperature is 0°C (32°F), the partial pressure of water in the air is 3.0 mm Hg. Calculate the relative humidity of the outside air and the relative humidity of the same air when it is brought into a house and warmed to 20°C (68°F).

Solution   The equilibrium vapor pressure of water is 4.6 mm Hg at 0°C and 19.8 mm Hg at 20°C (Appendix 1). Consequently

$$\text{R. H. outside} = \frac{3.0}{4.6} \times 100\% = 65\%$$

$$\text{R. H. inside} = \frac{3.0}{19.8} \times 100\% = 15\%$$

Exercise   At what temperature would the relative humidity become 10%? Answer: 29°C (84°F), where vp H_2O is 30.0 mm Hg.

Cloud Formation and Weather Modification

If a warm air mass is suddenly cooled, $P^0_{H_2O}$ in Equation 15.7 drops and the relative humidity rises. When it reaches 100%, liquid water condenses. This is precisely the way in which clouds are formed in the atmosphere. Clouds consist of many billions of tiny droplets of liquid water, on the average perhaps 0.01 mm in diameter. Droplets of this size are too small to fall to the earth's surface as rain. The growth of small water droplets at temperatures above 0°C is ordinarily a very slow process. The rate of growth is increased in the presence of dust particles, which act as nuclei upon which small droplets can condense. This explains why volcanic eruptions are often followed by rainstorms.

More frequently ice crystals formed in the colder upper regions of clouds act as nuclei for precipitation. In principle, ice crystals should form at 0°C; in practice, they seldom develop unless the temperature drops to at least −15°C. To stimulate the formation of ice crystals, clouds are sometimes seeded with dry ice (Fig. 15.1). The sublimation of solid carbon dioxide absorbs enough heat from the cloud to reduce the temperature below that required for ice crystal formation. Another substance that is frequently used is finely divided silver iodide, which has a crystal structure similar to that of ice. The presence of silver iodide tends to prevent supercooling and, hence, allows ice crystals to form at temperatures close to 0°C.

Silver iodide has also been used with some success in seeding hurricanes. The objective here is to add so much silver iodide that an enormous number of tiny ice crystals form. These crystals are too small to bring about precipitation, but the heat

Figure 15.1 Cloud seeding with Dry Ice. The particles of dry ice sublime rapidly in the cloud, cooling the fine water droplets to the point where they freeze and act as nuclei for condensation of other droplets in the cloud.

evolved in their formation tends to dissipate the storm clouds at the eye of the hurricane. One difficulty is that hurricane seeding may simply shift the storm off course without weakening it appreciably. For this reason, seeding experiments are limited to storms that are far removed from populated areas.

The Greenhouse Effect

The mean global temperature at the earth's surface is about 15°C (59°F). This temperature is determined by a delicate balance between:

—the energy which is absorbed from the sun. This energy covers a broad spectrum of wavelengths from the ultraviolet (<400 nm) through the visible (400 to 800 nm) into the infrared (>800 nm).

—the energy emitted back into space by the earth. This consists of infrared radiation, mostly at wavelengths between 5000 and 25,000 nm.

Any change in the amount of energy absorbed or emitted by the earth could upset this balance, affecting our climate. In this connection, we must consider what happens to the infrared radiation given off by the earth. Part of it is absorbed by the atmosphere rather than being lost to outer space. Two gases in the air, H_2O and CO_2, absorb infrared radiation (Fig. 15.2, p. 358). This way, they act as an insulating blanket to prevent heat from escaping; this is often referred to as a "greenhouse effect." Were it not for this effect, the mean global temperature would be −25°C (−13°F) rather than 15°C (59°F).

Of the two gases, water vapor absorbs more infrared radiation than carbon dioxide because its concentration is higher. This property of water vapor accounts for the fact that the temperature drops less on nights when there is a heavy cloud cover. In desert regions, where there is very little water vapor, large variations between day and night temperatures are common.

The concentration of CO_2 in the atmosphere is low (about 330 ppm) but is known

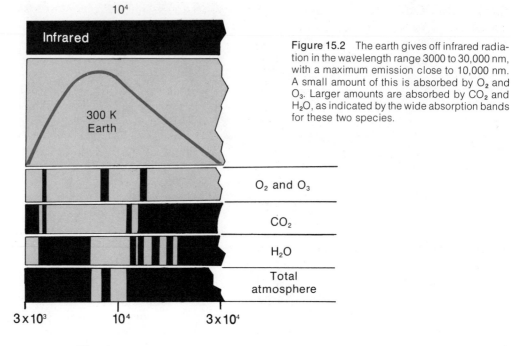

Figure 15.2 The earth gives off infrared radiation in the wavelength range 3000 to 30,000 nm, with a maximum emission close to 10,000 nm. A small amount of this is absorbed by O_2 and O_3. Larger amounts are absorbed by CO_2 and H_2O, as indicated by the wide absorption bands for these two species.

to be increasing. A century ago the CO_2 concentration was less than 300 ppm. The increase has been caused by man's activities. Increased consumption of fossil fuels is mainly responsible. Extensive land clearing, which reduces the amount of CO_2 consumed in photosynthesis, is also a factor. If this trend continues, the CO_2 level could approach 400 ppm by the year 2000. This could increase temperatures by as much as 1°C, producing important changes in our climate.

15.3   THE UPPER ATMOSPHERE

The species in air near the surface of the earth are mostly stable atoms and molecules such as N_2, O_2, Ar, CO_2, and H_2O. Above about 30 km, the situation is quite dif-

TABLE 15.3   FORMATION OF HIGH-ENERGY SPECIES IN THE UPPER ATMOSPHERE

MOLECULAR FRAGMENTS			CATIONS		
Reaction	ΔE (kJ/mol)	λ_{max}* (nm)	Reaction	ΔE (kJ/mol)	λ_{max}* (nm)
$NO_2 \rightarrow NO + O$	+305	392	$NO \rightarrow NO^+ + e^-$	+920	130
$O_2 \rightarrow O + O$	+494	242	$O_2 \rightarrow O_2^+ + e^-$	+1109	108
$H_2O \rightarrow H + OH$	+502	238	$O \rightarrow O^+ + e^-$	+1314	91
$NO \rightarrow N + O$	+632	189	$H \rightarrow H^+ + e^-$	+1314	91
$N_2 \rightarrow N + N$	+941	127	$N_2 \rightarrow N_2^+ + e^-$	+1510	79

*Longest wavelength of light which can supply the energy indicated.

ferent. Exploration of the upper atmosphere by rockets and satellites reveals additional species such as those listed in Table 15.3. Some of these are neutral particles (oxygen, nitrogen, and hydrogen atoms, and OH radicals). Others are positive ions formed from diatomic molecules (NO^+, O_2^+, N_2^+) or atoms (O^+, H^+). At high altitudes, some of these species make up an appreciable fraction of air. At a height of 120 km, there are about as many free O atoms as O_2 molecules.

All the reactions listed in Table 15.3 are endothermic. They require either the breaking of a chemical bond or the removal of an electron. In the upper atmosphere, sunlight is the source of energy for such reactions. The wavelengths of light given in Table 15.3 are calculated from the Einstein equation (Chap. 8) in the form

$$\lambda_{max} \text{ (nm)} = \frac{1.196 \times 10^5}{\Delta E(\text{kJ/mol})} \qquad (15.8)$$

Notice that all of the wavelengths listed are in the ultraviolet region, below 400 nm. Very little of this high-energy radiation reaches the surface of the earth, so the reactions in Table 15.3 are not very likely to occur in the air around us. Above 30 km, significant amounts of radiation in the near UV (200 to 400 nm) are available and species such as O, H, and OH begin to appear. At still higher altitudes, sunlight contains radiation in the far UV (<200 nm). This can bring about highly endothermic processes, forming such species as N, O_2^+, and N_2^+.

The UV is absorbed when the reactions in Table 15.3 occur

Example 15.4   The high-energy species shown in Table 15.3 react with one another or, more often, with stable molecules. A typical reaction is that between oxygen atoms and water molecules:

$$O(g) + H_2O(g) \rightarrow 2\ OH(g)$$

Kinetic studies show that the rate constant k for this reaction is 2×10^{-3} $\ell/(\text{mol} \cdot \text{s})$ at 25°C and the activation energy, E_a, is 77.4 kJ. What is the value of k at an altitude of 20 km, where the temperature is −55°C?

Solution   The appropriate equation, from Chapter 14, is

$$\log_{10} \frac{k_2}{k_1} = \frac{E_a(T_2 - T_1)}{19.14\ T_2 T_1}$$

Here, $E_a = 77,400$ J, $T_2 = 298$ K, and $T_1 = 218$ K. Thus,

$$\log_{10} \frac{k_2}{k_1} = \frac{77,400\ (80)}{19.14(298)(218)} = 5.0; \quad \frac{k_2}{k_1} = 1 \times 10^5$$

$$k_1 = \frac{k_2}{1 \times 10^5} = \frac{2 \times 10^{-3}}{1 \times 10^5} = 2 \times 10^{-8}\ \ell/(\text{mol} \cdot \text{s})$$

Note that the rate constant at this temperature is only 1/100,000 of that at room temperature.

Exercise   At an altitude of 50 km, the temperature is about 0°C. What is k for this reaction at that temperature? Answer: 1×10^{-4} $\ell/(\text{mol} \cdot \text{s})$.

Ozone

One component of the upper atmosphere has received more attention than any other in recent years. This is ozone, an allotropic form of oxygen, whose electronic structure was discussed in Chapter 12. Interest in O_3 centers upon its ability to absorb ultraviolet light. From 95 to 99% of sunlight in the wavelength range of 200 to 300 nm is absorbed by ozone in the upper atmosphere. If this radiation were to reach the surface of the earth, it could have several adverse effects. A decrease in O_3 concentration of only 5% could produce an additional 8000 cases of skin cancer annually in the United States.

The concentration of ozone in the earth's atmosphere passes through a maximum of about 10 ppm at an altitude of 30 km. The ozone in this region is formed by a two-step process. The first step involves the dissociation of an O_2 molecule:

$$O_2(g) \rightarrow 2\ O(g) \tag{15.9}$$

This is followed by a collision between an oxygen atom and an O_2 molecule:

$$O_2(g) + O(g) \rightarrow O_3(g) \tag{15.10}$$

Ozone molecules formed by Reaction 15.10 decompose by several mechanisms. One of the most important is

$$O_3(g) + O(g) \rightarrow 2\ O_2(g) \tag{15.11}$$

This reaction takes place at a rather slow rate by direct collision between an O_3 molecule and an O atom. It can occur more rapidly by a two-step process in which a trace component of the upper atmosphere acts as a catalyst. Two such catalysts which have received a great deal of attention are listed in Table 15.4.

Since NO acts as a catalyst for ozone decomposition, an increase in nitric oxide concentration in the upper atmosphere could cause the depletion of ozone. This was one factor which influenced the United States to abandon the development of the supersonic transport (SST). These planes burn jet fuel with air at high temperatures. It was suggested that combustion might produce significant amounts of NO by the reaction

$$N_2(g) + O_2(g) \rightarrow 2\ NO(g) \tag{15.12}$$

However, research has shown that SST's probably *decrease* the concentration of NO rather than increasing it. Other species, such as OH radicals, formed in high-altitude flights react with NO already in the upper atmosphere.

More recently, concern has focused on the Cl-catalyzed decomposition of ozone. At the time this mechanism was discovered, in 1973, it was believed to be unimportant. There was no known source of Cl atoms in the upper atmosphere. Less than a year later it was suggested that two organic compounds, $CFCl_3$ and CF_2Cl_2, might be a source of Cl. These compounds are widely used as refrigerants and aerosol propellants. They decompose to form chlorine atoms when exposed to UV radiation at 200 nm:

These compounds are called Freons; both boil below 0°C, which makes them good refrigerants

TABLE 15.4   MECHANISM FOR THE CATALYTIC DECOMPOSITION OF OZONE

	CATALYST	
	NO Molecule	**Cl Atom**
Mechanism	$NO + O_3 \rightarrow NO_2 + O_2$ $NO_2 + O \rightarrow NO + O_2$	$Cl + O_3 \rightarrow ClO + O_2$ $ClO + O \rightarrow Cl + O_2$
Overall reaction	$O_3 + O \rightarrow 2\ O_2$	$O_3 + O \rightarrow 2\ O_2$

$$CFCl_3(g) \rightarrow CFCl_2(g) + Cl(g) \qquad (15.13)$$

$$CF_2Cl_2(g) \rightarrow CF_2Cl(g) + Cl(g) \qquad (15.14)$$

Light at this wavelength is readily available at a height of 40 km. Significant amounts of $CFCl_3$, CF_2Cl_2, and Cl atoms have been detected at this altitude. So far, the evidence is mixed regarding the lowering of O_3 concentration by this mechanism. To ensure that depletion of the ozone layer does not occur in the future, the use of such compounds as aerosol propellants has been phased out in the United States, Canada, and Sweden. They are being replaced by other gases, including CO_2, C_3H_8, and C_4H_{10}.

Example 15.5   Consider the reaction mechanisms

$$O_3(g) + O(g) \rightarrow 2\ O_2(g); \ k_1 = 5 \times 10^6 \ \ell/(mol \cdot s)$$

$$O_3(g) + NO(g) \rightarrow NO_2(g) + O_2(g); \ k_2 = 1 \times 10^7 \ \ell/(mol \cdot s)$$

a. Write the rate expressions for the two mechanisms.
b. Calculate the ratio of the two rates (rate$_2$/rate$_1$) at an altitude of 40 km. Take the concentrations of O and NO to be 2×10^{-12} and 3×10^{-12} mol/ℓ, in that order.

Solution

a. rate$_1$ = k_1(conc. O_3) × (conc. O)
   rate$_2$ = k_2(conc. O_3) × (conc. NO)
b. Dividing: $\dfrac{\text{rate}_2}{\text{rate}_1} = \dfrac{k_2}{k_1} \times \dfrac{\text{conc. NO}}{\text{conc. O}} = \dfrac{1 \times 10^7}{5 \times 10^6} \times \dfrac{3 \times 10^{-12}}{2 \times 10^{-12}} = 3$

This calculation suggests that, under these conditions, the NO-catalyzed decomposition of O_3 is occurring three times as fast as the direct decomposition.

Exercise   Suppose the concentration of O were twice that of NO. How would the two rates compare in that case? Answer: Equal.

15.4 AIR POLLUTION

Since the discovery of fire, mankind has polluted the atmosphere with noxious gases and soot. When coal began to be used as a fuel in the fourteenth century, the problem became one of public concern. Increased fuel consumption by industry, concentration of population in urban areas, and the advent of motor vehicles have, over the years, made the problem worse. Today a major cause of pollution in our atmosphere is the gasoline engine.

But we did get rid of those sooty coal fires

Any substance whose addition to the atmosphere produces a measurable effect on man or his environment can be classified as a pollutant. A host of materials fit this broad definition. Radioactive species produced by nuclear fallout qualify, as do toxic gases released by accident. In this chapter we will limit our attention to five major types of pollutants. These are:

1. Suspended particles
2. Sulfur oxides (SO_2, SO_3)
3. Nitrogen oxides (NO, NO_2)
4. Hydrocarbons
5. Carbon monoxide (CO)

SUSPENDED PARTICLES. The finely divided solids and tiny drops of liquids suspended in the air vary greatly in size. Particles in cigarette smoke have diameters as small as 10^{-8} m. Dust particles produced by a cement kiln may be as large as 10^{-4} m. Industrial processes involving cutting, grinding, and spraying are major sources of suspended particles. The combustion of coal produces large amounts of unburned carbon (soot) and finely divided inorganic material (fly ash). Fuel oil is cleaner burning; natural gas is even better in this respect. A properly adjusted burner using natural gas produces virtually no smoke or soot.

Reduced visibility (Fig. 15.3) caused by absorption and scattering of light is one of the more obvious effects of suspended particles in the air. Soiled clothing and soot-covered buildings are other familiar symptoms. Many types of suspended particles are harmful to health. Black lung disease, common among coal miners, is caused by tiny particles of carbonaceous material settling in the lungs. Compounds of lead, one of the most toxic metals, enter the air when gasoline containing the antiknock additive tetra-ethyl lead, $Pb(C_2H_5)_4$, is burned.

SULFUR OXIDES. Most of the coal burned in heating and power plants in the United States contains from 1 to 3% sulfur, much of which is in the form of minerals such as pyrite, FeS_2. Upon combustion, the sulfur is converted to sulfur dioxide:

$$4\ FeS_2(s) + 11\ O_2(g) \rightarrow 2\ Fe_2O_3(s) + 8\ SO_2(g) \tag{15.15}$$

Most hydrocarbon fuels contain very little sulfur. An exception is the residual oil left after the distillation of petroleum. This tarry material, which is used in many industrial and municipal heating plants, contains from 1 to 2% sulfur. The sulfur, present largely as organic compounds, is converted to sulfur dioxide upon combustion. Metallurgical processes involving the roasting of sulfide ores (Chap. 5) are another major source of SO_2.

You can often smell SO_2 in the vicinity of volcanoes

Sulfur dioxide at concentrations as low as 0.3 ppm can cause acute injury to plants. Most healthy adults can tolerate considerably higher SO_2 levels without apparent adverse effects. However, individuals who suffer from chronic respiratory diseases such as bronchitis or asthma are much more sensitive. Increasing the concentration of SO_2

Figure 15.3　A smoggy vs. a clear day in New York City. The two photographs were taken at the same hour on successive days. *N.Y. Daily News photo.*

TABLE 15.5 EQUILIBRIUM CONSTANT FOR THE REACTION $SO_2(g) + \frac{1}{2}O_2(g) \rightleftharpoons SO_3(g); \Delta H = -99.1$ kJ							
t (°C)	25	200	400	500	600	700	800
K_c	9.2×10^{12}	5.0×10^6	2300	400	70	20	7

from 0.1 to 0.2 ppm can cause them to start coughing and experience severe difficulties in breathing.

From equilibrium considerations, we might expect sulfur dioxide in the air to be converted to sulfur trioxide (Table 15.5):

$$SO_2(g) + \frac{1}{2}O_2(g) \rightleftharpoons SO_3(g); \quad K_c = 9.2 \times 10^{12} \text{ at } 25°C$$

However, this reaction takes place very slowly in the absence of a catalyst. Ordinarily, less than 1% of the SO_2 in polluted air is converted to SO_3. Suspended solids in the atmosphere, such as those found in coal smoke, can raise this fraction to 5% or more by catalyzing this reaction.

Most of the adverse effects of sulfur oxides in the atmosphere are caused by the sulfuric acid produced when SO_3 reacts with water:

$$SO_3(g) + H_2O \rightarrow H_2SO_4(aq)$$

Sulfuric acid (and, to a lesser extent, nitric acid) is responsible for the high acidity of rainfall in the northeastern United States. Many remote lakes in the Adirondacks have

Acid rain is common near industrial areas

Figure 15.4 Marble statues sometimes become unrecognizable as a result of attack by oxides of sulfur. (From Wagner, R. H., *Environment and Man,* W. W. Norton & Co., New York, 1971.)

been depleted of trout because the water is too acidic. The sulfur oxides are coming from cities hundreds of miles away.

Sulfuric acid also attacks building materials such as limestone or marble (calcium carbonate):

$$CaCO_3(s) + H_2SO_4(aq) \rightarrow CaSO_4(s) + CO_2(g) + H_2O \qquad (15.16)$$

The calcium sulfate formed is soluble enough to be gradually washed away. This process is responsible for the deterioration of the Greek ruins on the Acropolis in Athens. These structures have suffered more damage in the twentieth century than in the preceding 2000 years. Another effect which is due to sulfuric acid is the deterioration of the paper in books and documents. Manuscripts printed before 1750 are almost immune to sulfur oxides. At about that time modern methods of papermaking were introduced. These leave traces of metal oxides which catalyze the conversion of SO_2 to SO_3 and hence to sulfuric acid.

Example 15.6   The reaction between SO_2 and O_2 is the key step in the industrial process for making sulfuric acid (see below). Commonly, the temperature used is about 600°C. If, in an equilibrium mixture at this temperature, $[SO_2] = [O_2] = 0.010$ mol/ℓ, calculate $[SO_3]$, using Table 15.5.

Solution   The expression for K_c is

$$K_c = \frac{[SO_3]}{[SO_2] \times [O_2]^{1/2}} = 70$$

Solving for $[SO_3]$:

$$[SO_3] = 70 \times [SO_2] \times [O_2]^{1/2}$$

But: $[SO_2] = 0.010; [O_2]^{1/2} = (0.010)^{1/2} = 0.10$

Hence, $[SO_3] = 70 \times 0.010 \times 0.10 = 0.070$ mol/ℓ

Exercise   Repeat this calculation at 700°C. Answer: $[SO_3] = 0.020$ mol/ℓ.

Commercially, sulfuric acid is made by the three-step process known as the *contact process:*

$$S(s) + O_2(g) \rightarrow SO_2(g) \qquad (15.17)$$

$$SO_2(g) + {}^1/_2\ O_2(g) \rightleftharpoons SO_3(g) \qquad (15.18)$$

$$SO_3(g) + H_2O(l) \rightarrow H_2SO_4(aq) \qquad (15.19)$$

Reaction 15.17 is carried out by burning sulfur in air. The sulfur dioxide formed is allowed to react with oxygen at a temperature of 600 to 700°C. A solid catalyst, usually vanadium pentoxide, V_2O_5, is used. The sulfur trioxide produced is passed into sulfuric acid solution, where Reaction 15.19 occurs.* If SO_3 is added directly to water, H_2SO_4 forms as a fog of tiny particles which are difficult to condense.

---

*This is the overall reaction. Sulfur trioxide first reacts with sulfuric acid to form pyrosulfuric acid, $H_2S_2O_7$. Subsequent addition of water to $H_2S_2O_7$ forms sulfuric acid, H_2SO_4.

NITROGEN OXIDES.   The high-temperature combustion of fuels is the principal source of this type of air pollutant. Detectable amounts of nitric oxide are produced by the reaction

$$N_2(g) + O_2(g) \rightarrow 2\ NO(g);\ \Delta H = +180.8\ kJ$$

In urban air, NO is converted to NO_2 by a mechanism which is poorly understood. Direct reaction with O_2 (Reaction 15.5) is too slow to account for the rapid build-up of NO_2 (see Example 15.7).

The oxides of nitrogen play a key role in the formation of *photochemical smog*. This type of air pollution was first noted in Los Angeles, but it is now common in cities as far apart as Honolulu, Denver, and Washington, D.C. Typically, it develops on bright, sunny mornings when the concentration of NO_2 is relatively high (Fig. 15.5). The key step in smog formation is the dissociation of nitrogen dioxide:

$$NO_2(g) \rightarrow NO(g) + O(g) \tag{15.20}$$

NO_2 is itself a noxious substance

This occurs on exposure to light at the edge of the visible region (392 nm; Table 15.3). The oxygen atoms produced react with O_2 molecules:

$$O_2(g) + O(g) \rightarrow O_3(g)$$

The product, ozone, is a major component of photochemical smog. Ozone molecules, oxygen atoms, and nitric oxide molecules attack organic compounds in the air. Unsaturated hydrocarbons, containing multiple carbon-carbon bonds, such as ethylene and propylene, are particularly reactive:

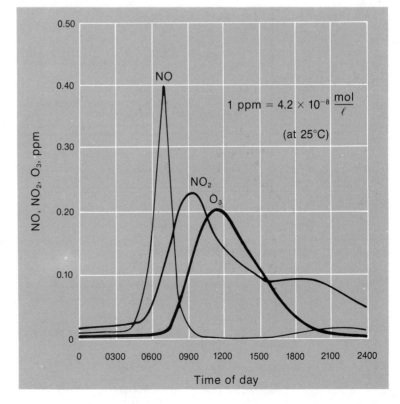

Figure 15.5 Average concentrations of the pollutants NO, NO_2, and O_3 on a smoggy day in Log Angeles. The concentration of NO from automobile exhaust builds up in the early morning rush hours. Later, NO_2 and O_3 are produced. The peak NO concentration of 0.40 ppm corresponds to about 1.7×10^{-8} mol/ℓ at 25°C.

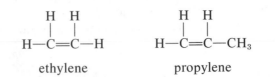

ethylene   propylene

Many different products are formed. Among these are acrolein and peroxyacetyl nitrate (PAN), which cause eye irritation.

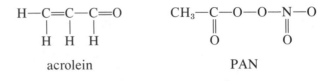

acrolein   PAN

Example 15.7   Consider the reaction $2 NO(g) + O_2(g) \rightarrow 2 NO_2(g)$, for which the rate expression is

$$\text{rate} = k(\text{conc. NO})^2 \times (\text{conc. } O_2), \text{ where } k = 2.5 \times 10^7 \, \ell^2/(\text{mol}^2 \cdot \text{h})$$

a. What is the rate of this reaction at the peak NO concentration in smog formation (Fig. 15.5), 1.7×10^{-8} mol/ℓ? The concentration of O_2 in the air is 8.8×10^{-3} mol/ℓ.
b. At this rate, how long would it take for the concentration of NO_2 to build up to its peak value, 1.0×10^{-8} mol/ℓ?

Solution

a. $\text{rate} = 2.5 \times 10^7 \, \dfrac{\ell^2}{\text{mol}^2 \cdot \text{h}} \times (1.7 \times 10^{-8} \text{ mol}/\ell)^2 \times (8.8 \times 10^{-3} \text{ mol}/\ell)$

$= 6.4 \times 10^{-11} \, \dfrac{\text{mol}}{\ell \cdot \text{h}}$

b. $\text{time} = \dfrac{\text{conc. } NO_2}{\text{rate}} = \dfrac{1.0 \times 10^{-8} \text{ mol}/\ell}{6.4 \times 10^{-11} \text{ mol}/(\ell \cdot \text{h})} = 160 \text{ h}$

Clearly, this cannot be the reaction by which NO_2 is produced from NO in smog formation. It is much too slow. From Figure 15.5, about 3 h is required for the concentration of NO_2 to build up to its peak value.

Exercise   Using the calculated rate, determine the concentration of NO_2 produced by this reaction in 3.0 h. Answer: 1.9×10^{-10} mol/ℓ.

HYDROCARBONS.   The major source of these organic compounds, at least in urban areas, is automobile exhaust. This contains significant amounts of unburned hydrocarbons unless special precautions are taken to ensure complete combustion. Most of these hydrocarbons are saturated (Chap. 12). However, a few unsaturated hydrocarbons are produced by reactions occurring within an automobile engine. A typical reaction involving octane, a major component of gasoline, is:

$$C_8H_{18}(l) \rightarrow 2 \, C_2H_4(g) + C_3H_6(g) + CH_4(g) \tag{15.21}$$

octane   ethylene   propylene   methane

As pointed out, unsaturated hydrocarbons such as ethylene and propylene produce some of the most noxious components of photochemical smog.

CARBON MONOXIDE. The incomplete combustion of fossil fuels produces carbon monoxide as well as unburned hydrocarbons. The principal culprit is again the automobile. If sufficient oxygen is present, all the carbon atoms in the fuel are converted to CO_2:

$$C_8H_{18}(l) + 12.5\ O_2(g) \rightarrow 8\ CO_2(g) + 9\ H_2O(l) \tag{15.22}$$

However, if the fuel mixture is too "rich," that is, contains too much fuel and too little air, considerable amounts of CO may be formed:

$$C_8H_{18}(l) + 11.5\ O_2(g) \rightarrow 6\ CO_2(g) + 2\ CO(g) + 9\ H_2O(l) \tag{15.23}$$

In principle, the carbon monoxide should be converted to CO_2 in the atmosphere

$$CO(g) + \tfrac{1}{2}\ O_2(g) \rightarrow CO_2(g);\ K_c = 3 \times 10^{45}\ \text{at } 25°C \tag{15.24}$$

but this reaction, like the conversion of SO_2 to SO_3, is ordinarily quite slow.

Carbon monoxide taken into the lungs reduces the ability of the blood to transport oxygen through the body. It does this by forming a complex with the hemoglobin of the blood which is more stable than that formed by oxygen (Chap. 19).

$$CO(g) + Hem \cdot O_2(aq) \rightleftharpoons O_2(g) + Hem \cdot CO(aq) \tag{15.25}$$

$$(Hem = hemoglobin)$$

$$K_c = \frac{[Hem \cdot CO] \times [O_2]}{[Hem \cdot O_2] \times [CO]} = 210$$

The equilibrium constant for Reaction 15.25 is large enough so that low levels of carbon monoxide can convert significant amounts of hemoglobin to the CO complex. Symptoms of carbon monoxide poisoning show up when 10% of the hemoglobin is tied up by CO. When the fraction rises to 20%, death can result unless the victim is removed from the poisonous atmosphere.

Example 15.8   Carbon monoxide is present in cigarette smoke. It is estimated that a heavy smoker is exposed to a CO concentration of 2.1×10^{-6} mol/ℓ (50 ppm). Taking the concentration of O_2 to be 8.8×10^{-3} mol/ℓ, calculate

    a. the ratio $[CO]/[O_2]$.
    b. the ratio $[Hem \cdot CO]/[Hem \cdot O_2]$.

Solution

    a. $\dfrac{[CO]}{[O_2]} = \dfrac{2.1 \times 10^{-6}\ \text{mol}/\ell}{8.8 \times 10^{-3}\ \text{mol}/\ell} = 2.4 \times 10^{-4}$

    b. $K_c = \dfrac{[Hem \cdot CO] \times [O_2]}{[Hem \cdot O_2] \times [CO]} = 210$

    Solving for the ratio $[Hem \cdot CO]/[Hem \cdot O_2]$:

$$\frac{[Hem \cdot CO]}{[Hem \cdot O_2]} = 210 \times \frac{[CO]}{[O_2]} = 210 \times 2.4 \times 10^{-4} = 0.050$$

This calculation suggests that in the blood stream of a heavy smoker about 5% of the hemoglobin is tied up as the CO complex. As a result, smokers are more susceptible to carbon monoxide poisoning than nonsmokers.

Exercise   The equilibrium constant of Reaction 15.25 differs from one mammal to another. In a rabbit, K_c is about 100. What is the ratio $[\text{Hem} \cdot \text{CO}]/[\text{Hem} \cdot \text{O}_2]$ in a rabbit who smokes two packs of cigarettes a day? Answer: 0.024.

Some rabbit

Reducing Air Pollution

HYDROCARBONS, CO, OXIDES OF NITROGEN.   These pollutants, as we have seen, come mainly from automobile exhaust. Since 1975, the principal method of reducing emissions of these gases has been to install catalytic converters in cars (Fig. 15.6). A catalytic converter contains from 1 to 3 g of a mixture of two heavy metals, platinum and palladium. These are embedded in a base of aluminum oxide, Al_2O_3. In the presence of the catalyst, CO and hydrocarbons in the exhaust are converted to carbon dioxide and water.

Cars equipped with catalytic converters cannot use "leaded" gasoline, which contains a mixture of $Pb(C_2H_5)_4$ and $C_2H_4Br_2$ to prevent knocking. As little as two tankfuls of gasoline containing this additive can "poison" the converter, making it ineffective. Unleaded gasoline is somewhat more expensive to make. It also contains relatively large amounts of aromatic hydrocarbons, some of which are carcinogenic.

Two different approaches have been studied for reducing exhaust emissions of oxides of nitrogen. The one currently in use involves recirculating from 10 to 20% of the exhaust gases. This lowers the combustion temperature and hence cuts down on the formation of NO from nitrogen and oxygen. Such a system has one inherent disadvantage. Unless the amount of exhaust gases recirculated is controlled very precisely, there are adverse effects on both performance and fuel economy. Another approach is to

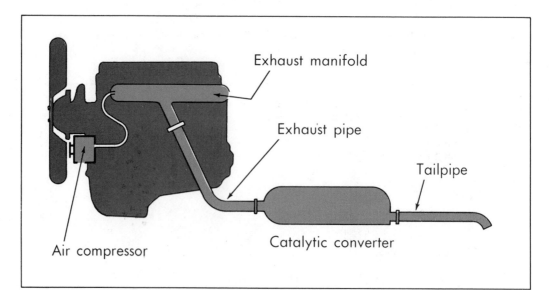

Figure 15.6   Schematic diagram of catalytic converters used in virtually all U.S. automobiles made since 1975. A mixture of Pt and Pd catalyzes the combustion of unburned hydrocarbons and the conversion of CO to CO_2.

introduce a second catalyst into the exhaust system to promote the decomposition of NO, perhaps by the reaction

$$CO(g) + NO(g) \rightarrow CO_2(g) + \frac{1}{2} N_2(g) \qquad (15.26)$$

As pointed out in Chapter 14, K_c for this reaction is very large, about 10^{60} at room temperature. In principle, it should go nearly to completion, removing nitric oxide and perhaps excess CO as well. However, the uncatalyzed reaction has a high activation energy and occurs very slowly under ordinary conditions. Several different catalysts capable of speeding up the reaction have been studied. The problem is that they are very susceptible to poisoning and become ineffective after a car has been driven a few thousand miles.

SULFUR OXIDES (SO_2, SO_3). The problem here is one of cleaning up emissions from power and heating plants. An obvious way to prevent SO_2 emissions is to remove sulfur compounds from fuel before it is burned. Pyrite, FeS_2, can be separated from coal by "washing" with a concentrated solution of calcium chloride ($d = 1.35$ g/cm³). The coal ($d = 1.2$ g/cm³) floats in this solution, while pyrite ($d = 4.9$ g/cm³) sinks to the bottom. This method is not effective with organic sulfur compounds, which account for nearly half the sulfur content of coal and nearly all of that in fuel oil.

A different approach is to add a chemical to react with the sulfur oxides after they are formed. A common method involves inserting a "scrubber" into the smoke stack. Here, a water solution of $Ca(OH)_2$ is used to bring about the reaction

$$Ca^{2+}(aq) + 2 OH^-(aq) + SO_2(g) + \frac{1}{2} O_2(g) \rightarrow CaSO_4(s) + H_2O \qquad (15.27)$$

It would be best if we could remove the SO_2, convert it to H_2SO_4, and sell it

This produces large amounts of calcium sulfate, which must be disposed of.

At present, most industries burning coal avoid the problem of the emission of sulfur oxides in one of two ways. Some use low-sulfur coal, which is somewhat more expensive. Others build extremely high stacks. These disperse gases into the air at a height of 1000 m or greater. This lowers the SO_2 concentration in the vicinity of the plant. Instead, the sulfur dioxide is spread out into neighboring states, where it produces acid rain.

SUSPENDED PARTICLES. Several different methods are used to remove smoke particles and dust from smokestack gases. Large particles (diameter $>5 \times 10^{-5}$ m) can

Figure 15.7 Schematic drawing of electrostatic precipitator. Particles pick up charge as they pass between plates and are precipitated on plates of opposite charge. The potential across the plates is about 50,000 V. (From Gordon, G., and Zoller, W., *Chemistry in Modern Perspective*, Addison-Wesley Publishing Co., Reading, Mass., 1975.)

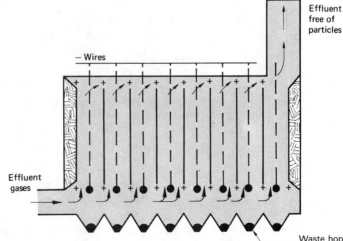

Effluent free of particles

— Wires

Effluent gases

Waste hoppers collect solid particle residue

be removed by allowing them to settle by gravity in a large chamber. This approach does not work with small particles, which settle more slowly. Particles as small as 1×10^{-6} m can be removed by passing stack gases through an electrostatic precipitator (Fig. 15.7). This device consists of a series of plates which are charged to high voltages, alternately + and −. Particles approaching a given plate tend to acquire its charge. They are then attracted to the next plate, from which they fall into the hopper below.

15.5 TWO HISTORICAL PERSPECTIVES

Fritz Haber (1868–1934)

The scientific career of Fritz Haber was characterized by theoretical studies in areas of emerging practical importance. Training as an organic chemist and self-development as a physical chemist permitted Haber to relate chemistry to engineering. His most important achievement was the synthesis of ammonia from nitrogen and hydrogen.

Haber's approach to the synthesis of ammonia was not original. The French chemist Le Châtelier (Chap. 13) first studied this reaction, but an explosion led him to abandon the project. The German chemist Nernst, a contemporary of Haber, was the first to accomplish the high-pressure synthesis. Disagreement with Nernst's data led Haber to seek the optimum temperature, pressure, and catalyst. After obtaining about 8% ammonia at 600°C and 200 atm, Haber turned the process over to engineers. In 1918 he received the Nobel Prize for his work.

During World War I, Haber directed chemical warfare in Germany, introducing the use of chlorine and mustard gas into warfare. After the war, he attempted unsuccessfully to recover gold from the oceans. In this way, he hoped to repay Germany's war debts (which were never paid).

In 1911, Haber became director of the new Kaiser Wilhelm Institute for Chemistry. Under his influence, the Institute became world famous, attracting outstanding students and professors. In 1933 Haber, who was Jewish, was removed from his directorship of the Institute by the Nazis. In his resignation letter he wrote, "For more than 40 years I have selected my collaborators on the basis of their intelligence and their character, not on the basis of their grandmothers. I am unwilling for the rest of my life to change this method which I have found so good." After a brief stay in England, Haber died enroute to a research directorship in Israel.

Friedrich Wilhelm Ostwald (1853–1932)

The conversion of ammonia to nitric acid was a minor contribution from this multifaceted individual. His early interests in chemistry, physics, painting, literature, music, and philosophy became life-long pursuits. With Van't Hoff, Arrhenius, and Gibbs, Ostwald helped to establish the fledgling discipline of physical chemistry. He speculated that he became a physical chemist because his chemical education was carried out in Russia rather than Germany, where the emphasis was on organic chemistry. Early work on catalysis and reaction rates (1881) earned him the 1909 Nobel Prize in chemistry.

At 53, Ostwald retired as director of the University of Leipzig Physical Chemistry Institute and spent his remaining years in research and writing. From 1890 on, his work and personal philosophy were organized about his science of "energetics" (he even named his house *Energie*). This point of view made him a vigorous opponent of the concept of atoms and molecules! Matter, he believed, was a combination of energies occurring simultaneously at one place: "In fact, energy is the unique, real entity in the world and matter is not a bearer but a manifestation of energy." Eventually (1909), in a preface to his general chemistry text, Ostwald grudgingly accepted the atomic theory, in part because of the work of Perrin in kinetic molecular theory and discoveries by Thomson and Rutherford.

A pacifist and an advocate of the conservation of natural energy sources, Ostwald was a visionary. He believed in international harmony and composed Ido, an international language. His love of painting led him to spend the last 20 years of his life developing the science of color. He created color standards, dyes, and a quantitative theory of color. An inspiring teacher, his textbooks for general and analytical chemistry revolutionized the teaching of these subjects. When Ostwald died, a former student wrote, "He was loved and followed by more people than any chemist of our time."

SUMMARY

This chapter describes selected topics in atmospheric chemistry, including:
—the fixation of nitrogen to form ammonia (Haber process) and the conversion of ammonia to nitric acid (Ostwald process).
—the effect of water and carbon dioxide upon our weather and climate (relative humidity, cloud formation, greenhouse effect).
—photochemical reactions taking place in the upper atmosphere (Tables 15.3 and 15.4).
—the sources, effects, and treatment of air pollution by suspended particles, sulfur oxides, nitrogen oxides, hydrocarbons, and carbon monoxide.

Several concepts from Chapters 13 and 14 were reviewed in this chapter. These include:
—equilibrium calculations (Examples 15.1, 15.6, and 15.8)
—effect of changes in conditions upon the position of an equilibrium and the rate at which it is reached (Example 15.2)
—activation energy and the effect of temperature upon reaction rate (Example 15.4)
—reaction mechanisms (Example 15.5)
—rate expressions (Example 15.7)

KEY WORDS AND CONCEPTS

mole fraction	Haber process	greenhouse effect	contact process
parts per million (ppm)	Ostwald process	wavelength	smog
parts per billion (ppb)	relative humidity	ozone	catalytic converter
nitrogen fixation			

QUESTIONS AND PROBLEMS

Catalog

General: 15.1–15.8, 15.23–15.29, 15.44–15.46
Calculations Involving K_c: 15.9–15.12, 15.30–15.33
Rate Expressions, Reaction Order: 15.13–15.17, 15.34–15.38

Activation Energy and Temperature: 15.18–15.20, 15.39–15.41
Mechanisms: 15.21, 15.22, 15.42, 15.43

15.1 Review and know the meaning of the key words and concepts in this chapter.

15.2 Criticize the following statements:

a. In the Haber process, a high pressure is used primarily to increase reaction rate.
b. Ice crystals form in clouds when the temperature drops below 0°C.
c. Ozone is found only in the upper atmosphere.
d. Sulfur dioxide is a more serious air pollutant than SO_3.

15.3 Write balanced equations (one or more) to represent the preparation of

a. ammonia by the Haber process.
b. sulfuric acid by the contact process.
c. ammonium sulfate.

15.4 Describe the effects which may result from

a. an increase in atmospheric CO_2 concentration.
b. a decrease in the concentration of O_3 in the upper atmosphere.
c. an increase in SO_3 concentration in the atmosphere.

15.5 Explain why

a. a catalytic converter is used in an automobile exhaust system.
b. catalytic converters are expensive.
c. even a low-level concentration of CO may be dangerous.

15.6 The 1980 USA emission standards goal for CO was 2.9 g CO/km (4.7 g/mi). In a large metropolitan area there are 100,000 automobiles driven an average of 32 km/day (20 mi/day).

a. Assuming that all the cars meet this standard, how many kg CO are emitted per day?
b. Assuming an air volume of 5000 km³ for the metropolitan area and a maximum allowable CO concentration of 4×10^{-7} mol/ℓ, does the daily CO output exceed this level?

15.7 A 20.0-ℓ sample of air contains 0.540 g of H_2O(g) at 30°C.

a. Using the Ideal Gas Law, calculate the partial pressure of water vapor in the air.
b. What is the relative humidity? (vp H_2O at 30°C = 31.8 mm Hg)

15.23 Classify the following as true or false. If the statement is false, correct it.

a. Fertilizer prices have risen in recent years because the price of N_2 has skyrocketed.
b. If the temperature of an air mass drops, relative humidity increases.
c. CO_2 in the atmosphere has a warming effect because it absorbs IR radiation from the sun.

15.24 Follow directions for Question 15.3 for the preparation of

a. nitric acid by the Ostwald process.
b. urea.
c. ammonium nitrate.

15.25 Describe the possible effects of

a. an increase in Cl concentration in the upper atmosphere.
b. an increase in NO concentration in the morning rush hour.
c. an increase in suspended particles in air.
d. continued use of leaded gasoline.

15.26 Explain

a. the greenhouse effect.
b. why chemical reactions which occur in the upper atmosphere (above 30 km) are not common near the surface of the earth.
c. the relation between SO_3 production in air and the contact process.

15.27 During the day, typical concentrations (ppm) of air pollutants in a metropolitan area are:

Time	NO	NO₂	O₃	Hydrocarbons
4 A.M.	0.07	0.06	0.02	0.28
6 A.M.	0.14	0.09	0.02	0.31
7 A.M.	0.14	0.16	0.02	0.36
8 A.M.	0.04	0.18	0.03	0.43
10 A.M.	0.01	0.06	0.08	0.33
Noon	0.01	0.04	0.19	0.25
2 P.M.	0.01	0.04	0.19	0.20
4 P.M.	0.02	0.05	0.12	0.15

Discuss these trends in light of the reactions mentioned in this chapter.

15.28 The dew point is the temperature at which water vapor in the air condenses upon cooling at constant pressure. On a day when the temperature is 20°C and the relative humidity is 60%, calculate the dew point. Use the table of water vapor pressures in Appendix 1.

15.8 The bond energy of the C—Cl bond is 331 kJ/mol. Use Equation 15.8 to calculate the maximum wavelength of radiation that can break this bond and hence lead to depletion of the ozone layer by Reactions 15.13 and 15.14.

15.9 One way to fix N_2 is to react it with O_2: $N_2(g) + O_2(g) \rightleftharpoons 2\ NO(g)$; $K_c = 1 \times 10^{-30}$ at 25°C.

 a. Comment on the yield of NO to be expected at 25°C and 1 atm.
 b. Would K_c increase, decrease, or remain the same if the temperature were increased? ($\Delta H = +181$ kJ)
 c. What would happen to K_c if the pressure were increased?

15.10 Consider the data in Table 15.2.

 a. Determine the number of moles of NH_3, N_2, and H_2 in an equilibrium mixture containing one mole of gas at 300 atm and 400°C. Note that $[H_2] = 3[N_2]$.
 b. Use the Ideal Gas Law to determine the volume of one mole of gas at these conditions.
 c. Combine your answers from (a) and (b) to calculate the equilibrium concentrations (mol/ℓ) of NH_3, N_2, and H_2.
 d. From your answers in (c), calculate K_c at 400°C. Compare your value with that given in the table.

15.11 Consider the Haber process:

$$N_2(g) + 3\ H_2(g) \rightleftharpoons 2\ NH_3(g)$$

At a certain temperature, 1.0 mol of N_2 and 3.0 mol of H_2 are placed in a 0.50-ℓ container. The equilibrium concentration of NH_3 is 2.0 mol/ℓ. Determine

 a. the equilibrium concentrations of N_2 and H_2.
 b. K_c.
 c. the approximate temperature (refer to Table 15.2).

15.12 Consider the equilibrium

$$CO(g) + Hem \cdot O_2(aq) \rightleftharpoons O_2(g) + Hem \cdot CO(aq)$$

$K_c = 210$. Calculate the ratio $[Hem \cdot CO]/[Hem \cdot O_2]$ when

 a. conc. CO = conc. O_2.
 b. conc. CO = 0.10 conc. O_2.

15.13 The reaction

$$NO(g) + N_2O(g) \rightarrow NO_2(g) + N_2(g)$$

is first order in each reactant. The rate is 1.23×10^{-15} mol/($\ell \cdot$ s) when the NO and N_2O concentrations are each 4.0×10^{-8} mol/ℓ.

 a. Calculate k for the forward reaction.
 b. At a conc. NO = 2.0×10^{-8} mol/ℓ, the rate is measured as 4.7×10^{-20} mol/($\ell \cdot$ s). What is conc. N_2O?

15.29 In the atmosphere, O_2 molecules are converted to oxygen atoms by UV radiation below 242 nm. Use Equation 15.8 to calculate the bond energy of O_2.

15.30 For the reaction in Problem 15.9, discuss the effect of temperature, pressure, and the use of a catalyst upon

 a. the position of the equilibrium.
 b. the rate at which equilibrium is reached.

15.31 Consider the data in Table 15.2.

 a. Calculate the mol % of all three gases at 500°C and 100 atm.
 b. Use the Ideal Gas Law to calculate the number of moles of gas in a 5.00-ℓ container at 100 atm and 500°C.
 c. Use the results from (a) and (b) to calculate the equilibrium concentrations of all three gases in a 5.00-ℓ vessel at 100 atm and 500°C.
 d. Use your answers in (c) to calculate K_c at 500°C. Compare your value with that in the table.

15.32 Consider the equilibrium

$$SO_2(g) + \tfrac{1}{2}\ O_2(g) \rightleftharpoons SO_3(g)$$

At a certain temperature, 1.00 mol of SO_2 and 0.50 mol of O_2 are placed in a 0.50-ℓ container. The concentration of SO_3 at equilibrium is 1.50 mol/ℓ. Determine

 a. the equilibrium concentrations of SO_2 and O_2.
 b. K_c.
 c. the approximate temperature, using Table 15.5

15.33 For the equilibrium in Problem 15.12, calculate the ratio $[CO]/[O_2]$ when 10.0% of the hemoglobin is in the form of the CO complex.

15.34 The reaction

$$2\ NO_2(g) \rightarrow N_2(g) + 2\ O_2(g)$$

is a second order reaction. The rate constant is 0.498 ℓ/(mol · s) at 592 K.

 a. What is the reaction rate at an initial conc. NO_2 = 0.050 mol/ℓ?
 b. What concentration of NO_2 is present when the reaction rate is reduced to half of that determined in (a)?

15.14 The following reaction is involved in smog formation:

$$O_3(g) + NO(g) \rightarrow O_2(g) + NO_2(g)$$

This reaction has been shown to be first order in both ozone and NO with a rate constant of 1.2×10^7 $\ell/(mol \cdot s)$. Calculate the concentration of NO_2 formed per second in polluted air where O_3 and NO concentrations are both 2×10^{-8} mol/ℓ. From the magnitude of your answer, would you expect the conversion of NO to be rapid or slow?

15.15 The rate constant for the decomposition of ozone by the mechanism $O_3(g) + O(g) \rightarrow 2\ O_2(g)$ is 7.8×10^5 $\ell/(mol \cdot s)$. Calculate the rate of ozone decomposition when the concentrations of O_3 and O are 3.0×10^{-8} and 1.2×10^{-14} mol/ℓ, respectively.

15.16 The rate constant for the NO-catalyzed decomposition of O_3 (Table 15.4) is 2.4×10^9 $\ell/(mol \cdot s)$, as compared to 7.8×10^5 $\ell/(mol \cdot s)$ for the direct decomposition (Problem 15.15). The second step in the catalyzed decomposition is rate-determining. If the concentration of NO_2 in the upper atmosphere is 1/2000 that of O_3, what is the ratio of the rate of the catalyzed reaction to that of the direct decomposition?

15.17 For the first order thermal decomposition of ozone

$$O_3(g) \rightarrow O_2(g) + O(g)$$

$k = 3 \times 10^{-26}$ s^{-1} at 25°C. What is the half-life for this reaction in years? Comment on the likelihood that this reaction contributes to the depletion of the ozone layer.

15.18 For the upper atmosphere reaction

$$OH(g) + H(g) \rightarrow H_2(g) + O(g)$$

E_a is 29 kJ. The heats of formation of O(g), H(g), and OH(g) are 248 kJ, 218 kJ, and 37 kJ, in that order. What is the activation energy for the reverse reaction?

15.19 For the reaction in Example 15.4, calculate the temperature at which k is one half its value at 25°C.

15.20 The rate constants are 4.35 and 36.3 $\ell/(mol \cdot s)$ at 650 and 750 K for the gas phase decomposition of NO_2. Calculate the activation energy for the reaction.

15.21 A possible mechanism for the reaction:

$$2\ NO(g) + O_2(g) \rightarrow 2\ NO_2(g)$$

is:

$NO(g) + NO(g) \rightleftharpoons N_2O_2(g)$ (fast)
$N_2O_2(g) + O_2(g) \rightarrow NO_2(g) + NO_2(g)$ (slow)

Derive the rate expression for this reaction, in terms of the concentrations of NO and O_2.

15.35 Formaldehyde, CH_2O, is a major eye irritant in smog. It may be formed by the reaction $O_3(g) + C_2H_4(g) \rightarrow 2\ CH_2O(g) + O(g)$, which is known to be first order in both O_3 and C_2H_4 with $k = 2 \times 10^3$ $\ell/(mol \cdot s)$. The concentrations of O_3 and C_2H_4 in heavily polluted air are estimated to be about 5×10^{-8} and 1×10^{-8} mol/ℓ, respectively. What is the rate of production of formaldehyde in $mol/(\ell \cdot s)$? How long will it take to build up a formaldehyde concentration of 1×10^{-8} mol/ℓ? This is the threshold above which eye irritation becomes noticeable. (Assume constant concentrations of O_3 and C_2H_4.)

15.36 If ozone is being produced by the mechanism $O_2(g) + O(g) \rightarrow O_3(g)$ at the same rate it is decomposing by the reaction in Problem 15.15, and if the concentrations of O_2 and O are 3.9×10^{-4} and 1.2×10^{-14} mol/ℓ, respectively, what is the rate constant for the above reaction?

15.37 It is estimated that the amount of O_3 in the upper atmosphere decomposing by the NO-catalyzed reaction is four times that decomposing by the direct reaction. Using the rate constants given in Problem 15.16, estimate the ratio of the concentrations of NO_2 and O_3.

15.38 For the reaction in Problem 15.17, how long would it take to lower the ozone concentration by 0.10%?

15.39 In the upper atmosphere, water molecules react with oxygen atoms:

$$H_2O(g) + O(g) \rightarrow 2\ OH(g)$$

For this reaction, E_a is 77 kJ. For the reverse reaction, E_a' is 9 kJ.

    a. What is ΔH for the forward reaction?
    b. Given $\Delta H_f\ H_2O(g) = -242$ kJ, and $\Delta H_f\ O(g) = 248$ kJ, calculate $\Delta H_f\ OH(g)$.

15.40 For the reaction in Example 15.4, calculate k at 50°C.

15.41 For the reaction given in Problem 15.18, $k = 6 \times 10^1$ $\ell/(mol \cdot s)$ at 25°C. Determine the value of k at -25°C.

15.42 Experimentally, it is found that the rate constant for the formation of NO_2 from NO *decreases* as temperature increases. Using the rate expression derived in Problem 15.21, explain how this can happen. (ΔH for the formation of N_2O_2 from NO is a negative quantity.)

15.22 Consider the following suggested mechanism:

$$NO_2(g) + NO_2(g) \rightleftharpoons N_2O_4(g) \text{ (fast)}$$
$$N_2O_4(g) + F_2(g) \rightarrow 2 NO_2F(g) \text{ (slow)}$$

a. Write the equation for the overall reaction.
b. Write the rate law for the overall reaction in terms of the concentrations of NO_2 and F_2.

15.43 The following mechanism has been proposed for the uncatalyzed decomposition of O_3:

$$O_3(g) \rightleftharpoons O_2(g) + O(g) \text{ (fast)}$$
$$O(g) + O_3(g) \rightarrow 2 O_2(g) \text{ (slow)}$$

Find the rate law associated with this mechanism. Express it in terms of the concentrations of O_3 and O_2.

*15.44 Two moles of SO_2 and one mole of O_2 are placed in a $1.00\text{-}\ell$ container at 900°C. At this temperature, $K_c = 2.4$ for the system $SO_2(g) + \frac{1}{2} O_2(g) \rightleftharpoons SO_3(g)$. Calculate the concentration of SO_3 at equilibrium.

*15.45 Most of the ozone in the upper atmosphere exists in a layer at an altitude between 20 and 40 km.

a. Calculate the total volume in liters of this layer, taking the radius of the earth to be 6.4×10^3 km and using the following relation for the difference in volume between two concentric spheres of radii r_1 and r_2: $\Delta V \approx 4 \pi r_1^2(r_2 - r_1)$.
b. Taking the average pressure of air in this layer to be 0.013 atm and the temperature to be −20°C, calculate the total number of moles of gas in the layer.
c. Assuming the average concentration of ozone in the layer to be 7 ppm, what is the total mass in grams of the ozone in the upper atmosphere?

*15.46 The ozone present in polluted air is formed by the two-step mechanism

$$NO_2(g) \xrightarrow{k_1} NO(g) + O(g) \qquad \text{(first order)}$$

$$O(g) + O_2(g) \xrightarrow{k_2} O_3(g) \qquad \text{(second order)}$$

It is known that $k_1 = 6.0 \times 10^{-3} \text{ s}^{-1}$ and $k_2 = 1.0 \times 10^6 \text{ } \ell/(\text{mol} \cdot \text{s})$. The concentrations of NO_2 and O_2 in polluted air are about 3.0×10^{-9} and 1.0×10^{-2} mol/ℓ, respectively. One can assume that the concentration of atomic oxygen reaches a "steady state," a low constant concentration at which point it is being consumed in the second reaction at the same rate as it is being produced in the first.

a. Calculate the steady rate concentration of O(g) in polluted air.
b. Calculate the rate of formation of O_3 in polluted air.
c. If the rate of formation of O_3 remains constant, how long would it take for its concentration to build up to one part per million in air at 25°C and 1 atm? Under such conditions air contains about 0.041 mol/ℓ.

16

SOLUTES IN WATER _____

16.1   INTRODUCTION

A solution is a homogeneous mixture of a solute (substance being dissolved) distributed uniformly throughout a solvent (substance doing the dissolving). Solutions may exist among any of the three states of matter: gas, solid, or liquid. Air, the most common gaseous solution, is a mixture of nitrogen, oxygen, and lesser amounts of other gases (Chap. 15). Many metal alloys are solid solutions. Examples include brass (Cu and Zn), bronze (Cu, Zn, and Sn), and the U.S. "nickel" coin (25% Ni, 75% Cu). Solutions in the liquid state are perhaps most familiar. They include gasoline, seawater, and alcoholic beverages. Of these, solutions in water are most important for our purposes in chemistry and will be emphasized in this chapter.

In most solutions, the identity of the *solute* in relation to the *solvent* is apparent. Solids or gases are solutes when dissolved in liquids—e.g., sugar in a soft drink or carbon dioxide in beer. When both components are liquids, the more abundant liquid is usually called the solvent. In a solution of 1 g of ethyl alcohol in 100 g of water, we would refer to water as the solvent and the alcohol as the solute. If the amounts are more nearly equal, the choice of terms is arbitrary and must be specified.

Several different adjectives can be used to indicate the relative amounts of solute and solvent in a solution. We may describe a solution containing a small amount of solute relative to solvent as being "dilute." Another solution containing more of that solute in the same amount of solvent might be called "concentrated." In a few cases these terms, through tradition, have taken on a quantitative meaning. Solutions of certain common reagents, labeled dilute or concentrated, have the concentrations specified in Table 16.1. These concentrations are given in *molarity* (moles of solute per liter of solution).

Most reactions in living organisms occur in, or in contact with, solutions

The terms conc. and dilute in this context are gradually disappearing

TABLE 16.1 CONCENTRATIONS OF WATER SOLUTIONS OF ACIDS AND BASES			
	Solute		**Molarity**
Hydrochloric acid	HCl	conc.	12.0
		dilute	6.0
Nitric acid	HNO_3	conc.	16.0
		dilute	6.0
Sulfuric acid	H_2SO_4	conc.	18.0
		dilute	3.0
Ammonia	NH_3*	conc.	15.0
		dilute	6.0

*Often labeled "NH_4OH."

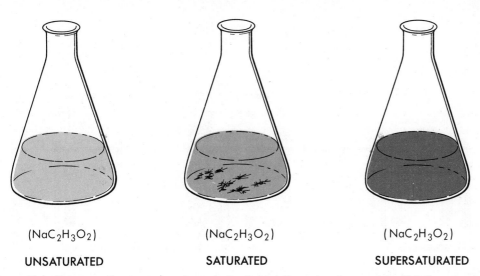

(NaC$_2$H$_3$O$_2$)   (NaC$_2$H$_3$O$_2$)   (NaC$_2$H$_3$O$_2$)

UNSATURATED   **SATURATED**   **SUPERSATURATED**

Figure 16.1 Unsaturated, saturated, and supersaturated solutions of sodium acetate. Only the saturated solution is in equilibrium with undissolved solute. If the supersaturated solution is disturbed, say by adding a crystal of sodium acetate, crystallization of excess solute will occur rapidly until equilibrium between solid and solution is established.

Relative concentrations of solutions are often expressed in a different way by using the terms "saturated," "unsaturated," and "supersaturated." A **saturated** solution is one which is in equilibrium with undissolved solute. An **unsaturated** solution contains a lower concentration of solute than the saturated solution. The unsaturated solution is not at equilibrium. If solute is added, it dissolves to approach saturation. A **supersaturated** solution contains more than the equilibrium concentration of solute. It too is unstable in the presence of excess solute.

To illustrate the meaning of these terms, consider solid sodium acetate, NaC$_2$H$_3$O$_2$ (Fig. 16.1). A saturated aqueous solution of sodium acetate at 20°C contains 46.5 g of NaC$_2$H$_3$O$_2$ in 100 g of water. Any solution in which the concentration of sodium acetate at 20°C is less than this value is unsaturated. If solid sodium acetate is added to such a solution, it will dissolve until the concentration reaches the saturation value. A supersaturated solution contains more than 46.5 g of NaC$_2$H$_3$O$_2$ per 100 g of water at 20°C. To prepare such a solution, we take advantage of the fact that the solubility of sodium acetate increases with temperature. At 50°C we can dissolve 80 g of NaC$_2$H$_3$O$_2$ in 100 g of water. If this solution is cooled carefully to 20°C, without shaking or stirring, the excess solute stays in solution. This produces a supersaturated solution. If a small seed crystal of sodium acetate is now added, crystallization quickly takes place. The excess solute (80 g − 46.5 g) comes out of solution, establishing equilibrium with the saturated solution.

Honey is often a super-
saturated solution

16.2   TYPES OF SOLUTES IN WATER: ELECTROLYTES
AND NONELECTROLYTES

Solutes in water can be classified according to the electrical conductivity of the solutions they form (Fig. 16.2). We can distinguish between two types of solutes:

1. **Nonelectrolytes,** whose water solutions do not conduct an electric current. These substances dissolve as molecules. Since the molecules carry no charge, they do not move in an electrical field and hence do not carry a current. Methyl alcohol is one such non-

Light bulb

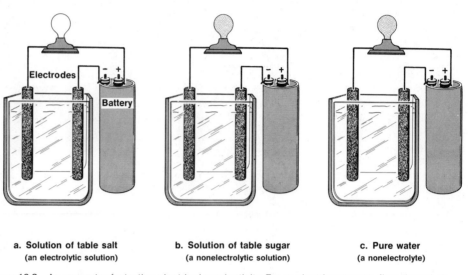

a. Solution of table salt
(an electrolytic solution)

b. Solution of table sugar
(a nonelectrolytic solution)

c. Pure water
(a nonelectrolyte)

Figure 16.2   An apparatus for testing electrical conductivity. For an electric current to flow, the solution must contain ions, which serve as carriers of electric charge. Pure water contains very few ions, so is not a conductor. If NaCl is dissolved in water, it forms ions. These conduct the current and the light bulb glows. Dissolving sugar in water produces only sugar molecules in the solution, so it does not become a conductor.

electrolyte. The process by which it dissolves in water may be represented by the simple equation

Roughly speaking, organic solutes tend to be nonelectrolytes, while inorganic are electrolytes

$$CH_3OH(l) \rightarrow CH_3OH(aq) \qquad (16.1)$$

2. **Electrolytes,** whose water solutions conduct an electric current. These substances exist as ions in solution. The charged ions migrate in an electrical field, thereby carrying a current. Sodium chloride is a familiar example of an electrolyte. In the solid, it consists of Na^+ and Cl^- ions. Upon dissolving in water, these ions are set free. The solution process may be represented as

$$NaCl(s) \rightarrow Na^+(aq) + Cl^-(aq) \qquad (16.2)$$

Other ionic solutes behave in a similar manner (Example 16.1). As a result of dissociation, ions are able to move about in solution.

Example 16.1   Write equations for the dissociation in water of each of the following solid electrolytes:

   a. KCl   b. Na_2SO_4 c. $Fe(NO_3)_3$

Solution   In each case, the "reactant" is simply the ionic solid. The "products" are the ions formed in water solution by dissociation.

   a. $KCl(s) \rightarrow K^+(aq) + Cl^-(aq)$

   b. $Na_2SO_4(s) \rightarrow 2\ Na^+(aq) + SO_4^{2-}(aq)$

   c. $Fe(NO_3)_3(s) \rightarrow Fe^{3+}(aq) + 3\ NO_3^-(aq)$

Exercise   What is the total number of moles of ions formed when one mole of KCl dissolves in water? one mole of Na_2SO_4? one mole of $Fe(NO_3)_3$? Answer: Two, three, four.

Hydrogen chloride, HCl, is typical of a certain class of covalent compounds which, *in water solution,* act as electrolytes. Pure hydrogen chloride as a solid, liquid, or gas has the properties of a molecular substance. For example, liquid HCl does not conduct an electric current. However, when put into water, HCl dissolves to form a conducting solution typical of an electrolyte. A reaction occurs with water to produce aqueous H^+ ions and Cl^- ions:

$$HCl(aq) \rightarrow H^+(aq) + Cl^-(aq) \qquad (16.3)$$

Actually, water is involved in this reaction (Ch. 7)

In water, the covalent bond holding the HCl molecule together breaks. This electron pair is transferred to chlorine, forming the Cl^- ion and leaving an H^+ ion. The presence of these two ions in solution makes HCl an electrolyte.

Table 16.2 summarizes the solution behavior of different kinds of solutes. Note from the table that a mole of a nonelectrolyte produces *one mole* of particles in solution. On the other hand, electrolytes provide *more than one mole* of particles per mole of solute dissolved.

Certain solutes dissolve in water to form a mixture containing both ions and molecules. An example is hydrogen fluoride, HF. In a water solution of HF, we have an equilibrium mixture of HF molecules, H^+ ions, and F^- ions:

$$HF(aq) \rightleftharpoons H^+(aq) + F^-(aq)$$

Compounds such as HF are often referred to as *weak electrolytes,* to distinguish them from *strong electrolytes* such as HCl and NaCl, which are completely dissociated to ions in solution. Water solutions of weak electrolytes have a measurable but low electrical conductivity. We will have more to say about this type of substance in Chapters 17 and 18.

TABLE 16.2   SOLUTION BEHAVIOR OF SOME AQUEOUS SOLUTES

SOLUTE	TYPE	SOLUTION EQUATION	MOLES OF SOLUTE PARTICLES PER MOLE SOLUTE
$I_2(s)$	Nonelectrolyte	$I_2(s) \rightarrow I_2(aq)$	1
$C_6H_{12}O_6(s)$ (glucose)	Nonelectrolyte	$C_6H_{12}O_6(s) \rightarrow C_6H_{12}O_6(aq)$	1
$C_{12}H_{22}O_{11}(s)$ (sucrose)	Nonelectrolyte	$C_{12}H_{22}O_{11}(s) \rightarrow C_{12}H_{22}O_{11}(aq)$	1
$HCl(g)$	Electrolyte	$HCl(g) \rightarrow H^+(aq) + Cl^-(aq)$	2 (1 H^+ and 1 Cl^-)
$LiBr(s)$	Electrolyte	$LiBr(s) \rightarrow Li^+(aq) + Br^-(aq)$	2 (1 Li^+ and 1 Br^-)
$K_2SO_4(s)$	Electrolyte	$K_2SO_4(s) \rightarrow 2 K^+(aq) + SO_4^{2-}(aq)$	3 (2 K^+ and 1 SO_4^{2-})
$LaBr_3(s)$	Electrolyte	$LaBr_3(s) \rightarrow La^{3+}(aq) + 3 Br^-(aq)$	4 (1 La^{3+} and 3 Br^-)

16.3 PRINCIPLES OF WATER SOLUBILITY

Solutes differ greatly in their water solubility. At one extreme, we recognize substances which are virtually insoluble. These include many gases (He, H_2, . . .), liquid hydrocarbons such as those found in gasoline, and a variety of solids such as iodine, I_2, and calcium carbonate, $CaCO_3$. In contrast, many other solutes have very high water solubilities. Methyl alcohol, CH_3OH, is soluble in water in all proportions; we might describe it as being infinitely soluble. Many ionic solids, including NaCl, KNO_3, and NaOH, are very soluble in water.

In this section, we will develop rules for predicting relative solubilities of solutes in water. We will also be concerned with the effect of temperature and pressure changes upon these solubilities.

Solubility of Nonelectrolytes

As a general rule, substances which dissolve in water as molecules do so to a very small extent. This is true of such species as iodine, methane, and carbon tetrachloride:

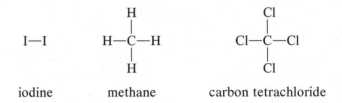

iodine   methane   carbon tetrachloride

We can readily understand why these nonpolar substances have such low water solubilities. To dissolve appreciable amounts of I_2, CH_4, or CCl_4 in water, it would be necessary to break the hydrogen bonds holding H_2O molecules together. There is no compensating attractive force to promote the solution process and supply the energy required to break into the water structure.

A few nonelectrolytes have high water solubilities. Among the species in this category are hydrogen peroxide, methyl alcohol, and ethylene glycol.

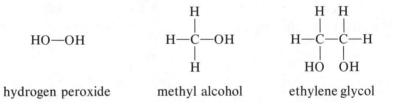

hydrogen peroxide   methyl alcohol   ethylene glycol

These compounds, all of which are liquids, are soluble in water in all proportions. They form hydrogen bonds with H_2O molecules (Fig. 16.3), and hence can readily dissolve by breaking into the water structure.

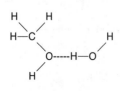

Figure 16.3 Methyl alcohol readily forms hydrogen bonds (dotted lines) with water. This explains why CH_3OH is infinitely soluble in water.

The process by which an ionic solid such as NaCl dissolves in water is shown in Figure 16.4. Solubility comes about because of an interaction between polar H_2O molecules and the ions making up the crystal. The extent to which solution occurs depends upon a balance between two forces, both electrical in nature.

1. The force of attraction between H_2O molecules and the ions of the solid, which tends to bring the solid into solution. If this factor predominates, we expect the compound to be very soluble in water, as is the case with NaCl, NaOH, and many other ionic solids.

2. The force of attraction between oppositely charged ions, which tends to keep them in the solid state. If this is the major factor, we expect the water solubility to be very low. The fact that $CaCO_3$ and $BaSO_4$ are almost insoluble in water implies that interionic attractive forces predominate with these ionic solids.

In $CaCO_3$ and $BaSO_4$ the ionic charges are higher than in NaCl and NaOH

Unfortunately, we cannot estimate from first principles the relative magnitudes of these two forces with a particular ionic solid. For this reason, among others, we cannot predict in a quantitative way the water solubilities of electrolytes. The best we can do is to write down a set of generalizations based upon experiment. These generalizations, called "solubility rules," are listed for common ionic solids in Table 16.3.

The solubility rules are rather simple to interpret. From Table 16.3, it should be clear that sodium chloride, potassium chloride, and sodium carbonate are soluble in water. The table also tells us that barium sulfate, calcium carbonate, and silver sulfide do not dissolve in water to an appreciable extent. Example 16.2 further illustrates the use of the data in Table 16.3.

Example 16.2   Using the solubility rules, indicate whether each of the following ionic compounds is soluble in water:
     a. NH_4NO_3 b. $BaCl_2$ c. $PbCO_3$

Solution
     a. Soluble (All nitrates are soluble.)
     b. Soluble ($BaCl_2$ is not one of the three insoluble chlorides listed.)
     c. Insoluble (Pb is not a Group 1 element.)

Exercise   Is HgS water-soluble? LiOH? Answer: HgS is insoluble; LiOH is soluble.

TABLE 16.3 SOLUBILITY RULES	
NO_3^-	All nitrates are soluble.
Cl^-	All chlorides are soluble except AgCl, Hg_2Cl_2, and $PbCl_2$.*
SO_4^{2-}	All sulfates are soluble except $CaSO_4$,* $SrSO_4$, $BaSO_4$, Hg_2SO_4, $HgSO_4$, $PbSO_4$, and Ag_2SO_4.*
CO_3^{2-}	All carbonates are insoluble except those of the Group 1 elements and NH_4^+.
OH^-	All hydroxides are insoluble except those of the Group 1 elements, $Sr(OH)_2$ and $Ba(OH)_2$. ($Ca(OH)_2$ is slightly soluble.)
S^{2-}	All sulfides except those of the Groups 1 and 2 elements and NH_4^+ are insoluble.

These rules are useful. Learn them

*Insoluble compounds are those which precipitate upon mixing equal volumes of solutions 0.1 M in the corresponding ions. Compounds which fail to precipitate at concentrations slightly below 0.1 M are starred.

Effect of Temperature upon Solubility

From experience, we know that more sugar dissolves in hot coffee than in cold coffee. In general, the solubility of a solid in a liquid rises with increasing temperature. We can explain this behavior in terms of Le Châtelier's Principle. Consider temperature change as a stress on the equilibrium between a solid and its saturated solution:

$$\text{solid + solvent} \rightleftharpoons \text{solution}; \ \Delta H > 0 \tag{16.4}$$

Ordinarily, dissolving a solid is an endothermic process ($\Delta H > 0$). This is the case because heat must be absorbed to break down the crystal lattice. Hence an increase in temperature shifts the equilibrium in Reaction 16.4 to the right and increases the solubility. In this way, heat is absorbed, partially counteracting the temperature increase.

Figure 16.5 shows the effect of temperature upon the solubility of several ionic solids in water. As you can see, solubility ordinarily increases with temperature. This explains why a supersaturated solution of such a solute can be formed by cooling a hot, saturated solution (Section 16.1). The process of fractional crystallization, discussed in Chapter 1, also takes advantage of this effect. In that case, the excess solute comes out of solution upon cooling.

Dissolving a gas in a liquid usually evolves heat ($\Delta H < 0$):

$$\text{gas + liquid} \rightleftharpoons \text{solution}; \ \Delta H < 0 \tag{16.5}$$

This means that the reverse process (gas coming out of solution) is endothermic. Hence, it is favored by an increase in temperature; gases become less soluble as the temperature rises. This rule is followed by all gases in water. You have probably noticed this effect when heating water in an open pan or beaker. Bubbles of air are driven out of the water by an increase in temperature. The reduced solubility of oxygen in water at high temperatures is a major factor in the "thermal pollution" of water supplies. In the warm water discharged by large power plants, the low concentration of oxygen makes it difficult for fish and other aquatic life to survive.

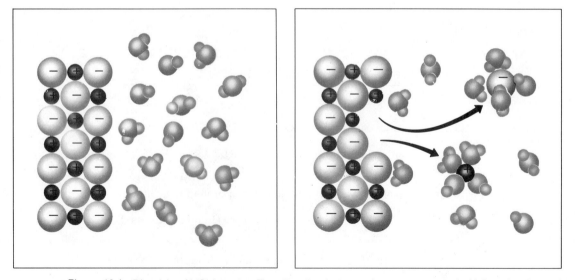

Figure 16.4 Dissolving NaCl in water. The attraction between the oxygen atoms in H_2O molecules and Na^+ ions in NaCl tends to bring the Na^+ ions into solution. At the same time, Cl^- ions, attracted to the hydrogen atoms of H_2O molecules, also enter the solution.

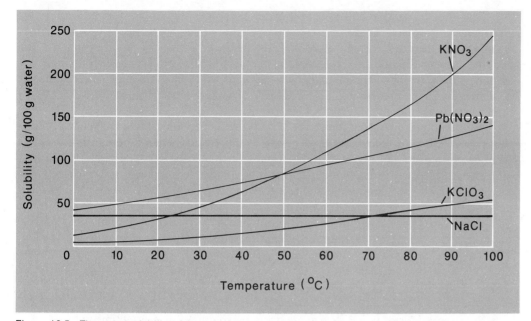

Figure 16.5 The water solubility of these four ionic compounds, and indeed that of most solids, increases with temperature. This reflects the fact that ΔH of solution of a solid is usually a positive quantity. With NaCl, where ΔH is very small (ΔH = +3.9 kJ/mol), solubility increases only slightly with temperature.

Effect of Pressure upon Solubility

Pressure has a major effect on solubility only for gas-liquid systems. At a given temperature, a rise in pressure increases the solubility of a gas. Indeed, at low to moderate pressures, gas solubility is directly proportional to pressure (Henry's Law). The solubility of O_2 in water is doubled by increasing the pressure from 1 to 2 atm. This effect is readily explained in terms of Le Châtelier's Principle (Fig. 16.6). The increase in pressure raises the concentration of molecules in the gas phase. To counteract this change, more molecules enter the solution. The equilibrium in 16.5 shifts to the right, and solubility goes up as pressure increases.

Figure 16.6 Effect of pressure on gas solubility. Doubling the pressure doubles the concentration of solute molecules in the gas phase. To maintain equilibrium, the concentration of solute molecules in the liquid must also double. More generally, solubility is directly proportional to the pressure of the gas.

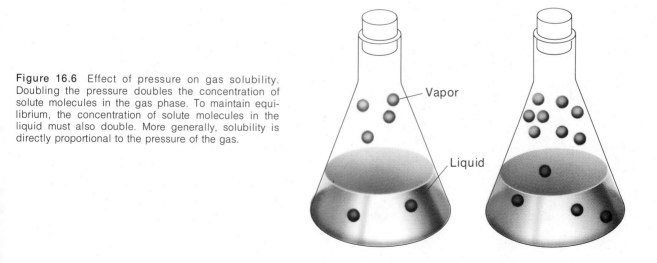

When the container is
opened, the CO_2 solu-
tion becomes super-
saturated

This effect is used in bottling carbonated beverages such as beer, sparkling wines, and many soft drinks. These beverages are bottled under pressures of CO_2 as high as 4 atm. When the bottle or can is opened, the pressure above the liquid drops to 1 atm and the carbon dioxide bubbles rapidly out of solution. Pressurized containers for shaving cream, whipped cream, and cheese spreads work on a similar principle. Pressing a valve reduces the pressure on dissolved gas, causing it to rush from solution, carrying liquid with it as a foam.

Another consequence of the effect of pressure on gas solubility is the painful, sometimes fatal, affliction known as the "bends." This occurs when a person goes rapidly from deep water (high pressure) to the surface (lower pressure). The rapid decompression causes air, dissolved in blood and other body fluids, to bubble out of solution. These bubbles impair blood circulation and affect nerve impulses. To minimize these effects, deep-sea divers and aquanauts breathe a helium-oxygen mixture rather than compressed air (nitrogen-oxygen). Helium is only about one fifth as soluble as nitrogen. Hence, much less gas comes out of solution upon decompression.

Even scuba divers
should be aware of this
problem

16.4 PRECIPITATION REACTIONS

When solutions of two different electrolytes are mixed, we sometimes observe that an insoluble solid comes out of solution. This solid is referred to as a *precipitate*. The chemical reaction by which it is formed is called a *precipitation reaction*. As an example, consider what happens when solutions of barium chloride and potassium sulfate are mixed. We observe that a white precipitate forms (Fig. 16.7). Using the solubility rules given in Table 16.3, it is possible to identify the precipitate.

We start by noting that in solutions of these two ionic solids the following ions are available:

$BaCl_2$ solution: Ba^{2+} and Cl^-
K_2SO_4 solution: K^+ and SO_4^{2-}

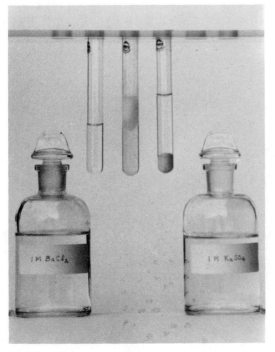

Figure 16.7 When solutions of $BaCl_2$ and Na_2SO_4 are mixed, a white precipitate forms (center test tube). This solid, which slowly settles out (test tube at right), can be identified as $BaSO_4$, which is very insoluble in water. It is formed when Ba^{2+} ions from the $BaCl_2$ solution combine with SO_4^{2-} ions from the Na_2SO_4 solution.

The precipitate must result from an interaction between two ions. We expect the cations to repel each other and the anions to do likewise. The interaction of Ba^{2+} with Cl^- ions or K^+ with SO_4^{2-} ions would simply produce the *soluble* starting materials. Two other ion pairings are possible:

$$K^+ \text{ and } Cl^-$$
$$Ba^{2+} \text{ and } SO_4^{2-}$$

Potassium chloride, KCl, is soluble in water, so we deduce that K^+ and Cl^- ions do not combine with each other. On the other hand, barium sulfate, $BaSO_4$, is insoluble. This means that Ba^{2+} and SO_4^{2-} ions cannot be present together in solution at appreciable concentrations. We conclude that Ba^{2+} and SO_4^{2+} ions must combine to form a precipitate of $BaSO_4$. The white, insoluble solid shown in Figure 16.7 is barium sulfate.

Equations for Precipitation Reactions

The precipitation reaction that occurs when solutions of $BaCl_2$ and K_2SO_4 are mixed can be represented by a simple equation. We have seen that the product is solid $BaSO_4$. This is formed by the reaction between Ba^{2+} and SO_4^{2-} ions in aqueous solution. The equation is

$$Ba^{2+}(aq) + SO_4^{2-}(aq) \rightarrow BaSO_4(s) \qquad (16.6)$$

Notice that we include in this equation only those ions which participate in the reaction. To be specific, K^+ and Cl^- ions do not appear in Equation 16.6. They are "spectator" ions, which are present in the solution before and after the precipitation of barium sulfate. Equations such as this, which involve ions and exclude any species which do not take part in the reaction, are often referred to as *net ionic equations*. They do not differ in any essential way from the equations we have been writing all along. Indeed, many of the equations written in earlier chapters qualify as net ionic equations.

If you put in K^+ and Cl^-, they appear as reactants and products, and so cancel out

Other examples of net ionic equations written to represent precipitation reactions include:

—formation of a precipitate of AgCl by mixing solutions of $AgNO_3$ and HCl:

$$Ag^+(aq) + Cl^-(aq) \rightarrow AgCl(s) \qquad (16.7)$$

—formation of a precipitate of Ag_2CrO_4 by mixing solutions of $AgNO_3$ and K_2CrO_4:

$$2\ Ag^+(aq) + CrO_4^{2-}(aq) \rightarrow Ag_2CrO_4(s) \qquad (16.8)$$

Using Table 16.3, we can readily predict whether or not a precipitation reaction will occur when two electrolyte solutions are mixed. To do this, we follow the reasoning discussed above in the $BaCl_2$–K_2SO_4 case. We start by considering what precipitates are possible ($BaSO_4$, KCl). Then, using the solubility rules, we identify the precipitate, if any, that forms ($BaSO_4$). Finally, we write a simple equation, analogous to 16.6 through 16.8, to describe the precipitation reaction.

Example 16.3   Write a (net ionic) equation for any precipitation reaction that occurs when solutions of the following ionic compounds are mixed:

   a. $Pb(NO_3)_2$ and NaCl   b. $Ba(OH)_2$ and $NiSO_4$ c. NaCl and KOH

Solution In each case we first decide what the two possible products are. Then, using the solubility rules we decide what compound, if any, precipitates. Finally, we write an equation for the precipitation reaction.

a. *Possible products:* $NaNO_3$ (Na^+ ions from the NaCl solution, NO_3^- ions from the $Pb(NO_3)_2$ solution), and $PbCl_2$ (Pb^{2+} ions from the $Pb(NO_3)_2$ solution, Cl^- ions from the NaCl solution).

Identity of precipitate: Table 16.3 tells us that $NaNO_3$ is soluble and $PbCl_2$ is insoluble. We conclude that a precipitate of $PbCl_2$ should form.

Equation: $Pb^{2+}(aq) + 2\ Cl^-(aq) \rightarrow PbCl_2(s)$

b. The possible products are barium sulfate, $BaSO_4$, and nickel hydroxide, $Ni(OH)_2$. $BaSO_4$ is listed in Table 16.3 as being insoluble. Since Ni is not a Group 1 element, we deduce that $Ni(OH)_2$ is insoluble. Thus we predict that both of these compounds will precipitate. We write two equations, since two entirely different precipitation reactions take place:

$$Ba^{2+}(aq) + SO_4^{2-}(aq) \rightarrow BaSO_4(s)$$

$$Ni^{2+}(aq) + 2\ OH^-(aq) \rightarrow Ni(OH)_2(s)$$

Experiment confirms our deductions. If we look at the precipitate under a microscope, we see white $BaSO_4$ crystals and green particles of $Ni(OH)_2$.

c. Both possible products, NaOH and KCl, are soluble. No precipitation reaction occurs, so there is no equation to write.

Exercise Write net ionic equations for the precipitation reactions that occur when the following solutions are mixed: $MgCl_2$ and NaOH; $SbCl_3$ and Na_2S. Answer: $Mg^{2+}(aq) + 2\ OH^-(aq) \rightarrow Mg(OH)_2(s)$; $2\ Sb^{3+}(aq) + 3\ S^{2-}(aq) \rightarrow Sb_2S_3(s)$.

Mole Relations in Precipitation Reactions

For a precipitation reaction, the coefficients in the equation represent the number of moles of ions reacting and precipitate formed. We can use these coefficients to carry out calculations similar to those presented in Chapters 4 and 5. Example 16.4 illustrates this application.

Example 16.4 When a solution of $AgNO_3$ is added to a solution of NaCl, the following precipitation reaction occurs:

$$Ag^+(aq) + Cl^-(aq) \rightarrow AgCl(s)$$

What volume of 0.200 M $AgNO_3$ solution is required to react with 100 cm³ of 0.250 M NaCl solution?

Solution The equation tells us that one mole of Ag^+ ion reacts with one mole of Cl^-. This suggests that we could work the problem by: (1) calculating the number of moles of Cl^- in the NaCl solution; (2) equating this to the number of moles of Ag^+ required; and (3) calculating the volume of $AgNO_3$ solution needed to give that number of moles of Ag^+.

1. moles NaCl = $0.250\ \dfrac{\text{mol}}{\ell} \times 0.100\ \ell = 0.0250$ mol NaCl

Since 1 mol NaCl gives 1 mol Cl^-:

moles Cl^- = moles NaCl = 0.0250 mol Cl^-

Ions, like other chemical species, obey the laws of stoichiometry (mass balance)

2. Since 1 mol Ag^+ reacts with 1 mol Cl^-:

$$\text{moles } Ag^+ = \text{moles } Cl^- = 0.0250 \text{ mol } Ag^+$$

3. Since 1 mol $AgNO_3$ gives 1 mol Ag^+, we need 0.0250 mol of $AgNO_3$. To find the volume of $AgNO_3$ solution required:

$$\text{volume } AgNO_3 \text{ solution} = \frac{\text{moles } AgNO_3}{\text{mol}/\ell \; AgNO_3} = \frac{0.0250 \text{ mol}}{0.200 \text{ mol}/\ell} = 0.125 \; \ell = 125 \text{ cm}^3$$

Exercise   If 50.0 cm³ of 0.100 M $AgNO_3$ is required to react with 25.0 cm³ of NaCl solution, what is the concentration, in moles per liter, of the NaCl? Answer: 0.200 M.

The exercise following Example 16.4 suggests an important application of precipitation reactions. The concentration of Cl^- in a solution can be determined by titrating with $AgNO_3$ solution of known concentration. From the volumes of the two solutions required for complete reaction, it is possible to calculate the concentration of Cl^- ions.

16.5 COLLIGATIVE PROPERTIES OF SOLUTIONS

The properties of a solution differ considerably from those of the pure solvent. Those solution properties which depend primarily on the *concentration of solute particles* rather than their nature are called **colligative properties.** These properties include vapor pressure lowering, osmotic pressure, boiling point elevation, and freezing point depression. The fact that these properties are colligative means that all water solutions containing 0.1 mol of solute particles per liter should have about the same vapor pressure, osmotic pressure, boiling point, or freezing point. This should be true regardless of the type of solute particles present. They may be molecules, like glucose ($C_6H_{12}O_6$) or sucrose ($C_{12}H_{22}O_{11}$), or ions, as with sodium chloride (Na^+, Cl^-). In this section we will emphasize the colligative properties of nonelectrolytes, such as glucose and sucrose. Colligative properties of electrolytes will be discussed briefly at the end of this section.

With colligative properties, an ion is as effective as a molecule

The relationships among colligative properties and solute concentration are best regarded as limiting laws. They are approached most closely when the solution is very dilute. In practice, the relationships we will discuss are valid, for nonelectrolytes, to within a few per cent at concentrations as high as 1 M.

Vapor Pressure Lowering (Nonelectrolytes)

The rate at which water molecules escape from the surface is reduced in the presence of a nonvolatile solute. Concentrated aqueous solutions of nonelectrolytes such as glucose or sucrose evaporate more slowly than pure water. This reflects the fact that the vapor pressure of water in the solution is less than that of pure water. This decrease in vapor pressure is:

A nonvolatile solute will always decrease the vapor pressure of the solvent

1. Independent of the nature of the solute.
2. Directly proportional to the concentration of solute.

To illustrate these principles, consider the data in Table 16.4. Here we show the vapor pressures of water over glucose and sucrose solutions of various molarities. In the two columns at the right, we show the **vapor pressure lowering** in these solutions:

$$\text{vp lowering} = \text{vp pure solvent} - \text{vp solvent in solution} \qquad (16.9)$$

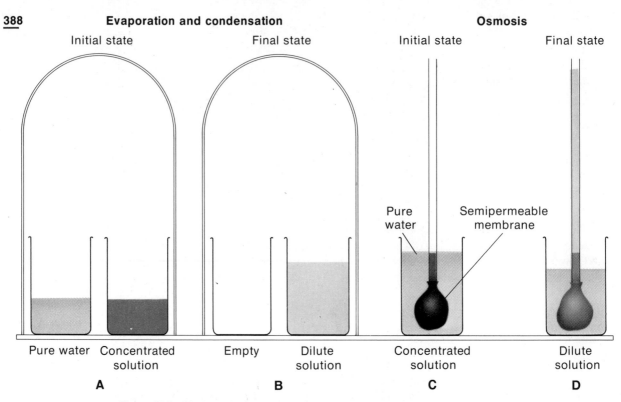

Figure 16.8 Water tends to move spontaneously from a region where its vapor pressure is high to a region where it is low. In (A) → (B), movement of water molecules occurs through the air trapped under the bell jar. In (C) → (D), water molecules move by osmosis through a semipermeable membrane. The driving force is the same in the two cases, although the mechanism differs.

Notice that, at a given molarity, the vapor pressure lowerings for glucose and sucrose are identical, at least to 0.001 mm Hg. Moreover, vapor pressure lowering is directly proportional to concentration. When the concentration of sucrose doubles from 0.10 to 0.20 M, the vapor pressure lowering becomes about twice as great, 0.017 mm Hg as opposed to 0.008 mm Hg.

One interesting effect of vapor pressure lowering is shown at the left of Figure 16.8. Here, we start with two beakers, one containing pure water and the other containing a sugar solution. These are placed next to each other, under a bell jar (Fig. 16.8A). As time passes, the liquid level in the beaker containing the solution rises. The level of

TABLE 16.4 VAPOR PRESSURE LOWERING IN AQUEOUS SOLUTIONS OF GLUCOSE AND SUCROSE AT 0°C

	VP H_2O (mm Hg)		VP LOWERING (mm Hg)	
MOLARITY	**Glucose Solution**	**Sucrose Solution**	**Glucose Solution**	**Sucrose Solution**
0.00	4.579	4.579		
0.05	4.575	4.575	0.004	0.004
0.10	4.571	4.571	0.008	0.008
0.20	4.562	4.562	0.017	0.017
0.30	4.554	4.554	0.025	0.025

pure water in the other beaker falls. Eventually, by evaporation and condensation, all the water is transferred to the solution (Fig. 16.8B). At the end of the experiment, the beaker that contained pure water is empty. The driving force behind this process is It would admittedly take a long time the difference in vapor pressure of water in the two beakers. *Water moves from a region in which its vapor pressure is high* (pure water) *to one in which its vapor pressure is low* (sugar solution). This is a general tendency, followed by all liquids, and is responsible for a variety of natural processes, including osmosis.

Osmotic Pressure (Nonelectrolytes)

The apparatus shown at the right of Figure 16.8 (Fig. 16.8C and D) can be used to achieve a result similar to that found in the bell jar experiment. Here, a sugar solution is separated from water by a "semipermeable" membrane. This may be an animal bladder, a slice of vegetable tissue, or a piece of parchment. The membrane, by a mechanism which is not well understood, allows water molecules to pass through it, but not sugar molecules. Here, as before, water moves from a region where its vapor pressure is high (pure water) to a region where it is low (sugar solution). This process, taking place through a membrane permeable only to the solvent, is called **osmosis.** As a result of osmosis, the water level rises in the tube and drops in the beaker (Fig. 16.8D).

The passage of solvent molecules through an osmotic membrane can be prevented by applying pressure to the solution (Fig. 16.9). The external pressure, P, just sufficient to prevent osmosis is equal to the **osmotic pressure** of the solution, π. If P is less than π, osmosis takes place in the normal way and water moves through the membrane into the solution (Fig. 16.9A). By making the external pressure large enough, it is possible to reverse this process (Figure 16.9B). When P > π, water molecules move through the membrane from the solution to pure water. This process, called *reverse osmosis,* is used to obtain fresh water from seawater in arid regions of the world.

Osmotic pressure, like vapor pressure lowering, is a colligative property. For a nonelectrolyte, π is directly proportional to molarity, M. The equation relating the two quantities is

$$\pi = M R T$$

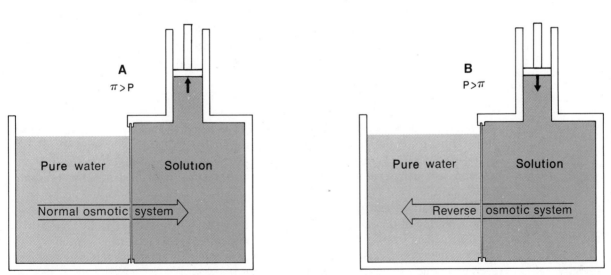

Figure 16.9 Osmosis can be prevented by applying to the solution a pressure P which just balances the osmotic pressure, π. If P < π, normal osmosis occurs. If P > π, water flows in the opposite direction. This process, called reverse osmosis, can be used to obtain fresh water from seawater.

where R is the gas law constant, $0.0821 \ \ell \cdot atm/(mol \cdot K)$, and T is the Kelvin temperature. Even in dilute solution, the osmotic pressure is quite large. Suppose, for example, we are dealing with a 0.10 M solution:

$$\pi = (0.10)(0.0821)(298) \ atm = 2.4 \ atm$$

At a given concentration the osmotic pressure is much, much larger than the vapor pressure lowering

A pressure of 2.4 atm is equivalent to a column of water 25 m (more than 80 ft) high. This pressure is great enough to rupture the coating of a wet, germinating seed and allow the young plant to push through the soil.

A striking example of a natural osmotic process can be seen under a microscope when red blood cells are placed in pure water. Water flows through the cell membrane to dilute the solution inside the cell. The cell swells and eventually bursts (Fig. 16.10A). If the red blood cells are placed in a concentrated sugar solution instead, water moves in the reverse direction. It flows from the cells, and they shrink and shrivel (Figure 16.10B). To avoid these effects, solutions used in intravenous feeding are made up to have the same osmotic pressure as blood (about 7.7 atm). A 0.31 M glucose solution has this osmotic pressure and is said to be *isotonic* with blood.

Boiling Point Elevation and Freezing Point Depression (Nonelectrolytes)

A solution of a nonvolatile solute boils at a *higher* temperature and freezes at a *lower* temperature than the pure solvent. Consider, for example, a solution containing 18.0 g of glucose, $C_6H_{12}O_6$, in 100 g of water. This solution boils at 100.52°C at 1 atm pressure; the normal boiling point of pure water is 100.00°C. The glucose solution freezes at −1.86°C (fp pure water = 0.00°C).

In discussing the boiling points or freezing points of solutions, we use the terms boiling point elevation, ΔT_b, and freezing point depression, ΔT_f. These are defined so as to be positive quantities. Thus,

$$\Delta T_b = bp \ solution - bp \ pure \ solvent \qquad (16.10)$$

$$\Delta T_f = fp \ pure \ solvent - fp \ solution \qquad (16.11)$$

For the glucose solution just referred to:

Actually, all the colligative properties are the result of vapor pressure lowering

$$\Delta T_b = 100.52°C - 100.00°C = 0.52°C$$

$$\Delta T_f = 0.00°C - (-1.86°C) = 1.86°C$$

Boiling point elevation is a direct result of vapor pressure lowering. At any given temperature, a solution has a vapor pressure *lower* than that of the pure solvent. Hence, a *higher* temperature must be reached before the solution boils—that is, before its vapor pressure becomes equal to the external pressure. Figure 16.10 illustrates this reasoning graphically.

The freezing point depression, like the boiling point elevation, is a direct result of the lowering of the solvent vapor pressure by the solute. Notice from Figure 16.10 that the freezing point of the solution is the temperature at which the solvent in solution has the same vapor pressure as the pure solid solvent. This implies that it is pure solvent (e.g., ice) which separates when the solution freezes.

Boiling point elevation and freezing point depression, like vapor pressure lowering, are colligative properties. They are directly proportional to solute concentration. The concentration unit commonly used in expressions for ΔT_b and ΔT_f is one we have not used previously. It is called **molality** and is given the symbol **m**. Molality is defined as the number of moles of solute per kilogram (1000 g) of solvent:

$$m = \frac{\text{no. moles solute}}{\text{no. kg solvent}} \qquad (16.12)$$

The molality of a solution is readily calculated if the masses of solute and solvent are known. Consider, for example, a solution containing 13.0 g of glucose, $C_6H_{12}O_6$, in 98.0 g of water. Since the molecular mass of glucose is 180, we have

$$13.0 \text{ g} \times \frac{1 \text{ mol}}{180 \text{ g}} = 0.0722 \text{ mol glucose}$$

dissolved in

$$98.0 \text{ g} \times \frac{1 \text{ kg}}{1000 \text{ g}} = 0.0980 \text{ kg water}$$

Hence

$$m = \frac{0.0722 \text{ mol solute}}{0.0980 \text{ kg solvent}} = 0.737$$

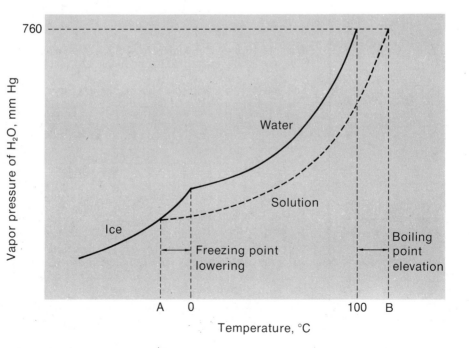

Figure 16.10 Since a nonvolatile solute lowers the vapor pressure of a solvent, the boiling point of a solution will be higher and the freezing point lower than the corresponding points for the pure solvent. Water solutions freeze *below* 0°C at A; and boil *above* 100°C at B.

The equations relating boiling point elevation and freezing point depression to molality are

$$\Delta T_b = k_b \times m \tag{16.13}$$

$$\Delta T_f = k_f \times m \tag{16.14}$$

The proportionality constants in these equations, k_b and k_f, are called the *molal boiling point constant* and the *molal freezing point constant*. Their magnitudes depend upon the nature of the solvent (Table 16.5). Note that when the solvent is water

$$k_b = 0.52°C; \quad k_f = 1.86°C$$

This tells us that a 1.00 m solution of a nonelectrolyte in water will boil at 100.52°C (at 1 atm) and freeze at −1.86°C. The organic solvents listed in Table 16.5 have values of k_f and k_b which are larger than those of water.

TABLE 16.5   MOLAL FREEZING POINT AND BOILING POINT CONSTANTS

SOLVENT	FREEZING POINT	k_f	BOILING POINT	k_b
Water	0°C	1.86	100°C	0.52
Acetic acid	17	3.90	118	2.93
Benzene	5.50	5.10	80.0	2.53
Cyclohexane	6.5	20.2	81	2.79
Camphor	178	40.0	208	5.95
p-Dichlorobenzene	53	7.1	—	—

The use of Equations 16.13 and 16.14 is illustrated in Example 16.5.

Example 16.5   An antifreeze solution is prepared containing 40.0 g of ethylene glycol (GMM = 62.0 g/mol) in 60.0 g of water. Calculate the freezing point of this solution.

Solution   We start by calculating the molality.

$$\text{moles ethylene glycol} = \frac{40.0 \text{ g}}{62.0 \text{ g/mol}} = 0.645 \text{ mol}$$

$$\text{molality} = \frac{\text{no. moles ethylene glycol}}{\text{no. kg water}} = \frac{0.645}{0.0600} = 10.8$$

For water, $k_f = 1.86°C$. Hence,

With concentrated solutions ΔT_f and ΔT_b can be quite large

$$\Delta T_f = k_f \times m = 1.86°C \times 10.8 = 20.1°C$$

We conclude that the freezing point of the solution should be 20.1°C below that of pure water (0°C). Hence the solution should freeze at −20.1°C. In practice, the freezing point is slightly lower, about −22°C. The deviation occurs because Equation 16.14 is a limiting law, strictly valid only at low concentrations.

Exercise   Estimate the boiling point of this solution at 1 atm. Answer: 105.6°C.

As you know, we take advantage of the freezing point depression when we add anti-freeze to automobile radiators in winter. Ethylene glycol is the solute commonly used in so-called "permanent" antifreezes. It has a high boiling point (197°C), is virtually nonvolatile at 100°C and, as we have seen, raises the boiling point of water. Hence, antifreeze containing ethylene glycol does not boil away in summer driving. Earlier antifreezes contained methyl alcohol. It is cheaper than ethylene glycol but has one disadvantage. It is volatile (bp = 65°C), lowers the boiling point of water,* and tends to boil out of solution. Ethyl alcohol behaves similarly.

Molecular Masses of Nonelectrolytes from Freezing Point Depression

The molecular mass of a gas or volatile liquid can be obtained from gas density measurements, as described in Chapter 7. This method is useless for nonvolatile solids or for substances that decompose on heating. Such substances do not ordinarily exist as vapors. It turns out that colligative properties, particularly freezing point depression, can be used to determine molecular masses of a wide variety of nonelectrolytes. The approach used is illustrated in Example 16.6.

---

*Volatile solutes ordinarily lower the boiling point because they contribute to the total vapor pressure of the solution.

Example 16.6   A student dissolves 1.50 g of a newly prepared compound in 75.0 g of cyclohexane. She measures the freezing point of the solution to be 2.70°C; that of pure cyclohexane is 6.50°C. Cyclohexane has a k_f of 20.2°C. Using these data, calculate the molecular mass of the compound.

Solution   We follow a three-step procedure. First, we obtain ΔT_f, using Equation 16.11. Then we use Equation 16.14 to calculate the molality. Finally, we obtain the molecular mass, using the defining equation for molality (Eq. 16.12).

(1) ΔT_f = fp pure cyclohexane − fp solution = 6.50°C − 2.70°C = 3.80°C

(2) $\Delta T_f = k_f \times m$; $m = \dfrac{\Delta T_f}{k_f} = \dfrac{3.80°C}{20.2°C} = 0.188$

(3) $m = \dfrac{\text{no. moles solute}}{\text{no. kg solvent}} = \dfrac{\text{no. g solute/GMM}}{\text{no. kg solvent}}$

where GMM is the gram molecular mass (g/mol). Solving the above equation for GMM:

$$\text{GMM} = \frac{\text{no. g solute}}{(m)(\text{no. kg solvent})}$$

The GMM is the mass of a mole in grams

All the quantities on the right side of this equation are known. We know that 1.50 g of solute is dissolved in 75.0 g of solvent (0.0750 kg); we calculated m to be 0.188. Hence:

$$\text{GMM} = \frac{1.50}{(0.188)(0.0750)} \frac{g}{\text{mol}} = 106 \text{ g/mol}$$

The molecular mass of the compound is 106.

Exercise   A solution of 5.00 g of a compound X in 60.0 g of water freezes at −1.00°C. What is the molecular mass of X? Answer: 155.

In carrying out a molecular mass determination by freezing point depression, we must choose a solvent in which the solute is readily soluble. Usually, several such solvents are available. Of these, we tend to pick one which has a large k_f. This makes ΔT_f large and so reduces the per cent error in the freezing point measurement. From this point of view, cyclohexane or other organic solvents are better choices than water, since their k_f values are larger.

Molecular masses can also be determined using other colligative properties. Osmotic pressure measurements are often used, particularly for solutes of high molecular mass where the concentration is likely to be quite low. The advantage of using osmotic pressure is that the effect is relatively large. A 0.1 M solution of a nonelectrolyte in water has an osmotic pressure of 2.4 atm, as compared to a freezing point depression of about 0.2°C, a boiling point elevation of 0.05°C, and a vapor pressure lowering of only 0.04 mm Hg at 25°C.

Colligative Properties of Electrolytes

As noted earlier, colligative properties of dilute solutions are directly proportional to the concentration of solute *particles*. On this basis, we would predict that, at a given concentration, an electrolyte would have a greater effect upon these properties than a nonelectrolyte. When one mole of a nonelectrolyte such as glucose dissolves in water, 1 mol of solute molecules is obtained. On the other hand, one mole of the electrolyte NaCl yields 2 mol of ions. With calcium chloride, $CaCl_2$, 3 mol of ions are produced per mole of solute.

This reasoning is confirmed experimentally. Suppose, for example, we compare the vapor pressure of 1 M solutions of glucose, sodium chloride, and calcium chloride. We find that the vapor pressure lowering is smallest for glucose and largest for calcium chloride. In other words,

vp pure water > vp glucose soln. > vp NaCl soln. > vp $CaCl_2$ soln.

Many electrolytes form saturated aqueous solutions whose vapor pressures are so low that the solids pick up water *(deliquesce)* when exposed to moist air. This occurs with calcium chloride, whose saturated solution has a vapor pressure only 20% that of pure water. If dry $CaCl_2$ is exposed to air in which the relative humidity is greater than 20%, it absorbs water and forms a saturated solution. Deliquescence continues until the vapor pressure of the solution becomes equal to that of the water in the air. Calcium chloride is sometimes spread on dirt roads in summer. This prevents them from becoming dusty and developing a rough, "washboard" surface.

The freezing points of electrolyte solutions, like their vapor pressures, are lower than those of nonelectrolytes at the same concentration. Sodium chloride and calcium chloride are used to melt ice on highways. Their aqueous solutions can have freezing points as low as −21°C and −55°C, in that order. The equation for the freezing point lowering of an electrolyte is similar to that for nonelectrolytes, except for the introduction of a multiplier i. For aqueous solution of electrolytes

$$\Delta T_f = 1.86°C \times m \times i \tag{16.15}$$

If we assume that the ions of an electrolyte behave independently, we would predict that i should be equal to *the number of moles of ions per mole of electrolyte*. Thus i should be 2 for NaCl and MgSO$_4$, 3 for CaCl$_2$ and Na$_2$SO$_4$, and so on.

Looking at the data in Table 16.6, we see that the situation is not so simple as our discussion might imply. The observed freezing point lowerings of NaCl and MgSO$_4$ are smaller than we would predict from Equation 16.15 with i = 2. In other words, at any finite concentration, the multiplier i is less than 2. It approaches a limiting value of 2 as the solution becomes more and more dilute. This behavior is generally typical of electrolytes. Interactions occur between positive and negative ions in solution. As a result, they do not behave as completely independent particles. The effect is ordinarily to make the observed freezing point depression or other colligative property less than we would expect from the number of ions in solution.

Despite the complications just mentioned, Equation 16.15 can be very useful. In particular, we can use it to decide how an electrolyte dissociates in water (Example 16.7).

At higher concentrations + and − ions tend to pair up to some extent

---

TABLE 16.6   FREEZING POINT LOWERINGS OF SOLUTIONS

	ΔT_f OBSERVED (°C)		i (CALC. FROM EQ. 16.15)	
MOLALITY	NaCl	MgSO$_4$	NaCl	MgSO$_4$
0.005	0.0182	0.0158	1.96	1.70
0.01	0.0360	0.0301	1.94	1.62
0.02	0.0714	0.0573	1.92	1.54
0.05	0.176	0.132	1.89	1.42
0.10	0.348	0.246	1.87	1.32
0.20	0.685	0.454	1.84	1.22
0.50	1.68	1.00	1.81	1.08

Example 16.7   The freezing point of a 0.010 m solution of NaHCO$_3$ is −0.038°C. On this basis, determine whether NaHCO$_3$ ionizes primarily as

$$NaHCO_3(s) \rightarrow Na^+(aq) + HCO_3^-(aq); i \rightarrow 2$$

or

$$NaHCO_3(s) \rightarrow Na^+(aq) + H^+(aq) + CO_3^{2-}(aq); i \rightarrow 3$$

Solution   Let us calculate the value of i in Equation 16.15:

$$0.038°C = 1.86°C \times 0.010 \times i$$

$$i = \frac{0.038}{0.0186} = 2.0$$

Clearly, the first mode of ionization is indicated.

Exercise   The freezing point of a 0.010 m solution of Na$_2$HPO$_4$ is −0.050°C. What ions are formed when this electrolyte dissociates? Answer: 2 Na$^+$, HPO$_4^{2-}$ (i = 2.7).

SUMMARY

A solution consists of solute(s) and solvent. In this chapter we have dealt primarily with water solutions. Solutes can be classified as nonelectrolytes or electrolytes, depending on their dissociation behavior in solution (Table 16.2 and Example 16.1). Nonelectrolytes tend to be insoluble in water unless they are capable of forming hydrogen bonds. Ionic solutes vary widely in their water solubility, as indicated by Table 16.3 and Example 16.2. The solubility rules are useful in predicting whether or not a precipitate will form when solutions of two ionic solutes are mixed. Precipitation reactions can be represented by net ionic equations (Example 16.3). Such equations can be used in the usual way to relate amounts of reactants and products (Example 16.4).

Solubilities of solutes in water are affected by temperature. If the solution process is endothermic, the solute becomes more soluble as temperature increases. If it is exothermic, solubility decreases with increasing temperature. The solubility of gaseous solutes is increased by increasing the pressure of the gas over the solution.

Solutes lower the vapor pressure of a solvent (Table 16.4, Fig. 16.8), depress the freezing point, and elevate the boiling point (Fig. 16.11). These properties, along with the osmotic pressure (Figs. 16.9 and 16.10), are colligative. For example, the extent to which the vapor pressure is lowered depends primarily upon the concentration rather than the type of solute particle. The same is true of freezing point depression and boiling point elevation. Here, the concentration unit used is molality, the number of moles of solute per kilogram of solvent (Table 16.5, Example 16.5). Freezing point lowering may be used to determine the molecular mass of an unknown (Example 16.6).

Since electrolytes dissociate in water, they have a greater effect on colligative properties than do nonelectrolytes. A 0.10 m solution of NaCl (Na^+, Cl^- ions) has a freezing point lowering about twice that of a 0.10 m solution of a nonelectrolyte. This effect makes it possible to determine the mode of ionization of a solute in water (Example 16.7).

KEY WORDS AND CONCEPTS

solution	saturated solution	solubility rules	osmotic pressure
solute	unsaturated solution	precipitate	boiling point elevation
solvent	supersaturated solution	net ionic equation	freezing point depression
dilute	nonelectrolyte	colligative property	molality
concentrated	electrolyte	vapor pressure lowering	reverse osmosis
molarity	weak electrolyte		

QUESTIONS AND PROBLEMS

Catalog

Solubility of Solutes: 16.2–16.6, 16.22–16.26
Precipitation Reactions: 16.7–16.10, 16.27–16.30
Molality and Molarity: 16.11–16.13, 16.31–16.33

Colligative Properties: 16.14–16.19, 16.34–16.39
General: 16.1, 16.20, 16.21, 16.40–16.45

16.1 Review and know the meaning of the key words and concepts in this chapter.

16.2 For each of the following pairs of solutes, state which one you would expect to be the more soluble in water. Explain your reasoning.

    a. NaCl or CCl_4
    b. CH_3—OH or CH_3—O—CH_3
    c. C_6H_6 or HO—OH
    d. CO_2 or SiO_2

16.3 Using Table 16.3, classify the following as "soluble," "slightly soluble," or "insoluble" in water.

    a. $CsNO_3$ b. $CaSO_4$ c. $NiCO_3$
    d. KOH   e. CuS

16.4 One mole of each of the following solutes is dissolved in water to form 500 cm³ of solution. Write a balanced equation for the dissociation of the solute and calculate the molarity of each ion in solution.

    a. $NaNO_3$ b. K_2CO_3 c. $FeBr_3$
    d. CaI_2

16.5 A certain gaseous solute dissolves in water, evolving 6.2 kJ/mol of heat. Its solubility at 25°C and 1.00 atm is 0.010 M. Would you expect the solubility to be greater or less than 0.010 M at

    a. 0°C and 1 atm?   b. 50°C and 1 atm?
    c. 25°C and 0.1 atm?   d. 15°C and 2 atm?

16.6 Referring to Figure 16.5,

    a. how much water is required to dissolve 50 g of KNO_3 at 50°C?
    b. if the solution in (a) is cooled to 20°C, how much KNO_3 comes out of solution?

16.7 Write net ionic equations to explain the formation of

    a. a blue precipitate when solutions of $Cu(NO_3)_2$ and sodium hydroxide are mixed.
    b. two different white precipitates when solutions of zinc sulfate and barium sulfide are mixed.

16.8 Using Table 16.3, decide whether a precipitate will form when the following solutions are mixed. If a precipitate forms, write a net ionic equation for the reaction.

    a. $CuCl_2$ and K_2CO_3 b. $AsCl_3$ and Na_2S
    c. $CuSO_4$ and KOH   d. $NaNO_3$ and $CaCl_2$

16.9 Zinc in a paint sample is converted to ZnS for analysis. If 0.132 g ZnS are obtained from a 5.00-g sample, what is the per cent zinc in the paint?

16.22 Choose the member of each set which you would expect to be the most soluble in water. Explain your reasoning.

    a. NH_3, CH_4, or H_2S
    b. ethane or ethylene glycol
    c. H_2, Br_2, or NaOH

16.23 Follow the directions of Question 16.3 for the following ionic solutes:

    a. $NiCl_2$ b. $Ca(OH)_2$
    c. $(NH_4)_2S$ d. $Fe(NO_3)_3$

16.24 0.50 mol of each of the following solutes is dissolved in water to form 2.0 ℓ of solution. Write a balanced equation to show the dissociation of the solute and calculate the molarity of each ion in solution.

    a. $KClO_4$ b. Na_2CrO_4 c. $CoCl_3$
    d. $Al_2(SO_4)_3$

16.25 Consider the process by which ammonium nitrate dissolves in water:

$$NH_4NO_3(s) \rightarrow NH_4^+(aq) + NO_3^-(aq)$$

    a. Using appropriate data from tables in Chapter 6, calculate ΔH.
    b. Would you expect the solubility of NH_4NO_3 to increase or decrease if the temperature is lowered?

16.26 Referring to Figure 16.5,

    a. how much $KClO_3$ can be dissolved in 29 g of water at 80°C?
    b. if the solution in (a) is cooled to 10°C, how much $KClO_3$ comes out of solution?

16.27 Write net ionic equations for the formation of

    a. a yellow precipitate when solutions of $Ba(NO_3)_2$ and K_2CrO_4 are mixed.
    b. a white precipitate, which is soluble in acid, when solutions of NaOH and $AlCl_3$ are mixed.

16.28 Follow the directions of Problem 16.8 for the following pairs of solutions:

    a. NaCl and $Pb(NO_3)_2$ b. BaS and NaOH
    c. $Ba(OH)_2$ and $Fe(NO_3)_3$ d. $Ni(NO_3)_2$ and KOH

16.29 Arsenic in an insecticide is converted to As_2S_3 for analysis. If the insecticide contains 37.6% As and an analyst must have at least 50 mg of As_2S_3, what is the minimum mass of sample which can be used?

16.10 The addition of Na_2SO_4 solution to one of $BaCl_2$ causes the following precipitation reaction:

$$Ba^{2+}(aq) + SO_4^{2-}(aq) \rightarrow BaSO_4(s)$$

a. What volume of 0.150 M Na_2SO_4 is needed to react with 10.0 cm³ of 0.200 M $BaCl_2$?
b. What mass in grams of $BaSO_4$ is formed?

16.11 Complete the following table:

Solute	g Solute	g H₂O	Molality
$C_6H_{12}O_6$	1.80	200	——
NaCl	5.85	50.0	——
C_2H_5OH	——	500	4.00
$MgBr_2$	49.0	——	0.30

16.12 The "proof" of an alcoholic beverage is twice the percentage by volume of ethyl alcohol, C_2H_5OH. For a 110 proof bourbon whiskey, what is the molality of ethyl alcohol? (d C_2H_5OH = 0.80 g/cm³; d H_2O = 1.00 g/cm³)

16.13 Concentrated phosphoric acid contains 85.0% H_3PO_4 by mass and has a density of 1.69 g/cm³.

a. Calculate the mass of H_3PO_4 in one liter of the acid solution.
b. What is the molarity of H_3PO_4?
c. What is the molality of H_3PO_4?

16.14 Calculate the freezing point and normal boiling point of each of the following solutions (see Table 16.5):

a. 19.4 g of glucose, $C_6H_{12}O_6$, in 100 g of water.
b. 4.00 g of urea $(NH_2)_2CO$, in 80.0 g of water.

16.15 A mixture of water and ethylene glycol, $C_2H_6O_2$, is the most common antifreeze.

a. Estimate the molality of ethylene glycol in such a mixture if it freezes at $-15°C$.
b. What volume (cm³) of ethylene glycol (d = 1.1 g/cm³) would have to be added to 8.0 kg of water to prepare this mixture?

16.16 Vitamin C contains 40.9% C, 4.58% H, and 54.5% O. A solution of 9.70 g of vitamin C in 50.0 g of water freezes at $-2.05°C$. What is the molecular formula of vitamin C?

16.17 A student dissolves 0.180 g of a nonelectrolyte in 50.0 g of benzene and determines the freezing point to be 5.20°C. Calculate the molecular mass of the solute (see Table 16.5).

16.30 Silver chromate precipitates when a solution containing Ag^+ ions is mixed with one containing CrO_4^{2-} ions:

$$2\ Ag^+(aq) + CrO_4^{2-}(aq) \rightarrow Ag_2CrO_4(s)$$

a. What volume of 0.050 M K_2CrO_4 is needed to react with 5.0 cm³ of 0.030 M $AgC_2H_3O_2$?
b. How many grams of Ag_2CrO_4 are formed?

16.31 Complete the following table:

Solute	g Solute	g H₂O	Molality
KCl	3.47	600	——
$(H_2N)_2CO$	3.00	250	——
$C_3H_8O_3$	——	750	10.2
CaI_2	29.4	——	0.70

16.32 Formalin is a preservative solution used in biology laboratories. It contains 40 cm³ of formaldehyde, CH_2O (d = 0.82 g/cm³) per 100 cm³ of water. What is the molality of formaldehyde in formalin?

16.33 Using the data in Table 16.1, calculate the molality of H_2SO_4 in

a. concentrated sulfuric acid. (d = 1.84 g/cm³)
b. dilute sulfuric acid. (d = 1.18 g/cm³)

16.34 How many grams of the following nonelectrolyte solutes would have to be dissolved in 50.0 g of water to give a solution freezing at $-1.50°C$? What would be the normal boiling point of each solution?

a. sucrose, $C_{12}H_{22}O_{11}$
b. glycerol, $C_3H_8O_3$

16.35 Isopropyl alcohol, C_3H_8O, and ethylene glycol, $C_2H_6O_2$, are used as antifreezes in automobile window washers and radiators, respectively.

a. How many grams of isopropyl alcohol lower the freezing point of water the same amount as 1.00 g of ethylene glycol?
b. If the prices of isopropyl alcohol and ethylene glycol are $2.70/kg and $1.80/kg, which is the cheaper antifreeze?

16.36 Nicotine has the empirical formula C_5H_7N. A solution of 0.80 g of nicotine in 12.0 g of water boils at 100.21°C at 760 mm Hg. What is the molecular formula of nicotine?

16.37 The freezing point of p-dichlorobenzene is 53.1°C; its k_f value is 7.10°C. A solution of 1.26 g of sulfanilamide (a sulfa drug) in 10.0 g of p-dichlorobenzene freezes at 47.9°C. What is the molecular mass of sulfanilamide?

16.18 Estimate the freezing points of 0.10 molal water solutions of

    a. KNO_3 b. $MgCl_2$ c. $Al(NO_3)_3$

16.19 The freezing point of 0.10 m HF is $-0.19°C$. Is HF primarily unionized in this solution (HF molecules), or is it dissociated to H^+ and F^- ions?

16.20 In your own words, explain

    a. why the concentrations of solutions used for intravenous feeding must be controlled carefully.
    b. why, when making fudge (a supersaturated sugar mixture), care must be taken to prevent it from getting "grainy."
    c. why fish in a lake seek deep, shaded places during summer afternoons.
    d. what causes the "bends" in divers.
    e. why champagne "fizzes" in a glass.

16.21 Explain, in your own words,

    a. how to determine experimentally whether a pure substance is an electrolyte or nonelectrolyte.
    b. why a cold glass of beer goes "flat" upon warming.
    c. the differences between molarity and molality.
    d. why the boiling point is raised by the presence of a solute.

16.38 Consider 0.10 m solutions of the solutes listed below. Arrange the solutes in order of increasing vapor pressure of their solutions.

    a. LiBr   b. $C_2H_6O_2$ c. $MgCl_2$

16.39 The freezing point of 0.050 m $KHSO_3$ is $-0.19°C$. Which of the following equations best represents what happens when $KHSO_3$ dissolves in water?

    a. $KHSO_3(s) \rightarrow KHSO_3(aq)$
    b. $KHSO_3(s) \rightarrow K^+(aq) + HSO_3^-(aq)$
    c. $KHSO_3(aq) \rightarrow K^+(aq) + H^+(aq) + SO_3^{2-}(aq)$

16.40 Criticize the following statements:

    a. A saturated solution is always a concentrated solution.
    b. The water solubility of a solid always decreases with a drop in temperature.
    c. For aqueous solutions, molarity and molality are equal.
    d. The freezing point depression of a 0.10 m $CaCl_2$ solution is twice that of a 0.10 m KCl solution.
    e. A 0.10 m sucrose solution and a 0.10 m NaCl solution have the same osmotic pressure.

16.41 In your own words explain why

    a. salt is added to ice in an ice cream maker to freeze the ice cream.
    b. seawater has a lower freezing point than fresh water.
    c. more ethyl alcohol, C_2H_5OH, dissolves in 100 g water than in 100 g of benzene, C_6H_6, at the same temperature.
    d. why we believe vapor pressure lowering is a colligative property.

*16.42 The vapor pressure lowering (vpL) in water solution is given by the relation: $vpL = X_{solute} \times P^0$, where X_{solute} is the mole fraction of solute and P^0 is the vapor pressure of pure water. A solution of a certain nonelectrolyte in water freezes at $-1.86°C$. What is its molality? the mole fraction of the solute? the vapor pressure lowering at 25°C, where $P^0 = 23.6$ mm Hg?

*16.43 The osmotic pressure, π, of a nonelectrolyte solution is given by the relation $\pi = MRT$, where M is the molarity of solute, R is the gas law constant, $0.0821 \ \ell \cdot atm/(mol \cdot K)$, and T is the absolute temperature. A solution of 1.00 g of a certain polymer in one liter of solution has an osmotic pressure of 2.1 mm Hg at 27°C. What is the molecular mass of the polymer?

*16.44 The water-soluble nonelectrolyte, X, has a molecular mass of 423. A 0.100-g sample of this substance mixed with sugar (MM = 342) is added to 1.00 g of water to give a solution freezing at $-0.500°C$. Estimate the mass % of X in the sample.

*16.45 A martini, weighing about 5.0 ounces (142 g), contains 30% by mass of alcohol. About 15% of the alcohol in the martini passes directly into the blood stream (7.0 ℓ for an adult). Estimate the concentration of alcohol in the blood (g/cm^3) of a person who drinks two martinis before dinner. (A concentration $\geq 0.0030 \ g/cm^3$ is frequently considered indicative of intoxication in a "normal" adult.)

ACIDS AND BASES

In previous chapters we have referred from time to time to water solutions of acids and bases. Such solutions are among the most useful of laboratory reagents. They are also common in the home. Vinegar, orange juice, and battery fluid are, to varying degrees, acidic. Oven and drain cleaners, whitewash (like Tom Sawyer used), and antacid tablets are basic.

This chapter examines the properties of acidic and basic water solutions. We start with working definitions of the terms acid and base. To begin, we take an *acid* to be a substance which, when added to water, produces *hydrogen ions* (**H⁺**). *Bases* produce *hydroxide ions* (**OH⁻**) in water solution.

All acidic water solutions have certain properties in common. They evolve hydrogen gas, H_2, with zinc or magnesium. They react with compounds containing the CO_3^{2-} ion to form CO_2 (Fig. 17.1). They affect the color of certain organic dyes, called indicators. For example, litmus turns red in acidic solution. These properties of acidic solutions are due to the presence of H⁺ ions.

Water solutions of bases also have identifying properties. They feel slippery and

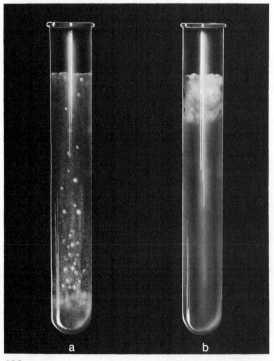

Figure 17.1 The H⁺ ions present in a water solution of an acid react with calcium carbonate to form carbon dioxide: $CaCO_3(s) + 2 H^+(aq) \rightarrow Ca^{2+}(aq) + CO_2(g) + H_2O$ (tube a). The OH⁻ ions present in a solution of a base precipitate $Mg(OH)_2$ from a solution containing Mg^{2+} ions: $Mg^{2+}(aq) + 2 OH^-(aq) \rightarrow Mg(OH)_2(s)$ (tube b).

turn litmus blue. In the presence of Mg^{2+} ions, they precipitate magnesium hydroxide, $Mg(OH)_2$. The characteristic properties of basic solutions are caused by OH^- ions.

17.1 WATER DISSOCIATION; ACIDIC, NEUTRAL, AND BASIC SOLUTIONS

The acidic and basic properties of aqueous solutions are dependent upon an equilibrium that involves the solvent, water. Water, when pure or as a solvent, tends to dissociate to some extent into hydrogen ions and hydroxide ions:

$$H_2O(l) \rightleftharpoons H^+(aq) + OH^-(aq) \qquad (17.1)$$

The forward reaction proceeds only slightly before equilibrium is reached. Only a small fraction of the total number of water molecules is dissociated.

Applying the general rules from Chapter 13 for equilibrium systems, we can write the equilibrium constant expression for Reaction 17.1 as

$$K_c = \frac{[H^+] \times [OH^-]}{[H_2O]}$$

In aqueous solutions, the concentration of H_2O is very high, typically about 55 M, and is essentially constant. Hence the term $[H_2O]$ can be combined with K_c to give a new constant, K_w, called the dissociation constant of water:

$$K_c \times [H_2O] = K_w = [H^+] \times [OH^-] = 1.0 \times 10^{-14} \qquad (17.2)$$

In the equation for K_w, $[H_2O]$ is always omitted

At 25°C, K_w is 1.0×10^{-14}. This small value reflects the slight dissociation of water into H^+ and OH^- ions. In any aqueous solution or in pure water, the product of $[H^+]$ times $[OH^-]$ at 25°C is always 1.0×10^{-14}.

We can readily calculate $[H^+]$ and $[OH^-]$ in pure water. From Equation 17.1, we see that equal amounts of these two ions form when water dissociates. Applying Equation 17.2 to pure water:

$$[H^+] = [OH^-]; [H^+] \times [OH^-] = [H^+]^2 = 1.0 \times 10^{-14}$$

$$[H^+] = 1.0 \times 10^{-7} \text{ M} = [OH^-]$$

Any aqueous solution in which $[H^+]$ *equals* $[OH^-]$ is called a *neutral* solution. It has a $[H^+]$ equal to 1.0×10^{-7} M at 25°C.

Ordinarily, the concentrations of H^+ and OH^- in a solution are not equal. Note from Equation 17.2 that as the concentration of one ion goes up, that of the other must go down so that the product $[H^+] \times [OH^-]$ stays constant. If the concentration of one ion is known, then that of the other can be calculated using Equation 17.2 (Example 17.1).

Example 17.1 An aqueous solution has a $[H^+]$ of 2.0×10^{-4} M. What is its $[OH^-]$?

Solution From Equation 17.2:

$$[OH^-] = \frac{K_w}{[H^+]} = \frac{1.0 \times 10^{-14}}{2.0 \times 10^{-4}} = 5.0 \times 10^{-11} \text{ M}$$

An aqueous solution, like that described in Example 17.1 where $[H^+]$ *is greater than* $[OH^-]$, is said to be *acidic*. An aqueous solution in which $[OH^-]$ *is greater than* $[H^+]$ is *basic*. Therefore,

if $[H^+] > 1.0 \times 10^{-7}$ M, $[OH^-] < 1.0 \times 10^{-7}$ M, **solution is acidic**

if $[OH^-] > 1.0 \times 10^{-7}$ M, $[H^+] < 1.0 \times 10^{-7}$ M, **solution is basic**

Table 17.1 indicates some possible combinations of concentrations of these ions. Solutions 1 through 4 are decreasingly acidic, 5 is neutral and 6 through 9 are progressively more basic.

17.2 pH

As we have seen, the acidity or basicity of a solution can be described by giving its H^+ ion concentration. Sorensen, in 1909, proposed an alternative method of accomplishing this purpose. He suggested using a term called pH, defined as

$$pH = -\log_{10}[H^+] = \log_{10} 1/[H^+] \tag{17.3}$$

Thus we have

$$[H^+] = 10^{-4} \text{ M; pH} = -\log_{10}(10^{-4}) = -(-4) = 4$$
$$[H^+] = 10^{-7} \text{ M; pH} = -\log_{10}(10^{-7}) = -(-7) = 7$$
$$[H^+] = 10^{-10} \text{ M; pH} = -\log_{10}(10^{-10}) = -(-10) = 10$$

A scientist will usually refer to a solution in terms of its pH rather than its $[H^+]$

Most aqueous solutions have hydrogen ion concentrations between 1 M and 10^{-14} M. By Equation 17.3, such solutions have pH's lying between 0 and 14. In this case, it is perhaps more convenient to express acidity in terms of pH rather than $[H^+]$. This avoids using small fractions or negative exponents.

Looking at the pH values in Table 17.1, we see that $[H^+]$ and pH are inversely

TABLE 17.1   RELATIONS BETWEEN $[H^+]$, $[OH^-]$, and pH IN
AQUEOUS SOLUTIONS

SOLUTION

	No. 1	No. 2	No. 3	No. 4	No. 5	No. 6	No. 7	No. 8	No. 9
$[H^+]$	10^0	10^{-2}	10^{-4}	10^{-6}	10^{-7}	10^{-8}	10^{-10}	10^{-12}	10^{-14}
$[OH^-]$	10^{-14}	10^{-12}	10^{-10}	10^{-8}	10^{-7}	10^{-6}	10^{-4}	10^{-2}	10^0
pH	0	2	4	6	7	8	10	12	14

Acidic   Neutral   Basic

TABLE 17.2 pH OF SOME COMMON LIQUIDS			
Lemon juice	2.2–2.4	Urine, human	4.8–8.4
Wine	2.8–3.8	Cow's milk	6.3–6.6
Vinegar	3.0	Saliva, human	6.5–7.5
Tomato juice	3.5	Drinking water	6.5–8.0
Beer	4–5	Blood, human	7.3–7.5
Cheese	4.8–6.4	Seawater	8.3

related. As $[H^+]$ decreases from 10^{-2} to 10^{-4} M, pH increases from 2 to 4. In general, **the higher the pH, the less acidic (more basic) the solution.** A solution of pH 4 has a lower concentration of H^+ and a higher concentration of OH^- than does a solution of pH 2. Notice also that when pH increases by one unit, $[H^+]$ decreases by a factor of 10. A solution of pH 3 has a hydrogen ion concentration one tenth that of a solution of pH 2 and ten times that of a solution of pH 4.

Previously, we used concentration of H^+ or OH^- to differentiate acidic, neutral, and basic solutions. pH can also be used for this purpose:

if pH < 7.0, solution is acidic
if pH = 7.0, solution is neutral
if pH > 7.0, solution is basic

Table 17.2 shows the pH of some common solutions. Example 17.2 shows how Equation 17.3 can be used to calculate pH from $[H^+]$, or vice versa.

Example 17.2   Calculate
  a. the pH of a solution in which $[H^+] = 4.0 \times 10^{-3}$ M.
  b. the $[H^+]$ of a beer with a pH of 4.7.

Solution   If you have a "scientific" calculator, you may be able to answer these questions instantly by pushing the proper button. In case your calculator does not do this, we will indicate briefly the procedure involved (see also Appendix 4 for a discussion of logarithms).

a. $pH = -\log_{10}[H^+] = -\log_{10}(4.0 \times 10^{-3}) = -(\log_{10}4.0 + \log_{10}10^{-3})$
$= -(0.60 - 3.00) = -(-2.40) = 2.40$

b. $pH = 4.7 = -\log_{10}[H^+]$. This means that $[H^+] = 10^{-4.7}$, which can be rewritten as $10^{0.3} \times 10^{-5}$. The antilog of 0.3 is 2; i.e., $10^{0.3} = 2$; that of 10^{-5} is -5. Therefore, a pH of 4.7 corresponds to a $[H^+]$ of 2×10^{-5} M.

Calculators are indeed helpful with pH problems

Exercise   Using Table 17.2, obtain the $[H^+]$ of seawater; the $[OH^-]$. Answer: 5.0×10^{-9} M; 2.0×10^{-6} M (seawater is basic).

Several experimental methods can be used to determine the pH of an aqueous solution. Most often a so-called "pH meter" is used (Figure 17.2). A more colorful (and time-consuming) way to determine pH is to use acid-base indicators. These substances undergo a color change within a narrow pH range, perhaps 1 to 2 pH units (see color plate 14, center of book). By using two or more of the indicators listed in Table 17.3, you can estimate quite accurately the pH of a solution. Suppose, for example, that a solution turns red when a drop of phenolphthalein is added. This means

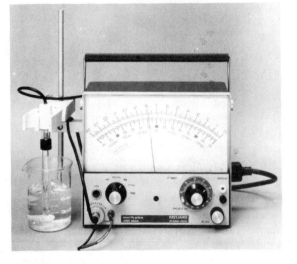

Figure 17.2 The simplest way to determine the pH of a solution is to measure it with an electrical device called a pH meter. Care must be taken in making such a measurement; the electrodes are easily broken. (Courtesy of the Arthur H. Thomas Co. and Corning Glass.)

that its pH is 10 or greater. If you find that this same solution gives a yellow color with alizarin yellow, its pH must be 10 or less. Putting these two observations together, the solution must have a pH of just about 10.

"pH paper" is coated with a mixture of indicators. Strips of pH paper are used widely to test the pH of biological fluids, soil, ground water, and foods. Depending upon the indicators used, a test strip can measure pH over a wide or narrow range.

17.3   STRONG AND WEAK ACIDS

For a species to act as an acid, it must supply H^+ ions to water. The acid, which may be a molecule or ion, contains hydrogen atoms. The H^+ ions are formed by the dissociation of the acid in water.

Strong acids dissociate completely in water, forming H^+ ions and anions. A typical strong acid is HCl. It undergoes the following reaction upon addition to water:

$$HCl(aq) \rightarrow H^+(aq) + Cl^-(aq) \tag{17.4}$$

This reaction goes to completion. In a dilute water solution of hydrochloric acid there are no HCl molecules, only H^+ ions and Cl^- ions. Consider, for example, a 0.10 M solu-

---

TABLE 17.3   TYPICAL ACID-BASE INDICATORS

INDICATOR	pH INTERVAL	ACID COLOR (LOWER pH)	BASE COLOR (HIGHER pH)
Methyl violet	0.0–1.6	yellow	violet
Methyl yellow	2.9–4.0	red	yellow
Methyl orange	3.1–4.4	red	yellow
Methyl red	4.8–6.2	red	yellow
Bromthymol blue	6.0–8.0	yellow	blue
Thymol blue	8.0–9.6	yellow	blue
Phenolphthalein	8.2–10.0	colorless	pink
Alizarin yellow	10.1–12.0	yellow	red

tion of HCl, prepared by adding 0.10 mol of HCl to water to form one liter of solution. The concentration of HCl molecules is virtually zero. The concentrations of H^+ and Cl^- are 0.10 M. The pH of the solution is 1.00. Any way you look at it, the HCl is completely dissociated into ions.

In contrast, **a weak acid is only partially dissociated in water.** A typical weak acid is HF. When hydrogen fluoride is added to water, the following *reversible* reaction occurs:

$$HF(aq) \rightleftharpoons H^+(aq) + F^-(aq) \tag{17.5}$$

In a solution of hydrofluoric acid, there are HF molecules as well as H^+ and F^- ions. In 0.10 M HF, prepared by adding 0.10 mol of HF to enough water to give one liter of solution, more than 90% of the HF molecules remain undissociated. The concentrations of H^+ and F^- ions are less than 0.01 M. The pH of the solution is slightly greater than 2.

As pointed out in Chapter 16, hydrogen fluoride behaves as a weak electrolyte in aqueous solution. A 0.10 M solution of HF has a small electrical conductivity due to the H^+ and F^- ions present. In contrast, a 0.10 M solution of HCl is a strong electrolyte, with a conductivity more than ten times that of 0.10 M HF. In its colligative properties, hydrogen fluoride is intermediate between a nonelectrolyte such as glucose and a strong electrolyte such as HCl. The freezing point of 0.10 M HF is −0.21°C as compared to −0.19°C for 0.10 M glucose and −0.37°C for 0.10 M HCl.

The Strong Acids

There are very few strong acids. For our purposes, we need consider only the six species listed in Table 17.4. You should learn the names and molecular formulas of these six strong acids. They will be referred to again and again, in this and following chapters.

We label the bottle 0.1 M HCl, but perhaps 0.1 M H^+, 0.1 M Cl^- would be better

H_2SO_4, HNO_3, and HCl rank among the most important industrial chemicals

		TABLE 17.4 COMMON STRONG ACIDS
STRONG ACID	**MOLECULAR FORMULA**	**MOLECULAR STRUCTURE**
Hydrochloric acid	HCl	H—Cl
Hydrobromic acid	HBr	H—Br
Hydriodic acid	HI	H—I
Nitric acid	HNO_3	H—O—N—O with =O below N
Sulfuric acid	H_2SO_4	H—O—S—O—H with O above and O below S
Perchloric acid	$HClO_4$	H—O—Cl—O with O above and O below Cl

Notice that all the substances that act as strong acids are molecular species when pure. In dilute water solution, they dissociate completely to form a H^+ ion and an anion. The dissociation reactions are similar to Reaction 17.4 for HCl. Examples include

$$HNO_3(aq) \rightarrow H^+(aq) + NO_3^-(aq) \qquad (17.6)$$

$$H_2SO_4(aq) \rightarrow H^+(aq) + HSO_4^-(aq) \qquad (17.7)$$

Species Which Act As Weak Acids

A wide variety of solutes behave as weak acids in water. For convenience, they can be classified into three categories: molecules, anions, and cations.

MOLECULES CONTAINING AN IONIZABLE PROTON. There are literally thousands of molecular weak acids. A few of the more common ones are listed in Table 17.5. All these species contain a hydrogen atom bonded covalently to a non-metal atom in the molecule. When placed in water, the weak acid molecule partially dissociates, forming H^+ ions and anions at low concentrations. The equations for dissociation are analogous to Reaction 17.5 for HF. Thus, for hypochlorous acid, HClO, and acetic acid, $HC_2H_3O_2$, we have

$$HClO(aq) \rightleftharpoons H^+(aq) + ClO^-(aq) \qquad (17.8)$$

$$HC_2H_3O_2(aq) \rightleftharpoons H^+(aq) + C_2H_3O_2^-(aq) \qquad (17.9)$$

TABLE 17.5   SOME COMMON MOLECULAR WEAK ACIDS

WEAK ACID	MOLECULAR FORMULA	MOLECULAR STRUCTURE	CONC. H^+ IN 0.10 M SOLN.	pH (0.10 M SOLN.)
Phosphoric acid	H_3PO_4	$\begin{matrix} & O & \\ & \| & \\ H-O-&P&-O-H \\ & \| & \\ & O-H & \end{matrix}$	0.024 M	1.62
Hydrofluoric acid	HF	H—F	0.0080 M	2.10
Nitrous acid	HNO_2	$\begin{matrix} H-O-N \\ \| \| \\ O \end{matrix}$	0.0065 M	2.19
Acetic acid	$HC_2H_3O_2$	$\begin{matrix} CH_3-C-O-H \\ \| \| \\ O \end{matrix}$	0.0013 M	2.87
Carbonic acid	H_2CO_3	$\begin{matrix} H-O-C-O-H \\ \| \| \\ O \end{matrix}$	0.00021 M	3.69
Hydrogen sulfide	H_2S	H—S—H	0.00010 M	4.00
Hypochlorous acid	HClO	H—O—Cl	0.000056 M	4.25
Hydrogen cyanide	HCN	H—C≡N	0.0000063 M	5.20

The properties of 0.10 M solutions of HClO or $HC_2H_3O_2$ resemble those of 0.10 M HF. All three species are weak electrolytes with small electrical conductivities. Each has a pH greater than 1 (Table 17.5).

Acid-base indicators, referred to previously, are weak acids. We might represent the dissociation of such an indicator by the equation:

$$HIn(aq) \rightleftharpoons H^+(aq) + In^-(aq)$$

The formula HIn stands for the weak acid molecule, *which has a color different from that of the In⁻ anion.* In the case of bromthymol blue (Table 17.3), the weak acid molecule is colored yellow while the anion is blue.

The position of the above equilibrium is sensitive to the concentration of H^+ ions. If $[H^+]$ is "high," the equilibrium lies far to the left and the principal species present is the HIn molecule. We see its color (yellow with bromthymol blue) when we look at a solution containing a few drops of the indicator. When $[H^+]$ is "low," the equilibrium shifts to the right, forming In⁻ ions. The solution takes on the color of In⁻ (blue with bromthymol blue).

Le Châtelier's Principle applies nicely here

ANIONS CONTAINING AN IONIZABLE PROTON. Let us refer back for a moment to Equation 17.7 for the dissociation of H_2SO_4. Notice that the anion formed, HSO_4^-, contains a hydrogen atom. In water, the HSO_4^- ion undergoes further dissociation, producing a H^+ ion and a SO_4^{2-} ion. The dissociation reaction is reversible, so we classify HSO_4^- as a weak acid:

$$HSO_4^-(aq) \rightleftharpoons H^+(aq) + SO_4^{2-}(aq) \qquad (17.10)$$

In practice, very few anions give acidic solutions when added to water. The only other anion of this type that we need be concerned with is the $H_2PO_4^-$ ion (see Example 17.3, p. 408).

CATIONS.   The ammonium ion, NH_4^+, behaves as a weak acid in water because of the following reversible reaction:

$$NH_4^+(aq) \rightleftharpoons H^+(aq) + NH_3(aq) \qquad (17.11)$$

The products are an ammonia molecule, NH_3, and a H^+ ion which makes the solution acidic. Note that the behavior of the NH_4^+ ion in water is very similar to that of the HF molecule (Eq. 17.5) or the HSO_4^- ion (Eq. 17.10). All three species contain hydrogen atoms which are converted to H^+ ions when the weak acid dissociates.

You may be surprised to learn that most metal cations, except those of Groups 1 and 2, are weak acids. At first, it is not at all obvious how a cation such as Zn^{2+} can make a water solution acidic. To understand how this is possible, we must realize that this cation and others like it are *hydrated* in water solution. When $ZnCl_2$ or $Zn(NO_3)_2$ is added to water, the cation formed is $Zn(H_2O)_4^{2+}$. Here, a Zn^{2+} ion is bonded to four water molecules. This complex cation is slightly dissociated in water, according to the following reaction:

$$Zn(H_2O)_4^{2+}(aq) \rightleftharpoons H^+(aq) + Zn(H_2O)_3(OH)^+(aq) \qquad (17.12)$$

The H^+ ion, which makes the solution acidic, comes from the ionization of one of the H_2O molecules bonded to Zn^{2+}. The OH⁻ ion formed at the same time remains bonded to Zn^{2+} and so does not directly affect the pH of the solution. (It does, however, affect the charge of the complex cation, reducing it from +2 to +1.)

Free H⁺ and OH⁻ ions determine pH

Example 17.3 Write dissociation equations for each of the following weak acids in water:

    a. H_2S b. $H_2PO_4^-$ c. $Al(H_2O)_6^{3+}$

Solution In each case, a proton (H^+ ion) is produced to make the solution acidic. The other species formed is the residue from the weak acid after removal of the proton. All the reactions go to equilibrium, indicated by a double arrow.

    a. $H_2S(aq) \rightleftharpoons H^+(aq) + HS^-(aq)$; compare Eqs. 17.5, 17.8, 17.9

    b. $H_2PO_4^-(aq) \rightleftharpoons H^+(aq) + HPO_4^{2-}(aq)$; compare Eq. 17.10

    c. $Al(H_2O)_6^{3+}(aq) \rightleftharpoons H^+(aq) + Al(H_2O)_5(OH)^{2+}(aq)$; compare Eq. 17.12

Exercise Consider the following ions: HS^-, HPO_4^{2-}, $Al(H_2O)_5(OH)^{2+}$. Each of these can dissociate in water to produce a H^+ ion and another species. In each case, give the formula of the "other species" formed. Answer: S^{2-}; PO_4^{3-}; $Al(H_2O)_4(OH)_2^+$.

17.4 STRONG AND WEAK BASES

For a species to act as a base, it must supply OH^- ions to water. Bases, like acids, are classified as "strong" or "weak." As with acids, there are only a few strong bases but a great many weak bases.

Strong Bases

A strong base dissociates completely in water to release OH^- ions. Sodium hydroxide, NaOH, is the most common strong base. It dissolves readily in water to give a solution containing Na^+ and OH^- ions:

$$NaOH(s) \rightarrow Na^+(aq) + OH^-(aq) \tag{17.13}$$

It may seem strange, but in 0.1 M NaOH, conc. NaOH = 0!

As with all strong bases, this reaction goes to completion. In a 0.10 M NaOH solution, prepared by dissolving 0.10 mol NaOH in enough water to give one liter of solution, the concentration of *undissociated* NaOH is virtually zero. In this solution, the concentrations of Na^+ and OH^- are 0.10 M. The pH is 13.00.

    Strong bases are limited to:

    1. **The hydroxides of the Group 1 metals** (LiOH, NaOH, KOH, RbOH, CsOH).

    2. **The hydroxides of the Group 2 metals,** $Mg(OH)_2$, $Ca(OH)_2$, $Sr(OH)_2$, $Ba(OH)_2$. With these compounds, two moles of OH^- are produced for every mole of solid that dissociates.

$$Ca(OH)_2(s) \rightarrow Ca^{2+}(aq) + 2\ OH^-(aq) \tag{17.14}$$

Of the strong bases, only NaOH and KOH are commonly used in the chemistry laboratory. All compounds of lithium, rubidium, and cesium, including the hydroxides, are expensive. The Group 2 hydroxides, particularly $Mg(OH)_2$, have limited solubilities. Calcium hydroxide is often used in industry when a strong base is needed and high solubility is not critical.

Weak bases furnish OH^- ions by a reversible reaction involving a water molecule. Perhaps the most common weak base is ammonia, NH_3. It reacts with water as follows:

$$NH_3(aq) + H_2O \rightleftharpoons NH_4^+(aq) + OH^-(aq) \qquad (17.15)$$

The forward reaction occurs to only a slight extent. In a 0.10 M solution of NH_3, pre-pared by adding 0.10 mol of NH_3 to enough water to form a liter of solution, nearly 99% of the NH_3 molecules remain unreacted. The concentration of NH_3 is about 0.099 M. In contrast, the concentrations of NH_4^+ and OH^- are only about 0.001 M. The Which makes it quite pH of the solution is about 11. basic

Most weak bases are anions. A typical example is the fluoride ion, F^-. It under-goes the following reversible reaction with water:

$$F^-(aq) + H_2O \rightleftharpoons HF(aq) + OH^-(aq) \qquad (17.16)$$

As with NH_3, the forward reaction occurs only to a slight extent. In a solution in which the F^- concentration is 0.10 M, the concentration of OH^- at equilibrium is only about 10^{-6} M. As small as this is, it is enough to make the solution basic, with a pH of about 8. A reaction similar to 17.16 occurs with the anion of any weak acid. With the acetate ion, $C_2H_3O_2^-$, the reaction is

$$C_2H_3O_2^-(aq) + H_2O \rightleftharpoons HC_2H_3O_2(aq) + OH^-(aq) \qquad (17.17)$$

Looking at Equations 17.15 through 17.17, we see that they resemble each other In a very real sense a very closely. In each case, a weak base (NH_3, F^-, $C_2H_3O_2^-$) picks up a proton (H^+ ion) weak base competes from a water molecule. The products are: with H_2O for H^+ ions

—a weak acid (NH_4^+, HF, $HC_2H_3O_2$).

—an OH^- ion, which makes the solution basic.

The general reaction is

$$\text{weak base(aq)} + H_2O \rightleftharpoons \text{weak acid(aq)} + OH^-(aq)$$

Since the reaction does not go to completion, relatively few OH^- ions are formed. This is why we refer to species such as NH_3, F^-, and $C_2H_3O_2^-$ as *weak* bases.

Example 17.4   Write equations for the reactions of the following weak bases with water:
   a. CN^- b. CO_3^{2-}

Solution   The equations are entirely similar to 17.16 and 17.17. In each case, the weak base picks up a proton (H^+ ion) from an H_2O molecule. This produces an OH^- ion, which makes the solution basic.

   a. $CN^-(aq) + H_2O \rightleftharpoons HCN(aq) + OH^-(aq)$

   b. $CO_3^{2-}(aq) + H_2O \rightleftharpoons HCO_3^-(aq) + OH^-(aq)$

Exercise   Write the formula for the weak acid formed by the reaction of each of the following weak bases with water: SO_3^{2-}; CH_3NH_2. Answer: HSO_3^-; $CH_3NH_3^+$ (analogous to NH_4^+).

17.5 ACID-BASE PROPERTIES OF SALT SOLUTIONS

After completing Sections 17.3 and 17.4, you should be able to predict correctly that an aqueous solution of HI or H_2SO_4 is acidic while a solution of NaOH or NH_3 is basic. Solutions of $NaNO_2$ or NH_4I might be more difficult for you to classify. These two compounds, and many others, such as NaCl, $Zn(NO_3)_2$, and $CuSO_4$ are **salts**. **A salt is an ionic compound containing a cation other than H^+, and an anion other than OH^- or O^{2-}.**

In dilute water solution, a salt is completely dissociated. Hence, the acid-base properties of its solution are determined by those of the ions present. A particular ion may be neutral, having no effect on the $[H^+]$ or $[OH^-]$ of water. Other ions are acidic; they increase $[H^+]$ to greater than 1.0×10^{-7} M. Still other ions are basic, since they raise $[OH^-]$ beyond 1.0×10^{-7} M. Table 17.6 summarizes the acid-base properties of ions commonly present in aqueous solutions.

Neutral Ions

A neutral ion does not react with water to produce H^+ or OH^- ions. Hence, it does not affect the pH. There are relatively few neutral ions. We see from Table 17.6 that:
—*the neutral anions are those derived from strong acids.*
—*the neutral cations are those derived from strong bases.*
A typical neutral anion is the chloride ion, produced by the dissociation of hydrochloric acid:

$$HCl(aq) \rightarrow H^+(aq) + Cl^-(aq)$$

You will recall from Section 17.3 that HCl is completely dissociated in water; it is a strong acid. This means that there is no tendency for the reverse of the above reaction to occur. Chloride ions, wherever they come from, do not combine with H^+ ions. In particular, Cl^- ions from a salt such as NaCl do not pick up H^+ ions from water. As a result, Cl^- ions, and the other neutral anions listed in Table 17.6 do not change the $[H^+]$ or pH of water.

A similar argument applies to cations like Na^+, produced by the dissociation of strong bases such as NaOH. Dissociation is complete:

$$NaOH(s) \rightarrow Na^+(aq) + OH^-(aq)$$

Explain in your own words why the NO_3^- ion is neutral

TABLE 17.6 ACID-BASE PROPERTIES OF SOME COMMON IONS IN WATER SOLUTION

	NEUTRAL		BASIC		ACIDIC
Anion	Cl^- Br^- I^-	NO_3^- ClO_4^- SO_4^{2-}	$C_2H_3O_2^-$ F^- CO_3^{2-} S^{2-} PO_4^{3-}	CN^- NO_2^- HCO_3^- HS^- HPO_4^{2-}	HSO_4^-, $H_2PO_4^-$
Cation	Li^+ Na^+ K^+	Mg^{2+} Ca^{2+} Ba^{2+}	none		Al^{3+} NH_4^+ transition metal ions

Hence, there is no tendency for the reverse reaction to occur. That is, Na^+ ions, regardless of their source, do not combine with OH^- ions in water. As a result, Na^+ and the other cations of the Group 1 and 2 metals are neutral.

Basic Anions

Recall from Section 17.4 that any anion derived from a weak acid acts as a weak base in water solution. There is a small army of such anions. Those listed in Table 17.6 are typical examples. In contrast, there are no common basic cations.

Acidic Ions

These include:
—**all cations except those of the metals in Groups 1 and 2.** Typical examples of acidic cations are listed in Table 17.6.
—the HSO_4^- and $H_2PO_4^-$ ions (see discussion on p. 407).

The principles just discussed can be applied to predict the acid-base properties of salt solutions. Here, two ions are present, a cation and an anion. To decide whether the solution is acidic, basic, or neutral, we have to consider the properties of both ions. The way this is done is shown in Example 17.5.

Example 17.5   Using Table 17.6, describe each of the following 0.1 M solutions as acidic, basic, or neutral:
   a. $KClO_4$ b. NH_4I c. Na_3PO_4 d. $Zn(NO_3)_2$

Solution

Solute	Cation	Anion	Aqueous Solution
a. $KClO_4$	K^+ (neutral)	ClO_4^- (neutral)	neutral
b. NH_4I	NH_4^+ (acidic)	I^- (neutral)	acidic
c. Na_3PO_4	Na^+ (neutral)	PO_4^{3-} (basic)	basic
d. $Zn(NO_3)_2$	Zn^{2+} (acidic)	NO_3^- (neutral)	acidic

acid + neutral → acidic

basic + neutral → basic

Experiment confirms these predictions; see color plate 15 (center of book).

Exercise   Write an equation to explain why 0.1 M Na_3PO_4 is basic. Answer: $PO_4^{3-}(aq) + H_2O \rightleftharpoons HPO_4^{2-}(aq) + OH^-(aq)$.

Anions containing ionizable protons can show both acidic and basic character. The hydrogen carbonate ion, HCO_3^-, is typical. It can dissociate via the reaction

$$HCO_3^-(aq) \rightleftharpoons H^+(aq) + CO_3^{2-}(aq) \qquad (17.18)$$

This reaction, by itself, would tend to make the solution acidic. However, since the HCO_3^- ion is the anion of a weak acid, H_2CO_3, it can also undergo the following reaction:

$$HCO_3^-(aq) + H_2O \rightleftharpoons H_2CO_3(aq) + OH^-(aq) \qquad (17.19)$$

This reaction, which tends to make the solution basic, occurs to a greater extent than

17.18. This explains why a water solution of $NaHCO_3$ ("bicarbonate of soda") is slightly basic, with a pH above 8.

Most anions of this type behave like the HCO_3^- ion. Solution containing HS^-, HPO_4^{2-}, and $H_2BO_3^-$ are all slightly basic. As previously mentioned, the only two acidic anions are HSO_4^- and $H_2PO_4^-$. With these two ions, a reaction similar to 17.18 predominates.

17.6 ACID-BASE REACTIONS

When an acidic water solution is mixed with a solution containing a base, a reaction occurs. The nature of this reaction, and the equation we write for it, depend upon whether the acid and base are strong or weak. In this section, we will look at several different types of acid-base reactions. All those considered have large equilibrium constants and, for all practical purposes, *go to completion*.

Reactions of Strong Acids with Strong Bases

Consider what happens when we add a solution of a strong acid like HNO_3 to a solution of a strong base such as NaOH. Both HNO_3 and NaOH are completely dissociated into ions:

$$\text{solution of } HNO_3\text{: } H^+, NO_3^- \text{ ions}$$
$$\text{solution of NaOH: } Na^+, OH^- \text{ ions}$$

The acid-base reaction that occurs involves the H^+ ion of the HNO_3 solution and the OH^- ion of the NaOH solution. The equation for the reaction is simply

$$H^+(aq) + OH^-(aq) \rightarrow H_2O \qquad (17.20)$$

This is the net reaction that occurs when *any strong acid reacts with any strong base*. Note that we do not include in the equation "spectator ions" such as Na^+ or NO_3^-, which do not take part in the reaction.

The reaction between a strong acid and a strong base is called **neutralization.** If just enough base is added to react with all the acid, the resulting solution is neutral. For example, if equal volumes of 0.10 M HNO_3 and 0.10 M NaOH are used, the final solution will contain only Na^+ and NO_3^- ions. Since both these ions are neutral, the solution will have a pH of 7.

Robert Boyle of gas law fame was probably the first to recognize that when an acid reacts with a base, they neutralize each other's properties. Nearly 150 years passed before the nature of this reaction was determined. The delay came because Lavoisier (1787) insisted that oxygen was the fundamental component of all acids. In 1811, Humphry Davy showed that hydrochloric acid contained no oxygen. Shortly thereafter (1814), Gay-Lussac concluded that it is the hydrogen in acids that neutralizes bases.

Reactions of Weak Acids with Strong Bases

When a strong base such as NaOH is added to a weak acid, a reaction similar in many ways to 17.20 occurs. The equation for the reaction, however, is different. To see why this is the case, let us consider the nature of the principal species present in the two solutions. The strong base NaOH is completely dissociated to Na^+ and OH^- ions. In a solution of a weak acid, HX, the situation is quite different. Here there are very few H^+ and X^- ions; the principal species is the undissociated HX molecule.

solution of NaOH: Na^+, OH^- ions
solution of weak acid HX: HX molecule

The acid-base reaction involves the HX molecule and the OH^- ion as reactants. The products are an H_2O molecule and an X^- ion in solution. The equation for the reaction is

$$HX(aq) + OH^-(aq) \rightarrow H_2O + X^-(aq) \qquad (17.21)$$

We write the equation in terms of the principal species in the solution

Here, HX stands for any weak acid, such as HF, $HC_2H_3O_2$, ... By the same token, X^- is the anion derived from that acid (F^-, $C_2H_3O_2^-$, . . .).

You will recall that anions derived from weak acids are themselves weak bases. This means that a solution prepared by reacting equal amounts of HX and OH^- will be slightly basic. Consider, for example, what happens when we mix equal volumes of 0.10 M acetic acid, $HC_2H_3O_2$, and 0.10 M NaOH. The final solution contains the Na^+ ion, which is neutral, and the acetate ion, $C_2H_3O_2^-$, which is basic. The pH of this solution is greater than 7 (about 9).

Reactions of Strong Acids with Weak Bases

As an example of this type of reaction, consider what happens when a strong acid such as HCl is added to a water solution of ammonia, NH_3. Here, the acid is completely dissociated; HCl is a strong acid. In contrast, since ammonia is a weak base, the principal species present in its solution is the NH_3 molecule:

solution of HCl: H^+, Cl^- ions
solution of NH_3: NH_3 molecule

The acid-base reaction involves the H^+ ion of the HCl solution and the NH_3 molecule of the ammonia solution. The product is the NH_4^+ ion:

$$H^+(aq) + NH_3(aq) \rightarrow NH_4^+(aq) \qquad (17.22)$$

Recall that the ammonium ion, NH_4^+, is a weak acid. Hence, if equal amounts of H^+ and NH_3 are used, the final solution is slightly acidic, with a pH less than 7 (about 5).

As we saw in Section 17.3, many anions act as weak bases. When a strong acid is added to a solution containing such an anion, a reaction entirely analogous to 17.22 occurs. A weak acid forms. With F^- ions, the product is HF:

$$H^+(aq) + F^-(aq) \rightarrow HF(aq) \qquad (17.23)$$

In the case of the carbonate ion, CO_3^{2-}, reaction with H^+ occurs in two steps. The first product is the hydrogen carbonate ion, HCO_3^-. This ion can acquire another proton to form carbonic acid, H_2CO_3, an unstable species which decomposes to CO_2 and H_2O:

$$H^+(aq) + CO_3^{2-}(aq) \rightarrow HCO_3^-(aq) \qquad (17.24)$$

This reaction goes first.

$$H^+(aq) + HCO_3^-(aq) \rightarrow H_2CO_3(aq) \rightarrow CO_2(g) + H_2O \qquad (17.25)$$

This one goes if there is enough H^+ present

Addition of excess strong acid to a solution containing CO_3^{2-} ions evolves CO_2 gas. The overall reaction is obtained by summing 17.24 and 17.25:

$$2 H^+(aq) + CO_3^{2-}(aq) \rightarrow CO_2(g) + H_2O \qquad (17.26)$$

Carbon dioxide is also formed when sodium hydrogen carbonate, $NaHCO_3$, is added to an acidic solution (Reaction 17.25). This is what happens when bicarbonate of soda is taken to relieve acid indigestion.

Example 17.6 Write equations for each of the following acid-base reactions in water solution (include only those species, ions or molecules, that take part in the reaction):
  a. nitrous acid, HNO_2, with NaOH.
  b. sodium acetate, $NaC_2H_3O_2$, with nitric acid, HNO_3.
  c. hydrobromic acid, HBr, with NaOH.

Solution

  a. Nitrous acid is a weak acid; sodium hydroxide is a strong base. Hence the principal species present are

$$\text{solution of } HNO_2: HNO_2 \text{ molecule}$$
$$\text{solution of NaOH: } Na^+, OH^- \text{ ions}$$

The acid-base reaction is analogous to 17.21:

$$HNO_2(aq) + OH^-(aq) \rightarrow H_2O + NO_2^-(aq)$$

  b. Sodium acetate is a salt which is completely dissociated to Na^+ and $C_2H_3O_2^-$ ions in solution. Nitric acid, a strong acid, is completely dissociated to H^+ and NO_3^- ions:

$$\text{solution of } NaC_2H_3O_2: Na^+, C_2H_3O_2^- \text{ ions}$$
$$\text{solution of } HNO_3: H^+, NO_3^- \text{ ions}$$

The H^+ ions of the HNO_3 solution react with the weakly basic $C_2H_3O_2^-$ ions. A reaction analogous to 17.23 occurs; the product is the weak acid $HC_2H_3O_2$.

$$H^+(aq) + C_2H_3O_2^-(aq) \rightarrow HC_2H_3O_2(aq)$$

  c. HBr is a strong acid; NaOH is a strong base. The reaction is a simple neutralization:

$$H^+(aq) + OH^-(aq) \rightarrow H_2O$$

Exercise   Classify the solutions formed in (a), (b), and (c) as acidic, basic, or neutral.
Answer: (a) is basic, (b) is acidic, (c) is neutral.

17.7 ACID-BASE TITRATIONS

You may recall that in Chapter 4 we discussed briefly the principle involved in acid-base titrations. To refresh your memory, we show in Figure 17.3 the steps involved in an experiment of this type. Here, we are interested in determining the concentration of acetic acid in vinegar. To do this, we titrate with a sodium hydroxide solution of known concentration, 0.100 M. By carefully measuring the volume of this solution required to react with a known volume of vinegar, it is possible to determine the concentration of $HC_2H_3O_2$ in the vinegar (Example 17.7).

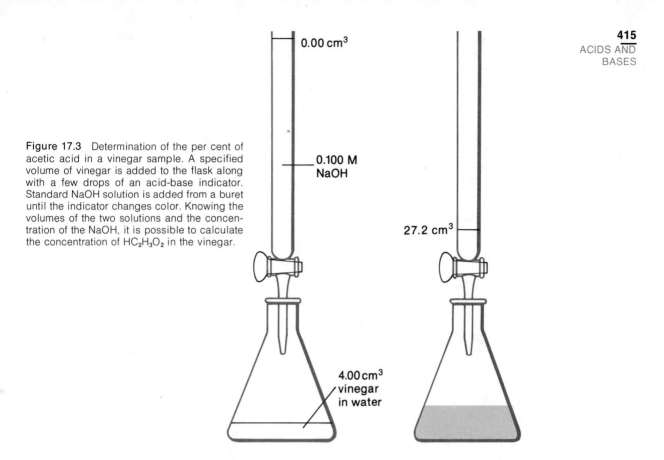

Figure 17.3 Determination of the per cent of acetic acid in a vinegar sample. A specified volume of vinegar is added to the flask along with a few drops of an acid-base indicator. Standard NaOH solution is added from a buret until the indicator changes color. Knowing the volumes of the two solutions and the concentration of the NaOH, it is possible to calculate the concentration of $HC_2H_3O_2$ in the vinegar.

0.00 cm³

0.100 M
NaOH

27.2 cm³

4.00 cm³
vinegar
in water

Example 17.7   By law, vinegar must contain at least 4.0% acetic acid, which corresponds to 0.67 M acetic acid. A 4.00-cm³ sample of Brand X vinegar is titrated with NaOH. It is found that 27.2 cm³ of 0.100 M NaOH is required for complete reaction. Does the concentration of acetic acid in the vinegar meet the legal standard?

Solution   To find M for $HC_2H_3O_2$, we start with the equation for the reaction

$$HC_2H_3O_2(aq) + OH^-(aq) \rightarrow H_2O + C_2H_3O_2^-(aq)$$

It is clear that the reactants combine in a 1:1 mole ratio:

no. moles OH^- used for titration = no. moles $HC_2H_3O_2$ in sample

But, no. moles = molarity × volume (in liters) = M × V
So: (M NaOH) × (V NaOH) = (M $HC_2H_3O_2$) × (V $HC_2H_3O_2$)

$$(0.100 \text{ mol}/\ell) \times (0.0272 \ \ell) = (M \ HC_2H_3O_2) \times (0.00400 \ \ell)$$

$$M \ H_2C_2H_3O_2 = \frac{(0.100 \text{ mol}/\ell) \times (0.0272 \ \ell)}{(0.00400 \ \ell)} = 0.680 \text{ M}$$

Brand X vinegar does comply with the law.

Exercise   How many cm³ of 0.0977 M HCl are required to titrate 30.0 cm³ of 0.103 M NaOH? Answer: 31.6 cm³.

Acid-base titrations can be used for a variety of purposes. We have just seen how they can be applied to find the concentration of a species in solution. They can also be used to find the percentage of an acidic or basic component of a solid mixture (Example 17.8).

Example 17.8   A research chemist isolates a sample of nicotinic acid, $HC_6H_4NO_2$ (MM = 123). To determine its purity, she titrates 0.450 g of the sample with 0.100 M NaOH. A volume of 36.2 cm³ of NaOH is required. The reaction is

$$HC_6H_4NO_2(aq) + OH^-(aq) \rightarrow H_2O + C_6H_4NO_2^-(aq)$$

Assuming any impurity present does not react with the base, calculate
    a. the mass in grams of nicotinic acid in the sample.
    b. the percentage of nicotinic acid in the sample.

Solution
    a. We first find the number of moles of NaOH, equate that to the number of moles of acid, and finally calculate the mass in grams of nicotinic acid.

$$\text{moles NaOH} = 0.100 \frac{\text{mol}}{\ell} \times 0.0362 \; \ell = 3.62 \times 10^{-3} \text{ mol}$$

$$\text{moles acid} = 3.62 \times 10^{-3} \text{ mol}$$

$$\text{mass acid} = 3.62 \times 10^{-3} \text{ mol} \times 123 \text{ g/mol} = 0.445 \text{ g}$$

    b. % acid $= \dfrac{0.445 \text{ g}}{0.450 \text{ g}} \times 100 = 98.9\%$

Exercise   Suppose that 35.8 cm³ of NaOH had been required, instead of 36.2 cm³. What percentage would you then calculate for nicotinic acid? Answer: 97.9%.

Indicators in Acid-Base Titrations

In an acid-base titration, we must know when to stop adding reagent. In other words, we must be able to tell at what point the reaction is complete. This is accomplished by adding an acid-base indicator. The indicator should change color when the reaction is complete—that is, when equivalent quantities of acid and base have been used. The point in the titration at which this occurs is called the *equivalence point*. If the indicator changes color before this point is reached, too little reagent will be added and the reaction will not be complete. If it changes too late, after the equivalence point has been passed, too much reagent will be added.

To choose the proper indicator for a particular titration, it is helpful to refer to what is known as a titration curve. Data for such a curve are obtained by determining the pH of the solution as a function of the volume of titrant added. Two such curves are shown in Figure 17.4. At the left, we show how pH changes when a sample of 1.00 M acetic acid, $HC_2H_3O_2$, is titrated with 1.00 M NaOH. The equation for the reaction is

$$HC_2H_3O_2(aq) + OH^-(aq) \rightarrow H_2O + C_2H_3O_2^-(aq)$$

At the right, we show what happens to pH when HCl is added to a water solution of ammonia, NH_3. Here the reaction is

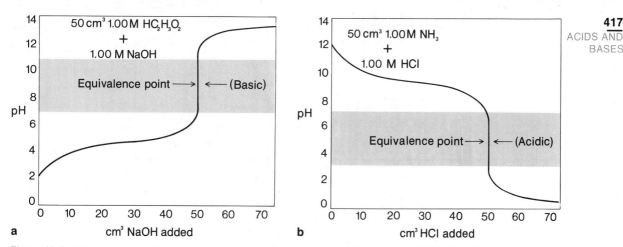

Figure 17.4 The curve at the left shows how pH changes as a solution of the weak acid $HC_2H_3O_2$ is titrated with a strong base, NaOH. The indicator used must change color close to pH 9, which is the pH of the solution of $NaC_2H_3O_2$ formed when the reaction is complete. At the right is the curve obtained when the weak base NH_3 is titrated with the strong acid HCl. The solution of NH_4Cl formed has a pH of 5, so we must use an indicator that changes color close to pH 5.

$$NH_3(aq) + H^+(aq) \rightarrow NH_4^+(aq)$$

Looking first at Figure 17.4a, we see that the pH starts off at about 2.3. This is the pH of 1.00 M $HC_2H_3O_2$, a weak acid. As OH^- ions are added the pH increases, rather slowly in the early stages of the titration. Near the equivalence point, the pH climbs more steeply, and then nearly levels off as excess NaOH is added. At the equivalence point, we have a solution of $NaC_2H_3O_2$. As noted earlier, this solution is basic, with a pH of about 9. Phenolphthalein, which changes from colorless to red at about pH 9, would be an excellent indicator for this titration. If we used methyl red (color change at pH 5), we would stop the titration much too early, when reaction is only about 65% complete. This situation is common in titrations of *weak acids* with *strong bases*. For such a titration we choose an indicator, such as phenolphthalein, which *changes color above pH 7.*

From Figure 17.4b, we see that in the titration of NH_3 with HCl, the pH starts off at about 11.7. This is the pH of 1.00 M NH_3, a weak base. As HCl is added the pH drops. At the equivalence point, we have a solution of NH_4Cl, with a pH on the acid side, about 5. Methyl red, which changes from yellow to red at pH 5, would be an ideal indicator for this titration. Phenolphthalein (color change at pH 9) would not be suitable. It would turn from pink to colorless too soon, before all the NH_3 had reacted. In general, in any titration of a *weak base* with a *strong acid,* we should choose an indicator which *changes color below pH 7.*

In Figure 17.5, we show a curve for the titration of a strong acid (1.00 M HCl) with a strong base (1.00 M NaOH). As we would expect, the pH at the equivalence point is 7, corresponding to a neutral solution of NaCl. This would suggest the use of an indicator such as bromthymol blue, which changes color at pH 7. However, looking more closely at the shape of this curve, we see that the choice of indicator is not as critical here as in the two cases just discussed. Notice that the steep portion of the curve is much longer than it is in the curves shown in Figure 17.4. Indeed, a single drop of 1.00 M NaOH (about 0.02 cm³) changes the pH from 4 to 10 (see Problem 17.49). Methyl red (color change at pH 5), or phenolphthalein (color change at pH 9) could be used. In general, for the titration of a strong acid with a strong base, we have

A good acid-base indicator for a titration changes color at the equivalence point of the titration

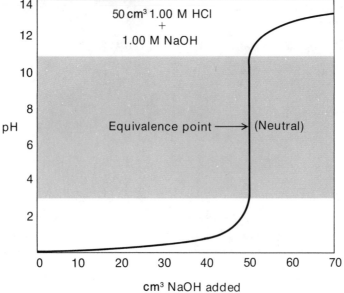

Figure 17.5 In the titration of HCl with NaOH, the pH is 7 at the equivalence point. However, the pH changes very rapidly near this point. Addition of a single drop of NaOH can change the pH by six units. As a result, almost any indicator is suitable for this titration.

a much wider choice of indicators than in other types of acid-base titrations. This is because the pH changes so rapidly near the equivalence point.

17.8 GENERAL MODELS OF ACIDS, BASES, AND ACID-BASE REACTIONS

So far in this chapter, we have considered an acid to be a substance which in water solution produces an excess of H^+ ions. A base was similarly defined to be a substance which, directly or indirectly, forms excess OH^- ions in water solution. This approach, first proposed by the Swedish chemist Svandte Arrhenius in 1884, is a very practical one. In particular, it allows us to predict whether such substances as HCl, $Ca(OH)_2$, NH_4Cl, NaCl, and K_2CO_3 will form acidic, basic, or neutral solutions.

The Arrhenius model does, however, have one disadvantage. It greatly restricts the number of reactions which can be considered to be of the acid-base type. Over the years, many other more general models of acids and bases have been proposed. In this section, we will discuss two such models. Curiously enough, they were both suggested in the same year, 1923. One of them was proposed independently by Brönsted in Denmark and Lowry in England. The other came from a man we have heard of before (Chap. 10), the American physical chemist, G. N. Lewis.

Brönsted-Lowry Concept

According to this model, an acid-base reaction is one in which there is a *proton transfer* from one species to another. The species which gives up or **donates the proton** is referred to as an **acid.** The molecule or ion which **accepts the proton** is a **base.**

A simple example of a Brönsted-Lowry acid-base reaction is that between acetic acid and hydroxide ions:

$$HC_2H_3O_2(aq) + OH^-(aq) \rightarrow C_2H_3O_2{}^-(aq) + H_2O \qquad (17.27)$$
$$\text{acid} \qquad\qquad \text{base}$$

Here, $HC_2H_3O_2$ gives up a proton to an OH^- ion. Hence, $HC_2H_3O_2$ is acting as an acid (proton donor) while OH^- is a base (proton acceptor). This conclusion is not particularly startling. We used almost the same words earlier in describing this reaction by the Arrhenius model.

Consider, however, the reverse of Reaction 17.27:

$$C_2H_3O_2^-(aq) + H_2O \rightarrow HC_2H_3O_2(aq) + OH^-(aq) \qquad (17.28)$$
$$\text{base} \qquad\quad \text{acid}$$

This is the equation written earlier to explain why a solution containing acetate ions is basic. At the time, we did not refer to it as an acid-base reaction. According to the Brönsted-Lowry model, it is. The $C_2H_3O_2^-$ accepts a proton from a water molecule and hence acts as a base. The H_2O molecule donates a proton to the acetate ion and so acts as a Brönsted-Lowry acid.

The Brönsted-Lowry model can be extended still further. Consider the equation written earlier for the dissociation of HCl in water:

$$HCl(aq) \rightarrow H^+(aq) + Cl^-(aq)$$

Here, although HCl appears to be the proton donor, there is not any obvious proton acceptor. According to Brönsted and Lowry, the proton acceptor is really a water molecule. They would rewrite this equation as

$$HCl(aq) + H_2O \rightarrow Cl^-(aq) + H_3O^+(aq) \qquad (17.29)$$
$$\text{acid} \qquad\quad \text{base}$$

By the Brönsted-Lowry model, the H^+ ion would always exist as H_3O^+

In this equation, HCl is donating a proton to H_2O (Fig. 17.6). So, HCl is acting as an acid and H_2O as a base. Equation 17.29 offers a plausible explanation as to why HCl dissociates in water. It does so by reacting with water to produce two more stable species, the Cl^- and H_3O^+ ions. In the Arrhenius model, the reason for the dissociation of HCl is not apparent.

You will note from this discussion that the H_2O molecule can act as either a

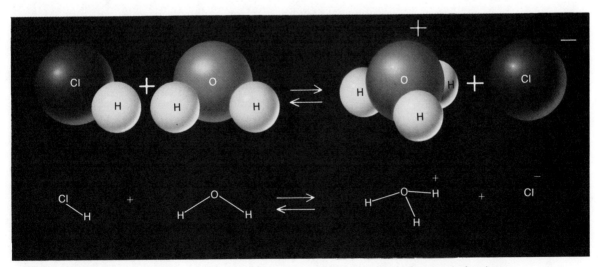

Figure 17.6   When HCl is added to water, there is a proton transfer from an HCl to an H_2O molecule, forming a Cl^- ion and a H_3O^+ ion. In this reaction, HCl acts as a Brönsted-Lowry acid, H_2O as a Brönsted-Lowry base.

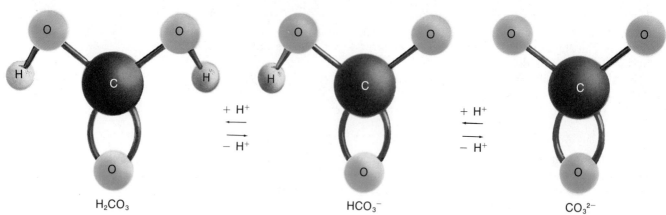

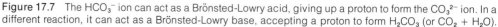

$$H_2CO_3 \qquad\qquad HCO_3^- \qquad\qquad CO_3^{2-}$$

Figure 17.7 The HCO_3^- ion can act as a Brönsted-Lowry acid, giving up a proton to form the CO_3^{2-} ion. In a different reaction, it can act as a Brönsted-Lowry base, accepting a proton to form H_2CO_3 (or $CO_2 + H_2O$).

Brönsted-Lowry acid or base. When it acts as an acid, it donates a proton to another species and is converted to an OH^- ion. When H_2O acts as a base, it accepts a proton, forming the *hydronium ion,* H_3O^+. Several other molecules and ions can behave as both acids and bases in the Brönsted-Lowry sense. Among these is the HCO_3^- ion:

$$\underset{\text{base}}{HCO_3^-(aq)} + \underset{\text{acid}}{H_2O} \rightarrow H_2CO_3(aq) + OH^-(aq) \qquad (17.30)$$

$$\underset{\text{acid}}{HCO_3^-(aq)} + \underset{\text{base}}{H_2O} \rightarrow CO_3^{2-}(aq) + H_3O^+(aq) \qquad (17.31)$$

In Reaction 17.31, CO_3^{2-} is the conjugate base of HCO_3^- and H_3O^+ the conjugate acid of H_2O

In discussing reactions such as 17.27 through 17.31, we often use a special terminology. A species formed from an acid by the loss of a proton is referred to as the *conjugate base* of that acid. Thus, we refer to the $C_2H_3O_2^-$ ion as the conjugate base of $HC_2H_3O_2$; the OH^- ion is the conjugate base of H_2O, and so on. A species formed from a base by gaining a proton is called the conjugate acid of that base. The *conjugate acid* of H_2O is the H_3O^+ ion; that of HCO_3^- is H_2CO_3, and so on.

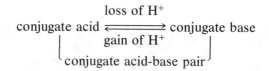

Example 17.9   Consider the reaction

$$H_2PO_4^-(aq) + C_2H_3O_2^-(aq) \rightleftharpoons HPO_4^{2-}(aq) + HC_2H_3O_2(aq)$$

Classify each of the four species involved as a Brönsted-Lowry acid or base.

Solution   In the forward reaction, the $H_2PO_4^-$ ion donates a proton to the $C_2H_3O_2^-$ ion. In the reverse reaction, $HC_2H_3O_2$ donates a proton to HPO_4^{2-}. Hence, $H_2PO_4^-$ and $HC_2H_3O_2$ are Brönsted-Lowry acids; $C_2H_3O_2^-$ and HPO_4^{2-} are Brönsted-Lowry bases.

Exercise   What is the conjugate base of the $H_2PO_4^-$ ion? the conjugate acid? Answer: HPO_4^{2-}; H_3PO_4.

We have seen that the Brönsted-Lowry model extends the Arrhenius picture of acid-base reactions considerably. However, the Brönsted-Lowry model is restricted in one important respect. It can be applied only to reactions involving a proton transfer. For a species to act as a Brönsted-Lowry acid, it must contain an ionizable hydrogen atom.

The Lewis acid-base model removes this restriction. A **Lewis acid** is a species which, in an acid-base reaction, **accepts** an **electron pair.** In this reaction, a **Lewis base donates** the **electron pair.**

From a structural point of view, the Lewis concept of a base does not differ in any essential way from the Brönsted-Lowry concept. In order for a species to accept a proton and thereby act as a Brönsted-Lowry base it must possess an unshared pair of electrons. Consider, for example, the NH_3 molecule, the H_2O molecule, and the F^- ion, all of which can act as Brönsted-Lowry bases:

$$\text{H---}\overset{\textstyle\cdot}{\text{N}}\text{---H,} \qquad \text{H---}\overset{..}{\underset{..}{\text{O}}}\text{---H,} \qquad (:\overset{..}{\underset{..}{\text{F}}}:)^-$$
$$\quad\ \ |$$
$$\quad\ \ \text{H}$$

Each of these species contains an unshared pair of electrons that is utilized in accepting a proton to form the NH_4^+ ion, the H_3O^+ ion, or the HF molecule.

$$\left[\begin{array}{c}\text{H}\\\text{H:}\overset{\textstyle\cdot}{\underset{\textstyle\cdot}{\text{N}}}\text{:H}\\\text{H}\end{array}\right]^+ \qquad \left[\begin{array}{c}..\\\text{H:}\overset{}{\underset{}{\text{O}}}\text{:H}\\\text{H}\end{array}\right]^+ \qquad \text{H:}\overset{..}{\underset{..}{\text{F}}}:$$

Clearly, NH_3, H_2O, and F^- can also be Lewis bases, since they possess an unshared electron pair which can be donated to an acid. We see then that the Lewis concept does not significantly change the number of species which can behave as bases.

On the other hand, the Lewis concept greatly increases the number of species which can be considered to be acids. The substance which accepts an electron pair and therefore acts as a Lewis acid can be a proton:

$$\underset{\text{acid}}{H^+(aq)} + \underset{\text{base}}{H_2O} \rightarrow H_3O^+(aq)$$

$$\underset{\text{acid}}{H^+(aq)} + \underset{\text{base}}{NH_3(aq)} \rightarrow NH_4^+(aq)$$

It can equally well be a cation, such as Zn^{2+}, which can accept electron pairs from a Lewis base:

$$\underset{\text{acid}}{Zn^{2+}(aq)} + 4\ \underset{\text{base}}{H_2O} \rightarrow Zn(H_2O)_4{}^{2+}(aq)$$

$$\underset{\text{acid}}{Zn^{2+}(aq)} + 4\ \underset{\text{base}}{NH_3} \rightarrow Zn(NH_3)_4{}^{2+}(aq)$$

We will discuss reactions of this type in greater detail in Chapter 19.

Another important class of Lewis acids comprises molecules containing an incomplete octet of electrons. A classic example is boron trifluoride, BF_3, which reacts readily with ammonia, accepting a pair of electrons:

MODEL	ACID	BASE
Arrhenius	supplies H^+ to water	supplies OH^- to water
Brönsted-Lowry	H^+ donor	H^+ acceptor
Lewis	electron pair acceptor	electron pair donor

TABLE 17.7 ALTERNATIVE DEFINITIONS OF ACIDS AND BASES

Clearly, a Lewis acid does not need to contain an H^+ ion

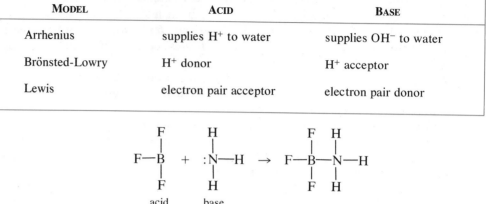

acid   base

The Lewis model is commonly used in organic chemistry to consider the catalytic behavior of such Lewis acids as $ZnCl_2$ and BF_3. In general, when proton transfer reactions are involved, most chemists use the Arrhenius or Brönsted-Lowry concepts. Table 17.7 summarizes the acid-base models we have discussed. As you might guess, other models have been proposed since 1923. However, the three listed will suffice for our purposes.

17.9 HISTORICAL PERSPECTIVE

Gilbert Newton Lewis (1875–1946)

The Lewis concept of acids and bases, like the Lewis structures discussed in Chapter 10, was the product of the American physical chemist, G. N. Lewis. Born in Massachusetts, Lewis grew up in Nebraska, then came back East to obtain his B.S. (1896) and Ph.D. (1899) at Harvard. Although he stayed on for a few years as an instructor, Lewis seems never to have been happy at Harvard. A precocious student and an intellectual rebel, he was repelled by the highly traditional atmosphere that prevailed in the chemistry department there in his time. Many years later, he refused an honorary degree from his alma mater.

After leaving Harvard, Lewis made his reputation at M.I.T., where he was promoted to full professor in only four years. In 1912, he moved across the country to the University of California at Berkeley as Dean of the College of Chemistry and department chairman. He remained there for the rest of his life. Under his guidance, the chemistry department at Berkeley became perhaps the most prestigious in the country. Among the faculty and graduate students that he attracted were five future Nobel Prize winners: Harold Urey in 1934, William Giauque in 1949, Glenn Seaborg in 1951, Willard Libby in 1960, and Melvin Calvin in 1961.

In administering the chemistry department at Berkeley, Lewis demanded excellence in both research and teaching. Virtually the entire staff was involved in the general chemistry program; at one time eight full professors carried freshman sections. Several department members became leaders of chemical education in America. Among them is Joel Hildebrand, who came to California in 1913 and was still active in teaching and research 60 years later.

Like so many physical chemists, G. N. Lewis maintained throughout his life a fascination with chemical thermodynamics. His Ph.D. thesis was in this area, as were all his early publications. Many of the standard electrode potentials given in Table 21.1, Chapter 21, are based on data obtained by Lewis and his students. In 1923 he published with Merle Randall a text entitled *Thermodynamics and the Free Energy of Chemical Substances*. More than 50 years later, a revised edition of that text is still widely used in graduate courses in chemistry.

Lewis' interest in chemical bonding and structure dates from 1902. In attempting to explain "valence" to a class at Harvard, he devised an atomic model to rationalize the octet rule. His model was deficient in many respects; for one thing, Lewis visualized cubic atoms with electrons located at the corners. Perhaps this explains why his ideas of atomic structure were not published until 1916. In that year, Lewis conceived of the electron-pair bond. This concept and its implications were elaborated upon in a book that he published in 1923, *Valence and the Structure of Atoms and Molecules* (recently reprinted by Dover Publications). Here, in Lewis' characteristically lucid style, we find many of the basic principles of covalent bonding that are accepted today. Here too is the Lewis definition of acids and bases as electron-pair acceptors and donors. Curiously enough, this general approach to acid-base reactions seems to have been virtually ignored for 15 years until revived in a paper published by Lewis in 1938.

The years from 1923 to 1938 were relatively unproductive for G. N. Lewis so far as his own research was concerned. The applications of the electron-pair bond came largely in the areas of organic and quantum chemistry; in neither of these fields did Lewis feel at home. In the early 1930's, he published a series of relatively minor papers dealing with the properties of deuterium. Then, in 1939, he began to publish in the field of photochemistry. Of approximately 20 papers in this area, several were of fundamental importance, comparable in quality to the best work of his early years. Retired officially in 1945, Lewis died a year later while carrying out an experiment on fluorescence.

SUMMARY

The acidity or basicity of a water solution can be expressed in terms of $[H^+]$, $[OH^-]$, or pH. These quantities are related by the equations:

$$[H^+] \times [OH^-] = K_w = 1.0 \times 10^{-14} \qquad \text{(Example 17.1)}$$

$$pH = -\log_{10} [H^+] \qquad \text{(Example 17.2)}$$

acidic solution:	$[H^+] > [OH^-]$; $[H^+] > 10^{-7}$ M	pH < 7
basic solution:	$[OH^-] > [H^+]$; $[OH^-] > 10^{-7}$ M	pH > 7
neutral solution:	$[H^+] = [OH^-] = 10^{-7}$ M	pH $= 7$

Strong acids are completely dissociated in dilute water solution. There are only six common strong acids (Table 17.4): HCl, HBr, HI, HNO_3, $HClO_4$, H_2SO_4. All other acids are weak; that is, they are partially dissociated to H^+ ions in water. Species that act as weak acids may be molecules (H_2S), cations (NH_4^+, Al^{3+}) or, in rare cases, anions ($H_2PO_4^-$) (Example 17.3). The only strong bases, completely dissociated to OH^- ions in water, are the hydroxides of the metals in Groups 1 and 2. Weak bases include ammonia (NH_3) and anions such as CN^- and CO_3^{2-} that are derived from weak acids. These species furnish OH^- ions by reacting reversibly with water:

$$\text{weak base} + H_2O \rightleftharpoons \text{weak acid} + OH^-$$

An aqueous solution of a salt may be acidic, basic, or neutral, depending upon the nature of the cation and anion (Table 17.6, Example 17.5).

Acidic and basic water solutions react with each other. The nature of the reaction and the equation written to represent it depend upon the strength of both the acid and the base (Example 17.6). The solution formed at the equivalence point may be neutral (strong acid, strong base), acidic (strong acid, weak base), or basic (weak acid, strong base). The pH at the equivalence point determines the choice of indicator. The concentration of an acidic or basic species can be determined by an acid-base titration (Examples 17.7 and 17.8).

The model of acid-base reactions described above is that of Arrhenius. Other more general models have been proposed (Table 17.7). One of the most useful of these is the Brönsted-Lowry model (Example 17.9).

KEY TERMS AND CONCEPTS

acid	pH	salt	Brönsted-Lowry acid, base
base	acid-base indicator	neutralization	Lewis acid, base
K_w	strong acid	acid-base titration	hydronium ion
acidic solution	weak acid	equivalence point	conjugate base
basic solution	strong base	Arrhenius model, acids and bases	conjugate acid
neutral solution	weak base		

QUESTIONS AND PROBLEMS

Catalog

$[H^+]$, $[OH^-]$, and pH: 17.2–17.6, 17.24–17.28
Strong and Weak Acids: 17.7, 17.8, 17.29, 17.30
Strong and Weak Bases: 17.9, 17.10, 17.31, 17.32
Salt Solutions: 17.11, 17.12, 17.33, 17.34

Acid-Base Reactions: 17.13–17.17, 17.35–17.39
Brönsted-Lowry, Lewis: 17.18–17.20, 17.40–17.42
General: 17.1, 17.21–17.23, 17.43–17.49

17.1 Review and know the meaning of the key words and concepts in this chapter.

17.2 Calculate the $[OH^-]$ and pH of a solution in which $[H^+] = 3.0 \times 10^{-3}$ M.

17.3 Calculate the pH of solutions with the following $[H^+]$:

a. 1×10^{-3} M   b. 10 M
c. 0.0000015 M   d. 4.3×10^{-8} M

Classify each of these solutions as acidic or basic.

17.4 Calculate the $[H^+]$ and $[OH^-]$ of solutions with the following pH values:

a. 8.0   b. 6.40   c. 0.00   d. −0.80

Classify each solution as acidic or basic.

17.24 The $[OH^-]$ in a certain solution is 5×10^{-5} M. Calculate $[H^+]$ and pH.

17.25 Apply the directions of Problem 17.3 to solutions in which $[H^+]$ is

a. 5×10^{-5} M.   b. 1.0 M.
c. 7.5×10^{-9} M.   d. 0.000028 M.

17.26 Find the $[H^+]$ and $[OH^-]$ in solutions having pH values of

a. 2.0   b. −1.00   c. 4.7   d. 6.90

Classify each solution as acidic or basic.

17.5 The most acidic rainfall ever measured occurred in 1974 in Scotland. The pH of the rain was 2.4.

    a. Calculate its $[H^+]$.
    b. Approximately how many times greater was its $[H^+]$ than its $[OH^-]$?

17.6 Find $[H^+]$, $[OH^-]$, and the pH of the following solutions of strong acids and bases:

    a. 0.50 M HI
    b. 0.22 M RbOH
    c. a solution made by dissolving 10.2 g KOH in water to make 200 cm³ of solution.
    d. a solution made by diluting 10.0 cm³ of 12 M HCl with water to make 200 cm³ of solution.

17.7 Classify each of the following as a strong or weak acid, and write an equation for its dissociation in water:

    a. $HClO_4$ b. $Zn(H_2O)_4^{2+}$ c. HI
    d. $H_2PO_4^-$ e. $Al(H_2O)_6^{3+}$

17.8 Write a balanced equation to explain why each of the following species gives an acidic water solution:

    a. HBr   b. H_2S c. $Zn(NO_3)_2$
    d. NH_4^+

17.9 Classify each of the following as a weak base, strong base, or neutral species:

    a. NaOH   b. NH_3 c. CO_3^{2-}
    d. Cl^- e. $C_2H_3O_2^-$

17.10 Each of the following acts as a weak base in water. Write balanced equations to show why this happens:

    a. BO_3^{3-} b. HCO_3^- c. S^{2-}

17.11 Use the term acidic, basic, or neutral to describe saturated solutions of the following salts in water:

    a. CsCl   b. NH_4Br c. NaCN
    d. $AlCl_3$ e. KNO_2

17.12 Explain, using balanced equations where appropriate, your answers to Question 17.11.

17.13 Write balanced equations for the reactions that occur when the following solutions are mixed. In each case, state whether the solution at the equivalence point is neutral, acidic, or basic.

    a. HBr and RbOH
    b. NH_3 and $HClO_4$
    c. $Ba(OH)_2$ and $HC_2H_3O_2$

17.14 20.0 cm³ of 0.144 M HNO_3 were required to neutralize a 15.0 cm³ sample of KOH exactly. Find the molarity of the KOH.

17.27 The pH of human urine can be as high as 8.40. What is its $[H^+]$? $[OH^-]$?

17.28 Determine the $[H^+]$, $[OH^-]$, and pH of the following solutions of strong acids and bases:

    a. 0.030 M CsOH
    b. 0.020 M $HClO_4$
    c. a solution made by dissolving 12.0 g LiOH in water to make 500 cm³ of solution.
    d. a solution made by diluting 4.0 cm³ of 2.0 M NaOH with water to a volume of 250 cm³.

17.29 Follow the directions of Question 17.7 for each of the following species:

    a. H_2SO_4 b. NH_4^+ c. $HC_2H_3O_2$
    d. $Cu(H_2O)_4^{2+}$ e. H_2CO_3

17.30 Explain by means of a balanced equation why water solutions of the following species are acidic:

    a. H_3PO_4 b. $N_2H_5^+$ c. $NaHSO_4$
    d. $CuCl_2$

17.31 Classify each of the following as a weak base, strong base, or neutral species.

    a. CN^- b. Br^- c. $Ba(OH)_2$
    d. NO_2^- e. NO_3^-

17.32 Follow the directions of Question 17.10 for

    a. F^- b. CH_3NH_2 c. HPO_4^{2-}
    d. PO_4^{3-}

17.33 Each of the following salts is dissolved in water to form a 0.5 M solution. Indicate whether the solution is acidic, basic, or neutral.

    a. KCl   b. $ZnCl_2$ c. $Ba(C_2H_3O_2)_2$
    d. NH_4I e. $NaNO_3$

17.34 For each salt in Question 17.33 that is acidic or basic, write a balanced equation to explain where the H^+ or OH^- ions come from.

17.35 Follow the directions of Question 17.13 for the following pairs of solutions:

    a. $Ba(OH)_2$ and NaH_2PO_4
    b. $Ba(OH)_2$ and HNO_3
    c. HCl and KCN

17.36 How many cm³ of 0.223 M NaOH are required to neutralize 36.2 cm³ of 0.168 M HCl exactly?

17.15 In titrating a $Ca(OH)_2$ solution, a student determined its molarity to be 0.0200 M by using exactly 9.52 cm³ of 0.1050 M HCl to neutralize the $Ca(OH)_2$ solution. What was the initial volume of the $Ca(OH)_2$ solution?

17.16 A vitamin C capsule is analyzed by titrating with 0.100 M NaOH. It is found that 24.4 cm³ of base is required to react with a capsule weighing 0.450 g. What is the percentage of vitamin C, $C_6H_8O_6$, in the capsule? (One mole of vitamin C reacts with one mole of OH^-.)

17.17 Three acid-base indicators and their pH color changes are: methyl red (5), bromthymol blue (7), and phenolphthalein (9). Which should be used for the following titrations? Explain.

    a. HNO_3 with KOH
    b. NH_3 with HBr
    c. HNO_2 with NaOH
    d. $NaC_2H_3O_2$ with HCl

17.18 For each of the following reactions, indicate the Brönsted-Lowry acids and bases:

    a. $H_2O + HF(aq) \rightarrow H_3O^+(aq) + F^-(aq)$
    b. $NH_3(aq) + H_2O \rightarrow NH_4^+(aq) + OH^-(aq)$
    c. $OH^-(aq) + NH_4^+(aq) \rightarrow H_2O + NH_3(aq)$

What are the conjugate acid-base pairs?

17.19 Which of the following species can act as Brönsted-Lowry acids? Brönsted-Lowry bases? Lewis acids? Lewis bases?

    a. H_3O^+ b. NH_3 c. HPO_4^{2-} d. Ni^{2+}

17.20 What is the conjugate base of

    a. $HClO_2$?   b. HS^-?   c. $Cu(H_2O)_4^{2+}$?
    d. H_3PO_4?

17.21 Which of the following are true regarding a 1 M aqueous solution of a strong acid, HX?

    a. The X^- concentration is 1 M.
    b. The HX concentration is 1 M.
    c. The sum of $[H^+]$ and $[X^-]$ is 2 M.
    d. The pH is zero.

17.22 Classify each of the following solutions as acidic, basic, or neutral:

    a. $Ca(OH)_2$ b. $Ca(C_2H_3O_2)_2$
    c. $Ca(NO_3)_2$ d. $HClO_4$ e. $KClO_4$
    f. $Zn(ClO_4)_2$ g. Na_2S h. H_2S

17.37 In titrating an $HClO_4$ solution, 22.4 cm³ of 0.215 M NaOH were required to neutralize 19.3 cm³ of the acid exactly. Find the molarity of the $HClO_4$.

17.38 The percentage of $NaHCO_3$ in a powder used for stomach upsets is found by titrating with 0.100 M HCl. 16.5 cm³ of the HCl is required to react with 0.302 g of the powder. What is the percentage of $NaHCO_3$?

17.39 Which of the acid-base indicators in Question 17.17 would be useful in the following titrations? Explain.

    a. KOH with $HClO_4$
    b. HF with KOH
    c. KCN with HCl
    d. $HC_2H_3O_2$ with KOH

17.40 For each of the following reactions, indicate the Brönsted-Lowry acids and bases:

    a. $HNO_2(aq) + H_2O \rightarrow H_3O^+(aq) + NO_2^-(aq)$
    b. $H_2O + S^{2-}(aq) \rightarrow HS^-(aq) + OH^-(aq)$
    c. $CN^-(aq) + HC_2H_3O_2(aq) \rightarrow C_2H_3O_2^-(aq) + HCN(aq)$

What are the conjugate acid-base pairs?

17.41 Which of the following species can act as Brönsted-Lowry acids? Brönsted-Lowry bases? Lewis acids? Lewis bases?

    a. H_2O b. Cu^{2+} c. HCO_3^- d. Cl^-

17.42 What is the conjugate acid of

    a. HS^-?   b. PO_4^{3-}?   c. H_2O?
    d. $Zn(H_2O)_3(OH)^+$?

17.43 Which of the following are true regarding a 0.10 M solution of a weak acid, HA?

    a. The A^- conc. is 0.10 M.
    b. The $[H^+] \approx [A^-]$.
    c. The $[HA] \gg [A^-]$.
    d. The pH is 1.

17.44 Classify each of the following solutions as acidic, basic, or neutral:

    a. Na_3PO_4 b. Na_2HPO_4 c. $Ca_2(PO_4)_3$
    d. $KHCO_3$ e. K_2CO_3 f. H_2CO_3
    g. $(HO)_2SO_2$ h. HOH

17.23 Give an example of

    a. a strong acid containing two ionizable protons.
    b. an indicator which changes color at pH 7.
    c. a salt containing a cation with a +1 charge which gives an acidic water solution.
    d. a neutral molecule which is a Lewis base.

17.45 Give an example of

    a. a weak acid that does not contain oxygen atoms.
    b. a salt containing C atoms which gives a basic solution.
    c. an indicator which changes color in acidic solution.
    d. a conjugate acid-base pair, both of which react with H^+ ions to give $CO_2(g)$.

*17.46 A chemistry student needs an aqueous solution of pH 8.0. To prepare it, he decides to dilute 1.0 M HCl with water until $[H^+]$ becomes 1.0×10^{-8} M. Will this work? Explain. Would dilution of 1.0 M NaOH work?

*17.47 Silver hydroxide, AgOH, in insoluble in water. Describe a simple qualitative experiment that would enable you to determine whether AgOH is a strong or weak base.

*17.48 Show by calculation that $[H_2O]$ in pure water is about 55 M.

*17.49 25.00 cm³ of 1.000 M HCl is titrated with 1.000 M NaOH. Calculate the $[H^+]$ and pH of the solution after the following volumes (cm³) of NaOH have been added:

    a. 0.00   b. 10.00   c. 20.00   d. 24.90   e. 24.99
    f. 25.00   g. 25.01   h. 25.10   i. 30.00

Use your data to construct a plot of pH vs. volume of NaOH added, similar to that shown in Figure 17.5.

18

IONIC EQUILIBRIA _____

In Chapters 16 and 17 we referred to three important types of equilibria involving ions in water solution. These are:

1. The dissolving of a slightly soluble ionic solid, such as silver chloride, AgCl:

$$AgCl(s) \rightleftharpoons Ag^+(aq) + Cl^-(aq)$$

2. The dissociation of a weak acid, such as hydrofluoric acid, HF:

$$HF(aq) \rightleftharpoons H^+(aq) + F^-(aq)$$

3. The reaction with water of a weak base, such as ammonia, NH_3:

$$NH_3(aq) + H_2O \rightleftharpoons NH_4^+(aq) + OH^-(aq)$$

In each of these processes, the reaction that occurs goes to an equilibrium state. For each reaction we can write an equilibrium constant expression, similar to those we worked with in gaseous equilibria (Chap. 13). In this chapter, we will examine the three types of ionic equilibria listed above. They are described by the equilibrium constants K_{sp}, K_a, and K_b, in that order. We will be interested in using these constants to deal with the chemical properties of slightly soluble electrolytes, weak acids, and weak bases.

18.1 THE EQUILIBRIUM CONSTANT OF A SLIGHTLY SOLUBLE IONIC SOLID, K_{sp}

In Chapter 16, we classified certain ionic solids as being "insoluble" in water. These solids do, however, dissolve to a slight extent; we might better refer to them as "very slightly soluble." A typical example is strontium chromate, $SrCrO_4$. When this solid is shaken with water, it establishes equilibrium with its ions in solution:

Only a small amount of solid dissolves

$$SrCrO_4(s) \rightleftharpoons Sr^{2+}(aq) + CrO_4^{2-}(aq) \tag{18.1}$$

The solution shows the characteristic yellow color of the aquated chromate ion, $CrO_4^{2-}(aq)$.

The chromate ion absorbs visible light at a wavelength of 370 nm, in the violet region of the spectrum. We can follow the establishment of equilibrium for 18.1 by measuring absorption as a function of time (Fig. 18.1). At first, the forward reaction dominates and absorption increases with time. When equilibrium is reached, the concentration of $CrO_4^{2-}(aq)$ and hence the absorption becomes constant.

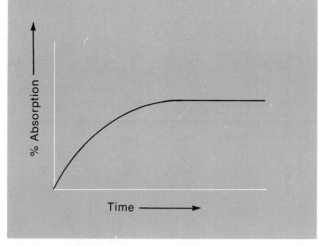

Figure 18.1 Establishment of equilibrium in the system: $SrCrO_4(s) \rightarrow Sr^{2+}(aq) + CrO_4^{2-}(aq)$. As $SrCrO_4$ dissolves, yellow chromate ions enter the solution. These ions absorb visible light at 370 nm; the per cent absorption increases with the concentration of CrO_4^{2-}. Eventually, this concentration levels off at its equilibrium value about 6×10^{-3} M.

The Solubility Product Constant, K_{sp}

For Reaction 18.1, the equilibrium constant expression is

$$K_{sp} = [Sr^{2+}] \times [CrO_4^{2-}]$$

For the reasons cited in Chapter 13, the concentration of solid does not appear in the expression. This equation tells us that, in a solution in equilibrium with solid $SrCrO_4$, the product $[Sr^{2+}] \times [CrO_4^{2-}]$ has a fixed value, given by K_{sp}. The constant in the equation, K_{sp}, is called a solubility product constant. The square brackets, as always, refer to equilibrium concentrations in moles per liter.

We can write a K_{sp} expression similar to that for $SrCrO_4$ for all solids of the type MX, where there is one mole of cations per mole of anions.

Thus, for AgCl we have

$$AgCl(s) \rightleftharpoons Ag^+(aq) + Cl^-(aq); \quad K_{sp} = [Ag^+] \times [Cl^-] \tag{18.2}$$

For slightly soluble ionic solids of the types MX_2 (e.g., PbI_2, CaF_2) or M_2X (Ag_2CrO_4), we have

$$MX_2(s) \rightleftharpoons M^{2+}(aq) + 2\ X^-(aq); \quad K_{sp} = [M^{2+}] \times [X^-]^2 \tag{18.3}$$
$$PbI_2(s) \rightleftharpoons Pb^{2+}(aq) + 2\ I^-(aq); \quad K_{sp} = [Pb^{2+}] \times [I^-]^2$$

and
$$M_2X(s) \rightleftharpoons 2\ M^+(aq) + X^{2-}(aq); \quad K_{sp} = [M^+]^2 \times [X^{2-}] \tag{18.4}$$
$$Ag_2CrO_4(s) \rightleftharpoons 2\ Ag^+(aq) + CrO_4^{2-}(aq); \quad K_{sp} = [Ag^+]^2 \times [CrO_4^{2-}]$$

More generally, we can express the solubility product principle as follows:

In any water solution in equilibrium with a slightly soluble ionic compound, the product of the concentrations of its ions, each raised to a power equal to its coefficient in the solubility equation, is a constant. This constant, K_{sp}, has a fixed value at a given temperature, independent of the concentrations of the individual ions. Values of K_{sp} for a variety of slightly soluble ionic solids are given in Table 18.1.

K_{sp} is really just a special case of K_c

TABLE 18.1   SOLUBILITY PRODUCT CONSTANTS AT 25°C

Acetates	$AgC_2H_3O_2$	2×10^{-3}	Iodides	AgI	1×10^{-16}
				PbI_2	1×10^{-8}
Bromides	$AgBr$	5×10^{-13}			
	$PbBr_2$	5×10^{-6}	Sulfates	$BaSO_4$	1.4×10^{-9}
				$CaSO_4$	3×10^{-5}
Carbonates	$BaCO_3$	2×10^{-9}		$PbSO_4$	1×10^{-8}
	$CaCO_3$	5×10^{-9}			
	$MgCO_3$	2×10^{-8}	Sulfides	Ag_2S	1×10^{-49}
				CdS	1×10^{-26}
Chlorides	$AgCl$	1.6×10^{-10}		CoS	1×10^{-20}
	Hg_2Cl_2	1×10^{-18}		CuS	1×10^{-35}
	$PbCl_2$	1.7×10^{-5}		FeS	1×10^{-17}
				HgS	1×10^{-52}
Chromates	Ag_2CrO_4	2×10^{-12}		MnS	1×10^{-15}
	$BaCrO_4$	2×10^{-10}		NiS	1×10^{-19}
	$PbCrO_4$	1×10^{-16}		PbS	1×10^{-27}
				ZnS	1×10^{-20}
Fluorides	BaF_2	2×10^{-6}			
	CaF_2	2×10^{-10}			
	PbF_2	4×10^{-8}			
Hydroxides	$Al(OH)_3$	5×10^{-33}			
	$Cr(OH)_3$	4×10^{-38}			
	$Fe(OH)_2$	1×10^{-15}			
	$Fe(OH)_3$	5×10^{-38}			
	$Mg(OH)_2$	1×10^{-11}			
	$Zn(OH)_2$	5×10^{-17}			

Uses of K_{sp}

There are many possible applications of the solubility product constant. The first one we consider is very simple. Suppose we have a water solution in equilibrium with a slightly soluble ionic solid. If the concentration of one of the ions in solution is known, then that of the other ion is readily calculated (Example 18.1).

Example 18.1

    a. Hydrochloric acid is added to a solution of $AgNO_3$. A precipitate of AgCl forms ($K_{sp} = 1.6 \times 10^{-10}$). If the final equilibrium concentration of Cl^- is 2.0×10^{-3} M, what is $[Ag^+]$?

    b. In a solution in equilibrium with PbI_2 ($K_{sp} = 1 \times 10^{-8}$), $[Pb^{2+}] = 1 \times 10^{-2}$ M. Calculate $[I^-]$.

Solution

    a. The expression for K_{sp} is

$$[Ag^+] \times [Cl^-] = 1.6 \times 10^{-10}$$

Solving for $[Ag^+]$ and substituting $[Cl^-] = 2.0 \times 10^{-3}$ M:

$$[Ag^+] = \frac{1.6 \times 10^{-10}}{[Cl^-]} = \frac{1.6 \times 10^{-10}}{2.0 \times 10^{-3}} = 0.80 \times 10^{-7} = 8.0 \times 10^{-8} \text{ M}$$

    b. Here we have $[Pb^{2+}] \times [I^-]^2 = 1 \times 10^{-8}$.

$$[I^-]^2 = \frac{1 \times 10^{-8}}{[Pb^{2+}]} = \frac{1 \times 10^{-8}}{1 \times 10^{-2}} = 1 \times 10^{-6}$$

$$[I^-] = (1 \times 10^{-6})^{1/2} = 1 \times 10^{-3} \text{ M}$$

Exercise For $Mg(OH)_2$, $K_{sp} = 1 \times 10^{-11}$. Calculate $[Mg^{2+}]$ in a solution in which $[OH^-]$ = 1×10^{-4} M. Answer: 1×10^{-3} M.

In Chapter 16, we used the solubility rules (Table 16.3) to predict whether or not a precipitate will form when two solutions are mixed. That type of prediction, you will remember, is limited to the case where the ions involved are at a concentration of 0.1 M or greater. We can use K_{sp} values such as those given in Table 18.1 to make a more general type of prediction. Regardless of the concentrations of ions, the solubility product principle can be applied to determine whether precipitation will occur.

To make this prediction, we compare the *ion product, Q*, to the value of K_{sp}. The ion product is the product of the *actual concentrations* of ions in solution, each raised to a power equal to that in the K_{sp} expression. Thus

Q involves the initial concentrations; K_{sp}, the equilibrium concentrations

$$Q_{AgCl} = (\text{conc. } Ag^+) \times (\text{conc. } Cl^-)$$

$$Q_{PbI_2} = (\text{conc. } Pb^{2+}) \times (\text{conc. } I^-)^2$$

where the concentrations are those in the mixture under study, which is ordinarily not at equilibrium.

We can distinguish three cases:

1. If $Q < K_{sp}$, the solution is unsaturated and no precipitate forms.

2. If, by chance, $Q = K_{sp}$, the solution is just saturated.

3. If $Q > K_{sp}$, the solution is temporarily supersaturated. Excess solid will precipitate until the ion product becomes equal to K_{sp}, giving a saturated solution.

Example 18.2 Chromate ions, CrO_4^{2-}, are added to a solution in which the original concentration of Sr^{2+} is 1.2×10^{-3} M. Will a precipitate of $SrCrO_4$ form ($K_{sp} = 3.6 \times 10^{-5}$) when

a. conc. $CrO_4^{2-} = 2.0 \times 10^{-2}$ M?   b. conc. $CrO_4^{2-} = 4.0 \times 10^{-2}$ M?

Solution Here the ion product is (conc. Sr^{2+}) × (conc. CrO_4^{2-}). This is to be compared to K_{sp}, 3.6×10^{-5}.

a. (conc. Sr^{2+}) × (conc. CrO_4^{2-}) = $(1.2 \times 10^{-3}) \times (2.0 \times 10^{-2}) = 2.4 \times 10^{-5}$

The ion product is less than 3.6×10^{-5}. The solution is unsaturated and no precipitate forms.

b. (conc. Sr^{2+}) × (conc. CrO_4^{2-}) = $(1.2 \times 10^{-3}) \times (4.0 \times 10^{-2}) = 4.8 \times 10^{-5}$

The ion product is greater than 3.6×10^{-5}. A precipitate forms, reducing the concentrations of the ions until their product becomes equal to K_{sp}.

Exercise Will PbI_2 ($K_{sp} = 1 \times 10^{-8}$) precipitate from a solution in which the concentrations of Pb^{2+} and I^- are both 1×10^{-3} M? Answer: No.

In principle, at least,* the solubility of a slightly soluble ionic compound in pure water can be calculated from its K_{sp} value, or vice versa (Example 18.3).

Example 18.3
   a. The measured solubility of $AgC_2H_3O_2$ in water at 20°C is 0.045 mol/ℓ. Calculate K_{sp} for silver acetate.
   b. Calculate the solubility, in moles per liter, of $CaCO_3$ ($K_{sp} = 5 \times 10^{-9}$).

Solution
   a. Silver acetate dissolves in water according to the equation

$$AgC_2H_3O_2(s) \rightleftharpoons Ag^+(aq) + C_2H_3O_2^-(aq)$$

Note that the solubility product and the solubility are *not* equal

For every mole of $AgC_2H_3O_2$ that dissolves, one mole of Ag^+ and one mole of $C_2H_3O_2^-$ enter the solution. Hence, when 0.045 mol/ℓ of solid dissolves, the equilibrium concentrations of Ag^+ and $C_2H_3O_2^-$ must both be 0.045 M.

$$K_{sp} = [Ag^+] \times [C_2H_3O_2^-] = (0.045) \times (0.045) = 2.0 \times 10^{-3}$$

   b. Calcium carbonate dissolves in water according to the equation

$$CaCO_3(s) \rightleftharpoons Ca^{2+}(aq) + CO_3^{2-}(aq)$$

For every mole of $CaCO_3$ that dissolves, one mole of Ca^{2+} and one mole of CO_3^{2-} are formed. Hence, if we let s be the solubility of $CaCO_3$ (that is, s = no. moles $CaCO_3$ dissolving per liter), then $[Ca^{2+}] = [CO_3^{2-}] = s$.

Hence, $K_{sp} = [Ca^{2+}] \times [CO_3^{2-}] = (s) \times (s) = s^2$.

Solving for s:

$$s = (K_{sp})^{1/2} = (5 \times 10^{-9})^{1/2} = (50 \times 10^{-10})^{1/2} = 7 \times 10^{-5} \text{ M}$$

Exercise   Suppose we represent by s the solubility, in moles per liter, of PbI_2:

$$PbI_2(s) \rightleftharpoons Pb^{2+}(aq) + 2 I^-(aq); K_{sp} = [Pb^{2+}] \times [I^-]^2$$

Express, in terms of s: $[Pb^{2+}]$; $[I^-]$; K_{sp} of PbI_2. Answer: s; 2s; $4s^3$.

In Example 18.3, we found that, for $CaCO_3$,

$$K_{sp} = s^2 \quad \text{or} \quad s = (K_{sp})^{1/2}$$

where s is the solubility in moles per liter. This same relation holds for any slightly soluble ionic compound of formula MX (one mole of cations per mole of anions). For example, K_{sp} and s are related in this way for AgCl, $SrCrO_4$, or $CaSO_4$.

---

*Experimentally, we usually find that the solubility is slightly greater than that predicted from K_{sp}. For example, the measured solubility of PbI_2 in water at 25°C is 1.7×10^{-3} M. This compares to a value of 1.3×10^{-3} M, calculated from the K_{sp} of PbI_2. The reason for this is that some of the lead in PbI_2 goes into solution in the form of species other than Pb^{2+}. For example, we can detect ions such as $Pb(OH)^+$ and PbI^+ in a water solution of lead iodide.

TABLE 18.2   RELATION BETWEEN K_{sp} AND SOLUBILITY, s

SOLID	K_{sp} EXPRESSION	EQUIL. CONC. IN SOLUTION		RELATION BETWEEN K_{sp} AND s
		Cation	Anion	
AgCl	$[Ag^+] \times [Cl^-]$	s	s	$K_{sp} = s^2; s = (K_{sp})^{1/2}$
CoS	$[Co^{2+}] \times [S^{2-}]$	s	s	$K_{sp} = s^2; s = (K_{sp})^{1/2}$
PbI_2	$[Pb^{2+}] \times [I^-]^2$	s	2s	$K_{sp} = 4s^3; s = \left(\dfrac{K_{sp}}{4}\right)^{1/3}$
Ag_2CrO_4	$[Ag^+]^2 \times [CrO_4^{2-}]$	2s	s	$K_{sp} = 4s^3; s = \left(\dfrac{K_{sp}}{4}\right)^{1/3}$
$Al(OH)_3$	$[Al^{3+}] \times [OH^-]^3$	s	3s	$K_{sp} = 27s^4; s = \left(\dfrac{K_{sp}}{27}\right)^{1/4}$

The nature of the K_{sp}-solubility relations changes when we go to a different type of electrolyte. In the exercise following Example 18.3, you found (we hope) that for PbI_2

$$K_{sp} = 4s^3 \quad \text{or} \quad s = \left(\frac{K_{sp}}{4}\right)^{1/3}$$

This relation holds for any slightly soluble ionic solid of the type MX_2 (PbI_2, $Mg(OH)_2$) or M_2X (Ag_2CrO_4). Table 18.2 summarizes relationships of this type.

The Common Ion Effect

As we have seen, in a solution saturated with calcium carbonate, the following equilibrium exists:

$$CaCO_3(s) \rightleftharpoons Ca^{2+}(aq) + CO_3^{2-}(aq) \tag{18.5}$$

It is possible to shift the position of this equilibrium in various ways. One way is to add a concentrated solution of Na_2CO_3. By Le Châtelier's Principle, the equilibrium shifts to the left to consume some of the added CO_3^{2-} ions. The same effect would be observed if we added Ca^{2+} ions from a concentrated solution of $CaCl_2$. In either case, we find that $CaCO_3$ precipitates out of solution. This phenomenon is known as the common ion effect (Fig. 18.2). A precipitate forms when we add an ion common to the slightly soluble solid, in this case Ca^{2+} or CO_3^{2-}.

Because of the common ion effect, the solubility of a compound such as $CaCO_3$ is lowered in a solution which originally contains Ca^{2+} or CO_3^{2-} ions. For example, in a solution 0.10 M in CO_3^{2-}, the equilibrium concentration of Ca^{2+} is only 5×10^{-8} M:

$$[Ca^{2+}] = \frac{K_{sp} \text{ of } CaCO_3}{[CO_3^{2-}]} = \frac{5 \times 10^{-9}}{1 \times 10^{-1}} = 5 \times 10^{-8} \text{ M}$$

This means that only about 5×10^{-8} mol of $CaCO_3$ can dissolve in 1 ℓ of 0.10 M Na_2CO_3 solution. We would say that, in 0.10 M Na_2CO_3, the solubility of $CaCO_3$ is 5×10^{-8} M. This is less than 0.1% of the solubility of $CaCO_3$ in pure water, 7×10^{-5} M.

A salt like $CaCO_3$ is much *less* soluble in a solution containing a common ion than it is in plain water

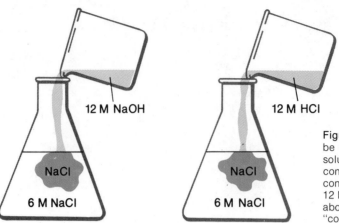

12 M NaOH

NaCl

6 M NaCl

12 M HCl

NaCl

6 M NaCl

Figure 18.2 Sodium chloride can be precipitated from its saturated solution (6 M) by adding a solution containing either Na^+ or Cl^- ions at a concentration greater than 6 M. Both 12 M NaOH and 12 M HCl will bring about this reaction, illustrating the "common ion" effect.

18.2 THE EQUILIBRIUM CONSTANT FOR DISSOCIATION OF A WEAK ACID, K_a

As noted in Chapter 17, most acids are "weak" in the sense that they are only partially dissociated in water. The dissociation of a weak acid, HB, is expressed by the general equation

$$HB(aq) \rightleftharpoons H^+(aq) + B^-(aq) \tag{18.6}$$

The products are a proton, H^+, and the conjugate base, B^-, of the weak acid.

Following the rules given in Chapter 13, we can write an expression for the equilibrium constant for Reaction 18.6:

$$K_a = \frac{[H^+] \times [B^-]}{[HB]} \tag{18.7}$$

The equilibrium constant, K_a, is called the *ionization constant* or *acid dissociation constant* of the weak acid HB.

To illustrate the form taken by K_a for various weak acids, consider the three species HF, NH_4^+, and HSO_3^-:

Weak acid		Conjugate base	

$$HF(aq) \rightleftharpoons H^+(aq) + F^-(aq) \qquad K_a = \frac{[H^+] \times [F^-]}{[HF]}$$

These equations follow from the general law of chemical equilibrium

$$NH_4^+(aq) \rightleftharpoons H^+(aq) + NH_3(aq) \qquad K_a = \frac{[H^+] \times [NH_3]}{[NH_4^+]}$$

$$HSO_3^-(aq) \rightleftharpoons H^+(aq) + SO_3^{2-}(aq) \qquad K_a = \frac{[H^+] \times [SO_3^{2-}]}{[HSO_3^-]}$$

The concentrations that appear in the expression for K_a are, as always, *equilibrium concentrations in moles per liter*.

The K_a values of some common weak acids are given in Table 18.3. These constants are a measure of the extent to which the acid dissociates in water solution. **The**

TABLE 18.3   DISSOCIATION CONSTANTS OF WEAK ACIDS AND BASES

	ACID		K_a	BASE	K_b
Sulfurous acid	H_2SO_3		1.7×10^{-2}	HSO_3^-	5.9×10^{-13}
Hydrogen sulfate ion	HSO_4^-		1.2×10^{-2}	SO_4^{2-}	8.3×10^{-13}
Phosphoric acid	H_3PO_4		7.5×10^{-3}	$H_2PO_4^-$	1.3×10^{-12}
Hydrofluoric acid	HF		7.0×10^{-4}	F^-	1.4×10^{-11}
Nitrous acid	HNO_2		4.5×10^{-4}	NO_2^-	2.2×10^{-11}
Formic acid	$HCHO_2$		1.8×10^{-4}	CHO_2^-	5.6×10^{-11}
Benzoic acid	$HC_7H_5O_2$		6.6×10^{-5}	$C_7H_5O_2^-$	1.5×10^{-10}
Acetic acid	$HC_2H_3O_2$		1.8×10^{-5}	$C_2H_3O_2^-$	5.6×10^{-10}
Propionic acid	$HC_3H_5O_2$		1.4×10^{-5}	$C_3H_5O_2^-$	7.1×10^{-10}
Carbonic acid	H_2CO_3		4.2×10^{-7}	HCO_3^-	2.4×10^{-8}
Hydrogen sulfide	H_2S		1×10^{-7}	HS^-	1×10^{-7}
Dihydrogen phosphate ion	$H_2PO_4^-$		6.2×10^{-8}	HPO_4^{2-}	1.6×10^{-7}
Hydrogen sulfite ion	HSO_3^-		5.6×10^{-8}	SO_3^{2-}	1.8×10^{-7}
Hypochlorous acid	HClO		3.2×10^{-8}	ClO^-	3.1×10^{-7}
Boric acid	H_3BO_3		5.8×10^{-10}	$H_2BO_3^-$	1.7×10^{-5}
Ammonium ion	NH_4^+		5.6×10^{-10}	NH_3	1.8×10^{-5}
Hydrocyanic acid	HCN		4.0×10^{-10}	CN^-	2.5×10^{-5}
Hydrogen carbonate ion	HCO_3^-		4.8×10^{-11}	CO_3^{2-}	2.1×10^{-4}
Hydrogen phosphate ion	HPO_4^{2-}		1.7×10^{-12}	PO_4^{3-}	5.9×10^{-3}
Hydrogen sulfide ion	HS^-		1×10^{-13}	S^{2-}	1×10^{-1}

Decreasing Acid Strength (left margin, downward)   *Increasing Basic Strength* (right margin, downward)

For the acids: $HB(aq) \rightleftarrows H^+(aq) + B^-(aq)$ $K_a = \dfrac{[H^+] \times [B^-]}{[HB]}$

For the bases: $B^-(aq) + H_2O \rightleftarrows HB(aq) + OH^-(aq)$ $K_b = \dfrac{[HB] \times [OH^-]}{[B^-]}$

smaller the dissociation constant, the weaker the acid. Notice that in the table the acids are arranged in decreasing order of K_a. For example, acetic acid, $HC_2H_3O_2$ ($K_a = 1.8 \times 10^{-5}$) lies below hydrofluoric acid, HF ($K_a = 7.0 \times 10^{-4}$). This means that at a given concentration, let us say 0.10 M:

—$[H^+]$ is lower for $HC_2H_3O_2$ than for HF (1.3×10^{-3} vs. 8.0×10^{-3} M).

—pH is higher for $HC_2H_3O_2$ than for HF (2.87 vs. 2.10).

Species which contain more than one ionizable hydrogen atom, such as phosphoric acid, H_3PO_4, dissociate stepwise. For each step, we can write a dissociation constant:

$$H_3PO_4(aq) \rightleftarrows H^+(aq) + H_2PO_4^-(aq); \quad K_{a_1} = 7.5 \times 10^{-3}$$

$$H_2PO_4^-(aq) \rightleftarrows H^+(aq) + HPO_4^{2-}(aq); \quad K_{a_2} = 6.2 \times 10^{-8}$$

$$HPO_4^{2-}(aq) \rightleftarrows H^+(aq) + PO_4^{3-}(aq); \quad K_{a_3} = 1.7 \times 10^{-12}$$

You will note from this example (and others listed in Table 18.3) that dissociation constants for such acids decrease with each successive step. Hence, for acids such as H_3PO_4 or H_2SO_3, virtually all the H^+ ions come from the first dissociation.

The value of K_a for a weak acid must be found experimentally. There are several ways to do this. Perhaps the most common involves measuring $[H^+]$ or pH in a solution prepared by dissolving a known amount of the weak acid to form a given volume of solution. The calculation is illustrated in Example 18.4.

Example 18.4   Lactic acid is found in sour milk (*L. lactis,* milk) and sore muscles. Its molecular formula is $HC_3H_5O_3$ (MM = 90.1). To save space, we will abbreviate this as HLac. A solution is prepared by adding 9.01 g (0.100 mol) of lactic acid, HLac, to enough water to give a liter of solution. In this solution, $[H^+]$ is measured to be 3.67×10^{-3} M. Calculate K_a for lactic acid.

Solution   For lactic acid, we have

$$HLac(aq) \rightleftharpoons H^+(aq) + Lac^-(aq); K_a = \frac{[H^+] \times [Lac^-]}{[HLac]}$$

To calculate K_a, we need the *equilibrium* concentrations of H^+, Lac^-, and *undissociated* HLac. The $[H^+]$ has been measured as 3.67×10^{-3} M. From the dissociation equation we note that one mole of lactate ion, Lac^-, is produced with every mole of H^+ ion. Therefore, in the acid solution,

$$[H^+] = [Lac^-] = 3.67 \times 10^{-3} \text{ M}$$

To obtain the equilibrium concentration of HLac, we note that a mole of HLac must be consumed for every mole of H^+ produced. Therefore, by subtracting $[H^+]$ from the original concentration of HLac, we obtain [HLac] at equilibrium:

$$[HLac] = \text{orig. conc. HLac} - [H^+]$$

$$= 0.100 \text{ M} - 0.00367 \text{ M} = 0.096 \text{ M}$$

Substituting in the expression for K_a:

$$K_a = \frac{[H^+] \times [Lac^-]}{[HLac]} = \frac{(3.67 \times 10^{-3}) \times (3.67 \times 10^{-3})}{0.096} = 1.4 \times 10^{-4}$$

Exercise   A 0.100 M HF solution has a $[H^+]$ of 8.0×10^{-3} M. Calculate the K_a of HF. Answer: 7.0×10^{-4}.

> The H⁺ comes from the HLac

18.3   SOME APPLICATIONS OF K_a

The dissociation constant of a weak acid can be used for several purposes. In this section, we will consider two of its more important uses:

1. The determination of $[H^+]$ in a buffer solution, where we know or can calculate both the concentration of the weak acid, HB, and that of its conjugate base, B^-.

2. The determination of $[H^+]$ in a solution prepared by dissolving the weak acid HB in water.

Buffers; pH Control Systems

Some solutions have a pH which is resistant to change upon addition of H^+ or OH^- ions. Such solutions are called buffers. Blood is an example of a buffer. Suppose we add 1 cm³ of 10 M HCl or NaOH (0.010 mol H^+ or OH^- ions) to a liter of blood. Its pH changes by less than 0.1 unit from its normal value of 7.4. In contrast, addition of these same quantities of acid or base to a liter of pure water changes its pH by about 5 units (from 7 to 2 with 0.010 mol HCl, from 7 to 12 with 0.010 mol NaOH).

A buffer ordinarily contains two species. One of these can react with H^+, the other with OH^- ions. The two species must not react with each other. Typically, a buffer is a mixture of a weak acid, HB (which reacts with OH$^-$ ions) and its conjugate base, B$^-$ (which reacts with H$^+$ ions). Thus *a buffer is a conjugate acid/base pair.*

A simple buffer can be made by adding acetic acid, $HC_2H_3O_2$, to a solution of sodium acetate (Na^+, $C_2H_3O_2^-$ ions). When a strong base such as NaOH is added, the OH$^-$ ions react with acetic acid molecules:

$$HC_2H_3O_2(aq) + OH^-(aq) \rightarrow C_2H_3O_2^-(aq) + H_2O$$

Addition of a strong acid such as HCl results in reaction of the H$^+$ ions of the acid with the acetate ions of the buffer:

A buffer solution will react with added H$^+$ and with added OH$^-$ ions

$$H^+(aq) + C_2H_3O_2^-(aq) \rightarrow HC_2H_3O_2(aq)$$

As pointed out in Chapter 17, both of these reactions go virtually to completion. Hence the added OH$^-$ or H$^+$ ions are consumed. They are unable to cause the drastic pH change that would occur if they were added to water or an unbuffered solution (see color plate 16, center of book).

To find [H$^+$] in a buffer, it is convenient to solve Equation 18.7,

$$K_a = \frac{[H^+] \times [B^-]}{[HB]}$$

for the concentration of H$^+$:

$$[H^+] = K_a \times \frac{[HB]}{[B^-]} \qquad (18.8)$$

This equation tells us that the concentration of H$^+$ depends upon two factors:
—the value for the dissociation constant of the weak acid, K_a.
—the ratio of the concentrations of the weak acid and its conjugate base, [HB]/[B$^-$].
Example 18.5 illustrates the use of Equation 18.8.

Example 18.5   A liter of lactic acid (HLac)–lactate (Lac$^-$) buffer is prepared using 1.00 mol HLac and 1.00 mol of NaLac. K_a of lactic acid is 1.4×10^{-4}. What is the [H$^+$] in this buffer?

Solution   In this case [HLac] = [Lac$^-$] = 1.00 M. Applying Equation 18.8:

The HLac and Lac$^-$ do not react with each other

$$[H^+] = K_a \times \frac{[HLac]}{[Lac^-]} = 1.4 \times 10^{-4} \times \frac{1.00}{1.00} = 1.4 \times 10^{-4} \text{ M}$$

Exercise   What is the pH of this buffer? Answer: 3.85.

The pH of a buffer changes slightly if a strong acid or strong base is added to it. By adding base, we convert a small amount of the weak acid HB to its conjugate base, B$^-$. Addition of strong acid converts a small amount of B$^-$ to HB. In both cases, the ratio [HB]/[B$^-$] changes; this in turn changes the [H$^+$] or pH of the buffer. The effect is ordinarily small, as indicated by Example 18.6.

Example 18.6   Consider the buffer described in Example 18.5, where [HLac] = [Lac⁻] = 1.00 M (K_a HLac = 1.4×10^{-4}). You will recall that in this buffer the pH is 3.85. Suppose now that we add a strong acid or strong base to this buffer. Calculate

a. [HLac], [Lac⁻], [H⁺], and pH after addition of 0.10 mol H⁺.

b. [HLac], [Lac⁻], [H⁺], and pH after addition of 0.10 mol OH⁻.

Solution

a. When H⁺ ions are added, the following reaction occurs:

$$H^+(aq) + Lac^-(aq) \rightarrow HLac(aq)$$

The addition of 0.10 mol H⁺ produces 0.10 mol HLac and consumes 0.10 mol Lac⁻. Originally, we had 1.00 mol of both HLac and Lac⁻.

Final = Initial + Change

$$[HLac] = 1.00 \text{ M} + 0.10 \text{ M} = 1.10 \text{ M}$$

$$[Lac^-] = 1.00 \text{ M} - 0.10 \text{ M} = 0.90 \text{ M}$$

$$[H^+] = K_a \times \frac{[HLac]}{[Lac^-]} = 1.4 \times 10^{-4} \times \frac{1.10}{0.90} = 1.7 \times 10^{-4}; \text{ pH} = 3.77$$

Note that the pH decreases by only 0.08 unit. Addition of the same amount of H⁺ to a liter of pure water would decrease the pH by 6 units, from 7 to 1.

b. This time, the reaction is

$$HLac(aq) + OH^-(aq) \rightarrow Lac^-(aq) + H_2O$$

Here, the concentration of HLac *decreases* by 0.10 M; that of Lac⁻ *increases* by 0.10 M.

$$[HLac] = 1.00 \text{ M} - 0.10 \text{ M} = 0.90 \text{ M}$$

$$[Lac^-] = 1.00 \text{ M} + 0.10 \text{ M} = 1.10 \text{ M}$$

$$[H^+] = K_a \times \frac{[HLac]}{[Lac^-]} = 1.4 \times 10^{-4} \times \frac{0.90}{1.10} = 1.2 \times 10^{-4}; \text{ pH} = 3.92$$

Again, the pH changes by less than 0.1 unit.

Exercise   Suppose that, in (a), we had added 0.50 mol H⁺ instead of 0.10 mol. What would be the final pH? Answer: 3.38.

In preparing buffer solutions, it is important to keep in mind two principles of buffer action:

1. *A buffer has a limited capacity to absorb H⁺ or OH⁻ ions.* To illustrate this effect, consider the exercise following Example 18.6. Addition of 0.50 mol of H⁺ to this buffer produced a rather large change in pH, decreasing it from 3.85 to 3.38. Addition of 1.00 mol or more of H⁺ would have an even more drastic effect. Indeed, it would consume all the Lac⁻ ions, thereby destroying the buffer (Fig. 18.3).

2. A buffer has its greatest capacity when the concentrations of weak acid and conjugate base are equal. *Hence, to make a buffer, we choose a weak acid whose K_a is close to the [H⁺] we want.* The lactic acid–sodium lactate buffer referred to in Example 18.6 works well if we wish to maintain [H⁺] around 10^{-4} (pH ≈ 4). If we wanted a pH of 7, this particular buffer system would be virtually useless. The concentration of lactic acid would be so low, about 0.001 M, that the system would have almost no capac-

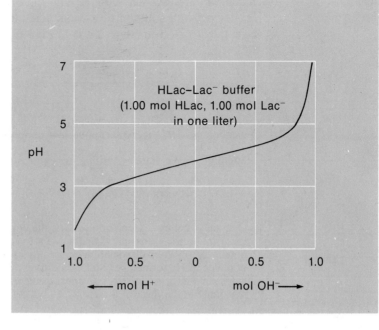

Figure 18.3 Effect of adding either acid or base to a lactic acid–sodium lactate buffer. The buffer itself has a pH of 3.85. Addition of OH^- ions produces a small increase in pH until most of the lactic acid is gone. Then the pH rises sharply (upper right of curve). The buffer behaves similarly toward addition of H^+ ions (lower left). The pH decreases slowly until nearly all the lactate ions are gone, then drops sharply.

ity for absorbing strong base. To maintain a pH of 7, we would choose a weak acid-conjugate base pair where K_a of the acid is around 10^{-7}. For example, $H_2PO_4^-$ and HPO_4^{2-}

Buffers are widely used in both chemistry and biology to maintain constant pH. Many natural systems are also buffered. We have mentioned blood, with a pH of 7.4. This value is maintained by a mixture of different buffers. One of the most important of these is the HCO_3^-–H_2CO_3 system. In blood, there are about 20 HCO_3^- ions for every H_2CO_3 molecule. The large amount of HCO_3^- gives blood a large capacity for absorbing strong acid without appreciable change in pH. This is essential because many cellular processes produce acidic waste products. Other crucial equilibria such as those involved in oxygen transport to cells and CO_2 transport to the lungs require close control of blood pH.

Biologists often calculate the pH of buffers by using a relation known as the Henderson-Hasselbalch equation. This is readily obtained from Equation 18.8. Taking the base 10 log of both sides gives

$$\log_{10} [H^+] = \log_{10} K_a + \log_{10} \frac{[HB]}{[B^-]}$$

Multiplying both sides by -1 and remembering that $-\log x = \log 1/x$, we obtain the equation:

$$pH = -\log_{10} K_a + \log_{10} \frac{[B^-]}{[HB]} = pK_a + \log_{10} \frac{[B^-]}{[HB]}$$

where pK_a is the negative base 10 logarithm of the dissociation constant of the weak acid in the buffer system.

Determination of H^+ in a Solution of a Weak Acid

In Example 18.4 we calculated K_a, knowing $[H^+]$ and the original concentration of HB in a solution prepared by adding a weak acid to water. It is possible to turn this

439

calculation around. Knowing K_a and the original concentration of HB, we can obtain $[H^+]$ or pH in a solution of a weak acid. Usually, the calculation is relatively simple (Example 18.7). Sometimes, it is more difficult (Example 18.8).

Example 18.7. Nicotinic acid, $C_6H_5O_2N$ ($K_a = 1.4 \times 10^{-5}$), is another name for niacin, an important vitamin. Determine $[H^+]$ in a solution prepared by dissolving 0.10 mol of nicotinic acid, HNic, to form one liter of solution.

Solution We start by writing the expression for the dissociation of this weak acid:

$$HNic(aq) \rightleftharpoons H^+(aq) + Nic^-(aq); \quad K_a = \frac{[H^+] \times [Nic^-]}{[HNic]} = 1.4 \times 10^{-5}$$

Originally, before dissociation occurs, the concentration of HNic is 0.10 M. At that point, there are no Nic^- ions and virtually no H^+ ions (neglecting the few produced by the dissociation of H_2O). When dissociation occurs, this situation changes. We will let x be the number of moles of H^+ formed per liter. According to the equation for dissociation, 1 mol of Nic^- is formed and 1 mol of HNic is consumed for every mole of H^+ formed. Summarizing this reasoning in the form of a table:

	Orig. conc. (M)	Change in conc. (M)	Equilibrium conc. (M)
H^+	0.00	$+x$	x
Nic^-	0.00	$+x$	x
HNic	0.10	$-x$	$0.10 - x$

The entries in the middle column follow from the dissociation equation

Substituting into the expression for K_a:

$$\frac{(x)(x)}{0.10 - x} = 1.4 \times 10^{-5}$$

This is a quadratic equation. It could be rearranged to the form $ax^2 + bx + c = 0$ and solved for x, using the quadratic formula. Such a procedure is time-consuming and, in this case, unnecessary. Remember that nicotinic acid is a weak acid, only slightly dissociated in water. The equilibrium concentration of HNic, $0.10 - x$, is probably only very slightly less than its original concentration, 0.10 M. We therefore make the approximation $0.10 - x \approx 0.10$. This simplifies the equation considerably:

When you can make things simple, you might as well do so

$$\frac{(x)(x)}{0.10} = 1.4 \times 10^{-5}; \quad x^2 = 1.4 \times 10^{-6}$$

Taking square roots, we have

$$x = 1.2 \times 10^{-3} \text{ M} = [H^+] = [Nic^-]$$

$$[HNic] = 0.10 - x = 0.10 - 0.0012 = 0.10 \text{ M (two sig. fig.)}$$

The fact that $[H^+]$, 0.0012 M, is so much less than the original concentration of nicotinic acid, 0.10 M, justifies the approximation that $0.10 - x \approx 0.10$.

Exercise In Example 18.7, what percentage of the original concentration of HNic, 0.10 M, is the value calculated for $[H^+]$, 0.0012 M? Answer: 1.2.

In general, the value of K_a is seldom known to better than $\pm 5\%$. Hence, in the expression

$$K_a = \frac{x^2}{a - x}$$

where $x = [H^+]$ and $a = $ *original* concentration of weak acid, you can neglect the x in the denominator if doing so does not introduce an error of more than 5%. In other words,

if $x \leq 0.05a$, then $a - x \approx a$

In most of the problems you will work, this condition holds and the simplifying approximation is valid. Sometimes, however, you will find that the $[H^+]$ you calculate is greater than 5% of the original concentration of weak acid. If this happens, you can follow either of the approaches suggested in Example 18.8.

Example 18.8   Calculate $[H^+]$ in a 0.100 M solution of nitrous acid, HNO_2, for which $K_a = 4.5 \times 10^{-4}$.

Solution   Proceeding as in Example 18.7, we arrive at the equation

$$\frac{x^2}{0.100 - x} = 4.5 \times 10^{-4}$$

Making the same approximation as before, $0.100 - x \approx 0.100$

$$x^2 = 0.100 \times 4.5 \times 10^{-4} = 4.5 \times 10^{-5} = 45 \times 10^{-6}$$

$$x = 6.7 \times 10^{-3} \approx [H^+]$$

In this case, x is more than 5% of a. That is, the calculated $[H^+]$ is more than 5% of the original concentration of undissociated acid.

$$\frac{6.7 \times 10^{-3}}{0.100} \times 100 = 6.7\%$$

To obtain a better value for x, we have a choice of two approaches. We can use:

1. *The method of successive approximations.* We know that

$$[HNO_2] = 0.100 - x$$

Our first approximation was to take $x = 0$. We can do better by making a second approximation. Here, we use the value just calculated for x, 6.7×10^{-3}, or 0.0067. Now,

$$[HNO_2] = 0.100 - 0.0067 = 0.093 \text{ M}$$

Substituting in the expression for K_a:

$$\frac{x^2}{0.093} = 4.5 \times 10^{-4}; \quad x^2 = 4.2 \times 10^{-5} = 42 \times 10^{-6}$$

$$x = 6.5 \times 10^{-3} \text{ M} \approx [H^+]$$

This value is closer to the true $[H^+]$, since 0.093 M is a better approximation for $[HNO_2]$ than was 0.100 M. If you're still not satisfied, you can go one step further. Using 6.5×10^{-3} for x instead of 6.7×10^{-3}, you can recalculate $[HNO_2]$ and solve again for x. If you do you will find that your answer does not change. In other words, "you have gone about as far as you can go." This is generally true in calculations involving $[H^+]$ in a solution of a weak acid. Usually, the first approximation is sufficient; almost never do you have to go beyond a second approximation.

2. *The quadratic formula.* This gives an exact solution for x, but is very time-consuming. Here, we would rewrite the equation

$$\frac{x^2}{0.100 - x} = 4.5 \times 10^{-4}$$

in the form $ax^2 + bx + c = 0$. Doing this, we obtain

$$x^2 + (4.5 \times 10^{-4} \, x) - (4.5 \times 10^{-5}) = 0$$

So, $a = 1$; $b = 4.5 \times 10^{-4}$; $c = -4.5 \times 10^{-5}$.

Applying the quadratic formula:

$$x = \frac{-b \pm \sqrt{b^2 - 4ac}}{2a} = \frac{-4.5 \times 10^{-4} \pm \sqrt{(4.5 \times 10^{-4})^2 + (18.0 \times 10^{-5})}}{2}$$

If you carry out the arithmetic properly, you should get two answers for x:

$$x = 6.5 \times 10^{-3} \text{ and } -6.9 \times 10^{-3}$$

The second answer is physically ridiculous; the concentration of H^+ cannot be a negative quantity. The first answer is the same one we obtained by the method of successive approximations.

> **Exercise** Suppose the original concentration of HNO_2 in this example had been 1.00 M instead of 0.100 M. Making the approximation $[HNO_2] = 1.00$ M, what would you calculate for $[H^+]$? Would you need to go beyond this first approximation? Answer: 0.021 M; no.

It is very easy to make arithmetic errors with the quadratic formula

18.4 THE EQUILIBRIUM CONSTANT FOR THE REACTION OF A WEAK BASE WITH WATER, K_b

As we saw in Chapter 17, weak bases remove a H^+ ion from water. The products are the conjugate acid of the weak base and an OH^- ion, which makes the solution basic. Perhaps the most common weak base is the NH_3 molecule:

$$NH_3(aq) + H_2O \rightleftharpoons NH_4^+(aq) + OH^-(aq) \tag{18.9}$$

We can write an equilibrium constant expression for Reaction 18.9 in the usual way:

$$K_b = \frac{[NH_4^+] \times [OH^-]}{[NH_3]} = 1.8 \times 10^{-5} \tag{18.10}$$

The concentration of water, which remains virtually constant at about 55 M, is incorporated into the equilibrium constant. The constant K_b is often called the *base dissociation constant* of NH_3. The fact that it is a small number means that Reaction 18.9 proceeds only to a small extent when ammonia is added to water.

Expressions for K_b for other weak bases are written in a very similar way. For example, consider the acetate ion, $C_2H_3O_2^-$:

$$C_2H_3O_2^-(aq) + H_2O \rightleftharpoons HC_2H_3O_2(aq) + OH^-(aq) \tag{18.11}$$

$$K_b = \frac{[HC_2H_3O_2] \times [OH^-]}{[C_2H_3O_2^-]} = 5.6 \times 10^{-10} \tag{18.12}$$

The fact that K_b for $C_2H_3O_2^-$ is a smaller number than K_b for NH_3 (5.6×10^{-10} vs. 1.8×10^{-5}) means that the acetate ion is a weaker base than the NH_3 molecule. Values of K_b for a variety of weak bases are listed in Table 18.3, p. 435, opposite the K_a value for the conjugate weak acids.

Relation Between K_a and K_b

Looking at Table 18.3, we see that there is an inverse relation between K_a of a weak acid and K_b of its conjugate base. As acid strength decreases (smaller K_a), base strength increases (larger K_b). We can derive a simple equation relating these two quantities. To do so, we start by writing general expressions for K_a and K_b. Using HB to represent the weak acid and B^- for the weak base:

$$HB(aq) \rightleftharpoons H^+(aq) + B^-(aq); \quad K_a = \frac{[H^+] \times [B^-]}{[HB]}$$

In a solution, both of these equations must be satisfied

$$B^-(aq) + H_2O \rightleftharpoons HB(aq) + OH^-(aq); \quad K_b = \frac{[HB] \times [OH^-]}{[B^-]}$$

Let us now multiply the two expressions together:

$$K_a \times K_b = \frac{[H^+] \times [B^-]}{[HB]} \times \frac{[HB] \times [OH^-]}{[B^-]}$$

Canceling terms and simplifying, we obtain

$$K_a \times K_b = [H^+] \times [OH^-]$$

But, as we saw in Chapter 17, the product $[H^+] \times [OH^-]$ is a constant (K_w), with a value of 1.0×10^{-14} at 25°C. Hence

$$K_a \times K_b = K_w = 1.0 \times 10^{-14} \tag{18.13}$$

Applying Equation 18.13 to several species from Table 18.3, we obtain

Acid	Base	K_a	K_b	$K_a \times K_b$
$HC_2H_3O_2$	$C_2H_3O_2^-$	1.8×10^{-5}	5.6×10^{-10}	1.0×10^{-14}
HCN	CN^-	4.0×10^{-10}	2.5×10^{-5}	1.0×10^{-14}
HCO_3^-	CO_3^{2-}	4.8×10^{-11}	2.1×10^{-4}	1.0×10^{-14}

Equation 18.13 allows us to calculate K_b for a weak base if K_a is known, or vice versa (Example 18.9).

Example 18.9   Butyric acid, HBut, has a K_a of 2.0×10^{-5}. Calculate K_b for the butyrate ion, But^-.

Solution   Applying Equation 18.13:

$$K_b \ But^- = \frac{1.0 \times 10^{-14}}{K_a \ HBut} = \frac{1.0 \times 10^{-14}}{2.0 \times 10^{-5}} = 0.50 \times 10^{-9} = 5.0 \times 10^{-10}$$

Exercise   Pyruvic acid, HPy, has a K_a of 3.3×10^{-3}. Which is the stronger acid, HBut or HPy? Which is the stronger base, But$^-$ or Py$^-$? Answer: HPy; But$^-$.

Determination of [OH$^-$] in a Solution of a Weak Base

We saw in Section 18.3 that the K_a of a weak acid can be used to calculate [H$^+$] in a solution of that acid. In a very similar way, we can use K_b to obtain [OH$^-$] in a solution of a weak base. If desired, we can go a step further and calculate [H$^+$] or pH (Example 18.10).

Example 18.10   For the butyrate ion, But$^-$, K_b is 5.0×10^{-10}. For a 1.0 M solution of sodium butyrate, NaBut, calculate
    a. [OH$^-$]   b. [H$^+$]   c. pH

Solution   When sodium butyrate dissolves in water, it dissociates completely into Na$^+$ and But$^-$ ions. The Na$^+$ ion does not react with water, but the butyrate ion does:

$$\text{But}^-(aq) + \text{H}_2\text{O} \rightleftharpoons \text{HBut}(aq) + \text{OH}^-(aq); \; K_b = \frac{[\text{HBut}] \times [\text{OH}^-]}{[\text{But}^-]} = 5.0 \times 10^{-10}$$

a. Let us represent [OH$^-$] by x. From the chemical equation, we see that when one mole of OH$^-$ is formed, one mole of HBut is also formed. At the same time, one mole of But$^-$ is consumed. Setting up a table, as in Example 18.7:

	Orig. conc.	Change	Equil. conc.
HBut	0.00	+x	x
OH$^-$	0.00	+x	x
But$^-$	1.0	$-$x	$1.0 - x$

Substituting into the expression for K_b:

$$\frac{x^2}{1.0 - x} = 5.0 \times 10^{-10}$$

Since K_b is so small, it seems safe to make the approximation that $1.0 - x \approx 1.0$. In this case $x^2 = 5.0 \times 10^{-10}$; $x = [\text{OH}^-] = 2.2 \times 10^{-5}$. (Note that x is far less than 5% of the original concentration of But$^-$, so the approximation is valid.)

b. Recall from Chapter 17 that $[\text{H}^+] \times [\text{OH}^-] = 1.0 \times 10^{-14}$.

So $[\text{H}^+] = \dfrac{1.0 \times 10^{-14}}{2.2 \times 10^{-5}} = 4.5 \times 10^{-10}$ M

c. pH $= -\log_{10} [\text{H}^+] = -\log_{10} (4.5 \times 10^{-10}) = -(0.65 - 10.00) = 9.35$

Exercise   What is [OH$^-$] in a 0.10 M solution of a weak base which has a K_b of 1.0×10^{-9}? Answer: 1.0×10^{-5} M.

18.5   RELATIONS BETWEEN EQUILIBRIUM CONSTANTS

In Section 18.4 we showed that for a conjugate acid-base pair

$$K_a \times K_b = 1.0 \times 10^{-14}$$

This is a very useful relationship. Knowing one of the quantities K_a or K_b, we can calculate the other directly (Example 18.9).

This kind of situation arises quite frequently in solution chemistry. Often we find that the equilibrium constant we need for a reaction can be determined from the known equilibrium constants for other, closely related reactions. In this section, we will look at two general relations which allow us to make calculations of this type.

Forward and Reverse Reactions

We can calculate the equilibrium constant, K, for a reaction if we know the equilibrium constant, K', for the reverse reaction. To do this, we use the general principle that K is the reciprocal of K',

$$K = \frac{1}{K'} \tag{18.14}$$

where K' is the equilibrium constant for the reaction written in the reverse direction.

To illustrate the use of this "reciprocal relation," suppose we want to know K for the neutralization reaction

$$H^+(aq) + OH^-(aq) \rightleftharpoons H_2O; \quad K = ? \tag{18.15}$$

Recall that, for the reverse reaction, the ionization of water

$$H_2O \rightleftharpoons H^+(aq) + OH^-(aq); \quad K_w = 1.0 \times 10^{-14} \tag{18.16}$$

Hence the equilibrium constant for Reaction 18.15 should be the reciprocal of that for 18.16. That is,

$$K = \frac{1}{1.0 \times 10^{-14}} = 1.0 \times 10^{14}$$

The validity of the reciprocal relation can be demonstrated by writing the expressions for K and K_w and multiplying:

$$K = \frac{1}{[H^+] \times [OH^-]}; \quad K_w = [H^+] \times [OH^-]$$

$$K \times K_w = \frac{[H^+] \times [OH^-]}{[H^+] \times [OH^-]} = 1; \quad K = 1/K_w$$

Example 18.11 Consider the reaction of the weak acid HF with a strong base:

$$HF(aq) + OH^-(aq) \rightleftharpoons F^-(aq) + H_2O$$

Using Table 18.3, find the equilibrium constant for the reverse reaction and then calculate K for this reaction.

You need to learn that
a reaction like this is
associated with K_b

Solution   Looking at this equation, we see that it is the reverse of the reaction of the weak base, F^-, with water:

$$F^-(aq) + H_2O \rightleftharpoons HF(aq) + OH^-(aq)$$

From Table 18.3, we see that $K_b\ F^- = 1.4 \times 10^{-11}$. Hence, for the reaction of HF with OH^-,

$$K = \frac{1}{K_b\ F^-} = \frac{1}{1.4 \times 10^{-11}} = 7.1 \times 10^{10}$$

Exercise   Given that K_a for NH_4^+ is 5.6×10^{-10}, calculate K for the reaction of NH_3 with a strong acid: $NH_3(aq) + H^+(aq) \rightleftharpoons NH_4^+(aq)$. Answer: 1.8×10^9.

We have just used the reciprocal relation to calculate K for three types of acid-base reactions discussed in Chapter 17:

strong acid–strong base: $H^+(aq) + OH^-(aq) \rightleftharpoons H_2O$; $K = 1.0 \times 10^{14}$

weak acid–strong base: $HF(aq) + OH^-(aq) \rightleftharpoons F^-(aq) + H_2O$; $K = 7.1 \times 10^{10}$

weak base–strong acid: $NH_3(aq) + H^+(aq) \rightleftharpoons NH_4^+(aq)$; $K = 1.8 \times 10^9$

Notice that, in each case, K is a very large number. This explains why these reactions, and others of the same type, go virtually to completion.

Multiple Equilibria

A relation which is very useful in dealing with solution equilibria states that:
If a reaction can be expressed as the sum of two other reactions, K for the overall reaction is the product of the equilibrium constants for the individual reactions. That is,

if   Reaction 3 = Reaction 1 + Reaction 2

then $K_3 = K_1 \times K_2$ (18.17)

To illustrate the use of this principle, often referred to as the "*rule of multiple equilibria*," consider the reactions for the stepwise dissociation of the weak acid H_2S:

$$H_2S(aq) \rightleftharpoons H^+(aq) + HS^-(aq); K_1 = 1 \times 10^{-7}$$

$$HS^-(aq) \rightleftharpoons H^+(aq) + S^{2-}(aq); K_2 = 1 \times 10^{-13}$$

If we add these two equations, the HS^- cancels and we obtain

$$H_2S(aq) \rightleftharpoons 2\ H^+(aq) + S^{2-}(aq); K = ?$$ (18.18)

Using the rule of multiple equilibria, we can calculate the equilibrium constant for Reaction 18.18:

$$K = K_1 \times K_2 = 1 \times 10^{-20}$$ (18.19)

The validity of this rule can be shown by working with the expressions for K_1, K_2, and K:

$$K_1 = \frac{[H^+] \times [HS^-]}{[H_2S]}; \quad K_2 = \frac{[H^+] \times [S^{2-}]}{[HS^-]}; \quad K = \frac{[H^+]^2 \times [S^{2-}]}{[H_2S]}$$

Multiplying K_1 by K_2, we obtain the expression just written for K:

$$K_1 \times K_2 = \frac{[H^+] \times [HS^-]}{[H_2S]} \times \frac{[H^+] \times [S^{2-}]}{[HS^-]} = \frac{[H^+]^2 \times [S^{2-}]}{[H_2S]} = K$$

Example 18.12   Given that the first and second dissociation constants of H_2CO_3 are 4.2×10^{-7} and 4.8×10^{-11}, calculate K for the reaction

$$H_2CO_3(aq) \rightleftharpoons 2\,H^+(aq) + CO_3{}^{2-}(aq)$$

Solution   The expression for the first and second dissociations are

Many complex equilibria can be dealt with by this approach

$$H_2CO_3(aq) \rightleftharpoons H^+(aq) + HCO_3{}^-(aq); \quad K_1 = 4.2 \times 10^{-7}$$

$$HCO_3{}^-(aq) \rightleftharpoons H^+(aq) + CO_3{}^{2-}(aq); \quad K_2 = 4.8 \times 10^{-11}$$

If we add these two equations, the $HCO_3{}^-$ ions cancel and we obtain

$$H_2CO_3(aq) \rightleftharpoons 2\,H^+(aq) + CO_3{}^{2-}(aq)$$

The equilibrium constant K for this reaction must then be the product of K_1 times K_2:

$$K = K_1 \times K_2 = 2.0 \times 10^{-17}$$

Exercise   Taking K for Reaction 18.18 to be 1×10^{-20}, calculate $[S^{2-}]$ in a solution in which $[H^+] = 0.3$ M, $[H_2S] = 0.1$ M. Answer: 1×10^{-20} M.

SUMMARY

In this chapter, we discussed three different types of equilibrium constants:

1. K_{sp}, the solubility product constant of a slightly soluble ionic solid:

$$M_xN_y(s) \rightleftharpoons x\,M^{y+}(aq) + y\,N^{x-}(aq); \quad K_{sp} = [M^{y+}]^x \times [N^{x-}]^y$$

K_{sp} can be determined from the solubility in moles per liter of the solid (Table 18.2). It can be used to calculate:
   a. The concentration of one ion in a solution in equilibrium with the solid, knowing that of the other ion (Example 18.1).
   b. Whether or not a precipitate will form (Example 18.2).
   c. The solubility of the solid in pure water (Example 18.3) or in a solution containing a common ion (p. 433).

2. K_a, the dissociation constant of a weak acid:

$$HB(aq) \rightleftharpoons H^+(aq) + B^-(aq); \quad K_a = \frac{[H^+] \times [B^-]}{[HB]}$$

K_a can be determined knowing $[H^+]$ in a solution prepared by dissolving the weak acid HB in water (Example 18.4). It can be used to calculate:

a. The $[H^+]$ or pH of a buffer: $[H^+] = K_a \times \dfrac{[HB]}{[B^-]}$; (Examples 18.5 and 18.6).

b. The $[H^+]$ or pH of a solution prepared by dissolving a weak acid in water: $[H^+]^2 = K_a \times [HB]$; (Examples 18.7 and 18.8).

3. **K_b,** the dissociation constant of a weak base:

$$B^-(aq) + H_2O \rightleftharpoons HB(aq) + OH^-(aq)$$

K_b can be calculated from the dissociation constant of the conjugate weak acid, using the relation $K_a \times K_b = 1.0 \times 10^{-14}$ (Example 18.9). It can be used to calculate $[OH^-]$ in a solution of a weak base: $[OH^-]^2 = K_b \times [B^-]$ (Example 18.10).

The equilibrium constant for a reaction is the reciprocal of the equilibrium constant for the reverse reaction: $K = 1/K'$ (Example 18.11). If Reaction 1 + Reaction 2 = Reaction 3, then $K_3 = K_1 \times K_2$ (Example 18.12).

KEY WORDS AND CONCEPTS

solubility product constant, K_{sp}
ion product, Q
acid dissociation constant, K_a
buffer

5% rule
successive approximations, method of
quadratic formula

base dissociation constant, K_b
reciprocal relation between K's
rule of multiple equilibria

QUESTIONS AND PROBLEMS

Catalog

K_{sp}: 18.2–18.7, 18.24–18.29
K_a, Determination of: 18.8, 18.30
Buffers: 18.9–18.12, 18.31–18.34
$[H^+]$ in Solutions of Weak Acids: 18.13–18.15, 18.35–18.37

K_b: 18.16–18.18, 18.38–18.40
Relations Between K's: 18.19, 18.20, 18.41, 18.42
General: 18.1, 18.21–18.23, 18.43–18.49

18.1 Review and know the meaning of the key words and concepts in this chapter.

18.2 Write a K_{sp} expression for each of the following slightly soluble electrolytes:

    a. NiS   b. CaF_2 c. $Fe(OH)_3$

18.3 Complete the following table. (K_{sp} of $PbCrO_4$ $= 1 \times 10^{-16}$)

$[Pb^{2+}]$	$[CrO_4^{2-}]$
1×10^{-9}	——
5×10^{-10}	——
——	0.05
——	0.1

18.24 Write the K_{sp} expression of each of the following sparingly soluble electrolytes:

    a. $BaCrO_4$ b. $Zn(OH)_2$ c. $Cr(OH)_3$

18.25 Complete the following table. (K_{sp} of CaF_2 $= 2 \times 10^{-10}$)

$[Ca^{2+}]$	$[F^-]$
2×10^{-4}	——
4×10^{-6}	——
——	2×10^{-2}
——	2×10^{-4}

18.4 Will $MgCO_3$ precipitate from a hard water solution containing 0.010 M Mg^{2+} and 0.0010 M CO_3^{2-}?

18.5 A sodium chloride solution is added to a solution containing 1.0×10^{-4} M Ag^+. (K_{sp} AgCl = 1.6×10^{-10})

a. What is the maximum conc. Cl^- that can be present without causing the precipitation of AgCl?

b. Sufficient NaCl is added to make the final conc. Cl^- 2.0×10^{-2} M. What is the equilibrium concentration of Ag^+ in the final solution?

18.6 The solubility of $CaCO_3$ in pure water is 7×10^{-5} mol/ℓ. Calculate

a. K_{sp} of $CaCO_3$.

b. the solubility of $CaCO_3$ in 0.05 M $CaCl_2$ solution.

18.7 Using K_{sp} values in Table 18.1, calculate the water solubility of

a. AgCl in grams per liter.

b. BaF_2 in moles per liter.

18.8 Aspirin is acetylsalicyclic acid (HAsp). A solution prepared by dissolving 0.0100 mol of HAsp to form a liter of solution has a pH of 3.30. What is the K_a of aspirin?

18.9 Which of the following pairs of solutions could be mixed to produce a buffer?

a. NaOH and HBr

b. HF and NaF

c. HClO and NaClO

d. HF and NaOH

18.10 You want to make a buffer with a pH of about 6. Which one of the following conjugate acid-base pairs would you use? Explain.

a. H_3PO_4–$H_2PO_4^-$ (K_a H_3PO_4 = 7.5×10^{-3})

b. H_2CO_3–HCO_3^- (K_a H_2CO_3 = 4.2×10^{-7})

c. NH_4^+–NH_3 (K_a NH_4^+ = 5.6×10^{-10})

18.11 One liter of a buffer is made using 0.10 mol acetic acid and 0.15 mol sodium acetate. (K_a $HC_2H_3O_2$ = 1.8×10^{-5})

a. Calculate the pH of the buffer.

b. What is the pH after 0.020 mol HCl is added?

c. What is the pH after 0.020 mol NaOH is added?

18.26 Fluoridation of water supplies may add about one part per million of F^- ion—that is, 1 g F^-/10^6 g of water.

a. What is the concentration of F^- in this water in moles per liter?

b. Will a precipitate form when hard water in which conc. Ca^{2+} = 2×10^{-4} M is fluoridated?

18.27 A solution contains 0.01 M Ba^{2+}. Enough CrO_4^{2-} is added to make $[CrO_4^{2-}]$ = 1×10^{-4} M. (K_{sp} $BaCrO_4$ = 1×10^{-10})

a. Will $BaCrO_4$ precipitate?

b. What is the minimum concentration of CrO_4^{2-} required to give a precipitate with 0.01 M Ba^{2+}?

18.28 The K_{sp} of $BaSO_4$ is 1.4×10^{-9}. How many moles of $BaSO_4$ will dissolve in

a. 200 cm^3 of pure water?

b. 200 cm^3 of 0.10 M Na_2SO_4?

18.29 Using data from Table 18.1, calculate the water solubility of

a. Ag_2CrO_4 in moles per liter.

b. $MgCO_3$ in grams per liter.

18.30 Propionic acid (HPro), in the form of the sodium salt, is added to bread to prevent molding. The $[H^+]$ in a 0.10 M solution of HPro is 1.2×10^{-3} M. Calculate the K_a of propionic acid.

18.31 Which of the following solutions could be mixed to form a buffer?

a. KCN and HCN

b. NaH_2PO_4 and Na_2HPO_4

c. NaOH and $HC_2H_3O_2$

d. NH_3 and NH_4Cl

18.32 To prepare a H_3PO_4–$H_2PO_4^-$ buffer, 0.10 mol of NaOH is added to 0.30 mol of H_3PO_4. Is this buffer more effective against the addition of H^+ ions or OH^- ions? Explain your reasoning.

18.33 A buffer is made up containing 0.010 mol KH_2PO_4 and 0.030 mol K_2HPO_4 per liter. (K_a $H_2PO_4^-$ = 6.2×10^{-8})

a. Calculate the pH of the buffer.

b. What is the pH after 0.005 mol of HBr is added?

c. What is the pH after 0.002 mol $Ca(OH)_2$ is added?

18.12 Blood is buffered mainly by the H_2CO_3–HCO_3^- system, in which the ratio of $[HCO_3^-]$ to $[H_2CO_3]$ is about 20:1.

    a. What would the pH of the blood be if this were the only buffer? (K_a $H_2CO_3 = 4.2 \times 10^{-7}$)

    b. What would the pH of the buffer become if the relative amount of H_2CO_3 were increased so that the concentration ratio was 2 HCO_3^-:1 H_2CO_3?

18.13 For a 0.20 M solution of benzoic acid ($K_a = 6.6 \times 10^{-5}$), calculate

    a. $[H^+]$ b. $[OH^-]$ c. pH

    d. % ionization $= 100 \times \dfrac{[H^+]}{\text{orig. conc. acid}}$

18.14 Consider acetic acid ($K_a = 1.8 \times 10^{-5}$). Calculate $[H^+]$ in solutions prepared by adding the following numbers of moles of acetic acid to one liter of water:

    a. 1.0   b. 0.50   c. 0.20   d. 0.10

For each solution, calculate:

$$\% \text{ ionization} = 100 \times \frac{[H^+]}{\text{orig. conc. acid}}$$

18.15 Chloroacetic acid, $ClCH_2COOH$, has a K_a of 1.4×10^{-3}. Calculate $[H^+]$ in a 0.100 M solution of this acid, using

    a. the quadratic formula.
    b. the method of successive approximations.

18.16 The K_a of pyridinium ion, $C_5H_5NH^+$, is 6.7×10^{-6}; K_a of HOCN is 2.2×10^{-4}.

    a. What is K_b of pyridine, C_5H_5N? of OCN^-?
    b. Which is the stronger acid, HOCN or $C_5H_5NH^+$?
    c. Which is the stronger base, OCN^- or C_5H_5N?

18.17 For the NO_2^- ion, $K_b = 2.2 \times 10^{-11}$.

    a. Write an equation for the reaction that makes the NO_2^- ion a weak base.
    b. Calculate $[OH^-]$ in a 0.10 M solution of $NaNO_2$.

18.18 Using Table 18.3, determine the pH of a 0.10 M solution of

    a. NH_3.
    b. PO_4^{3-} (use successive approximations).

18.19 Using Table 18.3 and the rule of multiple equilibria, calculate K for

    a. $H_2SO_3(aq) \rightleftharpoons 2\,H^+(aq) + SO_3^{2-}(aq)$
    b. $H_3PO_4(aq) \rightleftharpoons 3\,H^+(aq) + PO_4^{3-}(aq)$

18.34 A buffer is prepared in which the ratio $[HCO_3^-]/[CO_3^{2-}]$ is 5.0.

    a. What is the pH of this buffer? (K_a $HCO_3^- = 4.8 \times 10^{-11}$)
    b. Enough strong acid is added to make the pH of the buffer 9.30. What is the ratio $[HCO_3^-]/[CO_3^{2-}]$ at this point?

18.35 For a 0.10 M solution of propionic acid ($K_a = 1.4 \times 10^{-5}$), calculate

    a. $[H^+]$ b. $[OH^-]$ c. pH
    d. % ionization

18.36 Follow the directions of Problem 18.14 for $H_2PO_4^-$. ($K_a = 6.2 \times 10^{-8}$)

18.37 Dichloroacetic acid, $Cl_2CHCOOH$, has a K_a of 3.3×10^{-2}. Using the method of successive approximations, find $[H^+]$ in a 1.0 M solution of this acid.

18.38 The K_a of hypobromous acid, HOBr, is 2.1×10^{-9}; that of benzoic acid, $HC_7H_5O_2$, is 6.6×10^{-5}.

    a. What is K_b of the hypobromite ion, OBr^-? of the benzoate ion, $C_7H_5O_2^-$?
    b. Which is the stronger acid, hypobromous or benzoic?
    c. Which is the weaker base, OBr^- or $C_7H_5O_2^-$?

18.39 For the F^- ion, $K_b = 1.4 \times 10^{-11}$.

    a. What is K_a of HF?
    b. What is $[OH^-]$ in a 0.20 M solution of NaF?

18.40 Using Table 18.3, determine the pH of a solution that is 0.0100 M in

    a. ClO^-.
    b. CO_3^{2-} (use successive approximations).

18.41 Given that K_a for oxalic acid, $H_2C_2O_4$, is 6.5×10^{-2} and K_a for the $HC_2O_4^-$ ion is 6.1×10^{-5}, calculate K for

    a. $H_2C_2O_4(aq) \rightleftharpoons 2\,H^+(aq) + C_2O_4^{2-}(aq)$
    b. $H^+(aq) + C_2O_4^{2-}(aq) \rightleftharpoons HC_2O_4^-(aq)$

18.20 Using Table 18.3, obtain values of K for the following reactions. ($K_w = 1.0 \times 10^{-14}$)

    a. $H^+(aq) + C_2H_3O_2^-(aq) \rightleftharpoons HC_2H_3O_2(aq)$
    b. $HC_2H_3O_2(aq) + F^-(aq) \rightleftharpoons HF(aq)$
                     $+ C_2H_3O_2^-(aq)$
    c. $HC_2H_3O_2(aq) + OH^-(aq) \rightleftharpoons H_2O + C_2H_3O_2^-(aq)$

18.21 Using Table 18.3, arrange 0.10 M solutions of the following compounds in order of increasing pH:

    a. HCl   b. NaOH   c. NaCl
    d. HF   e. NH_4Cl f. NH_3

18.22 For each solution given below, indicate the main solute species present (0.10 M or greater) and calculate the pH of the solution:

    a. 0.20 M $HC_2H_3O_2$
    b. 0.20 M KCN
    c. 0.30 M KOH

18.23 K_a for butyric acid (HBut) is 2.0×10^{-5}. What is the pH of a solution that is 0.10 M in

    a. HBut?   b. But$^-$?
    c. both HBut and But$^-$?

18.42 Using Table 18.3, obtain values of K for the following reactions:

    a. $HNO_2(aq) + F^-(aq) \rightleftharpoons HF(aq) + NO_2^-(aq)$
    b. $HNO_2(aq) + OH^-(aq) \rightleftharpoons H_2O + NO_2^-(aq)$
    c. $H^+(aq) + NO_2^-(aq) \rightleftharpoons HNO_2(aq)$

18.43 Arrange 0.10 M solutions of the following compounds in order of decreasing pH:

    a. $HC_2H_3O_2$ b. $NaC_2H_3O_2$ c. Na_2CO_3
    d. NaOH   e. Na_2SO_3 f. HNO_3

18.44 Follow the directions of Problem 18.22 for the following solutions:

    a. 0.30 M NH_3
    b. 0.20 M HCl
    c. 0.40 M NH_4Cl

18.45 K_a for lactic acid (HLac) is 1.4×10^{-4}. What is the pH of a solution that is 1.0 M in

    a. HLac?   b. Lac$^-$?
    c. both HLac and Lac$^-$?

*18.46 The inside of a teakettle is coated with 10.0 g of $CaCO_3$. If the teakettle is washed with one liter of pure water, what fraction of the precipitate is removed? How many successive one-liter portions of water would have to be used to remove half of the $CaCO_3$?

*18.47 It is found that 0.20 M solutions of the three salts NaX, NaY, and NaZ have pH's of 7.0, 8.0, and 9.0, respectively. Arrange the acids HX, HY, and HZ in order of increasing acid strength. Where you can, find K_a for the acids.

*18.48 When CO_2 is bubbled into lime water, an aqueous solution of $Ca(OH)_2$, the solution becomes cloudy. If additional CO_2 is added, the solution again becomes colorless. Explain.

*18.49 50.00 cm³ of 1.000 M $HC_2H_3O_2$ ($K_a = 1.8 \times 10^{-5}$) is titrated with 1.000 M NaOH. Find the pH of the solution after the following volumes of NaOH have been added:

    a. 0.00 cm³   b. 25.00 cm³   c. 49.90 cm³   d. 50.00 cm³   e. 50.10 cm³   f. 100.00 cm³

Use your data to construct a plot similar to that shown in Figure 17.4 (pH vs. volume NaOH added).

COORDINATION COMPOUNDS; COMPLEX IONS

The brilliant colors of many materials are due to transition metal ions (color plate 17, center of book). These ions are responsible for the colors of such diverse materials as gems, blood, soil, plants, and oil paints. In this chapter, we will examine the origin of these colors. To do so, we will need to consider the structure and properties of an important class of compounds known as coordination compounds. These substances most often contain a transition metal in the form of a species called a complex ion. A large portion of this chapter will be devoted to the bonding and geometry in complex ions.

For over a century, chemists have studied the interaction of transition metal ions with other species. A mystery which intrigued chemists of the nineteenth century was how two stable compounds like NH_3 and $CuCl_2$ could react with each other. They indeed do so, forming a stable compound containing four moles of ammonia per mole of copper(II) chloride. We begin our study of coordination chemistry by examining the structure of this compound (color plate 18, center of book).

19.1 THE STRUCTURE OF COORDINATION COMPOUNDS

A water solution containing the Cu^{2+} ion has a pale blue color. When aqueous ammonia, NH_3, is added, the color changes to a deep, almost opaque, blue. The color change is due to a chemical reaction in which four NH_3 molecules combine with one Cu^{2+} ion:

$$Cu^{2+}(aq) + 4\ NH_3(aq) \rightarrow Cu(NH_3)_4{}^{2+}(aq) \tag{19.1}$$

light blue   deep blue

In electron-dot notation, we can represent this reaction as

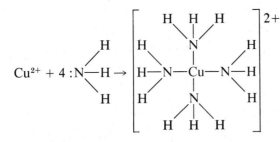

The nitrogen atom of each NH_3 molecule contributes a pair of unshared electrons to form a covalent bond with the Cu^{2+} ion. This bond and others like it, where both electrons are contributed by the same atom, is referred to as a **coordinate** covalent bond. There are four such bonds in the $Cu(NH_3)_4^{2+}$ ion.

The $Cu(NH_3)_4^{2+}$ ion is commonly referred to as a **complex ion.** We use the term complex ion to indicate a charged species in which *a metal ion is joined by coordinate covalent bonds to neutral molecules and/or negative ions.* Species such as $Al(H_2O)_6^{3+}$ and $Zn(H_2O)_3(OH)^+$, found in previous chapters, are further examples of complex ions. The metals which show the greatest tendency to form complex ions are those toward the right of the transition series (in the first transition series, $_{24}Cr$ through $_{30}Zn$). Nontransition metals, including Al, Sn and Pb, form a more limited number of stable complex ions.

In solution, species like Cu^{2+}, Cr^{3+}, and Zn^{2+} exist as complex ions

The metal cation in a complex is called the *central ion.* The molecules or anions bonded directly to it are called **ligands.** The number of bonds formed by the central ion is called its **coordination number.** In the $Cu(NH_3)_4^{2+}$ ion, the central ion is Cu^{2+}. The ligands are NH_3 molecules. Since the Cu^{2+} ion forms a total of four bonds, it has a coordination number of 4.

An ion such as $Cu(NH_3)_4^{2+}$ cannot exist by itself in the solid state. The $+2$ charge of this ion must be balanced by anions with a total charge of -2. A typical compound containing the $Cu(NH_3)_4^{2+}$ ion is

$$[Cu(NH_3)_4]Cl_2; \text{ one } Cu(NH_3)_4^{2+} \text{ ion, two } Cl^- \text{ ions}$$

Compounds such as this, which contain a complex ion, are referred to as **coordination compounds.** The formula of the complex ion is set off by brackets, [ ], to make the structure of the compound clear.

Table 19.1 gives the formulas of a series of coordination compounds formed by the Pt^{2+} ion, which has a coordination number of 4. Notice that:

1. Compounds 1 and 2 contain complex cations with charges of $+2$ and $+1$, in that order. These coordination compounds are analogous to the simple ionic compounds $CaCl_2$ and KCl. In both $[Pt(NH_3)_4]Cl_2$ and $CaCl_2$, two Cl^- ions are balanced by a $+2$ ion, $Pt(NH_3)_4^{2+}$ in one case, Ca^{2+} in the other.

2. Compounds 4 and 5 contain complex anions with charges of -1 and -2, in that order. In the solid, these are balanced by K^+ ions (1 K^+ per $Pt(NH_3)Cl_3^-$ ion, 2 K^+ per $PtCl_4^{2-}$ ion).

The term "complex" covers both charged and uncharged species

3. Compound 3 is a neutral complex, with zero charge. There are no ions present.

As we have pointed out, a coordination compound, like all compounds, must have a net charge of zero. The charge of a complex ion within such a compound is readily determined. It is the algebraic sum of the charges of the bare metal ion and the ligands. Applying this principle to the compounds in Table 19.1, where the species within the complexes are

$$Pt^{2+} \text{ (charge} = +2); NH_3 \text{ molecules (charge} = 0); Cl^- \text{ ions (charge} = -1)$$

TABLE 19.1   COORDINATION COMPOUNDS CONTAINING COMPLEXES OF Pt^{2+}

COORDINATION COMPOUND	COMPLEX	CHARGE OF COMPLEX	ANALOGOUS SIMPLE IONIC COMPOUND
1. $[Pt(NH_3)_4]Cl_2$	$Pt(NH_3)_4^{2+}$	$+2$	$CaCl_2$
2. $[Pt(NH_3)_3Cl]Cl$	$Pt(NH_3)_3Cl^+$	$+1$	KCl
3. $[Pt(NH_3)_2Cl_2]$	$Pt(NH_3)_2Cl_2$	0	
4. $K[Pt(NH_3)Cl_3]$	$Pt(NH_3)Cl_3^-$	-1	KNO_3
5. $K_2[PtCl_4]$	$PtCl_4^{2-}$	-2	K_2SO_4

we obtain

1. $+2 + 4(0) = +2$
2. $+2 + 3(0) + 1(-1) = +1$
3. $+2 + 2(0) + 2(-1) = 0$
4. $+2 + 1(0) + 3(-1) = -1$
5. $+2 + 4(-1) = -2$

Example 19.1   Determine the charge of the transition metal ion in each of the following species:

    a. $Cu(H_2O)_4^{2+}$ b. $Co(NH_3)_4Cl_2^+$ c. $K_2[PtCl_6]$

Solution

  a. The water molecules have no charge. The $+2$ charge of the complex ion must be the charge on the copper ion, Cu^{2+}.

  b. The charge of the complex ion is $+1$. Each chloride ion has a -1 charge. The NH_3 molecules have zero charge. Letting x be the charge on the cobalt:

$$+1 = x + 4(0) + 2(-1)$$

Solving for x: $x = +1 + 2 = +3$; Co^{3+} ion

  c. Here, the complex is a negative ion. Since it is balanced by two K^+ ions, the $PtCl_6^{2-}$ ion must have a charge of -2. Each Cl^- ion within the complex has a charge of -1:

$$-2 = x + 6(-1); \qquad x = +4; \qquad Pt^{4+}$$

Exercise   In a certain complex ion, a central Co^{3+} ion is bonded to four H_2O molecules and two Br^- ions. Give the formula and charge of the complex ion. Answer: $Co(H_2O)_4Br_2^+$.

The specialized nomenclature of coordination compounds is discussed in Appendix 3.

Ligands; Chelating Agents

In principle, any molecule or anion with an unshared pair of electrons can donate them to a metal ion to form a coordinate covalent bond. We expect species such as the ammonia molecule, the water molecule, or the fluoride ion to act as ligands.

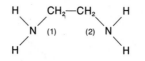

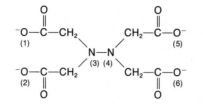

oxalate ion (bidentate)

ethylenediamine (bidentate)

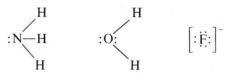

ethylenediaminetetraacetate ion (hexadentate)

Figure 19.1 Structures of three chelating agents, often abbreviated as "ox," "en," and EDTA. The numbers indicate the location of the unshared electron pairs that form coordinate covalent bonds with metal ions.

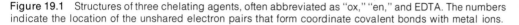

In practice, a ligand usually contains an atom of one of the more electronegative elements (C, N, O, S, F, Cl, Br, I). Several hundred different ligands are known. Those most commonly encountered in general chemistry are NH_3 and H_2O molecules and Cl^- and OH^- ions.

Ligands are classified according to the number of electron pairs donated. *Monodentate* ligands provide one electron pair per ligand molecule or ion. The species NH_3, H_2O, Cl^-, and OH^- are of this type. *Bidentate* (two-toothed) ligands furnish two electron pairs per molecule or ion, a tridentate ligand furnishes three pairs, and so on. The general term *polydentate* is used for any ligand that supplies more than one pair of electrons. The structures of three important polydentate ligands are shown in Figure 19.1. These are the ethylenediamine molecule, abbreviated "en," the oxalate ion ("ox"), and the ethylenediaminetetraacetate anion ("EDTA"). Note that ethylenediamine and the oxalate ion are both bidentate; EDTA is hexadentate, forming six bonds per ligand.

The complexes formed by polydentate ligands are often called **chelates** (Greek *chela*, meaning crab's claw). The structures of the chelates formed by Cu^{2+} with ethylenediamine and with the oxalate ion are shown in Figure 19.2, p. 456. Notice that in these complexes, as in the $Cu(NH_3)_4{}^{2+}$ ion (p. 453), Cu^{2+} has a coordination number of 4. When we write these complexes as $Cu(en)_2{}^{2+}$ and $Cu(ox)_2{}^{2-}$, we must remember that each ligand is forming two bonds. So, the coordination number of Cu^{2+} is $2 \times 2 = 4$.

In EDTA chelates there are three "claws"

Coordination Number

As shown in Table 19.2, p. 456, the most common coordination number is 6. A coordination number of 4 is less common. A value of 2 is restricted largely to Cu^+, Ag^+, and Au^+. Odd coordination numbers (1, 3, 5, 7, 9) are very rare.

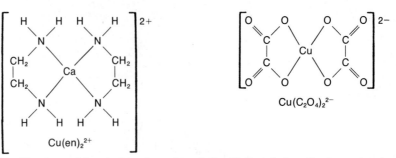

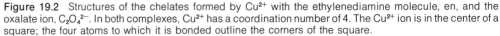

Figure 19.2 Structures of the chelates formed by Cu^{2+} with the ethylenediamine molecule, en, and the oxalate ion, $C_2O_4^{2-}$. In both complexes, Cu^{2+} has a coordination number of 4. The Cu^{2+} ion is in the center of a square; the four atoms to which it is bonded outline the corners of the square.

A few metal ions show only one coordination number in their complex ions. Thus Co^{3+} always shows a coordination number of 6, as in

$$Co(NH_3)_6^{3+}, \quad Co(NH_3)_4Cl_2^+, \quad \text{and} \quad Co(en)_3^{3+}$$

Other metal ions, such as Al^{3+} and Ni^{2+}, have variable coordination numbers, depending upon the nature of the ligand. With molecules or very small anions as ligands, we usually observe the higher coordination number, as in

It is not easy to predict coordination numbers for cations, beyond the fact that they are likely to be even

$$Al(H_2O)_6^{3+}, \quad AlF_6^{3-}, \quad Ni(H_2O)_6^{2+}, \quad Ni(NH_3)_6^{2+}$$

With larger anions, the lower coordination number is more common:

$$AlCl_4^-, \quad Al(OH)_4^-, \quad Ni(CN)_4^{2-}$$

TABLE 19.2   COORDINATION NUMBER AND GEOMETRY OF COMPLEX IONS*

METAL ION	COORDINATION NO.	GEOMETRY	EXAMPLE
Ag⁺, Au⁺, Cu⁺	2	linear	$Ag(NH_3)_2^+$
Cu²⁺, **Ni²⁺**, **Pd²⁺**, **Pt²⁺**	4	square planar	$Pt(NH_3)_4^{2+}$
Al³⁺, Au⁺, **Cd²⁺**, Co²⁺, Cu⁺, **Ni²⁺**, **Zn²⁺**	4	tetrahedral	$Zn(NH_3)_4^{2+}$
Al³⁺, **Cd²⁺**, **Co²⁺**, **Co³⁺**, **Cr³⁺**, **Cu²⁺**, **Fe²⁺**, **Fe³⁺**, **Ni²⁺**, **Pt⁴⁺**	6	octahedral	$Co(NH_3)_6^{3+}$

*Principal coordination number indicated by bold type.

19.2   GEOMETRY OF COMPLEX IONS

Coordination Number = 2

Complex ions in which the central metal ion forms only two bonds to ligands are *linear*. That is, the two bonds are directed at 180° angles. The structures of $CuCl_2^-$, $Ag(NH_3)_2^+$, and $Au(CN)_2^-$ may be represented as

$(Cl-Cu-Cl)^-$, $\begin{bmatrix} H & & & H \\ \diagdown & & & \diagup \\ H-N-Ag-N-H \\ \diagup & & & \diagdown \\ H & & & H \end{bmatrix}^+$, $(N\equiv C-Au-C\equiv N)^-$

Coordination Number = 4

Four-coordinate metal complexes may have either of two different geometries. The four bonds from the central metal ion may be directed toward the corners of a regular tetrahedron. Two common *tetrahedral* complexes are $Zn(NH_3)_4^{2+}$ and $CoCl_4^{2-}$. In the other case, known as *square planar,* the four bonds are directed toward the corners of a square. The $Pt(NH_3)_4^{2+}$ and $Ni(CN)_4^{2-}$ ions are of this type (Fig. 19.3). $Cu(NH_3)_4^{2+}$ is square planar

Certain square planar complexes occur in two different forms with quite different properties. The complex $[Pt(NH_3)_2Cl_2]$ has two forms differing in absorption spectrum, water solubility, melting point, and chemical reactivity. In one form, made by reacting NH_3 with the $PtCl_4^{2-}$ ion, the two NH_3 ligands are at adjacent corners of the square. In the other form, prepared by reacting $Pt(NH_3)_4^{2+}$ with HCl, the NH_3 molecules are at opposite corners of the square. These two forms are referred to as **geometric isomers.** Their structures differ only in the spatial arrangement of ligands about the central metal ion. The form in which like ligands are as close as possible is called the **cis** isomer. The **trans** isomer has these groups as far apart as possible.

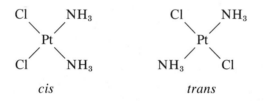

cis   trans

Geometric isomerism can occur with any square planar complex of the type Ma_2b_2 or Ma_2bc, where M refers to the central metal and a, b, c are different ligands. Because all positions in a tetrahedral complex are equidistant, geometric isomerism cannot occur with such complexes.

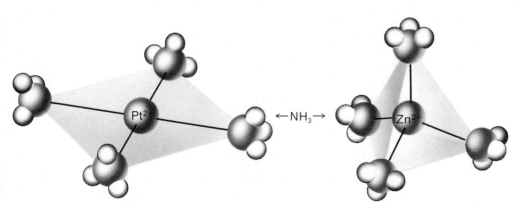

Figure 19.3 Structure of $Pt(NH_3)_4^{2+}$ (square planar) and $Zn(NH_3)_4^{2+}$ (tetrahedral). Geometrical isomers can exist for square planar complexes but not for tetrahedral complexes.

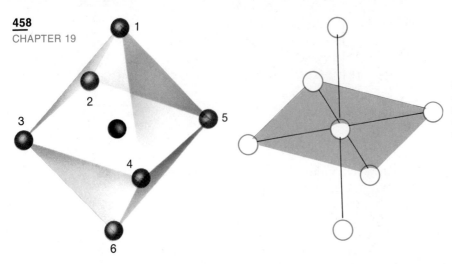

Figure 19.4 The drawing at the left shows six ligands at the corners of an octahedron with a metal ion at the center. A simpler way to represent an octahedral complex is shown at the right.

Coordination Number = 6

Octahedral geometry is characteristic of this coordination number. The six ligands surrounding the central metal ion in complexes such as $Fe(CN)_6^{3-}$ and $Co(NH_3)_6^{3+}$ are located at the corners of a regular octahedron. An octahedron is a geometric figure with six corners and eight faces, each of which is an equilateral triangle. The metal ion is located at the center of the octahedron (Fig. 19.4). The six ligands at the corners are *equidistant from the metal ion at the center.* The skeleton structure at the right of Figure 19.4 is easier to draw and more commonly used. From this skeleton, we see that an octahedral complex can be regarded as a derivative of a square planar complex. The two extra ligands are located above and below the square, on a line perpendicular to the square at its center.

Geometric isomerism can occur in octahedral complexes. To see how this is possible, consider Figure 19.4. For any given position of a ligand, four equivalent positions are equidistant while a fifth is at a greater distance. Suppose, for example, we choose position 1 as a point of reference. Groups at positions 2, 3, 4, and 5 will be equidistant from 1; 6 is farther away. We may refer to positions 1 and 2, 1 and 3, 1 and 4, or 1 and 5 as being *cis* to each other. On the other hand, positions 1 and 6 are *trans*. Hence, a complex ion like $Co(NH_3)_4Cl_2^+$ can exist in two isomeric forms. In the *cis* isomer, the two

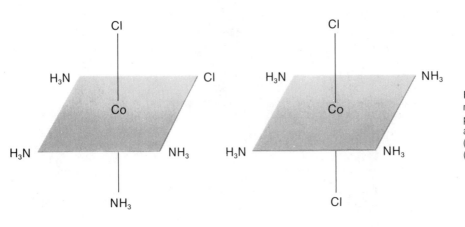

Figure 19.5 *Cis* and *trans* isomers of the $[Co(NH_3)_4Cl_2]^+$ complex ion. Note that the two Cl atoms are closer in the *cis* isomer (left) than in the *trans* isomer (right).

Cl⁻ ions are close together. In the *trans* form, they are far apart (Fig. 19.5). These two isomers differ in their physical and chemical properties. The most striking difference is color. Compounds of Co^{3+} containing the *cis* complex ion tend to be violet, while those containing the *trans* isomer are often green (see color plate 19, center of book).

Example 19.2   How many isomers are possible for the neutral complex $[Co(NH_3)_3Cl_3]$?

Solution   It is best to approach this problem systematically. We might start by putting two NH_3 molecules in *trans* positions, perhaps at the "top" and "bottom" of the octahedron (Fig. 19.6a). We then ask ourselves: In how many spatially different positions can we place the third NH_3 molecule? A moment's reflection should convince you that there is really only one choice. All four of the remaining positions are equivalent in that they are *cis* to the two groups that we have already located. Choosing one of these positions arbitrarily, we get our first isomer (Fig. 19.6b).

To see if there are other isomers we start again, this time locating two NH_3 molecules *cis* to each other (Fig. 19.6c). If we were to place the third NH_3 molecule at one of the other corners of the square, we would simply reproduce the first isomer. (Remember that the symmetry of a regular octahedron requires that the distance across the diagonal of the square be the same as that from "top" to "bottom.") We are left with two equivalent positions. Placing the third NH_3 molecule arbitrarily at the "top" we arrive at a second isomer (Fig. 19.6d), distinctly different from the first because all three NH_3 molecules are *cis* to one another.

We have now exhausted in a logical manner all the possibilities for geometric isomerism, finding two isomers. There are no others.

Different isomers will have different *cis-trans* geometries

Exercise   How many geometric isomers are there for $Cr(NH_3)_2Cl_4^-$? for the square planar complex $[Pt(NH_3)_2ClBr]$? Answer: Two; two.

Geometric isomerism can occur in chelated octahedral complexes (Fig. 19.7, p. 460).

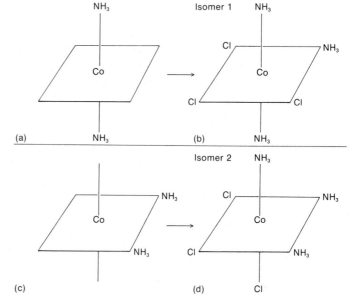

Figure 19.6   Isomers of $Co(NH_3)_3Cl_3$ (Example 19.2). You may be tempted to write down additional structures, but you will find that they are equivalent to one of the two isomers shown here.

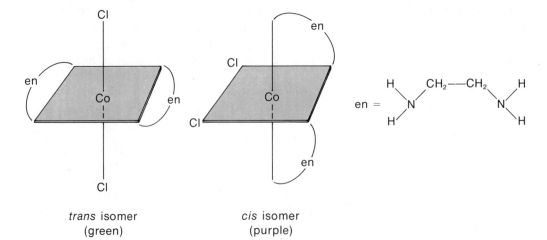

trans isomer
(green)

cis isomer
(purple)

Figure 19.7 There are two isomers of the complex ion $Co(en)_2Cl_2^+$ which differ from one another in color and in certain chemical properties. For example, the *cis* isomer reacts more readily with $C_2O_4^{2-}$ ions than does the *trans* isomer. (Why?)

19.3 USES OF COORDINATION COMPOUNDS

Coordination compounds are used for a variety of purposes inside and outside the laboratory. Perhaps their most important application is in bringing water-insoluble species into solution. In discussing the metallurgy of aluminum in Chapter 5, we pointed out that Al_2O_3 is separated from Fe_2O_3 and other impurities by heating with concentrated sodium hydroxide, NaOH. The reaction that takes place involves the formation of the stable $Al(OH)_4^-$ complex ion.

Fe^{3+} does not form complexes with OH^- ion

$$Al_2O_3(s) + 3\ H_2O + 2\ OH^-(aq) \rightarrow 2\ Al(OH)_4^-(aq) \tag{19.2}$$

In photography, developed film is "fixed" by washing with a solution of sodium thiosulfate ("hypo"), $Na_2S_2O_3$. Unreacted silver bromide on the film is brought into solution by formation of the $Ag(S_2O_3)_2^{3-}$ complex ion:

$$AgBr(s) + 2\ S_2O_3^{2-}(aq) \rightarrow Ag(S_2O_3)_2^{3-}(aq) + Br^-(aq) \tag{19.3}$$

It is even possible to bring unreactive metals such as silver and gold into solution by combining an oxidizing agent with a complexing agent. These metals are separated from low-grade ores by treatment with sodium cyanide solution in the presence of air. Oxygen in the air oxidizes the metals to cations (Ag^+ and Au^+). These are then converted to the complex ions $Ag(CN)_2^-$ and $Au(CN)_2^-$:

$$Ag^+(aq) + 2\ CN^-(aq) \rightarrow Ag(CN)_2^-(aq) \tag{19.4}$$

$$Au^+(aq) + 2\ CN^-(aq) \rightarrow Au(CN)_2^-(aq) \tag{19.5}$$

In this way, silver and gold enter the water solution while rocky impurities are left behind.

Another metal, aluminum, reacts with basic water solutions as a result of complex ion formation:

$$2\ Al(s) + 6\ H_2O(l) + 2\ OH^-(aq) \rightarrow 2\ Al(OH)_4^-(aq) + 3\ H_2(g) \tag{19.6}$$

This reaction has recently found practical application in the addition of finely divided aluminum to drain cleaners. The bubbles of hydrogen formed are supposed to loosen deposits of grease or dirt.

In many natural products, a transition metal ion is tied up in a coordination complex. Compounds of this type include hemoglobin (Fe^{2+}), chlorophyll (Mg^{2+}), and vitamin B_{12} (Co^{2+}). Other coordination compounds, made in the laboratory, are used in medicine. In the remainder of this section, we will look at a few of the biological uses of coordination compounds.

Hemoglobin and Oxygen Transport

The structure of heme, the pigment responsible for the color of blood, is shown in Figure 19.8. Notice that there is an Fe^{2+} ion at the center bonded to four nitrogen atoms at the corners of a square. Actually, the Fe^{2+} ion is part of an octahedral complex. A fifth ligand, not shown in the figure, is a large organic molecule called globin. In combination with heme, this gives the protein we refer to as hemoglobin. The sixth ligand, filling out the octahedron, is an H_2O molecule.

Even nature makes good use of complexes

The water molecule in hemoglobin can be replaced reversibly by an O_2 molecule. The product formed is called oxyhemoglobin; it has a bright red color. In contrast, hemoglobin itself is blue. This explains why arterial blood is red (high concentration of O_2), while that in the veins is bluish (low concentration of O_2).

$$\text{hemoglobin} \cdot H_2O(aq) + O_2(aq) \rightleftharpoons \text{hemoglobin} \cdot O_2(aq) + H_2O \qquad (19.7)$$

  blue   red

The equilibrium in Reaction 19.7 is sensitive to the concentration of oxygen. In the lungs, where there is a large amount of O_2 available, the equilibrium shifts to the right. Oxygen in the form of the hemoglobin complex is taken up by red blood cells and carried to the tissues. There, where the concentration of dissolved O_2 is low, the reverse reaction occurs. This supplies the oxygen required for metabolism.

Certain molecules and anions form complexes with hemoglobin that are more stable than that formed by O_2. One such species is the CO molecule. As pointed out in Chapter 15, carbon monoxide poisoning results from the formation of the hemoglobin·CO com-

Figure 19.8 Structure of heme. Fe^{2+} ion is at the center of an octahedron, surrounded by four nitrogen atoms, a globin molecule, and a water molecule.

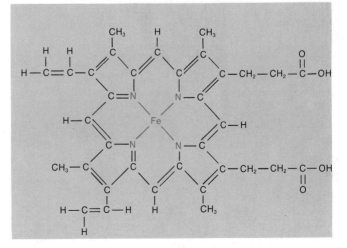

plex in preference to hemoglobin·O_2. Another species that behaves this way is the cyanide ion, CN^-, which has the same Lewis structure as CO:

$$:C\equiv O: \qquad (:C\equiv N:)^-$$

EDTA and Lead Poisoning

Lead compounds were used widely at one time as paint pigments. Children, by swallowing paint from an old crib or a windowsill in an old building, can develop lead poisoning. The Pb^{2+} ion interferes with hemoglobin production by deactivating an enzyme needed to make heme.

EDTA, whose structure was shown on page 455, is a very effective chelating agent. It forms complexes with a large number of cations, including those of some of the main group metals. The complex formed by calcium with EDTA is used to treat lead poisoning. When a solution containing the Ca-EDTA complex is given by injection, the calcium is displaced by lead. The more stable Pb-EDTA complex is eliminated in the urine. EDTA has also been used to remove radioactive isotopes of metals, notably plutonium, from body tissues.

EDTA therapy needs to be carefully controlled. Why?

Coordination Compounds in the Treatment of Cancer

The neutral complex *cis*-[Pt(NH$_3$)$_2$Cl$_2$] is an effective antitumor agent in cancer treatment (chemotherapy). The *trans* isomer is ineffective. The activity of the *cis* isomer is believed to reflect the ability of the two Cl atoms to interact with DNA, a molecule responsible for cell reproduction. A reaction occurs in which the Cl atoms are replaced by N atoms of the DNA molecule. The product is a Pt^{2+} complex that inhibits further cell growth (Fig. 19.9). This reaction cannot occur with the *trans* isomer, since the Cl atoms are too far apart.

19.4 ELECTRONIC STRUCTURES OF COMPLEX IONS

To develop the electron distribution in a complex ion, a logical starting point is the electronic structure of the central metal ion. The electron configurations of some of the cations in the first transition series are shown in Table 19.3. Notice that in these (and other) transition metal ions, **there are no outer s electrons.** One way to express this is

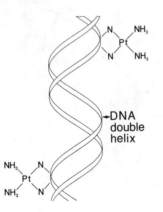

Figure 19.9 *cis*-Pt(NH$_3$)$_2$Cl$_2$ is effective in chemotherapy because of its ability to react with nitrogen atoms on DNA molecules, apparently causing the double helix to unwind. The *trans* isomer is ineffective. (For a more detailed discussion of this phenomenon, see *Chemical and Engineering News,* Jan. 21, 1980, p.35.)

TABLE 19.3 ELECTRON CONFIGURATIONS OF TRANSITION METAL ATOMS AND IONS

ATOMIC NUMBER	CONFIGURATION OF ATOM	CONFIGURATION OF ION
24	Cr $[_{18}Ar]3d^54s^1$	Cr^{3+} $[_{18}Ar]3d^3$
25	Mn $[_{18}Ar]3d^54s^2$	Mn^{2+} $[_{18}Ar]3d^5$
26	Fe $[_{18}Ar]3d^64s^2$	Fe^{3+} $[_{18}Ar]3d^5$
		Fe^{2+} $[_{18}Ar]3d^6$
27	Co $[_{18}Ar]3d^74s^2$	Co^{3+} $[_{18}Ar]3d^6$
		Co^{2+} $[_{18}Ar]3d^7$
28	Ni $[_{18}Ar]3d^84s^2$	Ni^{2+} $[_{18}Ar]3d^8$
29	Cu $[_{18}Ar]3d^{10}4s^1$	Cu^{2+} $[_{18}Ar]3d^9$
		Cu^+ $[_{18}Ar]3d^{10}$
30	Zn $[_{18}Ar]3d^{10}4s^2$	Zn^{2+} $[_{18}Ar]3d^{10}$

to say that when the ion is formed from the transition metal atom, the outer s (4s) electrons are lost first. From a slightly different point of view we might say that, in the transition metal *ion,* the 3d sublevel is lower in energy than the 4s.

Example 19.3 For the Mn^{3+} ion, write the
    a. electron configuration;   b. orbital diagram.

Solution
    a. There are 22 electrons in the Mn^{3+} ion, four beyond argon. These four electrons enter the 3d sublevel.

$$Mn^{3+} [_{18}Ar]3d^4$$

    b. Following Hund's rule (Chap. 8), the four 3d electrons enter singly into separate orbitals with parallel spins:

$$\begin{array}{c} 3d \\ Mn^{3+} [_{18}Ar] (\uparrow)(\uparrow)(\uparrow)(\uparrow)(\ ) \end{array}$$

Exercise How many unpaired electrons are there in the Co^{3+} ion? Answer: Four.

Remember, electrons remain unpaired when that doesn't take too much energy

---

With this background, we will look at two of the simpler models of the electronic structure of complex ions. One of these is the valence bond (atomic orbital) model, discussed in Chapter 10. The other is referred to as the crystal-field model.

Valence Bond (Atomic Orbital) Model

This model, for complex ions, starts with a simple assumption. It considers that electron pairs donated by ligands enter hybrid orbitals of the central metal ion. The particular orbitals which these electron pairs enter depend upon the coordination number and geometry of the complex (Table 19.4, p. 464). Note that the total number of hybrid orbitals occupied by ligand electrons is equal to the coordination number.

Of the types of hybrid orbitals listed in Table 19.4, the sp and sp³ sets were discussed earlier (Chap. 10). In square planar complexes, the four pairs of bonding electrons are accommodated by hybridizing a d orbital, an s orbital, and two p orbitals—hence,

TABLE 19.4 VALENCE BOND MODEL APPLIED TO COMPLEX IONS

COORDINATION NUMBER	GEOMETRY	HYBRID ORBITALS OCCUPIED BY LIGAND ELECTRONS	EXAMPLE
2	linear	sp	$Cu(NH_3)_2{}^+$
4	tetrahedral	sp^3	$Zn(NH_3)_4{}^{2+}$
4	square, planar	dsp^2	$Ni(CN)_4{}^{2-}$
6	octahedral	d^2sp^3	$Cr(NH_3)_6{}^{3+}$

"dsp²." For complex ions formed by metals in the first transition series, these would be one 3d, one 4s, and two 4p orbitals. In octahedral complexes, the six orbitals needed are obtained by hybridizing two d orbitals, an s orbital, and three p orbitals (d^2sp^3). In the first transition series, these are two 3d, one 4s, and three 4p orbitals.

Orbital diagrams for the four complex ions listed in Table 19.4 are shown in Figure 19.10. Here, the hybrid orbitals are enclosed by horizontal lines. The bonding electrons within these orbitals, all of which are contributed by the ligands, are shown as colored arrows. The electrons shown in black are those contributed by the metal ion itself; none of these are involved in bonding. To save space, only those electrons beyond the argon core are shown; the inner 18 electrons of each metal ion are omitted.

The orbital diagrams shown in Figure 19.10 can be derived in a series of simple steps:

Step 1: Determine the coordination number. This is the number of electron pairs (two, four or six) contributed by the ligands. Locate these electron pairs in the hybrid orbitals, one pair per orbital, using Table 19.4 as a guide. If the coordination number is 4, you must know (or be given enough information to deduce) the geometry. Only then can you decide whether the hybridization is dsp^2 or sp^3. For an octahedral complex such as $Cr(NH_3)_6{}^{3+}$, the six pairs of electrons are located in d^2sp^3 hybrid orbitals.

Step 2: Determine the charge of the central metal ion, using the discussion on page 453, if necessary. Now find the number of electrons beyond the argon shell in that ion. To do this, start with the atomic number of the metal, deduct the number of electrons lost in forming the ion and, finally, subtract 18. Thus, for Cr^{3+},

$$\text{no. of } e^- \text{ beyond Ar} = 24 - 3 - 18 = 3$$

Step 3: Locate the electrons from Step 2 in the lowest available orbitals, following Hund's rule. For Cr^{3+}, we have three 3d orbitals left vacant after locating the bonding electrons in Step 1. We put an electron into each of these orbitals, giving each electron the same spin. This gives the orbital diagram shown at the bottom of Figure 19.10 for $Cr(NH_3)_6{}^{3+}$.

Metal	Ion	Complex Ion	Orbital Diagram 3d	Orbital Diagram 4s	Orbital Diagram 4p	Hybridization
Cu^+	$3d^{10}$	$Cu(NH_3)_2{}^+$	(↑↓)(↑↓)(↑↓)(↑↓)(↑↓)	(↑↓)	(↑↓)( )( )	sp
Zn^{2+}	$3d^{10}$	$Zn(NH_3)_4{}^{2+}$	(↑↓)(↑↓)(↑↓)(↑↓)(↑↓)	(↑↓)	(↑↓)(↑↓)(↑↓)	sp^3
Ni^{2+}	$3d^8$	$Ni(CN)_4{}^{2-}$	(↑↓)(↑↓)(↑↓)(↑↓)(↑↓)	(↑↓)	(↑↓)(↑↓)( )	dsp^2
Cr^{3+}	$3d^3$	$Cr(NH_3)_6{}^{3+}$	(↑)(↑)(↑)(↑↓)(↑↓)	(↑↓)	(↑↓)(↑↓)(↑↓)	d^2sp^3

Figure 19.10 Orbital diagrams derived from the valence bond model for complexes of Cu^+ (linear), Zn^{2+} (tetrahedral), Ni^{2+} (square planar), and Cr^{3+} (octahedral). Electron pairs furnished by the ligands are shown in color.

Example 19.4   Write an orbital diagram for $Fe(CN)_6^{4-}$.

Solution   Since the coordination number is 6, we expect the bonding electrons from the CN^- ligands to enter d^2sp^3 hybrid orbitals. So, we put them there:

3d   4s   4p
( )( )( )(↑↓)(↑↓)   (↑↓)   (↑↓)(↑↓)(↑↓)

We must now decide how many electrons are contributed by the central metal ion. To do this, we must first deduce its charge. Since the complex ion has a charge of -4 and each CN^- ion has a charge of -1, we must be dealing with a $+2$ ion of iron. That is,

$$Fe^{2+} + 6\ CN^- \rightarrow Fe(CN)_6^{4-}$$

In the $+2$ ion of Fe (at. no. = 26), there must be 24 e^-; six of these are located beyond the argon core:

$$\text{total no. } e^- \text{ in } Fe^{2+} = 26 - 2 = 24$$
$$\text{no. of } e^- \text{ beyond [Ar] in } Fe^{2+} = 24 - 18 = 6$$

We have just enough room in the empty 3d orbitals to accommodate these six electrons, one pair to an orbital. Hence the completed orbital diagram is

3d   4s   4p
$Fe(CN)_6^{4-}$ [$_{18}$Ar] (↑↓)(↑↓)(↑↓)(↑↓)(↑↓)   (↑↓)   (↑↓)(↑↓)(↑↓)

Exercise   Write the orbital diagram for the $Co(NH_3)_6^{3+}$ ion. Answer: Same as $Fe(CN)_6^{4-}$ (why?).

Note that in this ion, which is very stable, Fe has the electronic structure of the noble gas Kr

The valence bond model has been quite successful in explaining certain of the properties of complex ions, notably their magnetic behavior. The structures shown in Figure 19.10 imply that the complex ions $Cu(NH_3)_2^+$, $Zn(NH_3)_4^{2+}$, and $Ni(CN)_4^{2-}$ should all be *diamagnetic;* they have no unpaired electrons. In contrast, the $Cr(NH_3)_6^{3+}$ ion, with three unpaired electrons, should be *paramagnetic*. These predictions are confirmed by experiment. The first three complexes are weakly repelled by a magnetic field. On the other hand, $Cr(NH_3)_6^{3+}$ is attracted into the field with a force that corresponds to three unpaired electrons.

With certain complex ions, however, the observed magnetic behavior does not agree with the structure predicted by valence bond theory. An example is the $Fe(H_2O)_6^{2+}$ ion. We might expect its orbital diagram to be identical with that of the $Fe(CN)_6^{4-}$ ion shown in Example 19.4. Hence, it should be diamagnetic, with no unpaired electrons. By experiment, the $Fe(H_2O)_6^{2+}$ ion is found to be paramagnetic, with four unpaired electrons. It is possible to rationalize this behavior within the framework of valence bond theory.* However, it is by no means obvious why the CN^- ion should form one type of complex with Fe^{2+} and the H_2O molecule a quite different type.

---

*One "explanation" is that the d orbitals occupied by ligand electrons in $Fe(H_2O)_6^{2+}$ are those in the fourth principal energy level rather than the third. The electrons in the Fe^{2+} ion (six beyond Ar) could then spread over all the 3d orbitals, giving the structure shown below

3d   4s   4p   4d
$Fe(H_2O)_6^{2+}$ [$_{18}$Ar] (↑↓)(↑)(↑)(↑)(↑)   (↑↓)   (↑↓)(↑↓)(↑↓)   (↑↓)(↑↓)( )( )( )

which does indeed have four unpaired electrons.

In another area, valence bond theory has been notably deficient. It does not explain the most striking property of coordination compounds, their brilliant colors. A quite different approach, known as the crystal-field model, has been much more successful dealing with this matter.

Crystal-Field Model

This model interprets the bonding between ligands and a metal ion as primarily electrostatic in nature. In this approach ligand electron pairs are *not* donated into orbitals of the central metal ion, as in atomic orbital theory. Instead, it is assumed that the only effect of the ligands is to create an electrostatic field around the d orbitals of the metal ion. **This field changes the relative energies of different d orbitals.** In the isolated metal ion, before interaction occurs, all the d orbitals in a sublevel such as 3d have the same energy. After interaction, the d orbitals are split into two groups with different energies.

To illustrate the crystal-field model, let us consider the formation of the octahedral $Fe(CN)_6^{4-}$ ion:

$$Fe^{2+} + 6\ CN^- \rightarrow Fe(CN)_6^{4-}$$

In this model, the metal ion and ligands do not share electrons

In the free $(3d^6)$ Fe^{2+} ion, all the 3d orbitals are equal in energy. Hence, the six 3d electrons are distributed in accordance with Hund's rule:

$$\begin{array}{c} 3d \\ Fe^{2+} \qquad (\uparrow\downarrow)(\uparrow)(\uparrow)(\uparrow)(\uparrow) \end{array}$$

To form the octahedral complex, the six CN^- ions must approach along the x, y, and z axes (Fig. 19.11*a*). This causes a splitting of the five 3d orbitals into two sets:

1. A higher energy pair, referred to as "$d_{x^2-y^2}$" and "d_{z^2}" orbitals.

2. A lower energy trio, called the d_{xy}, d_{yz}, and d_{xz} orbitals (Fig. 19.11*b*).

To understand why the splitting occurs, we look at the orientation of the electron density clouds associated with the five 3d orbitals (Fig. 19.12). Notice that the two

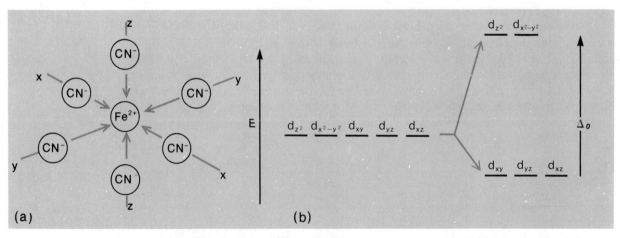

(a) (b)

Figure 19.11 When six ligands (CN^- ions) approach a central metal ion (Fe^{2+}) along the x, y, and z axes, the five d orbitals are split into two groups. Two of these are raised in energy, while the other three are lowered. The difference in energy between the two groups is the crystal-field splitting energy, Δ_o.

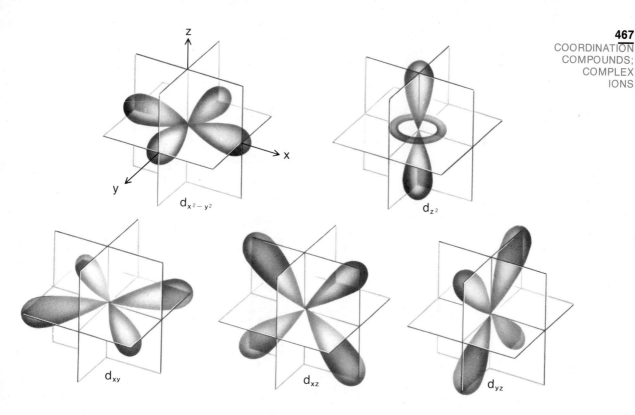

Figure 19.12 Spatial orientation of d orbitals. Note that the $d_{x^2-y^2}$ and d_{z^2} orbitals are oriented toward ligands approaching the corners of an octahedron.

orbitals called $d_{x^2-y^2}$ and d_{z^2} have their maximum electron density directly along the x, y, and z axes, respectively. In contrast, the other d orbitals have their electron densities concentrated between the axes rather than along them. Hence, as CN^- ligands approach along the x, y, and z axes, their electrostatic field is felt more strongly by electrons in the $d_{x^2-y^2}$ and d_{z^2} orbitals than by those in the d_{xy}, d_{yz}, and d_{xz} orbitals. The effect is to split the 3d orbitals into two sets of differing energy. The energy difference is known as the *crystal-field splitting energy*, Δ_0.

With $Fe(CN)_6^{4-}$, Δ_0 is *large enough* to cause the six 3d electrons of Fe^{2+} to *pair up* in the three lower energy orbitals, in "violation" of Hund's rule (Fig. 19.13a). In the $Fe(H_2O)_6^{2+}$ ion, the situation is quite different. Neutral water molecules repel electrons less than negatively charged CN^- ions. As a result, the splitting energy, Δ_0, is *smaller*. Indeed, it is *not* large enough to overcome the tendency for electrons to remain unpaired. In the $Fe(H_2O)_6^{2+}$ ion, as in the Fe^{2+} ion itself, the six 3d electrons are spread out over all the orbitals, in accordance with Hund's rule (Fig. 19.13b).

Looking at Figure 19.13, we see that the crystal field model explains why $Fe(CN)_6^{4-}$ is diamagnetic while $Fe(H_2O)_6^{2+}$ is paramagnetic. In the first complex there are *no* unpaired electrons; in the second there are *four* unpaired electrons. Pairs of complexes such as these are known for many other transition metal ions. The complex with the smaller number of unpaired electrons is known as the **"low-spin"** complex. It occurs, as with $Fe(CN)_6^{4-}$, when the splitting energy is relatively large. In contrast, the **"high-spin"** complex follows Hund's rule and has a larger number of unpaired electrons. Example 19.5 deals with the low-spin and high-spin complexes of Co^{2+}.

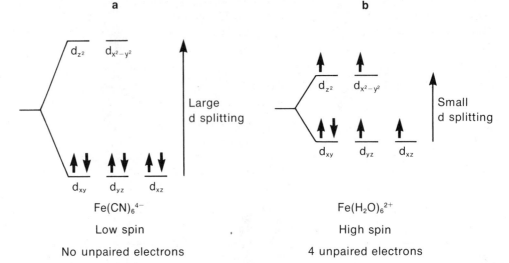

Figure 19.13 In the Fe(CN)$_6^{4-}$ ion, the energy difference between d orbitals, Δ_0, is large. Hence the six 3d electrons of the Fe^{2+} ion pair up in the three orbitals of lower energy. In Fe(H$_2$O)$_6^{2+}$, Δ_0 is small, and the electrons spread out among the five orbitals with a maximum number, 4, of unpaired electrons.

Example 19.5 Using the crystal-field model, derive the structure of the Co^{2+} ion in low-spin and high-spin octahedral complexes.

Solution We first determine the number of d electrons available. Since the atomic number of cobalt is 27, there are $27 - 2 = 25$ electrons in Co^{2+}. Of these, 18 are accounted for by the argon core, leaving seven in the 3d orbitals. In a low-spin complex, formed with strongly interacting ligands, six of these electrons are crowded into the lower three energy levels, leaving only one unpaired electron. In a high-spin complex, where the splitting is small, Hund's rule is followed and there are three unpaired electrons.

Small $\Delta_0 \rightarrow$ high spin
Large $\Delta_0 \rightarrow$ low spin

$(\uparrow\,)(\;\;)$ $(\uparrow\,)(\uparrow\,)$

$(\uparrow\downarrow)(\uparrow\downarrow)(\uparrow\downarrow)$ $(\uparrow\downarrow)(\uparrow\downarrow)(\uparrow\,)$

low-spin   high-spin

Both types of complexes are known. The Co(CN)$_6^{4-}$ ion is of the low-spin type; Co(H$_2$O)$_6^{2+}$ is a high-spin complex.

Exercise Consider a transition metal ion with four 3d electrons, such as Cr^{2+}. How many unpaired electrons are there in a high-spin complex of this ion? a low-spin complex? Answer: Four, two.

Bright line atomic spectra were explained in Chapter 8 in terms of electron transitions between energy levels. Crystal-field theory explains the color of complex ions in a similar way. The energy difference, Δ_0, between the two sets of d orbitals is, in many cases, equivalent to a wavelength in the visible region. Hence, by absorbing visible light, an electron may be able to move from the lower energy set of d orbitals to the higher one. This removes some of the component wavelengths of white light, so the light reflected or transmitted by the complex is colored.

The situation is particularly simple in $_{22}$Ti^{3+}, where there is only one 3d electron. Consider, for example, the Ti(H$_2$O)$_6^{3+}$ ion, which has an intense purple color. This ion absorbs at 510 nm, in the green region. The purple (red-blue) color that we see when we look at a solution of $Ti(H_2O)_6^{3+}$ is what is left when the green component is subtracted from the visible spectrum. Using Equation 8.2, Ch. 8, we can calculate the energy difference, ΔE, corresponding to a wavelength of 510 nm:

$$\Delta E = \frac{1.196 \times 10^5 \text{ kJ} \cdot \text{nm}}{\lambda} \frac{}{\text{mol}} = \frac{1.196 \times 10^5}{5.10 \times 10^2} \frac{\text{kJ}}{\text{mol}} = 234 \text{ kJ/mol}$$

We conclude that this is the energy absorbed to raise the 3d electron from a lower to a higher orbital. In other words, in the $Ti(H_2O)_6^{3+}$ ion, the two sets of d orbitals are separated by this amount of energy. The splitting energy, Δ_0, is 234 kJ/mol.

The magnitude of the splitting energy, Δ_0, determines the wavelength of the light absorbed by a complex and hence its color. This effect is shown in Table 19.5 and color plate 19 (center of book). Notice that when we substitute for NH_3 such ligands as NCS^-, H_2O, or Cl^-, which produce smaller values of Δ_0, the light absorbed shifts to longer wavelengths (lower energies). On the basis of these and other observations, we can arrange various ligands in order of decreasing tendency to split the d orbitals. A short version of such a *spectrochemical* series is

$$CN^- > NO_2^- > en > NH_3 > NCS^- > H_2O > F^- > Cl^-$$

strong field	$\xrightarrow{\text{decreasing } \Delta_0}$	weak field

Certain transition metal ions do not form colored complexes. Among these is Sc^{3+}, which has no d electrons. Another such ion is Zn^{2+}, where the d orbitals are completely filled:

$$Zn^{2+} \; [_{18}Ar]3d^{10}$$

Here, there is no possibility of an electron moving from one d orbital to another, so visible light is not absorbed. As a result, complexes such as $Zn(NH_3)_4^{2+}$ or $Zn(OH)_4^{2-}$ are colorless.

Would $Ag(NH_3)_2^+$ be colored?

Lest you suppose that crystal-field theory can explain all the properties of complex ions, we should point out one of its weaknesses. If the bonding in complex ions is primarily electrostatic, it is hard to see why certain molecules which are only slightly polar, such as CO, can act as ligands. To explain this, it is necessary to modify crystal-field theory to take into account covalent as well as ionic bonding. This is done in a more sophisticated approach known as *ligand-field theory*, which you may study in later chemistry courses.

TABLE 19.5   COLORS OF COMPLEX IONS OF Co^{3+}

COMPLEX	COLOR OBSERVED	COLOR ABSORBED	APPROXIMATE WAVELENGTH (NM) ABSORBED
$Co(NH_3)_6^{3+}$	yellow	violet	430
$Co(NH_3)_5NCS^{2+}$	orange	blue	470
$Co(NH_3)_5H_2O^{3+}$	red	blue-green	500
$Co(NH_3)_5Cl^{2+}$	purple	yellow-green	530
trans-$Co(NH_3)_4Cl_2^+$	green	red	680

19.5 COMPLEX IONS IN WATER; RATE AND EQUILIBRIUM CONSIDERATIONS

For reactions involving complex ions in water solution (as indeed with all reactions), we must be concerned with two factors:

1. *The rate of reaction.* Here we will consider only one type of reaction, that in which there is an exchange of ligands. An example is

$$Cu(H_2O)_4{}^{2+}(aq) + 4\ NH_3(aq) \rightarrow Cu(NH_3)_4{}^{2+}(aq) + 4\ H_2O \qquad (19.8)$$

Here, water molecules in the complex ion are replaced by NH_3 molecules.

2. *The position of the equilibrium involved.* Here we will consider the equilibrium constant for the dissociation of the complex ion. This is commonly called the dissociation constant and given the symbol K_d. A typical example is

Note the form of the equation for K_d

$$Cu(NH_3)_4{}^{2+}(aq) \rightleftharpoons Cu^{2+}(aq) + 4\ NH_3(aq);\ K_d = \frac{[Cu^{2+}] \times [NH_3]^4}{[Cu(NH_3)_4{}^{2+}]} \qquad (19.9)$$

Rate of Ligand Exchange

For reactions such as 19.8, we classify complex ions, more or less arbitrarily, into two types:

1. Complex ions which exchange ligands almost instantaneously are referred to as *labile*. Typically, they exchange ligands in water solution with a half-life of a minute or less. The $Cu(H_2O)_4{}^{2+}$ ion is of this type, as is $Ni(H_2O)_6{}^{2+}$. With the former, exchange occurs too rapidly to be measured. In the case of $Ni(H_2O)_6{}^{2+}$, the half-life for substitution by NH_3 or other ligands is of the order of 0.02 s at 25°C.

2. Complex ions which undergo ligand substitution slowly are called nonlabile or *inert* (color plate 20, center of book). A typical example of an inert complex is $Cr(H_2O)_6{}^{3+}$. Here, the half-life for substitution by NH_3 is about 40 h at 25°C. This means that, in the time required for one $Cr(H_2O)_6{}^{3+}$ ion to react with NH_3, several million labile $Ni(H_2O)_6{}^{2+}$ ions will have undergone reaction.

An interesting illustration of the difference between labile and inert species involves the complexes of iron(II) and iron(III) with the cyanide ion. The complex of Fe^{2+}, $Fe(CN)_6{}^{4-}$, is inert and nonpoisonous. In contrast, the complex of Fe^{3+}, $Fe(CN)_6{}^{3-}$, is labile. It is poisonous because the CN^- ions released in water solution are toxic.

Only a few transition metal ions form inert complexes consistently. Among these are Cr^{3+}, Co^{3+}, Pt^{2+}, and Pt^{4+}. Complexes of these ions were among the first to be studied in the laboratory. Once formed, they are relatively easy to separate from solution in the form of pure compounds. Moreover, inert complexes retain their identity in solution

With labile complexes, geometric isomers cannot be isolated

long enough for their physical and chemical properties to be measured. Much of our knowledge of the structures and properties of complex ions is based upon the behavior of inert complexes in water solution.

Dissociation Constants

Table 19.6 lists dissociation constants for several complex ions. In each case, K_d applies to the dissociation of the complex by a reaction similar to 19.9. The products are the simple cation, (e.g., Ag^+, Cu^{2+}) and ligand molecules or ions (e.g., NH_3, Cl^-). You will notice that in each case K_d is a small number, comparable to or even smaller

TABLE 19.6   DISSOCIATION CONSTANTS OF COMPLEX IONS

MC$_2$		MC$_4$		MC$_6$	
$AgCl_2^-$	1×10^{-5}	$Al(OH)_4^-$	1×10^{-33}	$Co(NH_3)_6^{2+}$	1×10^{-5}
$Ag(NH_3)_2^+$	4×10^{-8}	$Cd(NH_3)_4^{2+}$	1×10^{-7}	$Co(NH_3)_6^{3+}$	1×10^{-35}
$Ag(S_2O_3)_2^{3-}$	1×10^{-13}	$Cu(NH_3)_4^{2+}$	2×10^{-13}	$Fe(CN)_6^{4-}$	1×10^{-35}
$Ag(CN)_2^-$	1×10^{-21}	$Zn(CN)_4^{2-}$	1×10^{-17}	$Fe(CN)_6^{3-}$	1×10^{-42}
$CuCl_2^-$	1×10^{-5}	$Zn(NH_3)_4^{2+}$	1×10^{-9}	$Ni(NH_3)_6^{2+}$	2×10^{-9}
$Cu(NH_3)_2^+$	2×10^{-11}	$Zn(OH)_4^{2-}$	3×10^{-16}		

than the dissociation constants of weak acids (Chap. 18). This means that the complex ion is rather stable towards dissociation in water. Consider, for example, the $Ag(NH_3)_2^+$ ion:

$$Ag(NH_3)_2^+(aq) \rightleftharpoons Ag^+(aq) + 2\ NH_3(aq); \quad K_d = \frac{[Ag^+] \times [NH_3]^2}{[Ag(NH_3)_2^+]} \qquad (19.10)$$

The fact that K_d is very small, 4×10^{-8}, means that the $Ag(NH_3)_2^+$ ion is quite stable in water solution. Looking at it another way, even in dilute solutions of NH_3, most of the silver will be in the form of the complex ion rather than the free Ag^+ ion.

The stabilities of different complexes of the same cation are inversely related to their dissociation constants; the smaller the value of K_d, the more stable the complex. When we find that the equilibrium constant for the reaction

$$Ag(S_2O_3)_2^{3-}(aq) \rightleftharpoons Ag^+(aq) + 2\ S_2O_3^{2-}(aq); \quad K_d = \frac{[Ag^+] \times [S_2O_3^{2-}]^2}{[Ag(S_2O_3)_2^{3-}]}$$

is only 1×10^{-13}, we deduce that the $Ag(S_2O_3)_2^{3-}$ complex is even more stable than $Ag(NH_3)_2^+$ ($K_d = 4 \times 10^{-8}$).

$Ag(S_2O_3)_2^{3-}$ is so stable that AgBr will dissolve in $Na_2S_2O_3$ solution

19.6   HISTORICAL PERSPECTIVE

Alfred Werner and Sophus Mads Jorgensen

The basic ideas concerning the structure and geometry of complex ions presented in this chapter were developed by one of the most gifted individuals in the history of inorganic chemistry, Alfred Werner. His theory of coordination chemistry was published in 1893 when Werner was 26 years old, holding the equivalent of an associate professorship at the University of Zurich. In his paper Werner made the revolutionary suggestion that metal ions such as Co^{3+} could show two different kinds of valences. For the compound $Co(NH_3)_6Cl_3$, Werner postulated a central Co^{3+} ion joined by ''primary valences'' (ionic bonds) to three Cl^- ions and by ''secondary valences'' (coordinate covalent bonds) to six NH_3 molecules. Moreover, he made the inspired guess that the six secondary valences were directed toward the corners of a regular octahedron. In Table 19.7 we list the familiar Werner structures for the series of compounds $Co(NH_3)_xCl_3$, where x = 6, 5, 4, or 3. All these compounds were known at the time, and many of their properties had been established. In particular, it was known from conductivity studies and by precipitation with $AgNO_3$ that the first three members of the series yielded 3, 2,

TABLE 19.7   STRUCTURE AND PROPERTIES OF THE COMPOUNDS
$Co(NH_3)_xCl_3$

	STRUCTURE		MOLES Cl^- PER MOLE COMPOUND		
x	Werner	Jorgensen	Werner	Jorgensen	Observed
6	$[Co(NH_3)_6]^{3+}$, 3 Cl^-	Co with NH_3—Cl, NH_3—NH_3—NH_3—NH_3—Cl, NH_3—Cl	3	3	3
5	$[Co(NH_3)_5Cl]^{2+}$, 2 Cl^-	Co with Cl, NH_3—NH_3—NH_3—NH_3—Cl, NH_3—Cl	2	2	2
4	$[Co(NH_3)_4Cl_2]^+$, Cl^-	Co with Cl, NH_3—NH_3—NH_3—NH_3—Cl, Cl	1	1	1
3	$[Co(NH_3)_3Cl_3]^0$	Co with Cl, NH_3—NH_3—NH_3—NH_3—Cl, Cl	0	1	?

and 1 mol of Cl^-, respectively, when dissolved in water. This evidence was, of course, in complete agreement with Werner's theory.

Werner's structures, which seem so obvious to us today, aroused little enthusiasm among his contemporaries. The opposition was led by Sophus Mads Jorgensen, a 56-year-old professor of chemistry at the University of Copenhagen. Jorgensen was convinced that Co^{3+} could form no more than three bonds. To rationalize the existence of "addition compounds" of $CoCl_3$ containing six, five, four, or three NH_3 molecules, he invoked the chain structures shown in Table 19.7, where NH_3 molecules are linked together much like CH_2 groups in hydrocarbons. The differing extents of ionization of these compounds in water were explained by assuming that only those chlorine atoms bonded to NH_3 groups could ionize. Chlorines attached directly to cobalt were supposed to be held so tightly that they could not ionize in water.

From the data in Table 19.7 it would appear that the controversy between Werner and Jorgensen could have been settled quite simply by studying the behavior in water solution of the compound $Co(NH_3)_3Cl_3$. Werner's structure required that this species be a nonelectrolyte with no ionizable chlorine. In contrast, the chain structure of Jorgensen implied one ionizable chlorine, i.e., a 1:1 electrolyte similar to NaCl. Unfortunately the evidence was ambiguous; at 25°C, a water solution of $Co(NH_3)_3Cl_3$ has a conductivity intermediate between that of a nonelectrolyte and a 1:1 salt.

Werner's assumption of octahedral coordination around the Co^{3+} ion offered further opportunities for testing his ideas against those of Jorgensen. If Werner were correct, there should be two isomeric forms (cis and trans) of the compound $Co(NH_3)_4Cl_3$. At the time, only one compound of this formula was known. All of Werner's early attempts to prepare a second isomer failed, thereby weakening his position.

As the years passed, the weight of evidence began to shift toward Werner's structures. Studies at 0°C gave a very low value for the conductivity of $Co(NH_3)_3Cl_3$, which tended to support Werner's contention that the anomalous conductivity at 25°C was due to the reaction

$$Co(NH_3)_3Cl_3(s) + H_2O \rightarrow [Co(NH_3)_3(H_2O)Cl_2]^+(aq) + Cl^-(aq)$$

Moreover, Werner showed that the compound $Co(NH_3)_3(NO_2)_3$, which is entirely analogous to $Co(NH_3)_3Cl_3$, behaves as a true nonelectrolyte even at 25°C.

In 1907 Werner, after years of effort, finally succeeded in preparing a second isomer of the compound $Co(NH_3)_4Cl_3$. Jorgensen graciously accepted this new evidence as conclusive proof of Werner's structures, and the chain theory of coordination chemistry faded away. Six years later, in 1913, Alfred Werner received the Nobel Prize in chemistry.

SUMMARY

A complex ion consists of a central metal cation surrounded by ligands which may be molecules or anions. A ligand may form one or more bonds with the metal ion; if it forms more than one bond, it is called a chelating agent. The charge of the complex is the sum of the charges of the metal ion and the ligands (Example 19.1).

The coordination number of the metal ion is the number of bonds that it forms. Coordination numbers of 6, 4, and 2, in that order, are most common (Table 19.2). Complexes with a coordination number of 2 are linear. Those with a coordination number of 4 may be square planar or tetrahedral. A coordination number of 6 gives an octahedral complex (Fig. 19.4). Square planar and octahedral complexes can show cis-trans isomerism (Example 19.2).

Two different models of bonding in complex ions were presented in this chapter. Both start with the electron configuration of the central cation (Table 19.3 and Example 19.3). In the valence bond model, electron pairs from the ligands are fed into hybrid orbitals of the metal ion (sp, sp^3, dsp^2, or d^2sp^3). With this model, it is possible to derive orbital diagrams for complex ions (Example 19.4). In the crystal-field model, the only effect of the ligands is to change the relative energies of the d orbitals of the metal ion. In an octahedral complex, two of these orbitals are raised in energy as compared to the other three. This leads to the possibility of "high-spin" and "low-spin" complexes, differing in the number of unpaired electrons (Example 19.5).

In considering the reactivity of complex ions, we have to take into account two factors. One is the lability of the complex, which describes the rate at which it undergoes ligand exchange. The other is the dissociation constant of the complex, which describes the tendency of ligands to dissociate from the complex.

KEY WORDS AND CONCEPTS

coordinate covalent bond	chelating agent	orbital diagram	high-spin complex
complex ion	geometric *(cis, trans)* isomers	hybrid orbital	spectrochemical series
ligand	diamagnetic	crystal-field model	labile complex
coordination number	paramagnetic	crystal-field splitting energy	inert complex
polydentate ligand	atomic orbital model	low-spin complex	dissociation constant, K_d

QUESTIONS AND PROBLEMS

Catalog

Charge and Composition of Complex Ions: 19.2–19.4, 19.23–19.25

Chelating Agents; Coordination Number: 19.5, 19.6, 19.26, 19.27

Geometry of Complex Ions: 19.7–19.9, 19.28–19.30

Electron Configurations, Metal Ions: 19.10, 19.11, 19.31, 19.32

Valence Bond Model; Orbital Diagrams: 19.12–19.14, 19.33–19.35

Crystal-Field Model: 19.15–19.18, 19.36–19.39

Lability, Dissociation of Complex Ions: 19.19, 19.20, 19.40, 19.41

General: 19.1, 19.21, 19.22, 19.42–19.46

19.1 Review and know the meaning of the key words and concepts in this chapter.

19.2 For the complex ion $Co(NH_3)_3(H_2O)_2Cl^+$,

   a. identify the ligands and their charges.
   b. what is the charge of the central ion?
   c. what would be the formula of the nitrate salt of this ion?

19.3 What is the charge of the complex of Pt^{2+} in which the ligands are

   a. one NH_3 molecule, one water molecule, and two Cl^- ions?
   b. one ethylenediamine molecule, one OH^- ion, and one Cl^- ion?
   c. two ammonia molecules, one pyridine molecule, and one chloride ion.

19.4 1.00 mol of each of the following coordination compounds is dissolved in water. How many moles of ions are formed?

   a. $[Co(NH_3)_5Cl](NO_3)_2$
   b. $[Co(NH_3)_3(NO_2)_3]$
   c. $K_3[Co(NO_2)_6]$
   d. $[Co(en)_2Cl_2]Cl$

19.5 Which of the following would you expect to be effective chelating agents?

   a. CH_3CH_2OH
   b. $H_2N-(CH_2)_3-NH_2$
   c. $HO-\overset{\overset{O}{\|}}{C}-CH_2-\overset{\overset{O}{\|}}{C}-OH$
   d. PH_3

19.23 Given the complex ion $Cr(NH_3)_2(H_2O)_2Cl_2^+$,

   a. identify the ligands and give their charges.
   b. what is the charge on the central metal ion?
   c. what would be the charge of the complex ion if bromide ions replaced the H_2O?

19.24 State the charge on the central metal ion in each of the following:

   a. $MnCl_4^{2-}$
   b. $Cr(en)_2(H_2O)_2^{3+}$
   c. $Ru(H_2O)Cl_5^{2-}$

19.25 For each of the compounds in Question 19.4, state which of the following would most closely resemble it in colligative properties and conductivity: $CO(NH_2)_2$ (urea), KCl, K_2SO_4, K_3PO_4.

19.26 Classify the following ligands as monodentate, bidentate, etc.:

   a. $(CH_3)_3P$
   b. $N(CH_2CO_2)_3^{3-}$
   c. $H_2N-(CH_2)_2-NH-(CH_2)_2-NH_2$
   d. H_2O

19.6 What is the coordination number of the metal ion in

    a. $Co(H_2O)_6^{3+}$?   b. $Pt(NH_3)_2(C_2O_4)$?
    c. $Rh(en)_2Br_2^+$?   d. $Co(NH_3)_2(H_2O)_2Cl_2^+$?

19.7 Sketch the geometry of

    a. $Ni(NH_3)_5H_2O^{2+}$.
    b. $Zn(OH)_4^{2-}$ (tetrahedral).
    c. $Ni(en)_3^{2+}$.
    d. *trans*-$Ni(H_2O)_2Cl_2$.
    e. *cis*-$Pt(en)_2Cl_2^{2+}$.

19.8 Which of the following octahedral complexes show geometric isomerism? If isomers are possible, draw their structures.

    a. $Cr(NH_3)_5H_2O^{3+}$ b. $Cr(H_2O)_4Cl_2^+$
    c. $Cr(H_2O)_3Cl_3$ d. $Cr(H_2O)_2Cl_4^-$

19.9 Draw as many structural formulas as possible for octahedral complexes in compounds of the formula $Co(NH_3)_4Cl_2Br$.

19.10 Give the electron configuration of

    a. Cr^{2+} b. Mn^{3+} c. Fe^{2+} d. Ni^{2+}
    e. Ni^{3+}

19.11 Give the number of outer d electrons in

    a. Ag^+ b. Pd^{2+} c. Pt^{4+} d. Au^{3+}

19.12 Using the valence bond model, draw orbital diagrams to indicate the electron distribution around the central ion in

    a. square planar $Ni(CN)_4^{2-}$.
    b. $Cr(NH_3)_6^{3+}$.
    c. $Co(en)_3^{3+}$.
    d. $ZnCl_4^{2-}$ (tetrahedral).

19.13 Use the valence bond model and orbital diagrams to account for the fact that square planar Ni^{2+} complexes are diamagnetic while the tetrahedral complexes are paramagnetic.

19.14 State the number of unpaired electrons you would expect to find in the octahedral complexes of the following ions, using the valence bond model:

    a. Co^{3+} b. Cr^{3+} c. Mn^{2+} d. Fe^{2+}

19.15 Using the crystal-field model, give the electron distribution in the low-spin and high-spin complexes of

    a. Co^{3+} b. Mn^{3+}

19.16 Using crystal-field theory, explain why Ni^{2+} does not form high- and low-spin octahedral complexes.

19.17 $Cr(H_2O)_6^{3+}$ has a d orbital electronic transition where ΔE is 208 kJ/mol. Calculate the wavelength (nm) for this transition.

19.27 What is the coordination number of the metal ion in

    a. $W(OH)Cl_5$?   b. $Ru(NH_3)_4(C_2O_4)^+$?
    c. $Pt(NH_3)_2ClBr$?   d. $Cr(en)_3^{3+}$?

19.28 Sketch the geometry of

    a. $Pt(en)Br_4$.
    b. *trans*-$Pd(NH_3)_2(H_2O)Cl^+$.
    c. VCl_4^- (tetrahedral).
    d. *cis*-$Ni(H_2O)_2Cl_2$.
    e. *trans*-$Pt(en)_2Cl_2^{2+}$.

19.29 Follow the directions of Question 19.8 for

    a. $Fe(CN)_6^{4-}$.
    b. $Fe(H_2O)_2Cl_3Br^-$.
    c. $Fe(H_2O)_2Cl_2Br_2^{2-}$.

19.30 How many different octahedral complexes of Co^{3+} can you write using only ethylenediamine and/or Cl^- as ligands?

19.31 Give the electron configuration of

    a. Mn^+ b. Cr^{3+} c. Cu^+ d. V^{3+}

19.32 Give the number of outer d electrons in

    a. Rh^{3+} b. Zr^{4+} c. Re^{3+} d. Hg^{2+}

19.33 Follow directions for Question 19.12 for the following complexes:

    a. $NiCl_4^{2-}$ (tetrahedral)
    b. $Fe(CN)_6^{3-}$
    c. $Mn(H_2O)_6^{3+}$
    d. $Co(NH_3)_6^{3+}$

19.34 Use the valence bond model to show why the $CuCl_2^-$ ion is diamagnetic while the $CuCl_4^{2-}$ ion is paramagnetic.

19.35 State the number of unpaired electrons you would expect to find in the tetrahedral complexes of the following ions, using the valence bond model:

    a. V^{3+} b. Ni^{2+} c. Cu^+

19.36 Follow the directions of Question 19.15 for

    a. Fe^{2+} b. Mn^{2+}

19.37 V^{3+} does not form high- and low-spin octahedral complexes. Use crystal-field theory to explain why this is so.

19.38 The 460-nm absorption in MnF_6^{2-} corresponds to its crystal-field splitting energy, Δ_0. Find Δ_0 in kJ/mol.

19.18 Describe how the diamagnetic behavior of $Co(en)_3^{3+}$ is explained by the

 a. valence bond model.
 b. crystal-field theory.

19.19 Describe a simple experiment to demonstrate the lability of $Ni(H_2O)_6^{2+}$ compared to the inertness of $Cr(H_2O)_6^{3+}$.

19.20 Using Table 19.6 and the rule of multiple equilibria (Chap. 18), calculate K for the reaction

$$Zn(OH)_4^{2-}(aq) + 4\ CN^-(aq) \rightleftharpoons Zn(CN)_4^{2-}(aq) + 4\ OH^-(aq)$$

19.21 Explain why

 a. oxalic acid solution removes rust stains.
 b. EDTA is used in plant food.
 c. a pale green solution of nickel turns blue when NH_3 is added.
 d. the properties of Co^{3+} complexes are easier to study than those of Co^{2+} complexes.

19.22 A chemist synthesizes two coordination compounds. One compound decomposes at 210°C, the other at 240°C. Analysis of the compounds gives the same mass % data: 52.6% Pt, 7.6% N, 1.63% H, and 38.2% Cl. Both compounds contain a +4 central ion.

 a. What is the simplest formula of the compounds?
 b. Draw structural formulas for the complexes present.

19.39 $V(H_2O)_6^{3+}$ has two unpaired electrons per V^{3+}. Account for this using

 a. the valence bond model.
 b. crystal-field theory.

19.40 The *cis* and *trans* isomers of $Co(en)_2(H_2O)(NCS)^{2+}$ differ in color. Suggest an experimental technique which might be useful to measure the rate of *cis* to *trans* isomerization.

19.41 K_d for the $HgCl_4^{2-}$ ion is 1×10^{-16}. Calculate the ratio: $[Hg^{2+}]/[HgCl_4^{2-}]$ in a solution in which $[Cl^-] = 0.10$ M.

19.42 Indicate each of the following statements as true or false. If false, correct the statement to make it true.

 a. In $Pt(NH_3)_4Cl_4$, platinum has a +4 charge and a coordination number of 6.
 b. Complexes of Cu^{2+} are brightly colored while those of Zn^{2+} are colorless.
 c. The K_d value for the octahedral complexes A and B are 1×10^{-31} and 1×10^{-24}, respectively. This means that complex B is more stable than complex A.

19.43 Analysis of a coordination compound gives the following results: 22.0% Co, 31.4% N, 6.78% H, and 39.8% Cl. One mole of the compound dissociates in water to form four moles of ions.

 a. What is the formula of the compound?
 b. Write an equation for its dissociation in water.

*19.44 A certain coordination compound has the simplest formula $PtN_2H_6Cl_2$. It has a molar mass of about 600 g and contains both a complex cation and a complex anion. What is its structure?

*19.45 A child eats 10.0 g of paint containing 5.0% Pb. How many grams of the sodium salt of EDTA should he receive to bring the lead into solution?

*19.46 For the first order reaction

$$Cr(NH_3)_6^{3+}(aq) + H_2O \rightarrow Cr(NH_3)_5(H_2O)^{3+}(aq) + NH_3(aq)$$

$k = 8.0 \times 10^{-5}$/min at 40°C and E_a is 110 kJ. What is the half-life of this reaction at 25°C?

QUALITATIVE ANALYSIS

Ionic solutes are present in a wide variety of aqueous solutions. The solution may be a lake or stream sample, the body fluid of a hospital patient, or the waste water discharged from an industrial plant. The methods used to determine the ions present in such solutions make up the area of chemistry known as inorganic qualitative analysis.

Many different approaches can be used to test for the presence of ions in water solution. Ions can be identified on the basis of such physical properties as color, absorption spectra, or behavior in an electric or magnetic field. Physical methods such as these are rapid and can be adapted to detect ions at very low concentrations. Alternatively, we can use the chemical properties of ions to develop a scheme of qualitative analysis. This is the approach we will follow in this chapter, concentrating upon cations. We will consider how precipitation (Chap. 16), acid-base reactions (Chap. 17), and complex ion formation (Chap. 19) can be used in qualitative analysis.

Over the years, there has been developed a rather elaborate scheme of qualitative analysis, capable of testing for some 15 different anions and up to 25 different cations. The details of this scheme are described in standard texts on qualitative analysis.* Here our goal is less ambitious. In Section 20.1, we will consider the general features of this scheme, only from the viewpoint of separating cations into groups. Section 20.2 describes in detail the separation and identification of Group I cations. Section 20.3 shows how the principles and calculations of solution equilibria (Chap. 18) are applied to cation analysis.

20.1 ANALYSIS OF CATIONS; AN OVERVIEW

The analysis for cations in a mixture requires a systematic approach. Generally, the procedure is to remove successive groups of cations by precipitation from solution. The standard scheme of qualitative analysis separates cations into four groups (Table 20.1). Concentrations of reagents and solution pH are adjusted so that only one group is affected by a given precipitating agent. The group precipitate is removed from the solution by centrifuging the mixture and pouring the remaining solution from the solid. Within a given group, cations are separated and identified by selective chemical reactions.

In the qual scheme, the groups are separated one after the other from the solution of the sample

---

*See also *Chemical Principles with Qualitative Analysis,* W. L. Masterton and E. J. Slowinski, W. B. Saunders, 1978, and the accompanying laboratory manual.

TABLE 20.1   CATION GROUPS OF QUALITATIVE ANALYSIS

GROUP	CATIONS	PRECIPITATING REAGENT/CONDITIONS
I	Ag^+, Pb^{2+}, Hg_2^{2+}	6 M HCl
II	Cu^{2+}, Bi^{3+}, Hg^{2+}, Cd^{2+}, Sn^{4+}, Sb^{3+}	0.1 M H_2S at a pH of 0.5
III	Al^{3+}, Cr^{3+}, Co^{2+}, Fe^{2+}, Mn^{2+}, Ni^{2+}, Zn^{2+}	0.1 M H_2S at a pH of 9
IV	Ba^{2+}, Ca^{2+}, Mg^{2+}; Na^+, K^+, NH_4^+	0.2 M $(NH_4)_2CO_3$ at a pH of 9.5 No precipitates; separate tests for identification

Group I: Ag^+, Pb^{2+}, Hg_2^{2+}

To separate these three cations we add 6 M HCl. The members of Group I precipitate as white chlorides ($AgCl$, $PbCl_2$, Hg_2Cl_2).

Group II: Cu^{2+}, Bi^{3+}, Hg^{2+}, Cd^{2+}, Sn^{4+}, Sb^{3+}

Cations of Groups II, III, and IV do not form insoluble chlorides. Hence, they remain in solution after addition of 6 M HCl. To separate Group II from Groups III and IV, the H^+ ion concentration is first reduced to about 0.3 M (pH = 0.5). The solution is then saturated with hydrogen sulfide, H_2S. This precipitates the very insoluble sulfides of the Group II cations. The reaction of H_2S with Cu^{2+} is typical and may be considered to occur in two steps:

The sulfides of Group II have very low solubilities

H₂S dissociation: $H_2S(aq) \rightleftharpoons 2 H^+(aq) + S^{2-}(aq)$

CuS precipitation: $\dfrac{Cu^{2+}(aq) + S^{2-}(aq) \rightarrow CuS(s)}{Cu^{2+}(aq) + H_2S(aq) \rightarrow CuS(s) + 2 H^+(aq)}$ (20.1)

Several Group II sulfides have distinctive, identifying colors (see color plate 21, center of book).

Example 20.1   Write balanced equations for the precipitation of
  a. Pb^{2+} by HCl   b. Bi^{3+} by H_2S

Solution
  a. The product is $PbCl_2$. The Cl^- ions come from the strong acid HCl:

$$Pb^{2+}(aq) + 2 Cl^-(aq) \rightarrow PbCl_2(s)$$

  b. The product is Bi_2S_3. The sulfide ions come from the weak acid H_2S:

$$2 Bi^{3+}(aq) + 3 H_2S(aq) \rightarrow Bi_2S_3(s) + 6 H^+(aq)$$

HCl is 100% ionized in water solution. H_2S is not apprecialy ionized in solution

---

Exercise   Write a balanced equation for the precipitation of Hg_2^{2+} by HCl. Answer: $Hg_2^{2+}(aq) + 2 Cl^-(aq) \rightarrow Hg_2Cl_2(s)$.

The Group III cations form sulfides which are more soluble than those of Group II. Hence they do not precipitate at the very low S^{2-} concentration (10^{-20} M) present at $[H^+] = 0.3$ M. To bring the Group III cations out of solution, $[H^+]$ is reduced to about 10^{-9} M and the solution saturated with H_2S. Under these conditions, the equilibrium

$$H_2S(aq) \rightleftharpoons 2\ H^+(aq) + S^{2-}(aq)$$

shifts to the right. The concentration of S^{2-} becomes high enough to precipitate Co^{2+}, Fe^{2+}, Mn^{2+}, Ni^{2+}, and Zn^{2+} as the sulfides (see color plate 22, center of book). The reaction with Ni^{2+} is typical:

Lowering [H⁺] drives the reaction to the right

$$Ni^{2+}(aq) + H_2S(aq) \rightarrow NiS(aq) + 2\ H^+(aq) \qquad (20.2)$$

The two remaining ions in Group III—Al^{3+} and Cr^{3+}—precipitate as the hydroxides, $Al(OH)_3$ and $Cr(OH)_3$, rather than as sulfides.

Example 20.2   How would you analyze a solution which might contain Ag^+, Sn^{4+}, and Zn^{2+} but no other cations?

Solution Ag^+ is in Group I, Sn^{4+} in Group II, and Zn^{2+} in Group III. First add 6 M HCl; if a precipitate forms, it must be AgCl (white). Then adjust $[H^+]$ to 0.3 M and saturate with H_2S. If Sn^{4+} is present, it should precipitate as SnS_2 (yellow). Finally, adjust the pH to 9 and saturate again with H_2S; Zn^{2+} will precipitate as ZnS (white).

When a precipitate forms, separate it from the solution

Exercise   Write balanced equations for the reactions that occur when Ag^+, Sn^{4+}, and Zn^{2+} precipitate as just described. Answer:

$$Ag^+(aq) + Cl^-(aq) \rightarrow AgCl(s)$$
$$Sn^{4+}(aq) + 2\ H_2S(aq) \rightarrow SnS_2(s) + 4\ H^+(aq)$$
$$Zn^{2+}(aq) + H_2S(aq) \rightarrow ZnS(s) + 2\ H^+(aq)$$

Group IV: Ba²⁺, Ca²⁺, Mg²⁺, Na⁺, K⁺, NH₄⁺

Compounds of these ions are generally more soluble than those of ions in the preceding groups. The Group IV cations do not form precipitates with Cl^- or H_2S. The alkaline earth cations—Ba^{2+}, Ca^{2+}, and to a lesser extent Mg^{2+}—are precipitated as carbonates using 0.2 M $(NH_4)_2CO_3$. The precipitating solution is buffered with NH_3 to give a pH of about 9.5. The reaction of Ca^{2+} with CO_3^{2-} ions is typical:

$$Ca^{2+}(aq) + CO_3^{2-}(aq) \rightarrow CaCO_3(s) \qquad (20.3)$$

The three remaining cations in Group IV—Na^+, K^+, and NH_4^+—stay in solution, since their carbonates are soluble. These ions are tested for individually. The tests are carried out on the original solution, rather than one from which Groups I, II, and III have been removed. One reason for this is that reagents added in the separation scheme commonly contain Na^+ and NH_4^+ ions.

The two alkali metal ions, Na^+ and K^+, are determined by flame tests. The Na^+ ion gives an intense yellow color. The test for K^+ (violet) is much weaker. Usually, it is observed through a cobalt-blue glass plate to minimize interference by Na^+.

To test for NH$_4^+$, a small sample of the original solution is heated with 6 M NaOH. An acid-base reaction occurs between the OH$^-$ ion and the weak acid NH$_4^+$:

$$NH_4^+(aq) + OH^-(aq) \rightarrow NH_3(aq) + H_2O \qquad (20.4)$$

Ammonia gas can be detected by its odor or its basic properties. It is the only common gas that turns moist red litmus blue.

Example 20.3   A solution may contain any of the ions in Groups I through IV. Addition of 6 M HCl gives no precipitate. The H$^+$ concentration is adjusted to 0.3 M and the solution saturated with H$_2$S; no precipitate forms. However, when [H$^+$] is reduced to 10^{-9} M, and the solution again saturated with H$_2$S, a precipitate forms. This precipitate is later shown to be a Group III hydroxide. In the Group IV analysis, a precipitate is formed with (NH$_4$)$_2$CO$_3$. In the solution, what ions *may* be present from
    a. Group I?   b. Group II?   c. Group III?   d. Group IV?

Solution
    a. No Group I ions (would precipitate as chlorides).
    b. No Group II ions (would precipitate as sulfides).
    c. Al^{3+} or Cr^{3+}. These are the only ions in Group III which precipitate as hydroxides in the group separation.
    d. Any Group IV ion may be present. The carbonate precipitate could be BaCO$_3$, CaCO$_3$, or MgCO$_3$.

Exercise   How would you test the original solution for Na$^+$? NH$_4^+$? Answer: Flame test for Na$^+$ (yellow); heat with strong base and test for NH$_3$ to detect NH$_4^+$.

20.2   ANALYSIS FOR GROUP I

We will not attempt to describe the procedures used to separate and identify all the individual cations listed in Table 20.1. It will, however, be useful to illustrate the general approach followed by considering in some detail the analysis for Group I. Here, only three ions are involved: Ag$^+$, Pb^{2+}, and Hg$_2^{2+}$.*

A flow sheet for the analysis of the Group I cations is shown in Figure 20.1. After precipitation with HCl, the next step in the analysis involves heating the solids with water. Lead chloride, PbCl$_2$, is considerably more soluble than either AgCl or Hg$_2$Cl$_2$. Its solubility increases with temperature to the extent that it dissolves fairly readily at 100°C:

$$PbCl_2(s) \rightarrow Pb^{2+}(aq) + 2\ Cl^-(aq) \qquad (20.5)$$

The hot solution is quickly centrifuged to separate it from AgCl or Hg$_2$Cl$_2$. The presence of Pb^{2+} in the solution is detected by adding a solution of potassium chromate, K$_2$CrO$_4$. A bright yellow precipitate of lead chromate forms if lead is present:

$$Pb^{2+}(aq) + CrO_4^{2-}(aq) \rightarrow PbCrO_4(s) \qquad (20.6)$$
$$\text{yellow}$$

---

*The Hg$_2^{2+}$ ion has the structure [Hg–Hg]$^{2+}$.

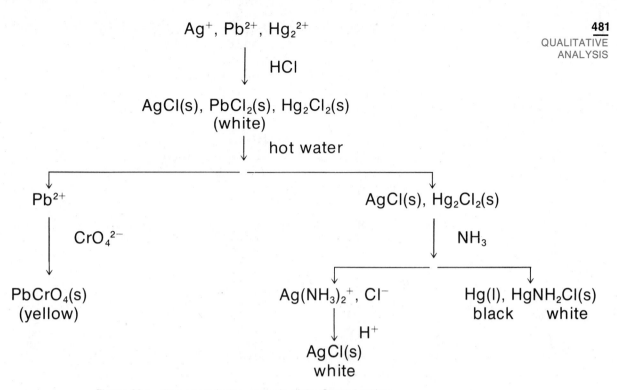

Figure 20.1 Flow sheet for the analysis of the Group I cations.

Any precipitate remaining after the removal of Pb^{2+} is treated with aqueous ammonia. If Hg_2Cl_2 is present, it undergoes the following reaction with NH_3:

$$Hg_2Cl_2(s) + 2\ NH_3(aq) \rightarrow HgNH_2Cl(s) + Hg(l) + NH_4^+(aq) + Cl^-(aq) \qquad (20.7)$$
$$\text{white} \qquad\qquad \text{black}$$

The gray precipitate formed is a mixture of finely divided mercury, which appears black, and $HgNH_2Cl$, which is white.

The addition of aqueous ammonia brings AgCl into solution. It does this by forming the stable $Ag(NH_3)_2^+$ complex ion. We might consider that this reaction occurs in two steps:

The silver and mercury are separated by addition of NH_3

1. $$AgCl(s) \rightarrow Ag^+(s) + Cl^-(aq)$$
2. $$\underline{Ag^+(aq) + 2\ NH_3(aq) \rightarrow Ag(NH_3)_2^+(aq)}$$
$$AgCl(s) + 2\ NH_3(aq) \rightarrow Ag(NH_3)_2^+(aq) + Cl^-(aq) \qquad (20.8)$$

In a sense, NH_3 molecules and Cl^- ions "compete" for Ag^+ ions. The NH_3 molecules "win" and the AgCl dissolves.

The solution formed by Reaction 20.8 is separated from any precipitate of Hg_2^{2+} by centrifuging. The solution must then be tested to establish the presence of silver. To do this, we add a strong acid, 6 M HNO_3. The H^+ ions of the acid destroy the $Ag(NH_3)_2^+$ complex ion. The free Ag^+ ions formed combine with the Cl^- ions already present to precipitate AgCl. Again, we can think of this as a two-step reaction. The first step is an acid-base reaction in which NH_3 molecules in the complex ion are converted to NH_4^+ ions:

$$\begin{array}{ll}
1. & Ag(NH_3)_2^+(aq) + 2\,H^+(aq) \rightarrow Ag^+(aq) + 2\,NH_4^+(aq) \\
2. & \underline{Ag^+(aq) + Cl^-(aq) \rightarrow AgCl(s)} \\
& Ag(NH_3)_2^+(aq) + 2\,H^+(aq) + Cl^-(aq) \rightarrow AgCl(s) + 2\,NH_4^+(aq) \quad (20.9)
\end{array}$$

Example 20.4   A Group I unknown gives a white precipitate with HCl. Treatment with hot water gives a solution to which K_2CrO_4 is added; a yellow precipitate forms. The white precipitate remaining from the hot water treatment dissolves completely in NH_3. What ions are present? absent?

Solution Pb^{2+} is present ($PbCl_2$ dissolves in hot water and then precipitates $PbCrO_4$). Hg_2^{2+} is absent (Hg_2Cl_2 would not dissolve in NH_3). Ag^+ is present (AgCl dissolves in NH_3).

Exercise   If HNO_3 is added to the solution formed with NH_3, a white precipitate forms. Write the equation for the reaction involved. Answer: Equation 20.9.

20.3   SOLUTION EQUILIBRIA IN CATION ANALYSIS

Cation analysis involves a wide variety of chemical reactions. A few of these are of the oxidation-reduction type (to be discussed in Chapters 21 and 22). Most of them, however, are of a type discussed in previous chapters (16, 17, 19). The extent to which these reactions occur can be determined using the principles of solution equilibria introduced in Chapter 18. Indeed, the entire scheme of cation analysis is based upon equilibrium considerations. In this section, we will look at the equilibria involved in the various types of reactions used in cation analysis.

Formation of Complex Ions

In one type of reaction in cation analysis, a simple cation in water solution is converted to a complex ion. Most often, the complexing agent (ligand) is an NH_3 molecule or an OH^- ion. Typical reactions of this type are

$$Cu^{2+}(aq) + 4\,NH_3(aq) \rightleftharpoons Cu(NH_3)_4^{2+}(aq) \quad (20.10)$$

$$Ni^{2+}(aq) + 6\,NH_3(aq) \rightleftharpoons Ni(NH_3)_6^{2+}(aq) \quad (20.11)$$

$$Al^{3+}(aq) + 4\,OH^-(aq) \rightleftharpoons Al(OH)_4^-(aq) \quad (20.12)$$

Cations in the qualitative analysis scheme which commonly form complex ions with NH_3 or OH^- are listed in Table 20.2. You will find this table very helpful in writing equations involving complex ions.

The extent to which complex ion formation takes place depends upon the magnitude of the equilibrium constant, K, for the reaction. This in turn can be calculated from the

TABLE 20.2   COMPLEXES OF CATIONS WITH NH_3 AND OH^-

CATION	NH_3 COMPLEX	OH^- COMPLEX
Ag^+	$Ag(NH_3)_2^+$	
Pb^{2+}		$Pb(OH)_3^-$
Cu^{2+}	$Cu(NH_3)_4^{2+}$ (blue)	
Cd^{2+}	$Cd(NH_3)_4^{2+}$	
Sn^{4+}		$Sn(OH)_6^{2-}$
Sb^{3+}		$Sb(OH)_4^-$
Al^{3+}		$Al(OH)_4^-$
Ni^{2+}	$Ni(NH_3)_6^{2+}$ (blue)	
Zn^{2+}	$Zn(NH_3)_4^{2+}$	$Zn(OH)_4^{2-}$

The stabilities of these complexes are high, so even insoluble species, like AgCl, will dissolve in the complexing agent, like 6 M NH_3

dissociation constant, K_d, of the complex ion. You will recall from Chapter 19 that for the reaction

$$Cu(NH_3)_4^{2+}(aq) \rightleftharpoons Cu^{2+}(aq) + 4\ NH_3(aq);\ K_d = 2 \times 10^{-13}$$

Looking at the equation just written, you can see that it is the reverse of Reaction 20.10. Hence we can apply the "reciprocal relation" discussed in Chapter 18 and write

$$K \text{ for Reaction } 20.10 = \frac{1}{K_d\ Cu(NH_3)_4^{2+}} = \frac{1}{2 \times 10^{-13}} = 5 \times 10^{12}$$

The fact that K for this reaction is so large means that Reaction 20.10 goes virtually to completion. Even at rather low concentrations of NH_3, virtually all the Cu^{2+} ions are converted to $Cu(NH_3)_4^{2+}$ (Example 20.5).

Example 20.5   For Reaction 20.10,
  a. write the expression for the equilibrium constant, K.
  b. determine the ratio $[Cu(NH_3)_4^{2+}]/[Cu^{2+}]$ in a solution in which $[NH_3] = 1$ M, given that $K = 5 \times 10^{12}$.

Solution
  a. $K = \dfrac{[Cu(NH_3)_4^{2+}]}{[Cu^{2+}] \times [NH_3]^4}$

  b. Solving for this ratio: $\dfrac{[Cu(NH_3)_4^{2+}]}{[Cu^{2+}]} = K \times [NH_3]^4 = (5 \times 10^{12}) \times 1^4$
$$= 5 \times 10^{12}$$

  Since this ratio is a very large number, virtually all the copper must be in the form of the $Cu(NH_3)_4^{2+}$ ion rather than the free Cu^{2+} ion.

Exercise   Calculate $[Cu(NH_3)_4^{2+}]/[Cu^{2+}]$ if $[NH_3] = 2$ M. Answer: 8×10^{13}.

The development we have just gone through for Reaction 20.10 can be applied equally well to other reactions of this type, such as 20.11 or 20.12. Here again, the calculated value of K is very large, which means that the reaction goes virtually to completion. This explains why complex ion formation is so useful in qualitative analysis. By adding an appropriate ligand, such as NH$_3$ or OH$^-$, we can convert an ion such as Cu^{2+}, Ni^{2+}, or Al^{3+} to a complex ion. In that way, it can often be separated from an ion which does not form a stable complex with the ligand. This is, for example, the basis of the separation of Cu^{2+} from Bi^{3+} in Group II. At one stage of the analysis these two ions are present together in solution. Addition of ammonia leaves copper in solution as the Cu(NH$_3$)$_4$$^{2+}$ complex. The Bi^{3+} ion does not form a stable complex with NH$_3$; instead, it precipitates as Bi(OH)$_3$.

Decomposition of Complex Ions

Often in qualitative analysis, we wish to destroy a complex ion formed in an earlier stage of the procedure. For example, in one step of the analysis for Group III, Al^{3+} is converted to Al(OH)$_4$$^-$. This separates aluminum from other ions in the group. To test for aluminum later in the analysis, it is necessary to convert the Al(OH)$_4$$^-$ ion back to Al^{3+}.

Complexes containing OH$^-$ or NH$_3$ ligands can be decomposed back to the free cation by adding a strong acid. The attraction of the H$^+$ ion for OH$^-$ or NH$_3$ is great enough to break up the complex. Consider, for example, the reaction of H$^+$ ions with Al(OH)$_4$$^-$:

$$Al(OH)_4^-(aq) + 4\ H^+(aq) \rightleftharpoons Al^{3+}(aq) + 4\ H_2O; \quad K = 10^{23}$$

The fact that K for this reaction is so large means that the reaction goes virtually to completion. A similar situation applies with the Ag(NH$_3$)$_2$$^+$ ion:

$$Ag(NH_3)_2^+(aq) + 2\ H^+(aq) \rightleftharpoons Ag^+(aq) + 2\ NH_4^+(aq); \quad K = 1.7 \times 10^{11}$$

Here again, K is a large number. This explains why, in the last step of the Group I analysis, addition of the strong acid HNO$_3$ destroys the Ag(NH$_3$)$_2$$^+$ complex. In general, we expect complexes containing the NH$_3$ molecule or OH$^-$ ion to be **stable in basic solution but to decompose in strong acid.**

Example 20.6   Write balanced equations for the reactions of H$^+$ with the complexes of Zn^{2+} given in Table 20.2.

Solution   With the OH$^-$ complex, reaction with H$^+$ forms water and the free Zn^{2+} ion:

$$Zn(OH)_4^{2-}(aq) + 4\ H^+(aq) \rightarrow Zn^{2+}(aq) + 4\ H_2O$$

The reaction with the ammonia complex is similar, except that NH$_4$$^+$ ions are formed rather than H$_2$O molecules:

$$Zn(NH_3)_4^{2+}(aq) + 4\ H^+(aq) \rightarrow Zn^{2+}(aq) + 4\ NH_4^+(aq)$$

Exercise   Write a balanced equation for the reaction of the hydroxide complex of Sn^{4+} with H$^+$. Answer: Sn(OH)$_6$$^{2-}$(aq) + 6 H$^+$(aq) $\rightarrow$ Sn^{4+}(aq) + 6 H$_2$O.

Formation of Precipitates

The most common type of reaction in qualitative analysis for cations is precipitation. Of the 22 cations listed in Table 20.1, 19 are separated from solution at one stage or another by precipitate formation. Sometimes, two or more such reactions are involved for a single cation. Recall, for example, that in Group I, lead is first precipitated as $PbCl_2$. Later, after separation from Ag^+ and Hg_2^{2+}, it is precipitated as $PbCrO_4$.

The simplest type of precipitation reaction is one in which the appropriate anion is added directly to a solution containing the cation being tested for. Thus, to separate the Group I cations from solution, we add the strong acid HCl to act as a source of Cl^- ions:

$$Ag^+(aq) + Cl^-(aq) \rightarrow AgCl(s)$$
$$Pb^{2+}(aq) + 2\ Cl^-(aq) \rightarrow PbCl_2(s)$$
$$Hg_2^{2+}(aq) + 2\ Cl^- \rightarrow Hg_2Cl_2(s)$$

For these reactions to be effective, they must lower the concentration of the cation to a very low value. Otherwise, it may cause problems in a later stage of the analysis (Example 20.7).

Example 20.7   In testing a Group I unknown, a student adds enough HCl to make $[Cl^-]$ = 0.10 M. A precipitate of $PbCl_2$ forms (K_{sp}= 1.7×10^{-5}). What is the concentration of Pb^{2+} remaining in solution?

Solution   Recall from Chapter 18 that

$$PbCl_2(s) \rightleftharpoons Pb^{2+}(aq) + 2\ Cl^-(aq);\ K_{sp} = [Pb^{2+}] \times [Cl^-]^2 = 1.7 \times 10^{-5}$$

Solving for Pb^{2+}:

$$[Pb^{2+}] = \frac{1.7 \times 10^{-5}}{[Cl^-]^2} \times \frac{1.7 \times 10^{-5}}{(1.0 \times 10^{-1})^2} = 1.7 \times 10^{-3}\ M$$

This is a relatively high concentration of Pb^{2+}, carried over from Group I analysis. Often, it is sufficient to form a precipitate of PbS ($K_{sp} = 1 \times 10^{-27}$) when H_2S is added in Group II. This explains why a black sulfide that doesn't behave like CuS or Bi_2S_3 sometimes shows up in Group II.

Exercise   Suppose the student adds more HCl, bringing $[Cl^-]$ up to 0.50 M. What would $[Pb^{2+}]$ be under these conditions? Answer: 6.8×10^{-5} M.

That would help clean out the Pb^{2+}

Frequently precipitation reactions are carried out in a somewhat different way. Instead of adding an anion directly, we add a molecule which forms that anion, at low concentration, in water solution. This procedure is used to precipitate hydroxides by adding NH_3. Consider, for example, what happens when we add ammonia to a solution containing Al^{3+}. A precipitate of $Al(OH)_3$ forms. We may consider that the reaction occurs in two steps:

$$3\ NH_3(aq) + 3\ H_2O \rightleftharpoons 3\ NH_4^+(aq) + 3\ OH^-(aq)$$
$$\underline{Al^{3+}(aq) + 3\ OH^-(aq) \rightleftharpoons Al(OH)_3(s)}$$
$$Al^{3+}(aq) + 3\ NH_3(aq) + 3\ H_2O \rightleftharpoons Al(OH)_3(s) + 3\ NH_4^+(aq) \qquad (20.13)$$

The ammonia furnishes the OH^- ions required to form $Al(OH)_3$. However, the concentration of OH^- ions is rather low (perhaps about 10^{-3} M), since NH_3 is a weak base. This, as a matter of fact, is highly desirable. If the concentration of OH^- were too high, the $Al(OH)_4^-$ ion would form (recall Eq. 20.12), and we wouldn't get any precipitate. That is why it would not be a good idea to try to precipitate $Al(OH)_3$ by using a strong base, such as NaOH. In general, if a cation is known to form both a precipitate and a stable complex with OH⁻, we should:

—add NaOH (high OH⁻ concentration) if we want to form the complex ion in solution.

—add NH_3 (low OH⁻ concentration) if we want to precipitate the hydroxide.

Example 20.8 Write a balanced equation for the precipitation of $Mn(OH)_2$ by addition of NH_3 to a solution containing Mn^{2+}.

Solution The equation is entirely analogous to Equation 20.13:

$$Mn^{2+}(aq) + 2\ NH_3(aq) + 2\ H_2O \rightarrow Mn(OH)_2(s) + 2\ NH_4^+(aq)$$

Exercise Of the cations listed in Table 20.2, which would you expect to precipitate as hydroxides upon addition of 6 M NaOH? Answer: Ag^+, Cu^{2+}, Cd^{2+}, and Ni^{2+} should precipitate; the other cations form complexes with OH⁻.

Another molecular precipitating agent is hydrogen sulfide. It acts as a source of sulfide ions in Group II analysis:

$$H_2S(aq) \rightleftharpoons 2\ H^+(aq) + S^{2-}(aq);\ K = \frac{[H^+]^2 \times [S^{2-}]}{[H_2S]} = 1 \times 10^{-20}$$

From the expression for the equilibrium constant for this reaction, we see that the concentration of S^{2-} is *inversely* related to that of H^+. That is,

$$[S^{2-}] = 1 \times 10^{-20} \times \frac{[H_2S]}{[H^+]^2}$$

If [H⁺] is high, [S²⁻] is very low

In the Group II analysis, $[H^+]$ is rather high, about 0.3 M; the concentration of H_2S is about 0.1 M. Under these conditions $[S^{2-}]$ is very low, about 1×10^{-20} M:

$$[S^{2-}] = 1 \times 10^{-20} \times \frac{0.1}{(0.3)^2} = 1 \times 10^{-20}\ M$$

This low concentration of sulfide ion is still high enough to precipitate CuS ($K_{sp} = 1 \times 10^{-35}$), but not NiS ($K_{sp} = 1 \times 10^{-19}$).

Example 20.9 Suppose that a solution is 0.1 M in both Cu^{2+} and Ni^{2+}. Under the conditions of Group II analysis, where $[S^{2-}] = 1 \times 10^{-20}$ M, will CuS precipitate? NiS?

Solution Recall from Chapter 18 that, to decide whether a precipitate forms, we compare the ion product, Q, to the solubility product constant, K_{sp}. Here:

$$Q = (conc.\ M^{2+}) \times (conc.\ S^{2-}) = 1 \times 10^{-1} \times 1 \times 10^{-20} = 1 \times 10^{-21}$$

where M^{2+} represents either Cu^{2+} or Ni^{2+}. We see that Q is *greater* than K_{sp} for CuS, but *less* than K_{sp} for NiS:

If Q > K_{sp} we get precipitation

$$CuS:\ 1 \times 10^{-21} > K_{sp}\ CuS\ (1 \times 10^{-35})$$
$$NiS:\ 1 \times 10^{-21} < K_{sp}\ NiS\ (1 \times 10^{-19})$$

We conclude that CuS should precipitate in Group II, which indeed it does. On the other hand, NiS should *not* precipitate under these conditions. Fortunately, it doesn't, since Ni^{2+} is supposed to be in Group III.

Exercise   The K_{sp} value of a certain sulfide MS is 1×10^{-26}. Would M^{2+} fall in Group II or III? Answer: Group II.

The argument we have just gone through can be applied generally to the cations in Groups II and III. The sulfides of all the Group II cations are very insoluble (low K_{sp} values). This means that they are precipitated by H_2S in Group II, even though the concentration of S^{2-} is very low, about 10^{-20} M. In contrast, the Group III sulfides are more soluble (larger K_{sp} values). As a result, the Group III ions stay in solution when Group II is precipitated.

Dissolving Precipitates

Frequently in qualitative analysis, we need to bring a water-insoluble compound into solution. For example, having precipitated AgCl in Group I, we need to bring the Ag^+ ion back into solution if we are to confirm its presence. Again, to separate the Group II ions from each other, we must bring their sulfides into solution.

There are many different ways to dissolve a precipitate. Sometimes, if it isn't too insoluble, simply raising the temperature will do the trick. This works with lead chloride, which has a rather large K_{sp} value:

$$PbCl_2(s) \rightleftharpoons Pb^{2+}(aq) + 2\ Cl^-(aq);\ K_{sp} = 1.7 \times 10^{-5} \text{ at } 25°C$$
$$= 5 \times 10^{-4} \text{ at } 100°C$$

Usually, however, we have to resort to a chemical reaction to bring about solution. Most often, we use either an acid-base reaction or complex ion formation. Thus, we may:

1. **Add a strong acid (H^+) to react with a basic anion in the precipitate.** In qualitative analysis, this reaction is perhaps most often used to bring water-insoluble hydroxides into solution. Looking at the values for K for the following reactions, you can guess that addition of H^+ ions should dissolve the hydroxides of Ni^{2+} and Fe^{3+}:

$$Ni(OH)_2(s) + 2\ H^+(aq) \rightleftharpoons Ni^{2+}(aq) + 2\ H_2O;\ K = 10^{13}$$

$$Fe(OH)_3(s) + 3\ H^+(aq) \rightleftharpoons Fe^{3+}(aq) + 3\ H_2O;\ K = 10^5$$

The H^+ ions react with OH^- ions from the insoluble hydroxide

Indeed, all metal hydroxides are soluble in strong acid. Carbonates, such as $BaCO_3$, also dissolve:

$$BaCO_3(s) + 2\ H^+(aq) \rightleftharpoons Ba^{2+}(aq) + CO_2(g) + H_2O;\ K = 10^8$$

2. **Add a ligand (NH_3, OH^-) to react with a cation in the precipitate.** The simplest example of this type of reaction is that used to bring silver chloride into solution in the Group I analysis.

1. $\qquad\qquad AgCl(s) \rightleftharpoons Ag^+(aq) + Cl^-(aq)$
2. $\qquad \dfrac{Ag^+(aq) + 2\ NH_3(aq) \rightleftharpoons Ag(NH_3)_2^+(aq)}{AgCl(s) + 2\ NH_3(aq) \rightleftharpoons Ag(NH_3)_2^+(aq) + Cl^-(aq)} \qquad\qquad (20.14)$

Example 20.10 Using the Rule of Multiple Equilibria, calculate K for Reaction 20.14. (K_{sp} AgCl = 1.7×10^{-10}; K_d Ag(NH$_3$)$_2^+$ = 4×10^{-8})

Solution Clearly, K = K$_1$ × K$_2$.

But K$_1$ = K_{sp} AgCl

$$K_2 = \frac{1}{K_d \text{ Ag(NH}_3)_2^+}$$ since this equation is the reverse of that for the dissociation of the complex ion.

$$K = \frac{K_{sp} \text{ AgCl}}{K_d \text{ Ag(NH}_3)_2^+} = \frac{1.7 \times 10^{-10}}{4 \times 10^{-8}} = 4 \times 10^{-3}$$

Notice that K is rather small. However, with excess 6 M NH$_3$, enough AgCl can be brought into solution to give a good test for Ag$^+$ (see Problem 20.22).

Exercise K_d for the Ag(S$_2$O$_3$)$_2^{3-}$ complex ion is 1.0×10^{-13}. Calculate K for the reaction AgCl(s) + 2 S$_2$O$_3^{2-}$(aq) $\rightleftharpoons$ Ag(S$_2$O$_3$)$_2^{3-}$(aq) + Cl$^-$(aq). Answer: 1.7×10^3.

Insoluble hydroxides can be dissolved by treatment with OH$^-$ ions, if the cation forms a stable hydroxo complex. Examples include Zn(OH)$_2$ and Al(OH)$_3$:

$$\text{Zn(OH)}_2(s) + 2 \text{ OH}^-(aq) \rightarrow \text{Zn(OH)}_4^{2-}(aq)$$

$$\text{Al(OH)}_3(s) + \text{OH}^-(aq) \rightarrow \text{Al(OH)}_4^-(aq)$$

These compounds, like all insoluble hydroxides, also dissolve in strong acid:

$$\text{Zn(OH)}_2(s) + 2 \text{ H}^+(aq) \rightarrow \text{Zn}^{2+}(aq) + 2 \text{ H}_2\text{O}$$

$$\text{Al(OH)}_3(s) + \text{H}^+(aq) \rightarrow \text{Al}^{3+}(aq) + 3 \text{ H}_2\text{O}$$

What are some other amphoteric hydroxides?

Hydroxides such as Zn(OH)$_2$ and Al(OH)$_3$ which dissolve in *either* strong base or strong acid are referred to as being **amphoteric.**

SUMMARY

Various aspects of the qualitative analysis of selected cations were discussed in this chapter. Principles of precipitate formation (Chap. 16), acid-base reactions (Chap. 17), solution equilibria (Chap. 18), and complex ion formation (Chap. 19) were applied to qualitative analysis.

The general scheme for separation of cations into groups is based upon selective precipitation (Table 20.1 and Example 20.1). Separation and identification of different cations from a mixture can be done based upon properties such as solubility and color (Example 20.2 and 20.3). Application of these ideas to Group I cations are noted in Figure 20.1 and Example 20.4.

Complex ions are formed from simple cations during qualitative analysis procedures (Table 20.2 and Example 20.5). Complex ions containing NH$_3$ or OH$^-$ are decomposed to the simple ions upon addition of strong acid (Example 20.6). Precipitate formation (Examples 20.8 and 20.9) and dissolving (Example 20.10) involve adjusting reagent concentrations and pH.

KEY WORDS AND CONCEPTS

qualitative analysis
Group I, II, III, IV (cations)

flow sheet
amphoteric

QUESTIONS AND PROBLEMS

Catalog

General: 20.1–20.6, 20.23–20.27, 20.44–20.46
Equations: 20.7–20.11, 20.28–20.32

Unknowns: 20.12–20.15, 20.33–20.36
Equilibrium Calculations: 20.16–20.22, 20.37–20.43

You will find that Tables 20.1 and 20.2 are helpful in working several of these questions and problems.

20.1 Review and know the meaning of the key words and concepts in this chapter.

20.2 Complete the following table for cations:

Cation	Analytical group	Precipitating agent	Ppt. formed
Hg_2^{2+}	____	____	____
Bi^{3+}	____	____	____
Fe^{2+}	____	____	____
Al^{3+}	____	____	____

20.3 Explain why

a. Mg^{2+} does not precipitate in Groups I, II, or III.
b. heating a solution of NH_4Cl with $NaOH$ gives a gas with a pungent odor.
c. gas bubbles form when $BaCO_3$ is treated with a strong acid.

20.4 Give the formula of

a. a Group I chloride which gives a precipitate with NH_3.
b. two ions in Group II which form complexes with NH_3.
c. an anion which gives a precipitate with Pb^{2+}.

20.5 Consider the $Al(OH)_4^-$ ion listed in Table 20.2.

a. What is the coordination number of Al^{3+} in this complex?
b. Write the electron configuration for an Al^{3+} ion (ground state).
c. Based on your answer to (b), what is the geometry of $Al(OH)_4^-$?

20.6 Explain why the concentration of S^{2-} in a solution of H_2S is pH-dependent. Why is $[H^+]$ kept rather high (0.3 M) in precipitating Group II?

20.23 Complete the following table for cations:

Species	Test reagent	Response to test reagent
K^+	Flame test	____
Ca^{2+}	____	____
____	Hot water	Precipitate dissolves
____	H_2S	White ppt. forms

20.24 What would happen if

a. an unknown which had been tested for Groups I through III were then tested for Na^+?
b. a solution containing Cu^{2+} and Zn^{2+} were treated with H_2S at pH 9?
c. acid were added to a solution containing the $Sn(OH)_6^{2-}$ ion?

20.25 Give the formulas of

a. an ion in Group I which forms a complex with NH_3.
b. two Group II cations which form complexes with OH^-.
c. a species which is a weak acid and gives a precipitate with Zn^{2+}.

20.26 In Group II analysis, tin is separated from antimony by the formation of an octahedral tin-oxalate complex ion. Draw the structural formula of this ion (see Fig. 19.2).

20.27 Explain why CuS precipitates in Group II, but ZnS does not. What would happen if $[H^+]$ were too low in the Group II precipitation?

20.7 Write a balanced equation for the reaction of each of the following with H^+:

    a. $Zn(OH)_2$ b. $BaCO_3$
    c. $Fe(OH)_3$ d. $Al(OH)_4^-$

20.8 Write balanced equations for the reaction with NH_3 by which

    a. AgCl dissolves.
    b. Al^{3+} precipitates as $Al(OH)_3$.
    c. Cu^{2+} forms a complex ion.

20.9 Write balanced equations for the reaction with OH^- by which

    a. Al^{3+} forms a complex ion.
    b. $Al(OH)_3$ dissolves.
    c. $Ni(OH)_2$ precipitates.

20.10 Write equations to illustrate that $Al(OH)_3$ and $Bi(OH)_3$ dissolve in strong acid but only the former dissolves in excess NaOH.

20.11 When aqueous NH_3 is added to a solution containing $Cu^{2+}(aq)$, a blue precipitate first forms. The precipitate dissolves in additional NH_3 to form a deep blue solution.

    a. Write an equation for the precipitate formation.
    b. Write an equation for the dissolving of the precipitate.

20.12 You are given two unlabeled white solids. One is $PbCl_2$, the other is AgCl. You may choose only one reagent to distinguish them from each other. What two reagents are acceptable?

20.13 An aqueous solution contains Ag^+ and Ni^{2+}. Devise a simple procedure for separating these ions from solution.

20.14 A Group I unknown gives a white precipitate with HCl which is completely soluble in hot water. Identify the unknown.

20.15 A solution gives no precipitate with any of the reagents listed in Table 20.1. Of the cations listed in the Table, which ones could be present?

20.16 A student adds enough HCl to a Group I unknown to make $[Cl^-] = 0.20$ M. Some $PbCl_2$ precipitates.

    a. What is $[Pb^{2+}]$ remaining? (K_{sp} $PbCl_2$ = 1.7 $\times$ 10^{-5})
    b. Under these conditions, will Pb^{2+} precipitate in Group II, where $[S^{2-}]$ is 1 $\times$ 10^{-20}? (K_{sp} PbS = 1 $\times$ 10^{-27})

20.17 For the $Ni(NH_3)_6^{2+}$ ion, K_d = 2 $\times$ 10^{-9}. What is the ratio $[Ni^{2+}]/[Ni(NH_3)_6^{2+}]$ in a solution 0.10 M in NH_3?

20.28 Write a balanced equation for the reaction of each of the following with strong acid:

    a. $CaCO_3$ b. $Al(OH)_3$
    c. $Ni(OH)_2$ d. $Ag(NH_3)_2^+$

20.29 Write balanced equations for the reaction with NH_3 by which

    a. Fe^{3+} forms a precipitate.
    b. Ag^+ forms a complex ion.
    c. $Ni(OH)_2$ dissolves.

20.30 Write balanced equations for the reaction with OH^- by which

    a. Zn^{2+} forms a complex ion.
    b. $Zn(OH)_2$ dissolves.
    c. $Mg(OH)_2$ precipitates.

20.31 $Ni(OH)_2$ and $Zn(OH)_2$ each dissolve in 6 M NH_3 but only the latter dissolves in 6 M NaOH. Write equations for each of these reactions.

20.32 Ni^{2+} behaves toward ammonia in much the same way as Cu^{2+}. Write equations for the two successive reactions involved with Ni^{2+} (see 20.11).

20.33 You are given unlabeled samples of NaCl and NH_4Cl (both are white) and asked to distinguish them from each other. You do it in a single step, adding only one reagent. What is the reagent?

20.34 Cu^{2+} and Zn^{2+} are present in a solution. Describe a short procedure separating these ions from solution.

20.35 A Group I unknown gives a white precipitate with HCl which is completely soluble in excess NH_3. Identify the unknown.

20.36 A solution upon treatment with HCl or with H_2S at pH 9 gives no precipitate. What, if anything, does this indicate about the cations that might be present in the solution?

20.37 At 100°C, K_{sp} $PbCl_2$ = 5 $\times$ 10^{-4}.

    a. What is the solubility, in moles per liter, of $PbCl_2$ at 100°C?
    b. How many grams of $PbCl_2$ can be dissolved by heating with 2.0 cm^3 of water at 100°C?

20.38 At what concentration of NH_3 will 99.9% of the Ni^{2+} ions in a solution be converted to $Ni(NH_3)_6^{2+}$? (K_d $Ni(NH_3)_6^{2+}$ = 2 $\times$ 10^{-9})

20.18 A solution is 0.010 M in H^+ and 0.10 M in H_2S. Taking $[H^+]^2 \times [S^{2-}]/[H_2S] = 1 \times 10^{-20}$, K_{sp} CuS = 1 $\times 10^{-35}$, and K_{sp} ZnS = 1 $\times 10^{-20}$, determine

    a. the concentration of S^{2-}.
    b. whether CuS will precipitate, taking conc. Cu^{2+} = 0.10 M.
    c. whether ZnS will precipitate, taking conc. Zn^{2+} = 0.10 M.

20.19 What is the lowest pH at which NiS (K_{sp} = 1 $\times 10^{-19}$) will precipitate from a 0.1 M Ni^{2+} solution when $[H_2S]$ = 0.1 M? Take $[H^+]^2 \times [S^{2-}]/[H_2S] = 1 \times 10^{-20}$.

20.20 A solution of 10 cm³ of 0.10 M HCl is mixed with 10 cm³ of a solution 0.010 M in Pb^{2+}, giving a total volume of 20 cm³.

    a. What is the concentration of Cl^- after mixing? the concentration of Pb^{2+}?
    b. Will a precipitate of $PbCl_2$ form? (K_{sp} $PbCl_2$ = 1.7 $\times 10^{-5}$)

20.21 Given that K_{sp} Al(OH)$_3$ = 5 $\times 10^{-33}$, K_d Al(OH)$_4^-$ = 1 $\times 10^{-33}$, use the Rule of Multiple Equilibria to obtain K for the reaction

$$Al(OH)_3(s) + OH^-(aq) \rightleftharpoons Al(OH)_4^-(aq)$$

20.22 In Example 20.10, we showed that for the reaction

$$AgCl(s) + 2\ NH_3(aq) \rightleftharpoons Ag(NH_3)_2^+(aq) + Cl^-(aq)$$

K = 4 $\times 10^{-3}$.

    a. Write the expression for K for this reaction.
    b. Note that for every mole of AgCl that dissolves, one mole of $Ag(NH_3)_2^+$ and one mole of Cl^- are formed. Calculate the number of moles per liter of AgCl that dissolves when $[NH_3]$ = 6 M.

20.39 In Group II precipitation, $[H^+]$ = 0.3 M, and $[H_2S]$ = 0.10 M. Under these conditions, show by calculations whether the following ions will precipitate, assuming they are present at a concentration of 0.10 M:

    a. Hg^{2+} (K_{sp} HgS = 1 $\times 10^{-52}$).
    b. Fe^{2+} (K_{sp} FeS = 1 $\times 10^{-17}$)

20.40 Repeat the calculation in Problem 20.19 for MnS. (K_{sp} = 1 $\times 10^{-15}$)

20.41 A Group I unknown is prepared by mixing 2.0-cm³ portions of three solutions, 0.12 M $AgNO_3$, 0.12 M $Pb(NO_3)_2$, and 0.12 M $Hg_2(NO_3)_2$, to give 6.0 cm³ of solution.

    a. What is the concentration of each cation in the unknown?
    b. At what $[Cl^-]$ will $PbCl_2$ start to precipitate from this solution?

20.42 Use the Rule of Multiple Equilibria to find K for

$$Al(OH)_3(s) + 3\ H^+(aq) \rightleftharpoons Al^{3+}(aq) + 3\ H_2O$$

given that K_{sp} Al(OH)$_3$ = 5 $\times 10^{-33}$ and $K_w = [H^+] \times [OH^-] = 1 \times 10^{-14}$.

20.43 Repeat the calculation of Problem 20.22 for AgI dissolving in NH_3. The reaction is

$$AgI(s) + 2\ NH_3(aq) \rightleftharpoons Ag(NH_3)_2^+(aq) + I^-$$

K for this reaction is 2.5 $\times 10^{-9}$.

*20.44 In Group IV analysis, a pH of 9.5 is established using an NH_3–NH_4^+ buffer. Calculate the $[NH_3]/[NH_4^+]$ ratio for this buffer. (K_a NH_4^+ = 5.6 $\times 10^{-10}$)

*20.45 Will Al(OH)$_3$ precipitate from an NH_3–NH_4^+ buffer in which $[NH_3]$ = $[NH_4^+]$, and $[Al^{3+}]$ = 0.10 M? (K_b NH_3 = 1.8 $\times 10^{-5}$; K_{sp} for Al(OH)$_3$ = 5 $\times 10^{-33}$)

*20.46 A Group III unknown contains only Ni^{2+} and Al^{3+}. It is treated with aqueous ammonia to give a colored precipitate. As more NH_3 is added, part of the precipitate dissolves to form a deep blue solution. The precipitate remaining goes into solution when treated with excess NaOH. If acid is slowly added to this solution, a white precipitate forms, which dissolves as more acid is added. Write balanced equations for each reaction that took place.

REDOX REACTIONS; STANDARD VOLTAGES

In Chapter 4 we discussed briefly a type of reaction known as oxidation-reduction. Such a process, commonly called a "redox reaction," involves an exchange of electrons. The species which *loses electrons* is said to be *oxidized*. The other species, which *gains electrons,* is *reduced*. A simple example of a redox reaction is that which occurs between the elements sodium and chlorine. The product of the reaction is the ionic compound sodium chloride (Na^+, Cl^- ions):

$$2 \ Na(s) + Cl_2(g) \rightarrow 2 \ NaCl(s) \tag{21.1}$$

Sodium atoms lose electrons and are oxidized to Na^+ ions:

$$2 \ Na \rightarrow 2 \ Na^+ + 2 \ e^- \qquad \text{oxidation}$$

Whenever one species is oxidized another is simultaneously reduced

At the same time, chlorine atoms in the Cl_2 molecule gain electrons and are reduced to Cl^- ions:

$$Cl_2 + 2 \ e^- \rightarrow 2 \ Cl^- \qquad \text{reduction}$$

In this chapter, we will examine more closely the processes of oxidation and reduction, as they apply to reactions taking place in water solution. To do this, we need to introduce a new concept, that of oxidation number (Section 21.1). Oxidation number will in turn be used to develop a general method of balancing redox equations (Section 21.2). Later (Section 21.3), we will examine a type of electrical cell known as a voltaic cell in which redox reactions serve as a source of electrical energy. From voltage measurements on such cells, principles can be developed to allow us to predict whether a redox reaction will take place when reactants are mixed directly in the laboratory (Sections 21.4 and 21.5).

21.1 OXIDATION NUMBER

The chemical equation for the reaction of hydrogen with chlorine

$$H_2(g) + Cl_2(g) \rightarrow 2 \ HCl(g) \tag{21.2}$$

resembles that for the reaction of sodium with chlorine

$$2 \text{ Na(s)} + \text{Cl}_2(g) \rightarrow 2 \text{ NaCl(s)}$$

Indeed, the two reactions themselves have much in common. In both there is an exchange of electrons between atoms. The major difference lies in the extent to which electrons are transferred. In Reaction 21.2 the product is a gaseous molecule, HCl, rather than a pair of ions (Na^+, Cl^-) in a solid. The valence electron of hydrogen is shared with chlorine instead of being transferred to it. This distinction, however, is one of degree rather than kind. The electrons in the H—Cl covalent bond are displaced strongly toward the more electronegative chlorine. So far as "electron bookkeeping" is concerned, we might assign these electrons to the chlorine atom:

$$H \bigg| \ddot{\ddot{\text{:Cl:}}}$$

By assigning electrons in this way we have, in a sense, given a -1 charge to chlorine. It now has one more valence electron, eight, than an isolated chlorine atom, which has seven valence electrons. The hydrogen atom, stripped of its valence electron by this assignment, has in effect acquired a $+1$ charge.

The accounting system we have just gone through is widely used in inorganic chemistry. The concept of oxidation number is used to refer to the charge an atom would have if the bonding electrons were assigned arbitrarily to the more electronegative element. In the HCl molecule, hydrogen is said to have an oxidation number of $+1$, and chlorine an oxidation number of -1. In water the bonding electrons are assigned to the more electronegative oxygen atom:

$$H \bigg| \ddot{\ddot{\text{:O:}}} \bigg| H$$

This gives oxygen an oxidation number of -2 (eight valence electrons vs. six in the neutral atom). A hydrogen atom in water has an oxidation number of $+1$ (zero valence electrons vs. one in the neutral atom). In a nonpolar covalent bond, the bonding electrons are split evenly between the two atoms:

$$\ddot{\ddot{\text{:Cl}}} \cdot \bigg| \cdot \ddot{\ddot{\text{Cl:}}} \qquad \text{oxidation number Cl} = 0$$

We should emphasize that the oxidation number of an atom in a molecule is an artificial concept. Unlike the charge of an ion, oxidation number cannot be determined experimentally. The hydrogen atom in the HCl or H_2O molecule does not carry a full positive charge. We might regard its oxidation number of $+1$ in these molecules as a "pseudocharge."

Oxidation numbers are useful even though they are really not experimental properties

Rules for Assigning Oxidation Numbers

In principle, oxidation numbers could be determined in any species by assigning bonding electrons in the manner just described. However, such a method would require that we know the Lewis structure of the species. In practice, oxidation numbers are ordinarily obtained in a much simpler way, applying certain arbitrary rules. There are four such rules:

1. **The oxidation number of an element in an elementary substance is 0.** For example, the oxidation number of chlorine in Cl_2 or of phosphorus in P_4 is 0.

2. **The oxidation number of an element in a monatomic ion is equal to the charge of**

that ion. In the ionic compound NaCl, sodium has an oxidation number of +1, chlorine an oxidation number of −1. The oxidation numbers of aluminum and oxygen in Al_2O_3 (Al^{3+}, O^{2-} ions) are +3 and −2, respectively.

3. **Certain elements have the same oxidation number in all or almost all their compounds.** The Group 1 metals always exist as +1 ions in their compounds and, hence, are assigned an oxidation number of +1. By the same token, Group 2 elements always have oxidation numbers of +2 in their compounds. Fluorine, the most electronegative of all elements, has an oxidation number of −1 in all of its compounds.

Oxygen is ordinarily assigned an oxidation number of −2 in its compounds. (An exception arises in compounds containing the peroxide ion, O_2^{2-}, where the oxidation number of oxygen is −1.)

Hydrogen in its compounds ordinarily has an oxidation number of +1. (The only exception is in metal hydrides such as NaH and CaH_2, where hydrogen is present as the H^- ion and hence is assigned an oxidation number of −1.)

4. **The sum of the oxidation numbers of all the atoms in a neutral species is 0; in an ion, it is equal to the charge of that ion.** The application of this very useful principle is illustrated in Example 21.1.

Example 21.1   What is the oxidation number of selenium in Na_2Se? of manganese in MnO_4^-?

Solution   For Na_2Se, knowing that the oxidation number of sodium must be +1, we have

$$2(+1) + \text{oxid no. Se} = 0; \quad \text{oxid no. Se} = -2$$

In the MnO_4^- ion, taking the oxidation number of oxygen to be −2 and realizing that the sum must be −1:

$$\text{oxid no. Mn} + 4(-2) = -1; \quad \text{oxid no. Mn} = +7$$

Exercise   What is the oxidation number of Cr in Cr_2O_3? in $Cr_2O_7^{2-}$? Answer: +3; +6.

This example shows how to find an oxidation number for an element not covered by Rules 1, 2, and 3

Oxidation and Reduction; A Working Definition

The concept of oxidation number leads directly to a working definition of the terms oxidation and reduction. **Oxidation** is defined as **an increase in oxidation number,** and **reduction** as a **decrease in oxidation number.** Examples include:

$$2\ Al(s) + 3\ Cl_2(g) \rightarrow 2\ AlCl_3(s) \qquad \begin{array}{l} \text{Al oxidized (oxid. no. } 0 \rightarrow +3) \\ \text{Cl reduced (oxid. no. } 0 \rightarrow -1) \end{array} \qquad (21.3)$$

$$4\ As(s) + 5\ O_2(g) \rightarrow 2\ As_2O_5(s) \qquad \begin{array}{l} \text{As oxidized (oxid. no. } 0 \rightarrow +5) \\ \text{O reduced (oxid. no. } 0 \rightarrow -2) \end{array} \qquad (21.4)$$

These definitions are compatible with our earlier interpretation of oxidation and reduction in terms of loss and gain of electrons. An element which loses electrons must increase in oxidation number. The gain of electrons always results in a decrease in oxidation number. However, defining oxidation and reduction in terms of changes in oxidation number has one distinct advantage. It greatly simplifies the electron bookkeeping in redox reactions. Consider, for example, the reaction

$$HCl(g) + HNO_3(l) \rightarrow NO_2(g) + \tfrac{1}{2}\ Cl_2(g) + H_2O(l) \qquad (21.5)$$

Analysis in terms of oxidation number reveals that chlorine is oxidized (oxid. no. = −1 in HCl, 0 in Cl_2). Nitrogen is reduced (oxid. no. = +5 in HNO_3, +4 in NO_2). It is much more difficult to decide precisely which atoms are "losing" or "gaining" electrons in this reaction.

Oxidizing and Reducing Agents

In a redox reaction, we usually have at least two reactants. One of these is referred to as the oxidizing agent, another as the reducing agent.

An oxidizing agent brings about the oxidation of another species. To do this, it must take electrons away from that species. Hence, the oxidizing agent is itself reduced in the reaction.

It may seem strange, but a good oxidizing agent is easily reduced

A reducing agent brings about the reduction of another species. To do this, it must give up electrons to that species. Hence, the reducing agent is itself oxidized in the reaction.

To illustrate what these statements mean, consider Reaction 21.3. Here, Cl_2 is the oxidizing agent. It oxidizes Al (0 → +3). The Cl_2 is itself reduced to Cl^- ions (0 → −1). The reducing agent is aluminum metal, Al. It reduces Cl_2 molecules to Cl^- ions. In the process, Al is oxidized to Al^{3+} ions.

Example 21.2   In Reaction 21.5, what is the oxidizing agent? the reducing agent?

Solution   Nitric acid is the oxidizing agent; it oxidizes chlorine from −1 in HCl to 0 in Cl_2. The reducing agent is HCl; it reduces nitrogen from +5 in HNO_3 to +4 in NO_2. Note that HNO_3 is itself reduced (to NO_2), while HCl is oxidized (to Cl_2).

Exercise   What is the oxidizing agent in Reaction 21.4? the reducing agent? Answer: O_2; As.

21.2 BALANCING REDOX EQUATIONS

Many of the equations written to represent redox reactions are simple enough to balance by inspection. This is the case, for example, with Equations 21.1 through 21.4. Frequently, however, you will be working with more complex redox reactions where the coefficients in the balanced equation are by no means obvious. In this section, we will discuss a general approach to balancing such equations. It is called the half-equation method and is perhaps the most straightforward way to balance a variety of redox equations for reactions in water solution.

The first step in this method involves breaking the overall equation down into two half-equations. One of these is an oxidation, the other is a reduction. The two half-equations are balanced separately. Finally, they are combined in such a way so as to obtain an overall equation in which there is no net change in the number of electrons. We will now consider several examples of the half-equation method of balancing redox equations.

Reaction Between Fe^{3+} and Cl^- Ions

To illustrate the half-equation method, we start with a simple example. This involves the reaction that occurs when a direct electric current is passed through a water solution of iron(III) chloride, $FeCl_3$, producing the two elements iron and chlorine. The unbalanced equation for the reaction is

$$Fe^{3+}(aq) + Cl^-(aq) \rightarrow Fe(s) + Cl_2(g)$$

To balance this equation, we proceed as follows:

1. *Split the equation into two half-equations,* one oxidation and one reduction:

reduction: $\qquad\qquad\qquad Fe^{3+}(aq) \rightarrow Fe(s) \qquad\qquad\qquad\qquad$ (1a)

oxidation: $\qquad\qquad\qquad Cl^-(aq) \rightarrow Cl_2(g) \qquad\qquad\qquad\qquad$ (1b)

2. *Balance these half-equations, first with respect to mass and then with respect to charge:*

Equation 1a is balanced insofar as mass is concerned, since there is one atom of Fe on both sides. The charges, however, are unbalanced: the Fe atom on the right has 0 charge while the Fe^{3+} ion on the left has a charge of $+3$. To correct this, we add three electrons to the left of 1a, arriving at

$$Fe^{3+}(aq) + 3\ e^- \rightarrow Fe(s) \qquad\qquad (2a)$$

We need to balance electrons as we do atoms

Equation 1b must first be balanced with respect to mass by providing two Cl^- ions to give one molecule of Cl_2:

$$2\ Cl^-(aq) \rightarrow Cl_2(g)$$

To balance charges, two electrons must be added to the right, giving a charge of -2 on both sides:

$$2\ Cl^-(aq) \rightarrow Cl_2(g) + 2\ e^- \qquad\qquad (2b)$$

3. Having arrived at two balanced half-equations, *combine them so as to make the number of electrons gained in reduction equal to the number lost in oxidation.*

In Equation 2a, three electrons are gained; in 2b, two electrons are given off. To arrive at a final equation in which no electrons appear, multiply 2a by 2, 2b by 3, and add:

$2 \times 2a$: $\qquad\qquad 2\ Fe^{3+}(aq) + \cancel{6}\ e^- \rightarrow 2\ Fe(s) \qquad\qquad$ (3a)

In any chemical reaction there is no net formation or consumption of electrons

$3 \times 2b$: $\qquad\qquad\qquad 6\ Cl^-(aq) \rightarrow 3\ Cl_2(g) + \cancel{6}\ e^- \qquad\quad$ (3b)

$$\overline{2\ Fe^{3+}(aq) + 6\ Cl^-(aq) \rightarrow 2\ Fe(s) + 3\ Cl_2(g)} \qquad (3)$$

Reaction Between MnO_4^- and Cl^- (Acidic Solution)

The equation just worked out was easy to balance because it involved only two species. One of these (Fe^{3+}) was reduced; the other (Cl^-) was oxidized. In many redox reactions, species other than those being reduced or oxidized take part in the reaction. Most commonly, they contain hydrogen (oxid. no. $= +1$) or oxygen (oxid. no. $= -2$). The presence of such species makes the redox equation a bit more difficult to balance. However, the half-equation method can still be applied.

To illustrate the approach used, consider the reaction between chloride ions (Cl^-)

and permanganate ions (MnO_4^-) in acidic solution (H^+). Experiment shows that this reaction is represented by the equation

$$MnO_4^-(aq) + H^+(aq) + Cl^-(aq) \rightarrow Mn^{2+}(aq) + Cl_2(g) + H_2O$$

Note that the two elements that undergo a change in oxidation number are manganese ($+7 \rightarrow +2$) and chlorine ($-1 \rightarrow 0$). Neither hydrogen nor oxygen change oxidation number, yet atoms of these elements participate in the reaction. The oxygen atoms tied up originally in the MnO_4^- ion end up as H_2O molecules; the H^+ ions meet the same fate.

Almost never do more than two elements change oxidation nos. in a reaction

To balance this equation, we proceed as follows:

1. Recognize which species undergo oxidation and reduction. Represent these by appropriate half-equations:

oxidation: $$Cl^-(aq) \rightarrow Cl_2(g) \tag{1a}$$

reduction: $$MnO_4^-(aq) + H^+(aq) \rightarrow Mn^{2+}(aq) + H_2O \tag{1b}$$

2. Balance the two half-equations with respect to mass (kinds of atoms). Half-equation (1a) is balanced by writing:

$$2\,Cl^-(aq) \rightarrow Cl_2(g) + 2\,e^- \tag{2a}$$

To balance (1b), we first make sure that there are the same number of Mn atoms on both sides—one. Next, oxygen is balanced by using H_2O. In this case a coefficient of 4 in front of H_2O on the right accounts for the four oxygens in a MnO_4^- ion:

$$MnO_4^-(aq) + H^+(aq) \rightarrow Mn^{2+}(aq) + 4\,H_2O$$

To complete the mass balance, the number of hydrogen atoms on the two sides must be equalized. The four H_2O molecules on the right contain eight hydrogen atoms; there must then be eight H^+ ions on the left:

$$MnO_4^-(aq) + 8\,H^+(aq) \rightarrow Mn^{2+}(aq) + 4\,H_2O$$

This half-equation now has a charge of $+2$ on the right and $+7$ on the left ($-1 + 8$). To balance each side with a $+2$ charge, five electrons are added to the left:

$$MnO_4^-(aq) + 8\,H^+(aq) + 5\,e^- \rightarrow Mn^{2+}(aq) + 4\,H_2O \tag{2b}$$

3. Finally, half-equations (2a) and (2b) are combined so as to eliminate electrons from the final equation. To do this, a least common multiple of 10 electrons is used. Thus, (2a) is multiplied by 5 and (2b) by 2, producing 10 e^- on each side.

$5 \times 2a$: $$10\,Cl^-(aq) \rightarrow 5\,Cl_2(g) + \cancel{10}\,e^- \tag{3a}$$

$2 \times 2b$: $$2\,MnO_4^-(aq) + 16\,H^+(aq) + \cancel{10}\,e^- \rightarrow 2\,Mn^{2+}(aq) + 8\,H_2O \tag{3b}$$

$$2\,MnO_4^-(aq) + 16\,H^+(aq) + 10\,Cl^-(aq) \rightarrow 2\,Mn^{2+}(aq) + 8\,H_2O + 5\,Cl_2(g) \tag{3}$$

Reaction Between MnO_4^- and I^- (Basic Solution)

Often we need to write balanced equations for redox reactions taking place in basic solution. Here, we should not have H^+ ions in the final equation, since their concentration in basic solution is very small ($<10^{-7}$ M). Instead, hydrogen in such equations should be in the form of OH^- ions or H_2O molecules. A simple way to accomplish this is to eliminate any H^+ ions appearing in the half-equations, "neutralizing" them by adding an equal number of OH^- ions to both sides. To illustrate, consider the oxidation, in basic solution, of iodide by permanganate ions:

$$I^-(aq) + MnO_4^-(aq) \rightarrow I_2(aq) + MnO_2(s) \text{ (basic solution)}$$

One can proceed exactly as in the foregoing example, to obtain the half-equations

oxidation: $\qquad\qquad\qquad 2\ I^-(aq) \rightarrow I_2(aq) + 2\ e^-$ (2a)

reduction: $\qquad\quad MnO_4^-(aq) + 4\ H^+(aq) + 3\ e^- \rightarrow MnO_2(s) + 2\ H_2O$ (2b)

The H^+ ions appearing in the reduction half-equation must now be removed to obtain an equation valid in basic solution. To do this, four OH^- ions are added to both sides, and water is formed on the left by combining H^+ with OH^- ions.

$$MnO_4^-(aq) + 4\ H^+(aq) + 3\ e^- \rightarrow MnO_2(s) + 2\ H_2O$$

$$+\ 4\ OH^-(aq) \rightarrow \qquad\qquad\qquad +\ 4\ OH^-(aq)$$

$$MnO_4^-(aq) + 4\ H_2O + 3\ e^- \rightarrow MnO_2(s) + 2\ H_2O + 4\ OH^-(aq)$$

Eliminating two water molecules from each side, we arrive at

$$MnO_4^-(aq) + 2\ H_2O + 3\ e^- \rightarrow MnO_2(s) + 4\ OH^-(aq)$$ (2b')

for the reduction half-reaction in basic solution. To obtain the overall equation, we proceed as before, combining 2a and 2b' in such a way as to make the electron gain equal the electron loss:

When you have a completed equation, check to see that all atoms balance and that charge is balanced

$3 \times 2a$: $\qquad\qquad\qquad\qquad 6\ I^-(aq) \rightarrow 3\ I_2(aq) + \cancel{6}\ e^-$ (3a)

$2 \times 2b'$: $\qquad 2\ MnO_4^-(aq) + 4\ H_2O + \cancel{6}\ e^- \rightarrow 2\ MnO_2(s) + 8\ OH^-(aq)$ (3b)

$6\ I^-(aq) + 2\ MnO_4^-(aq) + 4\ H_2O \rightarrow 3\ I_2(aq) + 2\ MnO_2(s) + 8\ OH^-(aq)$ (3)

21.3 VOLTAIC CELLS

In the most general sense, we can distinguish between two types of redox reactions. Certain reactions are **spontaneous,** in the sense that they take place directly when the reactants are mixed. A spontaneous reaction occurs "by itself" in the sense that we do not have to do work to make it go. An example of such a reaction is the formation of sodium chloride from the elements:

$$2\ Na(s) + Cl_2(g) \rightarrow 2\ NaCl(s)$$

This reaction takes place spontaneously when sodium metal is exposed to chlorine gas.

In contrast, certain redox reactions are **nonspontaneous.** They do not take place directly, no matter how long we wait. An example is the decomposition of sodium chloride to the elements:

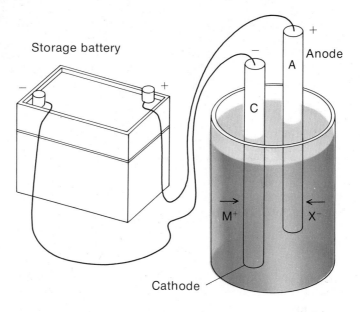

Figure 21.1 Schematic diagram of an electrolytic cell. Reduction occurs at the cathode (C). Oxidation occurs at the anode (A). During electrolysis the cations move toward the cathode, the anions toward the anode.

$$2 \text{ NaCl(s)} \rightarrow 2 \text{ Na(s)} + \text{Cl}_2(\text{g}) \qquad (21.6)$$

A sample of table salt does not spontaneously break down to sodium metal and chlorine gas. However, we saw in Chapter 5 that it is possible to bring about this reaction by electrolysis (recall Fig. 5.5, p. 90). In that case, we supply electrical energy from an outside source to make the reaction go. The device in which this is done is called an *electrolytic cell.* Figure 21.1 shows a general diagram of an electrolytic cell, suitable for bringing about a nonspontaneous reaction.

In this section we will discuss a different type of electrical cell known as a **voltaic cell.** A spontaneous redox reaction takes place within such a cell; it produces electrical energy. All of us are familiar with certain types of voltaic cells. They include "dry cells" used in flashlights and calculators, and the lead storage battery in an automobile. To understand how a voltaic cell works, we start with some simple cells that are easily constructed in the general chemistry laboratory.

Every voltaic cell depends for its voltage on a spontaneous redox reaction

The Zn–Cu²⁺ Cell (Zn/Zn²⁺ ‖ Cu²⁺/Cu)

When a piece of zinc is added to a water solution containing Cu^{2+} ions, the following redox reaction takes place:

$$\text{Zn(s)} + \text{Cu}^{2+}(\text{aq}) \rightarrow \text{Zn}^{2+}(\text{aq}) + \text{Cu(s)} \qquad (21.7)$$

Referring to color plate 23 (center of book), we see the result of this reaction. Copper metal plates out on the surface of the zinc. The blue color of the aqueous Cu^{2+} ion fades as it is placed by the colorless aqueous Zn^{2+} ion. From Equation 21.7, we see that the reaction amounts to electron transfer from a zinc atom to a Cu^{2+} ion.

To design an electrical cell using Reaction 21.7 as a source of electrical energy, the electron transfer must occur indirectly. That is, the electrons given off by zinc atoms must be made to pass through an external electric circuit before they reduce Cu^{2+} ions to copper atoms. One way to do this is shown in Figure 21.2. Let us trace the flow of electric current through this cell.

1. At the zinc *anode,* electrons are produced by the *oxidation* half-reaction

$$\text{Zn(s)} \rightarrow \text{Zn}^{2+}(\text{aq}) + 2 \text{ e}^- \qquad (21.7a)$$

This electrode, which "pumps" electrons into the external circuit, is ordinarily marked as the negative pole of the cell.

2. Electrons generated by Reaction 21.7a move through the external circuit (left to right in Figure 21.2). This part of the circuit may be a simple resistance wire, a light bulb, an electric motor, an electrolytic cell, or some other device that consumes electrical energy.

3. Electrons pass from the external circuit to the copper *cathode* where they are used in the *reduction* of Cu^{2+} ions in the surrounding solution:

$$Cu^{2+}(aq) + 2 e^- \rightarrow Cu(s) \tag{21.7b}$$

The copper electrode, which "pulls" electrons from the external circuit, is considered to be the positive pole of the cell.

4. To complete the circuit, ions must move through the aqueous solutions in the cell. As Reactions 21.7a and 21.7b proceed, a surplus of positive ions (Zn^{2+}) tends to build up around the zinc electrode. The region around the copper electrode tends to become deficient in positive ions as Cu^{2+} ions are consumed. To maintain electrical neutrality, cations must move toward the copper cathode or, alternatively, anions must move toward the zinc anode. In practice, both migrations occur.

In the cell shown in Figure 21.2, movement of ions occurs through a *salt bridge*. In its simplest form a salt bridge may consist of an inverted U-tube, plugged with glass wool at each end. The tube is filled with a solution of a salt which takes no part in the electrode reactions; potassium nitrate, KNO_3, is frequently used. As current is drawn from the cell, K^+ ions move from the salt bridge into the copper half-cell to compensate for the Cu^{2+} ions consumed at the cathode. At the same time, NO_3^- ions move into the zinc half-cell to compensate for the charge on the Zn^{2+} ions formed at the anode.

The cell shown in Figure 21.2 is often abbreviated as

$$Zn/Zn^{2+} \| Cu^{2+}/Cu$$

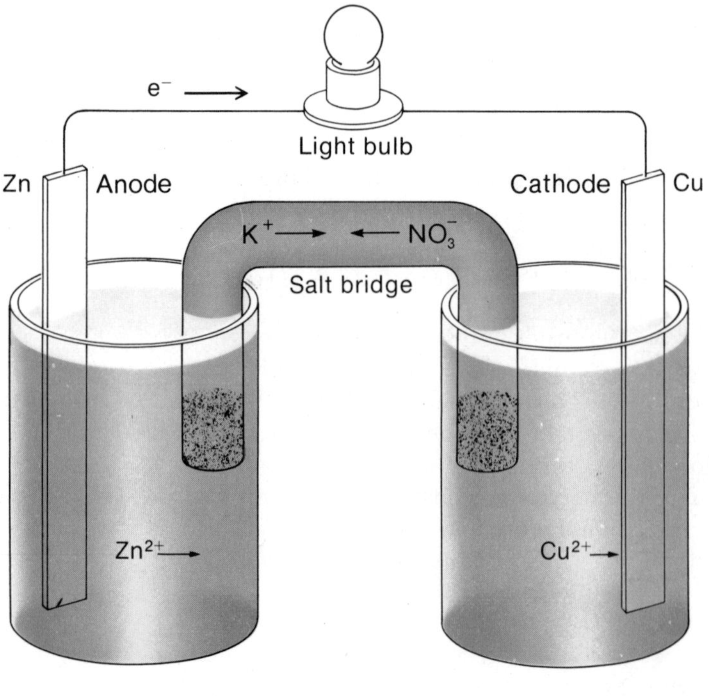

The salt bridge allows us to keep the solutions separated and still have a redox reaction

Figure 21.2 In this voltaic cell, the following spontaneous redox reaction takes place: $Zn(s) + Cu^{2+}(aq) \rightarrow Zn^{2+}(aq) + Cu(s)$. The salt bridge allows ions to pass from one solution to the other to complete the circuit. At the same time, it prevents direct contact between Zn atoms and Cu^{2+} ions.

In this notation:

—the **anode** reaction (**oxidation**) is shown at the left. Zn atoms are oxidized to Zn^{2+} ions.

—the salt bridge (or other means of separating the half-cells) is indicated by the symbol $\parallel$.

—the **cathode** reaction (**reduction**) is shown at the right. Cu^{2+} ions are reduced to Cu atoms.

Other Salt Bridge Cells

Cells similar to that shown in Figure 21.2 can be set up for many different spontaneous redox reactions. Consider, for example, the reaction

$$Ni(s) + Cu^{2+}(aq) \rightarrow Ni^{2+}(aq) + Cu(s) \qquad (21.8)$$

For this reaction the voltaic cell would closely resemble that in Figure 21.2. Indeed, the Cu^{2+}/Cu half-cell and the salt bridge would be identical to the one shown. The only difference would be in the oxidation half-cell. Here, we would use a nickel electrode, surrounded by a solution containing Ni^{2+} ions, such as $NiSO_4$ or $Ni(NO_3)_2$. The half-cell reactions would be

$$anode: Ni(s) \rightarrow Ni^{2+}(aq) + 2\ e^- \qquad (oxidation)$$
$$cathode: Cu^{2+}(aq) + 2\ e^- \rightarrow Cu(s) \qquad (reduction)$$

The cell notation would be $Ni/Ni^{2+} \parallel Cu^{2+}/Cu$.

Another spontaneous redox reaction that can serve as a source of electrical energy is that between zinc metal and H^+ ions:

$$Zn(s) + 2\ H^+(aq) \rightarrow Zn^{2+}(aq) + H_2(g) \qquad (21.9)$$

A voltaic cell using this reaction is shown in Figure 21.3. The Zn/Zn^{2+} half-cell and the salt bridge are the same as those in Figure 21.2. Since no metal is involved in the cathode half-reaction, we use an *inert* electrode. That is, the cathode is made of an unreactive

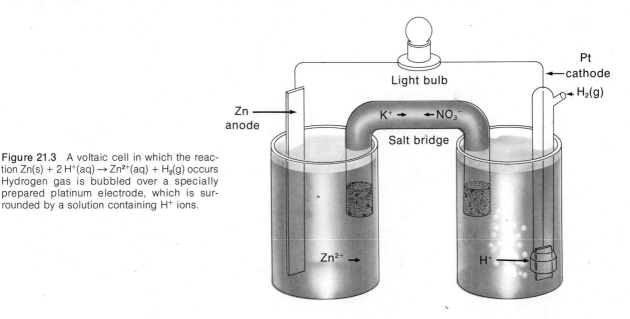

Figure 21.3 A voltaic cell in which the reaction $Zn(s) + 2\ H^+(aq) \rightarrow Zn^{2+}(aq) + H_2(g)$ occurs Hydrogen gas is bubbled over a specially prepared platinum electrode, which is surrounded by a solution containing H^+ ions.

material that conducts an electric current. In this case, it is convenient to use a special electrode made of platinum (graphite rods and Nichrome wires are often used as inert electrodes instead of platinum). Hydrogen gas is bubbled over the Pt electrode, which is surrounded by a solution containing H^+ ions (e.g., a solution of HCl).

The half-reactions occurring in the cell shown in Figure 21.3 are

$$\text{anode: } Zn(s) \rightarrow Zn^{2+}(aq) + 2\ e^- \qquad \text{(oxidation)}$$
$$\text{cathode: } 2\ H^+(aq) + 2\ e^- \rightarrow H_2(g) \qquad \text{(reduction)}$$

The cell notation is

$$Zn/Zn^{2+} \,||\, (Pt)H^+/H_2$$

The symbol (Pt) is used to indicate the presence of an inert platinum cathode.

Example 21.3   When chlorine gas is bubbled through a water solution of NaBr, a spontaneous redox reaction occurs:

$$Cl_2(g) + 2\ Br^-(aq) \rightarrow 2\ Cl^-(aq) + Br_2(l)$$

This reaction can serve as a source of electrical energy in the voltaic cell shown in Figure 21.4. In this cell
a. what is the cathode reaction? the anode reaction?
b. which way do electrons move in the external circuit?
c. which way do anions move within the cell? cations?

Solution
a. cathode: $Cl_2(g) + 2\ e^- \rightarrow 2\ Cl^-(aq)$ reduction
   anode: $2\ Br^-(aq) \rightarrow Br_2(l) + 2\ e^-$ oxidation
b. From anode to cathode (left to right in Fig. 21.4).
c. Anions move to the anode (right to left).
   Cations move to the cathode (left to right).

Exercise   What is the notation for the cell shown in Figure 21.4? Answer: $(Pt)Br^-/Br_2 \,||\, (Pt)Cl_2/Cl^-$.

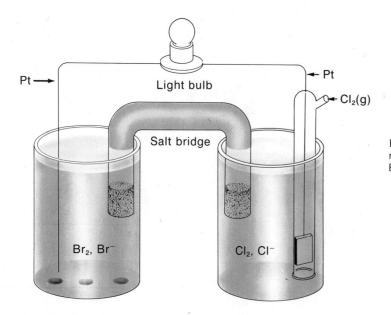

Figure 21.4 In this voltaic cell, the spontaneous redox reaction is: $Cl_2(g) + 2\ Br^-(aq) \rightarrow 2\ Cl^-(aq) + Br_2(l)$. Both electrodes are made of platinum.

Commercial Cells

503
REDOX
REACTIONS;
STANDARD
VOLTAGES

As we will see in Section 21.4, salt bridge cells yield valuable information about the spontaneity of redox reactions. However, they have too high an internal resistance to be used commercially. When an appreciable amount of current is drawn from a salt bridge cell, its voltage drops sharply. Commercial voltaic cells are designed to supply large amounts of current, at least for a short time. We will discuss two of the more familiar cells of this type.

DRY CELLS. The construction of the ordinary dry cell (Leclanché cell) used in flashlights is shown in Figure 21.5. The zinc wall of the cell is the anode. The graphite rod through the center of the cell is the cathode. The space between the electrodes is filled with a moist paste. This contains MnO_2, $ZnCl_2$, and NH_4Cl. When the cell operates, the half-reaction at the anode is

Dry cells are cheap, but deliver only small amounts of energy

$$Zn(s) \rightarrow Zn^{2+}(aq) + 2\ e^- \qquad (21.10a)$$

At the cathode, manganese dioxide is reduced to species in which Mn is in the $+3$ oxidation state, such as Mn_2O_3:

$$2\ MnO_2(s) + 2\ NH_4^+(aq) + 2\ e^- \rightarrow Mn_2O_3(s) + 2\ NH_3(aq) + H_2O \qquad (21.10b)$$

The overall reaction occurring in this voltaic cell is

$$Zn(s) + 2\ MnO_2(s) + 2\ NH_4^+(aq) \rightarrow Zn^{2+}(aq) + Mn_2O_3(s) + 2\ NH_3(aq) + H_2O \qquad (21.10)$$

If too large a current is drawn from a Leclanché cell, the ammonia forms a gaseous insulating layer around the carbon cathode. When this happens the voltage drops sharply, and then returns slowly to its normal value of 1.5 V. This problem can be avoided by using an "alkaline" dry cell, in which the paste between the electrodes contains KOH rather than NH_4Cl. In this case the overall cell reaction is simply

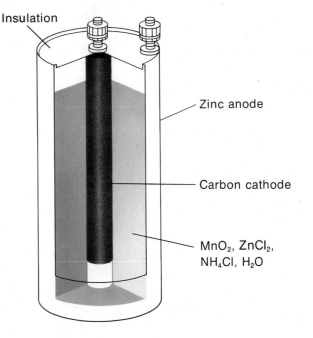

Insulation

Zinc anode

Carbon cathode

MnO_2, $ZnCl_2$, NH_4Cl, H_2O

Figure 21.5 Section of an ordinary Zn–MnO_2 dry cell. This cell produces 1.5 V and will deliver a current of about half an ampere for six hours.

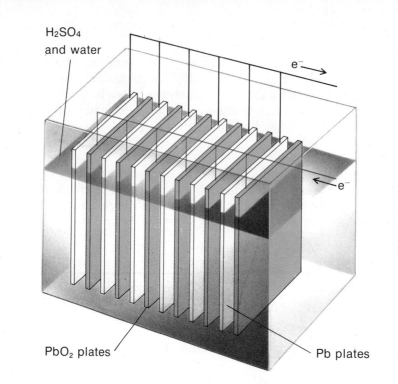

H₂SO₄ and water

e⁻ →

← e⁻

Figure 21.6 Lead storage battery. Two advantages of the lead storage battery are ability to deliver large amounts of energy for a short time and ability to be recharged. A disadvantage is its high mass/energy ratio.

PbO₂ plates

Pb plates

$$Zn(s) + 2\ MnO_2(s) \rightarrow ZnO(s) + Mn_2O_3(s) \tag{21.11}$$

No gas is produced. Moreover, an alkaline cell, unlike an ordinary dry cell, can be recharged several times. The alkaline dry cell, although more expensive than the Leclanché cell, is widely used in calculators and other small appliances.

LEAD STORAGE BATTERY.   The 12-volt storage battery used in automobiles consists of six voltaic cells of the type shown in Figure 21.6. A group of lead plates, the grids of which are filled with spongy gray lead, forms the anode of the cell. The multiple cathode consists of another group of plates of similar design filled with lead(IV) oxide, PbO_2. These two sets of plates alternate through the cell. They are immersed in a water solution of sulfuric acid, H_2SO_4, which acts as the electrolyte.

All commercial cells have only one solution and no salt bridges

When a lead storage battery is supplying current, the lead in the anode grids is oxidized to Pb^{2+} ions. These immediately react with SO_4^{2-} ions in the electrolyte, precipitating $PbSO_4$ (lead sulfate) on the plates. At the cathode, lead dioxide is reduced to Pb^{2+} ions, which also precipitate as $PbSO_4$.

$$Pb(s) + SO_4^{2-}(aq) \rightarrow PbSO_4(s) + 2\ e^- \tag{21.12a}$$
$$PbO_2(s) + 4\ H^+(aq) + SO_4^{2-}(aq) + 2\ e^- \rightarrow PbSO_4(s) + 2\ H_2O \tag{21.12b}$$
$$\overline{Pb(s) + PbO_2(s) + 4\ H^+(aq) + 2\ SO_4^{2-}(aq) \rightarrow 2\ PbSO_4(s) + 2\ H_2O} \tag{21.12}$$

Deposits of lead sulfate slowly build up on the plates, partially covering and replacing the lead and lead dioxide. As the cell discharges, the concentration of sulfuric acid decreases. For every mole of lead reacting, two moles of H_2SO_4 ($4\ H^+$, $2\ SO_4^{2-}$) are replaced by two moles of water. The state of charge of a storage battery can be checked by measuring the density of the electrolyte. When fully charged, the density is in the range of 1.25 to 1.30 g/cm³. A density below 1.20 g/cm³ indicates a low sulfuric acid concentration and hence a partially discharged cell.

A good storage battery can deliver over 100 amperes at 12 volts for 30 seconds

A lead storage battery can be recharged and thus restored to its original condition.

To do this, a direct current is passed through the cell in the reverse direction. While a storage battery is being charged, it acts as an electrolytic cell. The half-reactions 21.12a and 21.12b are reversed:

$$2\ PbSO_4(s) + 2\ H_2O \rightarrow Pb(s) + PbO_2(s) + 4\ H^+(aq) + 2\ SO_4^{2-}(aq) \qquad (21.13)$$

The electrical energy required to bring about this nonspontaneous reaction in an automobile is furnished by an alternator equipped with a rectifier to convert to direct current.

21.4   STANDARD VOLTAGES

Voltaic cells are of considerable interest in chemistry for one main reason: *A redox reaction which generates electrical energy in a voltaic cell will take place spontaneously outside the cell, in a beaker, test tube, or other container.*

To make the most effective use of this principle, we need to consider a property of a voltaic cell known as its *voltage.* The cell voltage is a measure of the "driving force" of the cell reaction—that is, its tendency to occur spontaneously. At a given temperature, let us say 25°C, the voltage depends upon two factors:

If voltage > 0, the reaction can occur spontaneously

1. The nature of the cell reaction.
2. The concentrations of the species taking part in the reaction.

Here we will be concerned only with the first factor; concentration effects will be considered in Section 21.5.

We will work with the **standard voltage, E^0,** for a given cell reaction. The standard voltage is that measured when *all ions or molecules in solution are at a concentration of 1 M and all gases are at a pressure of 1 atm.* To illustrate, consider the $Zn/Zn^{2+} \| (Pt)H^+/H_2$ cell shown in Figure 21.3. We find that when the pressure of hydrogen gas is 1 atm and the concentrations of Zn^{2+} and H^+ are 1 M, the cell voltage is $+0.76$ V. This quantity is referred to as the standard voltage for the reaction and is given the symbol E^0.

$$Zn(s) + 2\ H^+(aq,\ 1\ M) \rightarrow Zr.^{2+}(aq,\ 1\ M) + H_2(g,\ 1\ atm);\ E^0 = +0.76\ V \qquad (21.14)$$

E^0_{ox} and E^0_{red}

You will recall that any redox reaction can be split into two half-reactions. One of these is an oxidation, and the other a reduction. For Reaction 21.14:

oxidation: $Zn(s) \rightarrow Zn^{2+}(aq,\ 1\ M) + 2\ e^-$ $\qquad (21.14a)$

reduction: $2\ H^+(aq,\ 1\ M) + 2\ e^- \rightarrow H_2(g,\ 1\ atm)$ $\qquad (21.14b)$

It is possible to associate standard voltages with half-reactions such as these. Such a voltage is a measure of the driving force behind the half-reaction. The standard voltage for the oxidation half-reaction is given the symbol E^0_{ox}. That for the reduction half-reaction is written as E^0_{red}. The standard voltage for the cell reaction is the sum of these two quantities. Thus we have

For any cell $E^0 = E^0_{ox} + E^0_{red}$

$$Zn(s) \rightarrow Zn^{2+}(1\ M) + 2\ e^- \qquad\qquad E^0_{ox}\ (Zn \rightarrow Zn^{2+})$$

$$\underline{2\ H^+(1\ M) + 2\ e^- \rightarrow H_2(1\ atm)} \qquad \underline{E^0_{red}\ (H^+ \rightarrow H_2)}$$

$$Zn(s) + 2\ H^+(1\ M) \rightarrow Zn^{2+}(1\ M) + H_2(1\ atm) \qquad E^0 = E^0_{ox} + E^0_{red} = +0.76\ V$$

We would like to be able to calculate standard voltages for half-reactions such as those listed above. There is, however, a problem. As you can see from the equation just written, we have one known, E^0 (+0.76 V) and two unknowns, E^0_{ox} and E^0_{red}. No matter what we do, we cannot determine an individual half-reaction voltage experimentally. To resolve this dilemma, we make an arbitrary decision. We take the *standard voltage for the reduction of H^+ ions to H_2 gas to be zero.*

$$2 H^+(aq, 1 M) + 2 e^- \rightarrow H_2(g, 1 atm); E^0_{red} (H^+ \rightarrow H_2) = 0.00 V$$

Using this convention, and knowing that E^0 for the $Zn/Zn^{2+} \| (Pt)H^+/H_2$ cell is +0.76 V, it follows that the standard voltage for the oxidation of zinc must be +0.76 V. That is,

$$Zn(s) \rightarrow Zn^{2+}(aq, 1 M) + 2 e^-; E^0_{ox} (Zn \rightarrow Zn^{2+}) = +0.76 V$$

As soon as one voltage is established, others can be determined from measurements on appropriate cells. Suppose, for example, we want to determine the standard voltage for the reduction of Cu^{2+} to Cu. One way to do this is to set up the cell shown in Figure 21.2, using 1 M solutions of Zn^{2+} and Cu^{2+}, and measure the voltage. We find the standard voltage of this cell to be +1.10 V.

Cell voltages are determined experimentally

$$Zn(s) + Cu^{2+}(aq, 1 M) \rightarrow Zn^{2+}(aq, 1 M) + Cu(s); E^0 = +1.10 V$$

In this cell zinc is being oxidized and Cu^{2+} ions reduced. Hence

$$E^0_{ox} (Zn \rightarrow Zn^{2+}) + E^0_{red} (Cu^{2+} \rightarrow Cu) = +1.10 V$$

Since the standard voltage for the oxidation of zinc must be the same here as in the $Zn-H^+$ cell, +0.76 V,

$$+0.76 V + E^0_{red} (Cu^{2+} \rightarrow Cu) = +1.10 V$$

$$E^0_{red} (Cu^{2+} \rightarrow Cu) = +1.10 V - 0.76 V = +0.34 V$$

Standard half-cell voltages are ordinarily obtained from a list of *standard potentials* such as that given in Table 21.1. **The potentials listed give us directly the standard voltages for reduction half-reactions.** For example, since the standard potentials for $Zn^{2+} \rightarrow Zn$ and $Cu^{2+} \rightarrow Cu$ are −0.76 V and +0.34 V, respectively, we see immediately that

$$Zn^{2+}(aq) + 2 e^- \rightarrow Zn(s) \qquad E^0_{red} = -0.76 V$$

$$Cu^{2+}(aq) + 2 e^- \rightarrow Cu(s) \qquad E^0_{red} = +0.34 V$$

Standard voltages for oxidation half-reactions are obtained by changing the sign of the standard potential listed in Table 21.1. Thus, we have

$$Zn(s) \rightarrow Zn^{2+}(aq) + 2 e^- \qquad E^0_{ox} = +0.76 V$$

$$Cu(s) \rightarrow Cu^{2+}(aq) + 2 e^- \qquad E^0_{ox} = -0.34 V$$

In general, standard voltages for forward and reverse reactions (oxidation and reduction) are equal in magnitude but opposite in sign.

In the remainder of this section, we will consider some of the applications of standard voltages. We start by showing how they can be used to compare the strengths of

TABLE 21.1   STANDARD POTENTIALS IN WATER SOLUTION AT 25°C

OXIDIZING AGENT	REDUCING AGENT	E^0_{red} (V)
$Li^+(aq) + e^-$	$\rightarrow Li(s)$	−3.05
$K^+(aq) + e^-$	$\rightarrow K(s)$	−2.93
$Ba^{2+}(aq) + 2\ e^-$	$\rightarrow Ba(s)$	−2.90
$Ca^{2+}(aq) + 2\ e^-$	$\rightarrow Ca(s)$	−2.87
$Na^+(aq) + e^-$	$\rightarrow Na(s)$	−2.71
$Mg^{2+}(aq) + 2\ e^-$	$\rightarrow Mg(s)$	−2.37
$Al^{3+}(aq) + 3\ e^-$	$\rightarrow Al(s)$	−1.66
$Mn^{2+}(aq) + 2\ e^-$	$\rightarrow Mn(s)$	−1.18
$Zn^{2+}(aq) + 2\ e^-$	$\rightarrow Zn(s)$	−0.76
$Cr^{3+}(aq) + 3\ e^-$	$\rightarrow Cr(s)$	−0.74
$Fe^{2+}(aq) + 2\ e^-$	$\rightarrow Fe(s)$	−0.44
$Cr^{3+}(aq) + e^-$	$\rightarrow Cr^{2+}(aq)$	−0.41
$Cd^{2+}(aq) + 2\ e^-$	$\rightarrow Cd(s)$	−0.40
$PbSO_4(s) + 2\ e^-$	$\rightarrow Pb(s) + SO_4^{2-}(aq)$	−0.36
$Tl^+(aq) + e^-$	$\rightarrow Tl(s)$	−0.34
$Co^{2+}(aq) + 2\ e^-$	$\rightarrow Co(s)$	−0.28
$Ni^{2+}(aq) + 2\ e^-$	$\rightarrow Ni(s)$	−0.25
$AgI(s) + e^-$	$\rightarrow Ag(s) + I^-(aq)$	−0.15
$Sn^{2+}(aq) + 2\ e^-$	$\rightarrow Sn(s)$	−0.14
$Pb^{2+}(aq) + 2\ e^-$	$\rightarrow Pb(s)$	−0.13
$2\ H^+(aq) + 2\ e^-$	$\rightarrow H_2(g)$	0.00
$AgBr(s) + e^-$	$\rightarrow Ag(s) + Br^-(aq)$	0.07
$S(s) + 2\ H^+(aq) + 2\ e^-$	$\rightarrow H_2S(aq)$	0.14
$Sn^{4+}(aq) + 2\ e^-$	$\rightarrow Sn^{2+}(aq)$	0.15
$Cu^{2+}(aq) + e^-$	$\rightarrow Cu^+(aq)$	0.15
$SO_4^{2-}(aq) + 4\ H^+(aq) + 2\ e^-$	$\rightarrow SO_2(g) + 2\ H_2O$	0.20
$Cu^{2+}(aq) + 2\ e^-$	$\rightarrow Cu(s)$	0.34
$Cu^+(aq) + e^-$	$\rightarrow Cu(s)$	0.52
$I_2(s) + 2\ e^-$	$\rightarrow 2\ I^-(aq)$	0.53
$Fe^{3+}(aq) + e^-$	$\rightarrow Fe^{2+}(aq)$	0.77
$Hg_2^{2+}(aq) + 2\ e^-$	$\rightarrow 2\ Hg(l)$	0.79
$Ag^+(aq) + e^-$	$\rightarrow Ag(s)$	0.80
$2\ Hg^{2+}(aq) + 2\ e^-$	$\rightarrow Hg_2^{2+}(aq)$	0.92
$NO_3^-(aq) + 4\ H^+(aq) + 3\ e^-$	$\rightarrow NO(g) + 2\ H_2O$	0.96
$AuCl_4^-(aq) + 3\ e^-$	$\rightarrow Au(s) + 4\ Cl^-(aq)$	1.00
$Br_2(l) + 2\ e^-$	$\rightarrow 2\ Br^-(aq)$	1.07
$O_2(g) + 4\ H^+(aq) + 4\ e^-$	$\rightarrow 2\ H_2O$	1.23
$MnO_2(s) + 4\ H^+(aq) + 2\ e^-$	$\rightarrow Mn^{2+}(aq) + 2\ H_2O$	1.23
$Cr_2O_7^{2-}(aq) + 14\ H^+(aq) + 6\ e^-$	$\rightarrow 2\ Cr^{3+}(aq) + 7\ H_2O$	1.33
$Cl_2(g) + 2\ e^-$	$\rightarrow 2\ Cl^-(aq)$	1.36
$ClO_3^-(aq) + 6\ H^+(aq) + 5\ e^-$	$\rightarrow \frac{1}{2}\ Cl_2(g) + 3\ H_2O$	1.47
$Au^{3+}(aq) + 3\ e^-$	$\rightarrow Au(s)$	1.50
$MnO_4^-(aq) + 8\ H^+(aq) + 5\ e^-$	$\rightarrow Mn^{2+}(aq) + 4\ H_2O$	1.52
$PbO_2(s) + SO_4^{2-}(aq) + 4\ H^+(aq) + 2\ e^-$	$\rightarrow PbSO_4(s) + 2\ H_2O$	1.68
$H_2O_2(aq) + 2\ H^+(aq) + 2\ e^-$	$\rightarrow 2\ H_2O$	1.77
$Co^{3+}(aq) + e^-$	$\rightarrow Co^{2+}(aq)$	1.82
$F_2(g) + 2\ e^-$	$\rightarrow 2\ F^-(aq)$	2.87

BASIC SOLUTION

$Fe(OH)_2(s) + 2\ e^-$	$\rightarrow Fe(s) + 2\ OH^-(aq)$	−0.88
$2\ H_2O + 2\ e^-$	$\rightarrow H_2(g) + 2\ OH^-(aq)$	−0.83
$Fe(OH)_3(s) + e^-$	$\rightarrow Fe(OH)_2(s) + OH^-(aq)$	−0.56
$S(s) + 2\ e^-$	$\rightarrow S^{2-}(aq)$	−0.43
$Cu(OH)_2(s) + 2\ e^-$	$\rightarrow Cu(s) + 2\ OH^-(aq)$	−0.22
$CrO_4^{2-}(aq) + 4\ H_2O + 3\ e^-$	$\rightarrow Cr(OH)_3(s) + 5\ OH^-(aq)$	−0.12
$NO_3^-(aq) + H_2O + 2\ e^-$	$\rightarrow NO_2^-(aq) + 2\ OH^-(aq)$	0.01
$ClO_4^-(aq) + H_2O + 2\ e^-$	$\rightarrow ClO_3^-(aq) + 2\ OH^-(aq)$	0.36
$O_2(g) + 2\ H_2O + 4\ e^-$	$\rightarrow 4\ OH^-(aq)$	0.40
$ClO_3^-(aq) + 3\ H_2O + 6\ e^-$	$\rightarrow Cl^-(aq) + 6\ OH^-(aq)$	0.62
$ClO^-(aq) + H_2O + 2\ e^-$	$\rightarrow Cl^-(aq) + 2\ OH^-(aq)$	0.89

different oxidizing and reducing agents. Then we will consider their use in determining whether or not a given redox reaction will occur spontaneously in the laboratory.

Strength of Oxidizing and Reducing Agents

As pointed out earlier, an oxidizing agent gains electrons in a redox reaction. It follows that all the species listed in the column at the far left of Table 21.1 (Li^+, . . ., F_2) are, at least in principle, oxidizing agents. A *"strong"* oxidizing agent is one which has a strong attraction for electrons and hence *can readily oxidize other species*. In contrast, a "weak" oxidizing agent does not gain electrons readily. It is capable of reacting only with those species that are very easily oxidized.

The strength of an oxidizing agent is directly related to its value of E_{red}^0. **The more positive E_{red}^0 is, the stronger the oxidizing agent.** Looking at Table 21.1, we see that oxidizing strength increases as we move down the table. The Li^+ ion, at the top of the left column, is a very weak oxidizing agent. E_{red}^0 for Li^+ is a large *negative* number, which implies that it has little tendency to gain electrons:

$$Li^+(aq) + e^- \rightarrow Li(s); \; E_{red}^0 = -3.05 \text{ V}$$

In practice, cations of the Group 1 metals (Li^+, Na^+, K^+, . . .) and the Group 2 metals (Mg^{2+}, Ca^{2+}, . . .) never act as oxidizing agents in water solution. Further down the list, the H^+ ion has a greater tendency to gain electrons:

$$2 H^+(aq) + 2 e^- \rightarrow H_2(g); \; E_{red}^0 = 0.00 \text{ V}$$

It is capable of oxidizing metals such as Mg or Zn (to Mg^{2+}, Zn^{2+}). The strongest oxidizing agents are those at the bottom of the left column. Species such as $Cr_2O_7^{2-}$ ($E_{red}^0 = +1.33$ V), Cl_2 ($E_{red}^0 = +1.36$ V), and MnO_4^- ($E_{red}^0 = +1.52$ V) are commonly used as oxidizing agents in redox reactions. The fluorine molecule, F_2, is in principle the strongest of all oxidizing agents:

$$F_2(g) + 2 e^- \rightarrow 2 F^-(aq); \; E_{red}^0 = +2.87 \text{ V}$$

In practice, fluorine is seldom used as an oxidizing agent because it is too dangerous to work with. The F_2 molecule takes electrons away from just about anything, including water, often with explosive violence.

The argument we have just gone through can be applied, in reverse, to reducing agents. These species are listed in the column at the center of Table 21.1 (Li, . . ., F^-). In principle, at least, all of them can supply electrons to another species in a redox reaction. Their strength as reducing agents is directly related to their E_{ox}^0 values. **The more positive E_{ox}^0 is, the stronger the reducing agent.** Looking at the values, remembering that $E_{ox}^0 = -E_{red}^0$,

Best oxidizing agent:
F_2
F_2 has largest value of E_{red}^0

Best reducing agent:
Li
Li has the largest value of E_{ox}^0

$$Li(s) \rightarrow Li^+(aq) + e^-; \; E_{ox}^0 = +3.05 \text{ V}$$

$$H_2(g) \rightarrow 2 H^+(aq) + 2 e^-; \; E_{ox}^0 = 0.00 \text{ V}$$

$$2 F^-(aq) \rightarrow F_2(g) + 2 e^-; \; E_{ox}^0 = -2.87 \text{ V}$$

we conclude that reducing strength decreases as we move down the table. The strongest reducing agents are located at the upper right (Li, . . .), and the weakest at the lower right (. . ., F⁻).

Example 21.4   Consider the following species: MnO_4^-, I^-, NO_3^-, Fe^{2+}, Ca. Using Table 21.1, classify each species as an oxidizing or reducing agent. Arrange the oxidizing agents in order of increasing strength; do the same with the reducing agents.

Solution   We start by realizing that oxidizing agents (species that can be reduced) are located in the left column. Scanning that column, we find the following oxidizing agents:

$$Fe^{2+}: E^0_{red} (Fe^{2+} \rightarrow Fe) = -0.44 \text{ V}$$

$$NO_3^-: E^0_{red} (NO_3^- \rightarrow NO) = +0.96 \text{ V}$$

$$MnO_4^-: E^0_{red} (MnO_4^- \rightarrow Mn^{2+}) = +1.52 \text{ V}$$

Reducing agents (species that can be oxidized) are found in the right column. Here we locate

$$Ca: E^0_{ox} (Ca \rightarrow Ca^{2+}) = +2.87 \text{ V}$$

$$I^-: E^0_{ox} (I^- \rightarrow I_2) = -0.53 \text{ V}$$

$$Fe^{2+}: E^0_{ox} (Fe^{2+} \rightarrow Fe^{3+}) = -0.77 \text{ V}$$

(Note that Fe^{2+} can act as either an oxidizing agent, in which case it is reduced to Fe, or a reducing agent, where it is oxidized to Fe^{3+}.)
    Finally, noting that the more positive the voltage the stronger the oxidizing or reducing agent, we see that

$$\text{oxidizing agents: } Fe^{2+} < NO_3^- < MnO_4^-$$

$$\text{reducing agents: } Fe^{2+} < I^- < Ca$$

Exercise   Give the formula of a cation which is a stronger reducing agent than Fe^{2+}; a stronger oxidizing agent than Fe^{2+}. Answer: Sn^{2+}, Cr^{2+}; any cation below Fe^{2+} in the column at the left of Table 21.1.

Some species can be both oxidized and reduced

Calculation of E^0 from E^0_{red} and E^0_{ox}

As pointed out earlier, the standard voltage for a redox reaction is the sum of the standard voltages of the two half-reactions, reduction and oxidation. That is,

$$E^0 = E^0_{red} + E^0_{ox} \tag{21.15}$$

This simple relation makes it possible, using Table 21.1, to calculate standard voltages for more than 3000 different redox reactions. Example 21.5 illustrates how this is done.

Example 21.5   Using Table 21.1, calculate the standard voltage for the reaction

$$2 \text{ Ag}^+(aq) + \text{Cu}(s) \rightarrow 2 \text{ Ag}(s) + \text{Cu}^{2+}(aq)$$

Solution   Splitting the cell reaction into two half-reactions, finding the appropriate standard voltages from the table, and adding:

reduction: $2 \text{ Ag}^+(aq) + 2 \text{ e}^- \rightarrow 2 \text{ Ag}(s)$; $E^0_{red} = +0.80$ V

oxidation: $\text{Cu}(s) \rightarrow \text{Cu}^{2+}(aq) + 2 \text{ e}^-$; $E^0_{ox} = -0.34$ V

$$E^0 = +0.46 \text{ V}$$

Notice that E^0_{red} for Ag^+ is taken directly from Table 21.1, where we find

$$\text{Ag}^+(aq) + \text{e}^- \rightarrow \text{Ag}(s); \ E^0_{red} = +0.80 \text{ V}$$

We can't change a voltage by writing a number on a piece of paper

We do *not* multiply the voltage by two just because two Ag^+ ions appear in the balanced overall equation. E^0, E^0_{red}, or E^0_{ox} for a given reaction is independent of the number of electrons transferred.

Exercise   Calculate E^0 for the cell $Ni/Ni^{2+} \| Cu^{2+}/Cu$. Answer: $+0.59$ V.

You will notice that in all the examples we have worked, E^0 is a positive quantity. This is generally true for reactions taking place in a voltaic cell. A spontaneous reaction taking place within the cell generates a positive voltage. If the calculated voltage is negative, the reaction as written cannot serve as a source of energy in a voltaic cell. It may, however, be possible to carry out the reaction in an electrolytic cell. Consider, for example,

$$
\begin{array}{ll}
2 \text{ Cl}^-(aq) \rightarrow \text{Cl}_2(g) + 2 \text{ e}^- & E^0_{ox} = -1.36 \text{ V} \\
2 \text{ H}_2\text{O} + 2 \text{ e}^- \rightarrow \text{H}_2(g) + 2 \text{ OH}^-(aq) & E^0_{red} = -0.83 \text{ V} \\
\hline
2 \text{ Cl}^-(aq) + 2 \text{ H}_2\text{O} \rightarrow \text{Cl}_2(g) + \text{H}_2(g) + 2 \text{ OH}^-(aq) & E^0 = -2.19 \text{ V}
\end{array}
$$

You may recall (Chap. 5) that this reaction takes place when a water solution of sodium chloride is electrolyzed. A voltage of at least 2.19 V must be supplied, perhaps by a storage battery, to furnish the energy required for the electrolysis. A nonspontaneous reaction such as this one, with a negative E^0 value, can be made to occur in an electrolytic cell. To do so, we must apply an external voltage at least equal in magnitude to E^0.

Spontaneity of Redox Reactions

All we need to do is mix the reactants

As we have seen, the voltage of a cell in which a spontaneous redox reaction is taking place is always positive. Such a reaction will take place directly and spontaneously in the laboratory, perhaps in a test tube, beaker, or other container. This means that:

If the calculated voltage for a redox reaction is a positive quantity, the reaction will occur spontaneously in the laboratory. If the calculated voltage is negative, the reaction will not occur; instead, the reverse reaction will be spontaneous.

To illustrate this principle, consider the problem of reducing Ni^{2+} ions to nickel ($E^0_{red} = -0.25$ V). It should be possible to do this with zinc ($E^0_{ox} = +0.76$ V) but not with copper ($E^0_{ox} = -0.34$ V).

$$\text{Zn}(s) + \text{Ni}^{2+}(aq) \rightarrow \text{Zn}^{2+}(aq) + \text{Ni}(s)$$

$$E^0 = E^0_{ox} + E^0_{red} = +0.76 \text{ V} - 0.25 \text{ V} = +0.51 \text{ V} \quad \text{(spontaneous)}$$

$$Cu(s) + Ni^{2+}(aq) \rightarrow Cu^{2+}(aq) + Ni(s)$$

$$E^0 = E^0_{ox} + E^0_{red} = -0.34 \text{ V} - 0.25 \text{ V} = -0.59 \text{ V} \quad \text{(nonspontaneous)}$$

Sure enough, if we add zinc metal to a solution of $NiCl_2$, a reaction occurs. Nickel metal plates out on the zinc and the green color of the Ni^{2+} ion fades. If, on the other hand, we add copper to a solution of $NiCl_2$, there is no evidence of reaction. In contrast, the reverse reaction

$$Ni(s) + Cu^{2+}(aq) \rightarrow Ni^{2+}(aq) + Cu(s); E^0 = +0.59 \text{ V}$$

takes place spontaneously. Nickel metal reacts with a solution of $CuCl_2$ to plate out copper and form Ni^{2+} ions in solution.

Example 21.6 illustrates a somewhat more subtle application of this principle.

Example 21.6   Using the standard potentials listed in Table 21.1, decide whether
  a. Fe(s) will be oxidized to Fe^{2+} by treatment with 1 M hydrochloric acid (HCl).
  b. Cu(s) will be oxidized to Cu^{2+} by treatment with 1 M hydrochloric acid.
  c. Cu(s) will be oxidized to Cu^{2+} by treatment with 1 M nitric acid (HNO_3).

Solution
  a. In order for iron to be oxidized, some species must be reduced. In hydrochloric acid, the only reducible species is the H^+ ion. Looking up the appropriate potentials:

$$\begin{array}{ll} Fe(s) \rightarrow Fe^{2+}(aq) + 2 \text{ e}^- & E^0_{ox} = +0.44 \text{ V} \\ \underline{2 \text{ H}^+(aq) + 2 \text{ e}^- \rightarrow H_2(g)} & \underline{E^0_{red} = 0.00 \text{ V}} \\ Fe(s) + 2 \text{ H}^+(aq) \rightarrow Fe^{2+}(aq) + H_2(g) & E^0 = +0.44 \text{ V} \end{array}$$

Since the calculated voltage is positive, we deduce that the reaction should occur. In the laboratory, we find that it does. Iron filings dropped into hydrochloric acid dissolve, with the evolution of $H_2(g)$.
  b. Proceeding in the same way:

$$\begin{array}{ll} Cu(s) \rightarrow Cu^{2+}(aq) + 2 \text{ e}^- & E^0_{ox} = -0.34 \text{ V} \\ \underline{2 \text{ H}^+(aq) + 2 \text{ e}^- \rightarrow H_2(g)} & \underline{E^0_{red} = 0.00 \text{ V}} \\ Cu(s) + 2 \text{ H}^+(aq) \rightarrow Cu^{2+}(aq) + H_2(g) & E^0 = -0.34 \text{ V} \end{array}$$

We find, as predicted, that no reaction occurs when copper is added to 1 M hydrochloric acid.
  c. In HNO_3, there is another possible oxidizing agent, the NO_3^- ion. Combining the proper half-equations:

$$\begin{array}{ll} 3 [Cu(s) \rightarrow Cu^{2+}(aq) + 2 \text{ e}^-] & E^0_{ox} = -0.34 \text{ V} \\ \underline{2[NO_3^-(aq) + 4 \text{ H}^+(aq) + 3 \text{ e}^- \rightarrow NO(g) + 2 \text{ H}_2O]} & \underline{E^0_{red} = +0.96 \text{ V}} \\ 3 \text{ Cu(s)} + 2 \text{ NO}_3^- + 8 \text{ H}^+(aq) \rightarrow 3 \text{ Cu}^{2+}(aq) + 2 \text{ NO(g)} + 4 \text{ H}_2O & E^0 = +0.62 \text{ V} \end{array}$$

As predicted, nitric acid does indeed oxidize copper metal to Cu^{2+}; the reduction product is one of the lower oxidation states of nitrogen (e.g., NO or NO_2) rather than $H_2(g)$. The reaction is shown in color plate 24 (center of book).

Exercise   Which of the metals Zn, Cd, and Hg will react with HCl, based on E^0 values?
Answer: Zn ($E^0_{ox} = +0.76$ V) and Cd ($E^0_{ox} = +0.40$ V).

Strictly speaking, the conclusions reached in Example 21.6 apply only at standard concentrations (1 M for species in solution, 1 atm for gases).

21.5 EFFECT OF CONCENTRATION UPON VOLTAGE

To this point, we have dealt only with standard voltages: E^0, E^0_{red}, E^0_{ox}. These, you will recall, apply at standard concentrations (1 M for species in solution, 1 atm for gases). As we have seen, standard voltages are useful for many purposes. However, we often carry out redox reactions where the concentrations of one or more species are far removed from 1 M. Under these conditions, we need to consider the effect of concentration upon voltage.

Qualitatively, we can readily predict the direction in which voltage will shift when concentrations are changed. It is observed experimentally that:

1. A reaction becomes more spontaneous if the concentration of a reactant increases or that of a product decreases. Under these conditions, the voltage increases (becomes more positive).

2. A reaction becomes less spontaneous if the concentration of a reactant decreases or that of a product increases. Under these conditions, the voltage decreases (becomes less positive).

By Le Châtelier's Principle we drive the reaction to the right, so E goes up

Nernst Equation

Quantitatively, we use a relation called the Nernst equation to determine the effect of concentration upon voltage. Consider the general redox reaction at 25°C:

$$a \text{ A} + b \text{ B} \rightarrow c \text{ C} + d \text{ D}$$

Here, A, B, C, D are species whose concentrations can be varied. The small letters a, b, c, d refer to the coefficients in the balanced equation. The Nernst equation has the form, at 25°C,

$$E = E^0 - \frac{0.0591}{n} \log_{10} \frac{(\text{conc. C})^c \times (\text{conc. D})^d}{(\text{conc. A})^a \times (\text{conc. B})^b} \qquad (21.16)$$

In this equation, E is the voltage at a given concentration, E^0 is the standard voltage, and n is the number of moles of electrons transferred in the reaction.

To illustrate the use of Equation 21.16 consider the reaction

$$\text{Zn(s)} + \text{Cu}^{2+}\text{(aq)} \rightarrow \text{Zn}^{2+}\text{(aq)} + \text{Cu(s)}; E^0 = +1.10 \text{ V}$$

Here, n = 2 (two moles of electrons are produced by the oxidation of one mole of Zn and are consumed by the reduction of one mole of Cu^{2+}). Hence,

$$E = +1.10 \text{ V} - \frac{0.0591}{2} \log_{10} \frac{(\text{conc. Zn}^{2+})}{(\text{conc. Cu}^{2+})} \qquad (21.17)$$

Note that terms for solids (or pure liquids) do not appear in the expression for E.

In Table 21.2, we use Equation 21.17 to calculate E for various values of the ratio (conc. Zn^{2+})/(conc. Cu^{2+}). Note that:

—if (conc. Zn^{2+})/(conc. Cu^{2+}) is less than 1, E is larger than the standard voltage, +1.10 V. This agrees with the qualitative statement made earlier. To make this ratio

TABLE 21.2 VOLTAGE AT 25°C FOR THE REACTION $Zn(s) + Cu^{2+}(aq) \rightarrow Zn^{2+}(aq) + Cu(s)$								
(conc. Zn^{2+})/(conc. Cu^{2+})	10^{-10}	10^{-5}	10^{-1}	1	10^{1}	10^{5}	10^{10}	10^{37}
E (V)		1.40	1.25	1.13	1.10	1.07	0.95	0.80	0.00

less than 1, we could decrease the concentration of the product, Zn^{2+}, or increase that of the reactant, Cu^{2+}. Either of these changes should make the reaction more spontaneous and hence increase E.

—if (conc. Zn^{2+})/(conc. Cu^{2+}) is greater than 1, E is smaller than the standard voltage, +1.10 V. By increasing the concentration of the product, Zn^{2+}, relative to that of the reactant, Cu^{2+}, we make the reaction less spontaneous. Hence, the voltage decreases.

—only if the concentration ratio differs from 1 by several orders of magnitude does E differ appreciably from +1.10 V. As an extreme case, consider what happens when (conc. Zn^{2+})/(conc. Cu^{2+}) becomes 10^{37}. E drops to zero. This is the condition in a completely discharged $Zn/Zn^{2+} \| Cu^{2+}/Cu$ cell. Virtually all the reactant, Cu^{2+}, has been consumed. The system is at equilibrium and there is no further driving force for the reaction. The cell is "dead."

If you start with a standard cell, and use up 99.99% of the Cu^{2+}, the voltage is still about 1 V

Example 21.7 Consider a voltaic cell in which the following reaction occurs:

$$Zn(s) + 2\,H^+(aq) \rightarrow Zn^{2+}(aq) + H_2(g)$$

a. Set up the Nernst equation for this cell, relating E to E^0.
b. Taking E^0 = +0.76 V, calculate E when conc. Zn^{2+} = 1 M, P_{H_2} = 1 atm, conc. H^+ = 1×10^{-3} M.

Solution
a. The value of n is 2; two electrons are transferred from a Zn atom to two H^+ ions.

$$E = E^0 - \frac{0.0591}{2} \log_{10} \frac{(\text{conc. } Zn^{2+}) \times (P_{H_2})}{(\text{conc. } H^+)^2}$$

b. Substituting numbers:

$$E = +0.76\text{ V} - \frac{0.0591}{2} \log_{10} \frac{1 \times 1}{(1 \times 10^{-3})^2}$$

$$= +0.76\text{ V} - \frac{0.0591}{2} \log_{10} (10^6) = +0.76\text{ V} - \frac{0.0591}{2} \times 6.0 = +0.58\text{ V}$$

Exercise Calculate E for this cell under the same conditions except that conc. H^+ = 1×10^{-4} M. Answer: +0.52 V.

From Example 21.7 and the exercise that follows, it should be clear that the voltage of the $Zn/Zn^{2+} \| (Pt)H^+/H_2$ cell is directly related to the concentration of H^+. When conc. H^+ drops from 10^{-3} to 10^{-4} M, E decreases from 0.58 to 0.52 V. From a slightly different point of view, we might say that the cell voltage is inversely related to pH. E drops by 0.06 V when the pH increases by one unit.

Many other voltaic cells behave in this manner. Their voltage depends upon the pH of the solution in which one of the electrodes is immersed. The pH meter, referred to in Chapter 17, works on this principle. It contains a voltaic cell whose voltage is sensitive to pH. The measured voltage is translated into pH, which is read on the face of the meter.

21.6 ELECTROCHEMISTRY AND MICHAEL FARADAY

Throughout this chapter we have emphasized voltaic cells, in which a spontaneous redox reaction produces electrical energy. Historically, however, electrolytic cells played a more important role in developing the basic principles of electrochemistry. About 150 years ago, Michael Faraday studied several different types of electrolytic processes. He showed that *the amount of a substance produced by electrolysis is directly proportional to the quantity of electricity passed through the cell.* This becomes clear if we look at the half-equation for, let us say, the reduction of H^+ ions to H_2 gas:

$$2\ H^+(aq) + 2\ e^- \rightarrow H_2(g)$$

From this equation we see that 1 mol of H_2 gas is produced by 2 mol of electrons. More generally, we can say that

$$\text{no. moles } H_2 = \tfrac{1}{2} \text{ no. moles electrons}$$

or

$$\text{no. moles electrons} = 2 \times \text{no. moles } H_2 \text{ formed}$$

Similar relations can be written for any half-reaction, whether it occurs in an electrolytic cell, a voltaic cell, a beaker, or a test tube. Using n to represent number of moles

$$Ag^+(aq) + e^- \rightarrow Ag(s) \qquad n\ e^- = n\ Ag$$
$$Cu^{2+}(aq) + 2\ e^- \rightarrow Cu(s) \qquad n\ e^- = 2\ (n\ Cu)$$
$$Fe^{3+}(aq) + 3\ e^- \rightarrow Fe(s) \qquad n\ e^- = 3\ (n\ Fe)$$
$$2\ Cl^-(aq) \rightarrow Cl_2(g) + 2\ e^- \qquad n\ e^- = 2\ (n\ Cl_2)$$

The number in this equation (1, 2, 3, . . .) differs with the nature of the half-reaction and is equal to the number of electrons gained or lost in the half-reaction.

This simple relationship has several practical applications. Among other things, it offers an experimental approach to determining the charge of an ion in solution. When we find that one mole (6.022×10^{23}) of electrons is required to produce one mole of silver metal (107.87 g), we conclude that the charge of the silver cation must be +1. Again, knowing that 2 mol of electrons are required to produce 1 mol of Cu, it follows that the copper cation must have a charge of +2.

Michael Faraday (1791–1867) was perhaps the greatest experimentalist of the nineteenth century. He was born and lived almost all his life in what is now greater London. The son of a blacksmith, he had no formal education beyond the rudiments of reading, writing, and arithmetic. Apprenticed to a bookbinder at the age of 13, Faraday educated himself by reading almost every book that came into the shop. One that particularly impressed him was a textbook, *Conversations in Chemistry*, written by Mrs. Jane Marcet. Within a few years he was carrying out simple experiments in his home laboratory. He also attended lectures given by

Sir Humphry Davy at the Royal Institute (Davy was the leading English scientist of the period; among other things, he isolated six new elements, Na, K, Mg, Ca, Sr, and Ba). Anxious to escape a life of drudgery, Faraday wrote out a copy of these lectures and submitted it to Davy, asking to be employed. Shortly afterwards, a vacancy arose and he was hired.

In personality and temperament, Davy and Faraday were poles apart. To Sir Humphry Davy, science was a fascinating hobby which happened to bring him fame and fortune. He pursued it at his leisure, enjoying an active social life and writing mediocre poetry in his spare time. To Michael Faraday, science was an obsession; one of his biographers describes him as a "work maniac." An observer (Faraday had no students) said of him ". . . if he had to cross the laboratory for anything, he did not walk, he ran; the quickness of his perception was equalled by the calm rapidity of his movements." In 1839, he suffered a nervous breakdown, the result of overwork. For much of the rest of his life. Faraday was in poor health. He gradually gave up more and more of his social engagements but continued to do research at the same pace as before.

Faraday developed the laws of electrolysis between 1831 and 1834. In the summer of 1833, he showed that the amount of H_2 produced by electrolysis of H_2SO_4 was directly proportional to the amount of electricity passed through the solution. In mid-December of that same year, he began looking at the production of metals by electrolysis. These included, among others, tin, lead, and zinc. Despite taking a whole day off for Christmas, Faraday completed these experiments, wrote up 3 years' work, and published his paper in January, 1834. Here, Faraday introduced the basic vocabulary of electrochemistry. He used for the first time the terms "anode," "cathode," "ion," "electrolyte," and "electrolysis." One of the basic constants of electrochemistry was later named for him. The Faraday constant gives the number of coulombs, 96,500, in 1 mol of electrons. (The coulomb is the practical unit of electrical charge.)

Although we have emphasized Faraday's work in chemistry, his greatest contributions were to physics in the fields of electricity and magnetism. In 1821, he discovered that a wire through which a current is flowing will rotate about a magnetic pole, and so produced the first electric motor. Ten years later, he showed that an electric potential could be created by rotating a copper disc between the poles of a permanent magnet, and thereby discovered the principle governing all modern electrical generators. These two contributions made possible the controlled use of electricity by mankind and were crucial steps in the development of modern society. No less a scientist than Albert Einstein ranked Faraday, along with Newton, Galileo, and Maxwell, as one of the greatest physicists of the ages.

SUMMARY

This chapter dealt with redox reactions, in which:

1. One species is reduced (oxidation number decreases), and acts as an oxidizing agent.

2. Another species is oxidized (oxidation number increases), and acts as a reducing agent.

These concepts are illustrated in Example 21.2. Oxidation numbers are assigned according to arbitrary rules (Example 21.1). Equations for redox reactions can be balanced by the half-equation method, either in acidic solution (p. 497) or basic solution (p. 498).

In a voltaic cell, a spontaneous redox reaction is used to generate electrical energy.

Oxidation occurs at the anode, reduction at the cathode. Cations move to the cathode, and anions move to the anode. Voltaic cells can be used to measure standard voltages, E^0. By arbitrarily setting the standard reduction voltage (E_{red}^0) of H^+ equal to zero, it is possible to obtain standard potentials for the reduction (E_{red}^0) of various species (Table 21.1). Standard voltages for oxidation (E_{ox}^0) can be obtained by changing the sign of the E_{red}^0 value for the reverse half-reaction. Standard voltages for half-reactions can be used to:

1. Compare the strengths of oxidizing agents and reducing agents (Example 21.4). A strong oxidizing agent has a large positive E_{red}^0 value; a strong reducing agent has a large, positive E_{ox}^0 value.

2. Calculate the standard cell voltage (Example 21.5). $E^0 = E_{ox}^0 + E_{red}^0$.

3. Determine whether or not a redox reaction is spontaneous (Example 21.6). The reaction is spontaneous if $E^0 > 0$.

The voltage of a cell depends upon the concentrations of reactants and products. The effect can be calculated from the Nernst equation (Example 21.7).

KEY WORDS AND CONCEPTS

redox reaction	reducing agent	voltaic cell	standard voltage, E^0
oxidation number	half-equation	anode	standard reduction voltage, E_{red}^0
oxidation	spontaneous reaction	cathode	standard oxidation voltage, E_{ox}^0
reduction	nonspontaneous reaction	voltage	Nernst equation
oxidizing agent	electrolytic cell		

QUESTIONS AND PROBLEMS

Catalog

Oxidation Number, Oxidizing and Reducing Agents: 21.2, 21.3, 21.23, 21.24
Balancing Redox Equations: 21.4–21.9, 21.25–21.30
Voltaic Cell Diagrams: 21.10, 21.11, 21.31, 21.32
Strength of Oxidizing, Reducing Agents: 21.12, 21.13, 21.33, 21.34

Calculation of Standard Cell Voltage: 21.14, 21.15, 21.35, 21.36
Reaction Spontaneity: 21.16–21.18, 21.37–21.39
Nernst Equation: 21.19, 21.20, 21.40, 21.41
General: 21.1, 21.21, 21.22, 21.42–21.46

21.1   Review and know the meaning of the key words and concepts in this chapter.

21.2   Give the oxidation number of each atom in

    a. C_2H_6O b. $Cr_2O_7^{2-}$
    c. RuF_5 d. $Na_2Mo_2O_7$
    e. Cl_2O f. $HAsO_4^{2-}$

21.3   For each of the reactions given below (unbalanced equation)

    a. indicate the oxidation number of each atom.
    b. identify the oxidizing agent.
    c. identify the reducing agent.

$$Fe_2O_3(s) + Al(s) \rightarrow Fe(s) + Al_2O_3(s)$$

$$NO_3^-(aq) + Fe^{2+}(aq) + H^+(aq) \rightarrow NO(g) + Fe^{3+}(aq) + H_2O$$

21.23   Give the oxidation number of each atom in

    a. Sb_4O_{10} b. BrO_4^-
    c. $IrCl_6^-$ d. $Na_2Fe_2O_4$
    e. CaC_2O_4 f. HPO_3^{2-}

21.24   For the reaction (unbalanced equation)

$$CrO_4^{2-}(aq) + S^{2-}(aq) + H_2O \rightarrow$$
$$Cr(OH)_3(s) + S(s) + OH^-(aq)$$

identify

    a. the species oxidized.
    b. the species reduced.
    c. the oxidizing agent.
    d. the reducing agent.

21.4 Balance the equations in Question 21.3.

21.5 Balance the following equations:

a. $Al(s) + Pb^{2+}(aq) \rightarrow Al^{3+}(aq) + Pb(s)$
b. $Cu(s) + NO_3^-(aq) + H^+(aq) \rightarrow$
$Cu^{2+}(aq) + NO_2(g) + H_2O$
c. $UO_2^+(aq) + Fe^{2+}(aq) + H^+(aq) \rightarrow$
$Fe^{3+}(aq) + U^{4+}(aq) + H_2O$
d. $H_2O_2(aq) + Mn^{2+}(aq) \rightarrow$
$MnO_4^-(aq) + H^+(aq) + H_2O$

21.6 Write balanced equations for each of the following:

a. $MnO_4^-(aq) + AsO_3^{3-}(aq) \rightarrow$
$Mn^{2+}(aq) + AsO_4^{3-}$; acid
b. $MnO_4^-(aq) + IO_3^-(aq) \rightarrow Mn^{2+}(aq) + IO_4^-(aq)$;
base
c. $Re(s) + NO_3^-(aq) \rightarrow ReO_4^-(aq) + NO(g)$; acid

21.7 When moist air comes in contact with iron, corrosion takes place to form $Fe(OH)_3(s)$. Write balanced half-equations for the oxidation and reduction reactions and combine them to give the overall balanced equation.

21.8 Given the following half-equations for acidic solutions:

$$Cr(s) \rightarrow Cr^{3+}(aq)$$
$$NO(g) \rightarrow NO_3^-(aq)$$
$$Sn^{4+}(aq) \rightarrow Sn^{2+}(aq)$$
$$ClO_3^-(aq) \rightarrow Cl^-(aq)$$

a. Classify each as an oxidation or reduction reaction.
b. Balance each half-reaction equation.
c. Write as many balanced redox equations as possible by combining these half-equations.

21.9 Follow the directions for Question 21.8 for the following half-equations in basic solution:

$$S(s) \rightarrow S^{2-}(aq)$$
$$CrO_4^{2-}(aq) \rightarrow Cr(OH)_3(s)$$
$$ClO_3^-(aq) \rightarrow Cl^-(aq)$$
$$Fe(s) \rightarrow Fe(OH)_3(s)$$

21.10 Draw a diagram for salt bridge cells in which the following reactions occur. In each case, label the anode and cathode and indicate the direction of current flow in the circuit.

a. $Co(s) + Cu^{2+}(aq) \rightarrow Co^{2+}(aq) + Cu(s)$
b. $H_2(g) + I_2(s) + 2 OH^-(aq) \rightarrow 2 I^-(aq) + 2 H_2O$
c. $Al(s) + 3 H^+(aq) \rightarrow Al^{3+}(aq) + {}^3/_2 H_2(g)$

21.11 Write a balanced chemical equation for the overall cell reaction represented as

a. $Sn/Sn^{2+} \parallel Ag^+/Ag$
b. $Zn/Zn^{2+} \parallel (Pt)Cl_2/Cl^-$
c. $Ni/Ni^{2+} \parallel (Pt)Br_2/Br^-$

21.25 Balance the equation in Question 21.24.

21.26 Balance the following equations:

a. $Mn^{2+}(aq) + SO_4^{2-}(aq) + H_2O \rightarrow$
$MnO_4^-(aq) + SO_3^{2-}(aq) + H^+(aq)$
b. $Cr^{3+}(aq) + OH^-(aq) + H_2O_2(aq) \rightarrow$
$CrO_4^{2-}(aq) + H_2O$
c. $Fe^{2+}(aq) + Cr_2O_7^{2-}(aq) + H^+(aq) \rightarrow$
$Cr^{3+}(aq) + Fe^{3+}(aq) + H_2O$
d. $Zn(s) + NO_3^-(aq) + H^+(aq) \rightarrow$
$Zn^{2+}(aq) + NH_4^+(aq) + H_2O$

21.27 Write balanced equations for each of the following:

a. $Ce^{4+}(aq) + VO^{2+}(aq) \rightarrow Ce^{3+}(aq) + VO_3^-(aq)$;
acid
b. $Cr^{3+}(aq) + MnO_4^-(aq) \rightarrow$
$Cr_2O_7^{2-}(aq) + Mn^{2+}(aq)$; acid
c. $MnO_4^-(aq) + Mn^{2+}(aq) \rightarrow MnO_2(s)$; acid

21.28 In gold plating, metallic gold is formed from the $Au(CN)_2^-$ ion. At the same time, $O_2(g)$ is formed. The plating solution is basic. Write equations for the two half-reactions and the overall reaction.

21.29 Follow the instructions for Question 21.8 for the following half-equations (acidic solution):

$$Ni(s) \rightarrow Ni^{2+}(aq)$$
$$Cr_2O_7^{2-}(aq) \rightarrow Cr^{3+}(aq)$$
$$S(s) \rightarrow H_2S(g)$$
$$MnO_2(s) \rightarrow Mn^{2+}(aq)$$

21.30 Follow the directions for Question 21.8 for the following half-equations for reactions in basic solution.

$$H_2O \rightarrow H_2(g)$$
$$O_2(g) \rightarrow OH^-(aq)$$
$$NiO_2(s) \rightarrow Ni(OH)_2(s)$$
$$MnO(s) \rightarrow MnO_4^-(aq)$$

21.31 Follow the instructions for Question 21.10 for the following reactions:

a. $Cd(s) + Ni^{2+}(aq) \rightarrow Cd^{2+}(aq) + Ni(s)$
b. $2 Fe^{3+}(aq) + 2 I^-(aq) \rightarrow 2 Fe^{2+}(aq) + I_2(s)$
c. $3 Ag(s) + AuCl_4^-(aq) \rightarrow$
$3 Ag^+(aq) + Au(s) + 4 Cl^-(aq)$

21.32 Write a balanced chemical equation for the overall cell reaction represented as

a. $(Pt)H_2/H^+ \parallel (Pt)Cl_2/Cl^-$
b. $Sn/Sn^{2+} \parallel Pb^{2+}/Pb$
c. $(Pt)I^-/I_2 \parallel (Pt)Cl_2/Cl^-$

21.12 Use Table 21.1 to classify each of the following as a reducing or oxidizing agent. Indicate the strongest one in each category:

$$Co^{2+}, Al, Br_2, Sn^{2+}, Cu^{2+}$$

21.13 From Table 21.1 select a suitable species for each of the following (standard concentrations):

    a. an oxidizing agent to convert Fe to Fe^{2+} but not Sn to Sn^{2+}.
    b. a reducing agent capable of converting Co^{2+} to Co but not Mg^{2+} to Mg.
    c. an oxidizing agent capable of converting Cl^- to Cl_2 but not Co^{2+} to Co^{3+}.

21.14 Calculate E^0 for the following voltaic cells:

    a. $Sn^{4+}(aq) + Pb(s) \rightarrow Sn^{2+}(aq) + Pb^{2+}(aq)$
    b. $O_2(g) + 4 H^+(aq) + 4 Cr^{2+}(aq) \rightarrow$
    $2 H_2O + 4 Cr^{3+}(aq)$
    c. Mn-Mn^{2+} half-cell and Ni-Ni^{2+} half-cell

21.15 Calculate standard voltages for the cells in Question 21.11.

21.16 Which of the following reactions are spontaneous at standard concentrations?

    a. $I_2(s) + 2 Br^-(aq) \rightarrow Br_2(l) + 2 I^-(aq)$
    b. $Zn(s) + 2 Fe^{3+}(aq) \rightarrow Zn^{2+}(aq) + 2 Fe^{2+}(aq)$
    c. $O_2(g) + 4 H^+(aq) + 4 Cl^-(aq) \rightarrow 2 H_2O + 2 Cl_2(g)$

21.17 Which of the following metals will react with 1 M HCl?

    a. Ni   b. Au   c. Cu   d. Fe

21.18 Using Table 21.1, decide what reaction, if any, will occur when the following are mixed (standard concentrations):

    a. Co^{2+}, Co, Fe^{2+}
    b. Fe^{2+}, Fe^{3+}, Ag^+
    c. ClO_3^-, H^+, Hg(l)

21.19 Calculate the voltages of cells under the following conditions:

    a. $Fe(s) + Cu^{2+}(0.10 \text{ M}) \rightarrow Fe^{2+}(0.010 \text{ M}) + Cu(s)$
    b. $Cu(s) + 2 H^+(0.10 \text{ M}) \rightarrow$
    $Cu^{2+}(0.0010 \text{ M}) + H_2(1 \text{ atm})$

Are the cell reactions spontaneous? Explain.

21.20 At what concentration of H^+ is the voltage $+0.50$ V for the cell

$$Zn(s) + 2 H^+(aq) \rightarrow Zn^{2+}(1 \text{ M}) + H_2(g, 1 \text{ atm})?$$

21.33 Arrange the following species in order of increasing strength as reducing agents:

$$Au, SO_2, F^-, K, Fe^{2+}$$

21.34 Use Table 21.1 to select

    a. a reducing agent that will convert Pb^{2+} to Pb but not Cd^{2+} to Cd.
    b. an oxidizing agent that converts I^- to I_2 but not Br^- to Br_2.
    c. a reducing agent that converts Au^{3+} to Au but not $AuCl_4^-$ to Au.

21.35 Calculate E^0 for the following voltaic cells:

    a. $Al(s) + NO_3^-(aq) + 4 H^+(aq) \rightarrow$
    $NO(g) + 2 H_2O + Al^{3+}(aq)$
    b. $Cu(s) + NO_3^-(aq) + H_2O \rightarrow$
    $Cu(OH)_2(s) + NO_2^-(aq)$
    c. Cu-Cu^{2+} half-cell and I_2-I^- half-cell

21.36 Calculate E^0 for the cells in Question 21.32.

21.37 Which of the following reactions are spontaneous at standard concentrations?

    a. $AuCl_4^-(aq) + 3 Fe^{2+}(aq) \rightarrow$
    $Au(s) + 4 Cl^-(aq) + 3 Fe^{3+}(aq)$
    b. $2 NO_3^-(aq) + 8 H^+(aq) + 6 Cl^-(aq) \rightarrow$
    $2 NO(g) + 4 H_2O + 3 Cl_2(g)$
    c. $Cu(s) + 2 H^+(aq) \rightarrow Cu^{2+}(aq) + H_2(g)$

21.38 Which of the following species will react with 1 M HNO_3?

    a. Ag   b. Cl^- c. Cu   d. F^-

21.39 Predict what reaction, if any, will occur when Cl_2 gas is bubbled through an acidic aqueous solution of each of the following (standard concentrations):

    a. $Ca(NO_3)_2$
    b. FeI_2
    c. AgF

21.40 Calculate the voltages of cells at the following conditions:

    a. $Cd^{2+}(0.020 \text{ M}) + Zn(s) \rightarrow Zn^{2+}(0.50 \text{ M}) + Cd(s)$
    b. $Sn^{2+}(0.10 \text{ M}) + Cu^{2+}(1 \text{ M}) \rightarrow$
    $Cu(s) + Sn^{4+}(0.010 \text{ M})$

Are the cell reactions spontaneous? Explain.

21.41 At what concentration of Cu^{2+} is the voltage zero for the cell

$$Ni(s) + Cu^{2+}(aq) \rightarrow Ni^{2+}(1 \text{ M}) + Cu(s)?$$

21.21 Write balanced equations for

  a. the anode half-reaction in the ordinary dry cell.
  b. the anode half-reaction in the lead storage battery.
  c. the overall reaction in the alkaline dry cell.

21.22 Suppose E^0_{red} for $H^+ \rightarrow H_2$ were taken to be $+1.00$ V rather than zero. On this basis, what would be

  a. E^0_{ox} $H_2 \rightarrow H^+$?
  b. E^0_{red} $Cu^{2+} \rightarrow Cu$?
  c. E^0 for the $Zn/Zn^{2+} \parallel Ag^+/Ag$ cell?

21.42 Write balanced equations for

  a. the half-reaction at the cathode in the lead storage battery.
  b. the overall reaction in the lead storage battery.

21.43 Suppose E^0_{red} $Ag^+ \rightarrow Ag$ were taken to be zero. On this basis, what would be

  a. E^0_{ox} $Ag \rightarrow Ag^+$?
  b. E^0_{red} $H^+ \rightarrow H_2$?
  c. E^0 for the $Zn/Zn^{2+} \parallel Ag^+/Ag$ cell?

*21.44 Lead forms a series of oxides: PbO, PbO_2, Pb_2O_3, Pb_3O_4.

  a. Give the oxidation number of lead in each of these compounds.
  b. Lead occurs as either $+2$ or $+4$ ions in these compounds. Rationalize this with the data from (a).

*21.45 Given

$$Ag^+(aq) + e^- \rightarrow Ag(s); \quad E^0_{red} = +0.80 \text{ V}$$

$$AgCl(s) + e^- \rightarrow Ag(s) + Cl^-(aq); \quad E^0_{red} = +0.22 \text{ V}$$

Using these E^0 values and the Nernst equation, calculate the solubility product constant of $AgCl$.

*21.46 Using the Nernst equation, calculate E_{red} at pH 14.0 for the H^+ ion ($P_{N_2} = 1$ atm). To what standard potential in Table 21.1 does the voltage you have just calculated correspond?

REDOX REACTIONS; OXIDATION NUMBERS OF THE ELEMENTS

In Chapter 21, we introduced several new concepts. Among these were oxidation number (Section 21.1), standard voltages (21.4), and the Nernst equation (21.5). In this chapter, we will use these concepts to discuss the chemistry involved in a variety of redox reactions. We will concentrate upon those reactions which are likely to be most familiar to you. Many of the redox reactions discussed will be ones that you will carry out in the laboratory. Others take place in the world around you.

This chapter is organized according to the types of elements taking part in redox reactions. These include the transition metals (Section 22.2), oxygen (22.3), chlorine (22.4), nitrogen (22.5), and sulfur (22.6). It will be helpful to begin by surveying the common oxidation numbers of all the elements in the Periodic Table. Here we will be interested in trends as we move across or down the Table.

22.1 COMMON OXIDATION NUMBERS OF THE ELEMENTS

The oxidation numbers commonly shown by elements are given in Figure 22.1. We can extract several general principles from what appears at first glance to be a maze of numbers.

POSITIVE AND NEGATIVE OXIDATION NUMBERS. Metals commonly show only positive oxidation numbers in their stable compounds. Oxidation numbers of $+1$, $+2$, and $+3$ are derived from the charges of simple cations (e.g., Na^+, Mg^{2+}, Al^{3+}). Negative oxidation numbers in Figure 22.1 are restricted to nonmetals. For these elements, the lowest oxidation number is equal to the charge of the simple anion (e.g., N^{3-}, O^{2-}, F^-).

MULTIPLE OXIDATION NUMBERS. The metals at the far left of the Periodic Table show only one oxidation number in their compounds. These oxidation numbers reflect the charges of cations with noble-gas structures ($+1$ for Group 1, $+2$ for Group 2, ...). In contrast, most of the transition metals show various oxidation numbers (see color plate 25, center of book). The same is true of the elements in Groups 4 through 8. As an extreme example, nitrogen can have any oxidation number from $+5$ to -3.

Except for Groups 1 and 2, nearly all elements show more than one oxidation no.

520

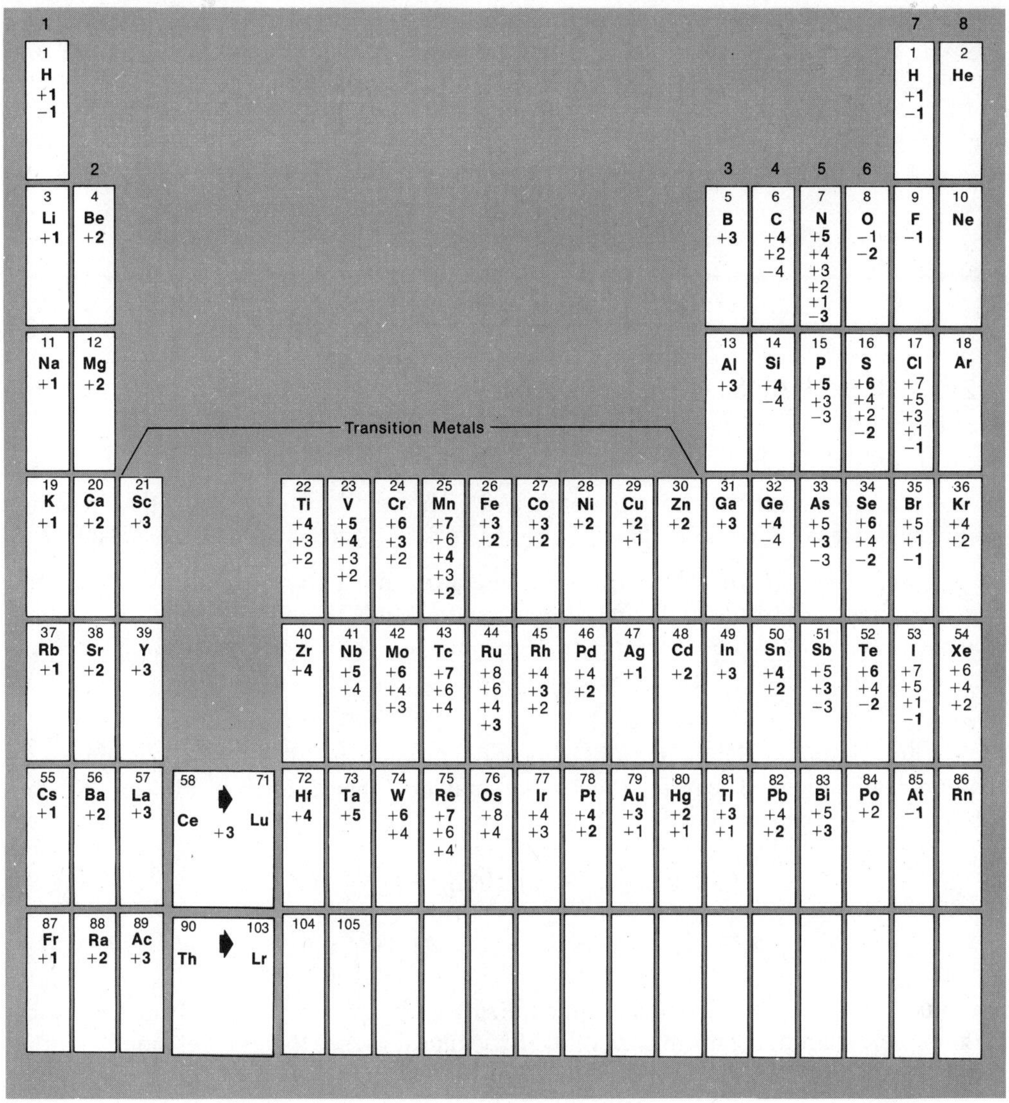

521

REDOX
REACTIONS;
OXIDATION
NUMBERS OF
THE ELEMENTS

Figure 22.1 Oxidation states of the elements in their compounds (a few rather rare oxidation numbers are omitted). The most common or stable states are shown in heavy type.

1	2	Transition Metals										3	4	5	6	7	8
1 H +1, −1																1 H +1, −1	2 He
3 Li +1	4 Be +2											5 B +3	6 C +4, +2, −4	7 N +5, +4, +3, +2, +1, −3	8 O −1, −2	9 F −1	10 Ne
11 Na +1	12 Mg +2											13 Al +3	14 Si +4, −4	15 P +5, +3, −3	16 S +6, +4, +2, −2	17 Cl +7, +5, +3, +1, −1	18 Ar
19 K +1	20 Ca +2	21 Sc +3	22 Ti +4, +3, +2	23 V +5, +4, +3, +2	24 Cr +6, +3, +2	25 Mn +7, +6, +4, +3, +2	26 Fe +3, +2	27 Co +3, +2	28 Ni +2	29 Cu +2, +1	30 Zn +2	31 Ga +3	32 Ge +4, −4	33 As +5, +3, −3	34 Se +6, +4, −2	35 Br +5, +1, −1	36 Kr +4, +2
37 Rb +1	38 Sr +2	39 Y +3	40 Zr +4	41 Nb +5, +4	42 Mo +6, +4, +3	43 Tc +7, +6, +4	44 Ru +8, +6, +4, +3	45 Rh +4, +3, +2	46 Pd +4, +2	47 Ag +1	48 Cd +2	49 In +3	50 Sn +4, +2	51 Sb +5, +3, −3	52 Te +6, +4, −2	53 I +7, +5, +1, −1	54 Xe +6, +4, +2
55 Cs +1	56 Ba +2	57 La +3	72 Hf +4	73 Ta +5	74 W +6, +4	75 Re +7, +6, +4	76 Os +8, +4	77 Ir +4, +3	78 Pt +4, +2	79 Au +3, +1	80 Hg +2, +1	81 Tl +3, +1	82 Pb +4, +2	83 Bi +5, +3	84 Po +2	85 At −1	86 Rn
87 Fr +1	88 Ra +2	89 Ac +3	104	105													

58 Ce → 71 Lu +3

90 Th → 103 Lr

RELATIVE STABILITIES OF OXIDATION NUMBERS OF AN ELEMENT.

Among the transition metals, *higher* oxidation numbers become more stable as we move *down* a given group. Consider, for example, the nickel subgroup. Nickel rarely shows an oxidation number higher than +2. In contrast, Pd and Pt form many stable compounds in which their oxidation number is +4.

Among the elements in Groups 4 through 8, *lower* oxidation numbers usually become more stable as we move *down* a group. For instance, in Group 4, the +4 state is the most common for the first three elements (C, Si, Ge). For tin, the +4 and the +2 states are about equally common (SnO_2, $SnCl_4$; SnO, $SnCl_2$). The last element in the group, lead, is most often found in the +2 state ($PbCl_2$, $PbSO_4$, . . .).

MAXIMUM OXIDATION NUMBER OF AN ELEMENT.

The highest oxidation number an element can have is its group number in the Periodic Table. When this number is +4 or greater, the element is covalently bonded to oxygen or another highly electronegative element. For example, sulfur has a +6 oxidation number in the species SO_3, H_2SO_4, SO_4^{2-}, and SF_6. We find +7 chlorine in the perchlorate ion, ClO_4^-. Here, the bonding electrons have all been assigned arbitrarily to oxygen:

$$\left[\begin{array}{c} :\ddot{O}: \\ :\ddot{O}: \boxed{Cl} :\ddot{O}: \\ :\ddot{O}: \end{array} \right]^-$$

22.2 THE TRANSITION METALS

Transition metals in their compounds are most often present as either:

—*cations,* which may be monatomic (Ag^+, Zn^{2+}, Cr^{3+}). More frequently, the cation is at the center of a complex ion such as $Ag(NH_3)_2^+$, $Zn(NH_3)_4^{2+}$, or $Cr(NH_3)_6^{3+}$.

—*oxyanions,* in which the transition metal is covalently bonded to oxygen. The most familiar ions of this type are permanganate (MnO_4^-), chromate (CrO_4^{2-}), and dichromate ($Cr_2O_7^{2-}$). The structures of these ions are shown in Figure 22.2.

Cations

A transition metal cation in the highest oxidation state of the metal, such as Fe^{3+}, can only be reduced. The reduction product may be either a cation in a lower oxidation state (Fe^{2+}) or the free metal (Fe):

$$Fe^{3+}(aq) + e^- \rightarrow Fe^{2+}(aq); \quad E^0_{red} = +0.77 \text{ V} \tag{22.1}$$

$$Fe^{3+}(aq) + 3 e^- \rightarrow Fe(s); \quad E^0_{red} = -0.04 \text{ V} \tag{22.2}$$

In either case, the Fe^{3+} ion acts as an oxidizing agent, picking up electrons from some other species. From the magnitude of the E^0_{red} values, we see that the Fe^{3+} ion is a rather strong oxidizing agent when it is reduced to Fe^{2+} ($E^0_{red} = +0.77$ V).

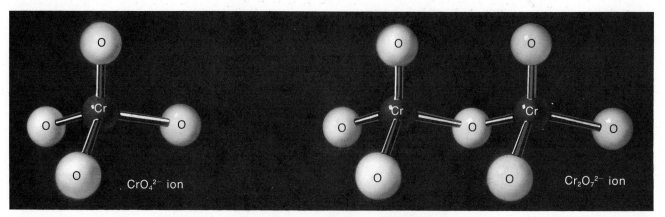

Figure 22.2 The structures of the chromate and dichromate ions. In the $Cr_2O_7^{2-}$ ion, two tetrahedra are linked through an oxygen atom. The oxidation number of chromium in both CrO_4^{2-} and $Cr_2O_7^{2-}$ is +6.

A transition metal cation in an intermediate oxidation state such as Fe^{2+} can act as either an oxidizing or reducing agent. When Fe^{2+} acts as an oxidizing agent, it is reduced to iron metal:

$$Fe^{2+}(aq) + 2 e^- \rightarrow Fe(s); \quad E^0_{red} = -0.44 \text{ V} \qquad (22.3)$$

Both Fe^{2+} and Fe^{3+} can exist in solution

When Fe^{2+} acts as a reducing agent, it is oxidized to Fe^{3+}. The half-reaction is the reverse of 22.1:

$$Fe^{2+}(aq) \rightarrow Fe^{3+}(aq) + e^-; \quad E^0_{ox} = -0.77 \text{ V} \qquad (22.4)$$

Many transition metal ions show this kind of behavior. Thus, Cr^{2+} can be reduced to Cr metal or oxidized to Cr^{3+}. Again, Cu^+ can be reduced to Cu metal or oxidized to Cu^{2+}. The behavior of the Cu^+ ion in water solution is somewhat unusual. It **disproportionates**—that is, it undergoes simultaneous oxidation and reduction. The products, formed together, are copper metal and the Cu^{2+} ion:

$$
\begin{aligned}
Cu^+(aq) + e^- &\rightarrow Cu(s); & E^0_{red} &= +0.52 \text{ V} \\
Cu^+(aq) &\rightarrow Cu^{2+}(aq) + e^-; & E^0_{ox} &= -0.15 \text{ V} \\
\hline
2\, Cu^+(aq) &\rightarrow Cu(s) + Cu^{2+}(aq); & E^0 &= +0.37 \text{ V}
\end{aligned}
\qquad (22.5)
$$

Since E^0 for this reaction is positive, the Cu^+ ion is unstable in solution. Indeed, the only known compounds of Cu(I) are ones which are very insoluble in water, such as CuCl ($K_{sp} = 1.0 \times 10^{-6}$).

Example 22.1   Would you expect the Fe^{2+} ion to disproportionate? That is, would the reaction $3\, Fe^{2+}(aq) \rightarrow 2\, Fe^{3+}(aq) + Fe(s)$ be spontaneous?

Solution   Taking the standard voltages quoted from Equations 22.3 and 22.4:

$$
\begin{aligned}
E^0 &= E^0_{red} + E^0_{ox} \\
&= -0.44 \text{ V} - 0.77 \text{ V} = -1.21 \text{ V}
\end{aligned}
$$

The Fe^{2+} ion should *not* disproportionate. Instead, the reverse reaction should be spontaneous. It is. Iron(II) compounds can be made by reacting iron metal with iron(III) compounds.

We sometimes add a little Fe to solutions of Fe^{2+} to reduce any Fe^{3+} that forms

---

Exercise   For $Au^{3+} \rightarrow Au^+$, $E^0_{red} = +1.40$ V; for $Au^+ \rightarrow Au$, $E^0_{red} = +1.69$ V. Would you expect the Au^+ ion to be stable in water solution? Answer: No; E^0 for disproportionation is +0.29 V.

Oxyanions

Oxyanions in which a transition metal is in its highest oxidation state are usually strong oxidizing agents in acidic solution. This is true of the permanganate ion (oxid. no. Mn = +7) and the dichromate ion (oxid. no. Cr = +6):

$$MnO_4^-(aq) + 8 H^+(aq) + 5 e^- \rightarrow Mn^{2+}(aq) + 4 H_2O; \quad E^0_{red} = +1.52 \text{ V} \qquad (22.6)$$

$$Cr_2O_7^{2-}(aq) + 14 H^+(aq) + 6 e^- \rightarrow 2 Cr^{3+}(aq) + 7 H_2O; \quad E^0_{red} = +1.33 \text{ V} \qquad (22.7)$$

The large positive values of E^0_{red} mean that the MnO_4^- and $Cr_2O_7^{2-}$ ions are very strong oxidizing agents. Indeed, the MnO_4^- ion is capable of oxidizing water to form O_2; aqueous solutions of $KMnO_4$ decompose slowly because of this reaction. A solution prepared by adding sulfuric acid to potassium dichromate ($K_2Cr_2O_7$) is often used as an oxidizing agent to clean laboratory glassware. It attacks greases, oils, and carbon deposits that are not affected by ordinary soaps or detergents.

Potassium permanganate, $KMnO_4$, is often used in redox titrations to determine the concentration of an oxidizable species such as Fe^{2+}. To carry out the titration, we start with a known volume of an acidified solution containing Fe^{2+} ions. A solution of $KMnO_4$ of known concentration is added from a buret. The end point of the titration is easily detected. When all the Fe^{2+} ions are gone, the addition of one or two drops of excess MnO_4^- gives the pink or purple color of that ion. Knowing:

— the volumes of the two reagents used
— the concentration of the $KMnO_4$ solution
— the balanced equation for the reaction

it is possible to calculate the concentration of Fe^{2+} ions in the solution (Example 22.2).

MnO_4^- ion is very highly colored

Example 22.2   A 20.0-cm³ sample containing Fe^{2+} is titrated with 0.100 M $KMnO_4$. A permanent pink color appears after addition of 18.0 cm³ of the $KMnO_4$ solution.
   a. Write a balanced equation for the redox reaction between Fe^{2+} and MnO_4^-.
   b. Calculate the concentration of Fe^{2+} ions in the solution.

Solution
   a. The half-equations are 22.4 and 22.6:

   oxidation: $Fe^{2+}(aq) \rightarrow Fe^{3+}(aq) + e^-$
   reduction: $MnO_4^-(aq) + 8\ H^+(aq) + 5\ e^- \rightarrow Mn^{2+}(aq) + 4\ H_2O$

   We multiply the oxidation half-equation by 5 to make the electron loss equal to the electron gain. The two half-equations are then added to obtain the balanced equation:

   $$MnO_4^-(aq) + 8\ H^+(aq) + 5\ Fe^{2+}(aq) \rightarrow Mn^{2+}(aq) + 5\ Fe^{3+}(aq) + 4\ H_2O$$

K_c for this reaction is enormous

   b. We first calculate the number of moles of MnO_4^- added. Then, using the equation just obtained, we can calculate the number of moles of Fe^{2+} in the sample. Finally, knowing the volume of the sample, we can calculate the concentration of Fe^{2+}.

   $$\text{moles } MnO_4^- = 0.100\ \frac{mol}{\ell} \times 0.0180\ \ell = 0.00180\ mol$$

   But: $1\ mol\ MnO_4^- \simeq 5\ mol\ Fe^{2+}$

   So: $\text{moles } Fe^{2+} = 0.00180\ mol\ MnO_4^- \times \dfrac{5\ mol\ Fe^{2+}}{1\ mol\ MnO_4^-} = 0.00900\ mol\ Fe^{2+}$

   $$\text{conc. } Fe^{2+} = \frac{0.00900\ mol}{0.0200\ \ell} = 0.450\ M$$

Exercise   Suppose 20.0 cm³ of 0.100 M $KMnO_4$ were required to titrate 18.0 cm³ of Fe^{2+}. What would be the concentration of Fe^{2+}? Answer: 0.556 M.

The oxidizing strength of an ion like MnO_4^- or $Cr_2O_7^{2-}$ depends upon the concentration of H^+ ion. From the half-equations for their reduction (Eqs. 22.6 and 22.7), we see that H^+ is a reactant. Hence, lowering the concentration of H^+ should make the reduction less spontaneous.

The effect of H^+ ion concentration upon the oxidizing strength of an oxyanion can be determined quantitatively, using the Nernst equation. Consider the half-reaction for the reduction of MnO_4^-:

Never add conc. H_2SO_4 to an oxyanion like MnO_4^-. The product may explode

$$MnO_4^-(aq) + 8\ H^+(aq) + 5\ e^- \rightarrow Mn^{2+}(aq) + 4\ H_2O$$

Applying the Nernst equation, we have

$$E_{red} = E_{red}^0 - \frac{0.0591}{5} \log_{10} \frac{(conc.\ Mn^{2+})}{(conc.\ MnO_4^-) \times (conc.\ H^+)^8} \qquad (22.8)$$

Reducing the concentration of H^+ from 1 M (pH = 0) to 1×10^{-7} M (pH = 7) lowers the reduction voltage by 0.66 V (Example 22.3).

Example 22.3   For the reduction of MnO_4^-, $E_{red}^0 = +1.52$ V. This value applies when conc. MnO_4^- = conc. Mn^{2+} = conc. H^+ = 1 M. Suppose we lower conc. H^+ to 1×10^{-7} M, but keep all the other concentrations the same. Using the Nernst equation, determine the reduction voltage, E_{red}, under these conditions.

Solution   Substituting concentrations into Equation 22.8:

$$E = +1.52\ V - \frac{0.0591}{5} \log_{10} \frac{1}{1 \times 10^{-56}}$$

$$= +1.52\ V - \frac{0.0591}{5}\ (56) = +1.52\ V - 0.66\ V = +0.86\ V$$

Exercise   Calculate E_{red} for this half-reaction at pH = 13.0. Answer: +0.29 V.

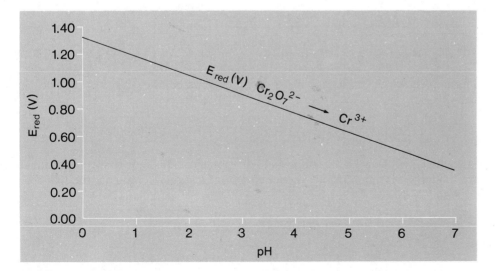

Figure 22.3   The oxidizing strength of the $Cr_2O_7^{2-}$ ion decreases with increasing pH (decreasing acidity). E_{red} drops by about 0.14 V when the pH increases by one unit.

We see from Example 22.3 that the concentration of H^+ ion has a large effect upon the oxidizing strength of the MnO_4^- ion. In 1 M acid, it is a very strong oxidizing agent. In neutral (conc. $H^+ = 10^{-7}$ M) or basic solution, the oxidizing strength of the MnO_4^- ion is greatly reduced. This effect is general for all oxyanions, since H^+ ions are involved as a reactant in their reduction. Decreasing the concentration of H^+ reduces their oxidizing strength, often to the point where they become ineffective. Figure 22.3 shows this effect with the $Cr_2O_7^{2-}$ ion.

22.3 OXYGEN

The redox chemistry of oxygen is relatively simple, at least in comparison with that of such nonmetals as chlorine, nitrogen, and sulfur. It shows only three common oxidation numbers: -2, -1, and 0. In most of its stable compounds, oxygen is present in the -2 oxidation state. This is the case, for example, in the

water molecule, H_2O
oxide ion, O^{2-}, and hydroxide ion, OH^-
molecules of oxyacids ($HClO_4$, HNO_3, H_2SO_4, . . .)
oxyanions (ClO_4^-, NO_3^-, HSO_4^-, SO_4^{2-}, CrO_4^{2-}, MnO_4^-, . . .)

-1 State (O_2^{2-}, H_2O_2)

In metal peroxides, such as Na_2O_2 and BaO_2, oxygen is in the -1 oxidation state. The same is true in hydrogen peroxide, H_2O_2

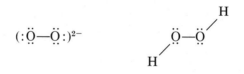

peroxide ion   hydrogen peroxide

Hydrogen peroxide can act as a strong oxidizing agent. In this case, it is reduced to H_2O (oxid. no. oxygen: $-1 \rightarrow -2$).

$$H_2O_2(aq) + 2\ H^+(aq) + 2\ e^- \rightarrow 2\ H_2O;\ E^0_{red} = +1.77\ V \qquad (22.9)$$

Alternatively, H_2O_2 can act as a weak reducing agent, being oxidized to O_2 (oxid. no. oxygen: $-1 \rightarrow 0$).

$$H_2O_2(aq) \rightarrow O_2(g) + 2\ H^+(aq) + 2\ e^-;\ E^0_{ox} = -0.68\ V \qquad (22.10)$$

As you might expect from these voltages, hydrogen peroxide tends to disproportionate in water, undergoing simultaneous reduction and oxidation. Combining the half-reactions 22.9 and 22.10:

$$2\ H_2O_2(aq) \rightarrow 2\ H_2O + O_2(g) \qquad (22.11)$$

$$E^0 = E^0_{red} + E^0_{ox} = +1.77\ V + (-0.68\ V) = +1.09\ V$$

This reaction occurs rather slowly in dilute solution (3% H_2O_2). It occurs more rapidly at higher concentrations and is catalyzed by many different species, including I^- ions and MnO_2. We would hardly consider hydrogen peroxide to be a "stable" species.

527

Laboratory, solutions of H_2O_2 should be freshly prepared

O_2 As an Oxidizing Agent

In the remainder of this section, we will consider redox reactions involving elementary oxygen (oxid. no. = 0). The O_2 molecule always acts as an oxidizing agent in these reactions. In acidic solution, O_2 is a strong oxidizing agent and is reduced to H_2O:

$$\tfrac{1}{2}\,O_2(g) + 2\,H^+(aq) + 2\,e^- \rightarrow H_2O; \quad E^0_{red} = +1.23\text{ V} \qquad (22.12)$$

In basic solution, O_2 is reduced to OH^- ions. The standard reduction voltage is considerably smaller:

$$\tfrac{1}{2}\,O_2(g) + H_2O + 2\,e^- \rightarrow 2\,OH^-(aq); \quad E^0_{red} = +0.40\text{ V} \qquad (22.13)$$

Of all oxidizing agents, elementary oxygen is the most abundant and, in many ways, the most important. Its presence in air insures that all water supplies including reagent solutions used in the laboratory will ordinarily be saturated with atmospheric oxygen. We often forget this and are puzzled by such phenomena as:

—the formation of white or yellow precipitates when hydrogen sulfide is used in qualitative analysis:

$$\tfrac{1}{2}\,O_2(g) + H_2S(g) \rightarrow S(s) + H_2O \qquad (22.14)$$

—the yellow color that solutions of NaI or KI acquire upon standing:

$$\tfrac{1}{2}\,O_2(g) + 2\,I^-(aq) + 2\,H^+(aq) \rightarrow I_2(aq) + H_2O \qquad (22.15)$$

—the cloudiness that develops in a solution of tin(II) chloride:

$$\tfrac{1}{2}\,O_2(g) + Sn^{2+}(aq) + H_2O \rightarrow SnO_2(s) + 2\,H^+(aq) \qquad (22.16)$$

The Corrosion of Iron

From an economic standpoint, the most important redox reaction involving dissolved oxygen is the corrosion of iron and steel. It is estimated that the annual cost to this country of corrosion of ferrous metals exceeds ten billion dollars. We see the results of corrosion all around us in junk piles and auto graveyards. Perhaps as much as 20 per cent of all the iron produced each year in this country goes to replace products whose usefulness has been destroyed by rust.

To understand how iron corrodes, consider what happens when a sheet of iron is exposed to a water solution containing dissolved oxygen. The iron tends to oxidize according to the half-reaction

Snow + ice + NaCl + cars → rust + disgruntled Minnesotans

$$Fe(s) \rightarrow Fe^{2+}(aq) + 2\,e^- \qquad (22.17a)$$

At the same time, oxygen molecules in the solution are reduced:

$$\tfrac{1}{2}\,O_2(g) + H_2O + 2\,e^- \rightarrow 2\,OH^-(aq) \qquad (22.17b)$$

Adding these two half-equations, and noting that iron(II) hydroxide is insoluble, we obtain for the primary corrosion reaction

$$Fe(s) + \tfrac{1}{2} O_2(g) + H_2O \rightarrow Fe(OH)_2(s) \qquad (22.17)$$

Reaction 22.17 appears to be the reaction involved in the first step of the corrosion of iron or steel. Ordinarily, iron(II) hydroxide is further oxidized in a second step:

O_2 can easily oxidize
Fe(II) to Fe(III)

$$2\ Fe(OH)_2(s) + \tfrac{1}{2} O_2(g) + H_2O \rightarrow 2\ Fe(OH)_3(s) \qquad (22.18)$$

The final product is the loose, flaky deposit that we call rust. It has the reddish brown color of iron(III) hydroxide, $Fe(OH)_3$.

Experimentally, we find that the two half-reactions 22.17a and 22.17b do not occur at the same location. The rust on a nail extracted from an old building is concentrated near the head, which has been in contact with moist air (Fig. 22.4). The most serious pitting, caused by oxidation, is found along the shank of the nail, which is embedded in the wood. These observations suggest that oxidation is occurring along a surface some distance away from the point where oxygen is being reduced.

The fact that oxidation and reduction half-reactions take place at different locations suggests that corrosion occurs by an electrochemical mechanism. The surface of a piece of corroding iron may be visualized as consisting of a series of tiny voltaic cells. At *anodic areas,* iron is oxidized to Fe^{2+} ions; at *cathodic areas,* elementary oxygen is reduced to OH^- ions. Electrons are transferred through the iron, which acts like the external conductor of an ordinary voltaic cell. The electrical circuit is completed by the flow of ions through the water solution or film covering the iron.

Many characteristics of corrosion are most readily explained in terms of an electrochemical mechanism. A perfectly dry metal surface is not attacked by oxygen; iron exposed to dry air does not corrode. This seems plausible if corrosion occurs through a voltaic cell, which requires a water solution through which ions can move to complete the circuit. The fact that corrosion occurs more readily in seawater than in fresh water has a similar explanation. The dissolved salts in seawater supply the ions necessary for the conduction of current.

If you can prevent formation of anodic and cathodic regions you can prevent corrosion

The existence of discrete cathodic and anodic areas on a piece of corroding iron requires that adjacent surface areas differ from each other chemically. There are several ways in which one small area on a piece of iron or steel can become anodic or cathodic with respect to an adjacent area. Two of the most important are:

1. *The presence of impurities at scattered locations along the metal surface.* A tiny crystal of a less active metal such as copper or tin embedded in the surface of the

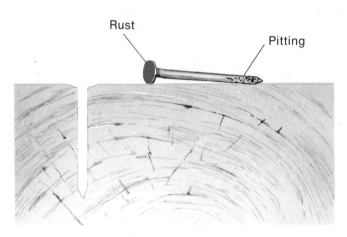

Rust

Pitting

Figure 22.4 Corrosion of an iron nail driven into wood. Rust collects near the head of the nail, but pitting occurs along its length.

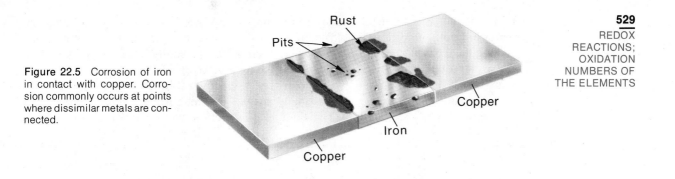

Figure 22.5 Corrosion of iron in contact with copper. Corrosion commonly occurs at points where dissimilar metals are connected.

529
REDOX
REACTIONS;
OXIDATION
NUMBERS OF
THE ELEMENTS

iron acts as a cathode at which oxygen molecules are reduced. The iron atoms in the vicinity of these impurities are anodic and undergo oxidation to Fe^{2+} ions. This effect can be demonstrated on a large scale by immersing in water an iron plate which has been partially copper-plated (Fig. 22.5). At the interface between the two metals, a voltaic cell is set up in which the iron is anodic and the copper cathodic. A thick deposit of rust forms at the interface. The formation of rust inside an automobile bumper where the chromium plate stops is another example of this phenomenon.

2. *Differences in oxygen concentration along the metal surface.* To illustrate this effect, consider what happens when a drop of water adheres to the surface of a piece of iron exposed to the air (Fig. 22.6). The metal around the edges of the drop is in contact with water containing a high concentration of dissolved oxygen. The water touching the metal beneath the center of the drop is depleted in oxygen, since it is cut off from contact with air. As a result, a small oxygen concentration cell is set up. The area around the edge of the drop, where the oxygen concentration is high, becomes cathodic; oxygen molecules are reduced there via Reaction 22.17b. Directly beneath the drop is an anodic area where the iron is oxidized. A particle of dirt on the surface of an iron object can act in much the same way as a drop of water to cut off the supply of oxygen to the area beneath it and thereby establish anodic and cathodic areas. This explains why garden tools left covered with soil are particularly susceptible to rusting.

22.4 CHLORINE

The element chlorine is a powerful oxidizing agent ($E^0_{red} = +1.36$ V). Perhaps the most familiar reactions in which Cl_2 acts as an oxidizing agent are those involving Br^- and I^- ions:

$$Cl_2(g) + 2\ Br^-(aq) \rightarrow 2\ Cl^-(aq) + Br_2(l);\ E^0 = +0.29\ V \qquad (22.19)$$

$$Cl_2(g) + 2\ I^-(aq) \rightarrow 2\ Cl^-(aq) + I_2(s);\ E^0 = +0.83\ V \qquad (22.20)$$

Figure 22.6 Corrosion of iron under a drop of water. The Fe^{2+} ions migrate toward the edge of the drop, where they precipitate as $Fe(OH)_2$, which is then further oxidized by air to $Fe(OH)_3$.

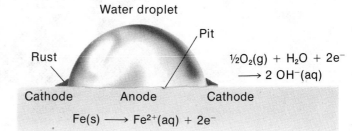

These reactions are often used to test for the presence of Br^- or I^- ions (see color plate 26, center of book). Addition of chlorine to a solution containing either of these ions gives the free halogens, Br_2 or I_2. After addition of chlorine, the solution is often shaken with an organic solvent. The free halogen enters the organic layer, in which it is more soluble. It gives that layer its characteristic color, reddish brown (bromine) or violet (iodine).

Chlorine, unlike oxygen, forms many species in which it has a positive oxidation number ($+7, +5, +3, +1$). In the remainder of this section, we will look at the properties of some of these species.

Oxyanions and Oxyacids of Chlorine

Table 22.1 lists the formulas and names of the oxyanions and corresponding oxygen acids of chlorine. Figure 22.7 shows the structures of the four oxyacids. Notice that, in each case, the *ionizable hydrogen atom is bonded to oxygen*. This is generally true of oxyacids of other elements as well.

Among the four oxyacids of chlorine, acid strength decreases as the number of oxygen atoms bonded to chlorine decreases (or as the oxidation number of chlorine decreases). Thus we have

<table>
<tr><td colspan="6" align="center">TABLE 22.1   OXYACIDS AND OXYANIONS OF CHLORINE</td></tr>
<tr><td rowspan="2">OXID. NO.
Cl</td><td colspan="2" align="center">OXYANION</td><td colspan="2" align="center">OXYACID</td></tr>
<tr><td>Formula</td><td>Name</td><td>Formula</td><td>Name</td></tr>
<tr><td>+7</td><td>ClO_4^-</td><td>perchlorate</td><td>$HClO_4$</td><td>perchloric acid</td></tr>
<tr><td>+5</td><td>ClO_3^-</td><td>chlorate</td><td>$HClO_3$</td><td>chloric acid</td></tr>
<tr><td>+3</td><td>ClO_2^-</td><td>chlorite</td><td>$HClO_2$</td><td>chlorous acid</td></tr>
<tr><td>+1</td><td>ClO^-</td><td>hypochlorite</td><td>$HClO$</td><td>hypochlorous acid</td></tr>
</table>

If you learn these names, you'll have a good guide to naming other similar species

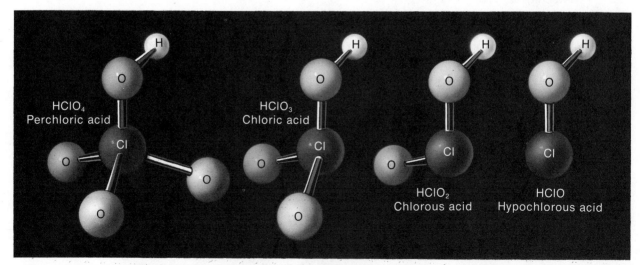

Figure 22.7   In perchloric acid, $HClO_4$, the Cl atom is at the center of a tetrahedron, bonded to four oxygen atoms; there is an H atom bonded to one of the oxygens. In $HClO_3$, $HClO_2$, and $HClO$, successive oxygen atoms are removed from corners of the tetrahedron.

$$HClO_4(aq) \rightarrow H^+(aq) + ClO_4^-(aq); \quad K_a \approx 10^7 \text{ (very strong)}$$
$$HClO_3(aq) \rightarrow H^+(aq) + ClO_3^-(aq); \quad K_a \approx 1 \text{ (strong)}$$
$$HClO_2(aq) \rightleftharpoons H^+(aq) + ClO_2^-(aq); \quad K_a = 1 \times 10^{-2} \text{ (weak)}$$
$$HClO(aq) \rightleftharpoons H^+(aq) + ClO^-(aq); \quad K_a = 3.2 \times 10^{-8} \text{ (very weak)}$$

531

REDOX
REACTIONS;
OXIDATION
NUMBERS OF
THE ELEMENTS

This behavior, as we will see, is typical of oxyacids of other nonmetals.

NOMENCLATURE. The names of the species listed in Table 22.1 follow a pattern typical of oxyanions and oxyacids in general. The rules followed are:

1. When a nonmetal forms two oxyanions, the suffix *ate* is used to denote the anion in which the nonmetal is in a higher oxidation state. The suffix *ite* is used for the anion in which the nonmetal is in a lower oxidation state. Thus we have chlor*ate*, ClO_3^-, and chlor*ite*, ClO_2^-; compare sulf*ate*, SO_4^{2-}, and sulf*ite*, SO_3^{2-}.

2. When an element forms more than two oxyanions, the prefixes *per* (highest oxidation number) and *hypo* (lowest oxidation number) are used as well. This is necessary with chlorine, where we have *per*chlor*ate*, ClO_4^-, and *hypo*chlor*ite*, ClO^-.

3. The name of an oxyacid is directly related to that of the corresponding oxyanion. The suffix *ate* (ClO_4^-, ClO_3^-) is replaced by *ic* ($HClO_4$, $HClO_3$). The suffix *ite* (ClO_2^-, ClO^-) is replaced by *ous* ($HClO_2$, $HClO$).

OXIDIZING STRENGTH. Figure 22.8 shows standard potentials (values of E_{red}^0) for several half-reactions involving the oxyanions or oxyacids of chlorine. Values are

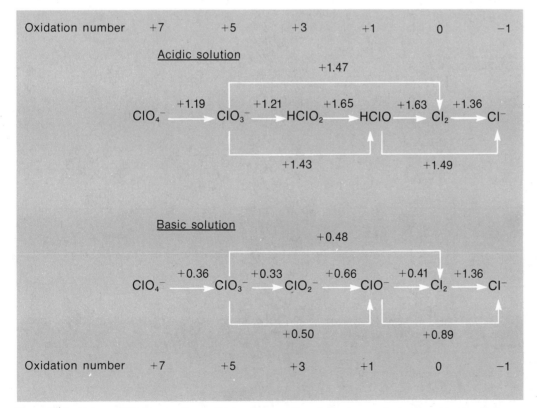

Figure 22.8 Values of E_{red}^0 for species containing Cl in acidic solution (upper diagram) and basic solution (lower diagram). Values of E_{ox}^0 can be obtained by changing the sign. Thus:

$$2\,ClO_3^-(aq) + 12\,H^+(aq) + 10\,e^- \rightarrow Cl_2(g) + 6\,H_2O; \quad E_{red}^0 = +1.47 \text{ V}$$

$$Cl_2(g) + 6\,H_2O \rightarrow 2\,ClO_3^-(aq) + 12\,H^+(aq) + 10\,e^-; \quad E_{ox}^0 = -1.47 \text{ V}$$

given for acidic solution (conc. $H^+ = 1$ M) and basic solution (conc. $OH^- = 1$ M; conc. $H^+ = 1 \times 10^{-14}$ M). You will note that in the +3 and +1 states, the species listed differ in acidic and basic solution. The weak acids $HClO_2$ and $HClO$ are the main species present at high H^+ concentration; in basic solution they are converted to the ClO_2^- and ClO^- ions.

Two points seem obvious from the figure:

1. All the species in which chlorine has a positive oxidation number are relatively strong oxidizing agents. All of them have positive values of E_{red}^0.

2. The oxidizing strength of these species is considerably greater in acidic than in basic solution. Compare, for example, the E_{red}^0 values for $ClO_4^- \rightarrow ClO_3^-$:

$$\text{acid: } ClO_4^-(aq) + 2\ H^+(aq) + 2\ e^- \rightarrow ClO_3^-(aq) + H_2O; \quad E_{red}^0 = +1.19\text{ V}$$

$$\text{base: } ClO_4^-(aq) + H_2O + 2\ e^- \rightarrow ClO_3^-(aq) + 2\ OH^-(aq); \quad E_{red}^0 = +0.36\text{ V}$$

The explanation here is the same as that offered earlier (Example 22.3). We see that H^+ ions are involved in the reduction of ClO_4^- ions. Hence, a decrease in their concentration makes the reduction less spontaneous and the voltage a smaller positive number.

The data in Figure 22.8 can be used to calculate standard voltages for a large number of redox reactions. Several of these involve disproportionation, in which the same species is both oxidized and reduced (Example 22.4).

Example 22.4 Consider the disproportionation of chlorous acid, $HClO_2$, to hypochlorous acid, $HClO$, and chlorate ion, ClO_3^-.

    a. Write a balanced equation for this reaction.

    b. Using Figure 22.8, calculate the standard voltage, E^0, for the reaction.

Solution

    a. The balanced half-equations, obtained in the usual way, are

      reduction: $HClO_2(aq) + 2\ H^+(aq) + 2\ e^- \rightarrow HClO(aq) + H_2O$
      oxidation: $HClO_2(aq) + H_2O \rightarrow ClO_3^-(aq) + 3\ H^+(aq) + 2\ e^-$

Adding the two half-equations gives, after cancelling out species that appear on both sides (2 e^-, 2 H^+, H_2O),

$$2\ HClO_2(aq) \rightarrow HClO(aq) + ClO_3^-(aq) + H^+(aq)$$

  b. $E^0 = E_{red}^0\ (HClO_2 \rightarrow HClO) + E_{ox}^0\ (HClO_2 \rightarrow ClO_3^-)$

From Figure 22.8, we read $E_{red}^0 = +1.65$ V directly. To obtain E_{ox}^0, we note that E_{red}^0 for the reverse reaction is +1.21 V. Changing the sign, we get $E_{ox}^0 = -1.21$ V.

$$E^0 = +1.65\text{ V} - 1.21\text{ V} = +0.44\text{ V}$$

The fact that E^0 is positive means that the $HClO_2$ molecule (oxid. no. Cl = +3) is unstable in water solution. $HClO_2$ disproportionates to $HClO$ (oxid. no. Cl = +1) and ClO_3^- (oxid. no. Cl = +5).

Exercise Determine E^0 for the disproportionation of the ClO_2^- ion (in basic solution) to give ClO^- and ClO_3^-. Answer: +0.33 V.

As we have just seen, the +3 state of chlorine is unstable in water solution. Both the $HClO_2$ molecule and ClO_2^- ion decompose spontaneously. Ordinarily, these reactions occur rather rapidly. This explains why you don't find compounds containing +3 chlorine in the general chemistry laboratory.

In the rest of this section, we will look at the properties of species where the oxidation number of chlorine is +1, +5, or +7. Typically, such compounds as NaClO, $KClO_3$, and $KClO_4$ are found in reagent bottles in the laboratory or storeroom. The acids HClO, $HClO_3$, and $HClO_4$ are less stable. Indeed, perchloric acid, $HClO_4$, is the only oxyacid of chlorine that can be isolated in the pure state.

533

REDOX
REACTIONS;
OXIDATION
NUMBERS OF
THE ELEMENTS

Hypochlorous Acid and the Hypochlorite Ion

When chlorine is added to water it undergoes the following reversible reaction:

$$Cl_2(g) + H_2O \rightleftharpoons HClO(aq) + H^+(aq) + Cl^-(aq) \qquad (22.21)$$

The resulting solution is called "chlorine water." It contains equimolar amounts of the weak acid HClO and the strong acid HCl. The concentrations of both acids in this solution are rather low, about 0.02 M.

The position of the equilibrium in Reaction 22.21 is strongly affected by the concentration of H^+ ions. In basic solution, where $[H^+]$ is low, chlorine is much more soluble than in pure water. The overall reaction that occurs when chlorine is bubbled through a solution of sodium hydroxide at room temperature is

$$Cl_2(g) + 2\ OH^-(aq) \rightarrow ClO^-(aq) + Cl^-(aq) + H_2O \qquad (22.22)$$

The solution formed in Reaction 22.22 is sold under various trade names as a household bleach and disinfectant. It is prepared commercially by electrolyzing a stirred water solution of sodium chloride. Recall that the electrolysis of an NaCl solution gives Cl_2 molecules and OH^- ions; stirring ensures that these species react with each other. The active ingredient of the resulting solution is the hypochlorite ion, which is a rather potent oxidizing agent:

$$ClO^-(aq) + H_2O + 2\ e^- \rightarrow Cl^-(aq) + 2\ OH^-(aq); \ E^0_{red} = +0.89\ V$$

Chlorates and Perchlorates

In hot, concentrated solution, the reaction between chlorine and OH^- is quite different from the one that occurs at room temperature. Any ClO^- ions formed disproportionate to ClO_3^- and Cl^- ions ($E^0 = +0.39\ V$). The net reaction is

$$3\ Cl_2(g) + 6\ OH^-(aq) \rightarrow ClO_3^-(aq) + 5\ Cl^-(aq) + 3\ H_2O \qquad (22.23)$$

Potassium chlorate is a powerful oxidizing agent in acidic solution (Fig. 22.8). It reacts violently with easily oxidized materials, including many organic compounds. It can be used as a laboratory source of oxygen if heated gently, using MnO_2 as a catalyst:

$$2\ KClO_3(s) \rightarrow 2\ KCl(s) + 3\ O_2(g) \qquad (22.24)$$

Without a catalyst, at 350 to 400°C, $KClO_3$ gives potassium perchlorate, $KClO_4$, and potassium chloride:

$$4\ KClO_3(s) \rightarrow 3\ KClO_4(s) + KCl(s) \qquad (22.25)$$

Perchloric acid can be prepared by heating (very carefully!) a metal perchlorate with sulfuric acid:

$$KClO_4(s) + H_2SO_4(l) \rightarrow KHSO_4(s) + HClO_4(l) \qquad (22.26)$$

The pure acid and its concentrated water solution (above 60% $HClO_4$) are explosive and unsafe to work with. In cold, dilute solution, perchloric acid is a stable, very strong acid.

22.5 NITROGEN

As pointed out earlier, nitrogen can have all possible oxidation numbers between $+5$ and -3. The species stable in acidic and basic solution for each of these oxidation states are listed in Table 22.2. Standard potentials (values of E^0_{red}) for some of the possible reductions are given in Figure 22.9.

Several of the species listed in Table 22.2 were discussed in previous chapters. In particular, the electronic structures and properties of the oxides of nitrogen were considered in Chapter 12. Ammonia, you will recall, is a weak base (Chap. 17) and an effective complexing agent (Chap. 19). It can also act as a reducing agent, since the nitrogen atom in NH_3 is in its lowest oxidation state, -3.

In the remainder of this section we will deal with the oxyacids and oxyanions of nitrogen in the $+5$ and $+3$ states. The Lewis structures of these species are shown on p. 535, below Table 22.2.

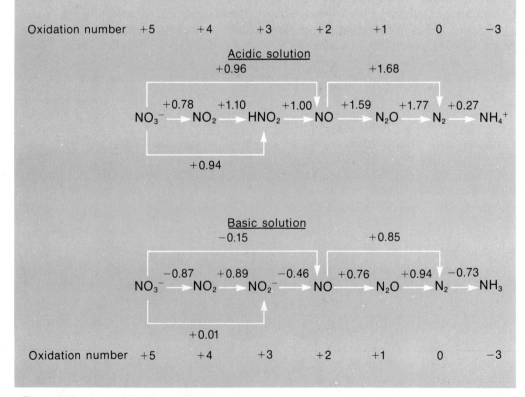

Figure 22.9 Values of E^0_{red} for species containing nitrogen. These voltages are interpreted in the same way as those listed in Figure 22.8.

535
REDOX
REACTIONS;
OXIDATION
NUMBERS OF
THE ELEMENTS

TABLE 22.2   OXIDATION STATES OF NITROGEN

Oxid. No. N	Acidic Solution	Basic Solution
+5	NO_3^-	NO_3^-
+4	$NO_2(g)$	$NO_2(g)$
+3	HNO_2*	NO_2^-
+2	$NO(g)$	$NO(g)$
+1	$N_2O(g)$	$N_2O(g)$
0	$N_2(g)$	$N_2(g)$
-1	NH_3OH^+	NH_2OH
-2	$N_2H_5^+$	N_2H_4
-3	NH_4^+	NH_3

*Slowly decomposes to NO and NO_3^-.

N wins the prize for multiple oxidation states

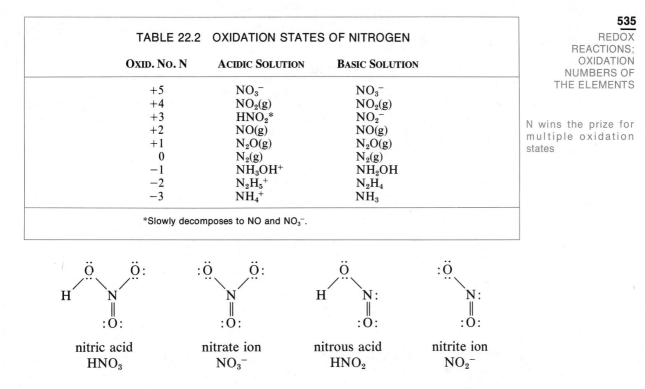

nitric acid	nitrate ion	nitrous acid	nitrite ion
HNO_3	NO_3^-	HNO_2	NO_2^-

+5 State (NO_3^-, HNO_3)

The preparation of nitric acid by the Ostwald process was discussed in Chapter 15. It is a strong acid, completely dissociated to H^+ and NO_3^- ions in dilute water solution:

$$HNO_3(aq) \rightarrow H^+(aq) + NO_3^-(aq)$$

Concentrated nitric acid (16 M) is colorless when pure. In sunlight, it turns yellow because of decomposition to NO_2:

$$4\ HNO_3(aq) \rightarrow 4\ NO_2(g) + 2\ H_2O + O_2(g)$$

The concentrated acid is a strong oxidizing agent. It can be reduced to any of the species listed below the NO_3^- ion in the "Acidic Solution" column of Table 22.2. The principal product is usually $NO_2(g)$. This is the case, for example, when 16 M HNO_3 reacts with copper or silver. The metals are oxidized and go into solution as cations:

$$Cu(s) + 2\ NO_3^-(aq) + 4\ H^+(aq) \rightarrow Cu^{2+}(aq) + 2\ NO_2(g) + 2\ H_2O \quad (22.27)$$

$$Ag(s) + NO_3^-(aq) + 2\ H^+(aq) \rightarrow Ag^+(aq) + NO_2(g) + H_2O \quad (22.28)$$

The 16 M HNO_3 solution can also be used to oxidize highly insoluble metal sulfides such as copper(II) sulfide, CuS:

$$CuS(s) + 2\ NO_3^-(aq) + 4\ H^+(aq) \rightarrow Cu^{2+}(aq) + S(s) + 2\ NO_2(g) + 2\ H_2O \quad (22.29)$$

This is about the only way to get CuS into solution

Here again the NO_3^- ion is reduced to NO_2. The sulfide ion, S^{2-}, is oxidized to sulfur and the cation, Cu^{2+}, goes into solution. Note that none of these reactions would take place with H^+ ions alone. The NO_3^- ion is the oxidizing agent in each case.

Example 22.5 Consider Reaction 22.28. Taking E^0_{ox} Ag $= -0.80$ V and E^0_{red} NO_3^- $= +0.78$ V,

a. calculate E^0 for the reaction.
b. write the Nernst equation as applied to this reaction.
c. calculate E for 10 M HNO_3, taking conc. $Ag^+ = 1$ M, $P_{NO_2} = 1$ atm.

Solution

a. $E = -0.80$ V $+ 0.78$ V $= -0.02$ V. The reaction should not go at standard concentrations.

b. $E = -0.02$ V $- \dfrac{0.0591}{1} \log_{10} \dfrac{(\text{conc. } Ag^+) \times (P_{NO_2})}{(\text{conc. } NO_3^-) \times (\text{conc. } H^+)^2}$

c. In 10 M HNO_3, assuming complete ionization, conc. $H^+ = $ conc. $NO_3^- = 10$ M.

$$E = -0.02 \text{ V} - \frac{0.0591}{1} \log_{10} \frac{1}{10^3}$$

$$= -0.02 \text{ V} - 0.0591 (-3.0) = +0.16 \text{ V}$$

In 16 M HNO_3, E should be a still larger positive number, so the reaction should, and does, go.

Exercise Calculate E under the same conditions except that conc. $HNO_3 = 0.10$ M. Answer: -0.20 V.

As you might suppose, dilute nitric acid (6 M) is a weaker oxidizing agent than 16 M HNO_3. It can also give a wider variety of products. Depending upon the reducing agent used, the NO_3^- ion may be reduced to HNO_2, NO, N_2O, or N_2. With very dilute acid and a strong reducing agent such as zinc ($E^0_{ox} = +0.76$ V), reduction goes all the way to the ammonium ion:

$$4 \text{ Zn(s)} + NO_3^-(aq) + 10 \text{ H}^+(aq) \rightarrow 4 \text{ Zn}^{2+}(aq) + NH_4^+(aq) + 3 \text{ H}_2O \qquad (22.30)$$

A few very inactive metals resist attack by nitric acid, either concentrated or dilute. Among these are gold and platinum. These metals can be brought into solution by treatment with "aqua regia," a mixture of 3 volumes of 12 M HCl with 1 volume of 16 M HNO_3. The reaction with gold is

$$\text{Au(s)} + 4 \text{ H}^+(aq) + 4 \text{ Cl}^-(aq) + NO_3^-(aq) \rightarrow AuCl_4^-(aq) + \text{NO(g)} + 2 \text{ H}_2O \qquad (22.31)$$

In this reaction, the nitrate ion is the oxidizing agent, taking gold from an oxidation number of 0 to +3. The function of the HCl is to furnish Cl^- ions to form the $AuCl_4^-$ complex ion. This drastically lowers the concentration of Au^{3+} and makes the reaction spontaneous.

Nitric acid is used in making explosives. Among these is nitroglycerine, $C_3H_5N_3O_9$, which is mixed with an absorbent solid, such as wood pulp, to give the explosive known as dynamite. One of the most common explosives nowadays is ammonium nitrate, NH_4NO_3. This is made by the reaction of ammonia with nitric acid:

The main use of HNO_3 is in making fertilizers, particularly NH_4NO_3

$$NH_3(aq) + \text{H}^+(aq) + NO_3^-(aq) \rightarrow NH_4NO_3(s)$$

Ammonium nitrate does not detonate as readily as dynamite and hence is safer to work with. A widely used type of blasting powder contains 95% NH_4NO_3 and 5% fuel oil.

+3 State (NO_2^- and HNO_2)

537
REDOX
REACTIONS;
OXIDATION
NUMBERS OF
THE ELEMENTS

Salts containing the nitrite ion are made from the corresponding nitrates. Thus sodium nitrite, $NaNO_2$, is prepared by heating sodium nitrate, $NaNO_3$:

$$2 \text{ NaNO}_3(s) \rightarrow 2 \text{ NaNO}_2(s) + O_2(g) \qquad (22.32)$$

Sodium nitrite is used as a food additive in hot dogs, ham, bacon, and cold cuts. It preserves the red color of the meat and also prevents the growth of an organism that causes botulism, an often fatal type of food poisoning. Recently, the use of $NaNO_2$ for this purpose has come under attack because it can produce compounds known as nitrosamines, which are carcinogenic (cancer-causing).

In August of 1980, the FDA decided *not* to ban nitrites in foods

A water solution of nitrous acid, HNO_2, is made by adding a strong acid to sodium nitrite, $NaNO_2$. Since HNO_2 is a weak acid ($K_a = 4.5 \times 10^{-4}$), the following acid-base reaction goes virtually to completion:

$$H^+(aq) + NO_2^-(aq) \rightarrow HNO_2(aq) \qquad (22.33)$$

At pH 0 (conc. $H^+ = 1$ M), about 99.95% of the NO_2^- ions are converted to HNO_2 molecules.

In HNO_2, nitrogen is in an intermediate oxidation state, +3. Hence, at least in principle, nitrous acid can act as either an oxidizing or reducing agent. As you can see from Figure 22.9, nitrous acid is a strong oxidizing agent

$$HNO_2(aq) + H^+(aq) + e^- \rightarrow NO(g) + H_2O; \; E^0_{red} = +1.00 \text{ V} \qquad (22.34)$$

but a weak reducing agent:

$$HNO_2(aq) + H_2O \rightarrow NO_3^-(aq) + 3 \text{ H}^+(aq) + 2 e^-; \; E^0_{ox} = -0.94 \text{ V} \qquad (22.35)$$

Looking at the equations just written, you might guess that HNO_2 would disproportionate in water solution. Indeed it does; the overall reaction, obtained by combining 22.34 and 22.35, is

$$3 \text{ HNO}_2(aq) \rightarrow 2 \text{ NO}(g) + NO_3^-(aq) + H_2O + H^+(aq); \; E^0 = +0.06 \text{ V} \qquad (22.36)$$

In basic solution nitrites are moderately stable

Ordinarily, this reaction occurs rather slowly, so the HNO_2 stays around long enough for its properties to be studied. However, the pure acid cannot be isolated.

22.6 SULFUR

In its compounds, sulfur shows oxidation numbers of +6, +4, +2, and −2. The species found in acidic and basic solution in these oxidation states are listed in Table 22.3, p. 539. Structures of the SO_4^{2-}, SO_3^{2-}, and $S_2O_3^{2-}$ anions are shown in Figure 22.10, p. 538. In the protonated species (HSO_4^-, H_2SO_4; HSO_3^-, H_2SO_3), H atoms are bonded to oxygen atoms of the oxyanions.

Figure 22.11, p. 538, lists standard potentials (E^0_{red} values) for the reduction of species in the several oxidation states of sulfur. Note that:

1. Species in the +6 state (H_2SO_4, HSO_4^-, SO_4^{2-}) can act only as oxidizing agents, never as reducing agents, in redox reactions.

2. Species in the +4 state (SO_2, H_2SO_3, HSO_3^-, SO_3^{2-}), the +2 state ($S_2O_3^{2-}$), or the 0 state can act as either oxidizing or reducing agents.

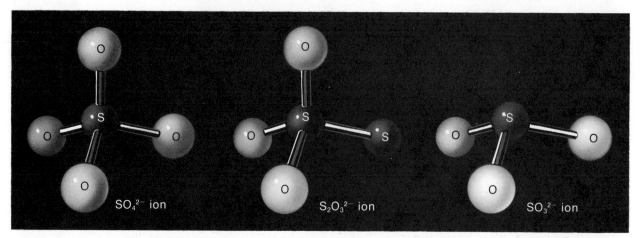

Figure 22.10 In the SO_4^{2-} ion, the sulfur atom is at the center of a tetrahedron, bonded to four oxygen atoms. Removing one of these oxygens gives the sulfite ion, SO_3^{2-}. Replacing an oxygen by a sulfur atom gives the thiosulfate ion, $S_2O_3^{2-}$.

3. Species in the -2 state (H_2S, HS^-, S^{2-}) can act only as reducing agents, never as oxidizing agents, in redox reactions.

You will also note from Figure 22.11 that reduction occurs more readily in acidic solution (E_{red}^0 positive) than in basic solution (E_{red}^0 negative). Hence, if we want to reduce, let us say SO_4^{2-} to SO_2, it would seem best to work in acidic solution ($E_{red}^0 = +0.20$ V). Conversely, oxidation of sulfur-containing species occurs more readily in basic solution. It is easier to oxidize S to $S_2O_3^{2-}$ in basic solution ($E_{ox}^0 = +0.74$ V) than in acidic solution. ($E_{ox}^0 = -0.50$ V).

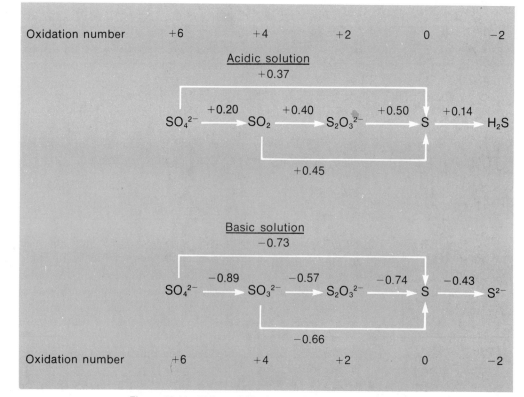

Figure 22.11 Values of E_{red}^0 for species containing sulfur.

539
REDOX
REACTIONS;
OXIDATION
NUMBERS OF
THE ELEMENTS

TABLE 22.3 OXIDATION STATES OF SULFUR

OXID. NO. S	ACIDIC SOLUTION	BASIC SOLUTION
+6	HSO_4^-, SO_4^{2-}	SO_4^{2-}
+4	$SO_2(g)$, H_2SO_3, HSO_3^-	SO_3^{2-}
+2	$*S_2O_3^{2-}$	$S_2O_3^{2-}$
0	$S(s)$	$S(s)$
−2	$H_2S(g)$	HS^-, S^{2-}

*Decomposes to S and SO_2.

+6 State (SO_4^{2-}, HSO_4^-, H_2SO_4)

The preparation of sulfuric acid by the contact process was discussed in Chapter 15. You will recall from Chapter 17 that H_2SO_4 is a strong acid, completely dissociated to H^+ and HSO_4^- ions in dilute water solution. The HSO_4^- ion dissociates further to give H^+ and SO_4^{2-} ions:

$$H_2SO_4(aq) \rightarrow H^+(aq) + HSO_4^-(aq)$$

$$HSO_4^-(aq) \rightleftharpoons H^+(aq) + SO_4^{2-}(aq); \quad K_a = 1.2 \times 10^{-2}$$

The dissociation constant of the HSO_4^- ion is relatively large. This explains why a solution made up by dissolving $NaHSO_4$ in water is acidic. Indeed, the pH of 0.10 M $NaHSO_4$ is about 1.5, only a little higher than that of 0.10 M HCl (pH = 1).

As you can see from Figure 22.11, sulfuric acid is not a very strong oxidizing agent. However, the concentrated acid (18 M) is capable of oxidizing copper metal:

$$Cu(s) + 4\, H^+(aq) + SO_4^{2-}(aq) \rightarrow Cu^{2+}(aq) + SO_2(g) + 2\, H_2O \qquad (22.37)$$

The SO_4^{2-} ion in neutral or basic solution is a very weak oxidizing agent (Example 22.6).

Example 22.6 Consider the reduction of SO_4^{2-} to SO_3^{2-} in basic solution.
  a. Write a balanced half-equation for the process.
  b. Taking $E_{red}^0 = -0.89$ V, calculate E when conc. $OH^- = 10^{-7}$ M and all other species are at unit concentrations.

Solution
  a. In acidic solution, the half-equation would be

$$SO_4^{2-}(aq) + 2\, H^+(aq) + 2\, e^- \rightarrow SO_3^{2-}(aq) + H_2O$$

Adding two OH^- ions to each side and simplifying:

$$SO_4^{2-}(aq) + H_2O + 2\, e^- \rightarrow SO_3^{2-}(aq) + 2\, OH^-(aq)$$

SO_4^{2-} in neutral or basic solution is a very stable species

  b. $E_{red} = E_{red}^0 - \dfrac{0.0591}{2} \log_{10} \dfrac{(\text{conc. } SO_3^{2-}) \times (\text{conc. } OH^-)^2}{(\text{conc. } SO_4^{2-})}$

$$= -0.89 \text{ V} - \frac{0.0591}{2} \log_{10} (10^{-7})^2$$

$$= -0.89 \text{ V} - \frac{0.0591}{2} (-14)$$

$$= -0.48 \text{ V}$$

Exercise What would be the value of E_{red} for this half-reaction at pH 10? Answer: -0.65 V.

Conc. H_2SO_4 is
1. an acid
2. an oxidizing agent
3. a dehydrating agent

Concentrated sulfuric acid is a reagent to be used only with great care. It gives off a large amount of heat when it dissolves in water, often enough to bring the solution to the boiling point. To prevent this and to avoid spattering, the acid should be added slowly to water, with constant stirring. If it comes in contact with the skin, concentrated sulfuric acid can cause painful chemical burns.

+4 State (SO_3^{2-}, HSO_3^-, H_2SO_3, SO_2)

Sulfur dioxide, SO_2, is very soluble in water. The concentration of its saturated solution at 25°C is about 1.3 mol/ℓ. The high solubility is explained in part by a reversible reaction with H_2O to form sulfurous acid, H_2SO_3:

$$SO_2(g) + H_2O \rightleftharpoons H_2SO_3(aq) \tag{22.38}$$

The equilibrium constant for this reaction is not known, but it appears that much of the SO_2 remains unreacted.

Sulfurous acid is a weak acid which ionizes in two steps:

$$H_2SO_3(aq) \rightleftharpoons H^+(aq) + HSO_3^-(aq); \quad K_a = 1.7 \times 10^{-2}$$

$$HSO_3^-(aq) \rightleftharpoons H^+(aq) + SO_3^{2-}(aq); \quad K_a = 5.6 \times 10^{-8}$$

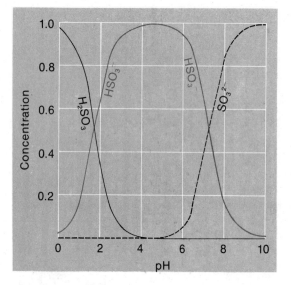

Figure 22.12 In the +4 state, sulfur can exist as H_2SO_3 (a water solution of SO_2), HSO_3^-, or SO_3^{2-}. The relative concentrations of these species change with pH. In strongly acidic solution, H_2SO_3 dominates. Close to pH 2, the concentration of HSO_3^- becomes equal to that of H_2SO_3; between pH 3 and pH 7, HSO_3^- is the major species present. In basic solution, above about pH 7, the SO_3^{2-} ion dominates.

The species present in the +4 state in water solution depends upon the acidity or pH of the solution. As you can see from Figure 22.12, H_2SO_3 and SO_2 are the main species below pH 2. Between pH 2 and 7, the major species is the hydrogen sulfite ion, HSO_3^-. In basic solution, above pH 7, the sulfite ion, SO_3^{2-}, is predominant.

541
REDOX
REACTIONS;
OXIDATION
NUMBERS OF
THE ELEMENTS

+2 State ($S_2O_3^{2-}$)

The thiosulfate ion, $S_2O_3^{2-}$, is formed by heating a basic solution containing the sulfite ion, SO_3^{2-}, with elemental sulfur:

$$SO_3^{2-}(aq) + S(s) \rightarrow S_2O_3^{2-}(aq) \qquad (22.39)$$

Sodium thiosulfate, $Na_2S_2O_3$, is made in this way from sodium sulfite, Na_2SO_3. The principal use of sodium thiosulfate ("hypo") is in photography (Chap. 19).

In acidic solution, the thiosulfate ion decomposes slowly as a result of the reaction

$$S_2O_3^{2-}(aq) + 2\ H^+(aq) \rightarrow S(s) + SO_2(g) + H_2O \qquad (22.40)$$

In basic solution the reaction is reversed

This is a disproportionation reaction; half the sulfur is reduced ($+2 \rightarrow 0$) and half is oxidized ($+2 \rightarrow +4$). From Figure 22.11, p. 538, we see that E^0 for this reaction in acidic solution is positive:

$$E^0 = E_{red}^0\ (S_2O_3^{2-} \rightarrow S) + E_{ox}^0\ (S_2O_3^{2-} \rightarrow SO_2)$$

$$= +0.50\ V - 0.40\ V = +0.10\ V$$

−2 State (S^{2-}, HS^-, H_2S)

Hydrogen sulfide is a poisonous gas with a foul odor, associated with rotten eggs. It is ordinarily prepared by an acid-base reaction between H^+ and S^{2-} ions. The reaction can be carried out in water solution:

$$S^{2-}(aq) + 2\ H^+(aq) \rightarrow H_2S(aq) \qquad (22.41)$$

H_2S is a key reagent in the standard qual analysis scheme

More commonly, an insoluble metal sulfide such as FeS is used as a source of sulfide ions:

$$FeS(s) + 2\ H^+(aq) \rightarrow Fe^{2+}(aq) + H_2S(g) \qquad (22.42)$$

Concentrated (12 M) HCl is most often used as the source of H^+ ions.

Hydrogen sulfide is a weak acid ($K_a = 1 \times 10^{-7}$; $K_a\ HS^- = 1 \times 10^{-13}$). In acid solution, H_2S is a relatively weak reducing agent:

$$H_2S(aq) \rightarrow S(s) + 2\ H^+(aq) + 2\ e^-; E_{ox}^0 = -0.14\ V \qquad (22.43)$$

In contrast, the S^{2-} ion in basic solution is a stronger reducing agent:

$$S^{2-}(aq) \rightarrow S(s) + 2\ e^-; E_{ox}^0 = +0.43\ V \qquad (22.44)$$

SUMMARY

Most elements show more than one oxidation number in their compounds. The metals of Groups 1 (+1) and 2 (+2) are exceptions. Positive oxidation numbers are most common. For the most part, negative oxidation numbers are limited to the strongly nonmetallic elements (N, O, S, and the halogens). The maximum oxidation number of a main group element is usually given by its group number in the Periodic Table (Fig. 22.1).

In their compounds, transition metals can exist as simple cations, complex ions, or oxyanions. The latter, such as CrO_4^{2-} and MnO_4^-, are often strong oxidizing agents, useful in redox titrations (Example 22.2). Their oxidizing strength increases with the concentration of H^+ ion (Example 22.3). Disproportionation occurs when a metal cation (Example 22.1) or other species (Example 22.4) is unstable with respect to a higher and a lower oxidation state.

Oxygen shows an oxidation number of -2 in most of its compounds; in the peroxide ion or H_2O_2, however, its oxidation number is -1. Elemental oxygen is the most common of all oxidizing agents, and is involved in the corrosion of iron and other metals.

Chlorine shows oxidation numbers of +7, +5, +3, and +1 in its oxyacids and oxyanions. The nomenclature of these species is typical of that of oxyacids and oxyanions in general. Acid strength increases with the number of oxygen atoms bonded to the central nonmetal atom. Standard potential diagrams such as Figure 22.8 allow us to calculate E^0 values for a variety of redox reactions.

Nitrogen can have oxidation numbers ranging from +5 to -3. The most important oxyanions and oxyacids of nitrogen are those in the +5 (NO_3^-; HNO_3) and +3 (NO_2^-; HNO_2) states.

Sulfur commonly shows oxidation numbers of +6, +4, +2 and -2. Species in the +6 state (H_2SO_4, HSO_4^-, SO_4^{2-}) act as oxidizing agents. Those in the -2 state (H_2S, HS^-, S^{2-}) are reducing agents. Species in the +4 state (SO_2, HSO_3^-, SO_3^{2-}) or +2 state ($S_2O_3^{2-}$) can act as both oxidizing and reducing agents.

KEY WORDS AND CONCEPTS

oxidation number	reducing agent	Nernst equation
oxyanion	redox titration	corrosion
oxidizing agent	disproportionation	oxyacid

QUESTIONS AND PROBLEMS

Catalog

General: 22.1–22.8, 22.22–22.28, 22.42–22.44
Oxidizing and Reducing Agents: 22.9, 22.10, 22.29, 22.30
Balancing Redox Equations: 22.11–22.13, 22.31–22.33

Redox Titrations: 22.14, 22.15, 22.34, 22.35
Cell Voltage and Spontaneity: 22.16–22.18, 22.36–22.38
Nernst Equation: 22.19–22.21, 22.39–22.41

22.1   Review and know the meaning of the key words and concepts in this chapter.

22.2   What is the oxidation number of Cl in

    a. ClO_3^-?   b. $HClO$?
    c. ClF_3?   d. $HClO_4$?

22.22   What is the oxidation number of N in

    a. NO_2^-?   b. NO_2?
    c. HNO_3?   d. NH_4^+?

22.3 Give the formula of

    a. an anion in which S has an oxidation number of +2.
    b. two anions in which S has an oxidation number of +4.
    c. three different acids of sulfur.

22.4 Give the formula(s) of the products formed when H^+ ions are added to a solution containing

    a. HSO_3^- b. $S_2O_3^{2-}$ c. ClO^-
    d. NO_2^- e. S^{2-}

22.5 Describe how you would convert

    a. Fe^{2+} to Fe^{3+}.
    b. Fe to $Fe(OH)_3$.
    c. Br^- to Br_2.
    d. $KClO_3$ to $KClO_4$.

22.6 Explain why corrosion

    a. occurs when iron is in contact with both H_2O and O_2.
    b. is prevented by washing garden tools thoroughly after use.
    c. usually occurs faster in seawater than in fresh water.
    d. occurs most rapidly near the water line on a bridge support.

22.7 Referring to Figure 22.1, identify all the main-group elements in which the maximum oxidation number is less than the group number in the Periodic Table; greater than the group number.

22.8 Which would you expect to be the stronger acid,

    a. $HBrO_3$ or $HBrO$?   b. HIO or HIO_2?
    c. HNO_2 or HNO_3?   d. H_2SO_4 or H_2SO_3?

22.9 Which of the following species can act both as oxidizing and reducing agents?

    a. H_2O_2 b. O_2 c. ClO_4^- d. ClO_3^-

22.10 Based on the potentials given in Figure 22.8, which species is

    a. the strongest oxidizing agent in acidic solution?
    b. the strongest oxidizing agent in basic solution?
    c. the strongest reducing agent in acidic solution?
    d. the strongest reducing agent in basic solution?

22.11 Write balanced equations for the disproportionation of

    a. ClO_3^- to ClO_4^- and ClO_2^-.
    b. HNO_2 to NO_3^- and NO (acid).
    c. NO_2^- to NO_3^- and NO (base).

22.23 Give the formula of a compound of nitrogen that is

    a. a weak base.
    b. a strong acid.
    c. a weak acid.
    d. capable of oxidizing copper.
    e. formed by heating $NaNO_3$.

22.24 Give the formula(s) of the products formed when OH^- ions are added to a solution containing

    a. $HClO$ b. HSO_3^- c. $HClO_4$
    d. NH_4^+

22.25 How would you prepare

    a. $HClO_4$ from $KClO_4$?
    b. NO_2 from HNO_3?
    c. $AuCl_4^-$ from Au?
    d. $S_2O_3^{2-}$ from S?

22.26 Explain why

    a. we believe corrosion occurs by an electrochemical mechanism.
    b. solid tin is often added to bottles containing Sn^{2+} ions.
    c. H_2O_2 can act as either an oxidizing or reducing agent.
    d. solutions of H_2S often turn cloudy on exposure to air.

22.27 Referring to Figure 22.1, identify all the elements that can show negative oxidation numbers; all the transition metals that have only one oxidation number listed.

22.28 Name all the *anions* derived from the acids listed in Question 22.8.

22.29 Which of the following species can act as both oxidizing and reducing agents?

    a. NO_2^- b. N_2H_4 c. S   d. SO_4^{2-}

22.30 Answer Question 22.10 for nitrogen compounds, using Figure 22.9.

22.31 Write balanced equations for the disproportionation of

    a. $HClO$ to $HClO_2$ and Cl_2 (acid).
    b. ClO^- to ClO_2^- and Cl_2 (base).
    c. N_2 to NO and NH_4^+ (acid).

22.12 Write balanced equations for the oxidation of the following species by $Cr_2O_7^{2-}$, which is reduced to Cr^{3+} in acidic solution:

a. I^- to I_2 b. Fe^{2+} to Fe^{3+}
c. HNO_2 to NO_3^- d. H_2S to S

22.13 Balance the following equations:

a. $Sn^{2+}(aq) + O_2(g) + H_2O \rightarrow SnO_2(s) + H^+(aq)$
b. $H_2O_2(aq) + I^-(aq) + H^+(aq) \rightarrow I_2(s) + H_2O$
c. $Fe(s) + O_2(g) + H_2O \rightarrow Fe(OH)_3(s)$ (base)

22.14 An acidic solution of potassium permanganate reacts with $C_2O_4^{2-}$ ions to form $CO_2(g)$.

a. Write a balanced equation for the reaction.
b. If 21.6 cm^3 of 0.100 M $KMnO_4$ are required to titrate 30.2 cm^3 of a sodium oxalate solution, what was conc. $C_2O_4^{2-}$ in the original solution?

22.15 Explain why no indicator needs to be added during the titration of Fe^{2+} by MnO_4^-.

22.16 A student analyzing an unknown reports that both Ag^+ and Sn^{2+} are present. Referring to Table 21.1 (p. 507) comment on the likelihood that this is correct.

22.17 Using Figures 22.8 and 22.9, decide whether each of the reactions in Question 22.11 will occur (standard concentrations).

22.18 Using Figure 22.11, decide whether each of the following reactions will occur at standard concentrations:

a. $SO_4^{2-}(aq) + S_2O_3^{2-}(aq) \rightarrow SO_3^{2-}(aq)$
b. $SO_4^{2-}(aq) + S_2O_3^{2-}(aq) \rightarrow SO_2(g)$
c. $S(s) + SO_2(g) \rightarrow S_2O_3^{2-}(aq)$

Balance any equation that represents a spontaneous reaction.

22.19 Consider the half-reaction

$$ClO_4^-(aq) + 2\ H^+(aq) + 2\ e^- \rightarrow ClO_3^-(aq) + H_2O$$

$E_{red}^0 = +1.19$ V. Calculate E_{red} when ClO_4^- and ClO_3^- are 1 M at pH

a. 0 b. 7 c. 14

22.20 In Example 22.4, we showed that for

$$2\ HClO_2(aq) \rightarrow HClO(aq) + ClO_3^-(aq) + H^+(aq)$$

$E^0 = +0.44$ V. Calculate E when conc. $HClO_2 = 0.01$ M, conc. $HClO = 0.10$ M, conc. $ClO_3^- = 0.10$ M, pH = 5.

22.32 Write balanced equations for the oxidation of the species in Question 22.12 by MnO_4^-, which is reduced to Mn^{2+}, in acidic solution.

22.33 Balance the following equations:

a. $Cl_2(g) + OH^-(aq) \rightarrow$
    $ClO_3^-(aq) + Cl^-(aq) + H_2O$
b. $CuS(s) + NO_3^-(aq) + H^+(aq) \rightarrow$
    $Cu^{2+}(aq) + S(s) + NO_2(g) + H_2O$
c. $Zn(s) + NO_3^-(aq) + H^+(aq) \rightarrow$
    $Zn^{2+}(aq) + NH_4^+(aq) + H_2O$

22.34 It is possible to determine I^- in solution by titrating with MnO_4^-.

a. Write a balanced equation for this reaction.
b. 27.6 cm^3 of 0.0800 M $KMnO_4$ are used to titrate 16.4 cm^3 of NaI solution to the end point. What was conc. I^- in the original solution?

22.35 The presence of Cl^- during the titration of Fe^{2+} with MnO_4^- causes an error in the titration data. Explain.

22.36 A classmate thoughtlessly uses an aluminum rod to stir a 1 M Cu^{2+} solution.

a. Describe what your classmate observes.
b. Write an equation to represent what occurs.

22.37 Using Figures 22.8 and 22.9, decide whether each of the reactions in Question 22.31 will occur (standard concentrations).

22.38 Using Table 21.1 (p. 507) decide whether the following species will disproportionate in acidic solution:

a. Cu^+ b. Sn^{2+} c. Hg_2^{2+} d. Co^{2+}

Write balanced equations for any disproportionation that is spontaneous.

22.39 Consider the half-reaction

$$NO_3^-(aq) + 2\ H^+(aq) + e^- \rightarrow NO_2(g) + H_2O$$

$E_{red}^0 = +0.78$ V. Calculate E_{red} when NO_3^- is 1 M, P_{NO_2} is 1 atm, and the pH is

a. 0 b. 7 c. 14

22.40 For the reaction

$$2\ ClO_2^-(aq) \rightarrow ClO^-(aq) + ClO_3^-(aq)$$

$E^0 = +0.33$ V. Calculate E when conc. $ClO_2^- =$ conc. $ClO^- =$ conc. $ClO_3^- = 0.10$ M.

22.21 For the reaction in Problem 22.20, taking all species to be 1 M except $HClO_2$, calculate the concentration of $HClO_2$ at which the reaction would no longer be spontaneous—that is, E = 0.

22.41 For the reaction in Problem 22.40, taking all species to be 1 M except ClO_2^-, calculate the concentration of ClO_2^- at which E = 0.

*22.42 Calcium in blood or urine can be determined by precipitation as CaC_2O_4. The precipitate is dissolved in strong acid and titrated with $KMnO_4$ (see Problem 22.14). A 24-h urine sample is collected from an adult patient, reduced to a small volume, and titrated with 27.2 cm³ of 0.0946 M $KMnO_4$. How many grams of CaC_2O_4 are in the sample? Normal range for Ca^{2+} output for an adult is 100 to 300 mg per 24 h. Is the sample within the normal range?

*22.43 Given that K for the reaction

$$Cl_2(g) + H_2O \rightleftharpoons Cl^-(aq) + H^+(aq) + HClO(aq)$$

is 3×10^{-5}, apply the Rule of Multiple Equilibria (Chapter 18) to evaluate K for

$$Cl_2(g) + 2\ OH^-(aq) \rightleftharpoons ClO^-(aq) + Cl^-(aq) + H_2O$$

*22.44 You may have noticed from Figures 22.8, 22.9, and 22.11 that E_{red}^0 values are not additive. For example, $E_{red}^0 (NO_3^- \rightarrow NO_2) + E_{red}^0 (NO_2 \rightarrow HNO_2)$ does *not* equal $E_{red}^0 (NO_3^- \rightarrow HNO_2)$. However, there is a simple general relation that applies in situations such as this. Can you discover this relation from the data in these three figures?

CHEMICAL THERMODYNAMICS; ΔH, ΔS, AND ΔG _____

A basic goal of chemistry is to predict whether or not a reaction will occur when reactants are brought together. Such a reaction is said to be *spontaneous;* it occurs "by itself" without the exertion of any outside force. All of us are familiar with certain spontaneous processes. Iron exposed to moist air rusts. A mixture of hydrogen and oxygen burns when we set a match to it. Certain other processes are clearly nonspontaneous. A hammer left out on the lawn all winter does not lose its coating of rust when spring comes. Water in a beaker in the laboratory does not decompose spontaneously into hydrogen and oxygen.

In several chapters of this text, we considered the problem of predicting whether or not a given reaction is spontaneous. We pointed out in Chapter 13 that a reaction between gases is likely to go nearly to completion if the equilibrium constant for that reaction is very large. In Chapter 16, we saw that mixing solutions of two ionic compounds is likely to yield a precipitate if one of the possible products is insoluble in water. More recently, in Chapter 21, we developed what might be called a "criterion of spontaneity" for redox reactions. If the calculated voltage, E, is positive, the reaction will take place.

Here, we look for a more general approach to the problem of predicting reaction spontaneity. We want to find a criterion which will allow us to decide whether any given reaction, at a specified temperature and pressure, will take place when the reactants are brought together. To do this, we work in that area of science called chemical thermodynamics. This considers, in a quantitative way, the energy changes which accompany a chemical reaction. Chemical thermodynamics deals with several different energy-related functions. One of these is the enthalpy change, ΔH. This quantity was first discussed in Chapter 6; it will be reviewed briefly in Section 23.1. We will introduce two new functions in this chapter. These are the entropy change, ΔS (Section 23.2) and the free energy change, ΔG (Section 23.3).

23.1 THE ENTHALPY CHANGE, ΔH

You will recall from Chapter 6 that, for a reaction carried out at constant pressure and temperature, the heat flow is equal to the difference in enthalpy between products and reactants, ΔH. That is,

$$\Delta H = H_{products} - H_{reactants} = \text{heat flow at constant T, P} \qquad (23.1)$$

So far as ΔH is concerned, we distinguish between two types of reactions. An *exothermic* reaction is one for which ΔH is a negative quantity ($\Delta H < 0$). The enthalpy of the products is less than that of the reactants; heat is given off by the reaction system to the surroundings. In contrast, for an *endothermic* reaction, ΔH is a positive quantity ($\Delta H > 0$). The enthalpy of the products is greater than that of the reactants. Heat must be absorbed from the surroundings for the reaction to take place.

Heat flow is with re-spect to the reaction mixture

The enthalpy change for a reaction can be calculated from molar heats of formation, ΔH_f. The general relation, given in Chapter 6, is

$$\Delta H = \Sigma \, \Delta H_f \text{ products} - \Sigma \, \Delta H_f \text{ reactants} \qquad (23.2)$$

Recall from Chapter 6 that:

—the molar heat of formation of a compound is the enthalpy change when one mole of the compound is formed from the elements in their stable forms.

—the molar heat of formation of an element is zero.

—molar heats of formation of ions in solution are established by setting $\Delta H_f \, H^+(aq)$ = 0.

Tables of heats of formation of compounds and of ions in aqueous solution were given in Chapter 6. They are repeated here (Tables 23.1 and 23.2) for convenience.

A hundred years ago many chemists felt that they had a general criterion for predicting reaction spontaneity. The prevailing idea, put forth by P. M. Berthelot in Paris and Julius Thomsen in Copenhagen, was that all spontaneous reactions are exothermic. If this were true, all we would have to do to predict reaction spontaneity would be to calculate the enthalpy change, ΔH, and look at its sign. If ΔH turned out to be negative, we could assume that the reaction must be spontaneous; if ΔH were positive, the reaction could not occur by itself.

It turns out that almost all exothermic *chemical reactions* are spontaneous *at 25°C and 1 atm*. Consider, for example,

TABLE 23.1 HEATS OF FORMATION (kJ/MOL) AT 25°C AND 1 ATM

AgBr(s)	−99.5	C₂H₂(g)	+226.7	H₂O(l)	−285.8	NH₄Cl(s)	−315.4
AgCl(s)	−127.0	C₂H₄(g)	+52.3	H₂O₂(l)	−187.6	NH₄NO₃(s)	−365.1
AgI(s)	−62.4	C₂H₆(g)	−84.7	H₂S(g)	−20.1	NO(g)	+90.4
Ag₂O(s)	−30.6	C₃H₈(g)	−103.8	H₂SO₄(l)	−811.3	NO₂(g)	+33.9
Ag₂S(s)	−31.8	n-C₄H₁₀(g)	−124.7	HgO(s)	−90.7	NiO(s)	−244.3
Al₂O₃(s)	−1669.8	n-C₅H₁₂(l)	−173.1	HgS(s)	−58.2	PbBr₂(s)	−277.0
BaCl₂(s)	−860.1	C₂H₅OH(l)	−277.6	KBr(s)	−392.2	PbCl₂(s)	−359.2
BaCO₃(s)	−1218.8	CoO(s)	−239.3	KCl(s)	−435.9	PbO(s)	−217.9
BaO(s)	−558.1	Cr₂O₃(s)	−1128.4	KClO₃(s)	−391.4	PbO₂(s)	−276.6
BaSO₄(s)	−1465.2	CuO(s)	−155.2	KF(s)	−562.6	Pb₃O₄(s)	−734.7
CaCl₂(s)	−795.0	Cu₂O(s)	−166.7	MgCl₂(s)	−641.8	PCl₃(g)	−306.4
CaCO₃(s)	−1207.0	CuS(s)	−48.5	MgCO₃(s)	−1113	PCl₅(g)	−398.9
CaO(s)	−635.5	CuSO₄(s)	−769.9	MgO(s)	−601.8	SiO₂(s)	−859.4
Ca(OH)₂(s)	−986.6	Fe₂O₃(s)	−822.2	Mg(OH)₂(s)	−924.7	SnCl₂(s)	−349.8
CaSO₄(s)	−1432.7	Fe₃O₄(s)	−1120.9	MgSO₄(s)	−1278.2	SnCl₄(l)	−545.2
CCl₄(l)	−139.5	HBr(g)	−36.2	MnO(s)	−384.9	SnO(s)	−286.2
CH₄(g)	−74.8	HCl(g)	−92.3	MnO₂(s)	−519.7	SnO₂(s)	−580.7
CHCl₃(l)	−131.8	HF(g)	−268.6	NaCl(s)	−411.0	SO₂(g)	−296.1
CH₃OH(l)	−238.6	HI(g)	+25.9	NaF(s)	−569.0	SO₃(g)	−395.2
CO(g)	−110.5	HNO₃(l)	−173.2	NaOH(s)	−426.7	ZnO(s)	−348.0
CO₂(g)	−393.5	H₂O(g)	−241.8	NH₃(g)	−46.2	ZnS(s)	−202.9

TABLE 23.2   HEATS OF FORMATION (kJ/mol) OF AQUEOUS IONS AT 25°C, 1 M

CATIONS				ANIONS			
$Ag^+(aq)$	+105.9	$K^+(aq)$	−251.2	$Br^-(aq)$	−120.9	$H_2PO_4^-(aq)$	−1302.5
$Al^{3+}(aq)$	−524.7	$Li^+(aq)$	−278.5	$Cl^-(aq)$	−167.4	$HPO_4^{2-}(aq)$	−1298.7
$Ba^{2+}(aq)$	−538.4	$Mg^{2+}(aq)$	−462.0	$ClO_3^-(aq)$	−98.3	$I^-(aq)$	−55.9
$Ca^{2+}(aq)$	−543.0	$Mn^{2+}(aq)$	−218.8	$ClO_4^-(aq)$	−131.4	$MnO_4^-(aq)$	−518.4
$Cd^{2+}(aq)$	−72.4	$Na^+(aq)$	−239.7	$CO_3^{2-}(aq)$	−676.3	$NO_3^-(aq)$	−206.6
$Cu^{2+}(aq)$	+64.4	$NH_4^+(aq)$	−132.8	$CrO_4^{2-}(aq)$	−863.2	$OH^-(aq)$	−229.9
$Fe^{2+}(aq)$	−87.9	$Ni^{2+}(aq)$	−64.0	$F^-(aq)$	−329.1	$PO_4^{3-}(aq)$	−1284.1
$Fe^{3+}(aq)$	−47.7	$Pb^{2+}(aq)$	+1.6	$HCO_3^-(aq)$	−691.1	$S^{2-}(aq)$	+41.8
$H^+(aq)$	0.0	$Sn^{2+}(aq)$	−10.0			$SO_4^{2-}(aq)$	−907.5
		$Zn^{2+}(aq)$	−152.4				

These numbers are based on the assumption that $\Delta H_f\ H^+ = 0$

—the combination of hydrogen and oxygen to give water:

$$2\ H_2(g) + O_2(g) \rightarrow 2\ H_2O(l) \tag{23.3}$$

—the neutralization of H^+ ions by OH^- ions:

$$H^+(aq) + OH^-(aq) \rightarrow H_2O(l) \tag{23.4}$$

Using Equation 23.2 with Tables 23.1 and 23.2 we can show that, in both these spontaneous reactions, ΔH is a negative quantity:

$$23.3:\ \Delta H = 2\ \Delta H_f\ H_2O(l) = -571.6\ \text{kJ}$$

$$23.4:\ \Delta H = \Delta H_f\ H_2O(l) - \Delta H_f\ OH^-(aq) = -55.9\ \text{kJ}$$

On the other hand, this simple rule fails for many familiar phase changes. An example is the melting of ice. This takes place spontaneously at 1 atm above 0°C, even though it is endothermic:

$$H_2O(s) \rightarrow H_2O(l); \Delta H = +6.0\ \text{kJ} \tag{23.5}$$

In another case we find that, above 100°C, liquid water vaporizes to steam at 1 atm. This process, like Reaction 23.5, absorbs heat:

$$H_2O(l) \rightarrow H_2O(g); \Delta H = +40.6\ \text{kJ} \tag{23.6}$$

There is still another basic objection to using the sign of ΔH as a general criterion for spontaneity. Endothermic reactions which are nonspontaneous at room temperature often become spontaneous when the temperature is raised. Consider, for example, the decomposition of limestone:

$$CaCO_3(s) \rightarrow CaO(s) + CO_2(g); \Delta H = +178.0\ \text{kJ} \tag{23.7}$$

At 25°C and 1 atm, this reaction is nonspontaneous. Witness the existence of the white cliffs of Dover and other limestone deposits over eons of time. However, if the temperature is raised to about 1100 K, the limestone decomposes to give off carbon dioxide gas at 1 atm. In other words, this endothermic reaction becomes spontaneous at high temperatures. This is true despite the fact that ΔH remains at $+178.0\,kJ$, nearly independent of temperature.

23.2 THE ENTROPY CHANGE, ΔS

To decide whether a given reaction will be spontaneous at a given temperature and pressure, we must consider another factor in addition to ΔH. This is the entropy change for the reaction, ΔS:

$$\Delta S = S_{products} - S_{reactants} \tag{23.8}$$

The entropy of a substance, like its enthalpy, is one of its characteristic properties. Entropy is a measure of disorder or randomness. Substances which are highly disordered have high entropies. Low entropy is associated with strongly ordered substances. By "order" we mean the extent to which particles of a substance are confined to a given region of space. In a crystal, where the atoms, molecules, or ions are fixed in position, the entropy is relatively low. When a solid melts, the particles are free to move through the entire liquid volume; the entropy of the substance increases. Upon vaporization the particles acquire still greater freedom to move about, and there is a large increase in entropy.

Volume is also a characteristic property

From what we have just said, it should be clear that for Reactions 23.5 and 23.6

$$H_2O(s) \rightarrow H_2O(l)$$
$$H_2O(l) \rightarrow H_2O(g)$$

the entropy change, ΔS, is a positive quantity. Liquid water has a higher entropy than ice; steam has a higher entropy than liquid water. Indeed, the increase in entropy is the driving force behind these processes. We might say that ice melts above 0°C and water vaporizes above 100°C *because* of the increase in randomness. This factor more than compensates for the fact that the processes are endothermic.

If we think about the matter carefully, we realize that there is an inherent tendency, common to inanimate substances (and probably to students and professors as well) to become disorganized. A beaker shatters when we drop it on a stone lab bench. Rivers become polluted and wilderness trails get cluttered with litter. All these processes involve an increase in entropy. They are all spontaneous, as anyone who tries to reverse them soon discovers.

Standard Molar Entropies of Elements and Compounds

Considering the rather vague way we have described entropy, you may be surprised to learn that we can determine absolute values for the molar entropies of pure substances. Actually, entropy can be defined very precisely and measured accurately. The details of how this is done are beyond the level of this text. The results, however, are useful for a variety of purposes. They are given in Table 23.3 in the form of *standard molar entropies* (S^0) at 25°C. These are the entropies per mole, at 1 atm, in the units of **joules per kelvin (J/K)**. You will note from the table that:

At T = 0 K, a pure substance is completely ordered. Its S = 0

—standard molar entropies of substances are always positive quantities ($S^0 > 0$).

—*elements as well as compounds are assigned standard entropies*. This is in contrast to the situation with heats of formation, where $\Delta H_f = 0$ for elements.

TABLE 23.3 STANDARD MOLAR ENTROPIES (J/K) OF ELEMENTS AND COMPOUNDS AT 25°C AND 1 ATM

ELEMENTS

Ag(s)	42.7	Co(s)	28.5	I_2(s)	116.7	O_2(g)	205.0
Al(s)	28.3	Cr(s)	23.8	K(s)	63.6	Pb(s)	64.9
Ba(s)	67	Cu(s)	33.3	Mg(s)	32.5	P_4(s)	177.4
Br_2(l)	152.3	F_2(g)	203.3	Mn(s)	31.8	S(s)	31.9
C(s)	5.7	Fe(s)	27.2	N_2(g)	191.5	Si(s)	18.7
Ca(s)	41.6	H_2(g)	130.6	Na(s)	51.0	Sn(s)	51.5
Cl_2(g)	222.9	Hg(l)	77.4	Ni(s)	30.1	Zn(s)	41.6

COMPOUNDS

AgBr(s)	107.1	C_2H_2(g)	200.8	H_2O(l)	69.9	NH_4Cl(s)	94.6
AgCl(s)	96.1	C_2H_4(s)	219.5	H_2O_2(l)	109.6	NH_4NO_3(s)	151.0
AgI(s)	114.2	C_2H_6(g)	229.5	H_2S(g)	205.6	NO(g)	210.6
Ag_2O(s)	121.7	C_3H_8(g)	269.9	H_2SO_4(l)	156.9	NO_2(g)	240.5
Ag_2S(s)	145.6	n-C_4H_{10}(g)	310.0	HgO(s)	72.0	NiO(s)	38.6
Al_2O_3(s)	51.0	n-C_5H_{12}(l)	262.8	HgS(s)	77.8	$PbBr_2$(s)	161.5
$BaCl_2$(s)	126	C_2H_5OH(l)	160.7	KBr(s)	96.4	$PbCl_2$(s)	136.4
$BaCO_3$(s)	112.1	CoO(s)	43.9	KCl(s)	82.7	PbO(s)	69.5
BaO(s)	70.3	Cr_2O_3(s)	81.2	$KClO_3$(s)	143.0	PbO_2(s)	76.6
$BaSO_4$(s)	132.2	CuO(s)	43.5	KF(s)	66.6	Pb_3O_4(s)	211.3
$CaCl_2$(s)	113.8	Cu_2O(s)	100.8	$MgCl_2$(s)	89.5	PCl_3(g)	311.7
$CaCO_3$(s)	92.9	CuS(s)	66.5	$MgCO_3$(s)	65.7	PCl_5(g)	352.7
CaO(s)	39.7	$CuSO_4$(s)	113.4	MgO(s)	26.8	SiO_2(s)	41.8
$Ca(OH)_2$(s)	76.1	Fe_2O_3(s)	90.0	$Mg(OH)_2$(s)	63.1	$SnCl_2$(s)	122.6
$CaSO_4$(s)	106.7	Fe_3O_4(s)	146.4	$MgSO_4$(s)	91.6	$SnCl_4$(l)	258.6
CCl_4(l)	214.4	HBr(g)	198.5	MnO(s)	60.2	SnO(s)	56.5
CH_4(g)	186.2	HCl(g)	186.7	MnO_2(s)	53.1	SnO_2(s)	52.3
$CHCl_3$(l)	202.9	HF(g)	173.5	NaCl(s)	72.4	SO_2(g)	248.5
CH_3OH(l)	126.8	HI(g)	206.3	NaF(s)	58.6	SO_3(g)	256.2
CO(g)	197.9	HNO_3(l)	155.6	NaOH(s)	52.3	ZnO(s)	43.9
CO_2(g)	213.6	H_2O(g)	188.7	NH_3(g)	192.5	ZnS(s)	57.7

—solids as a group have lower entropies than liquids, which in turn have lower entropies than gases. Compare, for example,

	S^0 (J/K)
Si(s)	18.7
H_2O(l)	69.9
SO_2(g)	248.5

Diamond has the lowest molar entropy of any pure substance. Can you suggest why?

Standard Molar Entropies of Ions in Solution

It is possible to assign standard molar entropies to ions in water solution. To do this, we adopt a convention similar to that followed with heats of formation. The standard molar entropy of the H^+ ion in water solution is taken to be zero.

$$S^0 \ H^+(aq) = 0$$

Standard entropies of ions based upon this convention are listed in Table 23.4. They are given in joules per kelvin at 25°C and a concentration of 1 M.

TABLE 23.4 STANDARD MOLAR ENTROPIES (J/K) OF AQUEOUS IONS AT 25°C, 1 M

CATIONS				ANIONS			
$Ag^+(aq)$	73.9	$K^+(aq)$	102.5	$Br^-(aq)$	80.7	$H_2PO_4^-(aq)$	89.1
$Al^{3+}(aq)$	−313.4	$Li^+(aq)$	14.2	$Cl^-(aq)$	55.1	$HPO_4^{2-}(aq)$	−36.0
$Ba^{2+}(aq)$	13	$Mg^{2+}(aq)$	−118.0	$ClO_3^-(aq)$	162.3	I^-	109.4
$Ca^{2+}(aq)$	−55.2	$Mn^{2+}(aq)$	−73.6	$ClO_4^-(aq)$	182.0	$MnO_4^-(aq)$	190.0
$Cd^{2+}(aq)$	−61.1	$Na^+(aq)$	60.2	$CO_3^{2-}(aq)$	−53.1	$NO_3^-(aq)$	146.4
$Cu^{2+}(aq)$	−98.7	$NH_4^+(aq)$	112.8	$CrO_4^{2-}(aq)$	38.5	$OH^-(aq)$	−10.5
$Fe^{2+}(aq)$	−113.4	$Ni^{2+}(aq)$	−159.4	$F^-(aq)$	−9.6	$PO_4^{3-}(aq)$	−218
$Fe^{3+}(aq)$	−293.3	$Pb^{2+}(aq)$	21.3	$HCO_3^-(aq)$	95.0	$S^{2-}(aq)$	22.2
$H^+(aq)$	0.0	$Sn^{2+}(aq)$	−24.7			$SO_4^{2-}(aq)$	17.2
		$Zn^{2+}(aq)$	−106.5				

ΔS⁰ for Reactions

Using Tables 23.3 and 23.4, it is possible to calculate the standard entropy change, ΔS^0, for a variety of reactions. The appropriate relation is

$$\Delta S^0 = \Sigma\ S^0_{products} - \Sigma\ S^0_{reactants} \qquad (23.9)$$

As an example, consider the reaction

$$CaCO_3(s) \rightarrow CaO(s) + CO_2(g)$$

$$\Delta S^0 = S^0\ CaO(s) + S^0\ CO_2(g) - S^0\ CaCO_3(s)$$

$$= 39.7\ J/K + 213.6\ J/K - 92.9\ J/K = +160.4\ J/K$$

You will notice that ΔS^0 for the decomposition of calcium carbonate is a positive quantity. This is reasonable since the gas formed, CO_2, has a much higher molar entropy than either of the solids, CaO or $CaCO_3$. As a matter of fact, we almost always find that *a reaction which results in an increase in the number of moles of gas is accompanied by an increase in entropy. Conversely, if the number of moles of gas decreases, we expect ΔS to be a negative quantity.* Consider, for example, the reaction

$$2\ H_2(g) + O_2(g) \rightarrow 2\ H_2O(l)$$

$$\Delta S^0 = 2\ S^0\ H_2O(l) - 2\ S^0\ H_2(g) - S^0\ O_2(g)$$

$$= 139.8\ J/K - 261.2\ J/K - 205.0\ J/K = -326.4\ J/K$$

Example 23.1 Calculate ΔS^0 for the reaction

$$HCl(g) \rightarrow H^+(aq) + Cl^-(aq)$$

Solution Using values of S^0 from Tables 23.3 and 23.4:

$$\Delta S^0 = S^0\ H^+(aq) + S^0\ Cl^-(aq) - S^0\ HCl(g)$$
$$= 0 + 55.1\ J/K - 186.7\ J/K = -131.6\ J/K$$

Exercise Calculate ΔS^0 for the process $NaCl(s) \rightarrow Na^+(aq) + Cl^-(aq)$. Answer: +42.9 J/K.

23.3 THE FREE ENERGY CHANGE, ΔG

We stated earlier that there are two thermodynamic quantities which affect the spontaneity of a reaction. One of these is the enthalpy, H. The other is the entropy, S. The problem is to put these two quantities together in such a way as to arrive at a single function which can be used to determine whether a reaction is spontaneous. This problem was first solved a century ago by J. Willard Gibbs, professor of mathematical physics at Yale. He introduced a new quantity into thermodynamics, now called the Gibbs free energy and given the symbol G. The free energy of a substance, like its enthalpy or entropy, is one of its characteristic properties. For a reaction the change in free energy, ΔG, is the difference between the free energies of products and reactants:

$$\Delta G = G_{products} - G_{reactants} \qquad (23.10)$$

Gibbs was able to show that the sign of ΔG can be used to determine whether or not a reaction is spontaneous. For a reaction carried out at constant temperature and pressure:

1. **If ΔG is negative, the reaction is spontaneous.**
2. **If ΔG is positive, the reaction will not occur spontaneously.** Instead, the reverse reaction will occur.
3. **If ΔG is 0, the reaction system is at equilibrium;** there is no tendency for it to occur in either direction.

Putting it another way, we can think of ΔG as a measure of the driving force of a reaction. **Reactions, at constant pressure and temperature, go in such a direction as to decrease the free energy of the system.** This means that the direction in which a reaction takes place is determined by the relative free energies of products and reactants. If the products have a lower free energy than the reactants ($G_{products} < G_{reactants}$), the forward reaction will occur (Fig. 23.1). If the reverse is true ($G_{reactants} < G_{products}$), the reverse reaction is spontaneous. Finally, if $G_{products} = G_{reactants}$, there is no driving force to make the reaction go in either direction.

Relation between ΔG, ΔH, and ΔS

The free energy change for a reaction, ΔG, is related in a simple way to the enthalpy change, ΔH, and the entropy change, ΔS. The equation relating these quantities, called the Gibbs-Helmholtz equation, is

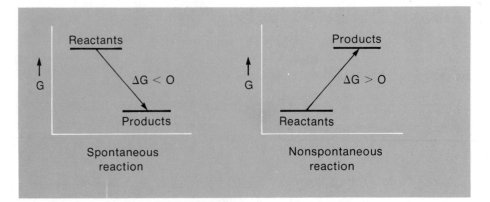

Figure 23.1 For a spontaneous reaction, the free energy of the products is less than that of the reactants: ΔG < 0. For a nonspontaneous reaction, the reverse is true: ΔG > 0.

$$\Delta G = \Delta H - T\Delta S \qquad (23.11)$$

where T is the temperature in K. In words, this equation tells us that the driving force for a reaction, ΔG, depends upon two quantities. One of these is the enthalpy change due to the making and breaking of bonds, ΔH. The other is the product of the change in randomness, ΔS, times the absolute temperature, T.

The two factors which tend to make ΔG negative and hence lead to a spontaneous reaction are:

1. **A negative value of ΔH.** Exothermic reactions ($\Delta H < 0$) tend to be spontaneous inasmuch as they contribute to a negative value of ΔG. On the molecular level, this means that there will be a tendency to form "strong" bonds at the expense of "weak" ones.

2. **A positive value of ΔS.** If the entropy change is positive ($\Delta S > 0$), the term $-T\Delta S$ will make a negative contribution to ΔG. Hence there will be a tendency for a reaction to be spontaneous if the products are less ordered than the reactants.

In many physical processes the entropy increase is the major driving force. An example is the formation of a solution. When nitrogen diffuses into oxygen or benzene dissolves in toluene, the enthalpy change is virtually zero, but ΔS is a positive quantity because the solution is less ordered than the pure substances from which it is formed.

In certain reactions, ΔS is nearly zero and ΔH is the only important component of the driving force for spontaneity. An example is the synthesis of hydrogen fluoride from the elements:

$$^1/_2 \ H_2(g) + \ ^1/_2 \ F_2(g) \rightarrow HF(g)$$

For this reaction, ΔH is a large negative number, -268.6 kJ. This reflects the fact that the bonds in HF are stronger than those in the H_2 and F_2 molecules. As we might expect for a gaseous reaction in which there is no change in the number of moles, ΔS is very small, about 0.0065 kJ/K. The free energy change, ΔG, at 1 atm is -270.5 kJ at 25°C, almost identical with ΔH. Even at very high temperatures, the difference between ΔG and ΔH is small, amounting to only about 13 kJ at 2000 K.

The more common case is one in which both ΔH and ΔS make significant contributions to ΔG. To determine the sign of ΔG, we must consider the values of both ΔH and ΔS as well as the temperature. The approach we will follow is described below.

Standard Free Energy Change

Using the Gibbs-Helmholtz equation in the form

$$\Delta G^0 = \Delta H - T\Delta S^0 \qquad (23.12)$$

it is possible to calculate what is known as the standard free energy change for a reaction, ΔG^0. This is the free energy change when all species involved in the reaction are at unit concentrations (1 atm for gases, 1 M for ions or molecules in aqueous solution).

To illustrate the use of Equation 23.12, let us apply it first at 25°C (T = 298 K). Consider the reaction

$$2 \ H_2(g) + O_2(g) \rightarrow 2 \ H_2O(l)$$

Earlier in this chapter, we showed that ΔH for this reaction is -571.6 kJ and that ΔS^0 is -326.4 J/K. Hence, at 25°C,

$$\Delta G^0 = -571.6 \text{ kJ} - 298 \text{ K} \times (-0.3264 \text{ kJ/K})$$

$$= -571.6 \text{ kJ} + 97.3 \text{ kJ} = -474.3 \text{ kJ}$$

Notice that:

    —in making a calculation of this type, the units must be consistent. If, as is ordinarily the case, we want ΔG^0 in kilojoules, ΔS^0 must be expressed in kilojoules per kelvin. (1 J/K $= 10^{-3}$ kJ/K)

    —the quantity ΔG^0 for this reaction is negative. This is consistent with the fact that $H_2(g)$ and $O_2(g)$, both at 1 atm, react spontaneously at 25°C to form liquid water.

Example 23.2   For the reaction $H^+(aq) + OH^-(aq) \rightarrow H_2O(l)$, ΔH is -55.9 kJ.
  a. Calculate ΔS^0.
  b. Calculate ΔG^0 at 25°C.
  c. Interpret the sign of ΔG^0 in terms of the spontaneity of the reaction.

Solution
  a. Using Tables 23.3 and 23.4:

$$\Delta S^0 = S^0\ H_2O(l) - S^0\ H^+(aq) - S^0\ OH^-(aq)$$
$$= 69.9 \text{ J/K} - 0 - (-10.5 \text{ J/K})$$
$$= 69.9 \text{ J/K} + 10.5 \text{ J/K} = 80.4 \text{ J/K} = 0.0804 \text{ kJ/K}$$

  b. $\Delta G^0 = -55.9$ kJ $- 298$ K $(0.0804$ kJ$) = -79.9$ kJ

  c. At 25°C, 1 M H^+ and 1 M OH^- react spontaneously to form H_2O. (This corresponds to the neutralization of, let us say, 1 M HCl with 1 M NaOH.)

Exercise   For the process $NaCl(s) \rightarrow Na^+(aq) + Cl^-(aq)$, $\Delta H = +3.9$ kJ, $\Delta S^0 = +42.9$ J/K. Calculate ΔG^0 at 25°C. Answer: -8.9 kJ (NaCl dissolves spontaneously at 25°C to give a 1 M solution).

ΔG^0 depends on T, even though ΔH and ΔS^0 don't

    We can also use Equation 23.12 to calculate ΔG^0 at temperatures other than 25°C. To do this, we neglect the variations of ΔH and ΔS^0 with temperature, which are ordinarily small.* In other words, we take the values of ΔH and ΔS^0 from Tables 23.1 through 23.4 and simply insert them in Equation 23.12, using the appropriate value of T. Suppose, for example, we want to know ΔG^0 at 1000°C for the reaction

$$CaCO_3(s) \rightarrow CaO(s) + CO_2(g)$$

For this reaction, we showed earlier that $\Delta H = +178.0$ kJ, $\Delta S^0 = +160.4$ J/K. Substituting T $= 1000 + 273 = 1273$ K, we have

$$\Delta G^0 = +178.0 \text{ kJ} - 1273 \text{ K} \times (0.1604 \text{ kJ/K})$$

$$= 178.0 \text{ kJ} - 204.2 \text{ kJ} = -26.2 \text{ kJ}$$

We conclude that this reaction is spontaneous at 1000°C, since ΔG^0 has a negative sign.

---

    *As far as Equation 23.12 is concerned, there is another reason for ignoring the temperature dependence of ΔH and ΔS. These two quantities always change in the same direction as the temperature changes (that is, if ΔH becomes more positive, so does ΔS). Hence the two effects tend to cancel each other. The true value of ΔG at any temperature is about the same as the one we calculate taking ΔH and ΔS to be constant.

    While we are on the subject, we should comment on the pressure or concentration dependence of ΔG, ΔH, and ΔS. Both ΔG and ΔS are ordinarily very sensitive to variations in pressure or concentration. That is why we have been careful to use the symbols ΔG^0 and ΔS^0, implying 1 atm for gases and 1 M for species in solution. On the other hand, ΔH is much less sensitive to pressure or concentration.

Example 23.3   Calculate ΔG for the reaction $Cu(s) + H_2O(g) \rightarrow CuO(s) + H_2(g)$ at 500 K.

Solution   From Table 23.1:

$$\Delta H = \Delta H_f\ CuO(s) - \Delta H_f\ H_2O(g) = -155.2\ kJ + 241.8\ kJ = +86.6\ kJ$$

From Table 23.3:

$$\Delta S^0 = S^0\ CuO(s) + S^0\ H_2(g) - S^0\ Cu(s) - S^0\ H_2O(g)$$
$$= 43.5\ J/K + 130.6\ J/K - 33.3\ J/K - 188.7\ J/K$$
$$= -47.9\ J/K = -0.0479\ kJ/K$$

Hence $\Delta G^0 = +86.6\ kJ - 500\ K\ (-0.0479\ kJ/K) = +110.6\ kJ$

We conclude that this reaction is not spontaneous, since ΔG^0 is positive. Instead, the reverse reaction occurs at 1 atm and 500 K (227°C). Under these conditions, CuO is reduced to copper by heating in a stream of H_2 gas.

This example shows how we determine reaction spontaneity at any temperature

Exercise   Calculate ΔG^0 for this reaction at 1000 K. Answer: +134.5 kJ (still nonspontaneous).

Effect of T Upon Reaction Spontaneity

When the temperature of a reaction system is increased, the direction in which the reaction proceeds spontaneously may or may not change. Whether it does or not depends upon the relative signs of ΔH and ΔS. The four possible situations, deduced from the Gibbs-Helmholtz equation, are summarized in Table 23.5.

If ΔH and ΔS have opposite signs (Table 23.5, I and II) it is impossible to reverse the direction of spontaneity by a change in temperature alone. The two terms ΔH and $-T\Delta S$ reinforce one another. Hence, ΔG has the same sign at all temperatures. Reactions of this type are rather uncommon. One such reaction is that discussed in Example 23.3:

$$Cu(s) + H_2O(g) \rightarrow CuO(s) + H_2(g) \qquad (23.13)$$

TABLE 23.5   EFFECT OF TEMPERATURE ON REACTION SPONTANEITY
AT A GIVEN PRESSURE

	ΔH	ΔS	$\Delta G = \Delta H - T\Delta S$	REMARKS
I	−	+	always −	Spontaneous at all T; reverse reaction always nonspontaneous
II	+	−	always +	Nonspontaneous at all T; reverse reaction occurs
III	+	+	+ at low T − at high T	Nonspontaneous at low T; becomes spontaneous as T is raised
IV	−	−	− at low T + at high T	Spontaneous at low T; at high T, reverse reaction becomes spontaneous

Here, ΔH is $+86.6$ kJ and ΔS^0 is -0.0479 kJ/K. Hence,

$$\Delta G^0 = \Delta H - T\Delta S^0$$
$$= +86.6 \text{ kJ} + T(0.0479 \text{ kJ/K})$$

Clearly, ΔG^0 is positive at all temperatures. The reaction cannot take place spontaneously at 1 atm regardless of temperature.

It is more common to find that ΔH and ΔS^0 have the same sign (Table 23.5, III and IV). When this happens, the enthalpy and entropy factors oppose each other. ΔG changes sign as temperature increases, and the direction of spontaneity reverses. At low temperatures ΔH predominates and the exothermic reaction occurs. As the temperature rises, the quantity $T\Delta S^0$ increases in magnitude and eventually exceeds ΔH. At high temperatures, the reaction which leads to an increase in entropy occurs. In most cases, *For most reactions, at low T, ΔH determines spontaneity. At high T, ΔS does* 25°C is a "low" temperature, at least at a pressure of 1 atm. This explains why exothermic reactions are usually spontaneous at room temperature and atmospheric pressure.

An example of a reaction for which ΔH and ΔS have the same sign is the decomposition of calcium carbonate:

$$CaCO_3(s) \rightarrow CaO(s) + CO_2(g)$$

Here, as pointed out previously, $\Delta H = +178.0$ kJ and $\Delta S^0 = +160.4$ J/K. Hence,

$$\Delta G^0 = +178.0 \text{ kJ} - T (0.1604 \text{ kJ/K})$$

Figure 23.2 shows a plot of ΔG^0 vs. T. Notice that:

—below about 1100 K (830°C), ΔH predominates, ΔG^0 is positive, and the reaction is nonspontaneous at 1 atm.

—above about 1100 K, $T\Delta S$ predominates, and the reaction is spontaneous at 1 atm.

—at about 1100 K, $\Delta H = T\Delta S^0$, so $\Delta G^0 = 0$. At this temperature, the system is at equilibrium at 1 atm. This means that if we put some $CaCO_3$ in a container and heat it to 1100 K, the pressure of CO_2 developed will be 1 atm.

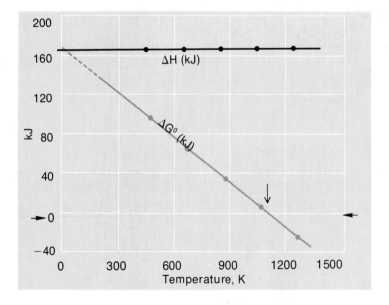

Figure 23.2 ΔH and ΔG^0 as a function of temperature for the reaction $CaCO_3(s) \rightarrow CaO(s) + CO_2(g)$. Below about 1110 K, ΔG^0 is positive and the reaction is nonspontaneous. Above about 1100 K, ΔG^0 is negative and $CaCO_3$ decomposes spontaneously to CaO and CO_2 at 1 atm.

Example 23.4   Consider the vaporization of water at 1 atm pressure:

$$H_2O(l) \rightarrow H_2O(g)$$

For this process, $\Delta H = +44.0$ kJ, $\Delta S^0 = +118.8$ J/K. Calculate the temperature at which liquid water and water vapor are at equilibrium at 1 atm ($\Delta G^0 = 0$).

Solution

$$\Delta G^0 = \Delta H - T\Delta S^0$$

For ΔG^0 to be zero, $\Delta H = T_e\Delta S^0$, where T_e is the equilibrium temperature.

Solving for T_e: $T_e = \dfrac{\Delta H}{\Delta S^0} = \dfrac{44.0 \text{ kJ}}{0.1188 \text{ kJ/K}} = 370$ K

From experience, we know that this answer is reasonable. Liquid water and water vapor are in equilibrium at 1 atm pressure at 100°C (373 K). Below this temperature, ΔG^0 is positive and water vapor condenses spontaneously. Above 100°C, ΔG^0 is negative and liquid water vaporizes at 1 atm. (The reason that we get 370 K rather than 373 K is that slight changes in ΔH and ΔS^0 between 25 and 100°C introduce a small error in the equilibrium temperature.)

Exercise   Calculate ΔG^0 for this process at 75°C; at 125°C. Answer: +2.7 kJ; −3.3 kJ.

23.4   THE FREE ENERGY CHANGE AND CELL VOLTAGE

We have just seen how the free energy change for a reaction can be calculated from the enthalpy and entropy changes. This is a general approach to determining ΔG, and is a very useful one. There is, however, a more direct way to obtain the free energy change, at least for redox reactions. It is based on a principle first developed by Gibbs. He showed that ΔG is a measure of the "useful work" that can be obtained from a reaction. In practice, the useful work is most easily obtained by measuring the electrical work associated with a cell in which the reaction is taking place. We can distinguish between two cases:

Electrical work is useful work

1. *A spontaneous reaction taking place in a voltaic cell.* Here, ΔG is a negative quantity. The maximum amount of electrical energy that can be obtained from the cell is numerically equal to ΔG. Consider, for example, the voltaic cell in Figure 23.3, p. 558. Here, the following spontaneous reaction is taking place at 25°C:

$$Zn(s) + Cu^{2+}(aq, 1 \text{ M}) \rightarrow Zn^{2+}(aq, 1 \text{ M}) + Cu(s); \quad \Delta G = -212.0 \text{ kJ}$$

$E^0_{cell} > 0$

The amount of electrical energy that we get from this reaction depends upon the way the cell is constructed and the nature of the external circuit. If we short-circuit the cell, we get no electrical energy at all; the energy difference between products and reactants is dissipated as heat. More typically, we might get 100 to 200 kJ of electrical energy per mole of zinc reacting. If we are clever enough, we can get 212.0 kJ of electrical energy, but no more. The value of ΔG sets an upper limit on the amount of electrical energy we can hope to obtain from the reaction.

2. *A nonspontaneous reaction taking place in an electrolytic cell.* Here, ΔG is a positive quantity. To make the reaction occur, we have to supply an amount of elec-

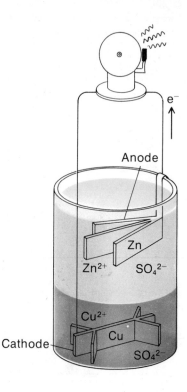

Figure 23.3 For the reaction $Zn(s) + Cu^{2+}(aq) \rightarrow Zn^{2+}(aq) + Cu(s)$, ΔG^0 is -212.0 kJ. Up to 212 kJ of useful electrical work can be obtained when one mole of zinc reacts with Cu^{2+} in a cell such as that shown here.

trical energy at least equal to ΔG. An example of such a reaction is the electrolysis of water:

$E^0_{cell} < 0$

$$H_2O(l) \rightarrow H_2(g, 1 \text{ atm}) + {}^1/_2 O_2(g, 1 \text{ atm}); \Delta G = +237.2 \text{ kJ}$$

To electrolyze one mole of water, we must supply at least 237.2 kJ of electrical energy, perhaps from batteries. In practice, we would probably use more energy, perhaps 250 or 300 kJ. The point is that, no matter what we do, we cannot reduce this quantity below 237.2 kJ.

From this discussion, it should be clear that we can obtain the value of ΔG from a reaction by measuring the amount of electrical energy produced or expended when the reaction is carried out in a cell. As it turns out, we do not have to actually make this measurement. All we need, to obtain the free energy change, is the cell voltage, which is directly proportional to the amount of electrical energy involved. The relationship is a simple one. At standard concentrations,

$$\Delta G^0 \text{ (in kJ)} = -96.5n \ E^0 \text{ (at 25°C)} \qquad (23.14)$$

Here ΔG^0 is the standard free energy change for the reaction, E^0 is the standard voltage (Chap. 21), and n is the number of moles of electrons transferred in the reaction. From Equation 23.14, we see that ΔG^0 and E^0 have opposite signs. A moment's reflection should convince you that this is reasonable. For a spontaneous reaction, ΔG^0 is *negative* and E^0 is *positive*. For a nonspontaneous reaction, ΔG^0 is *positive* and E^0 is *negative*.

To illustrate the use of Equation 23.14, consider the reaction

$$Zn(s) + Cu^{2+}(aq) \rightarrow Zn^{2+}(aq) + Cu(s)$$

We saw in Chapter 21 that E^0 for this spontaneous reaction is $+1.10$ V. The number of moles of electrons transferred is two (2 mol e$^-$ required to reduce 1 mol Cu^{2+}; 2 mol e$^-$ produced by the oxidation of 1 mol Zn). Hence

$$\Delta G^0 = -96.5(2)(+1.10) \text{ kJ} = -212 \text{ kJ}$$

For reactions that will occur in voltaic cells, it's easy to obtain ΔG^0 from E^0

Example 23.5   Consider the reaction

$$Cl_2(g) + 2 \; Br^-(aq) \rightarrow 2 \; Cl^-(aq) + Br_2(l)$$

a. Calculate E^0, using Table 21.1 (p. 507).
b. Calculate ΔG^0, using Equation 23.14.

Solution
a. $E^0 = E^0_{red} (Cl_2 \rightarrow Cl^-) + E^0_{ox} (Br^- \rightarrow Br_2)$
   $= +1.36 \text{ V} - 1.07 \text{ V} = +0.29 \text{ V}$

b. n = 2 (Two moles of e$^-$ are required to reduce one mole of Cl$_2$ or oxidize two moles of Br$^-$.)

$$\Delta G^0 = -96.5(2)(+0.29) \text{ kJ} = -56 \text{ kJ}$$

Exercise   What is ΔG^0 for the reaction 2 Cl$^-$(aq) + Br$_2$(l) $\rightarrow$ Cl$_2$(g) + 2 Br$^-$(aq)? Answer: +56 kJ.

23.5   THE FREE ENERGY CHANGE AND THE EQUILIBRIUM CONSTANT

In this chapter, we have emphasized that the free energy change, ΔG, is the basic criterion of spontaneity. A reaction occurs spontaneously if ΔG is a negative quantity. In earlier chapters, we discussed the extent to which reactions take place in terms of the equilibrium constant, K. As you might expect, these two quantities are intimately related. It can be shown by a thermodynamic argument, which we will not go into here, that

$$\Delta G^0 \text{ (in kJ)} = -0.0191 \; T \log_{10} K \qquad (23.15)$$

or, at 25°C (T = 298 K):

$$\Delta G^0 \text{ (in kJ)} = -5.71 \log_{10} K \text{ (at 25°C)} \qquad (23.16)$$

Here, ΔG^0 is the standard free energy change and K is the equilibrium constant for a reaction occurring in aqueous solution.

Looking at Equation 23.16, we can distinguish three possible situations:

1. If ΔG^0 is a negative quantity, $\log_{10} K$ must be positive, K is greater than one, and the reaction proceeds spontaneously in the forward direction when all species are at unit concentrations.

2. If ΔG^0 is positive, $\log_{10} K$ must be negative, K is less than one, and the reverse reaction is spontaneous when all species are at unit concentrations.

3. If, perchance, $\Delta G^0 = 0$, $\log_{10} K = 0$, K = 1, and the reaction is at equilibrium when all species are at unit concentrations.

TABLE 23.6   RELATION BETWEEN ΔG^0 AND K AT 25°C

ΔG^0 (kJ)	−120	−80	−40	0	+40	+80	+120
$\log_{10}$ K	+21.0	+14.0	+7.0	0	−7.0	−14.0	−21.0
K	1×10^{21}	1×10^{14}	1×10^{7}	1	1×10^{-7}	1×10^{-14}	1×10^{-21}

Equation 23.16 also allows us to give physical meaning to the magnitude of the free energy change. If ΔG^0 is a large negative number, K will be much greater than one and the forward reaction, under ordinary conditions, will go virtually to completion. Conversely, if ΔG^0 is a large positive number, K will be a small fraction and the reverse reaction will go nearly to completion.

In Table 23.6, we show values of K calculated from Equation 23.16 for various values of ΔG^0. Note that if ΔG^0 is more negative than −40 kJ, K is very large (K = 1×10^7 when ΔG^0 = −40 kJ). Such reactions will go virtually to completion under ordinary conditions. Conversely, if ΔG^0 is +40 kJ or greater, K is very small (K = 1×10^{-7} when ΔG^0 = +40 kJ). Reactions in this category will not occur to an appreciable extent. Only if ΔG^0 is rather close to zero, perhaps in the range −40 to +40 kJ, will the value of K be such that there will be appreciable amounts of both products and reactants at equilibrium.

Equation 23.16 can be used to calculate ΔG^0 for a reaction whose equilibrium constant is known (Example 23.6).

Example 23.6   Calculate ΔG^0 at 25°C for the reactions
   a. $AgCl(s) \rightarrow Ag^+(aq) + Cl^-(aq)$; $K_{sp} = 1.6 \times 10^{-10}$
   b. $H^+(aq) + OH^-(aq) \rightarrow H_2O(l)$; $K_w = 1.0 \times 10^{14}$

Solution   Using Equation 23.16:
   a. $\log_{10} (1.6 \times 10^{-10}) = 0.20 - 10 = -9.80$

   $\Delta G^0 = -5.71(-9.80)$ kJ = +56.0 kJ

The fact that ΔG^0 for this reaction is a positive quantity means that it does not occur at "standard" concentrations. That is, AgCl does not dissolve to give a solution 1 M in Ag^+ and Cl^- ions. This agrees with the fact that K_{sp} is a small number, which means that silver chloride is not very soluble in water. Indeed, at equilibrium, the concentrations of Ag^+ and Cl^- are very small, about 1.3×10^{-5} M.

   b. $\log_{10} (1.0 \times 10^{14}) = 14.00$; $\Delta G^0 = -5.71(14.00) = -79.9$ kJ

You may recall that we calculated ΔG^0 for this reaction by a rather different approach in Example 23.2. There we used the Gibbs-Helmholtz equation. The two answers are of course the same, about −80 kJ.

Exercise   The ionization constant of acetic acid is 1.8×10^{-5}. Calculate ΔG^0 at 25°C for the reaction: $HC_2H_3O_2(aq) \rightarrow H^+(aq) + C_2H_3O_2^-(aq)$. Answer: +27.1 kJ.

Equations 23.15 or 23.16 can be used to relate ΔG^0 to K for gas phase reactions of the type discussed in Chapter 13. There is, however, one complication. For such reactions, ΔG^0 is the free energy change when all gases involved are at a partial pressure of 1 atm. Under these conditions, the K in Equation 23.15 or 23.16 is not the equilibrium constant, K_c, used in Chapter 13. Instead, it is a quantity known as K_p, the equilibrium constant obtained when the concentrations of all gases are expressed as their partial pressures in atmospheres.

For reactions involving gases, we get K_p from the thermodynamic data

To illustrate the difference between K_c and K_p, consider the reaction

$$N_2(g) + 3 H_2(g) \rightleftharpoons 2 NH_3(g)$$

where $K_c = \dfrac{[NH_3]^2}{[N_2] \times [H_2]^3}$; $\qquad\qquad K_p = \dfrac{(P_{NH_3})^2}{(P_{N_2}) \times (P_{H_2})^3}$

In the expression for K_c the terms $[NH_3]$, $[N_2]$, and $[H_2]$ represent, as we have seen, equilibrium concentrations in moles/liter. In contrast, in the expression for K_p, the terms P_{NH_3}, P_{N_2}, and P_{H_2} represent equilibrium partial pressures in atmospheres.

If we know the value of K_c for a reaction, we can readily calculate K_p. To do this, we use the Ideal Gas Law, which tells us that

$$\text{conc. in mol}/\ell = \frac{n}{V} = \frac{P}{RT} = \frac{P}{0.0821\,T}$$

Using this relation, it is possible to show that

$$K_p = K_c(0.0821\,T)^{\Delta n_g}$$

We could use this equation to get K_c from K_p

The exponent in this equation, Δn_g, is the change in the number of moles of gas as the reaction proceeds in the forward direction. For the reaction referred to above,

$$\Delta n_g = 2 - 4 = -2$$

Hence: $K_p = K_c(0.0821\,T)^{-2}$

At 25°C (T = 298 K): $K_p = \dfrac{K_c}{(0.0821 \times 298)^2} = \dfrac{K_c}{600} = 0.00167 K_c$

23.6 HISTORICAL PERSPECTIVE: J. WILLARD GIBBS (1839–1903)

A century ago chemistry was primarily an empirical science. The outstanding chemists of that era were experimentalists who isolated and characterized new substances. The principles of chemistry were descriptive or correlative in nature, as illustrated by the atomic theory of Dalton and the Periodic Table of Mendeleev. Two theoreticians working in the latter half of the nineteenth century changed the very nature of chemistry by deriving the mathematical laws that govern the behavior of matter undergoing physical or chemical change. One of these was James Clerk Maxwell, whose contributions to kinetic theory were discussed in Chapter 7. The other was J. Willard Gibbs, professor of mathematical physics at Yale from 1871 until his death in 1903.

In 1876 Gibbs published the first portion of a remarkable paper in the *Transactions of the Connecticut Academy of Sciences* entitled ''On the Equilibrium of Heterogeneous Substances.'' When the paper was completed in 1878 (it was 323 pages long), the foundation was laid for the science of chemical thermodynamics. Here, for the first time, the concept of free energy appeared. Included as well were the basic principles of chemical equilibrium (Chap. 13), phase equilibrium (Chap. 11), and the relations governing energy changes in electrical cells (Chap. 23).

If Gibbs had never published another paper, this single contribution would have placed him among the greatest theoreticians in the history of science. Generations of experimental scientists have established their reputations by demonstrating in the laboratory the validity of the relationships that Gibbs derived at his desk. Many of these relationships were rediscovered by others; an example is the

Gibbs-Helmholtz equation developed in 1882 by Helmholtz, who was completely unaware of Gibbs' work.

In the 25 years that remained to him, Gibbs made substantial contributions in chemistry, astronomy, and mathematics. Among these were two papers published in 1881 and 1884 that established the discipline known today as vector analysis. His last work, published in 1901, was a book entitled *Elementary Principles in Statistical Mechanics*. Here Gibbs used the statistical principles that govern the behavior of systems to develop thermodynamic equations that he had derived from an entirely different point of view at the beginning of his career. Here, too, we find the "randomness" interpretation of entropy that has received so much attention in the social as well as the natural sciences.

J. Willard Gibbs is often cited as an example of the "prophet without honor in his own country." His colleagues in New Haven and elsewhere in the United States seem not to have realized the significance of his work until late in his life. During his first 10 years as a professor at Yale he received no salary. In 1920 when he was first proposed for the Hall of Fame of Distinguished Americans at New York University, he received nine votes out of a possible 100. Not until 1950 was he elected to that body. Even today the name of J. Willard Gibbs is generally unknown among educated Americans outside of those interested in the natural sciences.

Admittedly, Gibbs himself was largely responsible for the fact that for many years his work did not attract the attention it deserved. He made little effort to publicize it; the *Transactions of the Connecticut Academy of Sciences* was hardly the leading scientific journal of its day. Gibbs was one of those rare individuals who seem to have no inner need for recognition by contemporaries. His satisfaction came from solving a problem in his mind; having done so, he was ready to pass on to other problems without being concerned whether people understood what he had done. His papers are not easy to read; he seldom cites examples to illustrate his abstract reasoning. Frequently the implications of the laws that he derives are left for the reader to grasp on his own. One of his colleagues at Yale confessed many years later that none of the members of the Connecticut Academy of Sciences understood his paper on thermodynamics; as he put it, "We knew Gibbs and took his contributions on faith."

Gibbs achieved recognition in Europe long before his work was generally appreciated in this country. Maxwell read Gibbs' paper on thermodynamics, saw its significance, and referred to it repeatedly in his own publications. Wilhelm Ostwald, who said of Gibbs, "To physical chemistry, he gave form and content for a hundred years," translated the paper into German in 1892. Seven years later, Le Châtelier translated it into French.

SUMMARY

Two new functions, entropy (S) and free energy (G), were introduced in this chapter. We also made use of the enthalpy function (H), discussed in Chapter 6. Our interest is in how these quantities change in a reaction.

Change	Nature of Reaction	Example
$\Delta H = \Sigma \ \Delta H_f \text{ prod.} - \Sigma \ \Delta H_f \text{ react.}$	$\Delta H < 0$; exothermic $\Delta H > 0$; endothermic	

$$\Delta S = \Sigma\, S_{prod.} - \Sigma\, S_{react.}$$

$\Delta S > 0$; randomness increases   23.1, 23.2
$\Delta S < 0$; order increases

$$\Delta G = \Delta H - T\Delta S$$

$\Delta G < 0$; spontaneous   23.2, 23.3
$\Delta G > 0$; nonspontaneous
$\Delta G = 0$; equilibrium

Both ΔS and ΔG depend upon the pressures or concentrations of species taking part in the reaction. We dealt only with the standard entropy change, ΔS^0, and the standard free energy change, ΔG^0 (1 atm, 1 M). Both ΔH and ΔS^0 are taken to be independent of temperature; ΔG^0 is a linear function of T (Fig. 23.2). The fact that ΔG^0 becomes zero at equilibrium allows us to calculate the temperature at which a reaction is at equilibrium at standard pressures and concentrations (Example 23.4).

The standard free energy change, ΔG^0, is related to two other important quantities. These are the standard voltage, E^0, and the equilibrium constant, K. At 25°C:

1. ΔG^0 (in kilojoules) $= -96.5n\, E^0$; (Example 23.5)
2. ΔG^0 (in kilojoules) $= -5.71 \log_{10} K$; (Example 23.6)

Note that:

Spontaneous reaction (at standard concentrations): $\Delta G^0 < 0$; $E^0 > 0$; $K > 1$.
System at equilibrium (at standard concentrations): $\Delta G^0 = 0$; $E^0 = 0$; $K = 1$.
Nonspontaneous reaction (at standard concentrations): $\Delta G^0 > 0$; $E^0 < 0$; $K < 1$.

KEY WORDS AND CONCEPTS

chemical thermodynamics	enthalpy change, ΔH	standard molar entropy, S^0	standard free energy change, ΔG^0
spontaneous reaction	exothermic	entropy change, ΔS	Gibbs-Helmholtz equation
nonspontaneous reaction	endothermic	free energy, G	standard voltage, E^0
enthalpy, H	entropy, S	free energy change, ΔG	equilibrium constant, K

QUESTIONS AND PROBLEMS

Catalog

Enthalpy Change: 23.2, 23.22
Entropy Change: 23.3, 23.4, 23.23, 23.24
ΔG; Gibbs-Helmholtz Equation: 23.5–23.10, 23.25–23.30

ΔG^0, E^0, and K: 23.11–23.14, 23.31–23.34
General: 23.1, 23.15–23.21, 23.35–23.45

23.1   Review and know the meaning of the key words and concepts in this chapter.

23.2   Using Tables 23.1 and 23.2, calculate ΔH for

   a. $CaSO_4(s) \rightarrow CaO(s) + SO_3(g)$
   b. $H_2PO_4^-(aq) + 2\,OH^-(aq) \rightarrow$
      $PO_4^{3-}(aq) + 2\,H_2O(l)$
   c. $Cu(s) + 4\,H^+(aq) + 2\,NO_3^-(aq) \rightarrow$
      $Cu^{2+}(aq) + 2\,H_2O(l) + 2\,NO_2(g)$

23.22   Using Tables 23.1 and 23.2, calculate ΔH for

   a. $Ca(OH)_2(s) \rightarrow CaO(s) + H_2O(g)$
   b. $Ag(s) + 2\,H^+(aq) + NO_3^-(aq) \rightarrow$
      $Ag^+(aq) + H_2O(l) + NO_2(g)$
   c. $3\,Ag(s) + 4\,H^+(aq) + NO_3^-(aq) \rightarrow$
      $3\,Ag^+(aq) + 2\,H_2O(l) + NO(g)$

23.3 Predict the sign of ΔS for

a. dissolving tea in water.
b. firewood being stacked.
c. $I_2(s) \rightarrow I_2(g)$
d. $2 Cl(g) \rightarrow Cl_2(g)$

23.4 Using Tables 23.3 and 23.4, calculate ΔS^0 for each reaction in Problem 23.2.

23.5 Using the Gibbs-Helmholtz equation, calculate ΔG^0 at 25°C for each reaction in Problem 23.2. Which reactions are spontaneous at 25°C and standard concentrations?

23.6 Consider the reaction in Problem 23.2a. What is ΔG^0 for this reaction at

a. 100°C?   b. 200°C?   c. 500 K?

23.7 For the reaction

$$2 NO(g) + O_2(g) \rightarrow 2 NO_2(g)$$

calculate

a. ΔH b. ΔS^0
c. ΔG^0 at 298 K   d. ΔG^0 at 750 K

23.8 Consider the reaction

$$HgO(s) \rightarrow Hg(l) + \tfrac{1}{2} O_2(g)$$

a. Obtain an expression for ΔG^0 as a function of temperature. Use it to prepare a table of ΔG^0 values at 200 K intervals between 200 and 1000 K.
b. Calculate the temperature at which ΔG^0 becomes zero.

23.9 For the reaction

$$2 NaHCO_3(s) \rightarrow Na_2CO_3(s) + CO_2(g) + H_2O(g)$$

$\Delta G^0 = +129.3$ kJ at 0 K and +33.1 kJ at 298 K. Determine

a. ΔH b. ΔS^0
c. the temperature at which the reaction becomes spontaneous at 1 atm.

23.10 The standard free energy of formation, ΔG_f^0, of a compound is defined as ΔG^0 when one mole of the compound is formed from the elements in their stable states. Calculate ΔG_f^0 of $MgCl_2$ at 25°C, using data from Tables 23.1 and 23.3.

23.11 Use E^0 values (Table 21.1, p. 507) for the following reactions to determine ΔG^0 at 25°C:

a. $Ni(s) + Sn^{2+}(aq) \rightarrow Sn(s) + Ni^{2+}(aq)$
b. $Ni^{2+}(aq) + Sn^{2+}(aq) \rightarrow Ni(s) + Sn^{4+}(aq)$

Calculate K for each reaction.

23.23 Predict the sign of ΔS for

a. a log being split.
b. a log burning.
c. $Br_2(l) + 3 F_2(g) \rightarrow 2 BrF_3(l)$
d. $TiCl_4(g) + 2 Mg(s) \rightarrow Ti(s) + 2 MgCl_2(s)$

23.24 Using Tables 23.3 and 23.4, calculate ΔS^0 for each reaction in Problem 23.22.

23.25 Using the Gibbs-Helmholtz equation, calculate ΔG^0 at 25°C for each reaction in Problem 23.22. Which reactions are spontaneous at 25°C and standard concentrations?

23.26 Consider the reaction in Problem 23.22a. What is ΔG^0 for this reaction at

a. 100°C?   b. 200°C?   c. 500 K?

23.27 For the reaction

$$2 H_2S(g) + 3 O_2(g) \rightarrow 2 H_2O(g) + 2 SO_2(g)$$

calculate

a. ΔH b. ΔS^0
c. ΔG^0 at 298 K   d. ΔG^0 at 500 K

23.28 Earlier civilizations smelted iron from ore by heating with charcoal from a wood fire:

$$2 Fe_2O_3(s) + 3 C(s) \rightarrow 4 Fe(s) + 3 CO_2(g)$$

a. Obtain an expression for ΔG^0 as a function of temperature. Prepare a table of ΔG^0 values at 200 K intervals between 200 and 1000 K.
b. Calculate the lowest temperature at which the smelting could be carried out.

23.29 For the reaction

$$NH_4Cl(s) \rightarrow NH_3(g) + HCl(g)$$

$\Delta G^0 = +85.4$ kJ at 300 K and +34.6 kJ at 500 K. Determine

a. ΔH b. ΔS^0
c. the temperature at which $P_{NH_3} = P_{HCl} = 1$ atm.

23.30 Calculate the standard free energy of formation of NH_4NO_3 at 298 K (see 23.10).

23.31 Calculate E^0 and then ΔG^0 for the following reactions at 25°C:

a. $3 Mn(s) + 2 NO_3^-(aq) + 8 H^+(aq) \rightarrow$
   $2 NO(g) + 3 Mn^{2+}(aq) + 4 H_2O(l)$
b. $MnO_2(s) + 4 H^+(aq) + 2 Cl^-(aq) \rightarrow$
   $Mn^{2+}(aq) + Cl_2(g) + 2 H_2O(l)$

Calculate K for each reaction.

23.12 For the reaction

$$2 MnO_4^-(aq) + 5 H_2S(g) + 6 H^+(aq) \rightarrow$$
$$2 Mn^{2+}(aq) + 5 S(s) + 8 H_2O(l)$$

calculate

    a. ΔH b. ΔS^0
    c. ΔG^0 at 298 K   d. E^0 at 298 K

23.13 Calculate E^0, ΔG^0, and K at 298 K for the reaction

$$2 Fe^{3+}(aq) + Sn^{2+}(aq) \rightarrow Sn^{4+}(aq) + 2 Fe^{2+}(aq)$$

23.14 The equilibrium constant (solubility product constant) for the reaction

$$AgCl(s) \rightarrow Ag^+(aq) + Cl^-(aq)$$

is 1.6×10^{-10} at 25°C. Calculate ΔG^0 from K and compare to the value of ΔG^0 calculated from Tables 23.1 through 23.4.

23.15 How might the following nonspontaneous processes be accomplished?

    a. $NaCl(s) \rightarrow Na(s) + \frac{1}{2} Cl_2(g)$
    b. removing rust from a car's bumper.
    c. making a supersaturated solution of CO_2 gas in water.

23.16 Discuss the effect of temperature upon the spontaneity of the following reactions at 1 atm:

    a. $Al_2O_3(s) + 2 Fe(s) \rightarrow 2 Al(s) + Fe_2O_3(s)$
        $\Delta H = +847.6$ kJ; $\Delta S^0 = +41.2$ J/K
    b. $CO(g) \rightarrow C(s) + \frac{1}{2} O_2(g)$
        $\Delta H = +110.5$ kJ; $\Delta S^0 = -89.9$ J/K
    c. $SO_3(g) \rightarrow SO_2(g) + \frac{1}{2} O_2(g)$
        $\Delta H = +99.1$ kJ; $\Delta S^0 = +94.8$ J/K

23.17 Which of the following are spontaneous? nonspontaneous?

    a. tea dissolving in water.
    b. removing seeds from a watermelon.
    c. recharging a battery.
    d. burning gasoline in an automobile engine.

23.18 Consider the reaction

$$PCl_5(g) \rightarrow PCl_3(g) + Cl_2(g)$$

    a. Is this reaction spontaneous at 25°C and 1 atm?
    b. Will the reaction become spontaneous as temperature increases?
    c. At what temperature will the system be in equilibrium at 1 atm?

23.32 For the reaction

$$3 Cu(s) + 8 H^+(aq) + 2 NO_3^-(aq) \rightarrow$$
$$3 Cu^{2+}(aq) + 2 NO(g) + 4 H_2O(l)$$

calculate

    a. ΔH b. ΔS^0
    c. ΔG^0 at 298 K   d. E^0 at 298 K

23.33 For the cell Co | Co^{2+} ‖ Cu^{2+} | Cu calculate

    a. E^0 b. ΔG^0 at 298 K   c. K at 298 K

23.34 The equilibrium constant (acid dissociation constant) for the reaction

$$H_2PO_4^-(aq) \rightarrow H^+(aq) + HPO_4^{2-}(aq)$$

is 6.2×10^{-8} at 25°C. Calculate ΔG^0 from K and compare to the value of ΔG^0 calculated from Tables 23.2 and 23.4.

23.35 When a lead storage battery is used to start your car, a nonspontaneous process is being driven by a spontaneous process. Identify the two processes involved.

23.36 Comment on the *general* validity of each of the following statements:

    a. An exothermic reaction is spontaneous.
    b. A reaction for which ΔS is positive is spontaneous.
    c. ΔS is positive for a reaction in which there is an increase in the number of moles.
    d. If ΔH and ΔS are both positive, ΔG will decrease when the temperature rises.

23.37 a. Under what conditions is the vaporization of water at 25°C spontaneous? nonspontaneous? neither?
    b. Will a small amount of KNO_3 added to a saturated solution of KNO_3 dissolve spontaneously? Suggest a way to make the added KNO_3 dissolve.

23.38 Consider the reaction

$$2 KCl(s) + 3 O_2(g) \rightarrow 2 KClO_3(s)$$

    a. Is this reaction spontaneous at 25°C and 1 atm?
    b. Explain why $KClO_3$ does not decompose at 25°C and 1 atm.
    c. To make $KClO_3$ decompose to KCl and O_2, would you increase or decrease the temperature?

23.19 Consider the general reaction

$$A + B \rightleftharpoons C + D$$

a. How does the standard free energy change for the reverse reaction, ΔG_R^0, compare to that for the forward reaction, ΔG_F^0?
b. Using the relation in (a) and Equation 23.15, show that the equilibrium constant for the reverse reaction, K_R, must be the reciprocal of that for the forward reaction, K_F.

23.20 Explain in your own words

a. why all exothermic reactions are not spontaneous.
b. what a ΔS value <0 means.
c. why we say the reaction "stops" when $\Delta G = 0$.

23.21 Fill in the blanks.

a. As E^0 becomes more positive, ΔG^0 becomes _____ and K _____.
b. In general, if $T\Delta S^0$ becomes more positive with increasing temperature, ΔG^0 _____.
c. A reaction for which ΔH is positive and ΔS is _____ will be nonspontaneous at _____ temperatures.

23.39 Free energy changes, like enthalpy changes, are additive. Consider the reactions

$$A \rightarrow B;\ \Delta G_1^0$$

$$B + C \rightarrow D;\ \Delta G_2^0$$

a. Relate the standard free energy change, ΔG_3^0, for the reaction

$$A + C \rightarrow D$$

to ΔG_1^0 and ΔG_2^0.
b. Using the relation in (a) and Equation 23.15, demonstrate the validity of the Rule of Multiple Equilibria—that is, $K_3 = K_2 \times K_1$.

Explain in your own words

a. why operating voltaic cells have $\Delta G^0 < 0$.
b. how entropy is involved in environmental pollution.
c. what the phrase "a dead battery" means from a thermodynamic standpoint.

Write an equation relating E^0 and K for a reaction, at 25°C.

*23.42 The heat of fusion of ice is 333 J/g. For the process

$$H_2O(s) \rightarrow H_2O(l)$$

Determine

a. ΔH b. ΔG^0 at 0°C   c. ΔS^0 d. ΔG^0 at -10°C   e. ΔG^0 at 10°C

*23.43 The electrolysis of Al_2O_3 to form the elements is carried out at about 1000°C. Calculate ΔG^0 for the electrolysis of one kilogram of Al_2O_3 under these conditions. Calculate the cost of the electrolysis at 6.0¢ per kilowatt hour; 1 kW·h = 3600 kJ. (The actual cost is greater, because much of the electrical energy is dissipated as heat.)

*23.44 For the Haber process $3 H_2(g) + N_2(g) \rightarrow 2 NH_3(g)$, calculate, at 700 K

a. ΔH b. ΔS^0 c. ΔG^0 d. K_p e. K_c

*23.45 The overall reaction that occurs when sugar is metabolized is

$$C_{12}H_{22}O_{11}(s) + 12 O_2(g) \rightarrow 12 CO_2(g) + 11 H_2O(l)$$

For this reaction, ΔH is -5650 kJ and ΔG^0 is -5790 kJ at 25°C.

a. If 30% of the free energy change is actually converted to useful work, how many kJ of work could be obtained when one gram of sugar is metabolized at body temperature, 37°C?
b. How many grams of sugar would you have to eat to get the energy to climb a mountain 1610 m high? (W = 9.79×10^{-3} mh, where W = work in kJ, m is body mass in kilograms, and h is height in meters.)

NUCLEAR REACTIONS

The "ordinary chemical reactions" discussed to this point involve changes in the outer electronic structure of atoms or molecules. In contrast, nuclear reactions result from changes taking place within atomic nuclei. You will recall (Chap. 2) that atomic nuclei are represented by symbols, such as

$$^{12}_{6}C \qquad ^{14}_{6}C$$

Here the atomic number (number of protons in the nucleus) is shown as a subscript at the lower left. The mass number (number of protons + neutrons in the nucleus) appears as a superscript at the upper left. Nuclei with the same number of protons but different numbers of neutrons are called isotopes. The symbols written above represent two isotopes of the element carbon (at. no. = 6). One isotope has six neutrons in the nucleus and hence has a mass number of $6 + 6 = 12$. The heavier isotope has eight neutrons and a mass number of 14.

Nuclear reactions differ from ordinary chemical reactions in several important respects. In particular:

1. In ordinary reactions, different isotopes of an element show virtually identical chemical properties. In nuclear reactions, these isotopes behave quite differently. Consider, for example, $^{12}_{6}C$ and $^{14}_{6}C$. The chemical properties of these isotopes of carbon are very similar. Their nuclear properties are very different. The carbon-12 nucleus is extremely stable, while the carbon-14 nucleus decomposes spontaneously.

2. The nuclear reactivity of an element is nearly independent of its oxidation state. In the nuclear chemistry of radium, it makes little difference whether we deal with the element itself or one of its compounds. The Ra atom in the element and the Ra^{2+} ion in $RaCl_2$ behave similarly from a nuclear standpoint. In ordinary chemical reactions the Ra atom and the Ra^{2+} ion behave quite differently, since they have different outer electronic structures.

3. Nuclear reactions usually involve the conversion of one element to another. Whenever such a reaction results in a change in the number of protons in the nucleus, a new element of different atomic number is formed. In contrast, elements taking part in ordinary chemical reactions retain their identity.

Transmutation of elements does occur, but not by the methods of the alchemists

4. Nuclear reactions are accompanied by energy changes which greatly exceed those associated with ordinary chemical reactions. The energy evolved when one gram of radium undergoes radioactive decay (Section 24.1) is 500,000 times as great as that given off when the same amount of radium reacts with chlorine to form $RaCl_2$. Still larger amounts of energy are given off in nuclear fission (Section 24.4) and nuclear fusion (Section 24.5).

24.1 RADIOACTIVITY

Certain unstable nuclei decompose spontaneously, giving off energy in the process. A few such nuclei occur in nature; their decomposition is referred to as natural radioactivity. Many more unstable nuclei have been prepared in the laboratory. The process by which these nuclei decompose is called induced radioactivity. We will consider the nature of the reactions that occur when radioactive nuclei decompose ("decay"). We will also be concerned with the effects of the products of these reactions on the surroundings.

Natural Radioactivity

This process was discovered, almost accidentally, by a French scientist, Henri Becquerel, in 1896. While studying the fluorescence of uranium salts, he found that they gave off a new type of high-energy radiation, capable of blackening a photographic plate. This radiation seems never to have been detected before, even though the element uranium had been known for more than a century.

Becquerel showed that the rate of emission of radiation from a uranium salt was directly proportional to the amount of uranium present. There was one exception to this rule. A certain uranium ore called pitchblende gave off radiation at a rate four times as great as one would calculate on the basis of its uranium content. In 1898 Marie and Pierre Curie, colleagues of Becquerel at the Sorbonne, searched for the active ingredient of pitchblende. They isolated a fraction of a gram of a new element from a ton of ore. This element was more intensely radioactive than uranium. They named it polonium, after Marie Curie's native country. Six months later the Curies isolated another, intensely radioactive, new element, radium. The Nobel Prize for physics in 1903 was awarded jointly to Becquerel and the Curies. Eight years later Marie Curie received an unprecedented second Nobel Prize, this time in chemistry.

The polonium and radium had been produced by the decay of the uranium

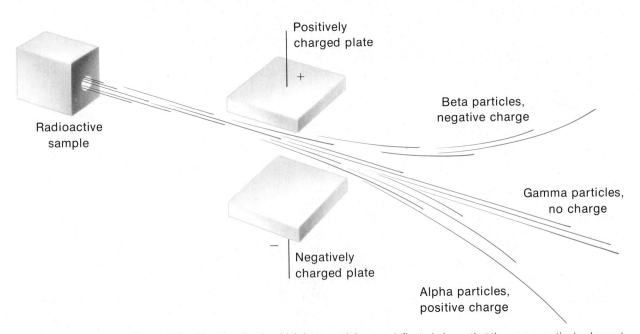

Figure 24.1 The direction in which beta particles are deflected shows that they are negatively charged. Alpha particles move so as to indicate that they carry a positive charge. Gamma rays are undeflected and so must be uncharged.

The radiation given off in natural radioactivity can be separated by an electrical or magnetic field into three distinct parts (Fig. 24.1).

1. **Alpha radiation** consists of a stream of positively charged particles (alpha particles) with a charge of +2 and a mass of 4 on the atomic mass scale. These particles are identical with the nuclei of ordinary helium atoms, $_2^4\text{He}$.

When an alpha particle is ejected from the nucleus, the atomic number decreases by two units; the mass number decreases by four units. Consider, for example, the loss of an alpha particle by a uranium atom with atomic number 92 and mass number 238. This gives an isotope of thorium with atomic number 90 and mass number 234. The transmutation of elements, long sought in vain by the alchemists, occurs by reactions of this sort. The nuclear reaction is represented by the equation

$$_{92}^{238}\text{U} \rightarrow {}_2^4\text{He} + {}_{90}^{234}\text{Th} \qquad (24.1)$$

Here, as in all nuclear equations, there is a balance of both atomic number ($90 + 2 = 92$) and mass number ($4 + 234 = 238$) on the two sides.

2. **Beta radiation** is made up of a stream of negatively charged particles (beta particles) identical in their properties to electrons. The ejection of a beta particle (mass ≈ 0, charge $= -1$) converts a neutron (mass $= 1$, charge $= 0$) in the nucleus into a proton (mass $= 1$, charge $= +1$). Hence, beta emission leaves the mass number unchanged but increases the atomic number by one unit. An example of beta emission is the radioactive decay of thorium-234 (90 protons, 144 neutrons) to protactinium-234 (91 protons, 143 neutrons):

$$_{90}^{234}\text{Th} \rightarrow {}_{-1}^{0}\text{e} + {}_{91}^{234}\text{Pa} \qquad (24.2)$$

The symbol $_{-1}^{0}\text{e}$ is written to stand for a beta particle (electron).

3. **Gamma radiation** consists of high-energy photons of very short wavelength ($\lambda = 0.0005$ to 0.1 nm). The emission of gamma radiation accompanies most nuclear reactions. It results from an energy change within the nucleus. An excited nucleus resulting from alpha or beta emission gives off a photon and drops to a lower, more stable energy state. Gamma emission changes neither the atomic number nor the mass number. For that reason, we shall frequently omit it in writing nuclear equations.

Reactions 24.1 and 24.2 represent the first two steps in a natural radioactive series which leads finally to a stable isotope of lead, $_{82}^{206}\text{Pb}$. There are a total of 14 steps in this process (see Problem 24.5). All the intermediate products are unstable and decay rapidly. The two steps that follow Reactions 24.1 and 24.2 are given in Example 24.1.

Example 24.1   The isotope formed in Reaction 24.2 is unstable, decomposing by emission of a beta particle (Reaction 1). The product of this reaction is also unstable; it decomposes by alpha emission (Reaction 2). Write balanced nuclear equations for Reactions 1 and 2.

Solution   The unbalanced equation for Reaction 1 is

$$_{91}^{234}\text{Pa} \rightarrow {}_{-1}^{0}\text{e} + \text{?}$$

where the question mark indicates the isotope formed by beta emission. We see that in order to balance mass and charge, this isotope must have a mass number of 234 and an atomic number of 92 (i.e., $92 - 1 = 91$). From the Periodic Table, we find this to be an isotope of uranium (at. no. = 92). Hence, the balanced equation is

$$_{91}^{234}\text{Pa} \rightarrow {}_{-1}^{0}\text{e} + {}_{92}^{234}\text{U} \qquad (1)$$

Both mass and charge balance in nuclear reactions

When $^{234}_{92}U$ loses an alpha particle, $^{4}_{2}He$, the mass number decreases by 4 and the atomic number by 2. The isotope produced must have a mass number of $234 - 4 = 230$ and an atomic number of $92 - 2 = 90$ (thorium, symbol Th). The balanced nuclear equation is

$$^{234}_{92}U \rightarrow {}^{4}_{2}He + {}^{230}_{90}Th \tag{2}$$

Exercise   The next step in this series involves α-emission by $^{230}_{90}Th$. Write a balanced nuclear equation for this reaction. Answer: $^{230}_{90}Th \rightarrow {}^{4}_{2}He + {}^{226}_{88}Ra$.

Induced Radioactivity. Bombardment Reactions

During the past 50 years, more than 1500 radioactive isotopes have been prepared in the laboratory. The number of such isotopes per element ranges from one (hydrogen and helium) to a maximum of 36 (indium). They are all prepared by bombardment reactions in which a stable nucleus is converted to one that is radioactive. A typical reaction is that which occurs when the stable isotope of aluminum, $^{27}_{13}Al$, absorbs a neutron to form $^{28}_{13}Al$. The latter is unstable, decaying by electron emission to give a stable isotope of silicon, $^{28}_{14}Si$. The two steps involved in the process are

$$\text{neutron bombardment: } {}^{27}_{13}Al + {}^{1}_{0}n \rightarrow {}^{28}_{13}Al \tag{24.3}$$

$$\text{radioactive decay: } \quad {}^{28}_{13}Al \rightarrow {}^{28}_{14}Si + {}^{0}_{-1}e \tag{24.4}$$

The first radioactive isotopes to be made in the laboratory were prepared in 1934 by Irene (daughter of Marie and Pierre) Curie and her husband, Frederic Joliot. They achieved this by bombarding certain stable isotopes with high-energy alpha particles. One reaction was

The α particles came from radium

$$^{27}_{13}Al + {}^{4}_{2}He \rightarrow {}^{30}_{15}P + {}^{1}_{0}n \tag{24.5}$$

The product, phosphorus-30, is radioactive. It decays by emitting a particle called a **positron,** which has the same mass as an electron, but a charge of $+1$ rather than -1.

$$^{30}_{15}P \rightarrow {}^{30}_{14}Si + {}^{0}_{1}e \tag{24.6}$$

Positron emission is never observed in natural radioactivity. However, it is a common mode of decay in induced radioactivity. Notice from Equation 24.6 that the result of positron emission is the conversion of a proton in the nucleus to a neutron (15 p, 15 n in P-30: 14 p, 16 n in Si-30). Positron emission occurs with "light" isotopes—that is, nuclei which have too few neutrons to be stable. An example is carbon-11 (6 p, 5 n), which decays by giving off a positron:

$$^{11}_{6}C \rightarrow {}^{11}_{5}B + {}^{0}_{1}e \tag{24.7}$$

Carbon-12 (6 p, 6 n) is neither "light" nor "heavy." It's just right

In contrast, the "heavy" isotope of the same element, carbon-14 (6 p, 8 n), decays by electron emission:

$$^{14}_{6}C \rightarrow {}^{14}_{7}N + {}^{0}_{-1}e \tag{24.8}$$

An interesting application of bombardment reactions is in the preparation of very heavy elements. During the past 40 years, 14 elements with atomic numbers greater than uranium (93 through 106) have been prepared. Much of this work was done by a group at the University of California at Berkeley, under the direction first of Glenn Seaborg and then Albert Ghiorso. In the past 15 years, a Russian group led by G. N. Flerov has made substantial contributions to the field.

Some of the reactions used to prepare elements beyond uranium are listed in Table 24.1. Neutron bombardment is effective for the lower members of the series. However, the yield of product decreases rapidly with increasing atomic number. To form very heavy elements, it is necessary to bombard appropriate targets with high-energy positive ions, accelerated to very high velocities. With heavy bombarding particles such as carbon-12, it is possible to achieve a large increase in atomic number.

Example 24.2   Dr. Seaborg has suggested that by using very heavy nuclei as bombarding particles, it may be possible to synthesize elements of atomic numbers much higher than any now known. One reaction that he has speculated upon is that of $^{48}_{20}Ca$ with $^{244}_{94}Pu$. Assuming that the product nucleus is an isotope of element 114 containing 174 neutrons, write a balanced nuclear equation for the reaction.

Solution   The unbalanced equation, from the information given, is

$$^{244}_{94}Pu + ^{48}_{20}Ca \rightarrow ^{288}_{114}X + \underline{\quad\quad}$$

Notice that atomic number is already balanced (114 on both sides). The other product must then be a neutron, ^1_0n. Since there is a deficiency of 4 in mass number on the right side (288 vs. 292), four neutrons must be formed. The balanced equation is

$$^{244}_{94}Pu + ^{48}_{20}Ca \rightarrow ^{288}_{114}X + 4\,^1_0n$$

Exercise   Suppose we wanted to obtain the same products using $^{54}_{22}Ti$ instead of $^{48}_{20}Ca$. What isotope should we substitute for $^{244}_{94}Pu$? Answer: $^{238}_{92}U$.

TABLE 24.1   SYNTHESIS OF TRANSURANIUM ELEMENTS

NEUTRON BOMBARDMENT

Neptunium, Plutonium	$^{238}_{92}U$ + ^1_0n	$\rightarrow$	$^{239}_{92}U$	$\rightarrow$	$^{239}_{93}Np$ + $^{\ 0}_{-1}e$	
	$^{239}_{93}Np$	$\rightarrow$	$^{239}_{94}Pu$ + $^{\ 0}_{-1}e$			
Americium	$^{239}_{94}Pu$ + $2\,^1_0n$	$\rightarrow$	$^{241}_{94}Pu$	$\rightarrow$	$^{241}_{95}Am$ + $^{\ 0}_{-1}e$	

POSITIVE ION BOMBARDMENT

Curium		$^{239}_{94}Pu$ + ^4_2He	$\rightarrow$	$^{242}_{96}Cm$ + ^1_0n	
Californium		$^{242}_{96}Cm$ + ^4_2He	$\rightarrow$	$^{245}_{98}Cf$ + ^1_0n	
	or:	$^{238}_{92}U$ + $^{12}_6C$	$\rightarrow$	$^{246}_{98}Cf$ + $4\,^1_0n$	
Element 104*		$^{249}_{98}Cf$ + $^{12}_6C$	$\rightarrow$	$^{257}_{104}?$ + $4\,^1_0n$	
Element 105*		$^{249}_{98}Cf$ + $^{15}_7N$	$\rightarrow$	$^{260}_{105}?$ + $4\,^1_0n$	
Element 106*		$^{249}_{98}Cf$ + $^{18}_8O$	$\rightarrow$	$^{263}_{106}?$ + $4\,^1_0n$	

*The equations given represent reactions used by the group at Berkeley. The names of elements 104 through 106 have not been established. For 104 and 105, the Berkeley group has suggested rutherfordium and hahnium, honoring Ernest Rutherford and Otto Hahn, discoverer of nuclear fission. The Russian group prefers the names bohrium and kurchatovium, after Niels Bohr and the Russian physicist I. V. Kurchatov.

It's hard to see what these elements might be useful for

Isotopes of the very heavy elements have very short half-lives. Moreover, most of them have been formed in very minute quantities, amounting in some cases to only a few atoms. One of the greatest achievements of scientists working in this field has been their ability to study the properties of these elements on submicrogram samples. Both chemical and physical evidence indicate that the elements of atomic numbers 90 through 103 are filling a second rare-earth series by completing the 5f sublevel. Element 104 is the first member of a new transition series, falling below hafnium in the Periodic Table.

Interaction of Radiation with Matter

The alpha, beta, and gamma rays given off in radioactive decay lose energy when they pass through matter. They do this by transferring energy to atoms, molecules, or ions with which they collide. These collisions may be elastic. That is, only kinetic energy may be transferred, raising the temperature of the impacted material. Eventually this kinetic energy is converted to heat, which is given off to the surroundings.

Frequently, interaction of radiation with matter results in inelastic collisions. An electron in the target species may be raised to a higher energy level by absorbing radiation. When it drops back to the ground level, energy is given off, often as visible light. An instrument used to detect and measure radiation, the *scintillation counter,* takes advantage of this effect. Light produced by radiation striking an organic solid or solution activates a photoelectric cell within the counter.

In another type of inelastic collision, an electron is removed from an atom or molecule to form a positively charged ion. The ionizing ability of alpha, beta, and gamma radiation is the basis of several methods used to study radioactivity. The Geiger-Müller counter (Fig. 24.2) amplifies the electric current produced by a flow of electrons and positive ions.

The harmful effect of radiation on human beings is caused by its ability to ionize and ultimately destroy the organic molecules of which body cells are composed. The extent of damage depends mainly upon two factors. These are the amount of radiation absorbed and the type of radiation. The former is commonly expressed in *rads*. A rad corresponds to the absorption of 10^{-2} J of energy per kilogram of tissue. The total bio-

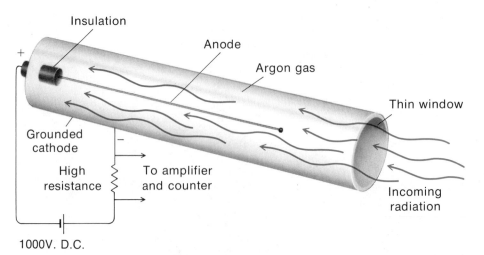

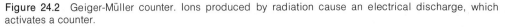

Figure 24.2 Geiger-Müller counter. Ions produced by radiation cause an electrical discharge, which activates a counter.

TABLE 24.2 EFFECT OF EXPOSURE TO A SINGLE DOSE OF RADIATION	
DOSE (REMS)	**PROBABLE EFFECT**
0 to 25	No observable effect
25 to 50	Small decrease in white blood cell count
50 to 100	Lesions, marked decrease in white blood cells
100 to 200	Nausea, vomiting, loss of hair
200 to 500	Hemorrhaging, ulcers, possible death
500+	Fatal

logical effect of radiation is expressed in *rems*. This is found by multiplying the number of rads by an appropriate factor, n, for the particular type of radiation:

$$\text{number of rems} = n(\text{number of rads})$$

Here, n is 1 for β, γ, and x-radiation and 10 for α radiation or high-energy neutrons. Table 24.2 lists some of the effects to be expected when a person is exposed to a single dose of radiation at various levels.

Small doses of radiation repeated over long periods of time can have very serious consequences. Many of the early workers in the field of radioactivity developed cancer in this way. Cases are known in which cancers developed as long as 40 years after initial exposure. Studies have shown an abnormally large number of cases of leukemia among the survivors of Hiroshima and Nagasaki.

Radiation can also have a genetic effect. That is, it can produce mutations in plants and animals by bringing about changes in chromosomes. There is every reason to suppose that similar effects can arise in human beings. Surveys of the children of radiologists show an increased frequency of congenital defects. This is confirmed by studies of children born to the survivors of Nagasaki and Hiroshima. A disturbing aspect of this problem is that there seems to be no lower limit, or "threshold," below which the genetic effects of radiation become negligible. Even a small increase in background radiation can be expected to produce a proportional increase in undesirable mutations.

It wasn't until after WW II that the dangers from radiation were recognized

Table 24.3 lists average exposures to radiation of people living in the United States. Several features of the data in this table are worthy of comment.

1. About two thirds of radiation exposure comes from natural sources. Here, the level depends upon the area in which a person lives. Cosmic radiation is much more intense at high elevations. A person living in Denver is exposed to about 100 millirems/year from this source. This is twice the national average.

2. The largest exposure to manmade sources of radiation is due to x-rays. The numbers quoted in Table 24.3 assume only occasional x-ray examinations. A single dental x-ray corresponds to 20 millirems, and a chest x-ray to 50 to 200 millirems. These amounts can be much higher if the operator of the x-ray machine is inexperienced or if the machine itself is defective. Clearly, excessive diagnostic use of x-rays should be avoided.

3. Exposure to radioactive fallout amounted to about 30 millirems/year prior to the nuclear test ban treaty in 1963. It has decreased steadily since then. Much of the present radiation from this source is coming from countries (France, China) which have continued atmospheric testing of nuclear weapons.

4. On the average, exposure to radiation from nuclear power plants is very low, about 0.2 millirem/year. This number is somewhat higher in the vicinity of the plant. Still, a person sitting on a fence at the boundary of a plant all year long would receive

TABLE 24.3 TYPICAL RADIATION EXPOSURES IN THE UNITED STATES (1 MILLIREM = 10^{-3} REM)	
NATURAL SOURCES	**MILLIREMS PER YEAR**
A. External to the body	
1. From cosmic radiation	50
2. From the earth	47
3. From building materials	3
B. Inside the body	
1. Inhalation of air	5
2. In human tissues (mostly $^{40}_{19}K$)	21
Total from natural sources	**126**
MANMADE SOURCES	
A. Medical procedures	
1. Diagnostic x-rays	50
2. Radiotherapy x-rays, radioisotopes	10
3. Internal diagnosis, therapy	1
B. Nuclear power industry	0.2
C. Luminous watch dials, TV tubes, industrial wastes	2
D. Radioactive fallout	4
Total from manmade sources	**67**
Total	**193**

only 5 millirems of radiation. The maximum possible exposure of a person standing at the gate of the Three Mile Island plant for the first 2 weeks of the March 1979 accident would have been 80 millirems. This is less than the radiation from a single chest x-ray.

Uses of Radioactive Isotopes

A large number of radioactive isotopes have been used in many areas of basic and applied research. A few of these are discussed below.

MEDICINE. The high-energy radiation given off by radium was used for many years in the treatment of cancer. Nowadays, cobalt-60, which is cheaper than radium and gives off even more powerful radiation, is used for this purpose. Certain types of cancer can be treated internally with radioactive isotopes. If a patient suffering from cancer of the thyroid drinks a solution of NaI containing radioactive iodide ions (^{131}I or ^{123}I), the iodine moves preferentially to the thyroid gland. There, the radiation destroys malignant cells without affecting the rest of the body.

The cobalt-60 is made in a nuclear reactor. The dosage used is large

Trace amounts of radioactive samples injected into the blood can be used to detect circulatory disorders. For example, a sodium chloride solution containing a small amount of radioactive sodium may be injected into the leg of a patient. By measuring the build-up of radiation in the foot, a physician can quickly find out whether the circulation in that area is abnormal.

INDUSTRY. The frictional wear of piston rings can be monitored by making a test ring slightly radioactive. The activity of the iron dust in the lubricating oil that circulates around the piston can then be measured. The rate of corrosion of steel can be measured by a similar technique.

The thickness of sheets of metal, paper, or plastic can be checked by putting them between a radioactive source and a Geiger counter. By measuring the fraction of radiation passing through the sheet, its thickness can be estimated quite accurately. Thin spots where the sheet might break down in use can also be detected.

CHEMISTRY. An early application of radioactive isotopes in chemistry was to study the nature of a dynamic equilibrium. In our discussion of chemical equilibria (Chap. 13), we defined the equilibrium state as one in which forward and reverse reactions are proceeding at the same rate. Consider, for example, a saturated solution of lead chloride:

$$PbCl_2(s) \rightleftharpoons Pb^{2+}(aq) + 2\ Cl^-(aq)$$

The fact that the concentration of Pb^{2+} does not change with time is explained by assuming that the two opposing processes, solution and precipitation, are occurring at the same rate. A radioactive isotope of lead, Pb-212, can be used to check this model. A small amount of this isotope, in the form of $Pb(NO_3)_2$, is injected into a solution saturated with $PbCl_2$. Within a very short time, some of the radioactive lead appears in the lead chloride. This indicates that exchange between the solid and the solution is indeed occurring.

Exchange of this sort occurs very rapidly

Radioactive isotopes are often used to trace the path of an element as it passes through various steps from reactant to final product. Organic chemists have learned a great deal about the mechanism of complex reactions by using carbon-14 as a tracer. One such reaction is the natural process of photosynthesis. The overall reaction can be represented as

$$6\ CO_2(g) + 6\ H_2O(l) \rightarrow C_6H_{12}O_6(s) + 6\ O_2(g) \qquad (24.9)$$

This reaction proceeds through a series of steps in which successively more complex organic molecules are formed. To study the path of reaction, plants are exposed to CO_2 containing carbon-14. At various time intervals, the plants are analyzed to determine which organic compounds contain carbon-14 and hence are early products of photosynthesis. Research along these lines by Melvin Calvin at the University of California at Berkeley led to a Nobel Prize in chemistry in 1961.

24.2 RATE OF RADIOACTIVE DECAY

The rate at which a radioactive sample decays can be measured by counting the number of particles given off in unit time. Instruments for measuring radioactivity, such as the Geiger counter, do this automatically. In using such instruments, it is necessary to correct for the "background" radiation given off by natural sources.

First Order Rate Law

Radioactive decay is a first order rate process. You will recall that for such reactions the rate equation (Chap. 14) is

$$\log_{10} \frac{X_0}{X} = \frac{kt}{2.30} \qquad (24.10)$$

Here, X_0 is the amount of radioactive substance at zero time (i.e., when the counting process starts) and X is the amount remaining after time t. The first order rate constant, k, is characteristic of the isotope undergoing radioactive decay.

Decay rates of radioactive isotopes are most often expressed in terms of their half-lives, $t_{1/2}$, rather than the first order rate constant, k. As noted in Chapter 14, these two quantities are related by the equation

$$k = \frac{0.693}{t_{1/2}} \qquad (24.11)$$

The application of these equations to radioactive decay processes is illustrated in Example 24.3.

Example 24.3 The common radioactive isotope of radium, $^{226}_{88}Ra$, has a half-life of 1620 yr. Calculate
a. the first order rate constant for the decay of radium-226.
b. the fraction of a sample of this isotope which will remain after 100 years.

Solution

a. $k = \dfrac{0.693}{1620 \text{ yr}} = 4.28 \times 10^{-4}/\text{yr}$

b. $\log \dfrac{X_0}{X} = \dfrac{(4.28 \times 10^{-4}/\text{yr})}{2.30} 100 \text{ yr} = 0.0186$

Taking antilogs: $X_0/X = 1.044$

The fraction *remaining* is

$$\frac{X}{X_0} = \frac{1}{1.044} = 0.958 \text{ or } 95.8\%$$

This problem illustrates a safety hazard inherent in any attempt to "dispose" of a sample of a relatively long-lived isotope. A sample of a radium salt cannot simply be discharged into the environment in the vain hope that it will decompose rapidly; the level of radiation would be virtually unchanged a century from now. This factor has to be taken into account in any proposed system of radioactive waste disposal.

Exercise How long does it take for one third of a Ra-226 sample to decay? Answer: 946 yr.

Half-lives can be interpreted in terms of the level of radiation of the corresponding isotopes. Since uranium-238 has a very long half-life (4.5×10^9 years), it gives off radiation very slowly. At the opposite extreme is polonium-214, which decays with a half-life of 1.6×10^{-4} s. Within a second, virtually all the radiation from this isotope is gone. Species such as this produce a very high level of radiation during their brief existence.

Isotopes with short half-lives tend to be "hot," with very dangerous radiation

Age of Rocks

Certain radioactive isotopes act as "natural clocks." That is, they help us to determine the time at which rock deposits solidified. The time elapsed since then is referred to as the "age" of the rock. To see how this information is obtained consider a uranium-bearing rock, formed billions of years ago by solidification from a molten mass. Once the rock became solid, the products of radioactive decay of uranium could no longer diffuse away. Hence, they were incorporated into the rock. Over time these products, all of which have short half-lives, were converted to lead-206. The overall equation for the decay process can be written:

$$^{238}_{92}U \rightarrow \, ^{206}_{82}Pb + 8 \quad ^{4}_{2}He + 6 \, ^{0}_{-1}e; \quad t_{1/2} = 4.5 \times 10^9 \text{ yr} \qquad (24.12)$$

Knowing the half-life for this process it should then be possible, by measuring the ratio of lead-206 to uranium-238 in the rock today, to calculate the time that has elapsed since the rock solidified. If we should find, for example, that equal numbers of atoms of these two isotopes were present, we would infer that the rock must be about 4.5×10^9 (4.5 billion) years old.

That's about the age of the earth

This method of estimating the age of mineral deposits assumes, among other things, that none of the lead-206 has become separated from the parent uranium-238. It is possible to check the validity of this assumption by referring to other "radioactive clocks" which operate in nature. One of these is the β-decay of rubidium-87, which has a half-life of 5.7×10^{10} yr.

$$^{87}_{37}Rb \rightarrow \, ^{87}_{38}Sr + \, ^{0}_{-1}e; \quad t_{1/2} = 5.7 \times 10^{10} \text{ yr} \qquad (24.13)$$

Ages of rocks determined by these methods range from (3 to 4.5) $\times 10^9$ years. The larger number is often taken as an approximate value for the age of the earth. Analyses of rock samples from the moon indicate ages in this same range. This argues against the once prevalent idea that the moon was torn from the earth's surface by a violent event a long time after the earth solidified.

Age of Organic Material

During the 1950's Professor W. F. Libby of the University of Chicago and others worked out a method for determining the age of organic material. It is based upon the decay rate of carbon-14. The method can be applied to objects from a few hundred up to 50,000 years old. It has been used to determine the authenticity of canvases of Renaissance painters and to check the ages of relics left by prehistoric cavemen.

Carbon-14 is produced in the atmosphere by the interaction of neutrons from cosmic radiation with ordinary nitrogen atoms:

$$^{14}_{7}N + \, ^{1}_{0}n \rightarrow \, ^{14}_{6}C + \, ^{1}_{1}H \qquad (24.14)$$

The carbon-14 formed by this nuclear reaction is eventually incorporated into the carbon dioxide of the air. A steady-state concentration, amounting to about one atom of carbon-14 for every 10^{12} atoms of carbon-12, is established in atmospheric CO_2. A living plant, taking in carbon dioxide, has this same $^{14}C/^{12}C$ ratio, as do plant-eating animals or human beings.

When a plant or animal dies, the intake of radioactive carbon stops. Consequently the radioactive decay of carbon-14

$$^{14}_{6}C \rightarrow \, ^{14}_{7}N + \, ^{0}_{-1}e \quad \text{(half-life} = 5720 \text{ yr)}$$

takes over and the ratio of $^{14}C/^{12}C$ drops. By measuring this ratio and comparing it to that in living plants, one can estimate the time at which the plant or animal died (Example 24.4).

There is a possible flaw in this approach. The $^{14}C/^{12}C$ ratio in the past may not have been what it is now

Example 24.4   A tiny piece of paper taken from the Dead Sea scrolls, believed to date back to the first century A.D., was found to have a $^{14}C/^{12}C$ ratio 0.795 times that in a plant living today. Estimate the age of the scrolls.

Solution   Knowing the half-life of carbon-14 ($t_{1/2} = 5720$ yr), we can calculate the first order rate constant from Equation 24.11. Then, using Equation 24.10, we can obtain the elapsed time.

$$k = \frac{0.693}{5720 \text{ yr}} = 1.21 \times 10^{-4}/\text{yr}$$

$$\log \frac{X_0}{X} = \frac{(1.21 \times 10^{-4}/\text{yr}) \times t}{2.30}$$

But, $X = 0.795\ X_0$, so $\log \dfrac{X_0}{X} = \log \dfrac{1.000}{0.795} = \log 1.26 = 0.100$

Hence, $0.100 = \dfrac{(1.21 \times 10^{-4}/\text{yr}) \times t}{2.30};$ $t = 1900$ yr

Exercise   What is the age of a piece of charcoal in which the $^{14}C/^{12}C$ ratio is 0.400 times that in a living plant? Answer: 7560 yr.

24.3 MASS-ENERGY RELATIONS

We pointed out at the beginning of this chapter that the energy change in nuclear reactions is much greater than that for ordinary chemical reactions. The energy change can be calculated from Einstein's equation:

$$\Delta E = \Delta mc^2 \qquad (24.15)$$

Here Δm is the change in mass,* ΔE is the change in energy, and c is the speed of light. If we substitute for c the value 3.00×10^8 m/s, Equation 24.15 gives the relation between the energy change in *joules* and the mass change in *kilograms* directly:

$$\Delta E \text{ (in joules)} = 9.00 \times 10^{16} \times \Delta m \text{ (in kilograms)}$$

In dealing with nuclear reactions, we usually want ΔE in *kilojoules* corresponding to a mass change in *grams*. Using the relations

$$1 \text{ kJ} = 10^3 \text{ J}; \ 1 \text{ kg} = 10^3 \text{ g}$$

we obtain

$$\Delta E \text{ (in kJ)} = 9.00 \times 10^{10} \Delta m \text{ (in grams)} \qquad (24.16)$$

---

*Specifically, Δm = mass of products − mass of reactants; ΔE = energy of products − energy of reactants. In spontaneous nuclear reactions, the products weigh less than the reactants (Δm negative). In this case, the energy of the products is less than that of the reactants (ΔE negative), and energy is evolved to the surroundings.

TABLE 24.4 NUCLEAR MASSES ON THE C-12 SCALE*

	At. No.	Mass No.	Mass		At. No.	Mass No.	Mass
n	0	1	1.00867	Br	35	79	78.8992
H	1	1	1.00728		35	81	80.8971
	1	2	2.01355		35	87	86.9028
	1	3	3.01550	Rb	37	89	88.8909
He	2	3	3.01493	Sr	38	90	89.8864
	2	4	4.00150	Mo	42	99	98.8849
Li	3	6	6.01348	Ru	44	106	105.8829
	3	7	7.01436	Ag	47	109	108.8789
Be	4	9	9.00999	Cd	48	109	108.8786
	4	10	10.01134		48	115	114.8793
B	5	10	10.01019	Sn	50	120	119.8747
	5	11	11.00656	Ce	58	144	143.8816
C	6	11	11.00814		58	146	145.8865
	6	12	11.99671	Pr	59	144	143.8807
	6	13	13.00006	Sm	62	152	151.8853
	6	14	13.99995	Eu	63	157	156.8914
O	8	16	15.99052	Er	68	168	167.8941
	8	17	16.99474	Hf	72	179	178.9048
	8	18	17.99477	W	74	186	185.9107
F	9	18	17.99601	Os	76	192	191.9187
	9	19	18.99346	Au	79	196	195.9231
Na	11	23	22.98373	Hg	80	196	195.9219
Mg	12	24	23.97845	Pb	82	206	205.9295
	12	25	24.97925		82	207	206.9309
	12	26	25.97600		82	208	207.9316
Al	13	26	25.97977	Po	84	210	209.9368
	13	27	26.97439		84	218	217.9628
	13	28	27.97477	Rn	86	222	221.9703
Si	14	28	27.96924	Ra	88	226	225.9771
S	16	32	31.96329	Th	90	230	229.9837
Cl	17	35	34.95952	Pa	91	234	233.9934
	17	37	36.95657	U	92	233	232.9890
Ar	18	40	39.95250		92	235	234.9934
K	19	39	38.95328		92	238	238.0003
	19	40	39.95358		92	239	239.0038
Ca	20	40	39.95162	Np	93	239	239.0019
Ti	22	48	47.93588	Pu	94	239	239.0006
Cr	24	52	51.92734		94	241	241.0051
Fe	26	56	55.92066	Am	95	241	241.0045
Co	27	59	58.91837	Cm	96	242	242.0061
Ni	28	59	58.91897	Bk	97	245	245.0129
Zn	30	64	63.91268	Cf	98	248	248.0186
	30	72	71.91128	Es	99	251	251.0255
Ge	32	76	75.90380	Fm	100	252	252.0278
As	33	79	78.90288		100	254	254.0331

*Note that these are *nuclear masses*. The masses of the corresponding atoms can be calculated by adding the masses of each extranuclear electron (0.000549). For example, for an *atom* of $_2^4$He we have

$$4.00150 + 2(0.000549) = 4.00260$$

Similarly, for an atom of $_6^{12}$C:

$$11.99671 + 6(0.000549) = 12.00000$$

Using Equation 24.16 along with the appropriate nuclear masses (Table 24.4), we can calculate the energy change accompanying a nuclear reaction (Example 24.5).

Example 24.5 For the radioactive decay of radium, $^{226}_{88}Ra \rightarrow ^{222}_{86}Rn + ^{4}_{2}He$, calculate ΔE in kJ when
  a. one mole of radium decays.
  b. one gram of radium decays.

Solution
  a. We first calculate Δm for the reaction and then obtain ΔE from Equation 24.16. For the decay of one mole of Ra, using Table 24.4

There is always a mass decrease in radioactive decay

Δm = mass of 1 mol $^{4}_{2}He$ + mass of 1 mol $^{222}_{86}Rn$ − mass of 1 mol $^{226}_{88}Ra$
= 4.0015 g + 221.9703 g − 225.9771 g
= −0.0053 g

(Note that since Δm is extremely small, it is necessary to know the masses of products and reactants very accurately to obtain the mass difference to two significant figures.)

$$\Delta E \text{ (in kJ)} = 9.00 \times 10^{10} \times (-0.0053) = -4.8 \times 10^{8} \text{ kJ}$$

  b. Since one mole of radium weighs 226 g, we have

And $\Delta E < 0$, so energy is always evolved

$$\Delta E = 1 \text{ g Ra} \times \frac{(-4.8 \times 10^{8} \text{ kJ})}{226 \text{ g Ra}} = -2.1 \times 10^{6} \text{ kJ}$$

Exercise Calculate ΔE for the decay of one mole of Pu-239: $^{239}_{94}Pu \rightarrow ^{4}_{2}He + ^{235}_{92}U$. Answer: -5.1×10^{8} kJ.

Energy changes in ordinary chemical reactions are of the order of 50 kJ/g or less. For example, in the combustion of petroleum, about 46 kJ of heat are evolved per gram of fuel burned. Looking at Example 24.5b, we see that ΔE for the radioactive decay of radium is about 50,000 times as great.

Using Table 24.4 it is possible to calculate, for various nuclei, what might be called an average mass per nuclear particle. To do this, we divide the mass of the nucleus by the mass number:

$$\text{average mass per nuclear particle} = \frac{\text{mass of nucleus on C-12 scale}}{\text{mass number}} \tag{24.17}$$

Thus, for a deuteron, $^{2}_{1}H$, we have

$$\text{average mass per nuclear particle} = \frac{2.01355}{2} = 1.00678$$

Figure 24.3 shows a plot of this quantity vs. mass number. Notice that the curve has a broad minimum in the vicinity of mass numbers 50 to 100. Consider, now, what would happen if a heavy nucleus such as $^{235}_{92}U$ were to split into smaller nuclei near this minimum. This process, referred to as *nuclear fission*, should result in a decrease in mass and hence an evolution of energy. The same effect would be obtained if very light nuclei such as $^{2}_{1}H$ were to combine with one another. Indeed, this process, called *nuclear fusion*, should evolve even more energy, since the average mass per nuclear particle drops off very sharply at the beginning of the curve.

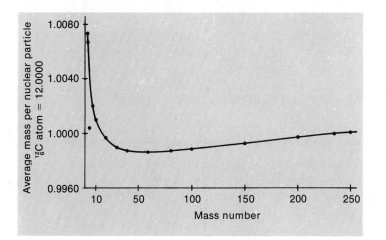

Figure 24.3 The average mass per nuclear particle goes through a minimum near mass number 50. This means that when very light nuclei combine (fusion) or very heavy nuclei split apart (fission) there is a decrease in mass and hence an evolution of energy.

24.4 NUCLEAR FISSION

Shortly before World War II several groups of scientists were studying the products obtained by bombarding uranium with neutrons. They were looking for new elements with atomic numbers greater than 92. In 1938 Hahn and Strassman in Germany isolated a compound of a Group 2 element, which they originally believed to be radium (at. no. 88). Later, they showed that this element was barium (at. no. 56), indicating that a uranium atom had been split into fragments. Hahn's first reaction to this discovery was one of disbelief. He later stated, in January of 1939, "As chemists, we should replace the symbol Ra . . . by Ba . . . [but], as nuclear chemists, closely associated with physics, we cannot decide to take this step in contradiction to all previous experience in nuclear physics."

If Hahn was reluctant to admit the possibility of an entirely new type of nuclear reaction, a former colleague of his, Lisa Meitner, was not. In a letter published with O. R. Frisch in January, 1939, she stated: "At first sight, this result seems very hard to understand. . . . On the basis, however, of present ideas about the behavior of heavy nuclei, an entirely different picture of these new disintegration processes suggests itself. . . . It seems possible that the uranium nucleus . . . may, after neutron capture, divide itself into nuclei of roughly equal size." This revolutionary suggestion was confirmed by experiments carried out all over the world.

With the outbreak of World War II, interest in nuclear fission focused on the enormous amount of energy released in the process. At Los Alamos, in the mountains of New Mexico, a group of scientists under J. Robert Oppenheimer worked feverishly to produce the fission, or "atomic," bomb. Many of the members of this group were exiles from Nazi Germany. They were spurred on by the fear that Hitler would obtain the bomb first. Their work led to the explosion of the first atomic bomb in the New Mexico desert at 5:30 A.M. on July 16, 1945. Less than a month later (August 6, 1945) the world learned of this new weapon when another bomb was exploded over Hiroshima. This bomb killed 70,000 people and completely devastated an area of ten square kilometers. Three days later Nagasaki and its inhabitants met a similar fate. On August 14th, Japan surrendered and World War II was over.

Right or wrong, we have to live with these events

The Fission Process (U-235)

Several isotopes of the heavy elements undergo fission if bombarded by neutrons of high enough energy. In practice, attention has centered upon two particular isotopes, $^{235}_{92}U$ and $^{239}_{94}Pu$. Both of these can be split into fragments by low-energy neutrons. The
Our discussion will concentrate upon the uranium-235 isotope. It makes up only

about 0.7% of naturally occurring uranium. The more abundant isotope, uranium-238, does not undergo the fission reaction. During World War II, several different processes were studied for the separation of these two isotopes. The most successful technique was that of gaseous effusion (Chap. 7), using the volatile compound UF_6 (bp = 56°C).

FISSION PRODUCTS. When a uranium-235 atom undergoes fission, it splits into two unequal fragments and a number of neutrons and beta particles. The fission process is complicated by the fact that different uranium-235 atoms split up in many different ways. For example, while one atom of $^{235}_{92}U$ is splitting to give isotopes of rubidium (at. no. 37) and cesium (at. no. 55), another may break up to give isotopes of bromine (at. no. 35) and lanthanum (at. no. 57), while still another atom yields isotopes of zinc (at. no. 30) and samarium (at. no. 62).

$$^{90}_{37}Rb + ^{144}_{55}Cs + 2\,^{1}_{0}n \tag{24.18}$$

$$^{1}_{0}n + ^{235}_{92}U \nearrow \rightarrow ^{87}_{35}Br + ^{146}_{57}La + 3\,^{1}_{0}n \tag{24.19}$$

$$\searrow ^{72}_{30}Zn + ^{160}_{62}Sm + 4\,^{1}_{0}n \tag{24.20}$$

Fission produces lots
of "hot" species

More than 200 isotopes of 35 different elements have been identified among the fission products of uranium-235.

The stable neutron-to-proton ratio near the middle of the Periodic Table, where the fission products are located, is considerably smaller (~1.2) than that of uranium-235 (1.55). Hence, the immediate products of the fission process contain too many neutrons for stability. In the case of rubidium-90, three steps are required to reach a stable nucleus:

$$^{90}_{37}Rb \rightarrow ^{90}_{38}Sr + ^{0}_{-1}e; \; t_{1/2} = 2.8 \text{ min}$$

$$^{90}_{38}Sr \rightarrow ^{90}_{39}Y + ^{0}_{-1}e; \; t_{1/2} = 29 \text{ yr}$$

$$^{90}_{39}Y \rightarrow ^{90}_{40}Zr + ^{0}_{-1}e; \; t_{1/2} = 64 \text{ h}$$

The radiation hazard associated with nuclear fallout arises from the formation of radioactive isotopes such as these. One of the most dangerous is strontium-90. In the form of strontium carbonate, $SrCO_3$, it is incorporated into the bones of animals and human beings.

You will notice from Equations 24.18 to 24.20 that two to four neutrons are produced by fission for every one consumed. Once a few atoms of uranium-235 split, the neutrons produced can bring about the fission of many more uranium-235 atoms. This creates the possibility of a chain reaction. This is precisely what happens in the atomic bomb. The energy evolved in successive fissions escalates to give, within a few seconds, a tremendous explosion.

For nuclear fission to result in a chain reaction, the sample must be large enough so that most of the neutrons are captured internally. If the sample is too small, most of the neutrons escape, breaking the chain. The *critical mass* of uranium-235 required to maintain a chain reaction in a bomb appears to be about 40 kg. In the Hiroshima bomb, the critical mass was achieved by using a conventional explosive to fire one piece of uranium-235 into another.

FISSION ENERGY. The evolution of energy in nuclear fission is directly related to the decrease in mass that takes place. About 80,000,000 kJ of energy is given off for every gram of U-235 that reacts. This is about 40 times as great as the energy change for simple nuclear reactions such as radioactive decay. The heat of combustion of coal is only about

30 kJ/g; the energy given off when TNT explodes is still smaller, about 2.8 kJ/g. Putting it another way, the fission of 1 g of U-235 produces as much energy as the combustion of 2700 kg of coal or the explosion of 30 metric tons (3×10^4 kg) of TNT.

The first atomic bomb was equivalent to about 18,000 tons of TNT

Nuclear Reactors

Even before the first atomic bomb exploded, scientists and political leaders began to speculate on the use of fission as a peacetime energy source. Nuclear reactors which convert the heat produced by the fission of uranium-235 into electrical energy are now a reality. About 70 such reactors supply 10% of the electrical energy used in the United States. Without them, meeting energy needs in the 1980's would be extremely difficult.

The type of reactor which is most common in the United States today is shown in Figure 24.4. Water at a pressure of 140 atm is passed through the reactor to absorb the heat given off by fission. The water, coming out of the reactor core at 320°C, circulates through a closed loop containing a heat exchanger. A second stream of water at a lower pressure passes through the heat exchanger and is converted to steam at 270°C. This steam is used to drive a turbogenerator which produces electrical energy.

The most serious problem posed by nuclear reactors involves the radioactive products of fission. The amounts of such lethal isotopes as Sr-90 and Cs-137 that accumulate in a few months are equal to those produced by an atomic bomb of the size dropped on Hiroshima. If the reactor were broken open by a fire, explosion, or earthquake, the fallout could raise the radiation level in the vicinity to deadly concentrations. Despite all the precautions taken to prevent such an accident, there can be no guarantee that it will not happen. This was shown dramatically by the "near-miss" at the Three Mile Island reactor in Pennsylvania in March of 1979. A set of valves in the steam generating system accidentally closed. Within seconds, the temperature and pressure within the reactor

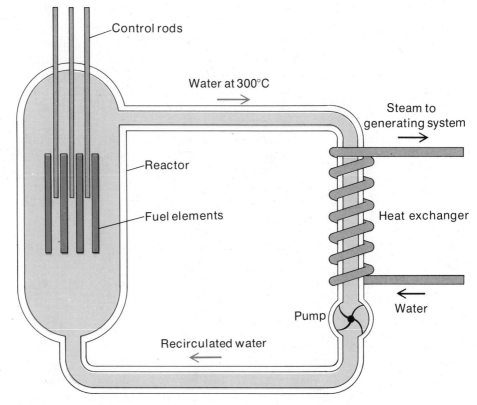

Figure 24.4 Nuclear reactor of the pressurized water type. The control rods are made of a material such as cadmium, which absorbs neutrons very effectively. The rate of fission is carefully monitored and controlled.

core increased to the danger point. The problem was compounded when the operators cut off the supply of emergency cooling water. For a considerable time, the fuel elements were uncovered and there was a very real danger that they might "melt down." Had that happened, the release of radiation from the plant could have been many times greater than the small amount that actually escaped.

With the large number of nuclear reactors now in use, it is imperative that we develop methods for safe, long-term storage of the radioactive wastes that they produce. So far these wastes have been stored, as concentrated aqueous solutions, in huge underground storage tanks. These tanks are heavily shielded and cooled to absorb the heat given off by radioactive decay. In some cases, this heat would be great enough to keep the solutions at the boiling point. At present, about $5 \times 10^8 \ \ell$ of liquid waste is in storage in the United States. The plan is to evaporate this material to obtain solids, which can be stored more safely. One proposal is to put the solids in abandoned salt mines deep beneath the surface of the earth.

Breeder Reactors

From a long-range standpoint, nuclear reactors using uranium-235 will not be able to satisfy our future power needs. Before the end of this century, our supplies of this relatively rare isotope will be depleted and we will have to turn to other sources of energy. One possibility is to convert the more abundant isotope, uranium-238, into plutonium-239

$$^{238}_{92}\text{U} + {}^{1}_{0}\text{n} \rightarrow {}^{239}_{94}\text{Pu} + 2 \ {}^{0}_{-1}\text{e} \tag{24.21}$$

which can then undergo fission by reactions such as

$$^{239}_{94}\text{Pu} + {}^{1}_{0}\text{n} \rightarrow {}^{90}_{38}\text{Sr} + {}^{147}_{56}\text{Ba} + 3 \ {}^{1}_{0}\text{n} \tag{24.22}$$

Notice that this two-step reaction sequence produces more neutrons than it consumes. Consequently, it leads to a chain reaction similar to the fission of uranium-235, producing comparable amounts of energy. By using this process, we could make use of nearly all the uranium found in nature rather than a tiny fraction of that element. So-called **breeder reactors** based on Reactions 24.21 and 24.22 could satisfy our energy needs for a century or more.

Energy-wise, breeder reactors are very attractive

The scientific feasibility of the U-238/Pu-239 cycle was shown many years ago. The bomb exploded over Nagasaki contained plutonium-239 made from uranium-238. The first nuclear reactor to produce electrical energy in 1951 was of the breeder type. However, 30 years later, we do not yet have a breeder reactor in commercial operation. A host of technical problems and safety hazards have held back the development of such reactors.

A major concern with breeder reactors is the behavior of plutonium-239. For one thing, it is highly radioactive. As little as ten micrograms (1×10^{-5} g) can cause cancer if it enters the body. Fears have also been expressed that plutonium from breeder reactors might be diverted to produce atomic bombs. Only about 5 kg of plutonium-239 would be required for such a bomb.

24.5 NUCLEAR FUSION

Recall (Fig. 24.3) that very light isotopes such as those of hydrogen are unstable with respect to fusion into heavier isotopes. Indeed, the energy available from nuclear fusion is considerably greater than that given off in the fission of an equal mass of a heavy element (Example 24.6).

Example 24.6 Calculate the amount of energy evolved, in kilojoules per gram of reactants, in
  a. a fusion reaction, $^2_1H + ^2_1H \rightarrow ^4_2He$.
  b. a fission reaction, $^{235}_{92}U \rightarrow ^{90}_{38}Sr + ^{144}_{58}Ce + ^1_0n + 4\,^{0}_{-1}e$.

Solution
  a. We first calculate the change in mass per mole of product, using Table 24.4.

$$\Delta m = 4.00150 \text{ g} - 2(2.01355 \text{ g}) = -0.02560 \text{ g}$$

Converting to kilojoules:

$$\Delta E = -2.56 \times 10^{-2} \text{ g} \times 9.00 \times 10^{10} \text{ kJ/g} = -2.30 \times 10^9 \text{ kJ}$$

Since 4.03 g of deuterium are involved, ΔE per gram of reactant is

$$\Delta E = \frac{-2.30 \times 10^9 \text{ kJ}}{4.03} = -5.71 \times 10^8 \text{ kJ}$$

  b. Proceeding as in (a), we find that, per mole of uranium reacting,

$$\Delta m = 89.8864 \text{ g} + 143.8816 \text{ g} + 1.0087 \text{ g} + 4(0.00055 \text{ g}) - 234.9934 \text{ g}$$
$$= -0.2145 \text{ g}$$

Hence, for one mole of U-235:

$$\Delta E = -0.2145 \text{ g} \times 9.00 \times 10^{10} \text{ kJ/g} = -1.93 \times 10^{10} \text{ kJ}$$

For one gram of U-235

$$\Delta E = \frac{-1.93 \times 10^{10} \text{ kJ}}{235} = -8.21 \times 10^7 \text{ kJ}$$

Comparing the answers to (a) and (b), we conclude that the fusion reaction produces about seven times as much energy per gram of starting material (57.1×10^7 vs. 8.21×10^7 kJ) as does the fission reaction. This factor varies from about three to ten, depending upon the particular reactions chosen to represent the fusion and fission processes.

Exercise Calculate ΔE per gram of reactant for $^2_1H + ^3_1H \rightarrow ^4_2He + ^1_0n$. Answer: -3.38×10^8 kJ.

As an energy source, nuclear fusion possesses several additional advantages over nuclear fission. For one thing, fusion is a "clean" process in the sense that the products are stable isotopes such as ^4_2He, rather than the hazardous radioactive isotopes formed by fission. Equally important, light isotopes suitable for fusion are far more abundant than the heavy isotopes required for fission. We can calculate, for example (Problem 24.45), that the fusion of only 2×10^{-13} per cent of the deuterium (^2_1H) in seawater would meet the total annual energy requirements of the world.

Unfortunately fusion processes, unlike neutron-induced fission, have very high activation energies. In order to overcome the electrostatic repulsion between two deuterium nuclei and cause them to react, they have to be accelerated to velocities of about 10^6 m/s, about 10,000 times greater than ordinary molecular velocities at room temperature. The corresponding temperature for fusion, as calculated from kinetic theory (Problem 24.44), is of the order of $10^9°C$. In the hydrogen bomb, temperatures of this magnitude were achieved by using a fission reaction to trigger nuclear fusion. If fusion

Unfortunately, we can't do fusion reactions in a nuclear reactor

reactions are to be used to generate electricity, it will be necessary to develop equipment in which very high temperatures can be maintained long enough to allow fusion to occur and give off energy. In any conventional container, the reactant nuclei would quickly lose their high kinetic energies by collisions with the walls.

One fusion reaction currently under study is a two-step process involving deuterium and lithium as the basic starting materials:

$$\frac{\begin{array}{l} {}^2_1H + {}^3_1H \rightarrow {}^4_2He + {}^1_0n \\[4pt] {}^6_3Li + {}^1_0n \rightarrow {}^4_2He + {}^3_1H \end{array}}{{}^2_1H + {}^6_3Li \rightarrow 2\ {}^4_2He} \tag{24.23}$$

This process is attractive because it has a lower activation energy than other fusion reactions.

Within the past few years, promising results have been obtained with Reaction 24.23 using "magnetic bottles" (Fig. 24.5) to confine the reactant nuclei. So far, prototype models using this principle have been able to sustain the reaction for a fraction of a second. To achieve a net evolution of energy, this time must be extended to at least one second. Scientists working in this area are confident this can be achieved by the mid-1980's. At least another 25 years will be required to develop commercial fusion reactors capable of making a significant contribution to our energy needs.

Another approach to nuclear fusion is shown in Figure 24.6. Tiny glass pellets (about 0.1 mm in diameter) filled with frozen deuterium and tritium are illuminated by a powerful laser beam. In principle at least, the neutrons produced should be able to react with lithium to complete Reaction 24.23 and give off energy. Unfortunately, the economic feasibility of this approach is extremely dubious. The glass pellets are expensive to produce and the laser beam has a very short life expectancy.

Perhaps the ultimate irony of our time is the fact that we have made so little use of the energy produced in a nuclear fusion process that has been going on since the universe was formed. The energy given off by the sun and other stars results from fusion reactions in which ordinary hydrogen is converted to helium. One mechanism which has been suggested for this process is

$$\frac{\begin{array}{l} {}^1_1H + {}^1_1H \rightarrow {}^2_1H + {}^0_1e \\[6pt] {}^2_1H + {}^1_1H \rightarrow {}^3_2He; \\[6pt] {}^3_2He + {}^1_1H \rightarrow {}^4_2He + {}^0_1e; \end{array}}{4\ {}^1_1H \rightarrow {}^4_2He + 2\ {}^0_1e;\quad \Delta E = -6.0 \times 10^8\ \text{kJ/g reactant}} \tag{24.24}$$

Figure 24.5 One way to carry out a controlled fusion reaction is to confine very light nuclei in a strong magnetic field. The problem is to extend the time during which the reaction occurs to one second or more.

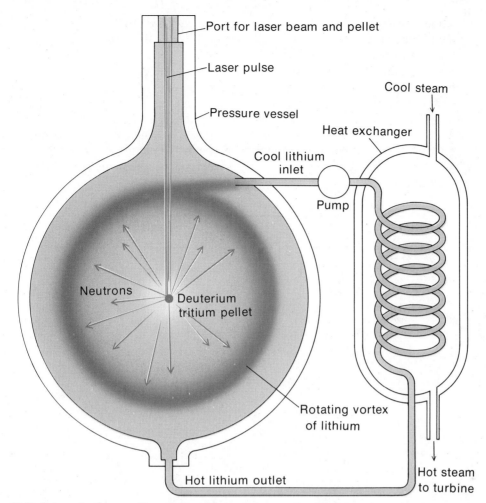

Figure 24.6 Schematic diagram of laser-induced fusion apparatus. A pellet of deuterium-tritium undergoes fusion on absorbing energy from a laser pulse. Neutrons from the fusion reaction react with molten lithium, liberating energy and regenerating tritium (Reaction 24.23).

Each day, processes such as this, occurring within the sun at a temperature of perhaps 10^9°C, produce enormous quantities of energy, of which about 6×10^{18} kJ reaches the surface of the earth. This is roughly equivalent to the total amount of energy that mankind has consumed since the beginning of time.

And the sun has been doing this for a long time

SUMMARY

Unstable radioactive nuclei decay by emitting:

—alpha particles ($_2^4$He nuclei). Emission of an α particle decreases the atomic number by two units and the mass number by four units.

—beta particles (electrons). Emission of a β particle increases the atomic number by one unit, leaving the mass number unchanged.

—gamma radiation.

Nuclear decay can be represented by a nuclear equation (Example 24.1). Stable nuclei can be made radioactive by bombardment with neutrons or high-energy, positively charged particles. Bombardment reactions are used to make isotopes of elements beyond uranium (Table 24.1 and Example 24.2).

Radioactive decay follows a first order rate law (Example 24.3). Measurements of decay rate can be used to estimate the age of rocks or dead organic matter (Example 24.4).

Changes in mass and energy associated with a nuclear reaction can be calculated using the Einstein equation (Example 24.5). In nuclear fission, energy is released when a heavy nucleus splits into lighter ones. A critical mass of fissionable material is required for a self-sustaining chain reaction. Such reactions in nuclear reactors evolve heat to produce steam for generating electricity (Fig. 24.4). Nuclear fusion releases energy when light nuclei combine to form heavier nuclei (Example 24.6). Current research efforts to achieve controlled fusion use magnetic containment (Fig. 24.5) or laser bombardment (Fig. 24.6).

KEY WORDS AND CONCEPTS

radioactivity, natural	gamma (γ) radiation	rad	Einstein equation
radioactivity, induced	positron	gray	fission
alpha (α) particle	scintillation counter	rem	nuclear reactor
beta (β) particle	Geiger-Müller counter	half-life	fusion

QUESTIONS AND PROBLEMS

Catalog

Nuclear Equations: 24.2–24.5, 24.22–24.25
Rate of Radioactive Decay: 24.6–24.11, 24.26–24.31

Mass-Energy Relations: 24.12–24.16, 24.32–24.36
General: 24.1, 24.17–24.21, 24.37–24.45

24.1   Review and know the meaning of the key words and concepts in this chapter.

24.2   The last three steps in the decay series of U-235 to Pb-207 are partially given below. Fill in the blanks.

$$\underline{\hspace{2cm}} \rightarrow {}^{0}_{-1}e + {}^{211}_{83}Bi$$

$$\underline{\hspace{2cm}} + {}^{211}_{84}Po$$

$${}^{207}_{82}Pb + \underline{\hspace{2cm}} \longleftarrow$$

24.3   Write balanced nuclear equations for

a. alpha emission by ${}^{196}_{83}Bi$.
b. beta emission by ${}^{210}_{83}Bi$.
c. positron emission by ${}^{206}_{83}Bi$.

24.4   Balance the following nuclear equations by filling the blanks:

a. ${}^{32}_{16}S + \underline{\hspace{1.5cm}} \rightarrow {}^{32}_{15}P + {}^{1}_{1}H$
b. ${}^{14}_{7}N + \underline{\hspace{1.5cm}} \rightarrow {}^{17}_{8}O + {}^{1}_{1}H$
c. ${}^{9}_{4}Be + {}^{4}_{2}He \rightarrow {}^{1}_{0}n + \underline{\hspace{1.5cm}}$

24.5   A certain natural radioactive series starts with U-238 and ends with Pb-206. Each step in the series involves the loss of either an alpha or a beta particle. In the entire series, how many alpha particles are given off? how many beta particles?

24.22   Th-232 is the beginning species in a decay series which ends with Pb-208. The first three emissions are alpha, beta, and beta, respectively. Write balanced nuclear equations which represent these changes.

24.23   Write balanced nuclear equations for

a. alpha emission by ${}^{142}_{58}Ce$.
b. beta emission by ${}^{144}_{58}Ce$.
c. positron emission by ${}^{131}_{58}Ce$.

24.24   Complete the following nuclear equations:

a. ${}^{27}_{13}Al + \underline{\hspace{1.5cm}} \rightarrow {}^{24}_{11}Na + {}^{4}_{2}He$
b. ${}^{24}_{12}Mg + \underline{\hspace{1.5cm}} \rightarrow {}^{27}_{14}Si + {}^{1}_{0}n$
c. ${}^{10}_{5}B + {}^{1}_{1}H \rightarrow {}^{4}_{2}He + \underline{\hspace{1.5cm}}$

24.25   A certain radioactive species starts with ${}^{237}_{93}Np$ and ends with ${}^{209}_{83}Bi$. How many alpha particles are emitted? how many beta particles?

24.6 Phosphorus-32 is a beta emitter with a half-life of 14.3 d.

    a. What is the rate constant, k, for this reaction?
    b. How long will it be before only 5.0% of the original isotope remains?

24.7 After one year, 98.0% of a certain radioactive isotope is left.

    a. What is the rate constant for the decay of this isotope?
    b. What is the half-life?

24.8 A scintillation counter is used to monitor the radioactivity level of a certain isotope. During a 24-h period the count rate drops from 400 counts/min to 240 counts/min.

    a. Calculate the rate constant for the decay of this isotope.
    b. What is its half-life?

24.9 The radioactive isotope tritium, ^3_1H, is produced in nature in much the same way as C-14. Its half-life is 12.3 yr. Estimate the age of a sample of Scotch whiskey which has a tritium content 0.61 times that of the water in the area where the whiskey was produced.

24.10 One way of dating rocks is to determine the relative amounts of K-40 and Ar-40; the decay of K-40 has a half-life of 1.27×10^9 yr. Analysis of a certain lunar sample gives the following results in mole ratios:

    Ar-40/K-40 = 4.13 $(t_{1/2}$ K-40 = 1.27×10^9 yr)
    Pb-206/U-238 = 0.66 $(t_{1/2}$ U-238 = 4.5×10^9 yr)
    Sr-87/Rb-87 = 0.041 $(t_{1/2}$ Rb-87 = 5.7×10^{10} yr)

Using these data, obtain the best possible value for the age of the sample. Can you suggest why the K-Ar method gives a low result?

24.11 Why would the U-238 to Pb-206 method be inappropriate for determining the age of a sample thought to be about 500 yr old?

24.12 For the reaction

$$^{230}_{90}Th \rightarrow \, ^{226}_{88}Ra + \, ^4_2He$$

    a. calculate Δm in grams when one mole of Th-230 decays.
    b. calculate ΔE in joules when one mole of Th-230 decays; one gram of Th-230 decays.

24.13 Determine, by calculation, whether the following nuclear reaction is spontaneous:

$$^{52}_{24}Cr \rightarrow \, ^{48}_{22}Ti + \, ^4_2He$$

24.26 The radioactive isotope Na-24 has a rate constant, k, of 0.0462/h. What fraction of an Na-24 sample remains after two days (48 h)?

24.27 After 3.0 h, 18% of a certain radioactive isotope has decayed.

    a. What is the rate constant for the decay?
    b. What is the half-life?

24.28 Iron-59 is a radioactive isotope (half-life = 45 d) used in anemia diagnosis. If the initial count rate of an Fe-59 sample is 680 counts/min, how long will it take for the count rate to drop to 500 counts/min?

24.29 An oil painting supposed to be by Rembrandt (1606–1669 A.D.) is checked by C-14 dating. The C-14 content ($t_{1/2}$ = 5720 yr) of the canvas is 0.959 times that in a living plant. Could the painting have been by Rembrandt?

24.30 The Rb-87 to Sr-87 method of dating rocks was used to analyze lunar samples from the Apollo-15 mission. Estimate the age of the lunar sample in which

    a. the mole ratio of Rb-87 to Sr-87 is 25.0 (see Eq. 24.13).
    b. the mole ratio of Rb-87 to Sr-87 is 20.0.

24.31 One objection to the C-14 method is that it assumes that the C-14/C-12 ratio in the atmosphere many years ago was the same as it is now. Suggest two ways in which the validity of this assumption might be checked.

24.32 Bk-245 decays by alpha emission. For one gram of Bk-245, calculate Δm (grams) and ΔE (kilojoules).

24.33 Will F-18 decay spontaneously by positron emission? B-10? Show by calculation.

24.14 For the fission reaction

$$\text{}^{1}_{0}\text{n} + \text{}^{235}_{92}\text{U} \rightarrow \text{}^{89}_{37}\text{Rb} + \text{}^{144}_{58}\text{Ce} + 3 \text{}^{0}_{-1}\text{e} + 3 \text{}^{1}_{0}\text{n}$$

a. How much energy (kJ) is given off per gram of U-235?
b. How many kilograms of TNT must be detonated to produce the same amount of energy? ($\Delta E = -2.76$ kJ/g)

24.15 Compare the energies given off per gram of reactant in the two fusion processes considered to occur during the formation of a star:

a. $\text{}^{2}_{1}\text{H} + \text{}^{1}_{1}\text{H} \rightarrow \text{}^{3}_{2}\text{He}$
b. $2 \text{}^{3}_{2}\text{He} \rightarrow \text{}^{4}_{2}\text{He} + 2 \text{}^{1}_{1}\text{H}$

24.16 Arrange the following isotopes in order of increasing average mass per nuclear particle:

a. $\text{}^{2}_{1}\text{H}$ b. $\text{}^{4}_{2}\text{He}$ c. $\text{}^{59}_{27}\text{Co}$ d. $\text{}^{238}_{92}\text{U}$

24.17 Consider element 114, referred to in Example 24.2.

a. What element would it fall beneath in the Periodic Table?
b. Would it be a metal, nonmetal, or metalloid?
c. How would its atomic radius compare to that of bismuth?

24.18 Explain why

a. I-131 and its -1 ion behave the same way in nuclear reactions.
b. a nonradioactive Na-23 $+1$ ion and a radioactive Na-24 $+1$ ion behave chemically alike.
c. alpha emission produces a nucleus having two less protons than the original nucleus.
d. mass changes are not observed in ordinary chemical reactions.

24.19 Predict what would happen if

a. a human being absorbed a single dose of 100 rads of high-energy neutron radiation during a nuclear accident.
b. the Rb-87 method were used to measure the age of a sample claimed to be 2000 yr old.

24.20 Using Table 24.3, estimate the percentage by which radiation exposure would be increased if atmospheric nuclear tests were still contributing 30 millirems per year.

24.21 Describe advantages and disadvantages of nuclear reactors as compared to conventional energy sources.

24.34 Consider the fission reaction

$$\text{}^{239}_{94}\text{Pu} + \text{}^{1}_{0}\text{n} \rightarrow \text{}^{146}_{58}\text{Ce} + \text{}^{90}_{38}\text{Sr} + 2 \text{}^{0}_{-1}\text{e} + 4 \text{}^{1}_{0}\text{n}$$

a. How many grams of Pu-239 would have to react to produce 1.00 kJ of energy?
b. How many atoms of Pu-239 would have to react to produce 1.00 kJ of energy?

24.35 Taking Equation 24.24 to represent the reaction that produces the sun's energy, how many grams of hydrogen would have to be fused to provide the 6.3×10^{18} kJ that reaches the earth each day?

24.36 Arrange the following isotopes in order of decreasing average mass per nuclear particle:

a. $\text{}^{40}_{20}\text{Ca}$ b. $\text{}^{72}_{30}\text{Zn}$ c. $\text{}^{109}_{47}\text{Ag}$

24.37 If element 116 were synthesized,

a. what would be its outer electron configuration?
b. what group of the Periodic Table would it fall in?

24.38 Explain how

a. alpha and beta radiation are separated by an electric field.
b. radioactive C-14 can be used as a tracer to study steps in photosynthesis.
c. elements of atomic number greater than 92 are prepared in the laboratory.
d. a self-sustaining chain reaction occurs in nuclear fission.

24.39 Describe an experiment to determine

a. the half-life of P-32 (about 14 d).
b. the age of a wooden utensil thought to be from an ancient Egyptian tomb (5000 B.C.).
c. where the O atom in H_2O comes from in:

$$CH_3COOH + CH_3OH \rightarrow H_2O + CH_3COOCH_3$$

using a radioactive isotope.
d. the solubility of water in benzene, which is very small.

24.40 Suppose a person living in the vicinity of the Three Mile Island accident were exposed to radiation of 0.010 rem. Estimate the percentage by which this would increase his or her total radiation exposure for 1979.

24.41 Why will it be necessary to develop breeder reactors if we are to continue to produce electricity by nuclear fission in the twenty-first century?

*24.42   How many alpha particles are given off in one second by a sample of uranium-238 weighing 1.00 mg ($t_{1/2}$ = 4.5 × 10^9 yr)?

*24.43   Plutonium-239 decays by the reaction $^{239}_{94}Pu \rightarrow {}^{235}_{92}U + {}^4_2He$, with a rate constant of 5.5 × 10^{-11}/min. In a one-gram sample of Pu-239,

        a. how many grams decompose in 10 min?
        b. how much energy in kilojoules is given off in 10 min?
        c. what radiation dosage in rems is received by a 70-kg man exposed to a gram of Pu-239 for 10 min?

*24.44   It is possible to estimate the activation energy for fusion by calculating the energy required to bring two deuterons close enough to form an alpha particle. This energy can be obtained by using Coulomb's Law: $E = \dfrac{q_1 \times q_2}{r}$, where q_1 and q_2 are the charges of the deuterons (4.8 × 10^{-10} esu), r is the radius of the helium nucleus (~1 × 10^{-12} cm), and E is the energy in ergs.

        a. Estimate E in ergs per alpha particle.
        b. Using the equation $E = mv^2/2$, estimate the velocity (cm/s) that a deuteron must have if a collision between two of them is to supply the activation energy for fusion (m is the mass of the deuteron in grams).
        c. Using the equation $v = (3RT/GMM)^{1/2}$, estimate the temperature that would have to be reached to achieve fusion $\left(R = 8.31 \times 10^7 \dfrac{erg}{mol \cdot K}\right)$.

*24.45   Consider the reaction $2\,{}^2_1H \rightarrow {}^4_2He$.

        a. Calculate ΔE in kilojoules per gram of deuterium fused.
        b. How much energy is potentially available from the fusion of all the deuterium atoms in seawater? The percentage of deuterium in water is about 0.015%. The total mass of water in the oceans is 1.3 × 10^{27} g.
        c. What fraction of the deuterium in the oceans would have to be consumed to supply the annual energy requirements of the world (2.3 × 10^{17} kJ)?

25

AN INTRODUCTION TO ORGANIC CHEMISTRY

Organic chemistry is the study of carbon compounds, especially those with hydrogen, oxygen, nitrogen, and the halogens. Organic molecules vary from simple ones like CH_4 to complex macromolecules such as plastics, proteins, and starch. More than a million organic compounds are known, far more than the sum of all other compounds.

Organic compounds are biologically important. Cells of your body consist mainly of organic molecules of great diversity. Many of these molecules are synthesized daily. In turn, they may be used to make other new molecules that aid and control life processes.

One important area of organic chemistry is concerned with the identification and synthesis of medicinal products. Developments in organic chemistry since 1940 have produced new compounds that have revolutionized the practice of medicine. Sulfa drugs, cortisones, steroids, tranquilizers, and other organic materials have been isolated or synthesized in the last four decades. These products have made it possible to treat illness more effectively than at any time in history. The ability to apply knowledge of organic chemistry to produce such medicines is one of the greatest contributions of science in this century.

In this chapter, we will look at the properties of a few types of organic compounds. Our goal is to present some of the basic principles that govern structure and reactivity in organic chemistry. Quite possibly, you will go on from here to take one or more courses in this subject area. There, these principles and others will be developed in much greater detail.

25.1 HYDROCARBONS

The simplest type of organic compound, called a *hydrocarbon*, contains only carbon and hydrogen atoms. Several compounds of this type were discussed in Chapter 12. You will recall that hydrocarbons are molecular in nature. Within the molecule, carbon atoms are joined by covalent bonds to other carbon atoms or to hydrogen atoms. In each case, a carbon atom forms a total of four bonds; a hydrogen atom forms one bond. Thus, we have

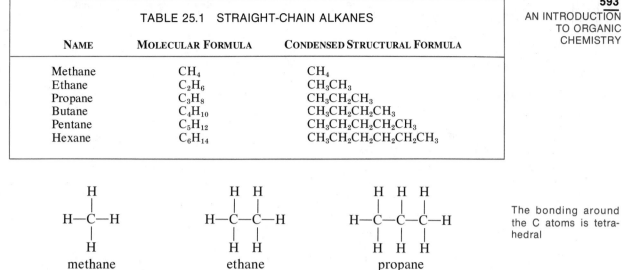

NAME	MOLECULAR FORMULA	CONDENSED STRUCTURAL FORMULA
Methane	CH_4	CH_4
Ethane	C_2H_6	CH_3CH_3
Propane	C_3H_8	$CH_3CH_2CH_3$
Butane	C_4H_{10}	$CH_3CH_2CH_2CH_3$
Pentane	C_5H_{12}	$CH_3CH_2CH_2CH_2CH_3$
Hexane	C_6H_{14}	$CH_3CH_2CH_2CH_2CH_2CH_3$

The bonding around the C atoms is tetrahedral

Saturated Hydrocarbons (Alkanes)

Hydrocarbons such as those just shown, in which all the carbon-carbon bonds are single bonds, are said to be *saturated*. They are called saturated hydrocarbons or **alkanes.** Their composition may be shown in any of several ways. Using propane as an example, we may cite its:

1. *Molecular formula,* C_3H_8. The general formula of an alkane is

$$C_nH_{2n+2}$$

where n is the number of carbon atoms in the molecule.

2. *Structural formula,* as shown in the diagrams above. Here we show all the bonds in the molecule, a total of ten in the case of propane. To save space, we often condense the structural formula. For propane, we write

$$CH_3CH_2CH_3$$

Here it is understood that each hydrogen atom is bonded to the carbon atom preceding it in the formula.

Table 25.1 gives the molecular formulas, condensed structural formulas, and names of six simple alkanes. In these molecules, the carbon atoms are bonded in a single continuous chain. These compounds are referred to as "unbranched" or *straight-chain* alkanes. (Actually, as we saw in Chapter 12, the carbon atoms are arranged in a zig-zag pattern with bond angles of 109°.)

Chains of C atoms can be very long

Example 25.1   For the straight-chain alkane containing seven carbon atoms, give the
a. molecular formula;   b. condensed structural formula.

Solution
a. Since n = 7, 2n + 2 = 14 + 2 = 16. C_7H_{16}
b. At each end of the chain, a carbon atom is bonded to three H atoms (CH_3). Within

the chain, the carbon atoms are bonded to two H atoms (CH_2). The condensed structural formula is

$$CH_3CH_2CH_2CH_2CH_2CH_2CH_3$$

This is often further abbreviated as

$$CH_3(CH_2)_5CH_3$$

Exercise   What is the molecular formula of an alkane containing 22 hydrogen atoms?
Answer: $C_{10}H_{22}$.

As pointed out in Chapter 12, alkanes containing four or more carbon atoms per molecule show *structural isomerism*. For example, there are two different alkanes with the molecular formula C_4H_{10}. One of these, the straight-chain compound called butane, is shown in Table 25.1. The other is 2-methylpropane, which has the structural formula

$$CH_3-\overset{\overset{\displaystyle H}{|}}{\underset{\underset{\displaystyle CH_3}{|}}{C}}-CH_3$$

2-methylpropane

In this isomer, the longest continuous chain contains only three carbon atoms. There is a —CH_3 (methyl) group bonded to the second carbon atom in this *branched chain* alkane.

As the number of carbon atoms per molecule increases, the number of structural

Isomerism is one reason we have so many organic compounds

isomers goes up rapidly. There are three such isomers for C_5H_{12} and five for C_6H_{14}. The carbon skeletons of the C_6H_{14} isomers are

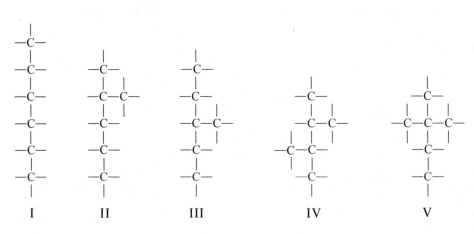

| I | II | III | IV | V |

These structural isomers have different properties. For example, their normal boiling points are:

I, 69°C; II, 60°C; III, 64°C; IV, 58°C; V, 50°C

The differences in boiling points reflect the fact that intermolecular attractive forces become weaker as chain branching increases and the molecules become more compact (Chap. 12). This principle is used in the seasonal blending of gasoline. A greater percentage of more volatile, highly branched alkanes is included in winter, making combustion occur more readily at low temperatures.

As a group, alkanes are relatively unreactive. Their principal reaction, which accounts for their use in fuels, is that of burning in air. Complete combustion gives carbon dioxide and water. A typical reaction is

$$C_3H_8(g) + 5\ O_2(g) \rightarrow 3\ CO_2(g) + 4\ H_2O(l) \tag{25.1}$$

Unsaturated Hydrocarbons

In an unsaturated hydrocarbon, at least one of the carbon-carbon bonds in the molecule is a multiple bond. In this section, we will discuss only two types of unsaturated hydrocarbons:

1. **Alkenes,** in which there is one carbon-carbon double bond in the molecule. The simplest alkene is ethylene, molecular formula C_2H_4. Its structural formula is

Planar; bond angles all equal 120°

ethylene

2. **Alkynes,** in which there is one carbon-carbon triple bond in the molecule. The simplest alkyne is acetylene, C_2H_2. Its structural formula is

$$H\!-\!C\!\equiv\!C\!-\!H$$

Linear

acetylene

When we introduce a double bond into a hydrocarbon molecule, two hydrogen atoms are eliminated. Hence, a molecule of an alkene contains two fewer H atoms than does an alkane molecule with the same number of carbon atoms (compare ethylene, C_2H_4, to ethane, C_2H_6). Recalling that the general formula of an alkane is C_nH_{2n+2}, we conclude that the general formula of an alkene must be

$$C_nH_{2n}$$

When we go from a double to a triple bond, the number of hydrogen atoms per molecule again decreases by two (compare acetylene, C_2H_2, to ethylene, C_2H_4). The general formula of an alkyne must then be

$$C_nH_{2n-2}$$

Example 25.2   For the alkene and the alkyne containing three carbon atoms, write the
   a. molecular formula;   b. structural formula.

Solution
   a. In both cases, n = 3.
      alkene: 2n = 6; molecular formula C_3H_6
      alkyne: 2n − 2 = 4; molecular formula C_3H_4
   b. Perhaps the simplest way to obtain the structural formulas is to replace a H atom of ethylene or acetylene with a —CH₃ group.

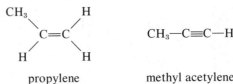

propylene   methyl acetylene

Exercise   Classify each of the following as an alkane, alkene, or alkyne: $C_{12}H_{22}$; C_9H_{20}; C_8H_{16}. Answer: Alkyne; alkane; alkene.

Unsaturated hydrocarbons are much more reactive than alkanes. Commonly, they undergo what is known as an *addition* reaction. Here, a small molecule such as H_2, Br_2, or HBr adds directly at the multiple bond. With an alkene, the product is a saturated compound in which all the bonds are single bonds:

$$C=C \quad + \quad Br-Br \rightarrow H-C-C-H \tag{25.2}$$

With an alkyne, such as acetylene, addition occurs in two steps. In the first step, a triple bond is converted to a double bond. Then the double bond is broken, forming a saturated compound:

$$H-C\equiv C-H + Br-Br \rightarrow C=C \tag{25.3}$$

$$C=C \quad + \quad Br-Br \rightarrow Br-C-C-Br \tag{25.4}$$

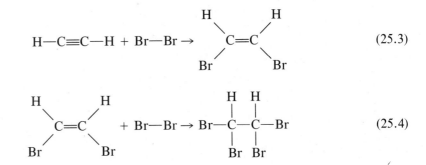

Aromatic Hydrocarbons

Hydrocarbons of this type can be considered to be derived from benzene, C_6H_6. Benzene, a liquid with a boiling point of 80°C, was discovered by Michael Faraday in 1825. Its formula, C_6H_6, indicates a high degree of unsaturation. Yet, its chemical reactivity is quite different from that of alkenes or alkynes. Consider, for example, its behavior toward bromine. Here, benzene reacts by *substitution* rather than addition:

C_6H_6 is much less reactive than C_2H_2

$$C_6H_6(l) + Br_2(l) \rightarrow C_6H_5Br(l) + HBr(g) \tag{25.5}$$

A bromine atom substitutes for a hydrogen in the C_6H_6 molecule, converting it to bromobenzene, C_6H_5Br.

Kekulé, a German chemist, was one of the first to study the chemical properties of benzene. He suggested a ring structure for the benzene molecule, with alternating single and double bonds. To explain the fact that the carbon-carbon bonds are all of the same length, we might regard benzene as a resonance hybrid of the two structures:

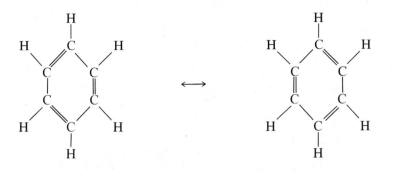

The C_6H_6 structure
would be the average
of these two

This model is consistent with many of the properties of benzene. The molecule is a planar hexagon with bond angles of 120°. The hybridization of each carbon atom is sp². However, this structure is misleading in one respect. If there were three double bonds in the molecule, we might expect benzene to react with bromine by addition rather than substitution.

A more satisfactory model of the electron distribution in the benzene molecule is suggested by modern bonding theory. Here we consider that:

—each carbon atom forms three sigma bonds, one to a hydrogen atom and two to adjacent carbon atoms.

—the three electron pairs remaining are not tied down *(localized)* to form three double bonds, as in the Kekulé structure. Instead, the six electrons are spread symmetrically over the entire molecule to form "delocalized" pi bonds. The orbitals occupied by these electrons are indicated in Figure 25.1.

This model is often represented by the structure

Here, it is understood that:

—there is a carbon atom at each corner of the hexagon.

—there is a H atom bonded to each carbon atom.

—the circle in the center of the molecule represents the six delocalized electrons.

A few typical aromatic hydrocarbons are shown in Figure 25.2, p. 598. Many of these are obtained from coal tar, produced by heating coal to about 1000°C in the absence of air. Notice that some aromatic hydrocarbons, such as naphthalene, contain two or more benzene rings fused together. Certain compounds of this type are potent carcinogens. One of the most dangerous is 3,4-benzpyrene, which has been detected in cigaret smoke. It is believed to be a cause of lung cancer, to which heavy smokers are susceptible.

Smokers who get lung cancer usually stop smoking

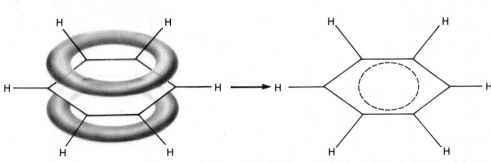

Figure 25.1   In benzene, three electron pairs are not localized on a particular carbon atom. Instead, they are spread out over two electron clouds of the shape shown, one above the plane of the benzene ring and the other below it.

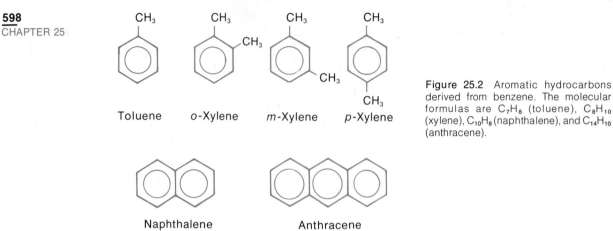

Figure 25.2 Aromatic hydrocarbons derived from benzene. The molecular formulas are C_7H_8 (toluene), C_8H_{10} (xylene), $C_{10}H_8$ (naphthalene), and $C_{14}H_{10}$ (anthracene).

Toluene   o-Xylene   m-Xylene   p-Xylene

Naphthalene   Anthracene

3,4-benzpyrene

Gasoline

Gasoline is a complex mixture of hydrocarbons (Fig. 25.3). Most of these are alkanes containing from four to ten carbon atoms per molecule. Smaller amounts of aromatics are present. There are virtually no alkenes or alkynes in gasoline.

Gasoline is produced by the fractional distillation of petroleum. The hot petroleum ("crude oil") is separated into fractions according to boiling point (Table 25.2). Distillation of a liter of crude oil gives about 250 cm³ of "straight-run" gasoline.

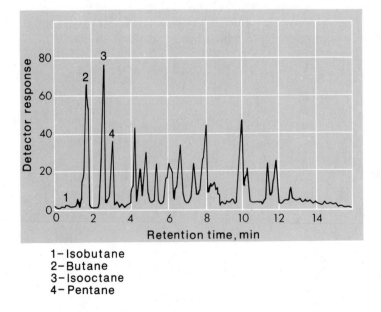

Figure 25.3 Separation of hydrocarbons in gasoline by vapor phase chromatography.

1 – Isobutane
2 – Butane
3 – Isooctane
4 – Pentane

FRACTION	BOILING RANGE (°C)	CARBON ATOM CONTENT	DIRECT USE
Gas	Below 20	C_1–C_4	Gas heating
Petroleum ether	20–60	C_5–C_6	Industrial solvent
Light naphtha	60–100	C_6–C_7	Industrial solvent
Straight-run gasoline	40–200	C_5–C_{10}	Motor vehicle fuel
Kerosene	175–325	C_{11}–C_{18}	Jet fuel
Gas oil	275–500		Diesel fuel
Lubricating oil	Above 400	C_{15}–C_{40}	Lubricant
Asphalt	Nonvolatile		Roofing and road construction

TABLE 25.2   FRACTIONS OBTAINED ON DISTILLATION OF PETROLEUM

It is possible to double the yield of gasoline from crude oil by converting higher or lower boiling fractions to hydrocarbons in the gasoline range. Several processes are used to do this. Two of the most important are *cracking* and *isomerization* (re-forming).

In cracking, high molecular mass fractions are heated, with catalysts, to 500 to 700°C. Under these conditions, carbon-carbon bonds are broken. The products include alkenes and alkanes of lower molecular mass than those originally present. A typical cracking reaction is

$$C_{13}H_{28}(l) \rightarrow C_8H_{18}(l) + C_2H_4(g) + C_3H_6(g) \tag{25.6}$$
$$\text{alkane} \qquad \text{alkane} \qquad \text{alkene} \qquad \text{alkene}$$

The alkanes formed are added to the straight-run gasoline, thus increasing the gasoline yield from crude oil. The alkenes are not used as a fuel. Rather, they serve as raw materials for making plastics (Chap. 26).

The light molecules can be separated by distillation

Isomerization further increases the yield and quality of gasoline. In this process straight-chain alkanes are converted to branched-chain isomers, which burn more efficiently. A typical reaction is that of pentane. When passed over an $AlCl_3$ catalyst at about 200°C, it isomerizes to produce the branched-chain compounds:

2-methylbutane   2,2-dimethylpropane

Some isomerization also occurs during cracking and further increases gasoline quality.

One problem with the internal combustion engine is that gasoline-air mixtures tend to ignite prematurely, or "knock," rather than burn smoothly. The *octane number* of a gasoline is a measure of its resistance to knock. It is determined by comparing the knocking characteristics of a gasoline sample to those of "isooctane" and heptane:

Diesels don't have this problem

```
      CH₃      H
       |       |
CH₃—C—CH₂—C—CH₃          CH₃—CH₂—CH₂—CH₂—CH₂—CH₂—CH₃
       |       |
      CH₃     CH₃
```

"isooctane"   heptane
(2,2,4-trimethylpentane)

Isooctane, which is highly branched, burns smoothly with little knocking and is assigned an octane number of 100. Heptane, being unbranched, knocks badly. It is given an octane number of zero. Gasoline with the same knocking properties as a mixture of 90% isooctane and 10% heptane is rated as "90 octane."

Octane numbers posted on gasoline pumps are an average of two different values. One of these is the "research octane number" (RON) determined with a test engine running at low speeds (600 rpm). The other is the "motor octane number" (MON), determined at a higher speed (900 rpm). Typically, a premium gasoline might have an RON of 99 and a MON of 91: the posted octane number would be 95.

Straight-run gasoline has an octane number of about 70. By blending it with products of cracking, isomerization, and other processes, regular gasoline of octane number 90 can be produced. Another way to boost octane rating is to use antiknock agents. From the 1920's to the mid-1970's, tetraethyl lead, $Pb(C_2H_5)_4$, was the major antiknock additive. A gallon of gasoline contained up to 2.4 g of this compound. Concern about lead pollution from auto exhaust and the poisoning of catalytic converters (Chap. 15) has reduced its usage. The switch to no-lead gasolines has required adding more expensive aromatics and highly branched alkanes to maintain high octane numbers.

Premium gasoline is fast becoming unavailable

The "energy crisis" of the 1970's has prompted scientists and engineers to look beyond petroleum for other sources of gasoline. The development of a program for making "synthetic" liquid fuels has become a national priority for the 1980's. The most promising source of such fuels is coal, of which the United States has abundant supplies. The technology for making gasoline from coal has been available for about 40 years. In the Lurgi process, developed in Germany, oxygen and steam are blown under pressure through hot coal. This produces a fuel called synthesis gas, which contains about 40 mol % H_2, 15 mol % CO, 15 mol % CH_4, and 30 mol % CO_2. In the presence of suitable catalysts, carbon monoxide and hydrogen undergo reactions such as

$$8 \; CO(g) + 17 \; H_2(g) \rightarrow C_8H_{18}(l) + 8 \; H_2O(l) \tag{25.7}$$

At 1 atm and 200°C, reactions of this type yield a variety of hydrocarbons, mostly unbranched alkanes and alkenes. These can be mixed with higher octane components to produce a suitable motor fuel. A large plant using the Lurgi process to make gasoline is in operation in South Africa. However, the product is about 1.5 times as expensive as gasoline from petroleum. This price ratio has stayed nearly constant for 30 years, despite the increase in the cost of petroleum.

We can make gasoline from coal, but it's expensive

Another possible future source of gasoline is oil shale. Huge deposits of this rocky material are found in Wyoming, Utah, and Colorado. The shale contains about 15% of an organic material called kerogen. When heated to about 500°C, kerogen breaks down to yield perhaps 100 ℓ of crude oil per metric ton of shale. The shale expands considerably upon heating, which means that large amounts of rocky residues must be disposed of. Moreover, it appears that significant government subsidies will be necessary to make the price of liquid fuels from oil shale competitive with that of ordinary gasoline.

25.2 ORGANIC COMPOUNDS CONTAINING CHLORINE, OXYGEN, AND NITROGEN

From one point of view, all organic compounds containing atoms other than C and H can be considered to be derived from hydrocarbons. In the simplest case, one or more H atoms in the hydrocarbon molecule are replaced by atoms of another nonmetal. This is true when the nonmetal is a halogen such as chlorine. The following compounds are derived from methane, CH_4, by replacing successive hydrogen atoms with chlorine:

Cl_2 will react with CH_4 at 300°C to produce these compounds

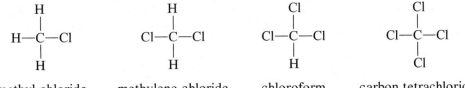

methyl chloride   methylene chloride   chloroform   carbon tetrachloride

These and many other organic chlorine compounds have a variety of commercial applications. They are employed as solvents, cooling fluids, pesticides, and herbicides (Table 25.3). Many of these compounds have been shown to be toxic or have other adverse effects upon our environment. Hardly a year passes without the use of an organic chlorine compound being banned or restricted.

Halogen atoms are able to substitute for hydrogen on a 1:1 basis because they, like hydrogen atoms, form one covalent bond to another atom. With oxygen *(two covalent bonds per atom)* or nitrogen *(three covalent bonds)* this is not possible. Organic compounds of oxygen or nitrogen can be considered to be derived from hydrocarbons by substituting a **functional group** for a hydrogen atom. A functional group is a small group of atoms which bonds to carbon in an organic compound. One such group is the —OH group found in alcohols such as

methyl alcohol   ethyl alcohol

Compounds containing a given functional group, such as —OH, resemble one another in their chemical properties. For that reason, the chemistry of organic com-

TABLE 25.3   SOME ORGANIC HALOGEN COMPOUNDS

NAME	FORMULA	USES
Dichlorodifluoromethane (a freon)	CCl_2F_2	Refrigerant, aerosol sprays
Dichlorodiphenyltrichloro-ethane (DDT)	Cl—◯—CH—◯—Cl CCl_3	Insecticide
2,4-dichlorophenoxyacetic acid (2,4-D)	Cl—◯—O—CH_2—COOH Cl	Herbicide
2,3,3′,4′,5-pentachloro-biphenyl (a typical poly-chlorinated biphenyl—PCB)	Cl Cl Cl \ \ / Cl—◯—◯ \Cl	Plasticizer, solvent, coolant
Trichloroethane	CH_3CCl_3	Solvent to replace CCl_4, $CHCl_3$
Vinyl chloride	$CH_2{=}CHCl$	Plastics (polyvinyl chloride)

It seems a bit strange, but there are very, very few halogen-containing compounds in living organisms

TABLE 25.4

Type of Compound	Functional Group	Examples	
Alcohol	—OH	CH_3—OH methyl alcohol	CH_3CH_2—OH ethyl alcohol
Aldehyde	$-\overset{\displaystyle \|\|}{\underset{O}{C}}-H$	$H-\overset{\displaystyle \|\|}{\underset{O}{C}}-H$ formaldehyde	$CH_3-\overset{\displaystyle \|\|}{\underset{O}{C}}-H$ acetaldehyde
Amide	$-\overset{\displaystyle \|\|}{\underset{O}{C}}-\overset{\|}{N}-$	$CH_3-\overset{\displaystyle \|\|}{\underset{O}{C}}-\overset{\|}{\underset{H}{N}}-H$ acetamide	$CH_3-\overset{\displaystyle \|\|}{\underset{O}{C}}-\overset{\|}{\underset{H}{N}}-CH_3$ N-methylacetamide
Amine	$-\overset{\|}{N}-$	$CH_3-\overset{\|}{\underset{H}{N}}-H$ methylamine	$CH_3-\overset{\|}{\underset{H}{N}}-CH_3$ dimethylamine
Carboxylic acid	$-\overset{\displaystyle \|\|}{\underset{O}{C}}-OH$	$H-\overset{\displaystyle \|\|}{\underset{O}{C}}-OH$ formic acid	$CH_3-\overset{\displaystyle \|\|}{\underset{O}{C}}-OH$ acetic acid
Ester	$-\overset{\displaystyle \|\|}{\underset{O}{C}}-O-$	$H-\overset{\displaystyle \|\|}{\underset{O}{C}}-O-CH_3$ methyl formate	$CH_3-\overset{\displaystyle \|\|}{\underset{O}{C}}-O-CH_2CH_3$ ethyl acetate
Ether	—O—	CH_3—O—CH_2CH_3 methyl ethyl ether	CH_3CH_2—O—CH_2CH_3 diethyl ether
Ketone	$-\overset{\displaystyle \|\|}{\underset{O}{C}}-$	$CH_3-\overset{\displaystyle \|\|}{\underset{O}{C}}-CH_3$ acetone	$CH_3-\overset{\displaystyle \|\|}{\underset{O}{C}}-CH_2-CH_3$ methyl ethyl ketone

There are a lot of organic compounds that contain oxygen or nitrogen

pounds of oxygen or nitrogen is often organized according to the type of functional group present (Table 25.4). In this section, we will look at the properties of some simple compounds containing a few different functional groups.

Alcohols. Functional Group —OH

Alcohols can be considered to be derived from hydrocarbons by replacing one or more H atoms by —OH groups. The simplest alcohols are methyl alcohol (methanol) and ethyl alcohol (ethanol), whose structures are indicated in Table 25.4.

The presence of an —OH group in the molecule makes it possible for an alcohol to form hydrogen bonds with itself or with water (Fig. 25.4). The formation of hydrogen bonds explains:

—the relatively high boiling points of alcohols. Methyl alcohol has about the same boiling point as hexane, C_6H_{14} (65°C vs. 69°C), even though its molecular mass is much lower (32 vs. 86).

Figure 25.4 Alcohol molecules, general formula ROH, can form hydrogen bonds (dotted lines) with other alcohol molecules or with water molecules.

—the high water solubility of alcohols. All the alcohols containing from one to three carbon atoms are soluble in water, in all proportions.

About 3×10^9 kg of methyl alcohol (methanol) are produced annually in the United States from water gas, a mixture of carbon monoxide and hydrogen:

$$CO(g) + 2\ H_2(g) \xrightarrow[\text{250 atm, 350°C}]{\text{ZnO, Cr}_2\text{O}_3} CH_3OH(g) \qquad (25.8)$$

It is also formed as a by-product when charcoal is made by heating wood in the absence of air. For this reason, methanol is sometimes called wood alcohol. Methanol is used in jet fuels and as a solvent, gasoline additive, and starting material for several industrial syntheses. It is poisonous, causing blindness or death. So, although it is an intoxicant, it must not be used in alcoholic beverages.

Ethyl alcohol (ethanol), the most common alcohol, can be prepared by the fermentation of grains or sugar. A typical reaction is that of glucose, $C_6H_{12}O_6$:

$$C_6H_{12}O_6(aq) \rightarrow 2\ C_2H_5OH(aq) + 2\ CO_2(g) \qquad (25.9)$$

A mixture of 95% ethanol and 5% water can be separated from the fermentation products by distillation. Ethanol, a colorless liquid, is the active ingredient of alcoholic beverages. There it is present in various concentrations (4 to 8% in beers, 12 to 15% in wine, and 40% or more in distilled spirits). The "proof" of an alcoholic beverage is twice the volume percentage of ethyl alcohol. Thus, an 86 proof bourbon whiskey contains 43% ethanol.

Most of the 1.5×10^9 kg of industrial ethanol made annually in the United States is produced synthetically. The starting material is ethylene, a by-product of the cracking of crude oil (recall Reaction 25.6). The reaction is

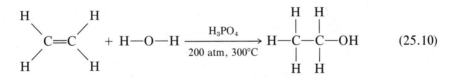

$$(25.10)$$

This is an addition reaction. The water molecule is split and the fragments add to the ethylene double bond.

Industrial ethanol is used in making many organic compounds

Recently, the use of ethanol as a gasoline additive has been heavily promoted. "Gasohol," a blend of 10% ethanol with 90% gasoline, can be burned in an automobile engine, apparently without adverse effects. In this way it may be possible to extend gasoline supplies and reduce petroleum imports. There are, however, a few problems. For one thing, the heat of combustion of ethanol, in joules per gram, is only about 60% that of gasoline. Moreover, the energy required to produce ethanol by fermentation and distillation is at least equal to that released when it burns.

Certain alcohols contain two or more —OH groups per molecule. Perhaps the most familiar compounds of this type are ethylene glycol and glycerol:

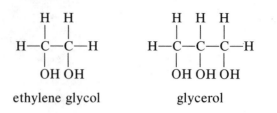

ethylene glycol   glycerol

Ethylene glycol is widely used as an antifreeze (Chap. 16). Glycerol is formed as a by-product in making soaps and detergents (p. 606). It is a viscous, sweet-tasting liquid, used in making drugs, antibiotics, plastics, and explosives (nitroglycerin).

Carboxylic Acids. Functional Group —C—OH ‖ O

Carboxylic acids can be considered to be derived from hydrocarbons by replacing one or more H atoms by a carboxyl group, —C—OH, often abbreviated —COOH. The ‖ O

most common carboxylic acid, at least in general chemistry, is acetic acid, whose formula may be written as:

$$H—\overset{\overset{\displaystyle H}{|}}{\underset{\underset{\displaystyle H}{|}}{C}}—\overset{}{\underset{\underset{\displaystyle O}{\|}}{C}}—OH \quad or \quad CH_3—\underset{\underset{\displaystyle O}{\|}}{C}—OH \quad or \quad CH_3—COOH$$

acetic acid

Acetic acid is the active ingredient of vinegar, responsible for its sour taste. A variety of other carboxylic acids are found in natural products (Table 25.5).

Carboxylic acids, like alcohols, show hydrogen bonding (Fig. 25.5). This accounts for their high water solubility and relatively high boiling points. Acetic acid is soluble in water, in all proportions, and boils at 118°C. As pointed out in Chapter 17, acetic acid (like other carboxylic acids) is slightly ionized in water solution. It is the H atom of the carboxyl group that ionizes:

$$CH_3—\underset{\underset{\displaystyle O}{\|}}{C}—OH(aq) \rightleftharpoons CH_3—\underset{\underset{\displaystyle O}{\|}}{C}—O^-(aq) + H^+(aq) \qquad (25.11)$$

acetic acid   acetate ion

Treatment of acetic acid or other organic acids with the strong base NaOH converts them to the corresponding sodium salts. In the case of acetic acid, the acid-base reaction is given by Equation 25.12:

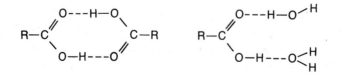

Figure 25.5 Carboxylic acids, like alcohols, can form hydrogen bonds (dotted lines).

TABLE 25.5   NATURALLY OCCURRING CARBOXYLIC ACIDS

NAME	STRUCTURE	NATURAL SOURCE
Acetic acid	CH_3—COOH	Vinegar
Citric acid	HOOC—CH_2—$\overset{\overset{\displaystyle OH}{\mid}}{\underset{\underset{\displaystyle COOH}{\mid}}{C}}$—$CH_2$—COOH	Citrus fruits
Lactic acid	CH_3—$\underset{\underset{\displaystyle OH}{\mid}}{CH}$—COOH	Sour milk
Malic acid	HOOC—CH_2—$\underset{\underset{\displaystyle OH}{\mid}}{CH}$—COOH	Apples
Oleic acid	$CH_3(CH_2)_7$—CH=CH—$(CH_2)_7$—COOH	Vegetable oils
Oxalic acid	HOOC—COOH	Rhubarb, spinach, cabbage, tomatoes
Stearic acid	$CH_3(CH_2)_{16}$—COOH	Animal fats
Tartaric acid	HOOC—$\underset{\underset{\displaystyle OH}{\mid}}{CH}$—$\underset{\underset{\displaystyle OH}{\mid}}{CH}$—COOH	Grape juice, wine

Organic acids are pres-
ent in many foods

$$CH_3COOH(aq) + OH^-(aq) \rightarrow CH_3COO^-(aq) + H_2O \qquad (25.12)$$

Sodium acetate is simi-
lar in its properties to
inorganic salts

Evaporation of the solution formed by this reaction gives the salt sodium acetate, which contains Na^+ and CH_3COO^- (or $C_2H_3O_2^-$) ions.

Esters. Functional Group $-\overset{\overset{\displaystyle }{}}{\underset{\underset{\displaystyle O}{\parallel}}{C}}-O-$

The reaction between a carboxylic acid and an alcohol forms an ester, containing the functional group $-\overset{}{\underset{\underset{\displaystyle O}{\parallel}}{C}}-O-$, often abbreviated —COO—. The reaction between acetic acid and methyl alcohol is typical:

$$CH_3-\underset{\underset{\displaystyle O}{\parallel}}{C}-OH(aq) + HO-CH_3(aq) \xrightarrow{H^+} CH_3-\underset{\underset{\displaystyle O}{\parallel}}{C}-O-CH_3(aq) + H_2O \qquad (25.13)$$

acetic acid   methyl alcohol   methyl acetate

This reaction is ordinarily carried out in dilute H_2SO_4 solution; the H^+ ion acts as a catalyst. Evidence from tracer studies, using the $^{18}_{8}O$ isotope, indicates that the —OH group comes from the acid rather than the alcohol.

Most esters have a pleasant odor. Butyl acetate gives bananas their odor. Ethyl acetate is one of at least five esters found in pineapples. In addition to being used in synthetic fragrances, esters are also industrial solvents and starting materials for plastics such as Plexiglas (Chap. 26).

Example 25.3   Give the structural formula of
  a. the three-carbon alcohol with an —OH group at the end of the chain.
  b. the three-carbon carboxylic acid.
  c. the ester formed when these two compounds react.

Solution

  a. $CH_3CH_2CH_2$—OH   b. CH_3CH_2—C—OH
                                         $\|$
                                         O

  c. CH_3CH_2—C—O—$CH_2CH_2CH_3$
                $\|$
                O

Exercise   Which one of the following is *neither* an alcohol, a carboxylic acid, nor an ester? CH_3CH_2—OH; H—C—OH; CH_3—O—CH_3; H—C—O—CH_3. Answer:
                                         $\|$ $\|$
                                         O   O
CH_3—O—CH_3.

Animal fats and vegetable oils are esters of long-chain carboxylic acids with glycerol. Like all esters, fats can be split (saponified) back to the parent alcohol and acid by heating with NaOH solution. The product is called a **soap.**

Glycerol is a valuable by-product of this reaction

$$\begin{array}{l} CH_2OC\overset{O}{\diagup}\text{—}(CH_2)_{14}CH_3 \\[2pt] \quad\quad\quad O \\ CHOC\overset{}{\diagup}\text{—}(CH_2)_{14}CH_3 + 3\ OH^- + 3\ Na^+ \rightarrow \\ \quad\quad\quad O \\ CH_2OC\overset{}{\diagup}\text{—}(CH_2)_{14}CH_3 \end{array}$$

a typical fat

$$\begin{array}{l} CH_2OH \\ CHOH + 3\ CH_3(CH_2)_{14}\overset{O}{\overset{\|}{C}}O^-, Na^+ \\ CH_2OH \quad\quad\quad \text{a soap} \\ \text{glycerol} \end{array}$$

(25.14)

A soap is the sodium salt of a long-chain carboxylic acid. It is a good cleansing agent because of the long hydrocarbon group, which is a good solvent for greases and oils, and the ionic COO^- group, which gives high water solubility. However, in hard water containing Ca^{2+} ions, the soap precipitates as an insoluble calcium salt. This causes various problems, particularly in laundering clothes.

Research to improve the solubility of cleaning agents resulted, in about 1950, in the development of substances called **detergents.** Detergents are also made from fats. However, following saponification, the acids themselves are recovered. These are reduced with hydrogen to alcohols, which are made into inorganic esters by reaction with sulfuric acid:

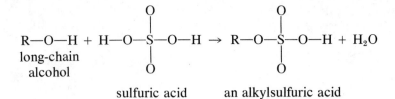

R—O—H + H—O—S(=O)(=O)—O—H → R—O—S(=O)(=O)—O—H + H_2O

long-chain alcohol   sulfuric acid   an alkylsulfuric acid

By reacting the remaining acidic hydrogen with NaOH, we obtain a detergent, the sodium salt of the alkylsulfuric acid:

$$R—O—\overset{O}{\underset{O}{S}}—O^-, \ Na^+$$

Detergents do not precipitate in hard water

The annual market for detergents is of the order of 2×10^9 kg, which may give you an idea of why companies like Colgate-Palmolive hire chemists.

Amines. Functional Group —N—

Amines are perhaps visualized most simply as derivatives of ammonia, NH_3,

$$H—\overset{H}{N}—H$$

in which one or more of the H atoms are replaced by hydrocarbon groups. We distinguish between:

—*primary* amines, in which one H atom of NH_3 is replaced.

—*secondary* amines, in which two H atoms of NH_3 are replaced.

—*tertiary* amines, in which all three H atoms of NH_3 are replaced.

Examples include

$$CH_3—\overset{H}{N}—H \text{ or } CH_3NH_2 \qquad CH_3—\overset{CH_3}{N}—H \text{ or } (CH_3)_2NH \qquad CH_3—\overset{CH_3}{N}—CH_3 \text{ or } (CH_3)_3N$$

methylamine   dimethylamine   trimethylamine
(primary)   (secondary)   (tertiary)

They also smell somewhat like NH_3

Amines act as weak bases in water solution. Their reaction with water is analogous to that of ammonia:

$$NH_3(aq) + H_2O \rightleftharpoons NH_4^+(aq) + OH^-(aq); \ K_b = 1.8 \times 10^{-5}$$

$$CH_3NH_2(aq) + H_2O \rightleftharpoons CH_3NH_3^+(aq) + OH^-(aq); \ K_b = 4.4 \times 10^{-4}$$

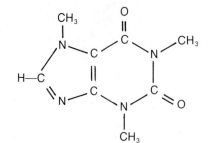

Caffeine

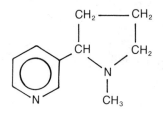

Nicotine

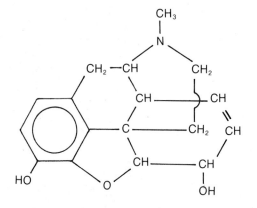

Morphine

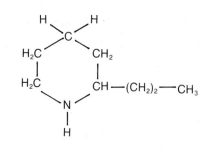

Coniine

Figure 25.6 Molecular structures of four alkaloids. Caffeine, nicotine, and morphine are tertiary amines; coniine is a secondary amine.

$$(CH_3)_2NH(aq) + H_2O \rightleftharpoons (CH_3)_2NH_2^+(aq) + OH^-(aq); \; K_b = 5.1 \times 10^{-4}$$

$$(CH_3)_3N(aq) + H_2O \rightleftharpoons (CH_3)_3NH^+(aq) + OH^-(aq); \; K_b = 5.3 \times 10^{-5}$$

Alkaloids such as caffeine, nicotine, morphine, and codeine (Fig. 25.6) form an important class of naturally occurring amines. Caffeine occurs in tea leaves, coffee beans, and cola nuts used to make cola soft drinks. Nicotine is the most abundant alkaloid in tobacco. The painkillers morphine and codeine are obtained from unripe opium poppy seed pods. They are narcotics which, with repeated usage, cause addiction. Coniine, extracted from hemlock, is the alkaloid that killed Socrates. He was sentenced to drink a brew made from hemlock because of his unconventional teaching methods and religious practices.

Teaching evaluations were tough in ancient Greece

25.3 ISOMERISM IN ORGANIC COMPOUNDS

Isomers are distinctly different compounds, with different properties, which have the same molecular formula. In Section 25.1, we briefly considered **structural isomers** of alkanes. You will recall that compounds such as butane and 2-methylpropane have the same molecular formula, C_4H_{10}, but different structural formulas (p. 594). In these, as in all structural isomers, the order in which the atoms are bonded to each other differs. Specifically, butane is a "straight-chain" alkane while 2-methylpropane has a branched carbon chain.

Structural isomerism is common among all types of organic compounds. Consider, for example:

1. The three structural isomers of the alkene C_4H_8:

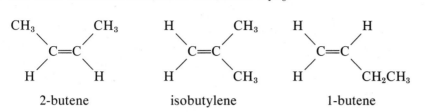

2-butene   isobutylene   1-butene

2. The two structural isomers of the three-carbon alcohol C_3H_7OH:

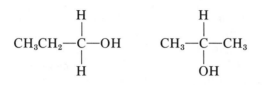

propyl alcohol   isopropyl alcohol

3. The two structural isomers dimethyl ether and ethyl alcohol, both of which have the molecular formula C_2H_6O:

$$CH_3-O-CH_3 \qquad CH_3-\overset{\overset{\displaystyle H}{|}}{\underset{\underset{\displaystyle H}{|}}{C}}-OH$$

dimethyl ether   ethyl alcohol

Example 25.4   Consider the molecule $C_3H_6Cl_2$, which is derived from propane by substituting two Cl atoms for H atoms. Draw the structural isomers of $C_3H_6Cl_2$.

Solution   There are four isomers. In one, both Cl atoms are bonded to a carbon atom at the end of the chain. In another, both Cl atoms are bonded to the central carbon atom:

$$CH_3-CH_2-\overset{\overset{\displaystyle Cl}{|}}{\underset{\underset{\displaystyle Cl}{|}}{C}}-H \qquad CH_3-\overset{\overset{\displaystyle Cl}{|}}{\underset{\underset{\displaystyle Cl}{|}}{C}}-CH_3$$

In the other two isomers, the Cl atoms are bonded to different carbons:

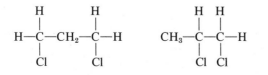

When drawing structural isomers, make sure the bonding is different in each one

---

Exercise   How many structural isomers are there of $C_3H_5Cl_3$? Answer: Five.

As we have seen, there are three structural isomers of the alkene C_4H_8. You may be surprised to learn that there are actually *four* different alkenes with this molecular formula. The "extra" compound arises because of a phenomenon called **geometric isomerism.** There are two different 2-butenes:

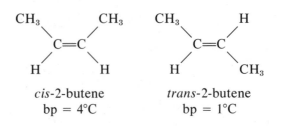

cis-2-butene
bp = 4°C

trans-2-butene
bp = 1°C

In the *cis* isomer, the two CH_3 groups (or the two H atoms) are as close to one another as possible. In the *trans* isomer, the two identical groups are farther apart. The two forms exist because there is no free rotation about the carbon-carbon double bond. The situation here is analogous to that with *cis-trans* isomers of square planar complexes (Chap. 19). In both cases, the difference in geometry is responsible for isomerism; the atoms are bonded to each other in the same way.

Geometric, or *cis-trans,* isomerism is common among alkenes. Indeed, it occurs with all alkenes *except those in which two identical atoms or groups are attached to one of the double-bonded carbons.* Thus, although 2-butene has *cis* and *trans* isomers, isobutylene and 1-butene (p. 609) do not.

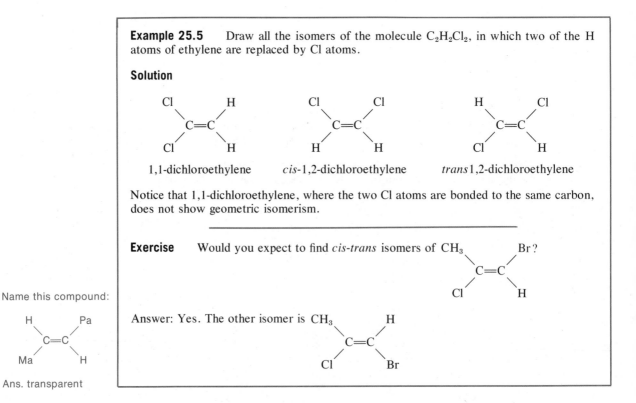

Example 25.5   Draw all the isomers of the molecule $C_2H_2Cl_2$, in which two of the H atoms of ethylene are replaced by Cl atoms.

Solution

1,1-dichloroethylene   *cis*-1,2-dichloroethylene   *trans*1,2-dichloroethylene

Notice that 1,1-dichloroethylene, where the two Cl atoms are bonded to the same carbon, does not show geometric isomerism.

Exercise   Would you expect to find *cis-trans* isomers of

Answer: Yes. The other isomer is

Name this compound:

Ans. transparent

This type of isomerism arises because of the tetrahedral nature of the bonding around a carbon atom. It occurs when at least one carbon atom in a molecule is bonded to four different atoms or groups. Consider, for example, the methane derivative CHClBrI. As you can see from Figure 25.7, there are two different forms of this molecule, which are mirror images of one another. The mirror images are not superimposable. That is, you cannot place one molecule over the other so that identical groups are touching. In this sense, the two isomers resemble right- and left-hand gloves. Optical isomers differ from geometric isomers in that the latter are not mirror images of one another.

A carbon atom with four different atoms or groups attached to it is referred to as a chiral center. A molecule containing such a carbon atom shows optical isomerism. It exists in two different forms which are nonsuperimposable mirror images. These forms are referred to as optical isomers or *enantiomers*. Molecules may contain more than one chiral center, in which case there can be more than two enantiomers.

Sometimes such a carbon atom is called optically active

Example 25.6   In the following structural formulas, locate each chiral carbon atom:

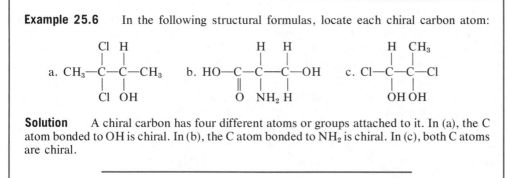

Solution   A chiral carbon has four different atoms or groups attached to it. In (a), the C atom bonded to OH is chiral. In (b), the C atom bonded to NH_2 is chiral. In (c), both C atoms are chiral.

Exercise   Consider the glycerol molecule, whose structural formula is shown on p. 604. Would you expect glycerol to show optical isomerism? Answer: No.

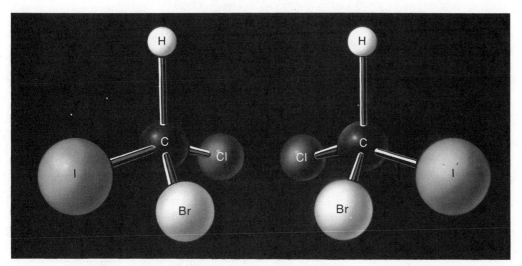

Figure 25.7   The two optical isomers of CHClBrI are mirror images of each other.

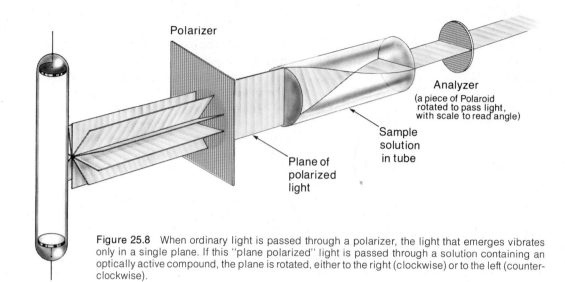

Figure 25.8 When ordinary light is passed through a polarizer, the light that emerges vibrates only in a single plane. If this "plane polarized" light is passed through a solution containing an optically active compound, the plane is rotated, either to the right (clockwise) or to the left (counterclockwise).

The term "optical isomerism" comes from the effect that enantiomers have upon plane polarized light, such as that produced by a Polaroid lens (Fig. 25.8). When this light is passed through a solution containing a single enantiomer, the plane is rotated from its original position. One isomer rotates it to the right (clockwise), and the other to the left (counterclockwise). If both isomers are present in equal amounts, we obtain what is known as a racemic mixture. In this case, the two rotations offset each other and there is no effect on plane polarized light.

In nature, usually only one of the isomers exists

Enantiomers ordinarily resemble each other closely in their physical and chemical properties. For example, the two forms of lactic acid have the same melting point (52°C), density (1.25 g/cm³), and acid dissociation constant ($K_a = 1.4 \times 10^{-4}$).

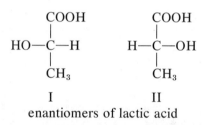

enantiomers of lactic acid

On the other hand, enantiomers frequently differ in their physiological activity. This was discovered by Louis Pasteur, the father of modern biochemistry. Working with a mixture of the optical isomers of lactic acid, he found that mold growth occurred only with enantiomer II. Apparently, the mold was unable to metabolize enantiomer I. A more modern example of this type involves amphetamine, often used illicitly as an "upper" or "pep pill." Amphetamine consists of two enantiomers:

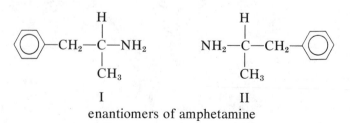

enantiomers of amphetamine

We don't know if there is a difference with left-handed people

Enantiomer I, called Dexedrine, is by far the stronger stimulant. It is about four times as active as Benzedrine, the racemic mixture of the two isomers.

SUMMARY

Hydrocarbons are classified as saturated, unsaturated, or aromatic. In saturated hydrocarbons (alkanes), all the carbon-carbon bonds are single bonds (Table 25.1, Example 25.1). The carbon atoms may form a "straight" (unbranched) chain or a branched chain. Unsaturated hydrocarbons contain carbon-carbon multiple bonds. There is one double bond in an alkene molecule and one triple bond in an alkyne molecule (Example 25.2). These compounds undergo addition reactions in which small molecules add directly to the multiple bond. Benzene, the simplest aromatic hydrocarbon, is a six-membered ring molecule with three pairs of delocalized electrons (Fig. 25.1). Gasoline is a complex mixture, containing mostly saturated hydrocarbons (Fig. 25.3).

Organic compounds may contain halogen, oxygen, and nitrogen atoms in addition to carbon and hydrogen. In such compounds, an atom or small group of atoms serves as a functional group (Table 25.4, Example 25.3). Compounds containing the same functional group show similar chemical properties.

Isomerism is common among organic molecules. Isomers have the same molecular formula but differ in their structures and properties. Three types of isomerism were considered in this chapter:

Type	Difference Between Isomers	Example
Structural	Bonding pattern of atoms	25.4
Geometric	Distance between groups	25.5
Optical	Molecules not superimposable	25.6

KEY WORDS AND CONCEPTS

hydrocarbon	addition reaction	aldehyde	fat
alkane	aromatic hydrocarbon	amine	soap
straight-chain alkane	substitution reaction	carboxylic acid	detergent
structural isomerism	cracking	ester	geometric isomerism
branched-chain alkane	isomerization	ether	optical isomerism
unsaturated hydrocarbon	octane number	ketone	chiral center
alkene	functional group		enantiomer
alkyne	alcohol		

QUESTIONS AND PROBLEMS

Catalog

Hydrocarbons: 25.2–25.6, 25.22–25.26
Functional Groups: 25.7–25.11, 25.27–25.31
Structural Isomerism: 25.12–25.14, 25.32–25.34

Geometric Isomerism: 25.15, 25.16, 25.35, 25.36
Optical Isomerism: 25.17, 25.18, 25.37, 25.38
General: 25.1, 25.19–25.21, 25.39–25.45

25.1   Review and know the meaning of the key words and concepts in this chapter.

25.2   Classify the following as alkanes, alkenes, or alkynes:

   a. C_5H_{10} b. C_5H_8
   c. C_8H_{18} d. C_7H_{12}

25.22   Write the molecular formula of a hydrocarbon containing six carbon atoms which is an

   a. alkane   b. alkene
   c. alkyne   d. aromatic

25.3 Write molecular formulas for

a. an alkane containing 20 hydrogen atoms.
b. a three-carbon alkene.
c. an alkyne with six hydrogen atoms.

25.4 Write the structural formula of the product formed when Br_2 adds to

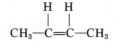

25.5 Fuels can have an octane number greater than 100. Explain why this is possible.

25.6 Give all the bond angles in the toluene molecule (see Fig. 25.2).

25.7 Classify each of the following as an alcohol, aldehyde, carboxylic acid, ester, ether, or ketone:

a. CH_3—O—CH_2—CH_3

b. CH_3—CH_2—$\underset{\underset{O}{\|}}{C}$—OH

c. CH_3—$\underset{\underset{O}{\|}}{C}$—O—$CH_3$

d. CH_3—CH_2—OH

e. CH_3—$\underset{\underset{O}{\|}}{C}$—$CH_2$—$CH_3$

25.8 Which of the following compounds would you expect to show hydrogen bonding?

a. CH_3—OH   b. CH_3—O—CH_3

c. H—$\underset{\underset{O}{\|}}{C}$—OH   d. H—$\underset{\underset{O}{\|}}{C}$—O—$CH_3$

25.9 Write the formulas of all the amines derived from NH_3 by replacing one or more H atoms with CH_3—CH_2— groups.

25.10 Write the structural formula of the ester formed by methyl alcohol with

a. formic acid   b. acetic acid
c. CH_3—$(CH_2)_6$—COOH

25.23 For the compounds in Question 25.3, write

a. the condensed structural formula for (a).
b. the structural formulas of (b) and (c).

25.24 Write the structural formula of the product formed when H_2 adds to

$$(CH_3)_2—C\overset{\overset{H}{|}}{=}C—CH_3$$

25.25 Straight-run gasoline has an octane number of about 70. What types of compounds are blended with it to raise the number to about 90?

25.26 Give all the bond angles in propylene and methyl acetylene (see Example 25.2).

25.27 Classify each of the following as an alcohol, aldehyde, carboxylic acid, ester, ether, or ketone:

a. CH_3—CH_2—$\underset{\underset{O}{\|}}{C}$—H

b. CH_3—$\underset{\underset{O}{\|}}{C}$—$CH_3$

c. CH_3—$(CH_2)_5$—OH

d. H—$\underset{\underset{O}{\|}}{C}$—$OCH_3$

25.28 Which of the following compounds would you expect to show hydrogen bonding?

a. CH_3—$\underset{\underset{O}{\|}}{C}$—$CH_3$ b. CH_3—NH_2

c. CH_3—$\underset{\underset{O}{\|}}{C}$—OH   d. $(CH_3)_3N$

25.29 Which of the following would form basic water solutions? acidic water solutions?

a. CH_3—CH_2—OH

b. CH_3—CH_2—$\underset{\underset{O}{\|}}{C}$—OH

c. CH_3—$\underset{\underset{H}{|}}{N}$—$CH_2$—$CH_3$

d. CH_3—CH_2—O—CH_3

25.30 Write the structural formulas of all the esters that can be formed by ethylene glycol with formic and acetic acids. (One or both of the —OH groups of ethylene glycol can react.)

25.11 Draw the structure of a fat which, upon treatment with NaOH, gives glycerol and the sodium salt of an organic acid containing 14 carbon atoms.

25.12 How many different dibromobenzenes could one hope to isolate?

25.13 Alkanes undergo substitution reactions with halogens to form many products. How many trichlorinated isomers could be made from propane, C_3H_8?

25.14 Draw skeleton structures of all the alcohols with the molecular formula $C_5H_{12}O$.

25.15 Malèic acid and fumaric acid are the *cis* and *trans* isomers, respectively, of $C_2H_2(COOH)_2$, a dicarboxylic acid. Draw and label their structural formulas.

25.16 Write structural formulas for all the isomers, structural or geometric, of the alkenes of molecular formula C_5H_{10}.

25.17 Locate the chiral carbon(s), if any, in the following molecules:

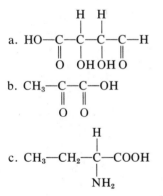

25.18 Which of the following would you expect to show optical isomerism?

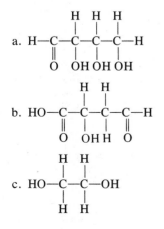

25.31 Draw the structure of a detergent (sodium salt) containing 14 carbon atoms.

25.32 How many different tribromobenzenes could there be?

25.33 How many isomers with the formula $C_4H_7Cl_3$ could one obtain by reacting Cl_2 with C_4H_{10}?

25.34 Draw skeleton structures for all the saturated carboxylic acids that contain four carbon atoms.

25.35 For which of the following is geometric isomerism possible?

    a. $(CH_3)_2C{=}CCl_2$
    b. $CH_3ClC{=}CCH_3Cl$
    c. $CH_3BrC{=}CCH_3Cl$

25.36 Write structural formulas for all the isomers, structural or geometric, of the alkene derivatives of molecular formula C_4H_6ClBr, in which Cl and Br are bonded to a double-bonded C.

25.37 Locate the chiral carbon(s), if any, in the following molecules:

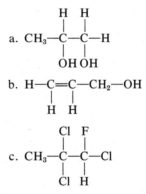

25.38 Which of the following would you expect to show optical isomerism?

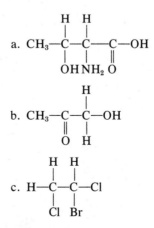

25.19 Noting that each carbon atom forms four bonds, put in the hydrogen atoms in the structure of 3,4-benzpyrene (p. 598) and give the molecular formula.

25.20 Write a balanced equation for the reaction that uses synthesis gas, a mixture of CO and H_2, to make C_4H_{10} and H_2O.

25.21 Calculate the mass % of carbon in a compound containing two carbon atoms which is an

    a. alkane   b. alkene
    c. acid   d. alcohol

25.39 Put H atoms in the proper places in the morphine molecule (Fig. 25.6) and give its molecular formula.

25.40 Write a balanced equation for the reaction that uses synthesis gas to make C_6H_{14} and H_2O.

25.41 Arrange the compounds acetaldehyde, acetic acid, acetylene, ethane, ethyl alcohol, and ethylene in order of decreasing mass % of carbon.

*25.42 Heroin is a diester derivative of morphine (Fig. 25.6). It can be considered to be the reaction product of two moles of acetic acid per mole of morphine. What are the percentages by mass of the elements present in heroin?

*25.43 A certain PCB (Table 25.3) contains 49.4% C, 48.6% Cl, and 2.07% H. What is its molecular formula? Draw structural formulas for at least five isomers of this compound.

*25.44 Noting the composition of the product gas in the Lurgi process (15% CO, 15% CH_4, 30% CO_2, 40% H_2), write a balanced equation for the overall reaction that occurs, starting with carbon, oxygen, and steam. What is ΔH for this reaction per mole of carbon consumed?

*25.45 Gasohol is a mixture of 10% ethyl alcohol and 90% gasoline by volume. The heat evolved in the combustion of gasoline is about 46 kJ/g and its density is 0.68 g/cm³. The density of ethyl alcohol is 0.79 g/cm³; its heat of combustion can be calculated from data in Table 23.1. Calculate ΔH for the combustion of one liter of gasohol and compare it to that for the combustion of one liter of gasoline.

ORGANIC POLYMERS: SYNTHETIC AND NATURAL

In previous chapters, we discussed the properties of a great many different substances. Almost all of these consisted of either small molecules or relatively simple ions. This chapter is concerned with *polymers*. These are solid substances, usually of high molecular mass, which are made by combining small molecular units called monomers. A typical polymer molecule may contain a chain of monomers several thousand units long. Some polymers are derived from a single monomer unit, repeated over and over again. Others may contain many different kinds of monomer units.

The properties of a polymer depend upon its molecular structure. Melting point, chemical reactivity, resistance to solvents, mechanical strength, and flexibility vary according to the type, number, and arrangement of monomer units. Polymers in which monomer units are arranged in nearly parallel chains form fibers which may be very strong and flexible. If the chains are tangled, the polymer may form as a strong film or bulk solid.

Natural polymers produced by plants and animals are essential to all forms of life. The cell walls of plants and the woody structures of trees are composed largely of cellulose, a polymeric carbohydrate (Section 26.3). Your body contains thousands of polymers, which make up tissues, blood, and skin. Many of these fall in the class of organic polymers known as proteins (Section 26.4). Natural polymers are also useful in our daily lives. For centuries, clothes have been made from cotton (largely cellulose) and wool, silk, and leather (protein materials).

Since about 1930, a wide variety of synthetic polymers (Sections 26.1 and 26.2) have become available. Examples include such well-known textile materials as Dacron and nylon. Most synthetic polymers, like natural polymers, are organic compounds of high molecular mass. They may contain, in addition to carbon and hydrogen, such elements as oxygen, nitrogen, and the halogens. Their molecular structures are much simpler than those of most natural polymers. Commonly, synthetic polymers are built from one or at most two different types of monomer units.

26.1 SYNTHETIC ADDITION POLYMERS

An addition polymer is formed by a reaction in which monomer units add to one another. The molecular mass of an addition polymer is the sum of the masses of the

TABLE 26.1 SOME COMMON ADDITION POLYMERS

MONOMER	NAME	POLYMER	USES
H₂C=CH₂ (H, H / C=C / H, H)	Ethylene	Polyethylene	Bags, coatings, toys
(H, H / C=C / H, CH₃)	Propylene	Polypropylene	Beakers, milk cartons
(H, H / C=C / H, Cl)	Vinyl chloride	Polyvinyl chloride, PVC	Floor tile, raincoats, pipe, phonograph records
(H, H / C=C / H, CN)	Acrylonitrile	Polyacrylonitrile, PAN	Rugs; Orlon and Acrilan are copolymers with other monomers
(H, H / C=C / H, C₆H₅)	Styrene	Polystyrene	Cast articles using a transparent plastic
(H, CH₃ / C=C / H, C—OCH₃ ‖ O)	Methyl methacrylate	Plexiglas, Lucite, acrylic resins	High quality transparent objects, latex paints
(F, F / C=C / F, F)	Tetrafluoroethylene	Teflon	Gaskets, insulation, bearings, pan coatings

These polymers were essentially unknown before WW II

monomer units from which it is formed. Table 26.1 lists some of the more familiar synthetic addition polymers. You will notice that each of these is derived from a monomer containing a carbon-carbon double bond. That is,

Upon polymerization, the double bond is converted to a single bond, and successive monomer units add to one another.

Polyethylene

The most common addition polymer is polyethylene, a solid derived from the monomer ethylene. We can represent the polymerization process as

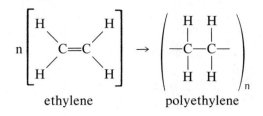

ethylene → polyethylene

$$(26.1)$$

All the monomer ends up in the polymer

Here, n is a very large number, of the order of 2000.

Example 26.1   What is
  a. the molecular mass of polyethylene, if $n = 2.0 \times 10^3$?
  b. the simplest formula of polyethylene?

Solution
  a. the molecular mass of ethylene is $2(12.0) + 4(1.0) = 28.0$
     MM polyethylene $= 2.0 \times 10^3 \times 28.0 = 5.6 \times 10^4$
  b. CH_2, the same as that of ethylene itself.

Exercise   What are the mass percentages of C and H in polyethylene? Answer: 85.6, 14.4.

There are several different ways to initiate Reaction 26.1. One method uses a small amount of a very reactive species called a *free radical,* which contains an unpaired electron. Such a species can be produced by adding the organic compound benzoyl peroxide to ethylene. Upon heating, the benzoyl peroxide decomposes to form a free radical:

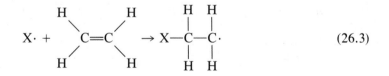

$$(26.2)$$

benzoyl peroxide   a free radical, X·

This free radical, which we will represent simply as X·, reacts rapidly with ethylene to form a species which still contains an unpaired electron:

$$\text{X·} + \text{C}=\text{C} \rightarrow \text{X}-\text{C}-\text{C·} \qquad (26.3)$$

The product of this reaction is able to add another ethylene molecule, and then another, and then another, ... Eventually, very long chains are formed, each with an unpaired electron at one end. Two of these chains can combine to terminate the polymerization process. The product formed by free radical polymerization of ethylene is one in which most of the chains are branched. Its structure might be shown as

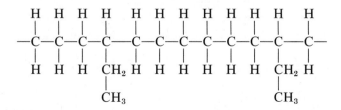

Figure 26.1 Two polyethylene bottles. The one on the left is made of semi-rigid, linear polyethylene. The other is made of more pliable, branched polyethylene. (From Chickos, J., et al.: *Chemistry; Its Role in Society.* New York, Heath, 1973, p. 94.)

In this polymer, known as *branched* polyethylene, neighboring chains are arranged in a somewhat random fashion, often overlapping each other. This produces a soft, flexible solid, used mostly in films and coatings where a pliable material is needed.

With a different type of catalyst (p. 622), it is possible to produce *linear* polyethylene. This consists almost entirely of unbranched chains. That is,

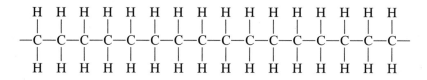

Linear P.E. is much harder and melts higher than branched P.E.

Neighboring chains in linear polyethylene line up nearly parallel to each other. This gives a polymer which approaches a crystalline material. It is used for bottles, toys, and other semirigid objects (Fig. 26.1).

Polyvinyl Chloride and Other Addition Polymers

Vinyl chloride, in contrast to ethylene, is an unsymmetrical molecule. We might refer to the CH_2 group in vinyl chloride as the "head" of the molecule and the CHCl group as the "tail":

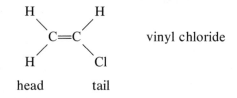

vinyl chloride

head   tail

In principle at least, vinyl chloride molecules can add to one another in any of three ways to form:

1. A *head-to-tail* polymer, in which there is a Cl atom on every other C atom in the chain:

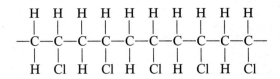

2. A *head-to-head, tail-to-tail* polymer, in which Cl atoms occur in pairs on adjacent carbon atoms in the chain:

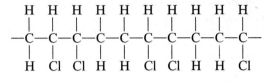

3. A *random* polymer:

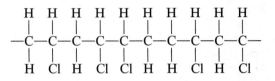

Similar arrangements are possible with polymers made from other unsymmetrical monomers (Example 26.2). In practice, the addition polymer is usually of the head-to-tail type. This is the case with polyvinyl chloride and polypropylene.

The monomer usually knows which end is up

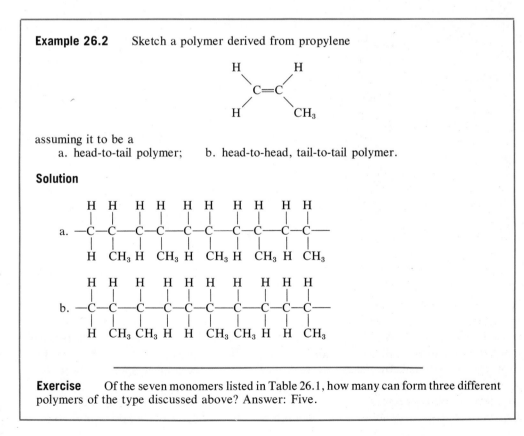

Example 26.2   Sketch a polymer derived from propylene

$$\underset{H}{\overset{H}{\diagdown}} C = C \underset{CH_3}{\overset{H}{\diagup}}$$

assuming it to be a
    a. head-to-tail polymer;   b. head-to-head, tail-to-tail polymer.

Solution

a.

b.

Exercise   Of the seven monomers listed in Table 26.1, how many can form three different polymers of the type discussed above? Answer: Five.

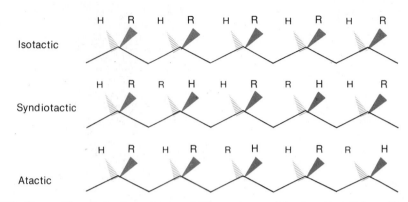

Figure 26.2 Three different structures for an addition polymer of the head-to-tail type. The solid line represents a carbon chain, with each lower vertex being a CH_2 group; the plane of the chain is that of the page. Depending on polymerization conditions, the polymer obtained can have any of the three structures shown.

You may be surprised to learn that the structure of polyvinyl chloride or polypropylene is not completely established by knowing that the monomers are arranged head-to-tail. Three different configurations of the polymer chain are possible. These are shown in Figure 26.2 for the general monomer CH_2=CHR (R = Cl for vinyl chloride, R = CH_3 for propylene).

In the isotactic configuration, the R group is always in the same direction with respect to the chain. That is, if one looks at the chain from above, all the R groups would be on the same side, say to the right. In the syndiotactic configuration, R groups alternate to the left and right down the chain. The structure in which R groups are randomly arranged is called atactic.

In about 1955, Giulio Natta, an Italian chemist, discovered some catalysts which allowed him to make both pure isotactic and pure syndiotactic polypropylene. Both of these have more desirable properties than the atactic (random) structure, which was the form produced by polymerization processes at that time. These catalysts contain a reducing agent and a transition metal salt; a common one is a mixture of $Al(C_2H_5)_3$ and $TiCl_4$. Natta catalysts proved to be extremely useful for controlling polymer structures. Karl Ziegler in Germany used them to make the first linear polyethylene. In 1963, these two men shared the Nobel Prize in chemistry for their discoveries.

The process used to make polyvinyl chloride is typical of that used with most addition polymers. Polymerization is carried out in a suspension of vinyl chloride in water at about 50°C, using an appropriate catalyst. The polymer forms granules as small as about 0.1 mm in diameter. A considerable amount of heat is liberated in the polymerization process and absorbed by the water:

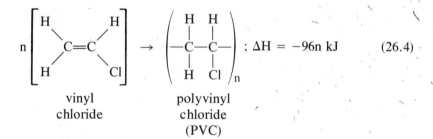

$$n \begin{bmatrix} H \\ \diagdown \\ C=C \diagup \\ H \diagup \diagdown Cl \end{bmatrix} \rightarrow \left(\begin{array}{cc} H & H \\ | & | \\ -C-C- \\ | & | \\ H & Cl \end{array} \right)_n \quad ; \Delta H = -96n \text{ kJ} \qquad (26.4)$$

vinyl chloride

polyvinyl chloride (PVC)

After drying, the PVC can be treated and formed into rigid articles like water pipe and steering wheels. Plasticizers are added to make it more flexible. These substances act by decreasing the dispersion forces between the chains of polyvinyl chloride. PVC can also be softened by polymerizing it with another monomer. Such "copolymers" containing PVC are used in garden hose, phonograph records, automobile seat covers, and swimming pool liners.

Most polymers decompose at high temperatures. For example, polyvinyl chloride starts to decompose at 100°C, liberating HCl. Teflon, a polymer of tetrafluoroethylene (Table 26.1) behaves quite differently. A hard waxy solid, Teflon is useful from about −70 to 250°C. Teflon is also extremely unreactive. It resists chemical attack by all known reagents except molten alkali metals. Its best known use is as a nonstick coating for frying pans. Teflon was discovered by accident at the Du Pont Company in 1938 during basic research on halogenated hydrocarbons. Louis Pasteur's maxim that "chance favors the prepared mind" was applied by Dr. Ray Plunkett and coworkers who discovered and developed Teflon.

For rugged gaskets, Teflon wins hands down

Rubber: Natural and Synthetic

Rubber is the most important natural addition polymer. It is obtained as a water suspension called latex, extracted from the *Hevea* (rubber) tree. The latex is collected and treated with acetic acid to precipitate the raw rubber.

Natural rubber is a polymer of isoprene, a molecule containing two carbon-to-carbon double bonds (see below). In the polymer, there is one double bond per isoprene unit. These units are joined head-to-tail with the CH_2 groups in a *cis* position relative to the double bond. On the average, rubber molecules contain about 6000 monomer units.

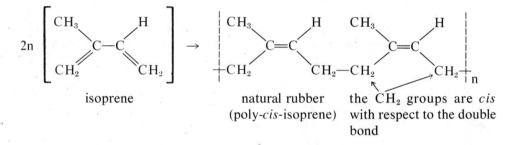

isoprene   natural rubber (poly-*cis*-isoprene)   the CH_2 groups are *cis* with respect to the double bond

Since raw rubber becomes sticky when warm and brittle when cold, it cannot be used directly. However, heating rubber with sulfur gives a product which is resilient over a wide temperature range. This reaction, called *vulcanization,* was discovered accidentally by Charles Goodyear in 1834. During vulcanization, the long hydrocarbon chains become cross-linked by short chains of one to six sulfur atoms. Cross-linking establishes a pattern in the solid which the chains, in a sense, "remember." They return to that pattern when the rubber is stretched and then released (Fig. 26.3). Ordinary rubber contains about 3% sulfur. Hard rubber, containing about 30% sulfur, has many more cross-links. This makes it rigid and brittle.

All rubbers need to be vulcanized

Since about 1930, several synthetic rubbers have been produced commercially. Wallace Carothers of Du Pont discovered **neoprene,** an addition polymer of chloroprene. Chloroprene (p. 624), like isoprene, has two carbon-to-carbon double bonds in the molecule.

Figure 26.3 Schematic drawing showing cross-linking in rubber. The sulfur cross-links are a few atoms long and prevent the flexible chains from moving with respect to each other. When tension is released, the chains assume their previous, more random, arrangement.

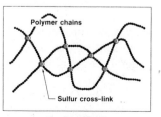

Unstretched

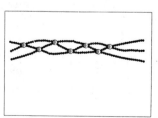

Stretched

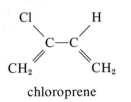

chloroprene

The structure of neoprene resembles that of natural rubber, with a Cl atom replacing the CH_3 group. Although more expensive than natural rubber, neoprene is superior in its resistance to attack by solvents and oxidizing agents. Shoe soles and gasoline fuel lines are commonly made from neoprene.

Another common synthetic rubber is **SBR**, a copolymer formed from the two monomers styrene and 1,3-butadiene:

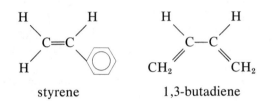

styrene   1,3-butadiene

SBR contains one part of styrene to three parts of 1,3-butadiene. It was first produced in the United States during World War II when natural rubber supplies from Southeast Asia were cut off. Currently, it is the most widely used synthetic rubber, mainly as tire tread material.

With the proper catalysts, 1,3-butadiene can be polymerized by itself to give a product very similar in its properties to natural rubber. The low cost of the process has made this synthetic rubber second only to SBR in annual production.

1,3-butadiene is easier to make than isoprene

26.2 SYNTHETIC CONDENSATION POLYMERS

In forming a condensation polymer, monomer units combine by splitting out a small molecule, most often water. Usually, two different monomers are involved. Each of these monomers has a functional group, such as

$$-\overset{\displaystyle O}{\underset{\displaystyle \parallel}{C}}-OH, \quad -OH, \quad -\overset{\displaystyle}{\underset{\displaystyle H}{N}}-H$$

at *both ends of the molecule.* We will consider two of the most common types of synthetic condensation polymers: polyesters and polyamides.

Polyesters

You will recall from Chapter 25 that an alcohol reacts with a carboxylic acid to form an ester. Representing the alcohol as R—OH and the acid as R′—COOH, we might write the general reaction as

$$R-OH + HO-\overset{\displaystyle O}{\underset{\displaystyle \parallel}{C}}-R' \rightarrow R-O-\overset{\displaystyle O}{\underset{\displaystyle \parallel}{C}}-R' + H_2O$$

Suppose now that the alcohol has two OH groups (HO—R—OH) and that the acid has two COOH groups (HOOC—R′—COOH). In this case, the ester formed still has a reactive group at each end of the molecule.

(26.5)

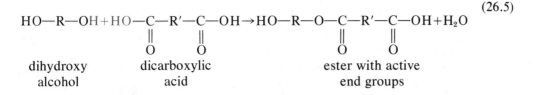

dihydroxy   dicarboxylic   ester with active
alcohol   acid   end groups

The COOH group at one end of the ester molecule can react with another alcohol molecule. The OH group at the other end can react with an acid molecule. This process can continue, leading eventually to a long-chain polymer containing 500 or more ester groups. The general structure of the **polyester** can be represented as

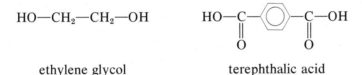

section of a polyester molecule

One of the most familiar polyesters is Dacron. Here, the monomers are ethylene glycol and terephthalic acid:

HO—CH$_2$—CH$_2$—OH

ethylene glycol   terephthalic acid

As a thin film the polymer is called Mylar, used in magnetic recording tape and weather balloons.

Example 26.3
    a. Write the structural formula of the ester formed when one molecule of ethylene glycol reacts with one molecule of terephthalic acid.
    b. Draw the structure of a section of the Dacron (Mylar) polymer.

Solution

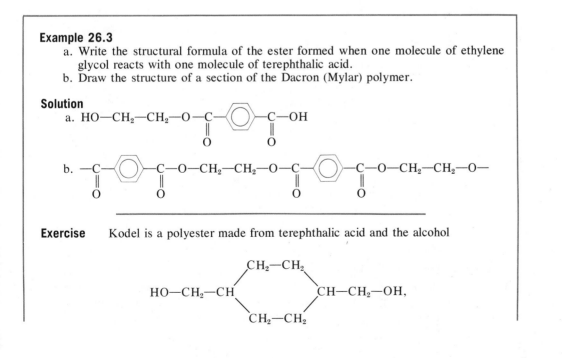

Exercise   Kodel is a polyester made from terephthalic acid and the alcohol

Draw the structure of a portion of the Kodel polymer. Answer:

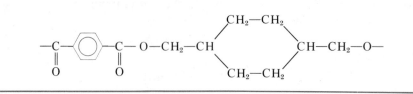

Polyamides

The reaction of an amine with a carboxylic acid leads to the formation of an *amide*:

$$\text{R}-\underset{\underset{\text{H}}{|}}{\text{N}}-\text{H} \quad + \quad \text{HO}-\underset{\underset{\text{O}}{\|}}{\text{C}}-\text{R}' \quad \rightarrow \quad \text{R}-\underset{\underset{\text{H}}{|}}{\text{N}}-\underset{\underset{\text{O}}{\|}}{\text{C}}-\text{R}' \quad + \quad \text{H}_2\text{O}$$

amine acid amide

This reaction is analogous to ester formation. The NH_2 group of the amine takes the place of the OH group of the alcohol.

When a diamine (molecule containing two NH_2 groups) reacts with a dicarboxylic acid (two COOH groups), a **polyamide** is formed. This condensation polymerization is entirely analogous to that used to make polyesters. Thus, we can write similar equations for the formation of a polyamide (Example 26.4).

Example 26.4   Consider the diamine $H_2N-(CH_2)_6-NH_2$ and the dicarboxylic acid $HOOC-(CH_2)_4-COOH$. Give the
  a. structural formula of the amide formed when one molecule of diamine reacts with one molecule of dicarboxylic acid.
  b. structure of a section of the polyamide formed by these two monomers.

Solution

  a. $H_2N-(CH_2)_6-\underset{\underset{H}{|}}{N}-\underset{\underset{O}{\|}}{C}-(CH_2)_4-\underset{\underset{O}{\|}}{C}-OH$

  b.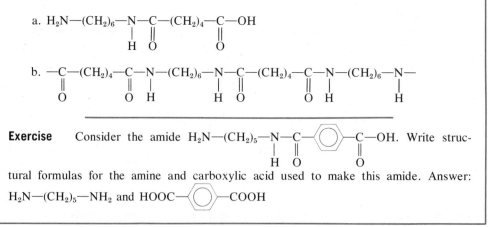

Exercise   Consider the amide $H_2N-(CH_2)_5-\underset{\underset{H}{|}}{N}-\underset{\underset{O}{\|}}{C}-\bigcirc-\underset{\underset{O}{\|}}{C}-OH$. Write structural formulas for the amine and carboxylic acid used to make this amide. Answer:
$H_2N-(CH_2)_5-NH_2$ and $HOOC-\bigcirc-COOH$

Modern industrial chemical research dates from the discovery of nylon

The polyamide in Example 26.4 is Nylon 66 (the numbers indicate the number of carbon atoms in the amine and acid monomers). This polyamide was first made in 1935 by Wallace Carothers, the discoverer of neoprene. Since then other nylons, all polyamides, have been synthesized. The earliest use of nylon was as a textile, especially

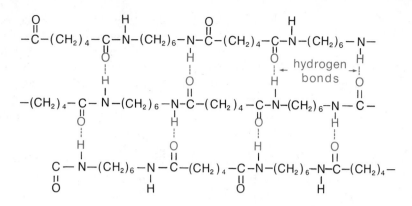

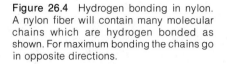

Figure 26.4 Hydrogen bonding in nylon. A nylon fiber will contain many molecular chains which are hydrogen bonded as shown. For maximum bonding the chains go in opposite directions.

for women's hosiery. Careful control of chain length during manufacturing gives nylon fibers of the desired sheerness and luster. The elasticity of nylon fibers is due in part to hydrogen bonds between adjacent polymer chains. These hydrogen bonds join carboxyl oxygen atoms on one chain to NH groups on adjacent chains (Fig. 26.4). Bulk nylon can be molded or cut into a variety of shapes. Its high wear resistance and slight slipperiness make bulk nylon ideal for door latches and gears.

26.3 CARBOHYDRATES

Carbohydrates, which comprise one of the three basic classes of foodstuffs, contain carbon, hydrogen, and oxygen atoms. Their general formula, $C_n(H_2O)_m$, is the basis for their name. They can be classified as:

Carbohydrates are not made by hydrating carbon

1. *Monosaccharides,* which cannot be broken down chemically to simpler carbohydrates. The most familiar monosaccharides, including glucose and fructose, contain six carbon atoms per molecule and have the molecular formula $C_6H_{12}O_6$.

2. *Disaccharides,* which are dimers formed when two monosaccharide units combine with the elimination of H_2O. The monosaccharides may be the same (two glucose units in maltose) or different (a glucose and fructose unit in sucrose).

3. *Polysaccharides,* which are condensation polymers containing from several hundred to several thousand monosaccharide units. Cellulose and starch are the most common polysaccharides.

Generally, mono- and disaccharides are sugars and have names ending in *-ose.* Glucose, sucrose (cane sugar), and lactose (milk sugar) are common examples. We will look at the structure and properties of one typical monosaccharide, glucose.

Glucose

The oxidation of glucose, $C_6H_{12}O_6$, is an important energy source for animals:

$$C_6H_{12}O_6(aq) + 6\ O_2(g) \rightarrow 6\ CO_2(g) + 6\ H_2O; \Delta H = -2880\ kJ \qquad (26.6)$$

In living cells, this reaction occurs in a complex series of over 20 steps, each catalyzed by a specific enzyme. Part of the energy released is stored in energy-rich compounds such as adenosine triphosphate, ATP. About 30% of the energy evolved in Reaction 26.6 can be converted into mechanical energy for muscular activities such as running or lifting

627

chemistry books. This energy conversion occurs at a constant temperature within the body, 37°C. No manmade machine could get any work at all from such a process.

Glucose in water solution exists primarily in the form of six-membered ring molecules, shown below. The rings are not planar but have a three-dimensional "chair" form. The carbon atoms at each corner of the ring are not shown in the structure; they are represented by the numbers 1 to 5. The sixth atom in the ring is oxygen. There is a CH_2OH group bonded to carbon atom 5. There is a H atom bonded to each of the ring carbons; an OH group is bonded to carbon atoms 1 through 4. The heavy lines indicate the front of the ring, projected toward you.

The bonds to the H and OH groups from the ring carbons are oriented either in the plane of the ring (*equatorial*) or perpendicular to it (*axial*). In β-glucose, all four OH groups are equatorial. This keeps these rather bulky groups out of each other's way. In α-glucose, the OH group on carbon atom 1 is axial; all the others are equatorial. In water solution, glucose exists as an equilibrium mixture, about 37% in the alpha form and 63% in the beta form.

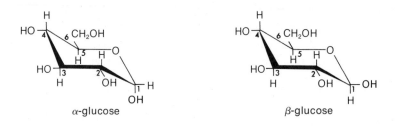

α-glucose   β-glucose

Although glucose can exist as a simple sugar, it is most often found in nature in combined form, as a disaccharide or polysaccharide. Several glucose-containing disaccharides are known. We will consider two of these, maltose and sucrose.

Maltose and Sucrose

Maltose is a dimer of two α-glucose molecules. These are combined head-to-tail; carbon atom 1 of one molecule is joined through an oxygen atom to carbon atom 4 of the second molecule. To form maltose, the two OH groups on these carbon atoms react, condensing out H_2O and leaving the O atom bridge.

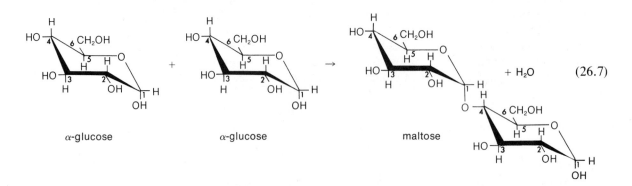

α-glucose   α-glucose   maltose $+ H_2O$ (26.7)

Reaction 26.7 can be reversed either by using the enzyme maltase as a catalyst or by heating with acid. Two molecules of α-glucose are formed when this happens.

Sucrose, the compound we call "sugar," is the most common disaccharide. One of the monomer units in sucrose is α-glucose. The other is fructose, a monosaccharide found in honey and natural fruit juices.

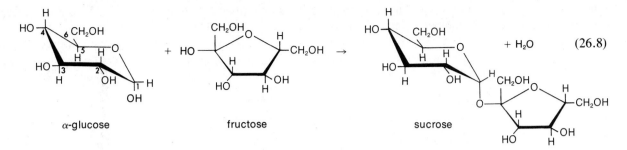

$$\alpha\text{-glucose} \qquad\qquad \text{fructose} \qquad\qquad \text{sucrose} \qquad\qquad + H_2O \qquad (26.8)$$

Reaction 26.8 can be reversed by acid or the enzyme sucrase. This occurs in digestion, and makes glucose and fructose available for absorption into the blood.

Example 26.5   Referring to the molecular structures just written, determine the molecular formulas of maltose and sucrose.

Solution   By counting atoms, we find that there are 12 C atoms, 22 H atoms, and 11 O atoms in each molecule. Both maltose and sucrose have the molecular formula $C_{12}H_{22}O_{11}$.

Exercise   Write a balanced chemical equation for the hydrolysis (reaction with water) of maltose. Answer: $C_{12}H_{22}O_{11}(aq) + H_2O \rightarrow 2\ C_6H_{12}O_6(aq)$

Disaccharide formation can be considered to be the first step in the polymerization of monosaccharides. An OH group in one disaccharide molecule can condense with an OH group in another molecule, eliminating H_2O. Eventually, a polymer is formed consisting of monosaccharide units bridged by oxygen atoms. Many natural polysaccharides have been isolated. The two most important and abundant are starch and cellulose.

Starch

Starch is a polysaccharide found in many plants, where it is stored in roots and seeds. It is particularly abundant in corn and potatoes, the major sources of commercial starch.

Starch is actually a mixture of two types of α-glucose polymers. One of these, called amylose, is insoluble in water and comprises about 20% of natural starch. It consists of long single chains of 1000 or more α-glucose units joined head-to-tail, as in maltose. The other glucose polymer found in starch is amylopectin, which is soluble in water. The linkage between α-glucose units is the same as in amylose. However, amylopectin has a highly branched structure. Short chains of 20 to 25 glucose units are linked through oxygen

Figure 26.5 In starch, glucose molecules are joined head-to-tail through oxygen atoms. A thousand or more glucose molecules may be linked in this way, either in long single chains (amylose) or branched chains (amylopectin).

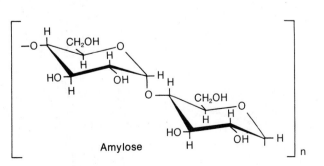

Amylose

629

atom bridges. The oxygen atom joins a number 1 carbon atom at the end of one chain to a number 6 carbon atom on an adjacent chain.

Plants use starch as their main source of glucose. In animals, glucose is stored mainly in the form of glycogen. This polysaccharide has a structure similar to that of amylopectin except that it is more highly branched. Glycogen is found principally in the liver and in muscle tissue.

Example 26.6   What is the simplest formula of amylose?

Solution   Amylose is a condensation polymer of glucose, whose formula is $C_6H_{12}O_6$. Every time a glucose unit is added to the polysaccharide chain a molecule of H_2O is eliminated. Hence, the unit added is not $C_6H_{12}O_6$, but rather $C_6H_{10}O_5$. This is the simplest formula of amylose.

Exercise   Amylose has a molecular mass of about 3.00×10^5. How many glucose units does an amylose molecule contain? Answer: 1850.

Cellulose

Cellulose contains long, unbranched chains of glucose units, about 10,000 per chain. It differs from starch in the way the glucose units are joined to each other. In starch, the oxygen bridge between the units is in the alpha position; in cellulose it is in the beta position (Fig. 26.6). Because humans lack the enzymes required to catalyze the hydrolysis of beta linkages, we cannot digest cellulose. The cellulose that we take in from fruits and vegetables remains undigested as "roughage."

The molecular structure of cellulose, unlike that of starch, allows for strong hydrogen bonding between polymer chains. This results in the formation of strong water-resistant fibers such as those found in cotton, which is 98% cellulose. Cotton actually has a tensile strength greater than that of steel. The major industrial source of cellulose is wood (~50% cellulose). Wood chips can be treated with hot NaOH solution, which dissolves some of the wood components and partially breaks down the polymer chains. The insoluble residue is impure cellulose in the form of wood pulp. It can be used directly to make paper or treated further to make various plastics. Although cellulose can be degraded completely to glucose, the process has not been practical economically. If it could be made so, the glucose could be used for food, or converted to ethanol by fermentation. Both such uses would be important, in view of world shortages of food and energy.

The world needs a good way to make glucose from cellulose

The OH groups of cellulose can be esterified. The most important cellulose esters are those formed with acetic acid and nitric acid. Cellulose acetate is used to make photographic film and synthetic fibers (Celanese, Estron). Cellulose nitrate is used to make explosives such as smokeless powder, the main propellant in shotgun and rifle shells.

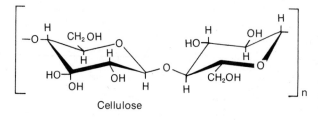

Cellulose

Figure 26.6 Cellulose structure. The bonding between glucose rings in cellulose is through oxygen bridges in the β position for each ring to the left of the bridge. This structure allows for ordered hydrogen bonding between chains and formation of long strong fibers. Cotton and wood fiber would have structures like that shown.

Celluloid, the first commercial plastic (1870) was made from cellulose nitrate. The motivation for its synthesis was a $10,000 prize offered for a product that could replace ivory in billiard balls. For about 50 years, celluloid was the only plastic made. The fact that celluloid has a noticeable odor, is highly flammable, and is attacked readily by many solvents encouraged chemists to look for substitutes. *But it took them a long time*

26.4 PROTEINS

The natural polymers known as proteins make up about 15% by mass of our bodies. They serve many functions. Fibrous proteins are the main components of hair, muscle, and skin. Other proteins found in body fluids transport oxygen, fats, and other substances needed for metabolism. Still others, like insulin and vasopressin, are hormones. Enzymes, which catalyze reactions in the body, are chiefly protein.

Protein is an important component of most foods. Nearly everything we eat contains at least a small amount of protein. Lean meats and vegetables such as peas and beans are particularly rich in protein. In our digestive system, proteins are broken down into small molecules called *α-amino acids*. These molecules can then be reassembled in cells to form other proteins required by the body.

α-Amino Acids

The monomers from which proteins are derived have the general structure shown below. These compounds are called α-amino acids. They have an NH_2 group attached to the carbon atom (the α carbon) adjacent to a COOH group.

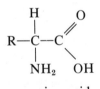

α-amino acid

Natural proteins can be broken down into about 20 different α-amino acids (Table 26.2). These molecules differ in the nature of the R group attached to the alpha carbon. As you can see from the table, R can be:

—a H atom (glycine).
—a simple hydrocarbon group (alanine, valine, . . .).
—a more complex group containing one or more atoms of oxygen (serine, aspartic acid, . . .), nitrogen (lysine, . . .), or sulfur (methionine, . . .).

Amino acids contain the COOH group, which is acidic, and the NH_2 group, which is basic. In the solid state, an amino acid ordinarily exists as a *zwitterion* formed by the transfer of a proton from a COOH group to an NH_2 group:

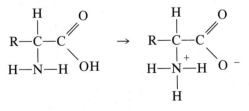

amino acid   zwitterion

TABLE 26.2 THE COMMON α-AMINO ACIDS
(R groups are indicated by shaded areas)

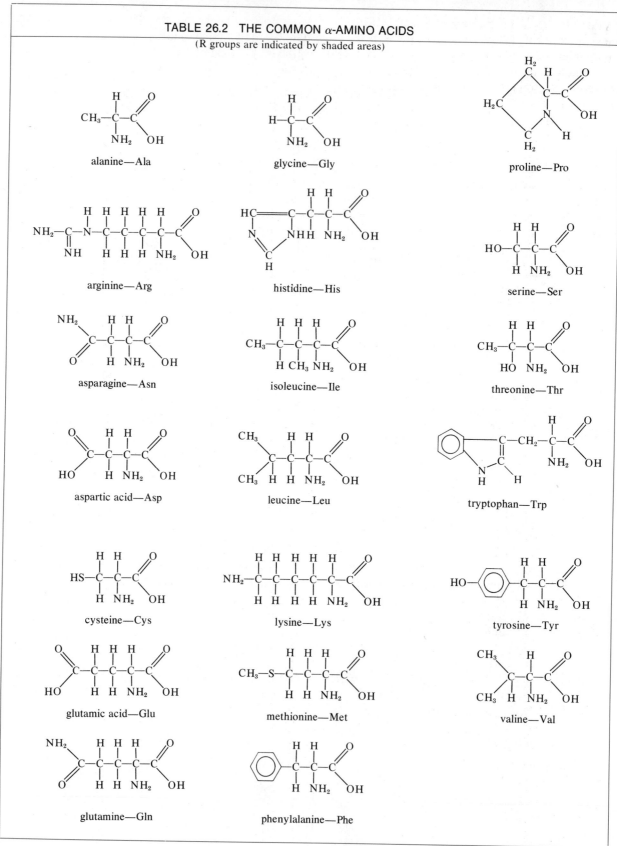

alanine—Ala

glycine—Gly

proline—Pro

arginine—Arg

histidine—His

serine—Ser

asparagine—Asn

isoleucine—Ile

threonine—Thr

aspartic acid—Asp

leucine—Leu

tryptophan—Trp

cysteine—Cys

lysine—Lys

tyrosine—Tyr

glutamic acid—Glu

methionine—Met

valine—Val

glutamine—Gln

phenylalanine—Phe

A zwitterion behaves like a polar molecule. Within it, there is a +1 charge at the nitrogen atom and a −1 charge at one of the oxygen atoms of the COOH group. Overall, the zwitterion has no net charge.

In water solution, zwitterions are stable only over a certain pH range. At low pH (high H^+ concentration), the COO^- group picks up a proton. The product, shown at the left below, has a +1 charge. At high pH (low H^+ concentration), the NH_3^+ group loses a proton. The species formed, shown at the right below, has a −1 charge.

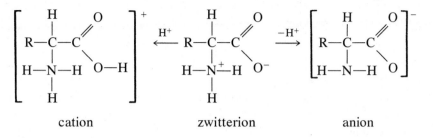

cation   zwitterion   anion

The pH range over which the zwitterion is stable depends upon the nature of the amino acid. With glycine, the zwitterion predominates between about pH 3 and pH 9. It has its maximum concentration, relative to the cation and anion, at pH 6. This is referred to as the isoelectric point of glycine; glycine has its lowest water solubility at this pH.

At pH 6, glycine is almost all zwitterion

Example 26.7   Using Table 26.2, write structural formulas for
    a. the zwitterion of alanine.
    b. the cation formed by alanine in acidic solution.
    c. the anion formed by alanine in basic solution.

Solution
    a. A proton is transferred from the COOH group to the NH_2 group.

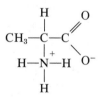

    b. The COO^- group of the zwitterion picks up a proton.

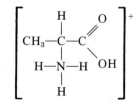

This acid is quite a bit stronger than acetic acid; $K_a = 2.5 \times 10^{-3}$

    c. The NH_3^+ group of the zwitterion loses a proton.

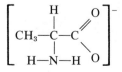

Exercise Write the structural formulas of the cation and anion derived from serine.
Answer:

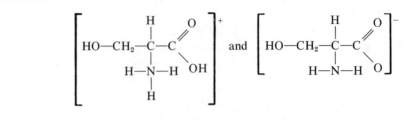

Polypeptides

As noted in Section 26.2, molecules containing NH_2 and $COOH$ groups can undergo condensation polymerization. Amino acids contain both groups in the same molecule. Hence, two amino acid molecules can combine by the reaction of the $COOH$ group in one molecule with the NH_2 group of the other molecule. If the acids involved are different, two different products are possible:

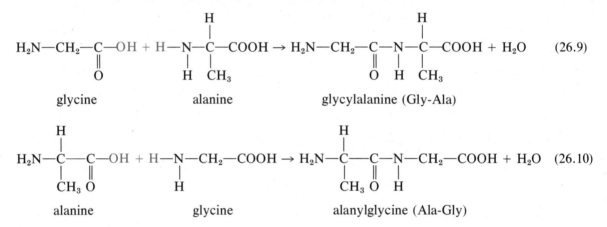

The products are referred to as *dipeptides;* they contain the group

commonly called a "peptide linkage." Dipeptides are named by writing first the amino acid which retains a free NH_2 group; the amino acid with a free $COOH$ group is written last.

Example 26.8 Draw the structural formula of glycylserine (Gly-Ser).

Solution The $COOH$ group of glycine condenses with the NH_2 group of serine:

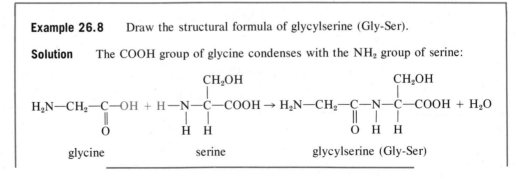

Exercise   Draw the structural formula of the other dipeptide that can be formed from glycine and serine. Answer:

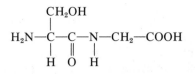

The dipeptides formed by Reactions 26.9 and 26.10 have reactive groups at both ends of the molecule (NH_2 at one end, COOH at the other). Hence they can combine further to give tripeptides (three amino acid units), tetrapeptides, . . ., and eventually long-chain polymers called polypeptides. Proteins are polypeptides containing a large number of amino acid units:

It's very possible that Carothers used protein structure as a guide for developing nylon

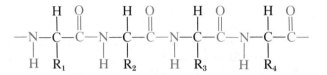

portion of a protein molecule (peptide linkages in color)

Proteins differ from the other polymers we have discussed in that they may contain up to 20 different monomer units. This means that there are a huge number of possible proteins. Using the amino acids listed in Table 26.2, we could make

$$20 \times 20 = 400 \text{ different dipeptides}$$
$$20 \times 20 \times 20 = 8000 \text{ different tripeptides}$$
$$20^n \text{ polypeptides containing n monomer units}$$

For a relatively simple protein containing only 50 monomer units, we have

$$20^{50} = 1 \times 10^{65} \text{ possibilities}$$

Amino Acid Sequence in Proteins

As you can imagine, it is not an easy task to identify all the amino acid units present in a protein chain. To go further and determine the order in which these units are arranged might seem next to impossible. However, in recent years, this type of analysis has become possible. The detailed amino acid sequence in some very complex proteins has been determined. The first protein for which this was done was insulin, which contains two chains linked by disulfide (—S—S—) bridges. One chain contains 21 amino acid units; the other has 30. The amino acid sequence in insulin is

Insulin is a small protein

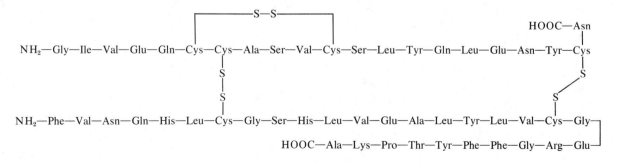

This structure was established by Frederick Sanger, who received the 1958 Nobel Prize in chemistry for his work.

In the body, proteins are built up by a series of reactions which in general produce a specific sequence of amino acids. Even tiny errors in this sequence may have serious effects. Among the genetic diseases known to be caused by improper sequencing are hemophilia, sickle cell anemia, and albinism. Sickle cell anemia is caused by the substitution of *one* valine unit for a glutamic acid unit in a chain containing 146 monomers.

Fortunately, nature doesn't make mistakes very often

Conformation of Protein Chains

The way in which protein chains are oriented in three dimensions is determined in large part by hydrogen bonding. Oxygen atoms on C=O groups can interact with H atoms in nearby N—H groups to form these bonds. This can occur within a single protein chain or between neighboring chains.

There are two ways in which a protein chain can be oriented to give maximum hydrogen bonding. When amino acid units with small R groups are present, as in glycine or alanine, hydrogen bonding leads to a pleated-sheet structure (Fig. 26.7). The sheet shown lies in the plane of the paper. It consists of many parallel chains held together by hydrogen bonds between peptide linkages in adjacent chains. By following the middle chain in Figure 26.7, you can see that hydrogen bonding occurs at each amino acid unit. This pleated-sheet structure is found in muscles and silk fibers.

Linus Pauling found the α-helix structure in proteins

Most proteins have amino acid units with R groups that are too bulky for a pleated-sheet structure. These proteins form a coil called an *α-helix* to maximize hydrogen bonding between peptide linkages. At the left of Figure 26.8, we show the outline of the protein chain forming the helix. The more complete structure is given at the right. This shows where the hydrogen bonds form between amino acid units in the protein chain. Notice that the bulky R groups are all on the outside of the helix, where they have the most room. The actual structure of the helix is nearly independent of the nature of the R groups. The dimensions of the helix correspond closely to those observed in such fibrous proteins as wool, hair, skin, feathers, and fingernails.

Many proteins have three-dimensional structures more complex than those shown in Figures 26.7 and 26.8. You may learn more about these structures if you take courses

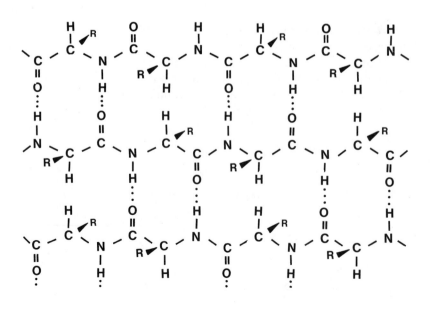

Figure 26.7 β-Keratin structure of silk fiber. When the R groups on the amino acid residues are small, the protein chains can hydrogen bond in the roughly planar structure shown. Since the bonding around the N and noncarbonyl carbon atoms is tetrahedral, the sheet formed has a pleated structure. In the drawing, the hydrogen bonds are shown by dots.

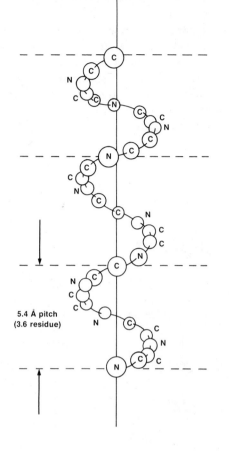

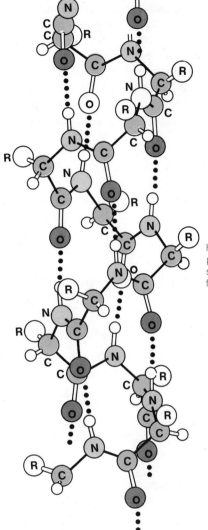

5.4 Å pitch
(3.6 residue)

Hydrogen bonding plays a large role in stabilizing protein conformations

Figure 26.8 α-Helix structure of proteins. The main atom chain in the helix is shown schematically on the left. The sketch on the right more nearly represents the actual positions of atoms and shows where intrachain hydrogen bonding occurs. Wool and many other fibrous proteins have the α-helix structure.

in organic chemistry and biochemistry. We should emphasize that the study of the structure and properties of proteins and related natural polymers is an area at the forefront of chemical research. It draws scientists from many disciplines, including biology, physics, and engineering. We can hope that this research will find solutions to such problems as hereditary diseases and cancer, and possibly even a cure for the common cold.

SUMMARY

An organic polymer contains monomer units joined in an extended chain. Addition polymers are made by the direct addition of monomer molecules containing double bonds (Table 26.1, Examples 26.1 and 26.2). Rubber is a natural addition polymer; synthetic rubbers are also available. A condensation polymer is made by splitting out a small molecule such as H_2O between two monomer molecules. Synthetic condensation polymers may be polyesters (Example 26.3) or polyamides (Example 26.4).

Carbohydrates can be classified as mono-, di-, and polysaccharides with increasing numbers of monomer units. The monosaccharide glucose is a monomer in the disaccharides maltose and sucrose. The monomer unit in starch is alpha-glucose (Fig. 26.5); in cellulose, the monomer unit is beta-glucose (Fig. 26.6).

Proteins are polymers of amino acid units (Table 26.2). The free amino acid can exist in solution as a cation, anion, or zwitterion, depending upon the pH (Example 26.7). Amino acid units condense to form dipeptides and eventually polypeptides or proteins by eliminating water (Example 26.8). Adjacent protein chains can be joined by hydrogen bonding to form a pleated-sheet structure (Fig. 26.7). Hydrogen bonding can also occur within a chain to form an alpha-helix (Fig. 26.8).

KEY WORDS AND CONCEPTS

polymer	vulcanization	monosaccharide	protein
monomer	condensation polymer	disaccharide	α-amino acid
addition polymer	polyester	polysaccharide	zwitterion
free radical	polyamide	equatorial	peptide linkage
head and tail polymerization	carbohydrate	axial	polypeptide

QUESTIONS AND PROBLEMS

Catalog

Addition Polymers: 26.2–26.5, 26.21–26.24
Condensation Polymers: 26.6–26.8, 26.25–26.27
Carbohydrates: 26.9–26.11, 26.28–26.30

Amino Acids, Proteins: 26.12–26.15, 26.31–26.34
General: 26.1, 26.16–26.20, 26.35–26.42

26.1   Review and know the meaning of the key words and concepts in this chapter.

26.2   Consider Teflon, the polymer made from tetra-fluoroethylene.

    a. Draw a portion of the Teflon molecule.
    b. Calculate the molecular mass of a Teflon polymer that contains 5.0×10^4 CF_2 units.
    c. What are the mass percentages of C and F in Teflon?

26.3   Styrene, $H_2C{=}C{-}\langle\bigcirc\rangle$, with an H on the C, forms a head-to-tail addition polymer. Sketch a portion of a polystyrene molecule.

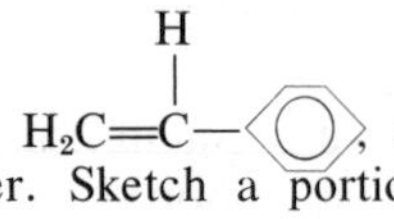

26.21   Consider a polymer made from tetrachloro-ethylene.

    a. Draw a portion of the polymer chain.
    b. What is the molecular mass of the polymer if it contains 3.2×10^3 tetrachloroethylene molecules?
    c. What are the mass percentages of C and Cl in the polymer?

26.22   Sketch a portion of the acrylonitrile polymer, assuming it is a

    a. head-to-tail polymer.
    b. head-to-head, tail-to-tail polymer.

26.4 Show the structure of the monomer used to make the following addition polymers:

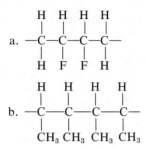

26.5 Neoprene, made from the chloroprene monomer, $H_2C=CH—CCl=CH_2$, is a rubber with a *trans* structure. Draw a section of a neoprene molecule.

26.6 A rather simple polymer can be made from ethylene glycol, $HO—CH_2—CH_2—OH$ and oxalic acid, $HO—\underset{\underset{O}{\|}}{C}—\underset{\underset{O}{\|}}{C}—OH$. Sketch a portion of the polymer chain obtained from these monomers.

26.7 Nylon 6 is made from a single monomer:

$$H_2N—(CH_2)_5—\underset{\underset{O}{\|}}{C}—OH$$

Sketch a section of the polymer chain in Nylon 6.

26.8 Identify the monomers from which the following condensation polymers are made:

a. $—\underset{\underset{H}{|}}{N}—CH_2—CH_2—\underset{\underset{H}{|}}{N}—\underset{\underset{O}{\|}}{C}—CH_2—\underset{\underset{O}{\|}}{C}—$

b. $—\underset{\underset{O}{\|}}{C}—\bigcirc—\underset{\underset{O}{\|}}{C}—O—CH_2—\underset{\underset{CH_3}{|}}{\overset{\overset{H}{|}}{C}}—O—$

26.9 Write a chemical equation, using molecular formulas, for the reaction of maltose with water to form glucose.

26.23 The polymer whose structure is shown below is made from two different monomers. Identify the monomers.

a.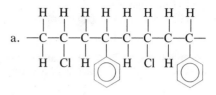

26.24 Gutta-percha, a rubber-type polymer used for golf ball covers, has the same monomer as in natural rubber, isoprene. However, in gutta-percha, the CH_2 groups are all *trans* with respect to the double bonds. Sketch a section of the gutta-percha chain.

26.25 Lexan is a very rugged polyester in which the monomers can be taken to be carbonic acid, $HO—\underset{\underset{O}{\|}}{C}—OH$, and

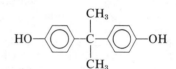

Sketch a section of the Lexan chain.

26.26 Para-aminobenzoic acid is an "essential vitamin" for many bacteria:

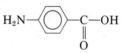

Sketch a portion of a polyamide polymer made from this molecule as the only monomer.

26.27 The following condensation polymer is made from a single monomer. Identify the monomer.

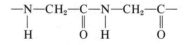

26.28 Write a chemical equation, using molecular formulas, for the reaction of sucrose with water to form glucose and fructose.

26.10 Cellulose consists of about 10,000 $C_6H_{10}O_5$ units linked together.

    a. What are the percentages by mass of C, H, and O in cellulose?
    b. What is the molecular mass of cellulose?

26.11 Explain why a plain cracker, if chewed for a long time without swallowing, starts to taste sweet.

26.12 Consider the cysteine molecule shown in Table 26.2. Write structural formulas for

    a. the zwitterion of cysteine.
    b. the cation formed in acid.
    c. the anion formed in base.

26.13 a. How many tripeptides can be made from glycine, alanine, and leucine, using each amino acid only once per tripeptide?
      b. Write the structural formulas of these tripeptides and name them in the shorthand abbreviation used for showing amino acid sequences.

26.14 On complete hydrolysis, a polypeptide gives two alanine, one leucine, one methionine, one phenylalanine, and one valine residue. Partial hydrolysis gives the following fragments: Ala-Phe, Leu-Met, Val-Ala, Phe-Leu. It is known that the first amino acid in the sequence is valine and the last one is methionine. What is the complete sequence of amino acids?

26.15 How would you explain the fact that the β-keratin structure present in silk is much less common in proteins than the α-helix structure?

26.16 Which of the following monomers could form an addition polymer? a condensation polymer?

    a. C_2H_6
    b. C_2H_4
    c. $HO-CH_2-CH_2-OH$
    d. $HO-CH_2-CH_3$

26.17 Consider a long-chain polypeptide made from glycine.

    a. What is the repeating unit in the polypeptide?
    b. What is the empirical formula of the polypeptide?

26.18 How many chiral carbon atoms are there in glucose? fructose?

26.19 Explain the difference between

    a. a head-to-tail and random polymer.
    b. natural rubber and neoprene.
    c. a polyester and a polyamide.
    d. α- and β-glucose.

26.29 Starch has the same empirical formula as cellulose and a molecular mass of about 1.0×10^5.

    a. What are the mass percentages of C, H and O in starch?
    b. How many $C_6H_{10}O_5$ units are linked together in a starch molecule?

26.30 Cellulose is useful for making paper, whereas starch is not. Suggest an explanation for this difference.

26.31 Follow the directions of Question 26.12 for serine.

26.32 A tripeptide contains valine, lysine, and phenylalanine residues.

    a. How many tripeptides are possible from these amino acids?
    b. Draw a structural formula for a possible form of the tripeptide and name it, using the shorthand form.

26.33 Suppose that, in the polypeptide referred to in Question 26.14, the first amino acid is alanine and the last one is also alanine. What is the complete sequence of amino acids?

26.34 In β-keratin, how many hydrogen bonds are present per amino acid residue (see Fig. 26.7)?

26.35 How would you explain to a young science student how to decide whether a given compound might be useful as a monomer for addition polymerization? condensation polymerization?

26.36 Consider the Dacron polymer shown in Example 26.3. Identify the repeating unit and give the empirical formula of Dacron.

26.37 Which of the following amino acids would you expect to show optical activity?

    a. glycine   b. alanine
    c. threonine   d. lysine

26.38 Explain the difference between

    a. linear and branched polyethylene.
    b. SBR and polybutadiene.
    c. a monosaccharide and a disaccharide.
    d. sucrose and maltose.

26.20 ΔH for Equation 26.4 is given as $-96n$ kJ. Explain where this number comes from, taking the C=C and C—C bond energies to be 598 and 347 kJ/mol, in that order.

26.39 An amino acid molecule is added to a polypeptide by splitting out water. List the number and type of bonds broken and formed. Using Table 6.3, calculate ΔH for this process.

*26.40 Glycerol (Chapter 25) and orthophthalic acid, —COOH, form a cross-linked polymer called an alkyd resin used in floor coverings and dentures.

  a. Write the structural formula for a portion of the polymer chain.
  b. Use your answer in (a) to show how cross-linking can occur between the polymer chains to form a water-insoluble, macromolecular solid.

*26.41 For alanine, the equilibrium constants relating the cation (C^+), zwitterion (Z), and anion (A^-) are as follows:

$$C^+(aq) \rightleftharpoons H^+(aq) + Z(aq); \; K_a = 5 \times 10^{-3}$$

$$Z(aq) \rightleftharpoons H^+(aq) + A^-(aq); \; K_a = 2 \times 10^{-10}$$

  a. Using the rule of multiple equilibria, show that

$$\frac{[H^+]^2 \times [A^-]}{[C^+]} = 1 \times 10^{-12}$$

  b. Determine the pH at which $[A^-] = [C^+]$. This is referred to as the isoelectric point. At this pH, there is no net charge on the alanine, almost all of which is in the form of the zwitterion.
  c. If the concentration of the zwitterion at the isoelectric point is 0.10 M, what is the concentration of C^+? A^-?

*26.42 Suppose solutions of alanine at various pH values are placed in an electrolysis cell. Using the equation in Problem 26.41a, determine whether there will be a net migration of alanine to the positive or the negative electrode at pH

  a. 2   b. 4   c. 7   d. 10

Appendix 1
CONSTANTS, REFERENCE DATA, AND SI UNITS

CONSTANTS

Acceleration of gravity = 9.806 m/s^2
Avogadro's number = 6.022×10^{23}
Electronic charge = 1.602×10^{-19} C
Faraday constant = 9.6485×10^4 J/V
Gas constant = 0.08206 $\ell \cdot$atm/(mol$\cdot$K)
    = 8.314 J/(mol$\cdot$K)
    = 8.314×10^7 g$\cdot$cm^2/(s$^2 \cdot$mol$\cdot$K)
Planck's constant = 6.626×10^{-34} J$\cdot$s
Velocity of light = 2.998×10^8 m/s
π = 3.142
e = 2.718
ln x = 2.3026 log x
2.3026 R = 19.14 J/(mol$\cdot$K)
2.3026 RT (at 25°C) = 5.708 kJ/mol

EQUATIONS INVOLVING CONSTANTS

Acid-base dissociation constants $\qquad$ $K_a \times K_b = 1.0 \times 10^{-14}$

Average velocity, gas molecule $\qquad$ $u^2 = \dfrac{3RT}{GMM}$

Bohr equation for H atom $\qquad$ $E = \dfrac{-1312}{n^2}$ kJ/mol

Energy and wavelength $\qquad$ $\Delta E = \dfrac{1.196 \times 10^5}{\lambda}$ kJ$\cdot$nm/mol

First order rate law $\qquad$ $\log_{10} X_0/X = kt/2.30$; $t_{1/2} = 0.693/k$

Free energy change and voltage	$\Delta G^0 = -96.5nE^0$ kJ/V
Free energy change and K	$\Delta G^0 = -0.0191T \log_{10} K$ kJ/K $= -5.71 \log_{10} K$ kJ (at 25°C)
Freezing point lowering, boiling point elevation, water	$\Delta T_f = 1.86$°C $\times$ m; $\Delta T_b = 0.52$°C $\times$ m
Gas Law	$PV = nRT$
Hydrogen sulfide equilibrium	$\dfrac{[H^+]^2 \times [S^{2-}]}{[H_2S]} = 1 \times 10^{-20}$
Ionization of water	$[H^+] \times [OH^-] = K_w = 1.0 \times 10^{-14}$ (at 25°C)
Mass-energy	$\Delta E = 9.00 \times 10^{10} \Delta m$ kJ/g
Nernst equation	$E = E^0 - \dfrac{0.0591}{n} \log \dfrac{(\text{conc. C})^c \times (\text{conc. D})^d}{(\text{conc. A})^a \times (\text{conc. B})^b}$
Rate constant and temperature	$\log_{10} \dfrac{k_2}{k_1} = \dfrac{E_a(T_2 - T_1)}{19.14\, T_2T_1}$
Temperature conversions	°F $= 1.8$(°C)$ + 32°$; K $=$ °C $+ 273$

TABLES AND FIGURES OF DATA

VAPOR PRESSURE OF WATER (mm Hg)

T(°C)	vp	T(°C)	vp	T(°C)	vp	T(°C)	vp
0	4.58	21	18.65	35	42.2	92	567.0
5	6.54	22	19.83	40	55.3	94	610.9
10	9.21	23	21.07	45	71.9	96	657.6
12	10.52	24	22.38	50	92.5	98	707.3
14	11.99	25	23.76	55	118.0	100	760.0
16	13.63	26	25.21	60	149.4	102	815.9
17	14.53	27	26.74	65	187.5	104	875.1
18	15.48	28	28.35	70	233.7	106	937.9
19	16.48	29	30.04	80	355.1	108	1004.4
20	17.54	30	31.82	90	525.8	110	1074.6

SI UNITS

Base Units

The International System of Units or *Système International* (SI), which represents an extension of the metric system, was adopted by the 11th General Conference of Weights and Measures in 1960. It is constructed from seven base units, each of which represents a particular physical quantity (Table I).

TABLE I SI BASE UNITS

PHYSICAL QUANTITY	NAME OF UNIT	SYMBOL
1. Length	metre	m
2. Mass	kilogram	kg
3. Time	second	s
4. Temperature	kelvin	K
5. Amount of substance	mole	mol
6. Electric current	ampere	A
7. Luminous intensity	candela	cd

Of the seven units listed in Table I, the first five are particularly useful in general chemistry. They are defined as follows:

1. The *metre* was redefined in 1960 to be equal to 1 650 763.73 wavelengths of a certain line in the orange-red region of the emission spectrum of krypton-86.

2. The *kilogram* represents the mass of a platinum-iridium block kept at the International Bureau of Weights and Measures at Sevres, France.

3. The *second* was redefined in 1967 as the duration of 9 192 631 770 periods of a certain line in the microwave spectrum of cesium-133.

4. The *kelvin* is 1/273.16 of the temperature interval between the absolute zero and the triple point of water (0.01°C = 273.16 K).

5. The *mole* is the amount of substance which contains as many entities as there are atoms in exactly 0.012 kg of carbon-12.

Prefixes Used with SI Units

Decimal fractions and multiples of SI units are designated by using the prefixes listed in Table II. Those which are most commonly used in general chemistry are underlined.

		TABLE II		SI PREFIXES		
FACTOR	**PREFIX**	**SYMBOL**	**FACTOR**	**PREFIX**	**SYMBOL**	
10^{12}	tera	T	10^{-1}	deci	d	
10^{9}	giga	G	10^{-2}	centi	c	
10^{6}	mega	M	10^{-3}	milli	m	
10^{3}	kilo	k	10^{-6}	micro	μ	
10^{2}	hecto	h	10^{-9}	nano	n	
10^{1}	deca	da	10^{-12}	pico	p	
			10^{-15}	femto	f	
			10^{-18}	atto	a	

Derived Units

In the International System of Units, all physical quantities are represented by appropriate combinations of the base units listed in Table I. To choose a particularly simple example, the SI unit for volume, the cubic metre, represents the volume of a cube one metre on an edge. Again, in SI, the density of a substance can be expressed by dividing its mass in kilograms by its volume in cubic metres. A list of the derived units most frequently used in general chemistry is given in Table III.

	TABLE III	SI DERIVED UNITS	
PHYSICAL QUANTITY	**NAME OF UNIT**	**SYMBOL**	**DEFINITION**
Area	square metre	m^2	
Volume	cubic metre	m^3	
Density	kilogram per cubic metre	kg/m^3	
Force	newton	N	$kg \cdot m/s^2$
Pressure	pascal	Pa	N/m^2
Energy	joule	J	$kg \cdot m^2/s^2$
Electric charge	coulomb	C	$A \cdot s$
Electric potential difference	volt	V	$J/(A \cdot s)$

Perhaps the least familiar of these units to the beginning chemistry student are the ones used to represent force, pressure, and energy.

The *newton* is defined as the force required to impart an acceleration of one metre per second squared to a mass of one kilogram (recall that Newton's second law can be stated as force = mass $\times$ acceleration).

The *pascal* is defined as the pressure exerted by a force of one newton acting upon an area of one square metre (recall that pressure = force/area). Commonly, pressures are expressed in kilopascals:

$$1 \text{ kPa} = 10^3 \text{ Pa}$$

Typically, atmospheric pressure near sea level is in the vicinity of 100 kPa.

The **joule** is defined as the work done when a force of one newton ($kg \cdot m/s^2$) acts through a distance of one metre (recall that work = force $\times$ distance). Commonly, energies are expressed in kilojoules:

$$1 \text{ kJ} = 10^3 \text{ J}$$

In terms of more familiar units, a kilowatt hour is 3600 kJ; a kilocalorie is 4.184 kJ.

Conversions Between SI and Other Units

In Table IV are listed appropriate conversion factors for translating units from other systems to the International System.

TABLE IV   CONVERSION FACTORS

QUANTITY	SI UNIT	OTHER UNIT	CONVERSION FACTOR
Energy	joule	calorie	$1 \text{ cal} = 4.184 \text{ J}$
		erg	$1 \text{ erg} = 10^{-7} \text{ J}$
Force	newton	dyne	$1 \text{ dyn} = 10^{-5} \text{ N}$
Length	metre	ångström	$1 \text{ Å} = 10^{-10} \text{ m} = 10^{-8} \text{ cm} = 10^{-1} \text{ nm}$
Mass	kilogram	pound	$1 \text{ lb} = 0.453\ 592\ 37 \text{ kg}$
Pressure	pascal	bar	$1 \text{ bar} = 10^{5} \text{ Pa}$
		atmosphere	$1 \text{ atm} = 1.01\ 325 \times 10^{5} \text{ Pa}$
		mmHg	$1 \text{ mmHg} = 133.322 \text{ Pa}$
		lb/in²	$1 \text{ lb/in}^2 = 6894.8 \text{ Pa}$
Temperature*	kelvin	Celsius	$1°\text{C} = 1 \text{ K}$
		Fahrenheit	$1°\text{F} = \dfrac{5}{9} \text{ K}$
Volume	cubic metre	litre	$1 \ \ell = 1 \text{ dm}^3 = 10^{-3} \text{ m}^3$
		gallon (U.S.)	$1 \text{ gal (U.S.)} = 3.7854 \times 10^{-3} \text{ m}^3$
		gallon (U.K.)	$1 \text{ gal (U.K.)} = 4.5641 \times 10^{-3} \text{ m}^3$
		cubic inch	$1 \text{ in}^3 = 1.6387 \times 10^{-6} \text{ m}^3$

*Temperatures in degrees Celsius (t_c) or degrees Fahrenheit (t_f) can be converted to temperatures in Kelvin (T) by using the equations:

$$t_c = T - 273.15; \qquad t_f = \frac{9}{5}T - 459.67$$

Appendix 2
ATOMIC AND IONIC RADII*

Element	Atomic Number	Atomic Radius in nm	Ionic Radius in nm
H	1	0.037	(−1) 0.208
He	2	0.05	
Li	3	0.152	(+1) 0.060
Be	4	0.111	(+2) 0.031
B	5	0.088	
C	6	0.077	
N	7	0.070	
O	8	0.066	(−2) 0.140
F	9	0.064	(−1) 0.136
Ne	10	0.070	
Na	11	0.186	(+1) 0.095
Mg	12	0.160	(+2) 0.065
Al	13	0.143	(+3) 0.050
Si	14	0.117	
P	15	0.110	
S	16	0.104	(−2) 0.184
Cl	17	0.099	(−1) 0.181
Ar	18	0.094	
K	19	0.231	(+1) 0.133
Ca	20	0.197	(+2) 0.099
Sc	21	0.160	(+3) 0.081
Ti	22	0.146	
V	23	0.131	
Cr	24	0.125	(+3) 0.064
Mn	25	0.129	(+2) 0.080
Fe	26	0.126	(+2) 0.075
Co	27	0.125	(+2) 0.072
Ni	28	0.124	(+2) 0.069
Cu	29	0.128	(+1) 0.096
Zn	30	0.133	(+2) 0.074
Ga	31	0.122	(+3) 0.062
Ge	32	0.122	
As	33	0.121	
Se	34	0.117	(−2) 0.198
Br	35	0.114	(−1) 0.195

*Radii are taken from Pauling, Linus: "The Nature of the Chemical Bond," 3rd ed., Ithaca, New York, Cornell University Press, 1960. Those of the noble gas atoms are from *Journal of Physical Chemistry, 69:* 596, 1965.

ELEMENT	ATOMIC NUMBER	ATOMIC RADIUS IN NM	IONIC RADIUS IN NM
Kr	36	0.109	
Rb	37	0.244	(+1) 0.148
Sr	38	0.215	(+2) 0.113
Y	39	0.180	(+3) 0.093
Zr	40	0.157	
Nb	41	0.143	
Mo	42	0.136	
Tc	43		
Ru	44	0.133	
Rh	45	0.134	
Pd	46	0.138	
Ag	47	0.144	(+1) 0.126
Cd	48	0.149	(+2) 0.097
In	49	0.162	(+3) 0.081
Sn	50	0.140	
Sb	51	0.141	
Te	52	0.137	(−2) 0.221
I	53	0.133	(−1) 0.216
Xe	54	0.130	
Cs	55	0.262	(+1) 0.169
Ba	56	0.217	(+2) 0.135
La	57	0.187	(+3) 0.115
Ce	58	0.182	(+3) 0.101
Pr	59	0.182	(+3) 0.100
Nd	60	0.182	(+3) 0.099
Pm	61		
Sm	62		
Eu	63	0.204	(+2) 0.097
Gd	64	0.179	(+3) 0.096
Tb	65	0.177	(+3) 0.095
Dy	66	0.177	(+3) 0.094
Ho	67	0.176	(+3) 0.093
Er	68	0.175	(+3) 0.092
Tm	69	0.174	(+3) 0.091
Yb	70	0.193	(+3) 0.089
Lu	71	0.174	(+3) 0.089
Hf	72	0.157	
Ta	73	0.143	
W	74	0.137	
Re	75	0.137	
Os	76	0.134	
Ir	77	0.135	
Pt	78	0.138	
Au	79	0.144	(+1) 0.137
Hg	80	0.155	(+2) 0.110
Tl	81	0.171	(+3) 0.095
Pb	82	0.175	
Bi	83	0.146	
Po	84	0.165	
At	85		
Rn	86	0.14	
Fr	87		
Ra	88	0.220	
Ac	89	0.20	
Th	90	0.180	
Pa	91		
U	92	0.14	

Appendix 3
NOMENCLATURE OF INORGANIC COMPOUNDS

In Chapter 3 we discussed briefly the system of nomenclature that is used with ionic compounds. Here we will review that system in somewhat greater detail and then examine how three other important classes of inorganic compounds (binary compounds of the nonmetals, oxyacids, and coordination compounds) are named.

IONIC COMPOUNDS

The names of ionic compounds are derived from those of the ions of which they are composed. We shall first consider the nomenclature of individual ions and then the names of the compounds they form.

Positive Ions

Monatomic positive ions take the names of the metal from which they are derived:

Na^+ sodium Ca^{2+} calcium Al^{3+} aluminum

When a metal forms more than one ion, it is necessary to distinguish between these ions. The accepted practice today is to indicate the charge of the ion by a Roman numeral in parentheses immediately following the name of the metal:

Fe^{2+}	iron(II)	Cu^+	copper(I)	Sn^{2+}	tin(II)
Fe^{3+}	iron(III)	Cu^{2+}	copper(II)	Sn^{4+}	tin(IV)

An earlier method, still widely used, adds to the stem of the Latin name of the metal the suffixes -*ous* or -*ic,* representing the lower and higher charges, respectively:

Fe^{2+}	ferrous	Cu^+	cuprous	Sn^{2+}	stannous
Fe^{3+}	ferric	Cu^{2+}	cupric	Sn^{4+}	stannic

The only polyatomic cations to be considered here are

NH_4^+ ammonium Hg_2^{2+} mercury(I) or mercurous

Negative Ions

Monatomic negative ions are named by adding the suffix *-ide* to the stem of the name of the nonmetal from which they are derived:

N^{3-}	nitride	O^{2-}	oxide	F^-	fluoride	H^-	hydride
		S^{2-}	sulfide	Cl^-	chloride		
		Se^{2-}	selenide	Br^-	bromide		
		Te^{2-}	telluride	I^-	iodide		

The nomenclature of polyatomic anions is more complex. The names of some of the more common oxyanions are

NO_3^-	nitrate	ClO_2^-	chlorite	MnO_4^-	permanganate
NO_2^-	nitrite	ClO^-	hypochlorite	CrO_4^{2-}	chromate
SO_4^{2-}	sulfate	CO_3^{2-}	carbonate	$Cr_2O_7^{2-}$	dichromate
SO_3^{2-}	sulfite	PO_4^{3-}	phosphate	OH^-	hydroxide
ClO_4^-	perchlorate	BO_3^{3-}	borate		
ClO_3^-	chlorate	SiO_4^{4-}	silicate		

Notice that when a nonmetal such as nitrogen or sulfur forms two different oxyanions, the suffix *-ate* is used with the anion containing the larger number of oxygen atoms (NO_3^-, SO_4^{2-}). The suffix *-ite* is used with the anion containing the smaller number of oxygen atoms (NO_2^-, SO_3^{2-}). Elements such as chlorine, which forms more than two oxyanions, use the prefixes *per-* (largest number of oxygen atoms) and *hypo-* (smallest number of oxygen atoms) as well.

Oxyanions that contain hydrogen as well as nonmetal and oxygen atoms are properly named as illustrated in the following examples:

HCO_3^-	hydrogen carbonate	HPO_4^{2-}	hydrogen phosphate
HSO_4^-	hydrogen sulfate	$H_2PO_4^-$	dihydrogen phosphate

Compounds

The name of the positive ion is given first, followed by the name of the negative ion. Examples are

$CaCl_2$	calcium chloride	$Fe(ClO_4)_3$	iron(III) perchlorate
$FeBr_2$	iron(II) bromide	$NaHCO_3$	sodium hydrogen carbonate
$(NH_4)_2SO_4$	ammonium sulfate		

In practice, compounds containing metal atoms, regardless of the type of bonding involved, are ordinarily named as if they were ionic. For example, in $AlCl_3$ and $SnCl_4$ the bonding is primarily covalent. However, they are named

$AlCl_3$ aluminum chloride $SnCl_4$ tin(IV) chloride

BINARY COMPOUNDS OF THE NONMETALS

When a pair of nonmetals form only one compound, that compound may be named quite simply. The name of the element whose symbol appears first in the formula (the element of lower electronegativity) is written first. The second portion of the name is formed by adding the suffix -*ide* to the stem of the name of the second nonmetal. Examples include

HCl	hydrogen chloride
H_2S	hydrogen sulfide
NF_3	nitrogen fluoride

If more than one binary compound is formed by a pair of nonmetals, as is often the case, the Greek prefixes, *di* = two, *tri* = three, *tetra* = four, *penta* = five, *hexa* = six, and so on, are used to designate the number of atoms of each element. Thus, for the oxides of nitrogen we have

N_2O_5	dinitrogen pentoxide*
N_2O_4	dinitrogen tetroxide*
NO_2	nitrogen dioxide
N_2O_3	dinitrogen trioxide
N_2O_2	dinitrogen dioxide
NO	nitrogen oxide
N_2O	dinitrogen oxide

A great many of the best-known binary compounds of the nonmetals have acquired common names which are widely and, in some cases, exclusively used. These include

H_2O	water	PH_3	phosphine
H_2O_2	hydrogen peroxide	AsH_3	arsine
NH_3	ammonia	NO	nitric oxide
N_2H_4	hydrazine	N_2O	nitrous oxide

OXYACIDS

The names of some of the more common oxygen acids are:

HNO_3	nitric acid	$HClO_4$	perchloric acid	H_2CO_3	carbonic acid
HNO_2	nitrous acid	$HClO_3$	chloric acid	H_3PO_4	phosphoric acid
H_2SO_4	sulfuric acid	$HClO_2$	chlorous acid	H_3BO_3	boric acid
H_2SO_3	sulfurous acid	$HClO$	hypochlorous acid	H_4SiO_4	silicic acid

It is of interest to compare the names of these oxyacids to those of the corresponding oxyanions listed previously. Note that oxyanions whose names end in -*ate* are derived from acids whose names end in -*ic*. Compare, for example, CO_3^{2-} (carbon*ate*) and H_2CO_3 (carbon*ic* acid); NO_3^- (nitr*ate*) and HNO_3 (nitr*ic* acid); ClO_4^- (perchlor*ate*) and $HClO_4$ (perchlor*ic* acid). Oxyanions whose names end in -*ite* are derived from acids whose names end in -*ous*. Thus we have NO_2^- (nitr*ite*) and HNO_2 (nitr*ous* acid); ClO^- (hypochlor*ite*) and $HClO$ (hypochlor*ous* acid).

---

*Note that in this case the *a* is dropped from the prefixes *penta* and *tetra* in the interests of euphony.

COORDINATION COMPOUNDS

The nomenclature of compounds containing complex ions (Chap. 19) is perhaps more involved than that of any other type of inorganic compound. Several rules are required, the more pertinent of which are as follows:

1. As in simple ionic compounds, the cation is named first, followed by the anion.

2. If there is more than one ligand of a particular type attached to the central atom, Greek prefixes are used to indicate the number of these ligands. Where the name of the ligand itself is complex (e.g., ethylenediamine), the number of such ligands is indicated by the prefixes *bis-* or *tris-* instead of *di-* or *tri-* and the name of the ligand is enclosed in parentheses.

3. In naming a complex ion, the names of anionic ligands are written first, followed by those of neutral ligands, and finally that of the central metal atom. This is exactly the reverse of the order in which the groups are listed in the formula of the complex ion: the symbol of the central atom is written first, followed by the formulas of neutral ligands and then those of negatively charged ligands. In writing the formula of a coordination compound, the complex ion is often set off by brackets.

4. The names of anionic ligands are modified by substituting the suffix *-o* for the usual ending. Thus we have

Cl^-	chloro	CO_3^{2-}	carbonato
OH^-	hydroxo	CN^-	cyano

The names of neutral ligands are ordinarily not changed. Two important exceptions are

H_2O	aquo	NH_3	ammine

5. The charge of the metal ion is indicated by a Roman numeral following the name of the metal. If the complex is an anion, the suffix *-ate* is added, often to the Latin stem of the name of the metal.

Applying these rules:

		Complex cation	Anion
$[Cr(NH_3)_4Cl_2]Cl$		dichlorotetramminechromium(III)	chloride
		2 Cl^- ligands 4 NH_3 ligands Cr^{3+}	

	Cation	Complex anion
$K_2[PtCl_4]$	potassium	tetrachloroplatinate(II)
		4 Cl^- ligands Pt^{2+}

Other examples are

$[Co(NH_3)_6]Cl_3$	hexamminecobalt(III) chloride
$[Co(en)_3](NO_3)_3$	tris(ethylenediamine)cobalt(III) nitrate
$[Pt(H_2O)_3Cl]Br$	chlorotriaquoplatinum(II) bromide
$K_3[Fe(CN)_6]$	potassium hexacyanoferrate(III)
$K_4[Fe(CN)_6]$	potassium hexacyanoferrate(II)

Appendix 4
REVIEW OF MATHEMATICS

The mathematics you will use in general chemistry is relatively simple. You will, however, be expected to:

—make calculations involving exponential numbers, such as 6.022×10^{23} or 1.6×10^{-10}.

—work with logarithms or antilogarithms, particularly in problems involving pH:

$$pH = -\log_{10} (\text{conc. H}^+)$$

In this appendix, we will review each of these topics briefly. To start with, it will be helpful to comment on electronic calculators, which are very useful for all kinds of calculations in general chemistry.

ELECTRONIC CALCULATORS

A so-called "scientific calculator," selling for between $20 and $50, is entirely adequate for general chemistry. Like any calculator, it will allow you to carry out such simple operations as addition, subtraction, multiplication, and division. Beyond that, make sure that the scientific calculator you buy can be used to:

—enter and perform operations on numbers expressed in exponential (scientific) notation.

—find a common (base 10) logarithm or antilogarithm (number corresponding to a given logarithm).

—raise a number to any power, n, or extract the nth root of a number.

The first thing you should do after buying a calculator is to learn how to use it. Read the instruction manual and work with the calculator until you become familiar with it. To get started, try carrying out the following operations:

a. $2.2 \times 6.1 = 13.42$
b. $8.1/2.7 = 3$
c. $(64)^{1/2} = 8$
d. $(27)^{1/3} = 3$
e. $3^4 = 81$

(In d and e, you will need to use the y^x or x^y key. Refer to your instruction manual for the sequence of operations, which differs depending upon the brand of calculator.)

f. $\dfrac{16 \times 9}{3 \times 8} = 6$

(This is carried out as a single operation; you do *not* solve for intermediate answers.)

In working with a calculator, you should be aware of one of its limitations. It does not indicate the number of significant figures in the answer. Consider, for example, the operations in a and b above. Assume that 2.2, 6.1, 8.1, and 2.7 represent experimentally measured quantities. Following the rules for significant figures (Chap. 1), the answers should be 13 and 3.0, in that order, *not* 13.42 and 3, the numbers appearing on the calculator. In another case, if you were asked to obtain the reciprocal of 3.68, your answer should be

$$1/3.68 = 0.272$$

not 0.2717391 . . ., or whatever other number appears on your calculator.

EXPONENTIAL NOTATION

In chemistry we frequently deal with very large or very small numbers. In one gram of the element carbon there are

$$50,150,000,000,000,000,000,000$$

atoms of carbon. At the opposite extreme, the mass of a single carbon atom is

$$0.00000000000000000000001994 \text{ g}$$

Numbers such as these are very awkward to work with. For example, neither of the numbers just written could be entered directly on a calculator. To simplify operations involving very large or very small numbers, we use what is known as **exponential** or **scientific** notation. To express a number in exponential notation, we write it in the form

$$C \times 10^n$$

where C is a number between 1 and 10 (e.g., 1, 2.62, 5.8) and n is a positive or negative integer (e.g., 1, -1, -3). To find n, we count the number of places that the decimal point must be moved to give the coefficient, C. If the decimal point must be moved to the *left,* n is a *positive* integer; if it must be moved to the *right,* n is a *negative* integer. Thus we have

$$26.23 = 2.623 \times 10^1 \quad \text{(decimal point moved 1 place to left)}$$
$$5609 = 5.609 \times 10^3 \quad \text{(decimal point moved 3 places to left)}$$
$$0.0918 = 9.18 \times 10^{-2} \quad \text{(decimal point moved 2 places to right)}$$

Numbers written in exponential notation can be given a very simple interpretation. Recognizing that $10^1 = 10$, $10^3 = 1000$, and $10^{-2} = 1/100 = 0.01$, we could express the three exponentials written above as

$$2.623 \times 10^1 \ = 2.623 \times 10$$
$$5.609 \times 10^3 \ = 5.609 \times 1000$$
$$9.18 \times 10^{-2} = 9.18 \ \times 0.01$$

The magnitude of a number written in exponential notation depends upon the values of both the coefficient, C, and the exponent, n. Suppose we compare two numbers which have the same value of n, such as 2.6×10^2 and 3.8×10^2. Here, the larger number is the one with the larger coefficient:

$$3.8 \times 10^2 > 2.6 \times 10^2; \ 380 > 260$$

Suppose now that we compare two exponential numbers with different values of n. Here *the larger number is the one which has the larger value of n:*

$$2.6 \times 10^2 > 4.8 \times 10^1; \ 260 > 48$$
$$3.2 \times 10^1 > 8.0 \times 10^{-1}; \ 32 > 0.80$$
$$2 \times 10^{-2} > 4 \times 10^{-3}; \ 0.02 > 0.004$$

MULTIPLICATION AND DIVISION. A major advantage of exponential notation is that it simplifies the processes of multiplication and division. To *multiply,* we *add exponents:*

$$10^1 \times 10^2 = 10^{1+2} = 10^3; \ 10^6 \times 10^{-4} = 10^{6+(-4)} = 10^2$$

To *divide,* we *subtract* exponents:

$$10^3/10^2 = 10^{3-2} = 10^1; \ 10^{-3}/10^6 = 10^{-3-6} = 10^{-9}$$

To multiply one exponential number by another, we can first multiply the coefficients in the usual manner and then add exponents. To divide one exponential number by another, we can find the quotient of the coefficients and then subtract exponents. For example,

$$(5.00 \times 10^4) \times (1.60 \times 10^2) = (5.00 \times 1.60) \times (10^4 \times 10^2) = 8.00 \times 10^6$$

$$(6.01 \times 10^{-3})/(5.23 \times 10^6) = \frac{6.01}{5.23} \times \frac{10^{-3}}{10^6} = 1.15 \times 10^{-9}$$

It often happens that multiplication or division yields an answer which is not in standard exponential notation. Thus we might have

$$(5.0 \times 10^4) \times (6.0 \times 10^3) = (5.0 \times 6.0) \times 10^4 \times 10^3 = 30 \times 10^7$$

The product is not in standard exponential notation since the coefficient, 30, does not lie between 1 and 10. To correct this situation, we could rewrite the coefficient as 3.0×10^1 and then add exponents:

$$30 \times 10^7 = (3.0 \times 10^1) \times 10^7 = 3.0 \times 10^8$$

In another case

$$0.526 \times 10^3 = (5.26 \times 10^{-1}) \times 10^3 = 5.26 \times 10^2$$

RAISING TO POWERS AND EXTRACTING ROOTS. To raise an exponential number to a power or extract a root, we make use of the rules

$$(10^n)^a = 10^{na}; (10^n)^{1/a} = 10^{n/a}$$

That is,

$$(10^{-2})^3 = 10^{-6}; (10^{-2})^{1/2} = 10^{-2/2} = 10^{-1}$$

Applying these rules to numbers expressed in exponential notation, we have

$$(2.0 \times 10^{-2})^3 = (2.0)^3 \times (10^{-2})^3 = 8.0 \times 10^{-6}$$

$$(4.0 \times 10^{-2})^{1/2} = (4.0)^{1/2} \times (10^{-2})^{1/2} = 2.0 \times 10^{-1}$$

Here, as in multiplication and division, we operate on the coefficient and exponential terms separately.

Extracting the square root of an exponential number is a bit more difficult when the exponent is not an even number. Consider, for example,

$$(4.0 \times 10^5)^{1/2}$$

Following the procedure described above, we would obtain

$$(4.0)^{1/2} \times (10^5)^{1/2} = 2.0 \times 10^{5/2}$$

The answer is not in standard exponential notation; indeed, $10^{5/2}$ is an awkward expression to work with because it does not readily translate into an ordinary number.

One way to handle cases of this type is to rewrite the exponential number to make the exponent an even number. To do this, we can divide the exponential by 10 and multiply the coefficient by 10:

$$(4.0 \times 10^5)^{1/2} = (40 \times 10^4)^{1/2}$$

Now, proceeding in the usual manner, we obtain

$$(40 \times 10^4)^{1/2} = (40)^{1/2} \times (10^4)^{1/2} = 6.3 \times 10^2$$

ADDITION AND SUBTRACTION. Occasionally we may find it necessary to add or subtract two exponential numbers. These processes are extremely simple if both exponents are the same. For example,

$$2.02 \times 10^7 + 3.16 \times 10^7 = (2.02 + 3.16) \times 10^7 = 5.18 \times 10^7$$

$$4.23 \times 10^{-5} - 1.61 \times 10^{-5} = (4.23 - 1.61) \times 10^{-5} = 2.62 \times 10^{-5}$$

If the exponents differ, one of the numbers can be rewritten to make the exponents the same. To add 5.04×10^8 to 3.0×10^7, we might express the latter as a number times 10^8:

$$5.04 \times 10^8 + 3.0 \times 10^7 = 5.04 \times 10^8 + 0.30 \times 10^8 = 5.34 \times 10^8$$

EXPONENTIAL NOTATION ON THE CALCULATOR. On all scientific calculators, it is possible to enter numbers in exponential notation. The method used depends upon the brand of calculator. Most often, it involves using a key labeled **EXP, EE,** or **EEX.** Check your instructor's manual for the procedure to be followed. To make sure you understand it, try entering the following numbers:

$$2.4 \times 10^6; \ 3.16 \times 10^{-8}; \ 6.2 \times 10^{-16}$$

All the operations we have discussed involving exponential numbers can be carried out directly on your calculator. Often, you will save considerable time that way. For example, the square root of 4.0×10^5 can be obtained without adjusting exponents (try it!). The same is true of finding the sum $5.04 \times 10^8 + 3.0 \times 10^7$. Try the following operations, first without using your calculator in exponential notation, to make sure you understand the principles involved. Then check your answers with the calculator set in exponential notation.

a. $(6.0 \times 10^2) \times (4.2 \times 10^{-4}) = ?$
b. $\dfrac{6.0 \times 10^2}{4.2 \times 10^{-4}} = ?$
c. $(2.50 \times 10^{-9})^{1/2} = ?$
d. $3.6 \times 10^{-4} + 4 \times 10^{-5} = ?$

The answers, expressed in exponential notation and following the rules of significant figures, are: a. 2.5×10^{-1}; b. 1.4×10^6; c. 5.00×10^{-5}; d. 4.0×10^{-4}.

LOGARITHMS AND ANTILOGARITHMS

The common (base 10) logarithm of a number is the power to which 10 must be raised to give that number. For example,

$$\text{since } 10^2 = 100, \ \log 100 = 2$$
$$\text{since } 10^0 = 1, \ \log 1 = 0$$
$$\text{since } 10^{-3} = 0.001, \ \log 0.001 = -3$$

Notice that numbers which are larger than 1 have a common logarithm greater than zero. Numbers smaller than one have a common logarithm which is less than zero.

An antilogarithm is, quite simply, the number corresponding to a given logarithm. Thus,

$$\text{since } \log 100 = 2, \ \text{antilog } 2 = 100$$
$$\text{since } \log 1 = 0, \ \text{antilog } 0 = 1$$
$$\text{since } \log 0.001 = -3, \ \text{antilog } -3 = 0.001$$

In other words, the numbers whose logarithms are 2, 0, and -3 are 100, 1, and 0.001, in that order. Notice that a positive antilogarithm signifies a number greater than 1, a negative antilogarithm a number less than 1.

The definitions we have just presented can be summarized in the following statement:

if $10^x = y$, log y = x, antilog x = y

FINDING LOGS AND ANTILOGS ON A CALCULATOR. Most numbers do not have logarithms that are simple integers like 2, 0, or -3. This is the case only for integral powers of 10 like 10^2, 10^0, or 10^{-3}. Logarithms of numbers such as 4.14 or 0.212 cannot be found by inspection. Until quite recently, finding logarithms of such numbers involved using a table which took some considerable practice to interpret correctly. Now, with the advent of the low-priced scientific calculator, "log tables" are almost obsolete.

To obtain a logarithm using a calculator, all you need do is enter the number and press the LOG **key.** This way you should find that

$$\log 4.14 = 0.617 \ldots$$
$$\log 0.526 = -0.279 \ldots$$

The same process can be used to find the logarithm of an exponential number. Thus,

$$\log (4.14 \times 10^2) = 2.617 \ldots$$

Notice that $\log (4.14 \times 10^2) = \log 4.14 + \log 10^2 = 0.617 + 2 = 2.617$
In the general case

$$\log (C \times 10^n) = \log C + n$$

This applies for negative as well as positive values of n. Thus,

$$\log (4.14 \times 10^{-1}) = 0.617 - 1 = -0.383$$

The process used to find an antilogarithm varies with the brand of calculator. Perhaps most commonly, you enter the number and press the **10^x** key to obtain the antilogarithm. If your calculator doesn't have such a key, refer to the instruction manual to find out how to do this operation. Then, find the numbers whose logarithms are 0.616, 2.415, and -1.057. You should find that

antilog $0.616 = 4.13 \ldots$, which means that $10^{0.616} = 4.13 \ldots$
antilog $2.415 = 260 \ldots$, which means that $10^{2.415} = 260 \ldots$
antilog $-1.057 = 0.0877 \ldots$, which means that $10^{-1.057} = 0.0877 \ldots$

SIGNIFICANT FIGURES IN LOGARITHMS AND ANTILOGARITHMS. As with other operations, a calculator does not indicate the number of significant figures to be retained in a logarithm or antilogarithm. In the examples just worked, you probably found several more digits in the calculator display, beyond those listed. To decide how many figures to retain, use the following rules:

1. In taking the logarithm of a number, retain after the decimal point in the log as many digits as there are significant figures in the number. (This part of the logarithm is often referred to as the *mantissa;* digits that precede the decimal point comprise the *characteristic* of the logarithm.) To illustrate this rule, consider the following:

$\log 2.00 = 0.301$	$\log (2.00 \times 10^3) = 3.301$
$\log 2.0 \ = 0.30$	$\log (2.0 \times 10^1) \ = 1.30$
$\log 2 \ \ \ = 0.3$	$\log (2 \times 10^{-3}) \ = 0.3 - 3 = -2.7$

2. In taking the antilogarithm of a number, retain as many significant figures in the antilogarithm as there are after the decimal point in the number. Thus,

antilog $0.301 = 2.00$	antilog $3.301 = 2.00 \times 10^3$
antilog $0.30 \ = 2.0$	antilog $1.30 \ = 2.0 \times 10^1$
antilog $0.3 \ \ \ = 2$	antilog $-2.7 = 2 \times 10^{-3}$

OPERATIONS INVOLVING LOGARITHMS. Since logarithms are exponents, the rules governing the use of exponents apply here as well.

MULTIPLICATION: $\log (xy) = \log x + \log y$

Example: $\log (6.02 \times 2.00) = \log 6.02 + \log 2.00$
$$1.081 = 0.780 + 0.301$$

DIVISION: $\log (x/y) = \log x - \log y$

Example: $\log (6.02/2.00) = \log 6.02 - \log 2.00$
$$0.479 = 0.780 - 0.301$$

RAISING TO A POWER: $\log (x^n) = n \log x$

Example: $\log (2.00^3) = 3 \log 2.00$
$$0.903 = 3(0.301)$$

EXTRACTING A ROOT: $\log (x^{1/n}) = \dfrac{1}{n} \log x$

Example: $\log (4.00^{1/2}) = \frac{1}{2} \log 4.00$
$$0.301 = \tfrac{1}{2}(0.602)$$

Another useful relationship involves logarithms of reciprocals. Since the logarithm of 1 is zero, it follows that

$$\log (1/x) = 0 - \log x = -\log x$$

Thus, $\log (1/2.00) = -\log 2.00 = -0.301$

NATURAL LOGARITHMS. For calculation purposes, common logarithms are most convenient since our number system is based upon multiples of 10. However, certain of the equations we use in general chemistry involve a different type of logarithm, taken to the base e, where, to four significant figures,

$$e = 2.718 \ldots$$

Logarithms to the base e are referred to as **natural** logarithms and are often written as "ln"; i.e.,

$$\log_e x \equiv \ln x; \qquad \log_{10} x \equiv \log x$$

To find the natural logarithm of a number on your calculator, enter the number and press the **LN** key. This way you should find that

$$\ln 10.0 = 2.303$$
$$\ln 2.00 = 0.693$$
$$\ln 1.00 = 0.000$$

There is a simple relationship between common and natural logarithms which you may be able to deduce from the examples just worked. Recalling that $\log 10.0 = 1.000$:

$$\frac{\ln 10.0}{\log 10.0} = \frac{2.303}{1.000} = 2.303$$

In the general case,

$$\ln x = 2.303 \log x$$

For example, $\ln 2.00 = 2.303 \log 2.00 = 2.303(0.301) = 0.693$

Sometimes you will need to evaluate an expression such as $e^{0.250}$, where the base of natural logarithms, e, is raised to a power. This can be done by entering the exponent, 0.250, on your calculator and pressing the e^x button:

$$e^{0.250} = 1.28$$

Appendix 5
ANSWERS TO PROBLEMS _____

CHAPTER 1

1.24 a. length   b. volume   c. density   d. pressure   e. energy
         f. mass   g. pressure

1.25 a. 50 m   b. 0.0500 g   c. 150 cm³   d. 1500 nm

1.26 a. $-89.8°C$ b. 183 K   **1.27** J·s

1.28 a. 3   b. 3   c. 3   d. 3   e. 5   f. 2

1.29 a. 2   b. 3   c. 2   d. 3   **1.30** 3.526×10^4 oz

1.31 a. 2.5 ℓ b. 1.5×10^2 in³   **1.32** 13.0 stone   **1.33** 19.5 km/h

1.34 a. 128   b. 1.4; 46   **1.35** 2.7×10^{13}

1.36 a. Ac, Al, Am, Ar, Cd, Ca, C, Cl, Cr, Co, Cm, Ho, H, La, Lr, Mn, Md, Mo, Nd,
         Np, N, No, O, Pd, P, Po, Pr, Pm, Pa, Ra, Rn, Rh, Na

1.37 a. N, P   b. As, Sb   c. Bi

1.38 metals: Mg, Li, Os;   nonmetals: S, Br;   metalloid: Sb

1.39 a. C   b. P   c. P   d. C   **1.40** 6.50×10^{-4} kg

1.41 1.1×10^3 kg   **1.42** a. T   b. T   c. F

1.43 a. add water, stir, filter   b. distill   c. fract. cryst.   d. fract. cryst.

1.44 use a different solvent or temperature range

1.45 a. 38 g; 22 g   b. ~25°C   c. ~20°C   **1.46** 574

1.47 1.2 km²   **1.48** 78.0 cm³   **1.49** 34% tryptophan, 0% alanine

CHAPTER 2

2.23 see discussion, Section 2.1

2.24 a. 0.2910, 0.1748   b. 0.2910/0.1748 = 5/3

2.25 a. 38   b. 52   c. 36   d. 38

2.26 0, 30, 34, 30; $^{31}_{15}P^{3-}$, 18; -1, 35, 44, 36; $^{56}_{26}Fe^{3+}$, 23

2.27 a. 16.0007   b. 12.0010   **2.28** 6.942

2.29 11% Mg-25; 10% Mg-26   **2.30** 5   **2.31** a. 183.18   b. lighter

2.32 1.2×10^{10} km³   **2.33** 6.0×10^8

2.34 a. 2.0×10^{-6} g   b. 8.667×10^{22}

2.35 a. 208   b. 1.253×10^{26} c. 3.13×10^{25}

2.36 a. 86.17 g   b. 11.6 mol

2.37 a. 1.19×10^{-2} b. 6.66×10^{-3} c. 2.15×10^{-2}

2.38 5.00×10^{-5}, 3.01×10^{19}, 9.03×10^{19}; 1.13, 7.53×10^{21}, 2.26×10^{22}; 4.5×10^8,
         5.0×10^6, 9.0×10^{30}; 6.0×10^{46}, 6.7×10^{44}, 4.0×10^{68}

2.39 b < a < d < c   **2.40** 2×10^{12} **2.41** 7.5×10^{15}

2.42 a. C atom > H_2 molecule   b. not $CaCl_2$
c. 6.022×10^{23} atoms weigh 12 g   d. no. molecules in NaCl
2.43 a. Find % of Sn, Cl; determine mass Cl per gram of Sn in both compounds.
b. Combine experiments of Thomson and Millikan.
c. Analyze different samples; show that % Na, Cl remain same.
2.44 a. 22.98922 g   b. 18.99895 g   c. 41.98817 g   **2.45** 6.03×10^{23}
2.46 6.01×10^{23} **2.47** 1.01

CHAPTER 3

3.19 a. 81.72   b. 40.00   c. 47.44   **3.20** 723.6 g
3.21 a. 53.2   b. 0.195 g   **3.22** 93.71, 6.29   **3.23** 47.2, 5.92, 46.8
3.24 a. Pb_2O_3 b. $C_7H_5O_6N_3$ c. $Cr_2S_3H_{30}O_{27}$ **3.25** 46.0
3.26 SnF_4 **3.27** Cu_2O **3.28** $C_3H_8O_3$
3.29 CHCl   **3.30** $C_{10}H_8$ **3.31** c, d
3.32 a. ammonium dichromate   b. calcium iodide   c. iron(II) carbonate
d. potassium perchlorate
3.33 a. MgF_2 b. cadmium telluride   c. Al_2S_3 d. iron(III) nitrate
e. $(NH_4)_2SO_4$ f. $Ca_3(PO_4)_2$ g. lead chromate
3.34 a. xenon tetrafluoride   b. xenon difluoride   c. ICl   d. ICl_3
e. tetrasulfur dinitride
3.35 a. 2   b. 0.750   c. 0.400   **3.36** a. V_2O_5, V_2O_3 b. 1.747 g
3.37 50.9   **3.38** 51.1% Cl, 12.3% Br   **3.39** 22

CHAPTER 4

4.18 a. $H_2O(g) + Cl_2O(g) \rightarrow 2\ HOCl(aq)$
b. $4\ NH_3(g) + 3\ O_2(g) \rightarrow 2\ N_2(g) + 6\ H_2O(l)$
c. $(NH_4)_2Cr_2O_7(s) \rightarrow Cr_2O_3(s) + 4\ H_2O(l) + N_2(g)$
d. $4\ KClO_3(s) \rightarrow 3\ KClO_4(s) + KCl(s)$
e. $Hg^{2+}(aq) + 2\ Cl^-(aq) \rightarrow HgCl_2(s)$
4.19 a. $C_{10}H_{16}(l) + 8\ Cl_2(g) \rightarrow 10\ C(s) + 16\ HCl(g)$
b. $2\ H_2(g) + O_2(g) \rightarrow 2\ H_2O(l)$
c. $3\ Ca^{2+}(aq) + 2\ PO_4^{3-}(aq) \rightarrow Ca_3(PO_4)_2(s)$
d. $2\ KClO_3(s) \rightarrow 2\ KCl(s) + 3\ O_2(g)$
e. $B_2O_3(s) + 6\ HF(l) \rightarrow 2\ BF_3(s) + 3\ H_2O(l)$
4.20 a. 1.35   b. 7.95   c. 6.03   d. 3.02
4.21 a. $2\ C_4H_{10}(g) + 13\ O_2(g) \rightarrow 8\ CO_2(g) + 10\ H_2O(l)$ b. 3.15 mol   c. 152 g
4.22 a. 190 g   b. 51 ℓ **4.23** a. 6.30 mol   b. 14.0 g   c. 4.0×10^9 g
4.24 12 ℓ **4.25** a. 0.166   b. 1.66×10^{-4} **4.26** 0.0915, 0.202, 0.0315, 362
4.27 155 cm^3 **4.28** a. 0.0614   b. 0.0614, 0.123
4.29 a. KCl, KCl   b. $MgBr_2$, $MgBr_2$ **4.30** 0.734 g
4.31 a. $Ba^{2+}(aq) + SO_4^{2-}(aq) \rightarrow BaSO_4(s)$ b. 1.6×10^{-3} c. 1.1 cm^3
4.32 50.0 cm^3 **4.33** a. 0.308 g   b. 21.8
4.34 a. $2\ Mg(s) + O_2(g) \rightarrow 2\ MgO(s)$ b. 2.387 g   c. 0.223 g
$MgO(s) + H_2O(l) \rightarrow Mg(OH)_2(s)$
$3\ Mg(s) + N_2(g) \rightarrow Mg_3N_2(s)$
4.35 1.6   **4.36** $C_6H_6(l) + 3\ HNO_3(l) \rightarrow C_6H_3O_6N_3(l) + 3\ H_2O(l)$ **4.37** 37

CHAPTER 5

5.24 S, N_2, Ar, (Cu); Cl_2, Mg, Al   **5.25** see text discussion
5.26 a. bp Ar closer to O_2 than N_2 b. less expensive
c. difference in reactivity   d. lower mp
5.27 a. steel contains carbon   b. Cu metal is formed   c. cathode
d. Cl_2 molecules oxidize Br^- ions to Br_2 molecules
5.28 a. $Cl_2(g) + 2\ I^-(aq) \rightarrow 2\ Cl^-(aq) + I_2(s)$

b. $2 Cl^-(aq) + 2 H_2O \rightarrow Cl_2(g) + H_2(g) + 2 OH^-(aq)$

c. $CaF_2(s) + H_2SO_4(l) \rightarrow CaSO_4(s) + 2 HF(l)$

5.29 a. $Mg^{2+}(aq) + 2 OH^-(aq) \rightarrow Mg(OH)_2(s)$

b. $Mg(OH)_2(s) + 2 H^+(aq) \rightarrow Mg^{2+}(aq) + 2 H_2O$

c. $MgCl_2(l) \rightarrow Mg(s) + Cl_2(g)$

d. $P_4(s) + 5 O_2(g) \rightarrow P_4O_{10}(s)$

5.30 a. $MgCO_3 \cdot CaCO_3(s) \rightarrow MgO(s) + CaO(s) + 2 CO_2(g)$

b. $HgS(s) + O_2(g) \rightarrow Hg(g) + SO_2(g)$

5.31 a. 32.85% Na, 12.85% Al, 54.30% F   b. 38.9 g   **5.32** MnO_2

5.33 a. $Cu_2CO_5H_2$ b. $CuCO_3 \cdot Cu(OH)_2$

5.34 a. $CaCO_3$; 1.00×10^{-2} b. $SrCO_3$; 6.77×10^{-3} c. $BaCO_3$; 5.07×10^{-3}

5.35 18.5, 370; 0.0375, 1.50; 0.0314, 0.0628   **5.36** a. 0.57   b. 13

5.37 $0.54 cm^3$ **5.38** a. 4.79 g   b. 13.4 g   c. 5.7 g

5.39 a. 6.28 mol   b. 154 ℓ c. 735 ℓ

5.40 a. $MnO_2(s) + 2 Cl^-(aq) + 4 H^+(aq) \rightarrow Mn^{2+}(aq) + Cl_2(g) + 2 H_2O$

b. 0.160 mol   c. 0.323 g   d. 1.97 ℓ **5.41** no

5.42 a. $Mg(OH)_2$ b. 3.5 mol   c. 1.4 M   **5.43** a. SiO_2 b. 1.93 g

5.44 a. $2 C_2H_2(g) + 5 O_2(g) \rightarrow 4 CO_2(g) + 2 H_2O(g)$

b. O_2 c. 33.8 g   d. 90.2   e. 106 g

5.45 22 g $[Co(NH_3)_5Cl]Cl_2$, 13 g KSCN   **5.46** 30 kg   **5.47** 2.46 M

5.48 $C_8H_{18}(l) + 8 H_2O(g) \rightarrow 8 CO(g) + 17 H_2(g)$; 13.26% H_2 **5.49** $2.1 cm^3$

CHAPTER 6

6.23 b. +1.78 kJ   c. 56.2 g

6.24 a. $2 Mg(s) + CO_2(g) \rightarrow 2 MgO(s) + C(s)$ b. −607 kJ   c. 8.74 kJ evolved

6.25 24 kJ   **6.26** −0.0827 kJ

6.27 a. +638.4 kJ   b. −2733.6 kJ   c. −904.4 kJ

6.28 a. −3.0 kJ   b. −216.8 kJ   c. +16.2 kJ

6.29 a. −1299.5 kJ   b. −1255.5 kJ   **6.30** −3012 kJ   **6.31** −296.1 kJ

6.32 a. +138 kJ   b. −885 kJ   c. +347 kJ   **6.33** −271 kJ

6.34 +4.03°C   **6.35** 3.2×10^3 J/°C   **6.36** −1360 kJ

6.37 see discussion, Section 6.6   **6.38** 33%

6.39 a. 1.3×10^5 kJ   b. $2.2   **6.40** a. +5 J   b. +47 J   **6.41** −1300 kJ

6.42 a. Use ΔH_f b. ΔH_f $H^+(aq)$ is zero, not $H^+(g)$ c. $Hg(l)$ is stable state

6.43 heat of sublimation of I_2 **6.44** a. 1.6×10^{16} kJ   b. 8.3 km/ℓ

6.45 3800 kJ

CHAPTER 7

7.23 $0.359 cm^3$ **7.24** 1.98×10^5 ℓ **7.25** $0.05 m^3$ **7.26** 215 kPa

7.27 1.1×10^2 atm   **7.28** 22.4 ℓ; 24.5 ℓ **7.29** a. 28.8   b. 1.17 g/ℓ

7.30 a. 99.1   b. $C_2H_4Cl_2$ **7.31** a. 53.1   b. B   **7.32** 6.3 ℓ

7.33 2.67 g   **7.34** a. 3.8×10^{-3} b. 0.032 g

7.35 a. 5×10^{-5} atm   b. 2×10^{-6} mol/ℓ **7.36** 1.02×10^3 mm Hg

7.37 a. 730 mm Hg   b. 0.240 g   **7.38** 170 g/ℓ vs 131 g/ℓ

7.39 a. lighter   b. 1.61   **7.40** a. 586 K   b. 1172 K

7.41 205 K; 462 K   **7.42** a. He   b. equal   c. O_2 d. O_2

7.43 a. straight line through origin   b. straight line through origin

c. hyperbola   d. straight line parallel to x axis

7.44 0.0456 $\ell \cdot$ atm/(mol $\cdot$ °R)   **7.45** 55%

7.46 a. $2.10 cm^3$ b. 0.0044   **7.47** 7.02 m

CHAPTER 8

8.24 1282 nm   **8.25** 97.24 nm   **8.26** a. 119.0 kJ/mol   b. 3

8.27 36.44 kJ/mol   **8.28** 8.716×10^{-18} J/atom; 524.9 kJ/mol

8.29 a. $1s^22s^22p^63s^23p^5$ b. $1s^22s^22p^63s^23p^6$ c. $1s^22s^22p^63s^23p^64s^23d^7$
8.30 a. Sc   b. He   c. P
8.31 a. excited   b. ground   c. excited   d. impossible   e. excited
    f. excited
8.32 a. 0   b. 1/2   c. 12/26
8.33   1s   2s   2p   3s   3p   4s   3d
    a. (↓↑) (↓↑) (↓↑)(↓↑)(↓↑)
    b. (↓↑) (↓↑) (↓↑)(↓↑)(↓↑)
    c. (↓↑) (↓↑) (↓↑)(↓↑)(↓↑) (↓↑) (↓↑)(↓↑)(↓↑) (↓↑) (↑)(↑)( )( )( )
8.34 Al   **8.35** a. 1   b. 0   c. 5   **8.36** f, 4, 7
8.37 a. 2   b. 5   c. 1
8.38 $4, 0, 0, +\frac{1}{2}$; $3, 2, 2, +\frac{1}{2}$; $3, 2, 1, +\frac{1}{2}$; $3, 2, 0, +\frac{1}{2}$; $3, 2, -1, +\frac{1}{2}$; $3, 2, -2, +\frac{1}{2}$
8.39 a. no 2d   b. no 3f   c. $m_s \neq 1$ d. $m_\ell \neq 2$ for s sublevel
8.40 $8s^2$ **8.41** a. 100   b. 96   c. 80   d. 62
8.42 a. Rb > Sr > Ca   b. Rb < Sr < Ca
8.43 a. N, P, As, Sb, Bi   b. Ne, Ar, Kr, Xe, Rn
8.44 a. one outer s electron and one electron short of noble gas
    b. extra proton in nucleus   c. only one electron in outer level
8.45 a. atomic radius *decreases*   b. *low* ionization energy
    c. in *same* principal level   d. *halving* the energy
8.46 Set up expression for $\Delta E = E_n - E_2$; then use Equation 8.2.   **8.47** $1s^21p^42s^2$
8.48 a. 3, 9   b. 12   c. $1s^32s^32p^2$; $1s^32s^32p^93s^2$

CHAPTER 9
9.21 LiOH; sodium peroxide; $Ca_3(PO_4)_2$; magnesium hydrogen carbonate
9.22 a. beryllium   b. sodium carbonate   c. lithium   d. beryllium
9.23 a. $MgBr_2$ b. MgO   c. Mg_3N_2 d. MgO
9.24 a. $K(s) + O_2(g) \rightarrow KO_2(s)$
    b. $2 Li(s) + 2 H_2O(l) \rightarrow 2 Li^+(aq) + 2 OH^-(aq) + H_2(g)$
    c. $2 OH^-(aq) + CO_2(g) \rightarrow CO_3^{2-}(aq) + H_2O$
    d. $2 NaHCO_3(s) \rightarrow Na_2CO_3(s) + H_2O(g) + CO_2(g)$
    e. $Mg^{2+}(aq) + 2 NaZ(s) \rightarrow MgZ_2(s) + 2 Na^+(aq)$
9.25 $Ba^{2+}(aq) + 2 Cl^-(aq) + 2 HR(s) + 2 R'OH(s) \rightarrow BaR_2(s) + 2 R'Cl(s) + 2 H_2O$
9.26 a. Bubble CO_2 through water solution of NH_3, NaCl at 0°C.
    b. Heat at 800°C; add CaO to water.
    c. Add to water.
9.27 661 nm   **9.28** 202.8 kJ/mol   **9.29** −643 kJ/mol
9.30 a. −38.9 kJ   b. +3890 kJ   **9.31** −347 kJ   **9.32** 0.517ℓ
9.33 a. 4.06ℓ b. 0.333   **9.34** MgF_2, $CaCl_2$, $SrBr_2$, BaI_2
9.35 a. Rb, Sr, Cs, Ba, Fr, Ra   b. Na, Mg, K, Ca, Rb, Sr, Cs, Ba, Fr, Ra
    c. Fr, Ra
9.36 a. Be > Mg > Na   b. Na > Mg > Be   c. Na > Mg > Be
9.37 $0.225 nm^3$ **9.38** 0.501 nm   **9.39** a. 0.151 nm   b. no
9.40 The body diagonal is the hypotenuse of a right triangle; one side of that triangle
    is the side of the cell, s, the other is a face diagonal, $\sqrt{2}$ s. Hence, the body diagonal
    has a length of $s\sqrt{3}$. Along body diagonal, there are four radii.
9.41 0.31ℓ **9.42** 0.414   **9.43** $3.62 g/cm^3$

CHAPTER 10
10.23 Al—Cl > Si—Cl > P—Cl > S—Cl   **10.24** I—F

10.25 a. :C̈l—Si—C̈l: b. H—P̈—H   c. $(:N\equiv O:)^+$ d. $:C\equiv O:$

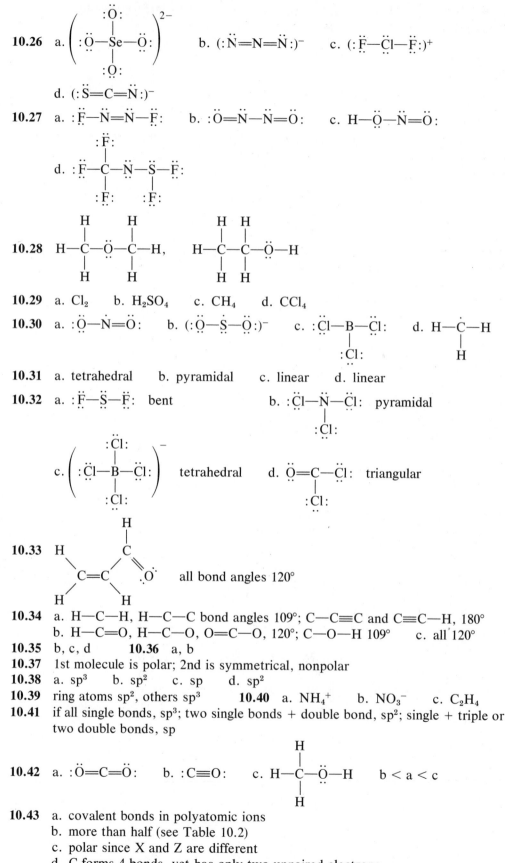

10.26 a. $\left(\begin{array}{c} :\ddot{O}: \\ | \\ :\ddot{O}-Se-\ddot{O}: \\ | \\ :\ddot{O}: \end{array}\right)^{2-}$ b. $(:\ddot{N}=N=\ddot{N}:)^-$ c. $(:\ddot{F}-\ddot{C}l-\ddot{F}:)^+$

d. $(:\ddot{S}=C=\ddot{N}:)^-$

10.27 a. $:\ddot{F}-\ddot{N}=\ddot{N}-\ddot{F}:$ b. $:\ddot{O}=\ddot{N}-\ddot{N}=\ddot{O}:$ c. $H-\ddot{O}-\ddot{N}=\ddot{O}:$

d. $:\ddot{F}-\overset{\displaystyle :\ddot{F}:}{\underset{\displaystyle :\ddot{F}:}{C}}-\ddot{N}-\overset{}{\underset{\displaystyle :\ddot{F}:}{\ddot{S}}}-\ddot{F}:$

10.28 $H-\overset{\displaystyle H}{\underset{\displaystyle H}{C}}-\ddot{O}-\overset{\displaystyle H}{\underset{\displaystyle H}{C}}-H,$ $H-\overset{\displaystyle H}{\underset{\displaystyle H}{C}}-\overset{\displaystyle H}{\underset{\displaystyle H}{C}}-\ddot{O}-H$

10.29 a. Cl_2 b. H_2SO_4 c. CH_4 d. CCl_4

10.30 a. $:\ddot{O}-\dot{N}=\ddot{O}:$ b. $(:\ddot{O}-\dot{S}-\ddot{O}:)^-$ c. $:\ddot{C}l-\overset{}{\underset{\displaystyle :\ddot{C}l:}{B}}-\ddot{C}l:$ d. $H-\overset{}{\underset{\displaystyle H}{\dot{C}}}-H$

10.31 a. tetrahedral   b. pyramidal   c. linear   d. linear

10.32 a. $:\ddot{F}-\ddot{S}-\ddot{F}:$ bent   b. $:\ddot{C}l-\overset{}{\underset{\displaystyle :\ddot{C}l:}{\ddot{N}}}-\ddot{C}l:$ pyramidal

c. $\left(\begin{array}{c} :\ddot{C}l: \\ | \\ :\ddot{C}l-B-\ddot{C}l: \\ | \\ :\ddot{C}l: \end{array}\right)^-$ tetrahedral   d. $\ddot{O}=C-\overset{}{\underset{\displaystyle :\ddot{C}l:}{\ddot{C}l}}:$ triangular

10.33

all bond angles 120°

10.34 a. H—C—H, H—C—C bond angles 109°; C—C≡C and C≡C—H, 180°
b. H—C=O, H—C—O, O=C—O, 120°; C—O—H 109°   c. all 120°
10.35 b, c, d   **10.36** a, b
10.37 1st molecule is polar; 2nd is symmetrical, nonpolar
10.38 a. sp^3 b. sp^2 c. sp   d. sp^2
10.39 ring atoms sp^2, others sp^3 **10.40** a. NH_4^+ b. NO_3^- c. C_2H_4
10.41 if all single bonds, sp^3; two single bonds + double bond, sp^2; single + triple or two double bonds, sp

10.42 a. $:\ddot{O}=C=\ddot{O}:$ b. $:C\equiv O:$ c. $H-\overset{\displaystyle H}{\underset{\displaystyle H}{C}}-\ddot{O}-H$ b < a < c

10.43 a. covalent bonds in polyatomic ions
b. more than half (see Table 10.2)
c. polar since X and Z are different
d. C forms 4 bonds, yet has only two unpaired electrons

10.44 H—N̈—N̈—H, two pyramids joined through N atoms; polar
      | |
      H H

10.45 $CH_3—CH_2—C—\ddot{O}—H$ and $CH_3—\ddot{O}—C—CH_3$ are two possibilities
            ‖   ‖
            :Ö: :Ö:

10.46 51

CHAPTER 11

11.23 a. 118 mm Hg   b. 118 mm Hg   c. 270 mm Hg

11.24 a. 4.00 ℓ b. 300 mm Hg   c. 240 mm Hg   **11.25** b. approx. 51°C

11.26 a. 89°C   b. 68°C   c. 26°C   **11.27** approx. 4.5 km

11.28 a. T   b. T   c. T   d. F   **11.29** dispersion: all; dipole: b, c, d

11.30 b, d   **11.31** $I_2 > Br_2 > Cl_2 > F_2$

11.32 a. urea, $CO(NH_2)_2$ b. NH_4Cl, KNO_3, . . .   c. BN

11.33 a. CaO, $CaCl_2$, . . .   b. $Ca(OH)_2$, $CaCl_2 \cdot 6 H_2O$ c. $CaCO_3$

11.34 a. ionic   b. molecular   c. metallic (Na), K . . .

11.35 a. AB   b. area bounded by AC, AD   c. A   **11.36** c

11.37 a. +50 kJ   b. +5.80 kJ   c. -6.06×10^2 kJ   **11.38** 11.8°C

11.39 a. small discrete molecules vs. network of covalent bonds
    b. forces between molecules vs. forces within molecules
    c. T at which solid, liquid, and gas are at equilibrium; highest T at which liquid
       can exist

11.40 a. cool below 0°C, evacuate to remove water   b. vapor pressure lower
    c. less air above   d. sublimation

11.41 a. dispersion forces   b. hydrogen bonds   c. ionic bonds
    d. dispersion forces

11.42 a. SiO_2 b. Mg   c. HF   d. SiH_4

11.43 a. 14.8 cm³   b. 1.5×10^{10} cm³   c. 9.39 cm³   d. 63.5; 6.3×10^{-8}

11.44 fibrous: per tetrahedron, there is 1 Si and $2 + 2(\frac{1}{2}) = 3$ oxygens. Hence SiO_3
    layer: per tetrahedron, there is 1 Si and $1 + 3(\frac{1}{2}) = 2.5$ oxygens. Hence Si_2O_5

11.45 −0.60°C; transfer of heat   **11.46** 50°C

CHAPTER 12

12.23 molecular: I_2, S_8, P_4; macromolecular: C, Si, P (red)

12.24 a, b: long chains   c. graphite-like   d. impurity atoms of Group 3 element

12.25 a. cool in air   b. heat NH_4NO_3 c. burn S   d. burn P_4 in limited O_2

12.26 a. S   b. Br   c. P

12.27 a. T   b. F; paramagnetic   c. F; reverse is true   d. F; As, not Ge
    e. F; different boiling points

12.28 a. (triangular, nonpolar)   b. H—S̈b—H, H (pyramidal, polar)   c. Ö—Ö with :F̈ and F̈: (polar)

    d. :C̈l—Si—C̈l: with :C̈l: above and :C̈l: below (tetrahedral, nonpolar)

12.29 120°; polar   **12.30** 90° bond angles; shorter bonds

12.31

	N_2O_5	N_2O_4	N_2O_3	N_2O_2	N_2O	NO_2	NO
π	2	2	2	2	2	1	1
σ	6	5	4	3	2	2	1

12.32 a. Si_4H_{10}; SiH_4 b. dispersion forces increase with molecular mass

12.33 dispersion forces weaker in compact molecule

12.34 a. dispersion force   b. H bond   c. covalent bond   d. dispersion force

12.35 a. see p. 288   b. see p. 276   c. similar to CO_2 (Example 12.1)
d. three resonance forms similar to those of SO_3

12.36 a. $N{\equiv}N{-}\ddot{N}$:   and $\ddot{N}{-}N{\equiv}N$:   b. no

H   H

12.37 similar to those of C_5H_{12}, Example 12.4

12.38 $CH_3{-}CH_2{-}CH_2{-}CH_2Cl$, $CH_3{-}\overset{\overset{\displaystyle H}{|}}{C}{-}CH_2{-}CH_3$, $CH_3{-}\overset{\overset{\displaystyle H}{|}}{\underset{\underset{\displaystyle CH_2Cl}{|}}{C}}{-}CH_3$,
                                                          Cl

$CH_3{-}\overset{\overset{\displaystyle Cl}{|}}{\underset{\underset{\displaystyle CH_3}{|}}{C}}{-}CH_3$

12.39 a. 6, sp^3d^2 b. 5, sp^3d c. 5, sp^3d d. 5, sp^3d

12.40 a. octahedral   b. linear   c, d. distorted, tetrahedral

12.41 a. N_2 b. NO_3^- c. NH_4^+ d. PF_5 e. PF_6^-

12.42 a. 3 bonds, 0 unpaired e^- b. $2\frac{1}{2}$ bonds, 1 unpaired e^-
c. $\frac{1}{2}$ bond, 1 unpaired e^-

12.43

	Li_2^+	Be_2^+	B_2^+	C_2^+	N_2^+	O_2^+	F_2^+
bonds	$\frac{1}{2}$	$\frac{1}{2}$	$\frac{1}{2}$	$\frac{3}{2}$	$\frac{5}{2}$	$\frac{5}{2}$	$\frac{3}{2}$
unpaired e^-	1	1	1	1	1	1	1

12.44 possible skeletons include, among many others:
$N{-}N{-}O{-}O{-}O$, $O{-}N{-}N{-}O{-}O$, $O{-}N{-}O{-}N{-}O$,
$N{-}O{-}N{-}O{-}O$, $N{-}O{-}N{-}O$, $N{-}O{-}O{-}N{-}O$,
                                                                                |
                                                                                O
$N{-}O{-}O{-}O{-}N$

12.45 MM = 57; suggests $(HF)_3$ trimer

12.46 XeF_4: distorted tetrahedron ClF_3: triangular, planar molecule

12.47 6; sp^3d^2; octahedral

CHAPTER 13

13.22 a. $\dfrac{[CCl_4] \times [S_2Cl_2]}{[CS_2] \times [Cl_2]^3}$ b. $\dfrac{[Cl_2] \times [NO]^2}{[NOCl]^2}$ c. $\dfrac{[Ni(CO)_4]}{[CO]^4}$ d. $\dfrac{[HCl]^4}{[H_2O]^2}$

13.23 a. $\frac{1}{2} N_2(g) + \frac{1}{2} O_2(g) \rightleftharpoons NO(g)$ b. $H_2(g) + I_2(g) \rightleftharpoons 2\,HI(g)$
c. $H_2(g) + I_2(s) \rightleftharpoons 2\,HI(g)$ d. $CuO(s) + H_2(g) \rightleftharpoons Cu(s) + H_2O(g)$

13.24 0.21   **13.25** 0.53

13.26 a. 0.0060 M   b. 0.0060 M   c. 1.2×10^{-5} **13.27** 7.6×10^{-4} M

13.28 1.3×10^{-2} M   **13.29** 0.010 M, 0.010 M, 0.080 M

13.30

	[IBr]	[I_2]	[Br_2]
a.	0.091	0.0045	0.0045
b.	0.090	0.005	0.005
c.	0.136	0.007	0.007

13.31 [CO] = 0.30 M; [Cl_2] = [$COCl_2$] = 0.10 M

13.32 n NO_2 = 0.68; n N_2O_4 = 0.66   **13.33** a. 0.0052 M   b. 2 × 10^{-17} M

13.34 a. 0.72   b. [CO_2] = [H_2] = 0.015 M; [CO] = [H_2O] = 0.017 M

13.35 appreciable amounts of both   **13.36** NO

13.37 a. decrease   b. increase   c. increase   d. increase

13.38 endothermic   **13.39** Δ conc. B = −0.2 M, Δ conc. C = 0.1 M

13.40 a. [H_2] = 3x   b. error in a, and some NH_3 reacts

   c. [NH_3] = 1 − 2x

   d. errors in b, c, and did not use coeff. as exponents

13.41 a, b, e   **13.42** 0.0122 M, 0.0063 M

13.43 59; [F_2] = 0.23 M; added 0.14 + 0.12 = 0.26 mol F_2

13.44 0.2   **13.45** 0.019

CHAPTER 14

14.21 a. rate = k(conc. A)^2 b. 4.8 × 10^2 ℓ/mol·min   c. 0.071 mol/ℓ

14.22 a. 3   b. 1.2 × 10^{-5} mol/ℓ·s

14.23 a. 0.030 mol/ℓ·min   b. 0.20 min^{-1} c. 1.3 ℓ/mol·min

14.24 a. 2   b. 1   c. 3   **14.25** a. 1   b. 0   c. 2

14.26 a. 3 × 10^{-6} mol/ℓ·s   b. 3.09 × 10^{-6} mol/ℓ·s   c. rate in b

14.27 a. 0.12 s^{-1} b. 19 s ∕   c. 6 × 10^{-25} M

14.28 a. 0.093 M   b. 13 min

14.29 Activated complex 50 kJ or 30 kJ (catalyzed reaction) above reactants. Products 70 kJ below reactants.

14.30 E_a' = 254 kJ   **14.31** 190 kJ   **14.32** 1.0 × 10^2 kJ

14.33 263 K   **14.34** 37 kJ

14.35 rate = k_2K_1 (conc. NO)^2 × (conc. O_2) (in both cases)

14.36 NO_2(g) $\rightleftharpoons$ NO(g) + O(g)   fast

   CO(g) + NO(g) + O(g) → CO_2(g) + NO(g)   slow

14.37 a. too slow   b. rate = k   c. see discussion p. 338

   d. see discussion p. 343

14.38 rate = k(conc. ICl) × (conc. H_2); k(conc. ICl)^2; k(conc. H_2)^2

14.39 a. Δ conc. H_2/Δt   b. $-\frac{1}{2}$ Δ conc. HI/Δt

14.40 a. $\frac{-dX}{dt} = kX^2; \frac{1}{X} - \frac{1}{X_0} = kt$

   b. $\frac{-dX}{dt} = kX^3; \frac{1}{X^2} - \frac{1}{X_0^2} = 2kt$

14.41 rate = k(conc. A)^2 × (conc. B) × (conc. C)

14.42 Cl_2(g) $\rightleftharpoons$ 2 Cl(g)

   CO(g) + Cl(g) $\rightleftharpoons$ COCl(g)

   COCl(g) + Cl_2(g) → $COCl_2$(g) + Cl(g)

14.43 rate of step 3 = k_3(conc. NO_2) × (conc. NO_3) = $k_3K_1K_2$(conc. N_2O_5)

CHAPTER 15

15.23 a. F; H_2, not N_2 b. T   c. F; earth, not sun

15.24 a. Eq. 15.4–15.6   b. Eq. 15.3

   c. NH_3(aq) + H^+(aq) → NH_4^+(aq)

   NH_4^+(aq) + NO_3^-(aq) → NH_4NO_3(s)

15.25 a. destruction O_3 b. smog formation   c. see discussion, p. 362

   d. pollution by Pb, destruction of catalytic converter

15.26 a. see p. 357   b. molecules absorb UV from sun

   c. SO_3 formed from SO_2 in similar way

15.27 In morning rush hour, NO and HC build up from auto exhaust. Reaction with air forms NO_2, O_3.

15.28 12°C   **15.29** 494 kJ

15.30 a. increase in T, →; P, catalyst have no effect
b. increase in T, P, use of catalyst increase rate
15.31 a. 11% NH_3, 22% N_2, 67% H_2
b. 7.88 mol
c. $[NH_3] = 0.17$ M, $[N_2] = 0.34$ M, $[H_2] = 1.05$ M
d. 0.073
15.32 a. $[SO_2] = 0.50$ M, $[O_2] = 0.25$ M   b. 6.0   c. slightly above 800°C
15.33 5.29×10^{-4} **15.34** a. 1.2×10^{-3} mol/$\ell \cdot$s   b. 0.035 M
15.35 1×10^{-12} mol/$\ell \cdot$s; 3 h   **15.36** 60 ℓ/mol·s   **15.37** 1.3×10^{-3}
15.38 3×10^{22} s   **15.39** a. 68 kJ   b. 37 kJ   **15.40** 2×10^{-2} ℓ/mol·s
15.41 5 ℓ/mol·s   **15.42** K decreases with T more rapidly than k increases
15.43 rate = $k_2 K_1 \times$ (conc. O_3)^2/(conc. O_2)   **15.44** 1.2 M
15.45 a. 1.0×10^{22} ℓ b. 6.3×10^{18} mol   c. 2×10^{15} g
15.46 a. 1.8×10^{-15} mol/ℓ b. 1.8×10^{-11} mol/$\ell \cdot$s   c. 38 min

CHAPTER 16

16.22 a. NH_3 (H bonding)   b. ethylene glycol (H bonding)   c. NaOH (ionic)
16.23 a. S   b. SS   c. S   d. S
16.24 a. $KClO_4(s) \rightarrow K^+(aq) + ClO_4^-(aq)$; 0.25 M K^+, 0.25 M ClO_4^-
b. $Na_2CrO_4(s) \rightarrow 2 Na^+(aq) + CrO_4^{2-}(aq)$; 0.50 M Na^+, 0.25 M CrO_4^{2-}
c. $CoCl_3(s) \rightarrow Co^{3+}(aq) + 3 Cl^-(aq)$; 0.25 M Co^{3+}, 0.75 M Cl^-
d. $Al_2(SO_4)_3(s) \rightarrow 2 Al^{3+}(aq) + 3 SO_4^{2-}(aq)$; 0.50 M Al^{3+}, 0.75 M SO_4^{2-}
16.25 a. +25.7 kJ   b. decrease   **16.26** a. 12 g   b. 9 g
16.27 a. $Ba^{2+}(aq) + CrO_4^{2-}(aq) \rightarrow BaCrO_4(s)$
b. $Al^{3+}(aq) + 3 OH^-(aq) \rightarrow Al(OH)_3(s)$
16.28 a. $Pb^{2+}(aq) + 2 Cl^-(aq) \rightarrow PbCl_2(s)$
c. $Fe^{3+}(aq) + 3 OH^-(aq) \rightarrow Fe(OH)_3(s)$
d. $Ni^{2+}(aq) + 2 OH^-(aq) \rightarrow Ni(OH)_2(s)$
16.29 81 mg   **16.30** a. 1.5 cm³   b. 0.025 g
16.31 0.0776, 0.200, 704, 140   **16.32** 11   **16.33** a. 2.6×10^2 b. 3.4
16.34 a. 13.8 g, 100.42°C   b. 3.71 g, 100.42°C
16.35 a. 0.968 g   b. ethylene glycol cheaper   **16.36** $C_{10}H_{14}N_2$
16.37 170   **16.38** c < a < b   **16.39** b
16.40 a. if solubility is low, concentration of saturated solution is low
b. not true if ΔH of solution is negative
c. true only in dilute solution
d. about 3/2 as great
e. NaCl has higher osmotic pressure
16.41 a. salt melts ice, absorbing heat from cream
b. contains dissolved salts, which lower fp
c. hydrogen bonding
d. directly proportional to concentration
16.42 1.00, 0.0177, 0.418 mm Hg   **16.43** 8.9×10^3
16.44 42%   **16.45** 1.9×10^{-3} g/cm³

CHAPTER 17

17.24 $[H^+] = 2 \times 10^{-10}$ M; pH = 9.7
17.25 a. 4.3   b. 0.00   c. 8.12   d. 4.55; all are acidic except c
17.26

	$[H^+]$	$[OH^-]$	
a.	1×10^{-2} M	1×10^{-12} M	acidic
b.	10 M	1.0×10^{-15} M	acidic
c.	2×10^{-5} M	5×10^{-10} M	acidic
d.	1.3×10^{-7} M	7.7×10^{-8} M	acidic

17.27 $[H^+] = 4.0 \times 10^{-9}$ M; $[OH^-] = 2.5 \times 10^{-6}$ M

17.28

	$[H^+]$	$[OH^-]$	pH
a.	3.3×10^{-13} M	3.0×10^{-2} M	12.48
b.	2.0×10^{-2} M	5.0×10^{-13} M	1.70
c.	1.00×10^{-14} M	1.0 M	14.00
d.	3.1×10^{-13} M	0.032 M	12.51

17.29 a. $H_2SO_4(aq) \rightarrow H^+(aq) + HSO_4^-(aq)$; strong
  b. $NH_4^+(aq) \rightleftharpoons H^+(aq) + NH_3(aq)$; weak
  c. $HC_2H_3O_2(aq) \rightleftharpoons H^+(aq) + C_2H_3O_2^-(aq)$; weak
  d. $Cu(H_2O)_4^{2+}(aq) \rightleftharpoons H^+(aq) + Cu(H_2O)_3(OH)^+(aq)$; weak
  e. $H_2CO_3(aq) \rightleftharpoons H^+(aq) + HCO_3^-(aq)$; weak

17.30 a. $H_3PO_4(aq) \rightleftharpoons H^+(aq) + H_2PO_4^-(aq)$
  b. $N_2H_5^+(aq) \rightleftharpoons H^+(aq) + N_2H_4(aq)$
  c. $HSO_4^-(aq) \rightleftharpoons H^+(aq) + SO_4^{2-}(aq)$
  d. $Cu(H_2O)_4^{2+}(aq) \rightleftharpoons H^+(aq) + Cu(H_2O)_3(OH)^+(aq)$

17.31 a. WB   b. N   c. SB   d. WB   e. N

17.32 a. $F^-(aq) + H_2O \rightleftharpoons HF(aq) + OH^-(aq)$
  b. $CH_3NH_2(aq) + H_2O \rightleftharpoons CH_3NH_3^+(aq) + OH^-(aq)$
  c. $HPO_4^{2-}(aq) + H_2O \rightleftharpoons H_2PO_4^-(aq) + OH^-(aq)$
  d. $PO_4^{3-}(aq) + H_2O \rightleftharpoons HPO_4^{2-}(aq) + OH^-(aq)$

17.33 a. N   b. A   c. B   d. A   e. N

17.34 b. $Zn(H_2O)_4^{2+}(aq) \rightleftharpoons H^+(aq) + Zn(H_2O)_3(OH)^+(aq)$
  c. $C_2H_3O_2^-(aq) + H_2O \rightleftharpoons HC_2H_3O_2(aq) + OH^-(aq)$
  d. $NH_4^+(aq) \rightleftharpoons H^+(aq) + NH_3(aq)$

17.35 a. $H_2PO_4^-(aq) + 2\,OH^-(aq) \rightarrow PO_4^{3-}(aq) + 2\,H_2O$; basic
  b. $H^+(aq) + OH^-(aq) \rightarrow H_2O$; neutral
  c. $H^+(aq) + CN^-(aq) \rightarrow HCN(aq)$; acidic

17.36 27.3 cm³   **17.37** 0.250 M   **17.38** 46.0

17.39 a. any indicator   b. PP   c. MR   d. PP

17.40 a. HNO_2 acid, H_2O base; HNO_2/NO_2^- and H_3O^+/H_2O
  b. H_2O acid, S^{2-} base; H_2O/OH^- and HS^-/S^{2-}
  c. $HC_2H_3O_2$ acid, CN^- base; $HC_2H_3O_2/C_2H_3O_2^-$ and HCN/CN^-

17.41 a. BL acid, BL base, Lewis base   b. Lewis acid
  c. BL acid, BL base, Lewis base   d. BL base, Lewis base

17.42 a. H_2S b. HPO_4^{2-} c. H_3O^+ d. $Zn(H_2O)_4^{2+}$ **17.43** b, c

17.44 a. B   b. B   c. B   d. B   e. B   f. A   g. A   h. N

17.45 a. HF, NH_4^+ b. Na_2CO_3, $NaHCO_3$, NaCN   c. methyl red
  d. HCO_3^-/CO_3^{2-}

17.46 from NaOH, but not from HCl   **17.47** test pH of $AgNO_3$ solution

17.48 1000/18.02 = 55.5

17.49 a. 0.00   b. 0.37   c. 0.96   d. 2.70   e. 3.70   f. 7.00   g. 10.30
  h. 11.30   i. 12.96

CHAPTER 18

18.24 a. $[Ba^{2+}] \times [CrO_4^{2-}]$ b. $[Zn^{2+}] \times [OH^-]^2$ c. $[Cr^{3+}] \times [OH^-]^3$

18.25 1×10^{-3}, 7×10^{-3}, 5×10^{-7}, 5×10^{-3} **18.26** a. 5×10^{-5} b. no

18.27 a. yes   b. 2×10^{-8} M   **18.28** a. 7.4×10^{-6} b. 2.8×10^{-9}

18.29 a. 8×10^{-5} b. 0.01   **18.30** 1.4×10^{-5} **18.31** a, b, c, d

18.32 OH^- **18.33** a. 7.68   b. 7.43   c. 7.96   **18.34** a. 9.62   b. 10

18.35 a. 1.2×10^{-3} M   b. 8.3×10^{-12} M   c. 2.92   d. 1.2

18.36 a. 2.5×10^{-4} M, 0.025   b. 1.8×10^{-4} M, 0.036
  c. 1.1×10^{-4} M, 0.055   d. 0.79×10^{-4} M, 0.079

18.37 0.17 M

18.38 a. 4.8×10^{-6}, 1.5×10^{-10} b. benzoic acid   c. benzoate ion
18.39 a. 7.1×10^{-4} b. 1.7×10^{-6} M   **18.40** a. 9.74   b. 11.11
18.41 a. 4.0×10^{-6} b. 1.6×10^4
18.42 a. 0.64   b. 4.5×10^{10} c. 2.2×10^3 **18.43** $d > c > e > b > a > f$
18.44 a. NH_3; 11.37   b. H^+, Cl^-; 0.70   c. NH_4^+, Cl^-; 4.82
18.45 a. 1.92   b. 8.92   c. 3.85   **18.46** 0.0007; 700
18.47 $HZ < HY < HX$; K_a HY $= 2 \times 10^{-3}$; K_a HZ $= 2 \times 10^{-5}$
18.48 $Ca^{2+}(aq) + 2\ OH^-(aq) + CO_2(g) \rightarrow CaCO_3(s) + H_2O$
$CaCO_3(s) + CO_2(g) + H_2O \rightarrow Ca^{2+}(aq) + 2\ HCO_3^-(aq)$
18.49 a. 2.38   b. 4.74   c. 7.44   d. 9.23   e. 11.00   f. 13.52

CHAPTER 19

19.23 a. $2\ NH_3$, $2\ H_2O$, $2\ Cl^-$ b. +3   c. −1
19.24 a. +2   b. +3   c. +3
19.25 a. K_2SO_4 b. $CO(NH_2)_2$ c. K_3PO_4 d. KCl
19.26 a. mono   b. tetra   c. tri   d. mono
19.27 a. 6   b. 6   c. 4   d. 6

19.28   a.   b.   c.

d.   e.

19.29 b. 3 isomers   c. 5 isomers   **19.30** 5
19.31 a. $[_{18}Ar]\ 3d^6$ b. $[_{18}Ar]\ 3d^3$ c. $[_{18}Ar]\ 3d^{10}$ d. $[_{18}Ar]\ 3d^2$
19.32 a. 6   b. 0   c. 4   d. 10
19.33

		3d	4s	4p
a.	$[_{18}Ar]$	(↑↓)(↑↓)(↑↓)(↑)(↑)	(↑↓)	(↑↓)(↑↓)(↑↓)
b.	$[_{18}Ar]$	(↑↓)(↑↓)(↑)(↑↓)(↑↓)	(↑↓)	(↑↓)(↑↓)(↑↓)
c.	$[_{18}Ar]$	(↑↓)(↑)(↑)(↑↓)(↑↓)	(↑↓)	(↑↓)(↑↓)(↑↓)
d.	$[_{18}Ar]$	(↑↓)(↑↓)(↑↓)(↑↓)(↑↓)	(↑↓)	(↑↓)(↑↓)(↑↓)

19.34 all orbitals filled in complexes of Cu^+; one orbital half-filled with complexes of Cu^{2+}
19.35 a. 2   b. 2   c. 0
19.36 a. ( )( ) and (↑)(↑)
(↑↓)(↑↓)(↑↓)   (↑↓)(↑)(↑)
b. ( )( ) and (↑)(↑)
(↑↓)(↑↓)(↑)   (↑)(↑)(↑)
19.37 only two d electrons; must remain unpaired   **19.38** 260 kJ/mol
19.39 a. $[_{18}Ar]$ (↑)(↑)( )(↑↓)(↑↓)   (↑↓)   (↑↓)(↑↓)(↑↓)
b. ( )( )
(↑)(↑)( )
19.40 study absorption as a function of time   **19.41** 1×10^{-12}
19.42 a. T   b. T   c. F; reverse is true
19.43 a. $CoN_6H_{18}Cl_3$ b. $[Co(NH_3)_6]Cl_3(s) \rightarrow Co(NH_3)_6^{3+}(aq) + 3\ Cl^-(aq)$
19.44 $[Pt(NH_3)_4][PtCl_4]$ or $[Pt(NH_3)_3Cl][Pt(NH_3)Cl_3]$ **19.45** 0.92 g
19.46 7.3×10^4 min

CHAPTER 20

20.23 violet color; $(NH_4)_2CO_3$, white ppt; $PbCl_2$; Zn^{2+}

20.24 a. gives positive test even if Na^+ is not present in unknown
b. precipitates both CuS and ZnS
c. forms Sn^{4+}

20.25 a. Ag^+ b. Sn^{4+}, Sb^{3+} c. H_2S

20.26 three oxalate ions, $\left(\begin{array}{c} \overset{\displaystyle O}{\overset{\displaystyle \|}{}} \ \overset{\displaystyle O}{\overset{\displaystyle \|}{}} \\ O-C-C-O \end{array} \right)^{2-}$, bonded at six positions

20.27 CuS much less soluble than ZnS; Group III would precipitate

20.28 a. $CaCO_3(s) + 2\ H^+(aq) \rightarrow Ca^{2+}(aq) + CO_2(g) + H_2O$
b. $Al(OH)_3(s) + 3\ H^+(aq) \rightarrow Al^{3+}(aq) + 3\ H_2O$
c. $Ni(OH)_2(s) + 2\ H^+(aq) \rightarrow Ni^{2+}(aq) + 2\ H_2O$
d. $Ag(NH_3)_2{}^+(aq) + 2\ H^+(aq) \rightarrow Ag^+(aq) + 2\ NH_4{}^+(aq)$

20.29 a. $Fe^{3+}(aq) + 3\ NH_3(aq) + 3\ H_2O \rightarrow Fe(OH)_3(s) + 3\ NH_4{}^+(aq)$
b. $Ag^+(aq) + 2\ NH_3(aq) \rightarrow Ag(NH_3)_2{}^+(aq)$
c. $Ni(OH)_2(s) + 6\ NH_3(aq) \rightarrow Ni(NH_3)_6{}^{2+}(aq) + 2\ OH^-(aq)$

20.30 a. $Zn^{2+}(aq) + 4\ OH^-(aq) \rightarrow Zn(OH)_4{}^{2-}(aq)$
b. $Zn(OH)_2(s) + 2\ OH^-(aq) \rightarrow Zn(OH)_4{}^{2-}(aq)$
c. $Mg^{2+}(aq) + 2\ OH^-(aq) \rightarrow Mg(OH)_2(s)$

20.31 $Ni(OH)_2(s) + 6\ NH_3(aq) \rightarrow Ni(NH_3)_6{}^{2+}(aq) + 2\ OH^-(aq)$
$Zn(OH)_2(s) + 4\ NH_3(aq) \rightarrow Zn(NH_3)_4{}^{2+}(aq) + 2\ OH^-(aq)$
$Zn(OH)_2(s) + 2\ OH^-(aq) \rightarrow Zn(OH)_4{}^{2-}(aq)$

20.32 $Ni^{2+}(aq) + 2\ NH_3(aq) + 2\ H_2O \rightarrow Ni(OH)_2(s) + 2\ NH_4{}^+(aq)$
$Ni(OH)_2(s) + 6\ NH_3(aq) \rightarrow Ni(NH_3)_6{}^{2+}(aq) + 2\ OH^-(aq)$

20.33 NaOH

20.34 Saturate with H_2S at pH = 0.5 to precipitate Cu^{2+}. Raise pH to 9 to precipitate Zn^{2+}.

20.35 Ag^+ **20.36** no ions in Groups I, II, III

20.37 a. 5×10^{-2} M b. 0.03 g **20.38** 0.1 M **20.39** a. yes b. no

20.40 3.5 **20.41** a. 0.040 M b. 0.021 M **20.42** 5×10^9

20.43 a. $K = \dfrac{[Ag(NH_3)_2{}^+] \times [I^-]}{[NH_3]^2} = 2.5 \times 10^{-9}$ b. 3×10^{-4}

20.44 2 **20.45** yes

20.46 $Ni^{2+}(aq) + 2\ NH_3(aq) + 2\ H_2O \rightarrow Ni(OH)_2(s) + 2\ NH_4{}^+(aq)$
$Al^{3+}(aq) + 3\ NH_3(aq) + 3\ H_2O \rightarrow Al(OH)_3(s) + 3\ NH_4{}^+(aq)$
$Ni(OH)_2(s) + 6\ NH_3(aq) \rightarrow Ni(NH_3)_6{}^{2+}(aq) + 2\ OH^-(aq)$
$Al(OH)_3(s) + OH^-(aq) \rightarrow Al(OH)_4{}^-(aq)$
$Al(OH)_4{}^-(aq) + H^+(aq) \rightarrow Al(OH)_3(s) + H_2O$
$Al(OH)_3(s) + 3\ H^+(aq) \rightarrow Al^{3+}(aq) + 3\ H_2O$

CHAPTER 21

21.23 a. +5, −2 b. +7, −2 c. +5, −1 d. +1, +3, −2 e. +2, +3, −2
f. +1, +3, −2

21.24 a. S^{2-} b. $CrO_4{}^{2-}$ c. $CrO_4{}^{2-}$ d. S^{2-}

21.25 $3\ S^{2-}(aq) + 2\ CrO_4{}^{2-}(aq) + 8\ H_2O \rightarrow 3\ S(s) + 2\ Cr(OH)_3(s) + 10\ OH^-(aq)$

21.26 a. $2\ Mn^{2+}(aq) + 3\ H_2O + 5\ SO_4{}^{2-}(aq) \rightarrow 2\ MnO_4{}^-(aq) + 6\ H^+(aq) + 5\ SO_3{}^{2-}(aq)$
b. $2\ Cr^{3+}(aq) + 10\ OH^-(aq) + 3\ H_2O_2(aq) \rightarrow 2\ CrO_4{}^{2-}(aq) + 8\ H_2O$
c. $6\ Fe^{2+}(aq) + Cr_2O_7{}^{2-}(aq) + 14\ H^+(aq) \rightarrow 6\ Fe^{3+}(aq) + 2\ Cr^{3+}(aq) + 7\ H_2O$
d. $4\ Zn(s) + NO_3{}^-(aq) + 10\ H^+(aq) \rightarrow 4\ Zn^{2+}(aq) + NH_4{}^+(aq) + 3\ H_2O$

21.27 a. $Ce^{4+}(aq) + VO^{2+}(aq) + 2\ H_2O \rightarrow Ce^{3+}(aq) + VO_3{}^-(aq) + 4\ H^+(aq)$
b. $10\ Cr^{3+}(aq) + 11\ H_2O + 6\ MnO_4{}^-(aq) \rightarrow 5\ Cr_2O_7{}^{2-}(aq) + 22\ H^+(aq) + 6\ Mn^{2+}(aq)$
c. $2\ MnO_4{}^-(aq) + 3\ Mn^{2+}(aq) + 2\ H_2O \rightarrow 5\ MnO_2(s) + 4\ H^+(aq)$

21.28 $4\ Au(CN)_2{}^-(aq) + 4\ OH^-(aq) \rightarrow 4\ Au(s) + 8\ CN^-(aq) + O_2(g) + 2\ H_2O$

21.29 a. O, R, R, R

b. $Ni(s) \rightarrow Ni^{2+}(aq) + 2 e^-$
$Cr_2O_7^{2-}(aq) + 14 H^+(aq) + 6 e^- \rightarrow 2 Cr^{3+}(aq) + 7 H_2O$
$S(s) + 2 H^+(aq) + 2 e^- \rightarrow H_2S(g)$
$MnO_2(s) + 4 H^+(aq) + 2 e^- \rightarrow Mn^{2+}(aq) + 2 H_2O$
c. $3 Ni(s) + Cr_2O_7^{2-}(aq) + 14 H^+(aq) \rightarrow 3 Ni^{2+}(aq) + 2 Cr^{3+}(aq) + 7 H_2O$
$Ni(s) + S(s) + 2 H^+(aq) \rightarrow Ni^{2+}(aq) + H_2S(g)$
$Ni(s) + MnO_2(s) + 4 H^+(aq) \rightarrow Ni^{2+}(aq) + Mn^{2+}(aq) + 2 H_2O$

21.30 a. R, R, R, O
b. $2 H_2O + 2 e^- \rightarrow H_2(g) + 2 OH^-(aq)$
$O_2(g) + 2 H_2O + 4 e^- \rightarrow 4 OH^-(aq)$
$NiO_2(s) + 2 H_2O + 2 e^- \rightarrow Ni(OH)_2(s) + 2 OH^-(aq)$
$MnO(s) + 6 OH^-(aq) \rightarrow MnO_4^-(aq) + 3 H_2O + 5 e^-$
c. $4 H_2O + 2 MnO(s) + 2 OH^-(aq) \rightarrow 5 H_2(g) + 2 MnO_4^-(aq)$
$5 O_2(g) + 4 MnO(s) + 4 OH^-(aq) \rightarrow 4 MnO_4^-(aq) + 2 H_2O$
$5 NiO_2(s) + 4 H_2O + 2 MnO(s) + 2 OH^-(aq) \rightarrow 5 Ni(OH)_2(s) + 2 MnO_4^-(aq)$

21.31 a. Cd anode, Ni cathode; e^- flow from Cd to Ni; + ions move to Ni, − ions to Cd
b. Pt anode, Pt cathode; e^- flow from anode to cathode; + ions move to cathode, − ions to anode
c. Ag anode, Au cathode; e^- flow from Ag to Au; + ions move to Au, − ions to Ag

21.32 a. $H_2(g) + Cl_2(g) \rightarrow 2 H^+(aq) + 2 Cl^-(aq)$
b. $Sn(s) + Pb^{2+}(aq) \rightarrow Sn^{2+}(aq) + Pb(s)$
c. $2 I^-(aq) + Cl_2(g) \rightarrow I_2(s) + 2 Cl^-(aq)$

21.33 $F^- < Au < Fe^{2+} < SO_2 < K$

21.34 a. Sn, Ag + I^-, Ni, Co, Tl, Pb + SO_4^{2-}
b. Fe^{3+}, Hg_2^{2+}, Ag^+, Hg^{2+}, NO_3^-, $AuCl_4^-$
c. Br^-, H_2O, Mn^{2+}, Cr^{3+}, Cl^-, Cl_2

21.35 a. +2.62 V b. +0.23 V c. +0.19 V

21.36 a. +1.36 V b. +0.01 V c. +0.83 V

21.37 (a) is spontaneous **21.38** Ag, Cu

21.39 a. no reaction
b. $2 Fe^{2+}(aq) + Cl_2(g) \rightarrow 2 Fe^{3+}(aq) + 2 Cl^-(aq)$
$2 I^-(aq) + Cl_2(g) \rightarrow I_2(s) + 2 Cl^-(aq)$
c. no reaction

21.40 a. +0.32 V; spontaneous b. +0.22 V; spontaneous

21.41 1×10^{-20} M **21.42** a. Eq. 21.12b b. Eq. 21.12

21.43 a. 0.00 V b. −0.80 V c. +1.56 V

21.44 a. +2, +4, +3, +8/3
b. Pb_2O_3: one Pb^{2+} for every Pb^{4+}
Pb_3O_4: two Pb^{2+} for every Pb^{4+}

21.45 2×10^{-10} **21.46** −0.83 V; $2 H_2O + 2 e^- \rightarrow H_2(g) + 2 OH^-(aq)$

CHAPTER 22

22.22 a. +3 b. +4 c. +5 d. −3

22.23 a. NH_3 b. HNO_3 c. HNO_2 d. HNO_3 e. $NaNO_2$

22.24 a. ClO^- b. SO_3^{2-} c. ClO_4^- d. NH_3

22.25 a. heat (carefully) with H_2SO_4 b. react with Cu
c. treat with HNO_3, HCl d. heat Na_2SO_3 solution with S

22.26 a. oxidation and reduction occur at two different locations
b. prevents oxidation to Sn^{4+} (SnO_2)
c. can be reduced to H_2O or oxidized to O_2
d. react with O_2 to produce S

22.27 H, C, Si, Ge, N, P, As, Sb, O, S, Se, Te, all halogens
Sc, Ni, Zn; Y, Zr, Ag, Cd; La, Hf, Ta; Ac

22.28 a. bromate, hypobromite   b. hypoiodite, iodite
c. nitrite, nitrate   d. sulfate, hydrogen sulfate; sulfite, hydrogen sulfite

22.29 a, b, c   **22.30** a. N_2O b. N_2O c. NH_4^+ d. NO_2

22.31 a. $3\ HClO(aq) \rightarrow HClO_2(aq) + Cl_2(g) + H_2O$
b. $3\ ClO^-(aq) + H_2O \rightarrow ClO_2^-(aq) + Cl_2(g) + 2\ OH^-(aq)$
c. $5\ N_2(g) + 6\ H_2O + 4\ H^+(aq) \rightarrow 6\ NO(g) + 4\ NH_4^+(aq)$

22.32 a. $2\ MnO_4^-(aq) + 16\ H^+(aq) + 10\ I^-(aq) \rightarrow 2\ Mn^{2+}(aq) + 8\ H_2O + 5\ I_2(s)$
b. $MnO_4^-(aq) + 8\ H^+(aq) + 5\ Fe^{2+}(aq) \rightarrow Mn^{2+}(aq) + 4\ H_2O + 5\ Fe^{3+}(aq)$
c. $2\ MnO_4^-(aq) + H^+(aq) + 5\ HNO_2(aq) \rightarrow 2\ Mn^{2+}(aq) + 3\ H_2O + 5\ NO_3^-(aq)$
d. $2\ MnO_4^-(aq) + 6\ H^+(aq) + 5\ H_2S(aq) \rightarrow 2\ Mn^{2+}(aq) + 8\ H_2O + 5\ S(s)$

22.33 a. $3\ Cl_2(g) + 6\ OH^-(aq) \rightarrow ClO_3^-(aq) + 3\ H_2O + 5\ Cl^-(aq)$
b. $CuS(s) + 4\ H^+(aq) + 2\ NO_3^-(aq) \rightarrow Cu^{2+}(aq) + S(s) + 2\ NO_2(g) + 2\ H_2O$
c. $4\ Zn(s) + 10\ H^+(aq) + NO_3^-(aq) \rightarrow 4\ Zn^{2+}(aq) + NH_4^+(aq) + 3\ H_2O$

22.34 a. $2\ MnO_4^-(aq) + 16\ H^+(aq) + 10\ I^-(aq) \rightarrow 2\ Mn^{2+}(aq) + 8\ H_2O + 5\ I_2(s)$
b. 0.671 M

22.35 some Cl^- reacts with MnO_4^- ($E^0 = +0.16$ V)

22.36 a. Al rod disappears; Cu^{2+} color fades
b. $2\ Al(s) + 3\ Cu^{2+}(aq) \rightarrow 2\ Al^{3+}(aq) + 3\ Cu(s)$; $E^0 = +2.00$ V

22.37 none   **22.38** a. $2\ Cu^+(aq) \rightarrow Cu^{2+}(aq) + Cu(s)$; $E^0 = +0.37$ V

22.39 a. +0.78 V   b. −0.05 V   c. −0.87 V   **22.40** +0.33 V

22.41 3×10^{-6} M   **22.42** 0.258 g; yes   **22.43** 9×10^{15}

22.44 $E^0 = (n_1E_1^0 + n_2E_2^0)/n$, where n is the number of electrons involved

CHAPTER 23

23.22 a. +109.3 kJ   b. +60.6 kJ   c. +43.1 kJ

23.23 a. +   b. +   c. −   d. −

23.24 a. +152.3 J/K   b. +195.2 J/K   c. +297.6 J/K

23.25 a. +63.9 kJ   b. +2.4 kJ   c. −45.6 kJ (spont)

23.26 a. +52.5 kJ   b. +37.3 kJ   c. +33.1 kJ

23.27 a. −1035.6 kJ   b. −151.8 J/K   c. −990.4 kJ   d. −959.7 kJ

23.28 a. 353.4 kJ, 242.9 kJ, 132.4 kJ, 21.9 kJ, −88.6 kJ   b. 840 K

23.29 a. +161.6 kJ   b. +0.254 kJ/K   c. 636 K   **23.30** −183.6 kJ

23.31 a. +2.14 V, −1240 kJ, 10^{217} b. −0.13 V, +25 kJ, 4×10^{-5}

23.32 a. −356.0 kJ   b. +12.0 J/K   c. −359.6 kJ   d. +0.621 V

23.33 a. +0.62 V   b. −120 kJ   c. 10^{21} **23.34** +41.2 kJ vs. +41.1 kJ

23.35 spontaneous reaction in battery starts motor (nonspontaneous)

23.36 a. only if $T\Delta S$ term can be neglected   b. only if ΔH term can be neglected
c. number of moles of gas   d. true

23.37 a. P below vapor pressure (23.8 mm Hg); P above vp; P = 23.8 mm Hg
b. no; raise T

23.38 a. no   b. too slow   c. raise T to make go faster

23.39 a. $\Delta G_3^0 = \Delta G_1^0 + \Delta G_2^0$ b. subst. Eq. 23.15 in (a); simplify

23.40 a. reaction must be spontaneous   b. pollutants dispersed over wide area
c. $E = 0$, $\Delta G = 0$

23.41 $E^0 = \dfrac{0.0591}{n} \log_{10} K$

23.42 a. +6.00 kJ   b. 0   c. +0.0220 J/K   d. +0.21 kJ   e. −0.23 kJ

23.43 $0.21

23.44 a. −92.4 kJ   b. −198.3 J/K   c. +46.4 kJ   d. 3.4×10^{-4} e. 1.1

23.45 a. 5.1 kJ   b. 3.1 g/kg of body mass

CHAPTER 24

24.22 $^{232}_{90}Th \rightarrow {}^{4}_{2}He + {}^{228}_{88}Ra$; ${}^{228}_{88}Ra \rightarrow {}^{0}_{-1}e + {}^{228}_{89}Ac$; ${}^{228}_{89}Ac \rightarrow {}^{0}_{-1}e + {}^{228}_{90}Th$

24.23 a. $^{142}_{58}Ce \rightarrow {}^{138}_{56}Ba + {}^{4}_{2}He$ b. $^{144}_{58}Ce \rightarrow {}^{144}_{59}Pr + {}^{0}_{-1}e$ c. $^{131}_{58}Ce \rightarrow {}^{131}_{57}La + {}^{0}_{1}e$

24.24 a. $^{1}_{0}n$ b. $^{4}_{2}He$ c. $^{7}_{4}Be$ **24.25** 7, 4 **24.26** 0.11

24.27 a. 0.066/h b. 11 h **24.28** 20 d **24.29** yes (345 yr old)

24.30 a. 3.2×10^9 yr b. 4.0×10^9 yr

24.31 check against historical records; use other radioactive "clocks"

24.32 -2.8×10^{-5} g; -2.5×10^6 kJ **24.33** yes; no

24.34 a. 1.32×10^{-8} g b. 3.33×10^{13} **24.35** 1.1×10^{10}

24.36 c > a > b **24.37** a. $7s^2 7p^4$ b. 6

24.38 a. see Figure 24.1
b. mix in known ratio with C-12; detect with Geiger counter
c. see Table 24.1
d. more neutrons produced than consumed

24.39 a. measure fraction left after few days
b. C-14 dating
c. label O atom in CH_3OH with radioactive isotope; see which product is radio-active
d. use water containing known fraction of tritium; measure tritium content of benzene saturated with water

24.40 5.2% **24.41** U-235 supply limited **24.42** 12

24.43 a. 5.5×10^{-10} g b. 1.2×10^{-3} kJ c. 17

24.44 a. 2.3×10^{-7} erg b. 2.6×10^8 cm/s c. 6×10^8 K

24.45 a. -5.72×10^8 kJ b. 1.1×10^{32} kJ c. 2.1×10^{-15}

CHAPTER 25

25.22 a. C_6H_{14} b. C_6H_{12} c. C_6H_{10} d. C_6H_6

25.23 a. $CH_3(CH_2)_7CH_3$ b.

c. $H-C \equiv C-CH_2CH_3$ or $CH_3-C \equiv C-CH_3$

24.24 $(CH_3)_2-\underset{\underset{H}{|}}{C}-CH_2-CH_3$

25.25 aromatics, branched alkanes

25.26 propylene: 120° around double bond, 109° in CH_3 group
methyl acetylene: 109° in CH_3 group, otherwise 180°

25.27 a. aldehyde b. ketone c. alcohol d. ester **25.28** b, c

25.29 acidic: b; basic: c

25.30 $H-\underset{\underset{O}{\|}}{C}-O-CH_2-CH_2-OH$, $H-\underset{\underset{O}{\|}}{C}-O-CH_2-CH_2-O-\underset{\underset{O}{\|}}{C}-H$

$CH_3-\underset{\underset{O}{\|}}{C}-O-CH_2-CH_2-OH$, $CH_3-\underset{\underset{O}{\|}}{C}-O-CH_2-CH_2-O-\underset{\underset{O}{\|}}{C}-CH_3$

$CH_3-\underset{\underset{O}{\|}}{C}-O-CH_2-CH_2-O-\underset{\underset{O}{\|}}{C}-H$

25.31 $CH_3(CH_2)_{13}-O-\underset{\underset{O}{|}}{\overset{\overset{O}{|}}{S}}-O^-$, Na^+ **25.32** 3 **25.33** 14

25.34 C—C—C—COOH, C—C—C   **25.35** b, c   **25.36** 9 in all

                    |
                    COOH

25.37 a. 2nd carbon from left   c. 3rd carbon   **25.38** a, c
25.39 $C_{17}H_{19}O_3N$ **25.40** $6\ CO(g) + 13\ H_2(g) \rightarrow C_6H_{14}(l) + 6\ H_2O(l)$
25.41 $C_2H_2 > C_2H_4 > C_2H_6 > C_2H_4O > C_2H_6O > C_2H_4O_2$
25.42 68.28% C, 6.28% H, 21.66% O, 3.79% N

25.43 $C_{12}Cl_4H_6$: 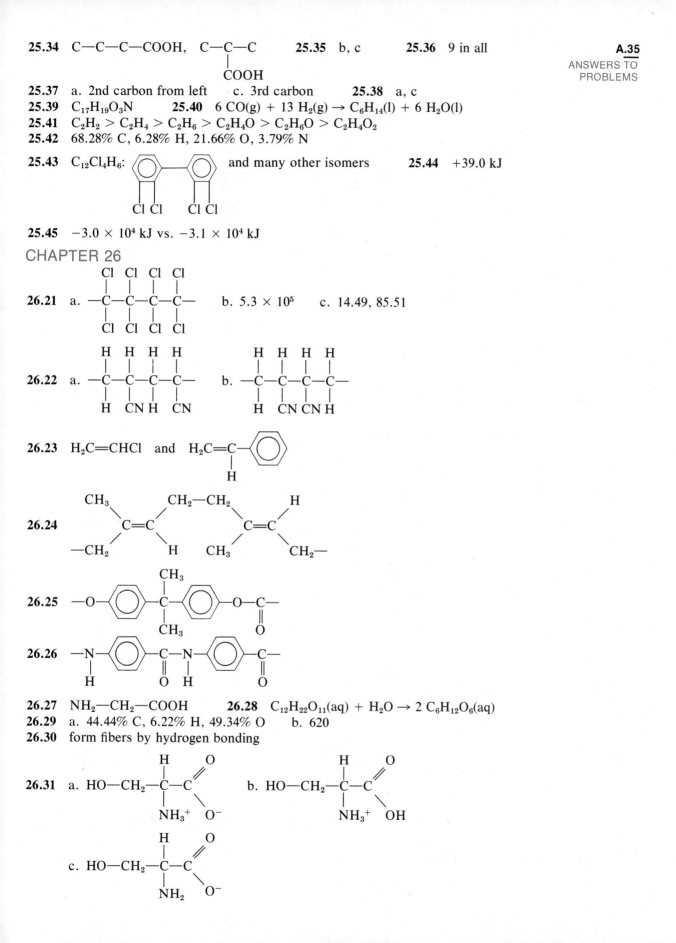   and many other isomers   **25.44** +39.0 kJ

25.45 -3.0×10^4 kJ vs. -3.1×10^4 kJ

CHAPTER 26

26.21 a. (structure)   b. 5.3×10^5 c. 14.49, 85.51

26.22 a. (structure)   b. (structure)

26.23 $H_2C{=}CHCl$ and $H_2C{=}C$—(ring)

26.24 (structure)

26.25 (structure)

26.26 (structure)

26.27 NH_2—CH_2—COOH   **26.28** $C_{12}H_{22}O_{11}(aq) + H_2O \rightarrow 2\ C_6H_{12}O_6(aq)$
26.29 a. 44.44% C, 6.22% H, 49.34% O   b. 620
26.30 form fibers by hydrogen bonding

26.31 a. HO—CH_2—C—C   b. HO—CH_2—C—C

c. HO—CH_2—C—C

26.32   a. 6   b. 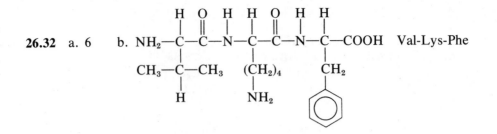   Val-Lys-Phe

26.33   Ala-Phe-Leu-Met-Val-Ala   **26.34**   1
26.35   Need double bond for addition, two functional groups for condensation

26.36 $-\overset{\|}{\underset{O}{C}}-\bigcirc-\overset{\|}{\underset{O}{C}}-O-CH_2-CH_2-O-$; $C_5H_4O_2$ **26.37**   b, c, d

26.38   a. see p. 620   b. see p. 624   c. see p. 627   d. see p. 628
26.39 -17 kJ

26.40   a. $-O-CH_2-\overset{H}{\underset{OH}{C}}-CH_2-O-\overset{\|}{\underset{O}{C}}-\bigcirc-\overset{\|}{\underset{O}{C}}-O-$

a. orthophthalic acid condenses with remaining OH groups in two adjacent
chains
26.41   b. 6.0   c. 2×10^{-5} **26.42**   a. $-$ b. $-$ c. $+$ d. $+$

GLOSSARY

Terms in this glossary are part of the basic chemical vocabulary; they are defined in the context used in this text. If you do not find the term you seek here, look for it in the Index, which will refer you to pages where the term is used.

A

absolute zero—the lowest possible temperature at which matter might exist; 0 K, −273.15°C.

absorption spectrum—graph showing the absorption of radiation by a substance over a range of wavelengths.

acid—a substance which on being dissolved in water produces a solution in which $[H^+]$ is greater than 10^{-7} M. Examples: HCl, HNO_3, H_2CO_3; CH_3COOH.

acid anhydride—a nonmetal oxide that reacts with water to form an acidic solution.

acid-base indicator—see *indicator, acid-base*.

acid-base titration—procedure used to determine the concentration of an acid or base. A sample is reacted to the equivalence point, using a measured volume of a base or acid of known concentration.

acid dissociation constant—K_a; the equilibrium constant for the following reaction of an acid HB: $HB(aq) \rightleftarrows H^+(aq) + B^-(aq)$.

$$K_a = \frac{[H^+][B^-]}{[HB]}$$

acidic solution—an aqueous solution with a pH < 7.0 ($[H^+] > 1.0 \times 10^{-7}$ M).

actinides—elements 90 (Th) through 103 (Lr).

activated complex—a species, formed by collision of energetic particles, which can react to form products or other intermediates.

activation energy, E_a—the minimum energy required for a reaction to occur.

actual yield—the amount of product obtained from reaction.

addition polymer—a polymer produced by reaction of a monomer, usually a derivative of ethylene, adding to itself; no other product is formed.

addition reaction—the insertion of a small molecule, such as H_2 or Cl_2, directly into a double or triple carbon-carbon bond.

alcohol—a substance containing an OH group attached to a hydrocarbon chain. Examples: C_2H_5OH, ethyl alcohol; C_4H_9OH, butyl alcohol.

aldehyde—a substance containing an $\overset{\displaystyle H}{\underset{\displaystyle H}{|}} C{=}O$ group at the end of a hydrocarbon chain. Example: $C_2H_5{-}\overset{\displaystyle |}{\underset{\displaystyle |}{C}}{=}O$, propionaldehyde.

alkali metal—a metal in Group 1. Examples: Li, Na, K.

alkaline (basic)—having a $[OH^-]$ which is greater than 10^{-7} M.

alkaline earth metal—a member of Group 2. Examples: Be, Mg, Ca, Ba.

alkane—a hydrocarbon containing only single carbon-carbon bonds. Examples: C_2H_6, C_6H_{14}.

alkene—a hydrocarbon which contains one carbon-carbon double bond. Examples: $CH_3CH{=}CH_2$; $H_2C{=}CH_2$.

alkyl group—a hydrocarbon group such as $CH_3{-}$, $C_2H_5{-}$, or $C_3H_7{-}$.

alkyne—a hydrocarbon containing one carbon-carbon triple bond. Example: $HC{\equiv}CH$.

allotrope—one of two or more forms of an elementary substance. Examples: O_2 and O_3 are allotropic forms of oxygen; graphite and diamond are allotropes of carbon.

alpha (α) particle—a helium nucleus. He^{2+} ion.

amalgam—a solution of a metal in mercury.

amine—organic compound containing the $-\overset{\displaystyle |}{N}-$ functional group. Examples include methylamine, CH_3NH_2, dimethylamine, $(CH_3)_2NH$, and trimethylamine, $(CH_3)_3N$.

α-amino acid—the monomer units of proteins. The amino acid contains an acid group ($-COOH$) and an amine group ($-NH_2$). The amine group is attached to the α carbon, the one adjacent to the $-COOH$ group.

amphoteric—capable of reacting with both H^+ and OH^- ions; usually an insoluble hydroxide. Examples: $Al(OH)_3$, $Zn(OH)_2$.

amplitude—the height of a standing wave.

anhydride—a substance derived from another by removal

of water. Examples: SO_3 is the anhydride of H_2SO_4; CaO is the anhydride of $Ca(OH)_2$.

anion—a species carrying a negative charge. Examples: Cl^-, CO_3^{2-}, $H_2PO_4^-$.

anode—an electrode at which oxidation occurs. Example: If, at a copper electrode, the reaction that occurs is $Cu(s) \rightarrow Cu^{2+}(aq) + 2\ e^-$, then the copper metal is behaving as an anode.

antibonding orbital—a molecular orbital which has decreased density between two proximate atoms. The energy of its two electrons is greater than that of those electrons in the separated atoms.

aqua regia—a mixture of concentrated hydrochloric and nitric acids.

aromatic substance—an organic compound containing a benzene ring. Examples: Benzene, C_6H_6, ⬡; toluene, C_7H_8, ⬡—CH_3; naphthalene, $C_{10}H_8$, ⬡⬡.

Arrhenius acid—a species which, upon addition to water, increases the concentration of H^+.

Arrhenius base—a species which, upon addition to water, increases the concentration of OH^-.

Arrhenius equation—the equation which expresses the temperature dependence of the rate constant: $\log_{10} k = A - E_a/2.30\ RT$.

atmosphere (atm)—standard unit of pressure; equal to 101.325 kPa; equivalent to the pressure exerted by a mercury column 760 mm high.

atom—smallest particle of an element; matter is composed of atoms in various chemical combinations. Example: The N atom is the smallest particle of the element nitrogen. Two nitrogen atoms combine to form an N_2 molecule, the smallest particle which has the properties of nitrogen as it is ordinarily found.

atomic mass—an averaged mass of atoms of one element relative to that of another element; based upon the atomic mass of a C-12 isotope taken to be exactly 12.

atomic number—a number equal to the number of electrons around the nucleus of an atom of an element; also the number of protons in the nucleus of that atom. Example: The atomic number of carbon is 6; there are six electrons outside the nucleus of a C atom and six protons in the nucleus of that atom.

atomic orbital model—see *valence bond model*.

atomic radius—the radius of an atom, taken to be one half the distance between two nuclei in the ordinary form of the elementary substance. Example: The radius of the Cl atom is 0.099 nm, since the internuclear distance in the Cl_2 molecule is 0.198 nm.

atomic spectrum—a spectrum containing lines characteristic of a particular element.

aufbau principle—the rule stating that electrons enter energy levels in an atom in order of increasing energy, filling one sublevel before moving into the next.

Avogadro's Law—a principle stating that equal volumes of gases at the same temperature and pressure contain equal numbers of molecules.

Avogadro's number—6.022×10^{23}; the number of units in a mole.

axial—adjective used to describe an atom or group that is perpendicular to the plane of a ring molecule.

B

balanced equation—an equation for a chemical reaction in which the reactants and products contain equal numbers of each kind of atom participating in the reaction. Example: The equation $CH_4(g) + 2\ O_2(g) \rightarrow CO_2(g) + 2\ H_2O(l)$ is balanced since both reactants and products contain one C, four H, and four O atoms.

base—a substance which on dissolving in water produces a solution in which $[OH^-]$ is greater than 10^{-7} M. Examples: NaOH, Na_2CO_3, NH_3.

base anhydride—a metal oxide that reacts with water to form a basic solution.

base dissociation constant—K_b; the equilibrium constant for the following reaction of the base B^-. $B^-(aq) + H_2O \rightleftharpoons HB(aq) + OH^-(aq)$.

$$K_b = \frac{[HB][OH^-]}{[B^-]}$$

basic solution—an aqueous solution having a pH > 7.0 ($[H^+] < 1.0 \times 10^{-7}$ M).

bent—adjective used to describe a molecule containing three atoms in which the bond angle is less than 180°. Examples include H_2O and SO_2.

beta radiation ($_{-1}^{0}e$)—one of the types of radiation emitted by unstable nuclei. Beta particles have properties identical to those of electrons.

body-centered cubic—a crystalline structure in which the unit cell is a cube with one atom at each of its corners and an atom at its center.

Bohr model—model of the hydrogen atom derived by Niels Bohr (see Section 8.2).

boiling point—that temperature of a liquid at which its vapor pressure equals the applied pressure; a liquid will tend to form bubbles and vaporize at its boiling point; usually reported at one atmosphere pressure.

boiling point elevation—increase in the boiling point of a liquid caused by addition of a nonvolatile solute. For a nonelectrolyte, the boiling point elevation, ΔT_b, is given by the equation: $\Delta T_b = k_b \times m$, where m is the molality and k_b is a constant for a given liquid (0.52°C for water).

bond—a linkage between two atoms.

bond energy—enthalpy change ΔH associated with a reaction in which a bond is broken. Example: For the reaction $HCl(g) \rightarrow H(g) + Cl(g)$, $\Delta H = 431$ kJ; the bond energy, B.E., of the H—Cl bond is 431 kJ/mol of bonds.

bonding orbital—an orbital associated with two atoms in which the energy of its two electrons is less than the energies of those electrons in the separated atoms. The presence of a populated bonding orbital between two atoms stabilizes the bond between them.

Boyle's Law—a relation stating that when a gas sample is compressed at a constant temperature, the product of the pressure and the volume remains constant.

branched-chain alkane—a saturated hydrocarbon in which not all of the carbon atoms are located in a single, continuous chain. The simplest branched-chain alkane is 2-methylpropane, which has the structure

$$CH_3-CH-CH_3.$$
$$\quad\quad\;\; |$$
$$\quad\quad CH_3$$

brine—a solution of a salt, usually NaCl, in water.

Brönsted-Lowry acid—a species which donates a proton to another species. Example: In the reaction $HB(aq) + H_2O \rightleftarrows H_3O^+(aq) + B^-(aq)$, HB behaves as a Brönsted-Lowry acid in that it donates a proton to H_2O.

Brönsted-Lowry base—a species that accepts a proton from another species. Example: In the reaction just given, H_2O behaves as a Brönsted-Lowry base, since it accepts a proton from HB.

buffer—a solution which resists change of pH more effectively than would a solution of a strong acid or base having the same pH; usually contains a weak acid and its conjugate base. Example: A solution containing 0.5 M H_2CO_3 and 0.5 M $NaHCO_3$ is a buffer with a pH of about 6.4; the pH is relatively resistant to change caused by addition of small amounts of either H^+ or OH^- ions.

C

calorie—a unit of thermal energy equal to 4.184 joules.

calorimeter—a device used to measure the heat flow in a chemical reaction or physical change.

calorimeter constant (C)—the product of the mass times the specific heat of a bomb calorimeter.

carbohydrate—a class of organic compounds in which the general formula is $C_m(H_2O)_n$. Examples include glucose, sucrose, starch, and cellulose.

carbonate ion—CO_3^{2-}.

carboxylic acid—an organic compound containing the functional group

$$-\overset{\displaystyle O}{\underset{\displaystyle \|}{C}}-OH.$$ An example is acetic

acid, $CH_3-\overset{\displaystyle O}{\underset{\displaystyle \|}{C}}-OH.$

catalyst—a substance which affects the rate of a reaction without being used up itself. Example: A piece of platinum foil can act as a catalyst for the combustion of methane in air.

catalytic converter—device inserted into the exhaust system of an automobile, containing finely divided Pt and Pd. This metal catalyst converts CO to CO_2 and unburned hydrocarbons to CO_2 and H_2O.

cathode—an electrode at which reduction occurs. Example: If, at a silver electrode, the reaction that occurs is $Ag^+(aq) + e^- \rightarrow Ag(s)$, then the silver metal is serving as a cathode.

cation—an ion having a positive charge. Examples: Fe^{2+}, K^+, and NH_4^+ are all cations.

cation exchange—process by which a cation in water solution is "exchanged" for a different cation, origi-

nally present in a solid resin. Used to soften water by exchanging Ca^{2+} for Na^+ ions.

Celsius degree—a unit of temperature, based on there being 100° between the freezing and boiling points of water; ultimately defined by means of a gas volume thermometer, the absolute temperature scale, and the 100° interval noted above.

centi—prefix on a metric unit indicating a multiple of 10^{-2}. Example: 1 centimeter = 10^{-2} meter.

chain reaction—a type of chemical reaction occurring in steps in which the product of a late step serves as a reactant in an earlier step, thereby allowing a reaction, when once begun, to continue.

Charles' Law—a relation stating that the volume of a gas sample at constant pressure is directly proportional to its absolute temperature.

chelating agent—a complexing ligand that can form more than one bond with a central ion. Example: Ethylenediamine, $H_2N-CH_2-CH_2-NH_2$, en, is a chelating agent which can form two bonds with a metal ion; its complex ion with Cu^{2+}, coordination number 4, has the formula $Cu(en)_2^{2+}$.

chemical equation—an expression which qualitatively and quantitatively describes the reactants and products of a chemical reaction as to their nature and amount. Example: $N_2(g) + 3 H_2(g) \rightarrow 2 NH_3(g)$, a chemical equation, tells us that one mole of nitrogen gas reacts with three moles of hydrogen gas to form two moles of ammonia gas.

chemical property—property of a substance related to its chemical changes.

chemical symbol—a one or two letter abbreviation of the name of an element. Example: Pb for lead (Latin, *plumbum*).

chemical thermodynamics—use of thermodynamic principles to predict whether a chemical reaction will be spontaneous; ordinarily involves calculation of ΔH, ΔS, and/or ΔG for the reaction.

chiral center—atom in a molecule which is bonded to four different groups; a source of optical isomerism.

chromatography—separation method in which the components of a solution are adsorbed at different locations on a solid surface.

cis **isomer**—a geometric isomer in which two identical bonded atoms or groups are relatively close to one another. Example: In square planar $Pt(NH_3)_2Cl_2$,

the *cis* isomer has the structure ; the

trans isomer has the structure

coefficient—a number preceding a symbol or formula in a chemical equation.

colligative property—a physical property of a solution which depends on the concentration, but not the kind, of solute particles. Example: The vapor pressure depression of a solution depends on the mole fraction of the solute but not on the nature of the solute, and so is a colligative property.

common ion effect—if, to a solution containing ions, a solute is added which furnishes one of the ions originally present, the common ion effect will change some of the properties of the solution. Example: The solubility of NaCl in water is decreased by addition of 6 M HCl.

complex ion—an ion containing a central metallic cation to which two or more groups are attached by coordinate covalent bonds. Example: In the $Ag(NH_3)_2{}^+$ complex ion the electrons in the coordinate covalent bonds between Ag^+ and NH_3 are furnished by the NH_3 molecules.

compound—a chemical substance containing more than one kind of atom.

concentrated—adjective used to describe a solution which contains a relatively high concentration of solute.

concentration—refers to relative amounts of solute and solvent in a solution; may be stated in many ways, such as per cent solute by mass, or mole fraction, but very often is given in terms of molarity, which is the number of moles of solute per liter of solution. Example: In 6 molar NaOH, 6 M NaOH, there are 6 mol of NaOH in a liter of solution.

condensation—conversion of a gas to a liquid or solid.

condensation polymer—a polymer formed from monomer units joined by splitting out a small molecule, usually water.

conductivity—a term referring to the relative ease with which a sample will transmit electricity or heat (should specify which). Example: Since a much larger electrical current will flow through an aluminum rod at a given voltage than through a glass rod of the same shape, the electrical conductivity of aluminum is much greater than that of glass.

configuration—a structure; a geometric arrangement. Example: Carbon tetrachloride molecules have a tetrahedral configuration.

conjugate—refers to related acids and bases, often in connection with Brönsted-Lowry description. Example: In the reaction $HNO_2(aq) + H_2O(aq) \rightleftharpoons H_3O^+(aq) + NO_2{}^-(aq)$, HNO_2 and H_3O^+ behave as acids, while H_2O and $NO_2{}^-$ act as bases. The $NO_2{}^-$ ion is the conjugate base of the acid HNO_2, and H_2O is the conjugate base of the H_3O^+ ion. Or, HNO_2 is the conjugate acid of the $NO_2{}^-$ ion and H_3O^+ ion is the conjugate acid of H_2O.

Conservation of Energy—law which states that energy can neither be created nor destroyed.

contact process—process used in the industrial preparation of sulfuric acid. SO_2 and O_2 are converted to SO_3 by bringing them into contact with a solid catalyst, which may be V_2O_5 or Pt.

continuous spectrum—a spectrum containing light of all wavelengths.

conversion factor—a ratio, numerically equal to one, by which a quantity is multiplied to obtain an equivalent quantity; often expressed as the equation from which the ratio can be obtained. Example: From the conversion factor, 12 in = 1 ft, one can set up the ratio 12 in/1 ft, which can be used to convert a distance in feet into its equivalent in inches.

coordinate covalent bond—a covalent bond in which the electrons are furnished by only one of the bonded atoms; most commonly encountered in complex ions. Example: In the $Zn(OH)_4{}^{2-}$ ion, the electrons in the bonds between Zn^{2+} and the OH^- ions are all furnished by the hydroxide ions; these bonds are, therefore, coordinate covalent bonds.

coordination number—the number of bonds formed from the central metal to the ligands in a coordination complex.

corrosion—a destructive chemical process, most often applied to the conversion of a metal to one of its compounds. An example is the corrosion of iron in contact with O_2 and H_2O to form first $Fe(OH)_2$ and eventually hydrated $Fe(OH)_3$.

Coulomb's Law—a relation expressing force between charged particles; $F = \dfrac{q_1 q_2}{Dr^2}$, where F is the force between two particles having charges q_1 and q_2 and r is the distance between them. If q_1 and q_2 have the same sign, force is repulsive; otherwise it is attractive. In a medium having dielectric constant D, the force is decreased by the factor D; D is 1 in a vacuum.

covalent bond—a chemical link between two atoms, produced by shared electrons in the region between the atoms. Example: In the H_2O molecule there is a covalent bond between the O atom and each H atom; each bond contains two electrons, one furnished by the H atom and one by the O atom; both atoms share the electrons in the bond.

critical pressure—the pressure at the critical temperature.

critical temperature—the highest temperature at which a substance can exhibit liquid-vapor equilibrium. Equilibrium pressure at that point is called the critical pressure. Above that temperature, liquid cannot be condensed from the vapor at any pressure. Example: Since water has a critical temperature of 374°C, above 374°C one cannot have liquid water in equilibrium with its vapor.

crystal—a sample of matter in which the component atoms or ions are arranged in a regular geometric pattern.

crystal field model—model of the bonding in complex ions. The bonding is considered to be essentially ionic; the only effect of the ligands is to change the relative energies of the d orbitals of the central metal ion.

crystal field splitting energy—the difference in energy between the two sets of d orbitals in a complex ion.

cubic centimeter (cm³)—a volume unit equal to the volume of a cube 1 cm on each edge; a milliliter.

D

Dalton's Law—a relation stating that the total pressure of a gas mixture is equal to the sum of the partial pressures of its components.

de Broglie relation—equation used to describe the wave properties of matter: $\lambda = h/mv$.

deionization—removal of ions from water. Can be achieved by passing through two successive columns. In one column, cations are exchanged for H^+ ions; in the other, anions are exchanged for OH^- ions. After reaction of H^+ with OH^- ions, no ions remain in solution.

density—a property of a sample equal to its mass per unit volume. Example: Since the density of mercury is 13.5 g/cm^3, 1.00 cm^3 of Hg weighs 13.5 g.

desiccant—a drying agent.

deuterium—a heavy isotope of hydrogen, ^2_1H.

diamagnetic—a descriptive term indicating that a substance does not contain unpaired electrons and so is not attracted into a magnetic field. Example: Since all of the electrons in NH_3 molecules are paired, NH_3 is diamagnetic.

diamond—one of the crystalline forms of carbon.

diffusion—a process by which one substance, by virtue of the kinetic properties of its particles, will gradually mix with another. Example: $H_2S(g)$ prepared in a test tube will slowly diffuse into the surrounding air.

dilute—refers to a solution containing a relatively small amount of solute; opposite of concentrated.

dipole—species in which there is a separation of charge, i.e., a positive charge at one point and a negative charge at a different point. Examples of dipoles are HF and H_2O.

dipole force—refers to attractive force between molecules possessing separate positive and negative poles. Example: Since the HCl molecule has positive and negative ends, there will be dipole forces between neighboring HCl molecules.

disaccharide—a dimer of two monosaccharide units. The units may be alike, as in maltose (both glucose), or different, as in sucrose (glucose and fructose).

dispersion force—an attractive force between molecules which arises because of the presence of temporary dipoles. Usually increases with MM.

disproportionation—a reaction in which a species undergoes oxidation and reduction simultaneously. Example: $2\ Cu^+(aq) \rightarrow Cu(s) + Cu^{2+}(aq)$.

dissociation—separation into two or more species; usually applied to weak acids or bases or complex ions. Example: The dissociation of acetic acid in water to form H^+ ions and acetate ions only occurs to a small extent.

dissociation constant, K_d—equilibrium constant for the dissociation of a complex ion. The expression for K_d of the $Cu(NH_3)_4^{2+}$ ion is $\dfrac{[Cu^{2+}] \times [NH_3]^4}{[Cu(NH_3)_4^{2+}]}$.

distillation—a procedure in which a liquid is vaporized under conditions where the evolved vapor is later condensed and collected.

double bond—two shared electron pairs between two bonded atoms.

ductility—ability of a solid to retain strength on being forced through an orifice; characteristic of metals.

E

E—see *energy*.

effusion—the movement of a gas through a capillary or porous solid into another gaseous region or vacuum.

Einstein's equation—the relation $\Delta E = \Delta mc^2$ relating mass and energy changes.

efflorescence—the loss of water of hydration from a hydrate.

electrode—a general name for anode or cathode.

electrolysis—the passage of a direct electric current through a solution containing ions, producing chemical changes at the electrodes.

electrolyte—a substance which exists as ions in water solution. Example: NaCl (Na^+ and Cl^- in water solution).

electrolytic cell—a cell in which the flow of electrical energy from an external source causes a redox reaction to occur.

electron—the negatively charged component of atoms; exists in a roughly spherical cloud around atomic nucleus; carries 1 unit of negative charge and has a very low mass.

electron cloud—a region of negative charge around an atomic nucleus; associated with an atomic orbital.

electron configuration—a statement of the populations of the electronic energy sublevels in an atom. Example: Since the electron configuration of the Li atom is $1s^22s$, there are two electrons in the 1s sublevel and one electron in the 2s sublevel.

electron pair repulsion—principle used to predict the geometry of a molecule or polyatomic ion. Electron pairs around a central atom tend to orient themselves so as to be as far apart as possible.

electron-sea model—model of metallic bonding in which cations are considered as fixed points in a mobile "sea" of electrons.

electron spin—a property of an electron loosely related to its spin around an axis. Only two spin states are allowed, usually described by quantum number m_s, which can assume the values $+1/2$ and $-1/2$.

electronegativity—a property of an atom which increases with its tendency to attract the electrons in a bond. Example: Since the Cl atom is more electronegative than the H atom, in the HCl molecule the bonding electrons will be closer to Cl.

electrostatic forces—the forces between particles caused by their electric charges.

element—a general name given to each of the 106 different atoms. Example: Sulfuric acid, H_2SO_4, contains three elements, hydrogen, sulfur, and oxygen; this is equivalent to saying that in H_2SO_4 there are three different kinds of atoms, H, S, and O.

elementary substance—ordinarily observed form of a chemical substance containing only one kind of atom. Examples: Elementary oxygen consists of O_2 molecules; elementary sodium is metallic Na.

empirical formula—an expression which furnishes relative numbers of atoms of the elements in a chemical substance; expressed as the lowest possible set of integers. Often called the simplest formula. Examples: NaCl, H_2SO_4, CH_2, Fe, HO (hydrogen peroxide).

enantiomer—one of a pair of optical isomers.

endothermic—describes a reaction during which heat must be furnished to the reacting mixture to maintain its temperature at the initial value; ΔH for the reaction is a positive quantity.

end point—the point during a reaction, usually in the course of a titration, at which a chemical indicator changes color. Example: The end point in a titration using phenolphthalein indicator occurs at a pH of 9.

energy—a property of a system which is related to its capacity to cause change; can be altered only by exchanging heat or work with the surroundings; given the symbol E.

energy level—the value of the **n** quantum number.

enthalpy—a property of a system which reflects its capacity to exchange heat Q with its surroundings, given the symbol H; defined so that $\Delta H = Q$ for changes in the system that occur at constant pressure.

enthalpy change, ΔH—difference in enthalpy between products and reactants.

enthalpy of formation (ΔH_f)—heat flow for the reaction in which a pure substance is formed at constant pressure from elementary substances.

entropy—a property of a system related to its degree of organization; highly ordered systems have low entropy; given the symbol S.

entropy change, ΔS—difference in entropy between products and reactants.

enzyme—a protein catalyst in biological systems.

equatorial—adjective used to describe an atom or group that is parallel to the plane of a ring molecule.

equilibrium—a state of dynamic balance, where rates of forward and reverse reactions are equal, so system does not change with time. Example: At 100°C, liquid water is in equilibrium with its vapor when the vapor is at a pressure of 1 atm.

equilibrium concentration—concentration, in moles per liter, of a species at equilibrium. Represented by the symbol [ ].

equilibrium constant—a number which imposes a condition on reactant and product concentrations in an equilibrium system; formulated according to Law of Chemical Equilibrium, given the symbol K_c. Example: For the reaction $PCl_3(g) + Cl_2(g) \rightleftharpoons PCl_5(g)$, $K_c = 20$ at 240°C; therefore, at equilibrium at 240°C, $\frac{[PCl_5]}{[PCl_3][Cl_2]} = 20$.

equivalence point—the point during a reaction between A and B, usually during a titration, when an amount of B has been added that is required to react exactly with the amount of A present. Example: The equivalence point in the reaction $H^+(aq) + OH^-(aq) \rightarrow H_2O(l)$ occurs when the number of moles of OH^- ion added to an acid solution equals the number of moles of H^+ ion in the solution.

ester—the product of the reaction between an alcohol and an acid. Example: When methanol, CH_3OH, reacts with acetic acid, CH_3COOH, the ester called methyl acetate, CH_3—O—CO—CH_3, is formed.

ether—an organic compound containing an oxygen atom connected to two alkyl groups. Examples: CH_3OCH_3, $C_2H_5OC_2H_5$.

excited state—an electronic state of a higher energy than the ground state.

exclusion principle—the rule stating that in an atom no two electrons can have the same set of four quantum numbers.

exothermic—describes a reaction during which heat must be removed from the reacting mixture to maintain its temperature at the initial value; ΔH for the reaction is a negative quantity.

expanded octet—more than four electron pairs about a central atom.

F

face-centered cubic—a type of crystal structure in which the unit cell is a cube with identical atoms at each corner and at the center of each face.

Fahrenheit degree—a degree based on the temperature scale on which water freezes at 32° and boils at 1 atm at 212°.

faraday—one mole of electrons; 96,485 coulombs of electric charge.

fat—an ester made from glycerol and a long-chain carboxylic acid; found in seeds and in fatty tissue of animals.

fatty acid—a long-chain carboxylic acid. Example: Stearic acid, $CH_3(CH_2)_{16}COOH$.

filtration—a process for separating a solid-liquid mixture by passing it through a barrier with fine pores, such as filter paper.

First Law of Thermodynamics—the statement that the change in energy, ΔE, of a system equals the heat flow, Q, into the system from the surroundings minus the work, W, done by the system on the surroundings ($\Delta E = Q - W$).

first order—a term describing a reaction whose rate depends on reactant concentration raised to the first power. Example: Since the rate of the reaction $2 N_2O_5(g) \rightarrow 2 N_2O_4(g) + O_2(g)$ is given by the equation rate = k(conc. N_2O_5), the reaction is first order.

fission—see *nuclear fission*.

five per cent rule—empirical rule that the approximation $a - x \approx a$ is valid if $x \leq 0.05$ a. The rule depends upon the generalization that the value of the constant in the equation in which x appears is seldom known to better than ±5%.

fixation of nitrogen—any process which converts $N_2(g)$ into a nitrogen-containing compound. Example: The fixation of nitrogen by the Haber process occurs via the reaction of N_2 with H_2 to make ammonia, NH_3.

flotation—a separation process used to free finely divided ore from rocky impurities. With a soapy emulsion of oil and water, the ore concentrates at the surface and can be skimmed off.

flow sheet—a diagram used to summarize a separation scheme in qualitative analysis.

formula—the expression used to describe the relative number of atoms of the different elements present in a substance; molecular formula is used with substances having molecules; empirical formula is used with nonmolecular substances.

formula mass—the sum of the atomic masses of the atoms in a formula.

fractional crystallization—process used to separate a pure solid from a mixture with another solid. The mixture is dissolved in the minimum amount of hot solvent. Upon cooling, one solid should crystallize from solution while the other remains in solution.

fractional distillation—a procedure used to separate components with different boiling points from a solution; based on passing vapors from a boiling solution up a column along which the temperature gradually decreases; higher boiling components condense on column and return to solution, lowest boiling component goes out of top of column, where it is condensed and collected.

free energy—a property of a system which reflects its capacity to do useful work, given the symbol G; ΔG for a reaction at constant temperature and pressure is equal to minus the amount of useful work the reaction can produce; spontaneous reactions are those for which ΔG is negative.

free radical—a species having an unpaired electron. Examples: The H atom, the NO molecule, and the CH_3 group are all free radicals.

freezing point—the temperature at which a solid and liquid phase can coexist at equilibrium; applies to both pure liquids and solutions. Example: The freezing point of a solution containing one mole NaCl in one liter of water is $-3.37°C$; at that point pure ice and the solution are in equilibrium.

freezing point depression—decrease in the freezing point of a liquid caused by addition of a solute. For a non-electrolyte, the freezing point depression ΔT_f is given by the equation: $\Delta T_f = k_f \times m$, where m is the molality and k_f is a constant for a given liquid ($1.86°C$ for water).

functional group—a small group of atoms in an organic molecule which give the molecule its distinctive chemical behavior.

fusion—the melting of a solid to a liquid; also refers to reaction between small atomic nuclei to form a larger one. Example: The fusion reaction $2\,{}^{2}_{1}H \rightarrow {}^{4}_{2}He$ would produce a large amount of energy.

G

G—see *free energy*.

gamma radiation—high energy photons emitted by radioactive nuclei.

gas constant (R)—the constant which appears in the Ideal Gas Law equation, $PV = nRT$; depends upon units of P, V, and T; equals $0.0821\ \dfrac{\ell \cdot atm}{mol \cdot K}$ in the units listed.

Gas Law—see *Ideal Gas Law*.

Geiger-Müller counter—device used to measure rate of radioactive decay (Figure 24.2).

geometric isomer—a species having the same kind and number of atoms as another species, but in which the geometric structure is different. Example: There are two geometric isomers with the molecular formula $Pt(NH_3)_2Cl_2$:

Cl   NH₃   Cl   NH₃

Pt   and   Pt

Cl   NH₃   H₃N   Cl

(structures are both planar).

Gibbs-Helmholtz equation—a relationship among ΔG, ΔH, and ΔS: $\Delta G = \Delta H - T\Delta S$.

Graham's Law—a relation stating that the rate of effusion of a gas is inversely proportional to the square root of its density or molecular mass.

gram—a unit of mass in the metric system; equal to mass of one cubic centimeter of water at $4°C$.

gram molecular mass—the mass in grams of a mole of a molecular substance. Example: Since the molecular mass of N_2 is 28, the gram molecular mass of nitrogen gas is 28 g.

gray—a unit of absorbed dose of radiation: one joule per kilogram of tissue.

greenhouse effect—phrase used to describe the effect of water and carbon dioxide in absorbing outgoing IR radiation, thereby raising the earth's temperature.

ground state—the lowest allowed energy state of an atom, ion, or molecule.

group—a vertical column of the Periodic Table.

Groups I, II, III, IV—cation groups in qualitative analysis. Roman numerals are used to distinguish from groups in the Periodic Table.

H

H—see *enthalpy*.

Haber process—industrial process used to make ammonia from nitrogen and hydrogen.

half-cell—half of a voltaic or electrolytic cell, at which either oxidation or reduction occurs. Example: The half-cell reaction at the anode is one of oxidation.

half-equation—equation written to describe a half-reaction of oxidation or reduction. An example of an oxidation half-equation is $Zn(s) \rightarrow Zn^{2+}(aq) + 2\ e^-$.

half-life ($t_{1/2}$)—the time required for a reaction to convert half of the initial reactant to product(s).

halide ion—F^-, Cl^-, Br^-, or I^-.

halogen—an elementary substance in Group 7. Examples: F_2, Cl_2, Br_2.

hard water—water containing excessive Ca^{2+} or Mg^{2+}.

head-to-tail polymerization—process leading to the formation of a polymer of the type:

heat—that form of energy which flows between two samples of matter because of their difference in temperature.

heat capacity—the amount of heat required to raise the temperature of a sample by 1°C.

heat flow—the amount of heat, Q, passing into or out of a system; Q is positive if flow is into system, negative if out of system.

heat of formation—see *enthalpy of formation.*

heat of fusion—ΔH for the conversion of unit amount (one gram or one mole) of a solid to a liquid at constant P, T.

heat of sublimation—ΔH for the conversion of unit amount (one gram or one mole) of a solid to a vapor at constant P, T.

heat of vaporization—ΔH for the conversion of unit amount (one gram or one mole) of a liquid to a vapor at constant P, T.

Hess's Law—a relation stating that the heat flow in a reaction which is the sum of two other reactions is equal to the sum of the two heat flows in those reactions.

heterogeneous—having nonuniform composition.

high-spin complex—complex which, for a particular metal ion, has the largest possible number of unpaired electrons.

homogeneous—having uniform composition.

Hund's rule—a relation stating that, ordinarily, electrons will not pair in an orbital until all orbitals of equal energy contain one electron.

hybrid atomic orbital—an orbital made from a mixture of s, p, d, or f orbitals. Example: An sp^2 hybrid orbital is derived from an s and two p orbitals.

hydrate—a substance containing bound water. Example: $BaCl_2 \cdot 2\,H_2O$ is a common hydrate.

hydrocarbon—a substance containing only hydrogen and carbon atoms.

hydrogen bonds—attractive forces between molecules, arising from interaction between a hydrogen atom in one molecule and a strongly electronegative atom (N, O, F) in a neighboring molecule. Example: Hydrogen bonding in water is due to interaction between the H atoms and O atoms on different H_2O molecules.

hydrolysis—a reaction in which a water molecule is split as a result of interaction with another species. Example: The hydrolysis of CN^- involves the reaction $CN^-(aq) + H_2O(l) \rightleftharpoons HCN(aq) + OH^-(aq)$.

hydronium ion—the H_3O^+ ion characteristic of acidic water solutions.

hydroxide ion—OH^-.

I

Ideal Gas Law—states the relationship between pressure, volume, temperature, and amount for any gas at moderate pressures; $PV = nRT$.

indicator, acid-base—a chemical substance which changes color with pH change; usually color change occurs over about two pH units.

inert complex—a complex ion which exchanges ligands very slowly.

inert gas—noble gas.

infrared—describes light having a wavelength greater than about 700 nm.

intermolecular—between molecules.

internal energy—see *energy.*

ion—a charged species.

ion pair—a species made up of a cation and anion held together by strong electrostatic forces. Examples: In solutions of magnesium sulfate one would find appreciable amounts of $(Mg^{2+}\,SO_4^{2-})$ ion pairs.

ion product (Q)—the product of the actual concentrations of ions, each raised to the appropriate power. This is in comparison to the product of the *equilibrium* concentrations of the ions.

ionic compound—a substance in which component species are cations and anions. Examples: Some common ionic compounds are NaCl, CaO, and NH_4NO_3.

ionic radius—the radius of an ion as based on the assumption that ions in a crystal are in contact with nearest neighbors.

ionization constant—a general term for dissociation constant of a weak acid or base; see *acid dissociation constant.*

ionization energy—the energy required to remove the outermost electron from a gaseous atom.

isoelectronic (with)—having the same number of electrons as.

isomer—a species having same number and kind of atoms as another species, but having different properties; structural, geometric, optical, and stereoisomers may occur. Example: Dimethyl ether, $CH_3—O—CH_3$ is a structural isomer of ethyl alcohol, $CH_3—CH_2—OH$.

isomerization—process in which straight-chain hydrocarbons are converted to their branched-chain isomers.

isotope—an atom having same number of nuclear protons as another, but with a different number of neutrons. Example: Ordinary oxygen has three isotopes, all with eight protons in the nucleus, but with eight, nine, and ten neutrons, respectively.

J

joule—the base SI unit of energy; equal to kinetic energy of a two-kilogram mass moving at a speed of one meter per second.

K

K_a, K_b, K_c, K_{sp}, K_w—see *acid dissociation constant, base dissociation constant, equilibrium constant, solubility product constant,* and *water dissociation constant,* respectively.

Kelvin temperature scale—an absolute temperature scale based on definition that the volume of a gas at constant (low) pressure is directly proportional to temperature, and that 100 degrees separate the freezing and normal boiling points of water.

ketone—an organic compound containing a nonterminal carbonyl, C=O, group. Example: The simplest ketone is acetone, $CH_3—CO—CH_3$.

kilo—a prefix on metric units indicating multiple of 1000. Example: One kilojoule equals 1000 J.

kilogram—basic unit of mass in SI: 1000 g.

kilojoule—unit of energy: 1000 J.

kilopascal (kPa)—a pressure unit: 1 kPa is approximately the pressure exerted by a 10-g mass resting on a 1-cm^2 area; 101.3 kPa = 1 atm.

kinetic—associated with motion. Example: The kinetic energy of a particle of mass m at speed v is equal to $\frac{1}{2}mv^2$.

kinetic theory—model of molecular motion used to explain many of the properties of gases.

kinetics—the study of rates of chemical reactions.

L

labile complex—a complex ion which rapidly reaches equilibrium with ligands in surrounding solution.

lanthanides—elements with atomic numbers 58–71; series in which 4f sublevel is being filled. Example: Europium, atomic number 63, is one of the lanthanides.

Law of Charles and Gay-Lussac—see *Charles' Law*.

Law of Chemical Equilibrium—a relation stating that in a reaction mixture at equilibrium there is a condition, given by the equilibrium constant, relating the concentrations of reactants and products; for the reaction $aA(g) + bB(g) \rightleftharpoons cC(g) + dD(g)$, $K_c = \frac{[C]^c[D]^d}{[A]^a[B]^b}$.

Law of Combining Volumes—a relation stating that relative volumes of gases in a chemical reaction are in the ratio of small integers (all gases at same T and P); also called Gay-Lussac's Law. Example: In the reaction $2 H_2(g) + O_2(g) \rightarrow 2 H_2O(g)$, 2 volumes H_2 react with 1 volume O_2 to produce 2 volumes H_2O.

Law of Conservation of Mass—a relation stating that in a chemical reaction the mass of the products equals the mass of the reactants.

Law of Constant Composition—a relation stating that the relative masses of the elements in a given chemical substance are fixed.

Law of Dulong and Petit—a relation stating that the heat capacity of one gram atomic mass of any metal is about 25 J/°C.

Law of Mass Action—same as Law of Chemical Equilibrium.

Law of Multiple Proportions—a relation stating that when two elements, A and B, form two compounds, the relative amounts of B which combine with a fixed amount of A are in a ratio of small integers. Example: In water and hydrogen peroxide, both of which contain hydrogen and oxygen, there are 8 and 16 g of oxygen, respectively, for each gram of hydrogen.

Le Châtelier's Principle—a relation stating that, when a system at equilibrium is disturbed, it will respond in such a way as to counteract the change.

Lewis acid—a species which can accept a pair of electrons. Example: In the reaction $BF_3 + NH_3 \rightarrow BF_3NH_3$, the BF_3 accepts a pair of electrons from NH_3 and so behaves as a Lewis acid.

Lewis base—a species which can donate a pair of electrons. Example: NH_3, in the above reaction.

Lewis structure—electronic structure of a molecule or ion in which electrons are shown by dots or dashes (electron pairs).

ligand—a molecule or anion bonded directly to the central metal in a complex ion.

limiting reagent—the least abundant reactant, based on the equation, in a chemical reaction; dictates the maximum amount of product which can be formed.

linear molecule—molecule containing three atoms in which the bond angle is 180°. Examples of linear molecules include BeF_2 and CO_2.

liter—1 dm = 1000 cm^3.

logarithm of a number—the exponent to which another number, usually 10, must be raised to give the number. Examples: The logarithm of 100 to the base of 10 is 2, since $10^2 = 100$; the logarithm of 3.00 is 0.477, since $10^{0.477} = 3.00$.

low-spin complex—complex which, for a particular metal ion, has the smallest possible number of unpaired electrons.

M

macromolecular—having a structure in which all the atoms in a crystal are linked by chemical bonds. Example: Since all the atoms in a silicon crystal are bonded chemically into a unit, silicon is macromolecular.

main group—a numbered group of the Periodic Table.

malleable—capable of being shaped, as by pounding with a hammer.

mass—a property reflecting the amount of matter in a sample.

mass number—an integer equal to the sum of the number of protons and neutrons in an atomic nucleus. Example: The mass number of a $^{37}_{17}Cl$ isotope is 37; the nucleus of that isotope contains 17 protons and 20 neutrons.

matter—a general term for any kind of material; the stuff of which pure substances are made.

Maxwellian distribution—a relation describing the way in which molecular speeds, or energies, are shared among the molecules in a gas.

mechanism—a sequence of steps that occurs during the course of a chemical reaction.

melting point—same as freezing point.

metal—a substance having characteristic luster, malleability, and high electrical conductivity; readily loses electrons to form positive ions.

metalloid—an element with properties intermediate between those usually associated with metals and nonmetals. Examples of metalloids are B, Si, Ge, As, Sb, and Te.

metallurgy—the science and processes of extracting metals from their ores.

meter—unit of length in the metric system.

milli—prefix on a metric unit indicating multiple of 1×10^{-3}. Example: One millimeter equals one one-thousandth of a meter, 0.001 m.

miscible (with)—soluble in.

mixture—two or more substances combined so that each substance retains its chemical identity.

mm Hg—a unit of pressure: 1 atm = 760 mm Hg.

molality—a concentration unit, defined equal to the number of moles of solute divided by number of kilograms of solvent. Example: A solution made by dissolving 0.10 mol of KNO_3 in 200 g of water would be 0.50 molal in KNO_3 (0.50 m KNO_3).

molarity—a concentration unit, defined equal to number of moles of solute divided by number of liters of solution. Example: In 6 molar HCl, 6 M HCl, there is 6 mol of HCl in one liter of solution.

molar mass—mass of one mole of a substance. Most often expressed in grams (GMM O_2 = 32.0 g), but can be expressed in kilograms (KMM O_2 = 0.032 kg).

mole—a convenient chemical mass unit; defined as containing 6.022×10^{23} molecules, atoms, or other units; the mass of the mole is equal to the gram formula mass of the substance. Examples: One mole of NH_3 contains 6.022×10^{23} molecules and weighs just about 17 g; one mole Cu contains 6.022×10^{23} atoms and weighs 63.54 g; one mole KNO_3 weighs 101.1 g.

mole fraction—a concentration unit, defined equal to number of moles of component divided by total number of moles in solution. Example: In a solution in which there is 1 mol benzene, 2 mol CCl_4, and 7 mol acetone, the mole fraction of the acetone is 0.7.

molecular formula—an expression stating the number and kind of each atom present in the molecule of a substance. Example: Since the molecular formula of hexane is C_6H_{14}, there are six C atoms and 14 H atoms in a hexane molecule.

molecular geometry—the shape of a molecule describing the relative positions of atomic nuclei.

molecular mass—a number equal to the sum of the atomic masses of all of the atoms in a molecule; tells the mass of the molecule relative to a ^{12}C atom, taken to have a mass of 12. Example: The molecular mass of C_2H_6 is about $30[(2 \times 12) + (6 \times 1)]$, so the molecule is about 2.5 times as heavy as a ^{12}C atom, and the same mass as an NO molecule, whose molecular mass is also 30 (i.e., 14 + 16 = 30).

molecular orbital—an orbital involved in the chemical bond between two atoms, and taken to be a linear combination of the orbitals on the two bonded atoms.

molecule—an aggregate of atoms, which is the characteristic component particle in all gases, many pure liquids, and some solids; often contains only a few atoms; has relatively little physical interaction with other molecules. Examples: In nitrogen gas, liquid benzene, and solid glucose, one finds N_2 molecules, C_6H_6 molecules, and $C_6H_{12}O_6$ molecules.

monomer—a small molecule which joins with other monomers to form a polymer.

monosaccharide—a class of sugars which cannot be broken down chemically to simpler sugars.

multiple equilibria, rule of—rule stating that if Equation 1 + Equation 2 = Equation 3, then $K_1 \times K_2 = K_3$.

N

nanometer—a unit of length equal to 10^{-9} m.

natural logarithm—a logarithm based upon the number e, 2.7182818 . . .; if $\log_e X = Y$, then $e^Y = X$, $\log_{10} X = \dfrac{\log_e X}{2.303}$; base e comes from the calculus, where certain derivatives and integrals are most easily expressed in terms of e.

Nernst equation—an equation relating the voltage of a cell to its standard voltage and the concentrations of reactants and products.

net ionic equation—a chemical equation for a reaction, in which only those species which actually react are listed. Example: When 1 M HCl and 1 M NaOH solutions are mixed, the net ionic equation for the reaction is $H^+(aq) + OH^-(aq) \rightarrow H_2O$; the Cl^- and Na^+ ions in the solution do not react and so are not in the equation.

neutral solution—an aqueous solution with a pH = 7.0 ($[H^+] = 1.0 \times 10^{-7}$ M).

neutralization—a reaction of an acid and a base to produce a neutral (pH 7) solution.

neutron—one of the particles in an atomic nucleus; mass = 1, charge = 0.

nitrogen fixation—see *fixation of nitrogen.*

noble gas—element in Group 8, at the far right of the Periodic Table.

noble gas structure—ns^2np^6 outer electron structure in an atom or ion; a particularly stable structure attained by atoms obeying the octet rule; sometimes called an inert gas structure.

nonelectrolyte—a substance that does not exist as ions in water solution. Example: Since ethyl alcohol does not ionize when dissolved in water, it is a nonelectrolyte.

nonmetal—one of the elements in the upper right corner of the Periodic Table that does not show metallic properties. Example: Nitrogen gas, N_2, is a nonmetal.

nonpolar bond—a chemical bond in which there are no positive and negative ends; found in homonuclear diatomic molecules such as H_2 and O_2.

nonpolar molecule—molecule in which there is no separation of charge and hence no negative and positive poles. Nonpolar molecules include H_2 and CO_2.

nonspontaneous reaction—a reaction which cannot occur by itself without input of work from an external source; $\Delta G > 0$ for nonspontaneous reactions at T and P.

normal boiling point—boiling point at one atmosphere pressure.

n-type semiconductor—semiconductor in which current is carried through a solid by "extra" electrons introduced by electron-rich impurity atoms.

nuclear fission—the splitting of a heavy nucleus by a

neutron into two lighter nuclei, accompanied by the release of energy.

nuclear fusion—the combining of light nuclei to form a heavier nucleus, accompanied by the release of energy.

nuclear reactor—device used to generate electrical energy using the heat given off by nuclear fission.

nucleus—the small, dense, positively charged region at the center of an atom.

O

octahedral—having the symmetry of a regular octahedron, a solid with six vertices and eight faces, each of which is an equilateral triangle.

octane number—number used to indicate the knocking resistance of a motor fuel. Based on a scale in which isooctane is rated 100 and heptane 0.

octet—group of eight valence electrons surrounding an atom. All the noble-gas atoms except He have an octet of valence electrons.

octet rule—the principle that bonded atoms tend to have a share in eight outermost electrons.

olefin—a hydrocarbon containing a carbon-carbon double bond. Same as alkene.

optical activity—the ability to rotate the plane of a beam of transmitted polarized light; a property possessed by substances having a chiral center.

optical isomerism—the phenomenon in which each member of a pair of molecules having the same molecular formula rotates a beam of plane polarized light in opposite directions. Such molecules have at least one chiral center.

orbital—an electron cloud with an energy state characterized by given values of n, ℓ, and m_ℓ quantum numbers; has a capacity for two electrons having paired spins; often associated with a particular region in the atom. Example: In an atom the electrons in the $2p_x$ orbital are in a dumbbell-shaped cloud concentrated along the x axis.

orbital diagram—a sketch showing electron populations of atomic orbitals, including electron spins.

order of reaction—an exponent to which concentration of a reactant needs to be raised to give observed dependence of reaction rate on concentration. Example: If, for the reaction A → products, rate = k(conc. A)², the reaction is second order.

ore—natural mineral deposit from which a metal can be extracted profitably.

organic—used to characterize any compound containing carbon, hydrogen, and possibly other elements. Example: Propionic acid, CH_3CH_2COOH, is an organic compound; SiO_2 is not.

osmotic pressure—the excess pressure which must be applied to a solution to prevent the pure solvent from diffusing into the solution through a semipermeable membrane.

Ostwald process—industrial process used to make nitric acid from ammonia.

overall order—number obtained by summing all the exponents in the rate expression. If rate = k(conc. A)^m × (conc. B)^n, then overall order = m + n.

oxidation—a half-reaction involving a loss of electrons or, more generally, an increase in oxidation number. Example: If, at an electrode, the reaction is Ag(s) → Ag^+(aq) + e$^-$, silver is undergoing oxidation, since its oxidation number is increasing from 0 to +1.

oxidation number—a number which can be assigned to an atom in a molecule or ion which reflects, qualitatively, its state of oxidation; the number is determined by applying a set of rules. Examples: In the NO_3^- ion the oxidation numbers of the N and O atoms are +5 and −2, respectively.

oxide—compound containing oxygen. The *oxide ion*, found in metal oxides, has the formula O^{2-}.

oxidizing agent—a species which accepts electrons from another. Example: In the reaction Cl_2(aq) + 2 Br$^-$(aq) → 2 Cl$^-$(aq) + Br_2(aq), Cl_2 serves as an oxidizing agent.

oxyacid—an acid containing oxygen. Examples: HNO_3, H_2SO_4, and HClO are oxyacids.

oxyanion—an anion containing oxygen. Examples: NO_3^-, SO_4^{2-}, and ClO$^-$ are oxyanions.

ozone—an allotropic form of oxygen, in which the molecule is O_3.

P

paired electrons—two electrons in the same orbital with spins equal to +½ and −½; an electron pair.

paraffin—a hydrocarbon in which all carbon-carbon bonds are single; an alkane.

paramagnetic—having magnetic properties caused by unpaired electrons. Examples: The NO molecule, H atom, and CH_3 radical are paramagnetic.

partial pressure of A—that part of the total pressure in a gaseous mixture which can be attributed to component A. The partial pressure of A is equal to the pressure A would exert in the container if A were there by itself. Example: In a mixture of 3 mol N_2 and 1 mol O_2 at a total pressure of 2 atm, the partial pressure of N_2 is 1.5 atm and that of O_2 is 0.5 atm.

parts per billion (ppb)—for gases, the number of molecules of solute per billion molecules of gas; for liquids and solids, the number of grams of solute per billion grams of sample.

parts per million (ppm)—for gases, number of molecules solute per million molecules of gas; for liquids and solids, number of grams solute per million grams of sample. Example: If a gas sample contains 6 parts per million CO, then in one mole of gas there would be 6×10^{-6} mol CO.

Pauli exclusion principle—see *exclusion principle*.

peptide linkage—the 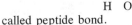 group in proteins; also called peptide bond.

per cent A—parts A in 100 parts of a sample; usually in mass per cent but may be in mole per cent or volume per cent. Example: A 5% NaOH solution contains 5 g NaOH in 100 g solution.

per cent composition—percentages by mass of the elements in a compound.

per cent yield—quantity equal to 100 × actual yield/theoretical yield.

period—a horizontal row of elements in the Periodic Table.

periodic—occurring in cycles.

Periodic Table—an arrangement of the elements into rows and columns in which those elements with similar properties occur in the same column.

peroxide—a binary compound containing an oxygen-oxygen single bond or the peroxide ion, O_2^{2-}.

pH—alternate way to express H^+ ion concentration; $pH = -\log_{10} [H^+]$.

phase diagram—for one-component systems, a graph of pressure vs. temperature, showing conditions under which the pure substance will exist as a liquid, solid, or gas and also the conditions under which two-phase and three-phase equilibria can exist.

photon—an individual quantum of radiant energy of wavelength λ.

photosynthesis—the process by which sunlight makes possible the synthesis of organic compounds from CO_2 and H_2O.

physical property—property of a substance related to its physical characteristics. Examples: Density, melting point.

pi (π) bond—a bond in which electrons are concentrated in orbitals which are located off the internuclear axis; one bond in a double bond is a pi bond, and there are two pi (π) bonds in a triple bond.

pi orbital—a molecular orbital in which electron density is concentrated in lobes which do not lie on the internuclear axis.

Planck's constant—constant in the equation: $E = h\nu = hc/\lambda$. $h = 6.626 \times 10^{-34}$ J·s.

polar bond—a chemical bond which has positive and negative ends; characteristic of all bonds between nonidentical atoms. Example: In CCl_4 the C—Cl bonds are polar, since the Cl atom tends to attract electrons more than the C atom does.

polarization—distortion of the electron distribution in a molecule, tending to produce positive and negative poles.

polar molecule—molecule in which there is a separation of charge and hence a positive and a negative pole. Polar molecules include HF and H_2O.

pollutant—a contaminant, or foreign species, present in a sample; usually has a deleterious effect on quality of sample as far as living things are concerned.

polyamide—a condensation polymer formed from carboxylic acid and amine units.

polyatomic ion—a charged species containing more than one atom.

polyamide—a condensation polymer formed from carboxylic acid and amine units.

polyatomic ion—a charged species containing more than one atom.

polydentate ligand—a ligand which forms two or more bonds to a central metal.

polyester—a condensation polymer made up of ester units. Dacron and Kodel are polyesters.

polymer—a molecule made up from many units which are linked together chemically. Example: In the polymer called polyethylene, many $H_2C=CH_2$ units become linked together by chemical reaction to form chains which have the structure

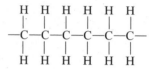

polypeptide—a condensation polymer in which the monomer units are α-amino acids; in the polymer the amino acid residues are linked by peptide bonds; another name for a protein.

polysaccharide—natural polymer (e.g., starch, cellulose) consisting of a large number of monomer units derived from simple sugars such as glucose.

positron ($^0_1$e)—a particle having the mass of an electron but a +1 charge.

post-transition metals—lower members of Periodic Table Groups 3, 4 and 5.

precipitate—a solid which forms when two solutions are mixed.

pressure—force per unit area; often expressed in mm Hg, atmospheres, or kilopascals.

principal energy level—energy level designated by the principal quantum number n. The first element in each period of the Periodic Table introduces a new principal energy level.

principal quantum number—quantum number **n**; the most important quantum number, since it has greatest effect on energy of electron; cited first in the set of four quantum numbers associated with an electron.

product—a substance formed as a result of a chemical reaction. Example: In the reaction $Ag^+(aq) + Cl^-(aq) \rightarrow AgCl(s)$, AgCl is the product.

"proof"—twice the volume percentage of ethyl alcohol in an alcoholic beverage.

property—a characteristic of a sample of matter that is fixed by its state. Example: The density and energy of a mole of H_2 at 100°C and 1 atm are properties of that sample of hydrogen gas.

protein—a polypeptide.

proton—the nucleus of a hydrogen atom, the H^+ ion; a component of atomic nuclei, with mass = 1, charge = +1.

psi (Ψ)—the amplitude (height) of an electron wave at various points in space.

p-type semiconductor—a semiconductor in which current is carried through a solid by electron flow into "positive holes" in the crystal, introduced by electron-deficient impurity atoms.

pyramidal—adjective used to describe the geometry of a molecule such as NH_3, in which one atom lies directly above the center of an equilateral triangle formed by three other atoms.

Q

Q—see *heat flow* and *ion product*.

quadratic formula—formula used to obtain the two roots of the general quadratic equation: $ax^2 + bx + c = 0$. The formula is $x = \dfrac{-b \pm \sqrt{b^2 - 4ac}}{2a}$.

qualitative analysis—the determination of the nature of the species present in a sample. Example: By qualitative analysis she found that the solution contained Cu^{2+}, Sn^{4+}, and Cl^- ions.

quantitative analysis—the determination of how much of a given component is present in a sample. Example: The students discovered, by quantitative analysis, that the ore contained 42.45% iron by mass.

quantum mechanics—approach used to calculate the energies and spatial distributions of small particles confined to very small regions of space.

quantum number—a number used in the description of the energy levels available to atoms and molecules; an electron in an atom or ion will have four quantum numbers to describe its state.

quantum theory—general theory which describes the allowed energies of atoms and molecules.

R

R—see *gas constant*.

rad—a unit of absorbed radiation equal to 10^{-2} J absorbed per kilogram of tissue.

radical—see *free radical*.

radioactivity—the ability possessed by some natural and synthetic isotopes (induced radioactivity) to undergo reactions involving nuclear transformations to other isotopes.

rare earth—the name sometimes given to members of the lanthanide series.

rate constant—the proportionality constant in the rate equation for a reaction. Example: If the rate equation is rate = k(conc. A)^n, then k is the rate constant.

rate expression (law)—a mathematical relationship describing the dependence of the reaction rate upon the concentration(s) of reactant(s) or product(s).

rate of a reaction—the magnitude of the change in concentration of a reactant or product divided by the time required for the change to occur (with both quantities relatively small). Example: For the reaction A → B, rate = $\dfrac{\Delta(\text{conc. B})}{\Delta t} = \dfrac{-\Delta(\text{conc. A})}{\Delta t}$.

rate of radioactive decay—the rate of emission of radiation by a radioactive species.

rate-determining step—the slowest step in a multistep reaction.

reactant—the starting material in a chemical reaction. Example: In the reaction $H_2(g) + \frac{1}{2} O_2(g) \rightarrow H_2O(l)$, H_2 and O_2 are both reactants.

reaction—a chemical change in which new substances are formed. Example: When aluminum burns in air, the chemical reaction that occurs is described by the equation $2\,Al(s) + \frac{3}{2} O_2(g) \rightarrow Al_2O_3(s)$.

reaction mechanism—see *mechanism*.

reciprocal relation—relation between equilibrium constants for forward (K_f) and reverse (K_r) reactions: $K_r = 1/K_f$.

redox reaction—a reaction involving oxidation and reduction.

redox titration—titration of an oxidizing agent by a reducing agent, or vice versa.

reducing agent—a species which furnishes electrons to another. Example: in the reaction $Zn(s) + 2\,H^+(aq) \rightarrow Zn^{2+}(aq) + H_2(g)$, the $Zn(s)$, metallic zinc, is the reducing agent.

reduction—a half-reaction in which a species gains electrons or, more generally, decreases in oxidation number. Example: In the reaction $Zn(s) + 2\,H^+(aq) \rightarrow Zn^{2+}(aq) + H_2(g)$, the H^+ ions (oxid. no. = +1) are reduced to H_2 (oxid. no. = 0).

relative humidity—$100 \times P/P_0$, where P = pressure of water vapor in air, P_0 = equilibrium vapor pressure of water at same T. Can also be thought of as a measure of the amount of water in the air divided by the amount that sample of air could hold. Example: When the relative humidity is 75%, there is three fourths as much water in the air as the air can hold at that temperature.

rem—a unit of absorbed radiation equal to n times the number of rads. The factor n depends upon the type of radiation absorbed (see p. 573).

resonance—used to rationalize properties of octet rule species for which one Lewis structure is inadequate; resonance structure is taken to be an average of two or more Lewis structures which differ only in positions of electrons; species are said to exhibit resonance.

reverse osmosis—process by which pure water is obtained from a salt solution. Under pressure, the water passes out of the salt solution through a semipermeable membrane.

rule of multiple equilibria—see *multiple equilibria*.

S

S—see *entropy*.

salt—a solid ionic compound made up from a cation other than H^+ and an anion other than OH^- or O^{2-}. Examples: $NaCl$, $CuSO_4$, NH_4NO_3.

saturated hydrocarbon—an alkane, a hydrocarbon in which all carbon-carbon bonds are single.

saturated solution—a solution containing as much solute as the amount of solvent can dissolve at a specific temperature.

Schrödinger equation—wave equation which relates mass, potential energy, kinetic energy, and coordinates of a particle.

scintillation counter—device used to measure the rate of radioactive decay by converting energy of radiation into light, which activates photoelectric cell.

second order reaction—a reaction whose rate depends on second power of reactant concentration; may be sum of exponents of two reactant concentrations. Examples: The two expressions, rate = k(conc. A)^2 and rate = k(conc. A) × (conc. B), are both associated with second order reactions.

semiconductor—a substance used in transistors and thermistors whose electrical conductivity depends on presence of tiny amounts of impurities such as As or B in a very pure crystal of an element such as silicon or germanium; conductance increases dramatically as temperature goes up.

semipermeable membrane—a film which allows passage of solvent molecules such as H_2O but does not pass solute molecules such as proteins or, in some cases, ions.

shielding—a term used to describe effect of inner electrons in decreasing the attraction of an atomic nucleus on outermost electrons.

SI unit—unit associated with the International System of Units; see Appendix 1.

sigma (σ) bond—a chemical bond in which electron density on internuclear axis is high, which is the case with all single bonds; double bonds contain one sigma and one pi bond, triple bonds contain one sigma and two pi bonds.

sigma (σ) orbitals—molecular orbitals associated with sigma bonds between atoms.

significant figures—meaningful digits in a measured quantity; number of digits in a number when expressed in exponential notation. Example: 1.035×10^3 has four significant figures (exponential doesn't count).

simple cubic cell—a unit cell containing atoms at each corner of a cube.

single bond—a pair of electrons shared between two bonded atoms.

smog—smoky fog containing harmful species such as SO_2, SO_3, NO_2, and O_3.

soap—sodium salt of a fatty acid.

solubility—the amount of a solute that will dissolve in a given amount of solvent at a specified temperature; may be stated in various ways, with moles solute per liter of solution being common.

solubility product constant (K_{sp})—equilibrium constant for the solution reaction of a relatively insoluble ionic compound. Example: For the reaction $Ca(OH)_2(s) \rightleftharpoons Ca^{2+}(aq) + 2\ OH^-(aq)$, $K_{sp} = [Ca^{2+}][OH^-]^2$.

solubility rules—rules used to classify ionic compounds as to their solubility in water (Table 16.3).

solute—the solution component present in smaller amount than the solvent.

solution—a liquid, gas, or solid phase containing two or more components dispersed uniformly throughout the phase.

Solvay process—industrial process used to make $NaHCO_3$ and Na_2CO_3 starting with sodium chloride, calcium carbonate, and water.

solvent—a substance, usually a liquid, in which another substance, called the solute, is dissolved.

species—a general term referring to a molecule, ion, or atom.

specific heat—the amount of heat required to raise the temperature of one gram of a substance by one degree Celsius.

spectrochemical series—arrangement of ligands in order of decreasing tendency to split d orbitals of transition metal cation. Substitution of a ligand for another higher in the series gives a smaller splitting and hence a longer wavelength of light absorbed.

spectrum—a pattern of characteristic wavelengths associated with excitation of an atom, molecule, or ion; also used as name of a pattern having a similar appearance obtained as a result of chromatographic or mass spectroscopic experiments.

spontaneous reaction—a reaction which can occur by itself, without input of work from outside; $\Delta G < 0$ for spontaneous reactions at T and P.

stable—will not change spontaneously. Nature of change should be specified. Example: Water is stable at 25°C with respect to thermal decomposition to hydrogen and oxygen.

standard free energy change, ΔG^0—ΔG when reactants and products are in their "standard state": 1 atm for a gas, effectively 1 M for a species in aqueous solution.

standard molar entropy, S^0—S of a substance in its standard state (1 atm for a gas, 1 M for an ion in water solution).

standard oxidation voltage (E_{ox}^0)—the voltage associated with an oxidation reaction at an electrode, when all solutes are 1 M (strictly speaking, unit activity) and all gases are at 1 atm.

standard potential—identical with the standard reduction voltage (described below).

standard reduction voltage (E_{red}^0)—the voltage associated with reduction reaction at an electrode, when all solutes are 1 M and all gases are at 1 atm. Given the E_{red}^0 for an electrode reaction, the voltage for the reaction in the opposite direction is equal to $-E_{red}^0$, and is the E_{ox}^0 for the latter reaction. Example: For the reaction $Cu^{2+}(aq) + 2\ e^- \rightarrow Cu(s)$, $E_{red}^0 = 0.34$ V; therefore, for the reaction $Cu(s) \rightarrow Cu^{2+}(aq) + 2\ e^-$, $E_{ox}^0 = -0.34$ V.

standard voltage (E^0)—voltage of a cell in which all species are in their standard states (solids and liquids are pure, solutes are at unit activity, often taken to be 1 M, and gases are at 1 atm).

state—condition of a system when its properties are fixed. Example: A mole of H_2O at 25°C and 1 atm is in a definite state in that all of its properties have values which are fixed.

state property—property of a system which is fixed when the temperature, pressure and composition are specified. One mole of water at 25°C and 1 atm has a fixed volume, enthalpy, and entropy; V, H, and S are state properties.

stoichiometric—having to do with masses (grams, moles) of reactants and products in a chemical equation.

straight-chain alkane—saturated hydrocarbon in which all the carbon atoms are arranged in a single, continuous chain.

strong—as applied to acids, bases, and electrolytes, indicates complete dissociation into ions when in water solution. Example: HCl is a strong acid, since it exists as H^+ and Cl^- ions in aqueous solution.

structural formula—a formula showing the arrangement of atoms in a molecule or polyatomic ion.

structural isomers—two or more isomers having the same molecular formula but different molecular structures.

sublevel—a subdivision of an energy level as designated by the quantum number ℓ.

sublimation—a change in state from solid to gas. Example: Iodine slowly sublimes in an open container; the sublimation is endothermic.

substitution reaction—reaction in which one atom or group is substituted for another. Example: $C_6H_6(l) + Br_2(l) \rightarrow C_6H_5Br(l) + HBr(g)$.

substrate—the reactant in an enzyme-catalyzed reaction.

sugar—a carbohydrate containing an aldehyde or ketone group and an OH and H group on all noncarbonyl group carbon atoms; sometimes, loosely, sucrose.

superoxide—an ionic oxygen compound containing the superoxide ion, O_2^-.

supersaturated—containing more solute than equilibrium conditions would allow; unstable to addition of solute crystal. Example: It is easy to make a supersaturated solution of sodium acetate by cooling a hot concentrated solution of the salt carefully to 25°C.

surroundings—everything outside the *system* being studied.

symbol—see *chemical symbol*.

system—the sample of matter under consideration.

T

temperature—a property of matter reflecting the amount of energy of motion of its component particles; measured value based on one of several possible scales.

tetrahedral—adjective describing the geometry of a molecule in which a central atom forms four bonds directed toward the corners of a regular tetrahedron (a solid with four vertices and four sides, all of which are equilateral triangles).

theoretical yield—the amount of product obtained from the complete conversion of the limiting reactant.

thermal—having to do with heat.

thermochemical equation—chemical equation in which the value of ΔH is specified.

thermodynamics—the study of heat, work, and the related properties of mechanical and chemical systems.

titration—a process in which a solution is added to another solution with which it reacts under conditions such that the volume of added solution can be accurately measured.

trans—as related to geometric isomers, referring to structures in which two identical groups are as far apart as possible, as opposed to *cis*, where they are as close as possible. Example: Cl and NH₃ would

be the structure of the *trans* isomer of the square planar $Pt(NH_3)_2Cl_2$ molecule.

transition metal—any one of the metals in the central groups in the fourth, fifth, and sixth periods in the Periodic Table. Examples: Fe, Zr, and W would all be classified as transition metal atoms.

translational energy—energy of motion through space. Example: A falling raindrop has translational energy.

triangular bipyramid—solid with five vertices and six sides; may be regarded as two pyramids fused through a base that is an equilateral triangle.

triple bond—three electron pairs shared between two bonded atoms.

triple point—that temperature and pressure at which the solid, liquid, and vapor of a pure substance can coexist in equilibrium.

U

ultraviolet radiation—light having a wavelength less than about 400 nm but greater than about 10 nm.

unit cell—the smallest unit of a crystal which, if repeated indefinitely, could generate the whole crystal.

unpaired electron—single electron occupying an orbital by itself.

unsaturated—referring to solutions, being able to dissolve more solute; referring to organic compounds, containing double or triple carbon-carbon bonds.

useful work—the work produced during a change in state in excess of that required to accomplish the change. Example: In the reaction $H_2O(l) \rightarrow H_2O(g)$, since the volume of the gas is larger than that of the liquid, work of expansion will be required to push back the surrounding air to make room for the water vapor; the useful work will be that work done during the change which is in excess of the work of expansion.

V

valence bond model—the theory that atoms tend to become bonded by pairing, and sharing, their outer, or valence, electrons; also referred to as the atomic orbital model.

valence electrons—those electrons in the outermost shell. Example: In the carbon atom, with electron configuration $1s^2 2s^2 2p^2$, there are four valence electrons, those in the 2s and 2p orbitals.

van der Waals equation—equation used to express the relation between P, V, and T of a real gas, used under conditions where the Ideal Gas Law is not sufficiently accurate.

van der Waals forces—a general name sometimes given to intermolecular forces.

vapor—a condensable gas.

vapor pressure—the pressure exerted by a vapor when it is in equilibrium with the liquid from which it is derived. Example: If liquid water is admitted to an evacuated container at 60°C, the pressure in the container when the liquid and vapor reach equilibrium becomes 149.4 mm Hg; therefore, the vapor pressure of water at 60°C is 149.4 mm Hg.

vapor pressure lowering—decrease in the vapor pressure of a liquid caused by addition of a nonvolatile solute.

visible light—radiation having wavelengths between 400 and 700 nm.

volatile—easily evaporated.

voltage—electric potential; a measure of tendency of a cell or other device to force electrons through an external circuit.

voltaic cell—a device in which a spontaneous chemical reaction is used to produce electrical work.

vulcanization—the cross-linking of rubber chains by short chains of sulfur atoms.

W

water dissociation constant (K_w)—equal to $[H^+][OH^-] = 1 \times 10^{-14}$ at 25°C.

water softening—the removal of ions, particularly Ca^{2+} and Mg^{2+}, from water.

wave function—the solution to Schrödinger wave equation, the square of whose magnitude at any point is proportional to probability of finding at that point the particle concerned; name arises from fact that solutions look roughly like waves when plotted on a graph.

wavelength—a characteristic property of light, similar to its color, and equal to the length of a full wave; often expressed in nanometers; can be measured with a spectroscope.

weak—as applied to acids and bases, being partially ionized in water solution. Example: Acetic acid, $HC_2H_3O_2$, is a weak acid because in water solution it is only slightly ionized to H^+ and $C_2H_3O_2^-$ ions.

weak electrolyte—species which, in water solution, forms an equilibrium mixture of molecules and ions. Examples include HF and NH_3.

work—one of the effects which may be associated with an energy change; during a change a system may do work on its surroundings, equivalent to the raising of a mass; work may be electrical, mechanical, or due to expansion or compression, and may be done by, or on, the system.

X

X—an unknown quantity.

x-rays—light rays having a wavelength from 0.01 to 1.0 nm.

Y

yield—the amount of product obtained from a reaction.

Z

zeolite—type of silicate mineral used to soften water by cation exchange.

zero order reaction—a reaction whose rate is independent of reactant concentration.

zwitterion—the dipolar form of an amino acid which occurs when an H^+ ion is transferred from an acid group to an amine group.

INDEX

Note: Italic page numbers indicate figures; t indicates tables.